Metal Ions in Life Sciences

Volume 16

Series editors
Astrid Sigel
Helmut Sigel
Roland K.O. Sigel

More information about this series at http://www.springer.com/series/8385

Astrid Sigel • Helmut Sigel • Roland K.O. Sigel
Editors

The Alkali Metal Ions: Their Role for Life

Springer

Editors
Astrid Sigel
Department of Chemistry
Inorganic Chemistry
University of Basel
Spitalstrasse 51
CH-4056 Basel, Switzerland
astrid.sigel@unibas.ch

Helmut Sigel
Department of Chemistry
Inorganic Chemistry
University of Basel
Spitalstrasse 51
CH-4056 Basel, Switzerland
helmut.sigel@unibas.ch

Roland K.O. Sigel
Department of Chemistry
University of Zürich
Winterthurerstrasse 190
CH-8057 Zürich, Switzerland
roland.sigel@chem.uzh.ch

ISSN 1559-0836 ISSN 1868-0402 (electronic)
Metal Ions in Life Sciences
ISBN 978-3-319-21755-0 ISBN 978-3-319-21756-7 (eBook)
DOI 10.1007/978-3-319-21756-7

Library of Congress Control Number: 2015958773

Springer Cham Heidelberg New York Dordrecht London

Cover illustration: Cover figure of the MILS series since Volume 11: RNA-protein interface of the Ile-tRNA synthetase complex held together by a string of Mg^{2+} ions, illustrating the importance of metal ions in both the protein and the nucleic acid world as well as connecting the two; hence, representing the role of *Metal Ions in Life Sciences*. tRNA synthetases are not only essential to life, but also serve as a target for novel classes of drugs making such RNA-protein complexes crucial also for the health sciences. The figure was prepared by Joachim Schnabl and Roland K. O. Sigel using the PDB coordinates 1FFY.

Printed on acid-free paper

Springer International Publishing AG Switzerland is part of Springer Science+Business Media (www.springer.com)

Historical Development and Perspectives of the Series

Metal Ions in Life Sciences*

It is an old wisdom that metals are indispensable for life. Indeed, several of them, like sodium, potassium, and calcium, are easily discovered in living matter. However, the role of metals and their impact on life remained largely hidden until inorganic chemistry and coordination chemistry experienced a pronounced revival in the 1950s. The experimental and theoretical tools created in this period and their application to biochemical problems led to the development of the field or discipline now known as *Bioinorganic Chemistry*, *Inorganic Biochemistry*, or, more recently also often addressed as, *Biological Inorganic Chemistry*.

By 1970 *Bioinorganic Chemistry* was established and further promoted by the book series *Metal Ions in Biological Systems* founded in 1973 (edited by H.S., who was soon joined by A.S.) and published by Marcel Dekker, Inc., New York, for more than 30 years. After this company ceased to be a family endeavor and its acquisition by another company, we decided, after having edited 44 volumes of the *MIBS* series (the last two together with R.K.O.S.), to launch a new and broader-minded series to cover today's needs in the *Life Sciences*. Therefore, the Sigels new series is entitled

Metal Ions in Life Sciences.

After publication of the first four volumes (2006–2008) with John Wiley & Sons, Ltd., Chichester, UK, and the next five volumes (2009–2011) with the Royal Society of Chemistry, Cambridge, UK, we are happy to join forces now in this still new endeavor with Springer Science & Business Media B.V., Dordrecht, the Netherlands, a most experienced Publisher in the *Sciences*.

---

*Reproduced with some alterations by permission of John Wiley & Sons, Ltd., Chichester, UK (copyright 2006), from pages v and vi of Volume 1 of the series *Metal Ions in Life Sciences* (MILS-1).

The development of *Biological Inorganic Chemistry* during the past 40 years was and still is driven by several factors; among these are (i) the attempts to reveal the interplay between metal ions and peptides, nucleotides, hormones or vitamins, etc.; (ii) the efforts regarding the understanding of accumulation, transport, metabolism, and toxicity of metal ions; (iii) the development and application of metal-based drugs; (iv) biomimetic syntheses with the aim to understand biological processes as well as to create efficient catalysts; (v) the determination of high-resolution structures of proteins, nucleic acids, and other biomolecules; (vi) the utilization of powerful spectroscopic tools allowing studies of structures and dynamics; and (vii), more recently, the widespread use of macromolecular engineering to create new biologically relevant structures at will. All this and more is and will be reflected in the volumes of the series *Metal Ions in Life Sciences*.

The importance of metal ions to the vital functions of living organisms, hence, to their health and well-being, is nowadays well accepted. However, in spite of all the progress made, we are still only at the brink of understanding these processes. Therefore, the series *Metal Ions in Life Sciences* will endeavor to link coordination chemistry and biochemistry in their widest sense. Despite the evident expectation that a great deal of future outstanding discoveries will be made in the interdisciplinary areas of science, there are still "language" barriers between the historically separate spheres of chemistry, biology, medicine, and physics. Thus, it is one of the aims of this series to catalyze mutual "understanding."

It is our hope that *Metal Ions in Life Sciences* proves a stimulus for new activities in the fascinating "field" of *Biological Inorganic Chemistry*. If so, it will well serve its purpose and be a rewarding result for the efforts spent by the authors.

Astrid Sigel and Helmut Sigel
Department of Chemistry, Inorganic Chemistry,
University of Basel, CH-4056 Basel, Switzerland

Roland K.O. Sigel
Department of Chemistry,
University of Zürich, CH-8057 Zürich, Switzerland

October 2005,
October 2008,
and August 2011

Preface to Volume 16

The Alkali Metal Ions: Their Role for Life

Sodium and potassium occur widely in the Earth's crust; they comprise 2.4 % and 2.6 %, respectively, of the mineral material. The content of sodium in seawater is 1.08 % (about 11 g/liter), whereas potassium is approximately 30 times less abundant with 0.04 %. Lithium, rubidium, and cesium, which belong to the same group in the periodic table, are rare, and the radioactive francium occurs only in traces; it is currently of little relevance. All these elements are metals with an ns^1 electron configuration (following a noble gas), and consequently, they occur in nature only as monocations. Indeed, Na^+ and K^+ represent the cationic electrolytes of living organisms. In human serum or plasma (extracellular fluid), the Na^+ concentration is close to 140 mM and that of K^+ amounts to about 4 mM, while in cellular fluid the concentrations are almost reversed, i.e., ca 10 mM for Na^+ and 150 mM for K^+. Potassium is accumulated within cells (in compartments) by the action of the sodium pump (Na^+/K^+-ATPase), which consumes adenosine 5′-triphosphate (ATP), that is, Na^+ is extruded from the cell during the uptake of K^+. The regulation of such metal ion flows is crucial for life. Such aspects are indicated in *Chapter 1*, where also the coordination chemistry of these cations is considered.

Chapter 2 provides an overview of the common methods for the determination of the alkali metal ions. They are drawn from the three principle branches of quantitative analysis and consist mainly of optical atomic spectrometric methods, ion-selective electrodes, and the separation methods of ion chromatography and capillary electrophoresis. Their main characteristics and performance parameters are discussed, and relevant specific applications are examined, namely, clinical and single cell analysis and the analysis of soil samples and nutrient solutions. Concerning the environment, important is the determination of the radioactive cesium isotopes ^{137}Cs and ^{134}Cs with half-lives of 30 and 2 years, which were released during the Chernobyl and Fukushima nuclear reactor disasters. This determination as well as, e.g., that of the naturally occurring ^{40}K, is carried out by highly sensitive γ-ray spectroscopy, which can detect extreme trace levels.

Solid-state structures of alkali metal ion complexes formed by low-molecular-weight ligands of biological relevance are summarized in *Chapter 3*. These ligands include amino acids and small peptides, nucleic acid constituents, simple carbohydrates, and naturally occurring antibiotic ionophores. Metal ion coordination and structural characteristics are described in detail. These insights are complemented in *Chapter 4*, which is devoted to the discriminating properties of alkali metal ions toward the constituents of proteins and nucleic acids based on gas-phase and theoretical studies. After considering the fundamentals of these techniques, the results for the complexes of alkali metal cations formed with amino acids, small peptides, and nucleobases are outlined, and periodic trends in how these interactions vary as the alkali cations get heavier are highlighted. In *Chapter 5* the situation in (aqueous) solution is reviewed, that is, complex formation between alkali metal ions and phosphates, nucleotides, amino acids, and related ligands of biological relevance. In general, the stability of alkali metal ion complexes of organic or inorganic ligands is rather low (usually $\log K < 2$) and depends on the charge of the ligand, owing to the (mainly) ionic nature of the interactions. However, the size of the cation is also important and influences the stability, which often follows the trend $Li^+ > Na^+ > K^+ > Rb^+ > Cs^+$.

Metal ions are essential cofactors for the structure and functions of nucleic acids; next to Mg^{2+} monovalent cations are of relevance as is emphasized in *Chapter 6*. For example, K^+ located in the core of a group I intron is important for folding and catalytic activity of the RNA. However, detection of K^+ and Na^+ is notoriously problematic, and the question about their specificity is recurrent. Therefore, the different methods that can be applied to detect K^+ and Na^+ ions in nucleic acid structures, namely, X-ray crystallography, nuclear magnetic resonance, or molecular dynamics simulations, are reviewed. The specific *versus* nonspecific binding of monovalent cations is also discussed. *Chapter 7* continues with G-quadruplexes that form with guanine-rich nucleic acids; these G-tetrads are stabilized by intra-quartet hydrogen bonds, inter-quartet stacking, and cation coordination. They are also detectable in human cells and are strongly suspected to be involved in important biological processes. The focus is on the essential and specific coordination of alkali metal cations, highlighting cation-dependent dissimilarities in stability, structure, formation, and interconversion of G-quadruplexes, considering also the kinetics of their formation.

Sodium and potassium ions also play important roles in the chemistry of proteins. They are not only nonspecific ionic buffering agents or mediators of solute exchange and transport. As pointed out in *Chapter 8*, molecular evolution and regulated high intracellular (K^+) and extracellular (Na^+) monovalent cation (M^+) concentrations led to incorporation of selective Na^+ and K^+ binding sites into enzymes to stabilize catalytic intermediates or to provide optimal positioning of substrates. The mechanism of M^+ activation, as derived from kinetic studies along with structural analysis, has led to the classification of cofactor-like (type I) or allosteric effector (type II) activated enzymes. In type I the mechanism substrate anchoring to the enzyme active site is mediated by M^+, often in tandem with a divalent cation like Mg^{2+}, Mn^{2+}, or

Zn^{2+}. In the allosteric type II mechanism, M^+ binding enhances enzyme activity through conformational transitions triggered upon binding to a distant site.

In dealing with plants, it is recalled in *Chapter 9* that Na^+ and K^+ have similar relative abundances in the Earth's crust, but display very different distributions in the biosphere. In all living organisms, K^+ is the major inorganic cation in the cytoplasm, where its concentration (ca. 0.1 M) is usually several times higher than that of Na^+. Accumulation of Na^+ at high concentrations in the cytoplasm results in deleterious effects on cell metabolism, e.g., on the photosynthetic activity of plants. Therefore, Na^+ is compartmentalized outside the cytoplasm. In fact, Na^+ is not an essential element for most plants, except for some halophytes. On the other hand, it can be beneficial by replacing K^+ as vacuolar osmoticum. In contrast, K^+ is an essential element; it is involved in electrical neutralization of inorganic and organic anions and macromolecules, pH homeostasis, control of membrane electrical potential, and the regulation of cell osmotic pressure. K^+ is involved in the activation of enzymes, protein synthesis, cell metabolism, and photosynthesis; it is taken up by roots from the soil solution and then distributed throughout the plant. Three families of ion channels and 3 families of transporters have been identified so far as contributing to K^+ and Na^+ transport across the plasmalemma and internal membranes.

Transport of Na^+ and K^+ ions across the cell membrane can be carried out by specialized pore-forming ion channel proteins, which discriminate with high fidelity between the two competing metal ions, as is described in *Chapter 10*. This striking ion selectivity of monovalent ion channels is astonishing in view of the close similarity between Na^+ and K^+: Both are spherical cations with the same charge, analogous chemical and physical properties, and relatively similar ionic radii. The monovalent ion channel selectivity filters (SFs), which dictate the selectivity of the channel, differ in oligomericity, composition, overall charge, pore size, and solvent accessibility. This diversity of SFs raises the intriguing questions addressed in this chapter: What factors govern the metal competition in these SFs? Which of these factors are exploited in achieving K^+ or Na^+ selectivity?

As pointed out in *Chapter 11*, Na^+ has an extraordinary role among the alkali cations in living cells since it serves to charge the "battery of life." To this end, sophisticated protein complexes in biological membranes convert chemical energy obtained from oxidation of NADH, or hydrolysis of ATP, into an electrochemical gradient of sodium ions. Cells use this so-called sodium-motive force stored in energy-converting membranes for important processes like uptake of nutrients, motility, or expulsion of toxic compounds. The Na^+ pumps act in concert with other enzymes embedded in the lipid membrane, and together they form the respiratory chain. The authors explain why Na^+ pumps are important model systems for the homologous, proton-translocating complexes and hope to convince the reader that studying the Na^+-translocating ATP synthase from the unimpressive bacterium *Ilyobacter tartaricus* has a big impact on our understanding of energy conversion by human ATP synthase. Of equal importance are the insights described in the very comprehensive *Chapter 12*, which is devoted to the transmembranal Na^+/H^+ antiporters. They transport sodium (or several other monovalent cations) in exchange for H^+ across lipid bilayers in all kingdoms of life. They are critical in pH homeostasis of the cytoplasm and/or organelles. The SLC9 gene family, which encodes Na^+/H^+ exchangers in many species from prokaryotes to eukaryotes, is a particularly notable example. Yet, the most extensively studied Na^+/H^+ antiporter is from *Escherichia coli*. A crystal structure determined at acidic pH has provided the first structural insights into the antiport mechanism and pH regulation, which is discussed in detail. It has enabled the structural modeling of three human plasmalemmal proteins that are members of the SLC9 family and which are involved in human pathophysiology. Remarkably, no crystal structure of any of the human Na^+/H^+ antiporters exists.

As a physiological phenomenon, acid secretion from the stomach was known already at least in the seventeenth century, yet its mechanism was elucidated only more recently as described in *Chapter 13*. Gastric H^+/K^+-ATPase in the parietal cells was found to be responsible for the final step of H^+ secretion in these cells. Lately, several Cl^--transporting proteins for gastric acid (hydrochloric acid) secretion have been found. As inhibitors of gastric acid secretion, histamine H_2 receptor antagonists (H_2 blockers) were developed in the 1970s. This discovery brought a great benefit allowing now to treat peptic ulcers with a drug. In the 1980s, proton pump inhibitors were developed, the target of which is gastric H^+/K^+-ATPase. In general, they exert more potent effects compared to H_2 blockers. Most recently, several K^+-competitive inhibitors of the ATPase are being developed.

Inspired by nature, scientists have created artificial synthetic transporting structures to mimic the natural systems. This is reviewed for Na^+ and K^+ ion channels in the very comprehensive *Chapter 14*. Amphipathic molecules spontaneously aggregate in water to form bilayers in which the polar groups are exposed to the aqueous medium, while the nonpolar chains self-organize by aggregating to each other to stay away from water. The insulating properties of such membranes are due to the formation of a hydrophobic bilayer with hydrophilic phosphate groups at its outer sides. Thus, lipophilic molecules can permeate the membrane freely, while charged or very hydrophilic molecules require the assistance of other membrane components to overcome the energetic costs needed for crossing the nonpolar region of the bilayer. Most of the large polar species (like oligosaccharides, polypeptides, or nucleic acids) cross into and out of the cell via endocytosis and exocytosis, respectively. For small hydrophilic molecules and ions, nature has created a series of systems (carriers and pores) to control the balance.

Chapter 15 deals in a sophisticated way with lithium salts in medicine, the pharmacology of Li^+, and a chemical perspective of its mechanism of action. It is described how lithium salts are used to treat mental illnesses, in particular bipolar disorders, but also other disease states. The authors highlight the application of chemical methodologies for the characterization of the cellular targets of lithium salts and their distribution in tissues. The concluding *Chapter 16* also dwells on medical problems; it deals with Na^+ and K^+ and their relation to Parkinson's disease (PD) and traumatic brain injury (TBI). Na^+ and K^+ are plentiful in biological systems, and they are vital, but still there is a limit in our understanding regarding their great ability to affect health and disease. The renin-angiotensin system, found at both peripheral and central levels, is the crucial regulator for physiological homeostasis. In particular, it displays direct or indirect interactions that are relevant in PD. Furthermore, some recent biochemical and clinical findings are discussed that help to describe the role of Na^+ and K^+ in the context of TBI, which is caused from a heavy striking of the head that strongly impacts ion flux in the affected tissue (brain) and damages cellular regulation systems.

Astrid Sigel
Helmut Sigel
Roland K.O. Sigel

Contents

Contributors to Volume 16

Numbers in parentheses indicate the pages on which the authors' contributions begin.

Fouad Razzaq Al Shiblawi Laboratory of Plant Biochemistry and Molecular Physiology, UMR BPMP CNRS/INRA/MontpellierSupAgro, University of Montpellier, INRA, Place Viala, F-34060 Montpellier cedex 1, France (291)

Manuel Amorín Department of Organic Chemistry, Singular Research Centre in Chemical Biology and Molecular Materials (CIQUS), University of Santiago de Compostela (USC), E-15782 Santiago de Compostela, Spain (485)

Katsuyuki Aoki Department of Environmental and Life Sciences, Toyohashi University of Technology, Tempaku-cho, Toyohashi 441-8580, Japan, ka003@edu.imc.tut.ac.jp (27)

Peter B. Armentrout Department of Chemistry, University of Utah, Salt Lake City, UT 84112, USA, armentrout@chem.utah.edu (103)

Pascal Auffinger Architecture et Réactivité de l'ARN, Université de Strasbourg, IBMC, CNRS, 15 rue René Descartes, F-67084 Strasbourg, France, p.auffinger@ibmc-cnrs.unistra.fr (167)

David G. Churchill Department of Chemistry, Korea Advanced Institute of Science and Technology (KAIST), 373-1 Guseong-dong, Yuseong-gu, Daejeon 305-701, Republic of Korea, dchurchill@kaist.ac.kr (1, 585)

Francesco Crea Dipartimento di Scienze Chimiche, Università di Messina, Viale F. Stagno d'Alcontres, 31, I-98166 Messina, Italy (133)

Luigi D'Ascenzo Architecture et Réactivité de l'ARN, Université de Strasbourg, IBMC, CNRS, 15 rue René Descartes, F-67084 Strasbourg, France, l.dascenzo@ibmc-cnrs.unistra.fr (167)

Concetta De Stefano Dipartimento di Scienze Chimiche, Università di Messina, Viale F. Stagno d'Alcontres, 31, I-98166 Messina, Italy (133)

Todor Dudev Faculty of Chemistry and Pharmacy, Sofia University, 1 James Bourchier Blvd., BG-1164 Sofia, Bulgaria, t.dudev@chem.uni-sofia.bg (325)

Eric Ennifar Architecture et Réactivité de l'ARN, Université de Strasbourg, IBMC, CNRS, 15 rue René Descartes, F-67084 Strasbourg, France, e.ennifar@ibmc-cnrs.unistra.fr (167)

Claudia Foti Dipartimento di Scienze Chimiche, Università di Messina, Viale F. Stagno d'Alcontres, 31, I-98166 Messina, Italy (133)

Günter Fritz Department of Neuropathology, University of Freiburg, Breisacherstrasse 64, D-79106 Freiburg, Germany, guenter.fritz@uniklinik-freiburg.de (349)

Alberto Fuertes Department of Organic Chemistry, Singular Research Centre in Chemical Biology and Molecular Materials (CIQUS), University of Santiago de Compostela (USC), E-15782 Santiago de Compostela, Spain (485)

Takuto Fujii Department of Pharmaceutical Physiology, Graduate School of Medicine and Pharmaceutical Sciences, University of Toyama, 2630 Sugitani, Toyama City, Toyama 930-0194, Japan (459)

Valérie Gabelica ARNA Laboratory, Université Bordeaux, IECB, 2, rue Robert Escarpit, F-33600 Pessac, France and ARNA Laboratory, INSERM, U869, F-33000 Bordeaux, France, valerie.gabelica@inserm.fr (203)

Jesse L. Goossens Department of Chemistry and Biochemistry, Loyola University Chicago, 1068 West Sheridan Road, Chicago, IL 60660, USA, jgoossens@luc.edu (557)

Juan R. Granja Department of Organic Chemistry, Singular Research Centre in Chemical Biology and Molecular Materials (CIQUS), University of Santiago de Compostela (USC), E-15782 Santiago de Compostela, Spain, juanr.granja@usc.es (485)

Yonghwang Ha Department of Chemistry, Korea Advanced Institute of Science and Technology (KAIST), 373-1 Guseong-dong, Yuseong-gu, Daejeon 305-701, Republic of Korea (585)

Peter C. Hauser Department of Chemistry, Analytical Sciences, University of Basel, Spitalstrasse 51, CH-4056 Basel, Switzerland, peter.hauser@unibas.ch (11)

Jeong A Jeong Department of Chemistry, Korea Advanced Institute of Science and Technology (KAIST), 373-1 Guseong-dong, Yuseong-gu, Daejeon 305-701, Republic of Korea (585)

Ning-Hai Hu Changchun Institute of Applied Chemistry, Chinese Academy of Sciences, Changchun 130022, China, hunh@ciac.ac.cn (27)

Youngsam Kim Department of Chemistry, Korea Advanced Institute of Science and Technology (KAIST), 373-1 Guseong-dong, Yuseong-gu, Daejeon 305-701, Republic of Korea (1, 585)

Meytal Landau Department of Biology, Technion-Israel Institute of Technology, Haifa, Israel, mlandau@technion.ac.il (391)

Gabriele Lando Dipartimento di Scienze Chimiche, Università di Messina, Viale F. Stagno d'Alcontres, 31, I-98166 Messina, Italy (133)

Eric Largy ARNA Laboratory, Université Bordeaux, IECB, 2, rue Robert Escarpit, F-33600 Pessac, France and ARNA Laboratory, INSERM, U869, F-33000 Bordeaux, France (203)

Brian D. Leverson Department of Chemistry and Biochemistry, Loyola University Chicago, 1068 West Sheridan Road, Chicago, IL 60660, USA, bleverson@luc.edu (557)

Carmay Lim Institute of Biomedical Sciences, Academia Sinica, Taipei 11529, Taiwan, and Department of Chemistry, National Tsing Hua University, Hsinchu 300, Taiwan, carmay@gate.sinica.edu.tw (325)

Jean-Louis Mergny ARNA Laboratory, Université Bordeaux, IECB, 2, rue Robert Escarpit, F-33600 Pessac, France and ARNA Laboratory, INSERM, U869, F-33000 Bordeaux, France, jean-louis.mergny@inserm.fr (203)

Demetrio Milea Dipartimento di Scienze Chimiche, Università di Messina, Viale F. Stagno d'Alcontres, 31, I-98166 Messina, Italy (133)

Duarte Mota de Freitas Department of Chemistry and Biochemistry, Loyola University Chicago, 1068 West Sheridan Road, Chicago, IL 60660, USA, dfreita@luc.edu (557)

Kazutaka Murayama Graduate School of Biomedical Engineering, Tohoku University, Aoba, Sendai 980-8575, Japan, kmura@bme.tohoku.ac.jp (27)

Thuy-Tien T. Nguyen Department of Chemistry, Korea Advanced Institute of Science and Technology (KAIST), 373-1 Guseong-dong, Yuseong-gu, Daejeon 305-701, Republic of Korea (1)

Manuel Nieves-Cordones Laboratory of Plant Biochemistry and Molecular Physiology, UMR BPMP CNRS/INRA/MontpellierSupAgro, University of Montpellier, INRA, Place Viala, F-34060 Montpellier cedex 1, France (291)

Etana Padan Alexander Silberman Institute of Life Sciences, Hebrew University of Jerusalem, 91904 Jerusalem, Israel, etana@vms.huji.ac.il (391)

Mary T. Rodgers Department of Chemistry, Wayne State University, Detroit, MI 48202, USA, mrodgers@chem.wayne.edu (103)

Nuria Rodríguez-Vázquez Department of Organic Chemistry, Singular Research Centre in Chemical Biology and Molecular Materials (CIQUS), University of Santiago de Compostela (USC), E-15782 Santiago de Compostela, Spain (485)

Hideki Sakai Department of Pharmaceutical Physiology, Graduate School of Medicine and Pharmaceutical Sciences, University of Toyama, 2630 Sugitani, Toyama City, Toyama 930-0194, Japan, sakaih@pha.u-toyama.ac.jp (459)

Silvio Sammartano Dipartimento di Scienze Chimiche, Università di Messina, Viale F. Stagno d'Alcontres, 31, I-98166 Messina, Italy, ssammartano@unime.it (133)

Joachim Schnabl Department of Chemistry B, University of Zürich, Winterthurerstrasse 190, CH-8057 Zürich, Switzerland, joachim.schnabl@chem.uzh.ch (259)

Hervé Sentenac Laboratory of Plant Biochemistry and Molecular Physiology, UMR BPMP CNRS/INRA/MontpellierSupAgro, University of Montpellier, INRA, Place Viala, F-34060 Montpellier cedex 1, France, sentenac@supagro.inra.fr (291)

Julia Steuber Institute of Microbiology, University of Hohenheim (Stuttgart), Garbenstrasse 30, D-70599 Stuttgart, Germany, julia.steuber@uni-hohenheim.de (349)

Noriaki Takeguchi Takeguchi Digestive Research, Toyama 930-0125, Japan, qqze72p9@tiara.ocn.ne.jp (459)

Milan Vašák Department of Chemistry B, University of Zürich, Winterthurerstrasse 190, CH-8057 Zürich, Switzerland, mvasak@bioc.uzh.ch (259)

Titles of Volumes 1–44 in the *Metal Ions in Biological Systems* Series

edited by the SIGELs
and published by Dekker/Taylor & Francis (1973–2005)

Volume 1:	**Simple Complexes**
Volume 2:	**Mixed-Ligand Complexes**
Volume 3:	**High Molecular Complexes**
Volume 4:	**Metal Ions as Probes**
Volume 5:	**Reactivity of Coordination Compounds**
Volume 6:	**Biological Action of Metal Ions**
Volume 7:	**Iron in Model and Natural Compounds**
Volume 8:	**Nucleotides and Derivatives: Their Ligating Ambivalency**
Volume 9:	**Amino Acids and Derivatives as Ambivalent Ligands**
Volume 10:	**Carcinogenicity and Metal Ions**
Volume 11:	**Metal Complexes as Anticancer Agents**
Volume 12:	**Properties of Copper**
Volume 13:	**Copper Proteins**
Volume 14:	**Inorganic Drugs in Deficiency and Disease**
Volume 15:	**Zinc and Its Role in Biology and Nutrition**
Volume 16:	**Methods Involving Metal Ions and Complexes in Clinical Chemistry**
Volume 17:	**Calcium and Its Role in Biology**
Volume 18:	**Circulation of Metals in the Environment**
Volume 19:	**Antibiotics and Their Complexes**
Volume 20:	**Concepts on Metal Ion Toxicity**
Volume 21:	**Applications of Nuclear Magnetic Resonance to Paramagnetic Species**
Volume 22:	**ENDOR, EPR, and Electron Spin Echo for Probing Coordination Spheres**
Volume 23:	**Nickel and Its Role in Biology**
Volume 24:	**Aluminum and Its Role in Biology**
Volume 25:	**Interrelations Among Metal Ions, Enzymes, and Gene Expression**
Volume 26:	**Compendium on Magnesium and Its Role in Biology, Nutrition, and Physiology**
Volume 27:	**Electron Transfer Reactions in Metalloproteins**

Contents of Volumes in the *Metal Ions in Life Sciences* Series

edited by the SIGELs

Volumes 1–4
published by John Wiley & Sons, Ltd., Chichester, UK (2006–2008)
<http://www.Wiley.com/go/mils>
<http://www.wiley.com/WileyCDA/Section/id-300350.html>

Volumes 5–9
by the Royal Society of Chemistry, Cambridge, UK (2009–2011)
<http://www.bioinorganic-chemistry.org/mils> <http://www.mils-WdG.com>

and from Volume 10 on
by Springer Science & Business Media BV, Dordrecht, The Netherlands (2012–2014) and by Springer International Publishing, Cham, Switzerland (since 2015)
<http://bioinorganic-chemistry.org.mils> <http://www.springer.com/series/8385>

Volume 1 Neurodegenerative Diseases and Metal Ions

Volume 2 Nickel and Its Surprising Impact in Nature

Volume 3 The Ubiquitous Roles of Cytochrome P450 Proteins

Volume 4 Biomineralization. From Nature to Application

Volume 5 Metallothioneins and Related Chelators

Volume 6 Metal-Carbon Bonds in Enzymes and Cofactors

Volume 7 Organometallics in Environment and Toxicology

Volume 8 Metal Ions in Toxicology: Effects, Interactions, Interdependencies

Volume 9 Structural and Catalytic Roles of Metal Ions in RNA

Volume 10 Interplay between Metal Ions and Nucleic Acids

Volume 11 Cadmium: From Toxicity to Essentiality

Volume 12 Metallomics and the Cell

Guest Editor: Lucia Banci

Volume 13 Interrelations between Essential Metal Ions and Human Diseases

Volume 14 The Metal-Driven Biogeochemistry of Gaseous Compounds in the Environment

Guest Editors: Peter M. H. Kroneck and Martha E. Sosa-Torres

Volume 15 Sustaining Life on Planet Earth: Metalloenzymes Mastering Dioxygen and Other Chewy Gases

Guest Editors: Peter M. H. Kroneck and Martha E. Sosa-Torres

Volume 16 The Alkali Metal Ions: Their Roles for Life (this book)

Astrid Sigel, Helmut Sigel, and Roland K.O. Sigel

Volume 17 Lead: Its Effects on Environment and Health (in preparation)

Comments and suggestions with regard to contents, topics, and the like for future volumes of the series are welcome.

Chapter 1
Bioinorganic Chemistry of the Alkali Metal Ions

Youngsam Kim, Thuy-Tien T. Nguyen, and David G. Churchill

Contents

Abstract The common Group 1 alkali metals are indeed ubiquitous on earth, in the oceans and in biological systems. In this introductory chapter, concepts involving aqueous chemistry and aspects of general coordination chemistry and oxygen atom donor chemistry are introduced. Also, there are nuclear isotopes of importance. A general discussion of Group 1 begins from the prevalence of the ions, and from a comparison of their ionic radii and ionization energies. While oxygen and water molecule binding have the most relevance to biology and in forming a detailed understanding between the elements, there is a wide range of basic chemistry that is potentially important, especially with respect to biological chelation and synthetic multi-dentate ligand design. The elements are widely distributed in life forms, in the terrestrial environment and in the oceans. The details about the workings in animal, as well as plant life are presented in this volume. Important biometallic aspects of human health and medicine are introduced as well. Seeing as the elements are widely present in biology, various particular endogenous molecules and enzymatic

Y. Kim • T.-T.T. Nguyen • D.G. Churchill (✉)
Department of Chemistry, Korea Advanced Institute of Science and Technology (KAIST), 373-1 Guseong-dong, Yuseong-gu, Daejeon 305-701, Republic of Korea
e-mail: dchurchill@kaist.ac.kr

A. Sigel, H. Sigel, and R.K.O. Sigel (eds.), *The Alkali Metal Ions: Their Role for Life*, Metal Ions in Life Sciences 16, DOI 10.1007/978-3-319-21756-7_1

systems can be studied. Sodium and potassium are by far the most important and central elements for consideration. Aspects of lithium, rubidium, cesium and francium chemistry are also included; they help in making important comparisons related to the coordination chemistry of Na^+ and K^+. Physical methods are also introduced.

Keywords Hydration sphere • Isotopes • Ionic radius • Lithium • Physical methods

Please cite as: *Met. Ions Life Sci.* 16 (2016) 1–10

1 Introduction

The alkali metals (Li^+, Na^+, K^+, Rb^+, Cs^+, Fr^+) are ubiquitous in nature and their biological relevance and importance is hard to ignore. These ns^1 ions are centrally important in aspects of life on earth and within biological systems [1–56], from the aspects of the salt water of the oceans, down to miniscule intracellular cellular compartments. In terms of elemental abundances, Li^+, Rb^+, and Cs^+ are far less abundant and important (Figures 1, 2, and 3). Fr, of course, is a radioactive trace element of little current relevance. The ions are commonly involved in osmotic systems, electrolyte balances, ion channels,

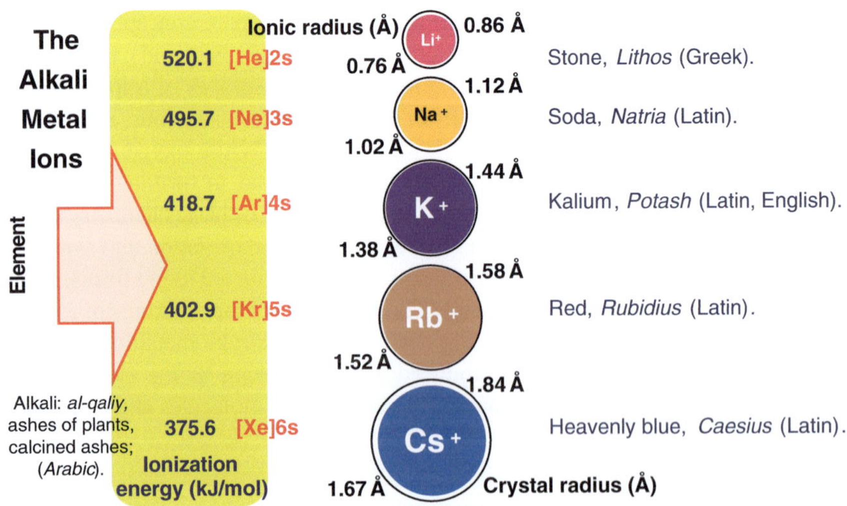

Figure 1 Electronic configurations, ionic radii, and ionization energy of 6-coordinate ions. Origin of the names of the elements. The color of the element is consistent with the color seen in the flame test [2, 11].

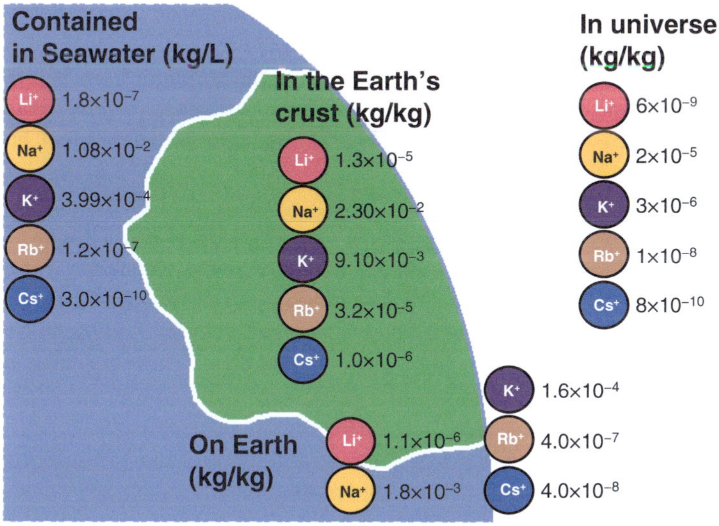

Figure 2 The occurrence of the elements on earth, in the oceans, and in the unverse. In all cases, francium is believed to be a trace element [6–10].

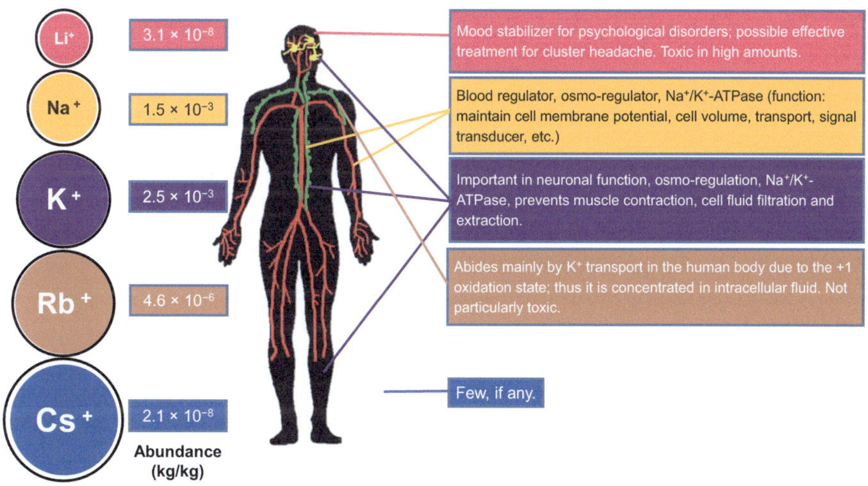

Figure 3 The abundance of alkali metals in the human body and their biological or medicinal roles [3].

and biological systems [1–5]. In the human body, sodium and potassium are by far the most abundant. These ions are often termed "spectator" ions.

This chapter is pertinent to ions in biological media and this inevitably means that fully-hydrated species should be considered. The polarization that the ionic species possess is also important. The inner hydration sphere, defined as the number of direct metal-oxygen-bound water molecule sites, is affected by cation size; as the cation radius increases (down the group) the number of inner sphere waters increases. The oxophilic nature allows for synthetic ionophore design [16]. Beyond the oxophilic character of the ions, they also have attraction to phosphates, and nitrogen-containing species. Predominant ionic, but also partly covalent bonding exists according to Fajan's rules [42].

Studies of natural and synthetic ligand designs for Group 1 metal ions continue to appear in the literature. These often relate to the biological implications of these metals. The basis for coordination chemistry in polydentate chelate design has often come from motifs also found in amino acids and nucleic acid bases (bioinspired) [2]. Furthermore, biologically there are sugar, vitamin, and hormone complexes that can be formed with the group I metal ions. There are also natural chelators and synthetic chelation motifs involving metal phosphate ligation as well. Ligation aspects are covered in various ways by authors of this volume: Chapter 3 (solid state), Chapter 4 (gas phase and theory) and the solution state (Chapter 5). Below, we will describe some points about the biological relevance of these elements. The basic coordination chemistry can be underscored to support that ionic interactions are predominant but are not the whole story.

2 Spectroscopic Techniques and Other Physical Methods

Various important physical techniques, including spectroscopy and electrochemistry, can be used analytically in the determination of amounts and speciation. Also, means of physical separation are available including ion chromatography and capillary electrophoresis. These techniques can be applied to issues in the chemical laboratory, in understanding biological systems, to issues in human health, within biological building blocks (living cells), and to environmental issues. Spectroscopy has commonly involved atomic absorption spectroscopy (AAS); this has been a powerful method to determine the presence and concentrations of certain elements in samples, especially trace elements in aqueous samples. This technique involves the thermal excitation of the ion through flame absorption spectrophotometry (flame photometry) whereby the species exhibits characteristic absorptions in nanometers

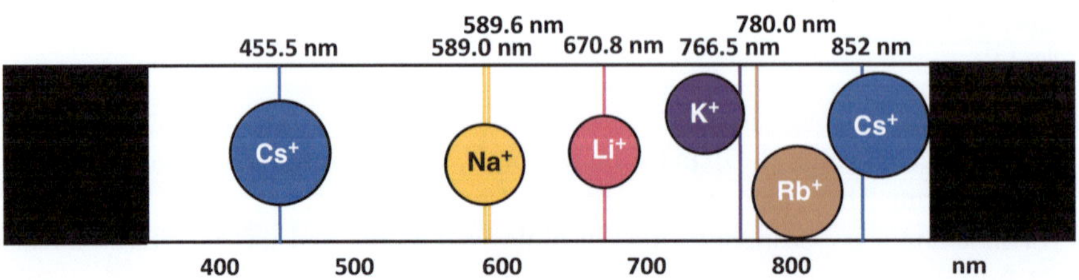

Figure 4 Select wavelengths and observed colors in atomic absorption spectroscopy analysis [15, 17].

[15, 17] (see also Chapter 2). Usually a principal absorption which gives rise to the observed flame color is monitored (Figure 4).

More recent efforts in chemosensing involve quantitative fluorescent methods which can also be used in the detection of metal ions in solution. Well-defined synthetic molecules containing an ionophore can selectively bind to a specific metal ion in solution. This technique is frequently used in biological studies for imaging common endogenous metal ions, such as Na^+ and K^+, within the workings of a single living cell [19, 20]. These cellular methods help probing spectroscopically silent metals [21]. Electrochemistry methods include the development and use of ion-selective electrodes (see also Chapter 2). These can also be used in targeting living cells and the concentrations within a single cell.

3 Lithium

Lithium exists in the lithosphere in various mineralogical forms and as a trace element in biology (Figures 1 and 2). Li^+ concentration is intimately connected to the physiological Na^+/K^+ balance. Thus, the element is important in clinical and pharmacological applications (Figure 3) [33, 34]. The lower bracket of Li^+ concentration in biological systems is not well defined; however, there is distinct Li^+ toxicity in human health often arising from prescription medication above a certain level; Li^+ has a narrow therapeutic index. The attention of lithium in science and society has also involved manufacturing of batteries [18, 22].

3.1 Introduction to the Coordination Chemistry of Li^+

In terms of its aqueous coordination chemistry, Li^+ bears an inner hydration sphere that is primarily tetrahedral. In solution, less ionic mobility is observed with Li^+; increasing ionic mobility is observed going down the group [23]. Li^+ is found "diagonally" (upper-left) to that of Mg^{2+} in the Periodic Table; such a "diagonal relationship" is sometimes helpful in understanding trends and tendencies and also building important relationships between metal ions and elements. When dissolved, Li^+ leads to a solution with very low vapor pressure; a markedly lowered melting point for water is also observed. Hydration sphere considerations can be extended to other solvent systems. The ion can crystallize with diethyl ether, or tetrahydrofuran. Sometimes nitrogen donors such as pyridine are present [1, 2].

In terms of solubility for Li^+, the carbonate and hydroxide are relatively insoluble (sparingly soluble) in water. The element's oxide chemistry contains important insights into the difference between it and heavier group members: upon heating, lithium hydroxide goes to an E_2O compound (Li_2O) analogous to water; MOH species of heavier congeners, however, undergo sublimation with no accompanying chemical change [1, 2]. The evaporation of LiOH gives the monohydrate compound which exists in a tetrahedral environment. Lithium carbonate is sparingly soluble and *anhydrous* as a solid, unlike LiOH. Lithium nitrate ($LiNO_3$) is hygroscopic, but can be prepared as an anhydrous crystalline sample as well.

3.2 *Recent Research Trends Regarding* Li^+

There is a wide range of applications for lithium [23–34]. Li^+ has been used reliably for decades to treat manic depressive disorder (see Chapter 15). This application relates to a large drug market. The common medicinal formulation, lithium carbonate ($LiCO_3$), exists commonly in nature. With the pharmacology as such, in recent years, lithium ion has been investigated as a possible treatment for Alzheimer's Disease and in other potentially important applications [23, 24, 34].

Current industrial applications involve extensive battery technology, grease production, metallic lithium alloys, and lithium carbonate glasses and porcelain manufacturing [25–32].

4 Sodium and Potassium

Na^+ and K^+ are grouped here; their chemical properties and abundances are relatively similar. The salts of both metals have various uses in biological as well as industrial processes. Sodium has an abundance of 1.08 % in oceans. It has only one stable isotope (^{23}Na). Sodium nitrate ($NaNO_3$), also known as Chile saltpeter or Peru saltpeter, is found in large deposits in the Atacama Desert of South America. Potassium (0.040 %) is about 30 times less abundant than sodium. Potassium has three naturally abundant isotopes ^{39}K, ^{40}K, and ^{41}K [13]. Particular foods such as bananas contain significant amounts of K^+.

The salts of Na^+ and K^+ were not differentiable until ca. 1807, at which point the method of electrolysis allowed for separation of the two species. Na^+ and K^+ are found in ionic salts in a wide variety of minerals such as halite, and zeolite for Na^+ and orthoclase, granite, and sylvite for K^+. The salts are commonly formed from evaporated portions of seawater in which the less soluble minerals deposit at the bottom. Potassium nitrate (KNO_3, ordinary saltpeter) can be used interchangeably for many purposes. Biologically, there is an interplay between the balance of Na^+ and K^+ and the balance between Mg^{2+} and Ca^{2+}.

4.1 *Introduction to the Coordination Chemistry of* Na^+ *and* K^+

Na^+ and K^+ have ionic radii that are small to medium considering the range found for the group I ions (Figure 1). The sodium chloride structure is used as a common model for discussions of unit cells [37]. Na^+ and K^+ are more capable of forming coordination complexes (multidentate binding) than the heavier alkali elements. Previous experimental and theoretical research suggests that the dominant coordination number for the primary hydration sphere for both K^+ and Na^+ is 6. A coordination number of 7 is also claimed for K^+ [14]. In a separate study, support for a primary coordination sphere of 4 for the metal ions has been made as well [16].

A natural selectivity for Na^+ or K^+ is manifested in biology. In spite of the similarity between Na^+ and K^+ (the Na^+ ion being smaller than K^+), only K^+ will pass along the potassium channels such as KcsA (found in *Streptomyces lividans*) for example, an apparatus in bacterial systems. This transport is due to effective solvation of a K^+ bound to eight peptidic oxygen donor atoms [17]. Artificial channels are discussed in Chapter 15. In terms of oxo chemistry, potassium superoxide (KO_2) is used extensively as a reagent in cell assays and as an analyte for molecular chemosensing.

The relatively small radii of Na^+ and K^+ allow for direct size fitting into ligand hosts. Coordination complex formation with crown ethers and other macrocyclic systems are well known. Primarily, electrostatic interactions exist between the monoatomic ion and the ligand donor atoms [4]. Host–guest chemistry is exemplified through various examples of synthetic supramolecular binding of Na^+ and K^+. The story of synthetic macrocyclic chemistry is fascinating and led to the Nobel Prize in 1987 [43]. Since host–guest chemistry may also include non-metallic cations, it is interesting to note that the K^+ ion has been stated to be similar to the ammonium ion (NH_4^+) [44]. Regarding natural systems, the selectivity in ion channels (K^+ *versus* Na^+) is discussed in Chapter 10.

The ions are also used ubiquitously in industry and in the household. Na^+ and K^+ are involved in processes involving curing meats [45], serve as counter ions in surfactants [46], in bleach, and in innumerable formulations in industry, [47–49, 55, 56], in salt baths, in KI which is used orally as a protectant for radioactive fallout [50], and in salt baths that are used for heat transfer. Azo-dye production also features sodium ion [51].

4.2 *Recent Research Trends Regarding Na^+ and K^+*

Contemporary research in the life sciences continues to focus on intricate biological pumps, enzymes and related aspects of cellular biology. The most important topics involving Na^+ and K^+ are covered in individual chapters in this volume.

5 Rubidium and Cesium

Rubidium is of more modest abundance (Figure 2) and does not feature prominently in biology. Rubidium has an abundance in seawater of 1.2×10^{-7} kg L^{-1}. It is likely that its primary hydration sphere consists of 6 coordinated water molecules. Rubidium appears with a deep red signal in the flame test (AAS).

Cesium is a relatively rare element with its concentration in seawater being approximately 3.0×10^{-6} kg L^{-1} (Figure 2). It has a golden appearance in the flame test (AAS). Regarding the Cs^+ ion, found in discussions regarding the cesium chloride unit cell with a coordination number of 8 (for both cesium and chloride), it is likely that the primary coordination sphere under aqueous conditions consists of 6 water molecules. It is the most polarizable of the congeners (excluding Fr^+) and is least disruptive to the aqueous environment. Also, it has the lowest overall *total* hydration number and the greatest ionic mobility in the series. There are some notable deposits in the World such as in Bernic lake (Tanco mine) in Canada. Radioactive forms of cesium (^{137}Cs, half-life=~30 years) is a main constituent of fallout (or nuclear fission) [52, 53]. Lastly, the element has found practical use in atomic clocks (stable isotope ^{133}Cs).

6 Francium

There is a dearth of studies relating to francium in biology because of its short radioisotopic lifetime (half-life $^{223}_{87}\text{Fr} = 21\,\text{min}$). Thus, francium is only of trace abundance. It is found in mineral deposits as ^{223}Fr alongside its progenitor isotopes uranium (^{235}U) and thorium (^{231}Th) (see the ^{235}U radioactive decay chain) [54]. As such, it is produced, but is then quickly lost to the formation of astatine (^{219}At), radium (^{223}Ra), and radon (^{219}Rn). While much less has been explored about it compared to the lighter congeners, it stands as an interesting frontier for future endeavors in the biomedical sciences.

7 Conclusions, Outlook, and Further Considerations for Future Studies

The alkali earth metals are a diverse group of elements that are essential for biology and which are found widely in mineral and electrolyte forms in nature – in the oceans, the earth's crust and in the universe (Figure 2). In particular, Na^+ and K^+ species are vital and important ions for life and the ecosystem.

The nuances in size, polarizability, and coordination chemistry among the elements is important to consider. Furthermore, additional donor atom chemistry such as nitrogen chemistry, fluoride chemistry, and phosphate chemistry, while not elaborated upon here, give an additional and fuller perspective of Group 1 ion chemistry in biology and in nature.

There are other singly-charged monoatomic, and even polyatomic species and ions that also behave like alkali metals; these surrogates can and should be studied in their relationship to biology.

A complete treatment of the differences between the Group 1 ions must involve a detailed analysis and expanded discussion of concepts that include ionic strength, ionic mobility, hydration energy, hydration number, hydrated radii, and crystal radii.

Further scientific instrumentation and methods may well focus on emergent spectroscopic techniques that can be more completely exploited, or derived, as well as to better explore issues in the frontiers of the chemical sciences and biomedical sciences. In particular, a thorough understanding of biological compartmentalization and transport of metal ions, it is hoped, can be much better elucidated in years to come.

Acknowledgment Prof. D. G. Churchill (D.G.C.) and Mr. Youngsam Kim acknowledge support from the NRF (National Research Foundation) of Korea (Grant 2011–0017280) for the operation of the Molecular Logic Gate Laboratory. D.G.C. acknowledges support from the Institute of Basic Science (IBS) of Korea.

References

1. For a general reference to alkali metal chemistry, refer to: N. N. Greenwood, A. Earnshaw, *Chemistry of the Elements*, Pergamon Press, Oxford, 1984, pp. 75–116.
2. For a general reference to alkali metal chemistry, refer to this source and references found therein: F. A. Cotton, G. Wilkinson, C. A. Murillo, M. Bochmann, *Advanced Inorganic Chemistry*, 6th edn., J. Wiley & Sons, New York, 1999, pp. 92–110.
3. E. J. Verspohl, *Lithium in Biology and Medicine*, Eds. G. N. Schrauzer, K. Klippel, VCH Verlagsges. mbH, Weinheim, **1991**, p. 209.
4. C. R. Fresenius, C. Remigius, *Manual Qualitative Chemical Analysis*, J. Wiley & Sons, New York, **1897**, p. 430.
5. R. H. Petrucci, F. G. Herring, J. D. Madura, C. Bissonnette, *General Chemistry, Principles & Modern Applications*, 9th edn., Macmillan Publishing Company, Toronto, **2007**, p. 877.
6. W. F. McDonough, *Compositional Model for the Earth's Core*, in *The Mantle and Core*, Elsevier Ltd., Oxford, **2005**, p. 554.
7. S. R. Taylor, S. M. McLennan, *The Continental Crust: Its Composition and Evolution*, Blackwell Sci. Publ., Oxford, **1985**, p. 330.
8. K. K. Turekian, *McGraw-Hill Encyclopedia of Science and Technology*, **1970**, *4*, p. 627.
9. Wolfram Research Inc., Mathematica, Version 10.0, http://reference.wolfram.com/language/ref/ElementData.html (accessed March 23rd, 2015).
10. C. K. Jorgensen, *Comments Astrophys.* **1993**, *17*, 49–101.
11. National Agricultural Library Digital Collections. http://handle.nal.usda.gov/10113/46493 (accessed Feb 2, 2015).
12. R. R. Crichton, R. Ward, *Metal-Based Neurodegeneration: From Molecular Mechanisms to Therapeutic Strategies*, John Wiley & Sons, Chichester, **2006**, pp. 24–30.
13. S. M. Blair, J. S. Brodbelt, A. P. Marchand, H.-S. Chong, S. Alihodzic, *J. Am. Soc. Mass Spectrom.* **2000**, *11*, 884–891.
14. C. M. Choi, J. Heo, N. J. Kim, *Chemistry Central Journal* [Online]. Published online June 1, **2012**. DOI: 10.1186/1752-153X-6-84 (accessed Feb. 2, 2015).
15. S. Svanberg, *Atomic and Molecular Spectroscopy: Basic Aspects and Practical Applications*, 4th edn., Springer-Verlag, Berlin, **2004**, p. 153.
16. A. E. H. Wheatley, *Chem. Soc. Rev.* **2001**, *30*, 265–273.
17. A. C. Menzies, *Anal. Chem.* **1960**, *32*, 898–904.
18. C. P. Grey, N. Dupre, *Chem. Rev.* **2004**, *104*, 4493–4512.
19. J. F. Birmingham, W. H. Wood, *J. Chem. Educ.* **1936**, *13*, 240–241.
20. M. H. Keefe, K. D. Benkstein, J. T. Hupp, *Chem. Rev.* **2000**, *205*, 201–228.
21. J. E. Penner-Hahn, *Coord. Chem. Rev.* **2005**, *249*, 161–177.

22. Global Strategic Metals Ltd. http://www.globalstrategicmetalsnl.com/_content/documents/405.pdf (accessed Jan. 30, 2015).
23. R. G. Keil, D. W. Johnson, M. A. Fryling, J. F. O'Brien, *Inorg. Chem.* **1989**, *28*, 2764–2766.
24. C. E. Kovacsics, I. I. Gottesman, T. D. Gould, *Annu. Rev. Pharmacol. Toxicol.* **2009**, *49*, 175–198.
25. R. J. Baldessarini, L. Tondo, P. Davis, M. Pompili, F. K. Goodwin, J. Hennen, *Bipolar Disord.* **2006,** *8*, 625–639.
26. Y. Huang, H. Liu, Y.-C. Lu, Y. Hou, Q. Li, *J. Power Sources* **2015**, *284*, 236–244.
27. S. Neuhold, D. J. Schroeder, J. T. Vaughey, *J. Power Sources* **2014**, *206*, 295–300.
28. T. M. Benedetti, E. Redston, W. G. Menezes, D. M. Reis, J. F. Soares, A. J.G. Zarbin, R. M. Torresi, *J. Power Sources* **2013**, *224*, 236–244.
29. A. R. Tuncdemir, E. Dilber, H. B. Kara, A. N. Ozturk, *Mater. Sci. Appl.* **2012**, *3*, 294–300.
30. K. Shikano, A. Mori, M. Shimizu, T. Ohtsuki, H. Yuki, K. Masumoto, *J. Radioanal. Nucl. Chem.* **2005**, *266*, 211–216.
31. A. Mohamed, A. A. Khattab, T. A. S. Osman, M. Zaki, *J. Nanotechnol.* **2013**, *2013*, Article ID 279090, 4 pages.
32. A. Mohamed, T. A. Osman, A. Khattab, M. Zaki, *J. Tribol.* **2015**, *137*, 011801–011805.
33. J. M. Tarascon, M. Armand, *Nature* **2001**, *414*, 359–367.
34. R. J. Baldessarini, L. Tondo, P. Davis, M. Pompili, F. K. Goodwin, J. Hennen, *Bipolar Disord.* **2007,** *9*, 314.
35. Y. Iwata, Y. Inoue, M. Minowa, *Jpn. J. Appl. Phys.* **2009**, *48*, 076505–076512.
36. C. N. Rowley, B. Roux, *J. Chem. Theory Comput.* **2012**, *8*, 3526–3535.
37. Bodner Research Web. Unit cells. http://chemed.chem.purdue.edu/genchem/topicreview/bp/ch13/unitcell.php#nacl (accessed Jan. 30, 2015).
38. T. Jun, O. G. Tsay, D. G. Churchill, in, *Radionuclides in the Environment*, Ed D. A. Atwood, John Wiley & Sons, Chichester, UK, 2011, pp. 65–72.
39. I. H. Shrivastava, D. P. Tieleman, P. C. Biggin, M. S. P. Sansom, *Biophys. J.* **2002**, *83*, 633–645.
40. Rensselaer Polytechnic Institute's Molecular Biochemistry I. K^+ channel. https://www.rpi.edu/dept/bcbp/molbiochem/MBWeb/mb1/part2/channel.htm#Selectivity (accessed Jan. 30, 2015).
41. J. D. Bradley, G. C. Gerrans, *J. Chem. Educ.* **1973**, *50*, 463–464.
42. K. Fajans, *Naturwiss.* **1923**, *11*, 165–72.
43. R. M. Izatt, *Chem. Soc. Rev.* **2007**, *36*, 143–147.
44. H. Amlal, M. Soleimani, *Biochim. Biophys. Acta, Biomembr.* **1997**, *1323*, 319–333.
45. D. C. Paik, D. V. Saborio, R. Oropeza, H. P. Freeman, *Int. J. Epidemiol.* **2001**, *30*, 181–182.
46. W. Y. Hwang, J. S. Shih, *J. Chin. Chem. Soc.*, **2000**, *47*, 1215–1222.
47. *Applications in Industry*, in *Adsorption and Its Application in Industry and Environmental Protection*, Vol. I, Ed. A. Dabrowski, Elsevier Science B. V., Amsterdam, **1999**, pp. 308–315.
48. N. Ediz, A. Yurdakul, *J. Ceram. Process Res.* **2009**, *10*, 758–769.
49. M. Sánchez, M. Navas, J. F. Ruggera, M. L. Casella, J. Aracil, M. Martínez, *Energy* **2014**, *73*, 661–669.
50. P. B. Zanzonico, D. V. Becker, *Health Phys.* **2000**, *78*, 660–667.
51. A. Reife, H. S. Freeman, *Environmental Chemistry of Dyes and Pigments*, John Wiley & Sons, New York, **1996**, pp. 273–274.
52. M. Sakai, T. Gomi, R. S. Naito, J. N. Negishi, M. Sasaki, H. Toda, M. Nunokawa, K. Murase, *J. Environ. Radioact.* **2015**, *144*, 15–20.
53. K. Oshita, H. Aoki, S. Fukutani, K. Shiota, T. Fujimori, M. Takaoka, *J. Environ. Radioact.* **2015**, *143*, 1–6.
54. S. S. Zumdahl, S. A. Zumdahl, *Chemistry: Media Enhanced Edition*, Houghton Mifflin, Boston, 2009, p. 327.
55. O. Çopuroğlu, A. L. A. Fraaij, J. M. J. M. Bijen, *Cem. Concr. Res.* **2006**, *36*, 1475–1482.
56. R. L. Frost, A. López, L. Wang, A. W. Romano, R. Scholz, *Spectrochim. Acta, Part A* **2015**, *137*, 70–74.

Chapter 2
Determination of Alkali Ions in Biological and Environmental Samples

Peter C. Hauser

Contents

Abstract An overview of the common methods for the determination of the alkali metals is given. These are drawn from all of the three principle branches of quantitative analysis and consist mainly of optical atomic spectrometric methods, ion-selective electrodes, and the separation methods of ion-chromatography and capillary electrophoresis. Their main characteristics and performance parameters are discussed. Important specific applications are also examined, namely clinical analysis, single cell analysis, the analysis of soil samples and hydroponic nutrient solutions, as well as the detection of the radioactive ^{137}Cs isotope.

Keywords Atomic spectroscopy • Clinical analysis • Ion-selective electrodes • Radioactive cesium • Single cell analysis • Soil analysis

Please cite as: *Met. Ions Life Sci.* 16 (2016) 11–25

P.C. Hauser (✉)
Department of Chemistry, Analytical Sciences, University of Basel,
Spitalstrasse 51, CH-4056 Basel, Switzerland
e-mail: peter.hauser@unibas.ch

A. Sigel, H. Sigel, and R.K.O. Sigel (eds.), *The Alkali Metal Ions: Their Role for Life*,
Metal Ions in Life Sciences 16, DOI 10.1007/978-3-319-21756-7_2

1 Introduction

For the quantitative analysis of the alkali metal ions practically no suitable classical gravimetric or titrimetric procedures are available due to a lack of selectivity. On the other hand, these elements are relatively easy to determine by instrumental methods. In contrast to their common ionic partners, i.e., the inorganic anions, a variety of methods exists, which include techniques from all of the three main groups of quantitative instrumental methods, i.e., spectroscopic and electrochemical methods as well as the separation techniques of ion chromatography and capillary electrophoresis. Usually the requirements on sensitivity are not very demanding, as the concentrations of sodium and potassium are generally high, and the analytical challenge may lie in achieving adequate precision. Note, that if concentrations below about 10^{-6} M must be determined, special measures are required to avoid contamination due to the ubiquity of the species, especially of sodium. Standard laboratory glassware has to be replaced by containers and pipettes made from plastic and the regular deionized laboratory water supply might not be adequate.

Sodium and potassium are routinely determined in clinical analysis. Together with other blood electrolytes these parameters are a well established diagnostic aid. The determination of sodium is of relevance in the food industry as it relates to table salt. Potassium as one of the essential growth factors and fertilizer components (the K in NPK fertilizers) needs to be frequently determined in soil samples when establishing their suitability for growing crops. In the life sciences it is sometimes desired to carry out the analysis in very small volumes, even inside single cells, and methods have been developed for these tasks. A special analytical endeavor is the determination of radioactive cesium as the main environmental contaminant from the Chernobyl and Fukushima disasters.

2 Spectrophotometry

Most metal ions can be determined by spectrophotometry (molecular absorption spectrometry in the ultraviolet and visible wavelength region) via the formation of colored complexes using chromogenic ligands. However, this is not true for alkali metals as ligands are generally not available. The situation improved somewhat with the introduction of the crown ethers in the 1960s. The most important examples are shown in Figure 1 [1].

12-crown-4 is well suitable for complexation of the lithium or sodium cation, while 15-crown-5 may preferentially be employed for sodium, and 18-crown-6 for potassium. As these ligands are not perfectly selective for just a single metal ion, 18-crown-6 as well as larger crown ethers may also be employed for complexation of rubidium and cesium. The respective sizes of metal ion and crown ether play a role in the selectivity but the crown ethers are not rigid molecules and complexation of a single ion by two

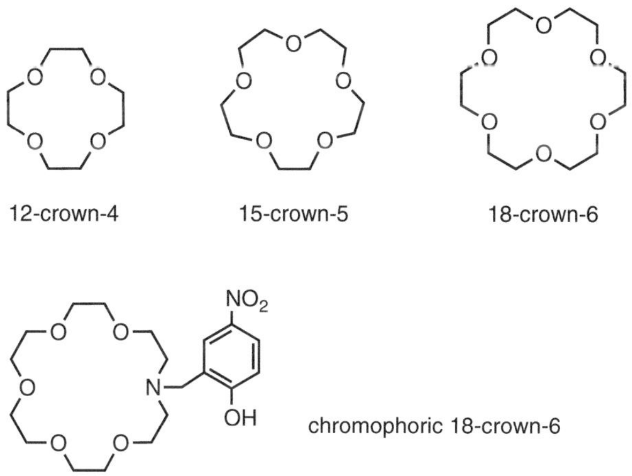

Figure 1 Structures of the three most common basic crown ethers which may be employed for solvent extraction of the alkali ions. The chromophoric crown ether looses a proton on complexation and thus changes color and is extracted as a zwitter-ionic complex [1].

rings is also possible. A large variety of derivatives of these basic crown ethers have been synthesized in order to alter their solubility properties or selectivity and some of these are available commercially. While these complexing agents do not directly induce a color for quantification by molecular absorption spectrometry, this can be achieved by using ion-pair extraction with an anionic dye into an organic solvent as described for example for sodium with 12-crown-4 and picrate into dichloromethane [2]. The selectivity of the method showed a preference for sodium, but Li^+, as well as the larger alkali ions K^+, Rb^+, and Cs^+ were also extracted to some extent.

The determination of potassium by extraction with dibenzo-18-crown-6 and bromothymol blue into chloroform was also described [3]. This was also not perfectly selective, as in particular Rb^+ and NH_4^+ interfered strongly, as did to a lesser extent sodium and cesium, as well as alkaline earth and transition metal ions. While Rb^+ is in practice not a problem, because it is normally not present in samples at high concentrations, this is not true for ammonium which generally cannot be ignored when employing crown ethers. The authors removed the interference by turning the sample solution alkaline with lithium hydroxide and boiling off the ammonium as ammonia.

The use of derivatives of crown-ethers which incorporate a chromophore or fluorophore, and therefore directly induce a color intensity change on complexation of an alkali ion as an analyte, has also been investigated intensively. An early example of such a 18-crown-6 derivative [1] reported is also shown in Figure 1. The phenyl group of this molecule is deprotonated on complexation of one of the alkali ions leading to a pronounced color change. The design of similar probe molecules, not just for alkali ions, has been a popular subject and discussions of more recent developments can be found in Valeur and Leray [4] and Callan et al. [5].

3 Atomic Spectroscopy

All alkali metals are readily determined by optical atomic spectroscopy. The narrowness of the atomic emission and absorption lines assure high selectivity and a lack of mutual interference of the alkali metals in these methods. The intense resonance lines, corresponding to the transition between the ground state electronic configuration and the first excited state, are all in the visible or near infrared range (Li: 670.8 nm; Na: 589.0 nm; K: 766.5 nm, Rb: 780.0 nm; Cs: 852.1 nm) and thus stimulation of light emission requires relatively little energy. Indeed, the flame spectroscope was instrumental in the discovery and differentiation of the alkali metals. Flame emission spectrometry (FES) is still a current method, in particular for the determination of Li, Na and K (and Ca as well). The instruments are relatively simple and inexpensive, employing a propane or butane flame, and optical filters for wavelength selection. A schematic representation of the fundamental arrangement of optical atomic spectrometers is given in Figure 2.

Samples are usually introduced in liquid form into a nebulizer. These make use of the Venturi effect created by the flow of fuel gas to aspirate the sample through a small tubing which on contact with the gas stream breaks up into an aerosol. The larger droplets are removed by impact on obstacles in the so-called aerosol chamber before arriving at the flame on the burner head. In the flame, first the solvent is evaporated, then the solutes are vaporized and atomized so that the metals end up as free and neutral atoms, which are thermally excited, leading to the emission of light.

As can be seen in Table 1 [6], limits of detection (LODs) of around 1 μM for the three alkali metals are achievable, and the working ranges cover about 3 orders of magnitude. As for other metals, the temperatures of flames are often not high enough to achieve the prerequisite for atomic emissions spectrometry, i.e.,

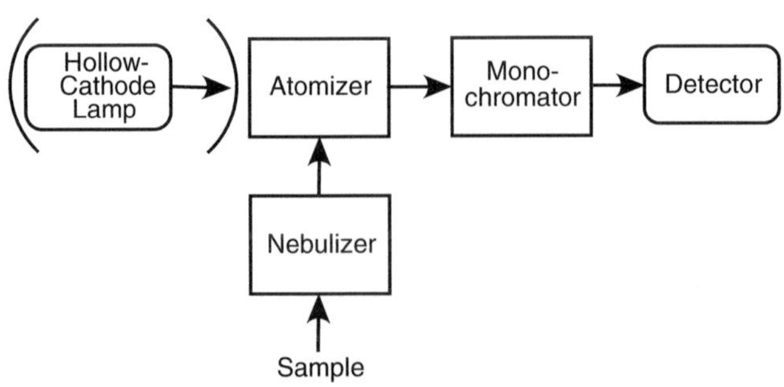

Figure 2 Block diagram of optical atomic spectrometers for emission and absorption measurements with sample introduction as aerosol. The hollow cathode lamp is only required for absorption measurement.

Table 1 Typical limits of detection (in nM) for the alkali metals achieved with the atomic spectroscopy methods [6].

Analyte	Flame Photometry	Flame-AAS	Furnace-AAS	ICP-OES	ICP-MS
Li	1000	71	7	130	14
Na	40	8.7	2.2	170	2.6
K	300	50	0.5	1280	256
Rb	3500	23	0.6	–	0.2
Cs	3800	60	0.4	–	0.2

adequate populations in the excited state. Nowadays, atomic absorption spectrometers (AAS) or inductively coupled plasma spectrometers (ICP) are more commonly found in the laboratory, as they allow the quantification of almost all metals. Therefore, often these instruments will also be employed for the determination of the alkali metals.

Atomic absorption requires neutral atoms in the ground state which are usually obtained in an acetylene flame. The sample is introduced again as an aqueous areosol produced from the sample in a similar fashion as in flame emission spectrometry. The fact that potassium, rubidium, and cesium are relatively easily ionized has to be considered in their determination by atomic absorption spectrometry as this leads to a loss of sensitivity and undesired non-linear calibration curves. Commonly an ionization suppressor is added in a relatively high concentration to the sample in the form of another easily ionizable alkali ion (e.g., Cs in the determination of K) which floods the flame with free electrons and thus reduces the ionization of the analyte. The use of propane or butane instead of acetylene in order to obtain a cooler flame (about 1900 °C instead of 2200 °C), and thus reduce ionization, is also of benefit. Alternatively, a graphite furnace atomizer may be employed, which has the advantage of requiring smaller samples volumes. Both types of instruments are more complex than the flame emission spectrometers as for the absorption measurements a monochromatic light source in form of a hollow cathode lamp is required. These lamps contain the same metal as the one to be determined in order to produce the correct lines, and therefore usually has to be changed when switching to a different analyte.

Dual element lamps containing Na and K are available, so that when changing between those analytes, which are frequently determined in the same samples, it is not necessary to change the lamp. The detection limits for the determination of the alkali metals by AAS are also given in Table 1. Note that for the furnace atomizer significantly lower detection limits are obtained. An important consideration is the limited dynamic range for absorption measurements which covers only about one order of magnitude. The reason for this is related to the fact that the absorbance parameter can only be obtained indirectly via a comparison of the intensities of the light beam before and after passage through the flame. Dynamic ranges at various higher levels as may be required for different analytical tasks can be obtained by employing different wavelengths.

The plasma spectrometers based on the ICP work with an argon plasma into which the sample is also normally introduced as aqueous aerosol. The ICP reaches temperatures of up to about 10’000 °C and thus all analytes are actually present as ions in the atomizer. Consequently, in contrast to flame emission spectrometers it is the emission from the ions, rather than the neutral atoms, which is employed in ICP-OES (OES = optical emission spectroscopy). Alternatively also ICP-MS (MS = mass spectrometry) is frequently employed. In comparison to atomic absorption spectroscopy both variants have the advantage of multi-analyte capability, and the mass spectrometric detection has generally much better limits of detection than the optical method. Typical values obtained for the alkali metals with both variants of ICP spectrometry are also given in Table 1. An advantage of ICP-MS is the fact that it allows the quantification of the isotopes. Both versions have also very wide dynamic ranges which cover many orders of magnitude. It has to be borne in mind, however, that the ICP-instruments are expensive, and as in the determination of the alkali metals the achievement of ultimate detection limits is not often required, the application of these instruments for the sole determination of these elements is usually not warranted.

4 Ion-Selective Electrodes

From the range of electrochemical methods it is potentiometry which provides an elegant solution for the determination of all of the alkali metal ions. The potentiometric determination of the ions is based on ion-selective electrodes (ISE) which employ a membrane which selectively takes up ions from the sample solution and thereby acquires a charge and hence a potential which is dependent on the activity of the ion in the solution. In order to create a defined interface between the backside of the membrane and the wire that leads to the electrometer, the ion-selective electrodes contain a filling solution. This contains the ion which is sensed by the membrane as well as chloride to create an internal silver/silver chloride electrode system of the second kind. A reference electrode is required to complete the measuring cell. The response behavior of an ion-selective electrode can be described by the Nicolsky equation, an extended form of the Nernst equation, which takes into account the fact that these electrodes are not perfectly selective. It is shown here in a simplified form:

$$E = E^0 + s \cdot \log\left(a_x + \sum K \cdot a_y\right) \tag{1}$$

The membrane potential, E, is logarithmically dependent on the activity, a_x, not the concentration, of the primary ion the membrane is designed for, as well as the sum of the products of the activity of any interfering ions, a_y, with their specific selectivity coefficient *K*. The activity is related to the concentration, but is also dependent on the ionic strength (which is a measure of the total concentration of all ions in the solution). The distinction is negligible if the total ion concentration is less than about 10^{-3} M, but becomes more significant for higher ion concentrations, and the activity of an ion is only about 80 % of the concentration for total ion concentrations of 0.1 M. In an evaluation, if the selectivity of the membrane is sufficient, the activity of the primary ion can be compared directly with the sum of the products of the activities of the interfering ion with their selectivity coefficients to obtain the relative error.

Current ion-selective electrodes for the alkali ions are of the polymeric membrane type. The membrane matrix typically consists of one third (by mass) of poly(vinyl chloride) (PVC) and two thirds of a plasticizer (such as bis(2-ethylhexyl) sebacate). The latter is essentially a water immiscible non-volatile organic solvent and allows dissolution of organic materials in the membrane. The selectivity of the membranes for the different alkali ions originates from an ionophore added in a small amount (approximately 1 % by weight) to the membrane. The potassium ion-selective electrode was one of the first polymeric membrane electrodes [7] and is based on valinomycin, a natural compound obtained from bacteria which is responsible for the selective transport of potassium through cell walls. Its macrocyclic structure is given in Figure 3. As can be seen from Table 2 this electrode has a high selectivity for potassium over Li^+ and Na^+ as well as other ions not shown in Table 2 (the more negative the log K value, the lower the interference by the species). Its response to Rb^+ is actually more pronounced than to K^+, but this is usually not a problem, as this potential interferent is not commonly present.

Ion-selective electrodes for the other alkali ions were developed subsequently. Crown-ethers may be employed as artificial ionophores for these ions, but higher selectivities are often obtained with other compounds, which notably may not be macrocyclic. Some structures and selectivity coefficients are also given in Figure 3 and Table 2 [8–10], respectively. Limits of detection of about 1 μM can be obtained with these ion-selective electrodes in the absence of significant interferences.

Based on these ionophores optical chemical sensors have also been developed. These generally employ plastiziced PVC membranes, which besides the ionophore contain a lipophilic pH-indicator. If the sample is buffered in pH at the appropriate value, the analyte cation, on being taken up into the membrane, will displace a proton from the membrane in order to maintain charge balance and hence lead to a deprotonation and a color change of the membrane. However, this and other types of optical chemical sensors have not found as wide use as the potentiometric sensors in form of ion-selective electrodes.

5 Ion Chromatography and Capillary Electrophoresis

The separation techniques of ion chromatography and capillary electrophoresis may also be employed for the quantification of the alkali metal ions. These methods are particularly useful when the concurrent determination of more than one of the species is required. There are several variants of ion chromatography, but the most common one is

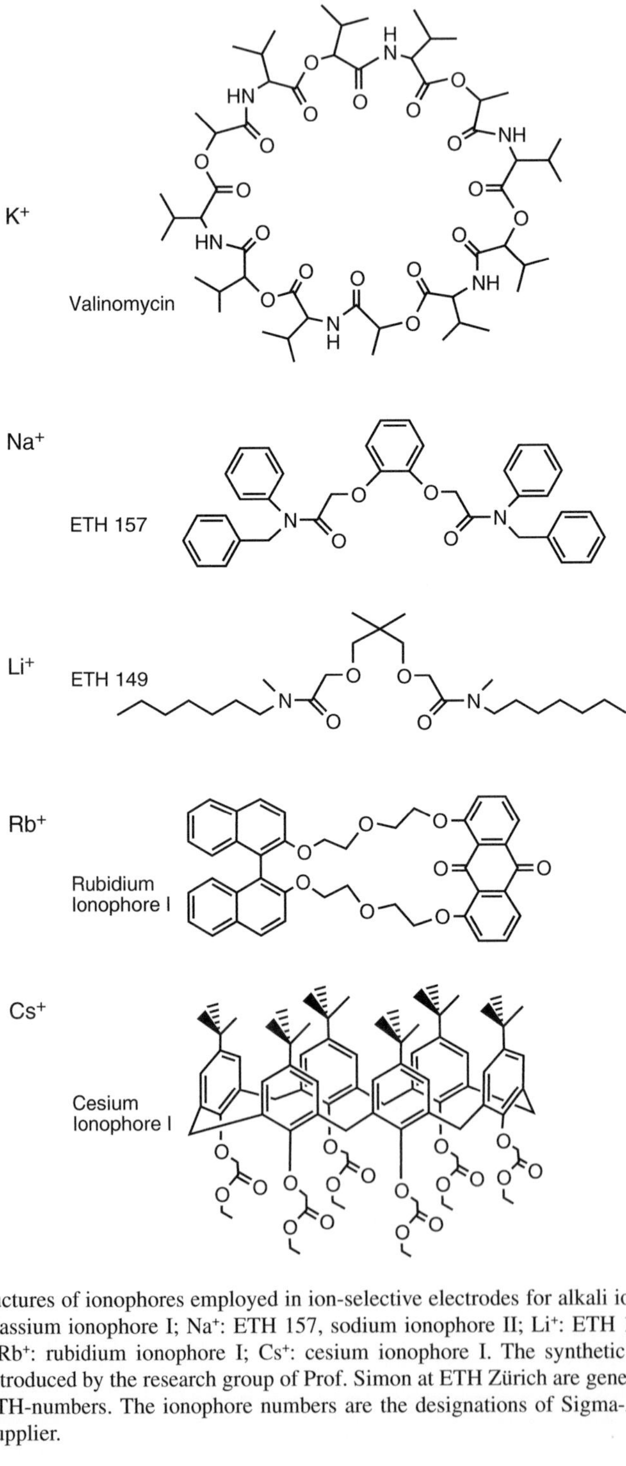

Figure 3 Structures of ionophores employed in ion-selective electrodes for alkali ions. K^+: valinomycin, potassium ionophore I; Na^+: ETH 157, sodium ionophore II; Li^+: ETH 149, lithium ionophore I; Rb^+: rubidium ionophore I; Cs^+: cesium ionophore I. The synthetic ionophores which were introduced by the research group of Prof. Simon at ETH Zürich are generally known under their ETH-numbers. The ionophore numbers are the designations of Sigma-Aldrich, the commercial supplier.

Table 2 Logarithmic selectivity coefficients (log *K*) for the alkali ions of ion-selective electrodes based on the ionophores of Figure 3.

Ion	Ionophore	Li^+	Na^+	K^+	Rb^+	Cs^+	
Li^+	ETH 149, lithium ionophore I	0	–1.4	–2.3	–2.2	–0.9	[8]
Na^+	ETH 157, sodium ionophore II	–1.7	0	–0.4	–	–	[8]
K^+	valinomycin, potassium ionophore I	–4.0	–3.5	0	0.5	–0.4	[8]
Rb^+	rubidium ionophore I	–2.5	–2.4	–1.0	0	–1.2	[9]
Cs^+	cesium ionophore I	–3.3	–2.1	–0.7	–1.2	0	[10]

ion-exchange chromatography which is based on a stationary phase having permanently charged sites, such as sulfonate groups, and an eluent containing a cation with appropriate affinity and concentration to obtain controlled distribution equilibra for the analyte cations between the stationary and mobile phases. The instruments are simpler and of lower cost than HPLC equipment and a separation of inorganic ions, including the alkali ions, can be achieved in about 10–15 min. The standard detection method is the conductivity measurement and detection limits of about 1 μM are achieved.

In capillary electrophoresis (CE) the separation of the alkali metal and other ions is usually achieved via the so-called zone electrophoresis, i.e., the separation due to their differences in the electrophoretic mobility. This is related to the speed of migration in the electric field which is dependent on the size of the ion (including the hydration sphere) and its charge. The separation has to be carried out in capillaries in order to limit the current as due to the high voltages needed otherwise the electrical power expansion leads to excessive Joule heating. The small internal diameters of the capillaries in the range from 10 to 100 μm lead to a challenge in detection. The alkali metal ions are again best quantified with conductivity detection. Due to the small dimensions and conflict with the separation voltage its implementation was difficult until the introduction of the capacitively coupled contactless conductivity detection method (commonly termed C^4D), which works with tubular electrodes fitted to the outside of the capillary. LODs for the alkali metal ions between about 0.1 to 1 μM are possible. Due to the relative simplicity of capillary electrophoresis (essential components are the inexpensive capillary and a high voltage power supply module) and C^4D it is possible to construct field-portable instruments. A further advantage of CE is the possibility to work with very small sample volumes down to about 10 μL, which can be of benefit for clinical and environmental applications. As illustration an electropherogram for the analysis of a saliva sample by CE-C^4D is given in Figure 4, showing prominently the peaks for sodium and potassium.

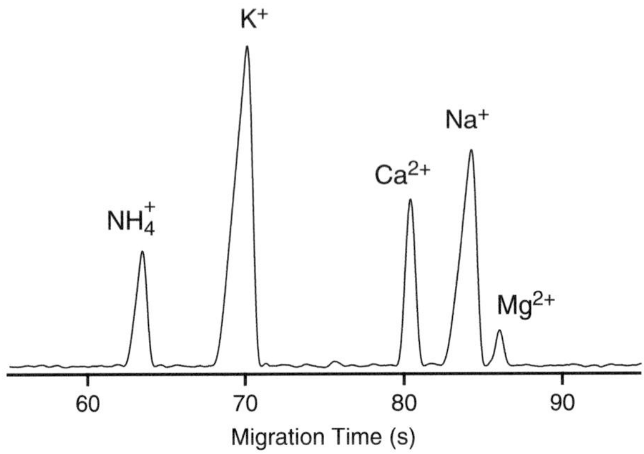

Figure 4 Electropherogram of a sample of saliva obtained with contactless conductivity detection.

6 Clinical Analysis

Sodium and potassium are routinely determined in clinical samples as these are the most abundant cations. Most often blood samples are analyzed, less frequently other fluids, such as urine, saliva, sweat or cerebrospinal fluid. When taking blood, the sampling is important, as differences may be found between venous and arterial blood, whether it is taken with a syringe or via a finger prick, and the rate of withdrawal [11]. In plasma the normal concentrations of sodium are in the range from 135 to 145 mM and of potassium from 3.5–4.5 mM (while in cellular fluid the concentrations are almost reversed, i.e., 10 mM for sodium and 150 mM for potassium). Deviations may be related to a variety of factors such as dehydration, renal dysfunction, hormonal problems or dietary deficiencies [12].

Lithium is determined when this is administered as a drug against bipolar disorders and some other health problems. During the initial treatment serum concentrations from 0.8 to 1.2 mM are recommended, while for long term maintenance levels of 0.6–1.0 mM are suggested [13]. Above 1.5 mM toxicity sets in [13]. As the therapeutic range, e.g., the useful range between effectiveness and poisoning, is thus very narrow, therapeutic drug monitoring (TDM) is essential in lithium therapy. Blood samples are sometimes analyzed immediately without further treatment, but if storage is required, these are centrifuged and stored at 4 °C or frozen for long term storage. Heparin is added to prevent coagulation.

The analysis is usually carried out with highly specialized flow through multispecies analyzers available from different suppliers. These systems are either large robotic systems designed for use in large hospitals or central service laboratories, or smaller instruments used directly at the point-of-care (POC), such as in a medical practitioner's surgery. Besides the alkali ions these robotic analyzers usually determine also other electrolytes such as calcium, chloride, phosphate, and pH, as well as a number of metabolites such as urea, uric acid, bilirubin, and creatinine. The standard method for the determination of the alkali ions in these clinical samples used to be atomic emission spectrometry with a flame photometer. However, this has largely been replaced by potentiometry, which is more suitable for automation. Flame photometers may still be used for occasional non-routine tests.

The ion-selective electrodes generally employ ionophore-doped PVC membranes, note however, that in some older systems glass membrane electrodes are used for the determination of sodium. Sodium, potassium as well as lithium ions in blood are free (in contrast to Ca^{2+}, of which about 60 % is bound to proteins) so that the concentrations for these ions determined by the potentiometric sensors agree with the distructive flame emission method. The discrepancy between activity, the parameter determined by the sensor, and concentration is eliminated by carrying out the calibration with standards matching the ionic composition of the blood samples. As the Nernst slope is dependent on temperature, the instrument and samples are carefully thermostatted and the instrument frequently calibrated and validated using certified samples in order to assure the high precision and accuracy required to give meaningful results for the clinical samples with the narrow dynamic ranges.

7 Single Cell Analysis

In the study of the physiological functions of sodium and potassium it is often desired to carry out investigations on the cellular level. In particular the transport of these ions across cell walls through ion channels under different conditions is of interest. The study of the concentrations of these ions on this microbiological level is possible by two different means, the use of fluorescent probe molecules or microscale ion-selective electrodes.

The structures of two fluorescent probes for intracellular studies, which are commercially available (from Invitrogen), are shown in Figure 5 [14]. Note, the two identical fluorophores bind to either a 15-crown-5 or a 18-crown-6 for the probes with preference for sodium and potassium, respectively. Cell studies are carried out on special fluorescence microscopes fitted with a light source for excitation, optical filters and a dichroic mirror to separate excitation and fluorescence wavelengths.

The use of microscale ion-selective electrodes is related to cell electrophysiology in which potentials are measured with micrometer-scale needle electrodes. The ion-selective microelectrodes are prepared by drawing out a glass capillary to a desired tip size of about one micrometer. The ionophore is dissolved in a viscous organic solvent (such as a plasticizer) which is then drawn into the tip and held there as the glass surface is made lipophilic through silanization with a chlorosilane. The back is filled with an aqueous electrolyte and the potential picked up with a miniature silver/silver chloride electrode as for a regular ISE. Similarly, miniature reference electrodes are employed. Details on this technique can be found in a monograph [8].

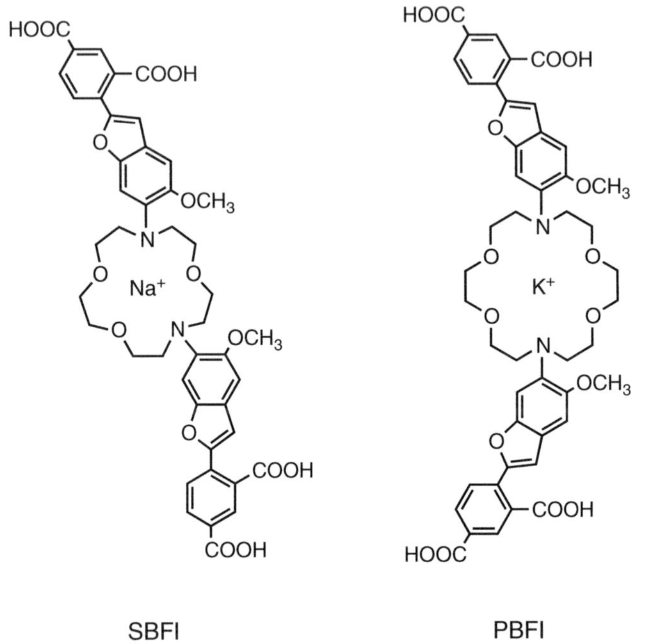

Figure 5 Structures of fluorescent molecular probes for sodium (SBFI) and potassium (PBFI) [14].

8 Environmental Samples

The foremost application of analysis of the alkali metal ions in environmental analysis is the determination of sodium and in particular of potassium in soil and plant material as the latter is one of the 3 essential nutrients, the K in the common NPK fertilizers. Sodium is not a nutrient but in high concentrations, as may occur in soils of costal or arid areas, has adverse affects on plants.

Potassium in soil samples is present in three fractions, dissolved, exchangeable, and bound [15], and it is usually of interest to determine the proportion which is available to the plant. For this reason the samples are commonly treated with an appropriate extractant. Frequently used are aqueous solutions of ammonium acetate [15]. To determine total potassium, a strong acid, such as HNO_3, may be employed as extractant [16]. Plant materials require homogenization and digestion with an acidic peroxide solution in presence of a catalyst such as Fe^{3+} or selenium [17]. The subsequent quantification may then be carried out by any available method. Both, soil and plant matter, is often dried before analysis for standardization.

Hydroponic nutrient solutions contain potassium. As the nutrients are consumed by the plants grown, their concentrations have to be determined frequently and the concentrations are corrected accordingly by addition of the fertilizer substances. To this end automated flow-through analyzers, which are usually based on ion-selective electrodes, have been reported (see for example [18]).

A very specific analysis concerning alkali metals in the environment is the determination of the radioactive cesium isotopes ^{137}Cs and ^{134}Cs. Due to their relative volatility, besides the iodine isotope ^{131}I these are the two main radioactive isotopes released into the environment after the Chernobyl and Fukushima nuclear reactor disasters. While the iodine isotope has a half-life of only 8 days and therefore does not pose a longterm problem, the cesium isotopes with half-lives of 30 years (^{137}Cs) and 2 years (^{131}Cs) are persistent. Do to its accumulation in the food chain in parts of Europe, 25 years after the event game meat was found to be still significantly contaminated with ^{137}Cs from the fallout of the Chernobyl disaster (see for example [19]). The determination of the cesium isotopes (as well as, for example, of naturally occurring ^{40}K) is carried out by γ-ray spectroscopy. The determination by this method is highly sensitive and extreme trace levels can be detected. ^{137}Cs was also produced in the nuclear bomb blasts and spread globally. A record of the presence of ^{137}Cs in the environment could be obtained by analysis of a series of vintage Bordeaux wines as shown in Figure 6 [20]. The graph clearly shows the production of the isotope during the bomb tests in the 1950s and 60s as well as the contamination from the Chernobyl incident. The data may be used to authenticate the vintage of a bottle of wine (the γ-ray spectra can be obtained without needing to open the bottle) and in particular detect fakes or resealing of old wines or other old containers as before 1945 no radioactive ^{137}Cs was present in the environment [20].

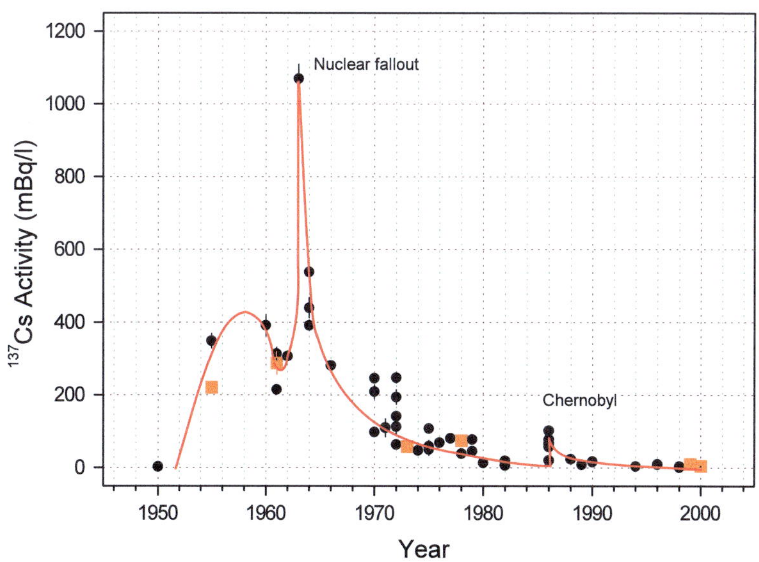

Figure 6 Activity of ^{137}Cs in Bordeaux wines of different vintage [20]. The orange squares represent measurements carried out on intact, unopened bottles. Reproduced with permission from Hubert et al., Radioactivity measurements applied to the dating and authentication of old wines. *C. R. Physique* **2009**, *10,* 622–629.

9 General Conclusions

For the determination of just one of the alkali metals, ion-selective electrodes are the most attractive option due to their simplicity, low cost, and the possibility of using them for field analysis. However, one has to carefully consider their selectivity, not just concerning the possible interference from other alkali ions, but from other ions present in the sample as well. The natural compound valinomycin as ionophore of the potassium electrode shows a low interference by sodium (more so by the larger alkali ions, which, however, is less critical), while the sodium electrode based on a synthetic compound is more prone to interference by potassium. If the required detection limits are modest, but the selectivity is a challenge, ion chromatography or capillary electrophoresis are not too expensive options, or possibly flame emission photometry. Capillary electrophoresis is suitable for field analysis as well. For high sensitivity, one of the other atomic spectrometric methods is required.

For quantitative analysis, generally the most simple methods available in most laboratories are the photometric ones, based on the formation of colored complexes with a suitable reagent. Unfortunately this is not readily possible for the alkali metal ions due to the lack of selective chromophoric reagents. The extraction with crown ethers is certainly a possibility, but a relatively cumbersome manual procedure. It also has limited specificity as the crown ethers have limited selectivity, despite their availability in different sizes. While their introduction was certainly of significant scientific value, unfortunately their analytical usefulness is limited. This also applies to the crown ether-based fluorophoric probe molecules for investigations at the single cell level, for which the microscale ion-selective electrodes appear to be better suited, even though they are perhaps more difficult to use.

Abbreviations and Definitions

AAS	atomic absorption spectrometry or spectrometer
C^4D	capacitively coupled contactless conductivity detection or detector
CE	capillary electrophoresis
FES	flame emission spectrometry or spectrometer
HPLC	high-performance liquid chromatography
ICP	inductively-coupled plasma spectrometry or spectrometer
ISE	ion-selective electrode
LOD	limit of detection
MS	mass spectrometry or spectrometer
OES	optical emission spectroscopy or spectrometer
POC	point-of-care
PVC	poly(vinyl chloride)
TDM	therapeutic drug monitoring

Acknowledgment The author thanks Dr. J. Sáiz (University of Alcalá, Alcalá de Henares, Spain) for the permission to use the electropherogram of Figure 4.

References

1. H. Nakamura, H. Sakka, M. Tagaki, K. Ueno, *Chem. Lett* **1981**, *10*, 1305–1306.
2. G. E. Pacey, Y. P. Wu, *Talanta* **1984**, *31*, 165–168.
3. R. Escobar, C. Lamoneda, F. de Pablos, A. Guiraúm, *Analyst* **1989**, *114*, 533–535.
4. B. Valeur, I. Leray, *Coord. Chem. Rev.* **2000**, *205*, 3–40.
5. J. F. Callan, A. P. de Silva, D. C. Magri, *Tetrahydron* **2005**, *61*, 8551–8588.
6. J. W. Robinson, *Atomic Spectroscopy*, Marcel Dekker, New York, 1996.
7. J. Pick, K. Tóth, E. Pungor, M. Vasák, W. Simon, *Anal. Chim. Acta* **1973**, *64*, 477–480.
8. D. Amman, *Ion-Selective Microelectrodes*, Springer, Berlin, 1986.
9. M. H. Hyun, M.-H. Piao, Y. J. Cho, Y.-B. Shim, *Electroanalysis* **2004**, *16*, 1785–1790.
10. A. Cadogan, D. Diamond, M. R. Smyth, G. Svehla, M. A. McKervey, E. M. Seward, S. J. Harris, *Analyst* **1990**, *115*, 1207–1210.
11. K. Camman, H. Galster, *Das Arbeiten mit ionenselektiven Elektroden*, Springer, Berlin, 1996.
12. H. G. J. Worth, *Analyst* **1988**, *113*, 373–384.
13. M. E. Cates, P. J. Sims, *Am. J. Pharm. Ed.* **2005**, *69*.
14. A. Minta, R. Y. Tsien, *J. Biol. Chem.* **1989**, *264*, 19449–19457.
15. W. Van Lierop, N. A. Gough, *Can. J. Soil Sci.* **1989**, *69*, 235–242.
16. L. Zhan, X. Li, J. Lu, Z. Liao, T. Ren, R. Cong, *Commun. Soil Sci. Plant Anal.* **2014**, *45*, 2921–2931.
17. K. Camman, *Instrumentelle Analytische Chemie*, Spektrum Akademischer Verlag, Heidelberg, 2001.
18. H.-J. Kim, W.-K. Kim, M.-Y. Roh, C.-I. Kang, J.-M. Park, K. A. Sudduth, *Comp. Electron. Agricul.* **2013**, *93*, 46–54.
19. N. Šprem, I. Babić, D. Barišić, D. Barišić, *J. Radioanal. Nucl. Chem.* **2013**, *298*, 513–517.
20. P. Hubert, F. Perrot, J. Gaye, B. Médina, M. S. Pravikoff, *C. R. Physique* **2009**, *10*, 622–629.

Chapter 3
Solid State Structures of Alkali Metal Ion Complexes Formed by Low-Molecular-Weight Ligands of Biological Relevance

Katsuyuki Aoki, Kazutaka Murayama, and Ning-Hai Hu

Contents

K. Aoki (✉)
Department of Environmental and Life Sciences, Toyohashi University of Technology, Tempaku-cho, Toyohashi 441-8580, Japan
e-mail: ka003@edu.imc.tut.ac.jp

K. Murayama
Graduate School of Biomedical Engineering, Tohoku University, Aoba, Sendai 980-8575, Japan
e-mail: kmura@bme.tohoku.ac.jp

N.-H. Hu
Changchun Institute of Applied Chemistry, Chinese Academy of Sciences, Changchun 130022, China
e-mail: hunh@ciac.ac.cn

A. Sigel, H. Sigel, and R.K.O. Sigel (eds.), *The Alkali Metal Ions: Their Role for Life*, Metal Ions in Life Sciences 16, DOI 10.1007/978-3-319-21756-7_3

Abstract This chapter provides structural data, mainly metal binding sites/modes, observed in crystal structures of alkali metal ion complexes containing low-molecular-weight ligands of biological relevance, mostly obtained from the Cambridge Structural Database (the CSD version 5.35 updated to February 2014). These ligands include (i) amino acids and small peptides, (ii) nucleic acid constituents (excluding quadruplexes and other oligonucleotides), (iii) simple carbohydrates, and (iv) naturally occurring antibiotic ionophores. For some representative complexes of these ligands, some details on the environment of the metal coordination and structural characteristics are described.

Keywords Alkali metal ions • Amino acids • Antibiotic ionophores • Carbohydrates • Crystal structures • Nucleic acid constituents • Peptides

Please cite as: *Met. Ions Life Sci.* 16 (2016) 27–101

1 Introduction

Alkali metal ions (Li^+, Na^+, K^+, Rb^+, and Cs^+) are involved in a variety of chemical and biological systems [1]. For example, because Na^+ and K^+ as well as Mg^{2+} and Ca^{2+} ions are present in the body in millimolar concentrations, negatively charged nucleic acids and nucleotides occur as complexes coordinated with metal ions [2]. The aim of this chapter is to provide structural information observed in crystal structures of alkali metal ion complexes formed with low-molecular-weight ligands of biological importance, mainly obtained from the Cambridge Structural Database (the CSD version 5.35 updated to February 2014), above all focusing on the metal binding sites/modes.

Today, single-crystal X-ray diffraction is still one of the most powerful methods to provide accurate information on the geometry of metal bonding. In this review, among a variety of biological substances, we deal with the ligands classified into the four categories: amino acids and peptides, nucleic acid constituents (excluding G-quadruplexes, for which see Chapter 7 of this volume), simple carbohydrates, and naturally occurring antibiotic ionophores.

To our knowledge, this is the first review on crystal structures of alkali metal ion complexes with amino acids and peptides, before only those for simple sugars were dealt with [3]. On the other hand, those for nucleic acid constituents [4] and for ionophore antibiotics [5–8] were repeatedly reviewed (containing not only main group metal ions but also transition metal ions), thus, the present Sections 3 and 5 rely heavily on these previous reviews. Here we treat exclusively the direct interaction between the metal ion and the ligand, that is, innersphere coordination, but no outersphere one, though hydrated metal ions could play important roles in biological systems. For example, metal cations interact with DNA most frequently via water molecules in their primary solvation shell as hydrogen bond donors in addition to point charges [9].

We also searched the Cambridge Structural Database (CSD) for water-soluble vitamins, major hormones, and neurotransmitters, but only a small number of alkali metal ion complexes were found, and these are not picked up here. For reference, they are: a Li^+ [CSD code WIKXUJ] and a K^+ [WIKYAQ] complex with vitamin B_{12} (cyanocobalamin) [10], a Li^+ complex of isoascorbate (derivative of vitamin C) [ZIBFAT] [11], and a K^+ complex of taurine (neurotransmitter) [SANSOQ] [12].

Alkali metal ions are spherical in shape with ionic radii (Å) of 0.60 for Li^+, 0.95 for Na^+, 1.33 for K^+, 1.48 for Rb^+, and 1.69 for Cs^+, and van der Waals radii (Å) of the ligating atoms are 1.40 for O and 1.50 for N [13].

2 Amino Acid and Small Peptide Complexes

2.1 Amino Acid Complexes

In this section, we are concerned with the twenty standard α-amino acids, which occur not only as amino acid residues in proteins but also as free amino acids within cells and blood plasma and which perform critical roles in processes such as neurotransmitter transport and biosynthesis. Amino acids are composed of amine (NH_2) and carboxylic acid (COOH) functional groups, along with a side-chain specific to each amino acid ($H_2NCHRCOOH$, where R is a side-chain). At physiological pH (6.7 ~ 7.4), they exist as zwitterions with the amino group protonated (for example, pK_a of 9.8 for glycine (R = H)) and the carboxyl group deprotonated (pK_a of 2.4 for glycine). The twenty amino acids can be classified here for convenience into four broad groups according to the properties of the side-chains: Group *1* amino acids with nonpolar and electrically neutral (at pH 7.4) side-chains; Group *2*, with polar and neutral side-chains; Group *3*, with basic polar side-chains; and Group *4*, with polar and acidic side-chains. Table 1 lists the total number of so far reported crystal structures of alkali metal ion complexes with the twenty amino acids [14–51]. No X-ray structures are available yet for half of the twenty amino acids, none for Rb^+ ion, and only one [39] for Cs^+ ion. Table 2 summarizes metal binding sites/modes observed in the individual complexes.

Group *1* amino acids with nonpolar and electrically neutral (at pH 7.4) side-chains are glycine (Gly), alanine (Ala), valine (Val), leucine (Leu), isoleucine (Ile), proline (Pro), phenylalanine (Phe), tryptophan (Trp), and methionine (Met). Among the group *1* amino acids, a total of thirty-four crystal structures of alkali metal ion complexes are reported but only for Gly, Ala, and Pro; most (twenty) are Gly complexes. In most of the complexes, the amino acid ligand takes the zwitterionic form for which the carboxyl oxygens are the only choice for metal bonding. When the triply protonated ammonium group loses a proton, the amino group becomes a possible metal binding site. This is the case for [Li(Gly–H)] [14], which was synthesized

Table 1 Total number of alkali metal ion complexes with the standard twenty amino acids in the solid state.[a,b]

Amino acid[c]	Li^+	Na^+	K^+	Rb^+	Cs^+	Total
Glycine (Gly)	8[d] (14)	8[e] (47)	4[f] (27)	0	0	20 (88)
Alanine (Ala)	0	5[g] (13)	1[h] (8)	0	0	6 (21)
Valine (Val)	0	0	0 (1)	0	0	0 (1)
Leucine (Leu)	0	0	0 (3)	0	0	0 (3)
Isoleucine (Ile)	0	0	0	0	0	0
Proline (Pro)	5[i] (5)	2[j] (9)	1[k] (8)	0	0	8 (22)
Phenylalanine (Phe)	0	0 (2)	0 (3)	0	0	0 (5)
Tryptophan (Trp)	0	0	0 (1)	0	0	0 (1)
Methionine (Met)	0	0 (1)	0 (1)	0	0	0 (2)
Serine (Ser)	0	3[l] (3)	0	0 (1)	1[m] (1)	4 (5)
Threonine (Thr)	0	1[n] (1)	0	0	0	1 (1)
Tyrosine (Tyr)	0	0 (1)	0	0	0	0 (1)
Cysteine (Cys)	0	2[o] (16)	1[p] (5)	0	0	3 (21)
Asparagine (Asn)	0	2[q] (2)	0	0	0	2 (2)
Glutamine (Gln)	0	0	0	0	0	0
Histidine (His)	0	0	0	0	0	0
Lysine (Lys)	0	0 (5)	0	0	0 (5)	0 (10)
Arginine (Arg)	0	0 (1)	0	0	0 (1)	0 (2)
Aspartic acid (Asp)	2[r] (2)	0 (3)	3[s] (3)	0	0 (1)	5 (9)
Glutamic acid (Glu)	1[t] (1)	1[u] (5)	1[v] (2)	0	0	3 (8)
Total	16 (22)	24 (109)	11 (62)	0 (1)	1 (8)	52 (202)

[a]Obtained from the Cambridge Structural Database (the CSD version 5.35 updated to February 2014).

[b]The number in the parenthesis denotes the sum of the number of the alkali metal ion complexes and that of the alkali metal ion salts, where the *complex* involves *direct* metal–ligand bonding while the *salt* does not.

[c]*Substituted* amino acids are excluded.

[d][14–19]. [e][20–23]. [f] [25, 28–30]. [g] [31–33]. [h] [34]. [i] [35, 36]. [j] [37]. [k] [38]. [l] [39]. [m] [39]. [n] [40]. [o] [41, 42]. [p] [43]. [q] [44]. [r] [45, 46]. [s] [45, 47, 48]. [t] [49]. [u] [50]. [v] [51].

using KOH, but somewhat surprisingly, this is the only case among the 52 complexes in Table 1. In the complexes in which transition metal ions coexist, the amino nitrogen usually coordinates to transition metal ions. Among such complexes, of special interest are a series of clusters of the type $[NaCu_6(aa)_8]^{5+}$ (aa: Gly–H [24, 27], Ala–H [31], Pro–H [37], and Thr–H [40]) or of a type $[NaCu_5(\text{L-Ala–H})_6]^{5+}$ [31], where the Cu^{2+} metal ions of the hexanuclear (or pentanuclear) units are held together by eight (or six) chelating aa ligands, through the carboxylate oxygen and the amino nitrogen, which in turn provide the remaining eight (or six) oxygen donors trapping a Na^+ ion in the center (see Figure 1), thus the $[Cu_6(aa)_8]^{4+}$ and $[Cu_5(aa)_6]^{4+}$ cages behave as 'metallacryptates' [40]. It has been demonstrated that, even under excess of Li^+ and/or K^+, only the Na^+ cluster could be crystallized,

Table 2 Alkali metal ion binding sites/modes on the amino acids in the solid state.[a,b]

Complex		Coordination geometry or number	Metal binding sites on amino acid ligands and other ligands/comments, if any	CSD code or CCDC number	Ref
Glycine complexes					
$\{[Li(Gly)(H_2O)]\cdot Cl\}_n$		Tet.	3O [α-CO_2^- (Mol.1-Mol.3): 1.951-1.953][b]	HEFWUK	[14]
[Li(Gly-H)]		Tet.	3O [α-CO_2^- (Mol.1-Mol.3): 1.943-1.967], N [NH_2 (Mol.4): 2.046]	HEFXAR	[14]
$\{[Li(Gly)_2(H_2O)]\cdot Br\}_n$		Tet	3O [α-CO_2^- (Mol.1-Mol.3): -[c]]	NEPWUC	[15]
$[Li_2(Gly)(SO_4)]_n$	Li1	Tet.	O [α-CO_2^- (Mol.1): 1.900], 3O [solvent: 1.956-1.965]	EVONAE01	[16]
	Li2	Tet.	O [α-CO_2^- (Mol.1): 1.884], 3O [solvent: 1.940-1.972]		
$[Li(Gly)(NO_3)]_n$		Tet.	3O [α-CO_2^- (Mol.1-Mol.3): 1.938-1.997], O [solvent:1.975]	ALUNEA	[17]
$\{[Li(Gly)_2]\cdot NO_3\}_n$		Tet.	4O [α-CO_2^- (Mol.1-Mol.4): 1.9018-1.9399]	ROZTUW	[18]
$\{[Li_2(Gly)_2(H_2O)(CrO_4)]_n$	Li1	Dist. tet.	4O [2O α-CO_2^- (Mol.1, Mol.2), 2O (CrO_4): 1.980-2.080]	UCIYOV	[19]
	Li2	Dist. tet.	4O [3O α-CO_2^- (Mol.2-Mol.4), 1O aqua: 1.936-2.104]		
$\{[Li_2(Gly)_2(MoO_4)]_n$	Li1	Dist. tet.	4O [2O α-CO_2^- (Mol.1, Mol.2), 2O (MoO_4): 1.904-1.959]	UCIYUB	[19]
	Li2	Dist. tet.	4O [2O α-CO_2^- (Mol.3, Mol.4), 2O (MoO_4): 1.906-1.967]		
$[Na_4(Gly\text{-}H)_2(H_2O)_{14}Cu_2TeMo_6O_{24}]\cdot$-$0.5H_2O\}_n$	Na1	Oct.	O [α-CO_2^- (Mol.1): 2.334], 3O [aqua: 2.318-2.476], 2O [$(TeMo_6O_{24})^{6-}$: 2.346, 2.347]	ULONUF	[20]
	Na2	Oct.	O [α-CO_2^- (Mol.1): 2.886], 5O [aqua: 2.290-2.458]		
$[Na(Gly)(NO_3)]_n$		8-Coord.	2O [α-CO_2^- (Mol.1, Mol.2): 2.317, 2.415], 6O [NO_3 (Mol.1-Mol3): 2.608-2.720]	TIJDAR01	[21]
$[Na(Gly)(NO_3)]_n$		8-Coord.	2O [α-CO_2^- (Mol.1, Mol.2): 2.322, 2.410], 6O [NO_3 (Mol.1-Mol.3]: 2.617-2.670]	TIJDAR02	[22]
$\{[\{Na_3(Gly\text{-}H)_2(H_2O)_{10}\}\{CuCr(OH)_6$-$Mo_6O_{18}\}]\cdot 13H_2O\}_n$	Na1	Dist. Oct.	O [α-CO_2^-: -], 5O [aqua: -]	XODGAZ	[23]
	Na2	Oct.	4O [aqua: -], 2O [$\{CuCr(OH)_6Mo_6O_{18}\}^{3-}$: -]		
$\{[Na(Gly\text{-}H)_6Co_2]\cdot ClO_4\}_n$		Oct.	6O [α-CO_2^- (Mol.1-Mol.6): -]	XODGED	[23]

(continued)

Table 2 (continued)

$\{[Na(Gly\text{-}H)_8Cu_6(ClO_4)_3(H_2O)]\cdot 2ClO_4\cdot 3H_2O\}_n$		8-Coord.	8O [α-CO_2^- (Mol.1-Mol.8): 2.543-2.621] Na^+ trapping in the $[Cu_6(aa)_8]^{4+}$ cage. See Figure 1.	POLKUX	[24]
$[Na_4(Gly)(S_2O_3)_2(H_2O)_2]_n$	Na1	Oct.	2O [α-CO_2^- (Mol.1): 2.420, 2.693], O [aqua: 2.389], S [$S_2O_3^{2-}$ (Mol.1): 3.014], 2O[$S_2O_3^{2-}$ (Mol.1, Mol.2): 2.428, 2.482]	MAZVOZ	[25]
	Na2	6-Coord.	2O [α-CO_2^- (Mol.1, Mol.2): 2.477, 2.419], 3O [aqua: 2.336-2.431], S [$S_2O_3^{2-}$ (Mol.2): 2.899]		
	Na3	6-Coord.	O [α-CO_2^- (Mol.1): 2.469], O [$S_2O_3^{2-}$ (Mol.1): 2.423], 2O [$S_2O_3^{2-}$ (Mol.3): 2.469, 2.960], 2O [$S_2O_3^{2-}$ (Mol.4, Mol.5): 2.333, 2.380]		
	Na4	6-Coord.	O [α-CO_2^- (Mol.3): 2.353], 3O [$S_2O_3^{2-}$ (Mol.1, Mol.5, Mol.6): 2.363-2.533], 2O [$S_2O_3^{2-}$ (Mol.4): 2.429, 2.674]		
$\{[Na(Gly\text{-}H)_6Cr_2]\cdot ClO_4\}_n$		Dist. Oct.	6O [α-CO_2^- (Mol.1-Mol.6): 2.414] Na^+ ion rides on a threefold axis.	EYOZAU	[26]
$[Na(Gly\text{-}H)_8Cu_6(H_2O)_2]\cdot[\{Na(Gly\text{-}H)_8Cu_6(H_2O)_2\}\{Mo_8(Gly)_2(O)_{26}Cu\text{-}(Gly\text{-}H)(H_2O)\}_2]\cdot[Mo_8(Gly)_2(O)_{26}]\cdot 16H_2O$	Na1	6-Coord.	6O [α-CO_2^- (Mol.1-Mol.6): -]	UZILEV	[27]
	Na2	6-Coord.	6O [α-CO_2^- (Mol.1-Mol.6): -] Na^+ trapping in the $[Na(aa)_8Cu_6]^{5+}$ cage. See Figure 1.		
$[K_2(Gly)(S_2O_3)]_n$	K1	6-Coord.	2O [α-CO_2^- (Mol.1, Mol.2): -] 2S [$S_2O_3^{2-}$ (Mol.1, Mol.2): -], 2O [$S_2O_3^{2-}$ (Mol.1, Mol.2): -]	MAZVUF	[25]
	K2	8-Coord.	2O [α-CO_2^- (Mol.2, Mol.3): -], 2O [$S_2O_3^{2-}$ (Mol.1): -], 2O [$S_2O_3^{2-}$ (Mol.2): -], 2O [$S_2O_3^{2-}$ (Mol.3, Mol.4): -]		
$\{[K_2(Gly)\{(VO)(O_2)_2\}_2]\cdot(H_2O)\}_n$	K	9-Coord.	O [α-CO_2^- : -], 6O [$\{VO(O_2)_2\}^-$: -], 2O [aqua: -]	QOQFOS	[28]
$\{[K_3(Gly\text{-}H)_6(H_2O)_6Cu_3(BW_{12}O_{40})]\cdot 4H_2O\}_n$	K1	7-Coord.	O [α-CO_2^- : 2.812], 3O [$(BW_{12}O_{40})^{5-}$: ave. 2.914], 3O [aqua: ave. 2.843]	QISXAS	[29]

	K2	8-Coord.	4O [α-CO_2^- (Mol.1-Mol.3): ave. 2.744], 2O [$(BW_{12}O_{40})^{5-}$: ave. 2.724], 2O [aqua: ave. 2.743]		
	K3	7-Coord.	3O [α-CO_2^- (Mol.1-Mol.3): ave. 2.689], 2O [$(BW_{12}O_{40})^{5-}$: ave. 2.891], 2O [aqua: ave. 2.967]		
{[K_2(Gly)$_2Mo_4O_{13}(H_2O)_2$]$_n$		11-Coord.	2O [α-CO_2^- (Mol.1, Mol.2): -], 7O [$(Mo_4O_{13})^{2-}$: -], 2O [aqua: -]	VEVXUQ	[30]
Alanine complexes					
[Na{Cu_6(L-Ala-H)$_8(ClO_4)(H_2O)$}]·- $4ClO_4·4H_2O$		8-Coord.	8O [α-CO_2^- (Mol.1-Mol.8): ave. 2.59]	QIYDIM	[31]
[Na{Cu_6(L-Ala-H)$_8(ClO_4)$}]$_2$·- [$Na(H_2O)_4(ClO_4)_2$]·$7ClO_4·4H_2O$	Na1	8-Coord.	8O [α-CO_2^- (Mol.1-Mol.8): -]	QIYDOS	[31]
	Na2	8-Coord.	8O [α-CO_2^- (Mol.9-Mol.17): -] Na^+ trapping within the [Cu_6(aa)$_8$]$^{4+}$ cage. See Figure 1.		
{[$Na(H_2O)${Cu_5(L-Ala-H)$_6(H_2O)_2$}]·- $5ClO_4·3H_2O$}$_n$		7-Coord.	6O [α-CO_2^- (Mol.1-Mol.6): -] O [aqua: -] Na^+ trapping within the [Cu_5(aa)$_6$]$^{4+}$ cage.	QIYDUY	[31]
[Na(L-Ala)(NO_3)]$_n$		8-Coord.	2O [α-CO_2^- (Mol.1, Mol.2):], 6O [nitrate (Mol.1-Mol.3): -]	KIKCAJ	[32]
{[{$Na(H_2O)_2$(DL-Ala-H)$_2$Cu}$_2$- {$Cr(OH)_6Mo_6O_{18}$}$_2$]·$23.5H_2O$}$_n$		Dist. Oct.	O [α-CO_2^- : -], 2O [aqua: -], 3O [{$Cr(OH)_6Mo_6O_{18}$}$^{3-}$: -]	HISVIP	[33]
[$K(H_2O)_3${Ni(L-Ala-H)$_3$}]$_n$		Oct.	3O [α-CO_2^- (Mol.1-Mol.3): 2.759], 3O [aqua: 2.785] K^+ ion rides on a threefold axis.	XOSQOM	[34]
Proline complexes					
{[Li(L-Pro)$_2$]·Br}$_n$		Tet.	4O [α-CO_2^- (Mol.1-Mol.4): 1.931-1.973]	EVUVIB	[35]
{[Li(L-Pro)$_2$]·Cl}$_n$		Tet.	4O [α-CO_2^- (Mol.1-Mol.4): 1.905-1.966]	EVUVUN	[35]
{[Li_2(L-Pro)$_4$]·$2NO_3$}$_n$	Li1	Tet.	4O [α-CO_2^- (Mol.1-Mol.4): -]	EVUWIC	[35]
	Li2	Tet.	4O [α-CO_2^- (Mol.1-Mol.4): -]		
{[Li(L-Pro)$_2$]·NO_3}$_n$	Li1	Tet.	4O [α-CO_2^- (Mol.1-Mol.4): -]	EVUWOI	[35]

(continued)

Table 2 (continued)

	Li2	Tet.	4O [α-CO_2^- (Mol.1-Mol.4): -]		
{[Li(L-Pro)(H_2O)]·Cl}$_n$		Tet.	3O [α-CO_2^- (Mol.1-Mol.3): -] O [aqua: -]	YOXBET	[36]
[Na(L-Pro-H)$_8$Cu$_6$(ClO_4)$_5$]·4H_2O		8-Coord.	8O [α-CO_2^- (Mol.1-Mol.8): 2.511-2.608] Na^+ ion rides on a fourfold axis. Na^+ trapping within the [$Cu_6(aa)_8$]$^{4+}$ cage. See Figure 1.	179092[d]	[37]
[Na(L-Pro-H)$_8$Cu$_6$(ClO_4)$_5$(MeOH)]·-H_2O		8-Coord.	8O [α-CO_2^- (Mol.1-Mol.8): 2.543, 2.549] Na^+ ion rides on a fourfold axis. Na^+ trapping within the [$Cu_6(aa)_8$]$^{4+}$ cage. See Figure 1.	UJAPOK	[37]
[K{PtCl$_2$(L-Pro-H)}]$_n$	K	Dist. tet.	4O [α-CO_2^- (Mol.1-Mol.4): 2.691-2.819]	EWOYUL	[38]
Serine complexes					
[Na$_2$(L-SerH)$_2$(H_2O)$_5$(tetra-*p*-sulfanatocalix[4]arene)$_2$]·12H_2O	Na1	6-Coord.	O [α-CO_2H (Mol.1): 2.359], 2O [sulfonate (Mol.1, Mol.2): 2.442, 2.439], 3O [aqua: 2.319-2.469]	KAFPEN	[39]
	Na2	6-Coord.	O [α-CO_2H (Mol.2): 2.407], O [alcohol OH (Mol.1): 2.537], 2O [sulfonate (Mol.1, Mol.2): 2.432, 2.436], 2O [aqua: 2.385, 2.506]		
{[Na$_5$(L-SerH)$_2$(H_2O)$_5$(tetra-*p*-sulfanatocalix[4]arene)$_2$]·10H_2O}$_n$	Na1	6-Coord.	O [α-CO_2H (Mol.1): 2.508]	KAFPIR	[39]
	Na2	6-Coord.	O [α-CO_2H (Mol.2): 2.583]		
	Na3	6-Coord.	–		
	Na4, Na5	6-Coord.	–		
{[Na$_2$(L-SerH)$_2$Cs$_2$(Cl)(H_2O)$_5$(tetra-*p*-sulfanatocalix[4]arene)$_2$]·6H_2O}$_n$	Na1	6-Coord.	2O [α-CO_2H (Mol.1, Mol.2): 2.426, 2.467), 2O [sulfonate: 2.302, 2.431], 2O [aqua: 2.377, 2.391]	KAFPOX	[39]
	Na2	6-Coord.	4O [sulfonate: 2.290-2.474], 2O [aqua: 2.355, 2.449]		
{[Cs$_2$Na$_2$(L-SerH)$_2$(Cl)(H_2O)$_5$(tetra-*p*-sulfanatocalix[4]arene)$_2$]·6H_2O}$_n$	Cs1	8-Coord.	O [α-CO_2H: 3.202], 5O [sulfonate: 2.950-3.148), 2O [aqua: 3.343, 3.457]	KAFPOX	[39]

	Cs2	6-Coord.	4O [sulfonate: 3.009-3.242], O [aqua: 2/945], Cl [2.905]		
Threonine complexes					
[Na(L-Thr-H)$_8$Cu$_6$(ClO$_4$)$_4$]·ClO$_4$·5H$_2$O		8-Coord.	8O [α-CO$_2^-$ (Mol.1-Mol.8): 2.543-2.616] Na$^+$ trapping within the [Cu$_6$(aa)$_8$]$^{4+}$ cage. See Figure 1.	MARWUY	[40]
Cysteine Complexes					
{[{Na(H$_2$O)$_3$(L-Cys-2H)Co(en)$_2$}-Ag{(L-Cys-H)Co(en)$_2$}$_2$]·ClO·4NO$_3$}$_n$		6-Coord.	3O [α-CO$_2^-$ (Mol.1-Mol.3): 2.376] 3O [aqua: -] Na$^+$ ion rides on a threefold axis.	XOZJAX	[41]
[Na$_2$(H$_2$O)$_5$(L-Cys-2H)$_2${Mo$_2$(μ-O)$_2$-O$_2$}]	Na1	6-Coord.	2O [α-CO$_2^-$ (Mol.1, Mol.2: 2.32, 2.37], O [Mo-O-Mo: 2.51], 3O [aqua: 2.37-2.45]	NACTMO10	[42]
	Na2	5-Coord.	5O [aqua: 2.26-2.53] Mo–S$^-$ bonding.		
K$_2$(H$_2$O)$_4${(L-Cys-2H)$_2$(Mo$_2$S$_2$O$_2$)}]·-H$_2$O}$_n$	K1	5-Coord.	2O [α-CO$_2^-$ (Mol.1, Mol.2: 2.729, 2.802], 3O [aqua: 2.654-2.806]	NARKAU	[43]
	K2	6-Coord.	O [α-CO$_2^-$ (Mol.3): 2.702], 2O [Mo=O: 2.791, 2.830], 3O [aqua: 2.653-2.865]		
Asparagine complexes					
{[Na$_2$(L-Asn)$_2$(H$_2$O)$_7${Zn(H$_2$O)$_2$}-{Cr(OH)$_6$Mo$_6$O$_{18}$}]·0.5(L-Asp)·-3.5H$_2$O}$_n$	Na1	6-Coord.	O [Asn α-CO$_2^-$ (Mol.1): -], O [MoO ({Cr(OH)$_6$Mo$_6$O$_{18}$}$^{3-}$): -], 4O [aqua: -]	VODLAC	[44]
	Na2	6-Coord.	O [Asn α-CO$_2^-$ (Mol.2): -], O [MoO (polyoxoanion): -], 4O [aqua: -]		
{[Na$_2$(D-Asn)$_2$(H$_2$O)$_7${Zn(H$_2$O)$_2$}-{Cr(OH)$_6$Mo$_6$O$_{18}$}]·0.5(D-Asp).-3.5H$_2$O}$_n$	Na1	6-Coord.	O [Asn α-CO$_2^-$ (Mol.1): -], O [MoO (polyoxoanion): -], 4O [aqua: -]	VOJLEM	[44]
	Na2	6-Coord.	O [Asn α-CO$_2^-$ (Mol.1): -], O [MoO (polyoxoanion): -], 4O [aqua: -]		
Aspartic acid complexes					
{[Li(L-Asp)]·H$_2$O}$_n$		Tet.	3O [α-CO$_2^-$ (Mol.1-Mol.3): 1.943-1.990],	SEBMOB	[45]

(continued)

Table 2 (continued)

			O [β-CO_2^- (Mol.4): 1.938]		
[Li_2(L-Asp-H)$_2$(H_2O)$_4$(Cp*Co)$_2$]·2Cl·-2H_2O	Li1	Tet.	O [α-CO_2^- (Mol.1): 2.05], O [β-CO_2^- (Mol.2): 1.95], 2O [aqua: 1.91, 1.99]	YEDSUV	[46]
	Li2	Tet.	O [α-CO_2^- (Mol.1): 1.99], O [β-CO_2^- (Mol.2): 1.95], 2O [aqua: 1.92, 1.95]		
[K(L-Asp)(H_2O)$_2$]$_n$		7-Coord.	2O [β-CO_2^- (Mo1, Mol.2): 2.698, 2.771], 2O [β-CO_2^- (Mol.3): 3.185, 3.193], 3O [aqua: 2.689-2.855] No K^+ bonding to α-CO_2^-.	SEBMUH	[45]
{[K_2{Cp*Co(L-Asp-H)$_2$}]·2I}$_n$	K1	5-Coord.	2O [β-CO_2^- (Mol.1): 2.76, 2.90], 2O [β-CO_2^- (Mol.2): 2.74, 3.16], O [α-CO_2^- (Mol.3): 2.63]	WEDJIY	[47]
	K2	5-Coord.	O [β-CO_2^- (Mol.1): 2.62], 2O [α-CO_2^- (Mol.3): 2.76, 2.97], 2O [α-CO_2^- (Mol.4): 2.86, 2.96]		
{[K(H_2O)$_2${Co(DL-Asp-H)$_2$}]·-1.5H_2O}$_n$		7-Coord.	2O [α-CO_2^- (Mol.1): 3.092; β-CO_2^- (Mol.1): 3.078], O [β-CO_2^- (Mol.2): 2.759], 2O [α-CO_2^- (Mol.3, Mol.4): 2.821, 2.862], 2O [aqua: 2.766, 2.799]	XAZVAW	[48]
Glutamic acid complexes					
[Li_2(L-Glu)$_2$]$_n$	Li1	Tet.	2O [α-CO_2^- (Mol.1, Mol.2): -], 2O [γ-CO_2^- (Mol.3, Mol.4): -]	IJECUV	[49]
	Li2	Tet.	2O [α-CO_2^- (Mol.2, Mol.5): -], 2O [γ-CO_2^- (Mol.1, Mol.6): -]		
[Na_2(L-Glu-H)$_2$(H_2O)$_2$]$_n$	Na1	6-Coord.	3O [α-CO_2^- (Mol.1-Mol.3): 2.311-2.442], O [γ-CO_2^- (Mol.4): 2.400], 2O [aqua: 2.396, 2.545]	SGLUTM01	[50]
	Na2	6-Coord.	O [α-CO_2^- (Mol.): 2.394], 2O [α-CO_2^- (Mol.5): 2.358, 2.461], O [γ-CO_2^- (Mol.6): 2.326], 2O [aqua: 2.339, 2.579]		
[K(L-Glu)(H_2O)]$_n$		6-Coord.	2O [α-CO_2^- (Mol.1): 2.822; γ-CO_2^- (Mol.1): 2.659], 3O [α-CO_2^- (Mol.1-Mol.3): 2.730-2.971], O [aqua: 2.756] See Figure 2.	KENKET	[51]

[a]Abbreviations: aqua: water ligand; Ala, zwitterionic alanine; Ala-H, alaninate anion with $-COO^-$ and $-NH_2$; Asn, zwitterionic asparagine with γ-$C(O)NH_2$; Asp, aspartate anion with α-COO^-, β-COO^-, and $-NH_3^+$; Asp-H, aspartate dianion with α-COO^-, β-COO^-, and $-NH_2$; Cp*, pentamethylcyclopentadienyl; Cys-2H, cysteinate dianion with $-COO^-$, $-NH_2$, and $-S^-$; Gly, glycine zwitterion with $-COO^-$ and $-NH_3^+$; Gly-H, glycinate anion with $-COO^-$ and $-NH_2$; Glu, glutaminate anion with α-COO^-, γ-COO^-, and $-NH_3^+$; Glu-H, glutaminate dianion with α-COO^-, β-COO^-, and $-NH_2$; Pro: zwitterionic proline; Pro-H, prolinate anion with $-COO^-$ and –N(H)–; Ser, zwitterionic serine with –OH; SerH, serinium cation with –COOH, $-NH_3^+$, and –OH; Thr, zwitterionic threonine with –OH; Thr-H, threoninate anion with $-COO^-$, $-NH_2$ and –OH; Tet., tetrahedral; Dist. tet., distorted tetrahedral; Oct. octahedral; Dist. oct., distorted octahedral.

[b]For example, '3O[α-CO_2^- (Mol.1–Mol.3): 1.951-1.953]' denotes that the metal ion binds to three α-carboxyl oxygens from three amino acid molecules (Gly in this case), Mol.1, Mol.2, and Mol.3, and their bond distances (Å).

[c]Not reported.

[d]CCDC number.

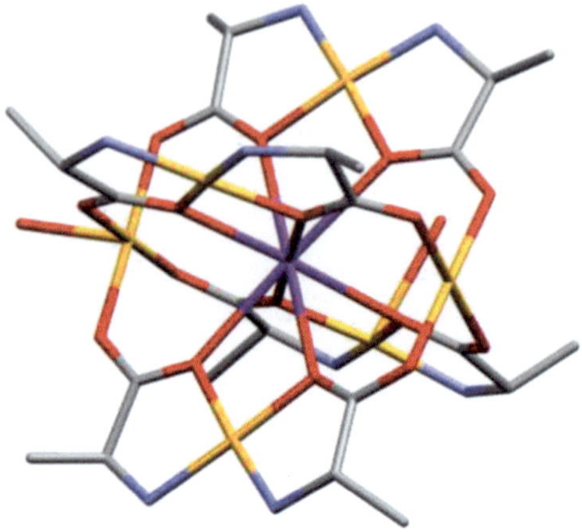

Figure 1 Structure of the $[NaCu_6(L\text{-}Ala\text{–}H)_8]^{5+}$ cluster [31] shown with color coded atoms (*red*, oxygen; *blue*, nitrogen; *grey*, carbon; *purple*, alkali metal ion; *ochre*, transition metal ion). The residues bound to the metal ion are depicted as 'sticks'. Note that the Na^+ ion is trapped in the center of the hexanuclear $[Cu_6(L\text{-}Ala\text{–}H)_8]^{4+}$ cage. Analogous clusters of the type $[NaCu_6(aa)_8]^{5+}$ (aa: amino acid; Gly–H [24, 27], Pro–H [37], and Thr–H [40]) or of a type $[NaCu_5(L\text{-}Ala\text{–}H)_6]^{5+}$ [31] are reported. Similar color codes are also used in subsequent diagrams of molecular structures.

implying that Li^+ is too small to be accommodated in the cluster whereas K^+ is too large [37].

Group *2*, amino acids with polar and neutral side-chains are serine (Ser), threonine (Thr), tyrosine (Tyr), cysteine (Cys), asparagine (Asn), and glutamine (Gln). A total number of nine complexes are available: four (three for Na^+ and one for Cs^+) for serine, one (for Na^+) for threonine, three (two for Na^+ and one for K^+) for cysteine, and two (for Na^+) for asparagine. The four Ser complexes form 'capsules' of tetra-*p*-sulfonatocalix[4]arene, in which chiral pairs of L-Ser molecules are confined in an overall bilayer arrangement, containing differing amounts of Na^+ and Cs^+ ions. The Na^+ complex of Thr is the 'Na^+-trapping' Cu^{2+} cluster noted above. The OH polar group of Ser and Thr and the $CONH_2$ of Asn are not involved in metal bonding neither to alkali metal ions nor to transition metal ions that coexist in the complexes. In each of the three Cys complexes, the SH group is deprotonated and coordinates to co-existing transition metal ions but not to alkali metal ions.

Group *3* amino acids with basic polar side-chains are lysine (Lys), arginine (Arg), and histidine (His) with 10 % basic and 90 % neutral species at pH 7.0 (pK_a of 6.0 for the imidazole NH of His). Unfortunately, no complex is reported.

Group *4* amino acids with acidic polar side-chains are aspartic acid (Asp) (pK_a 3.9 for β-COOH) and glutamic acid (Glu) (pK_a 4.1 for γ-COOH). Two Li^+ [45, 46] and three K^+ [45, 47, 48] complexes are reported for Asp, and three complexes for Glu (each one for Li^+ [49], Na^+ [50], and K^+ [51]). In all the reported complexes, the β-carboxyl group of Asp or γ-carboxyl group of Glu, which are deprotonated, are always involved in metal bonding, usually in addition to the α-carboxylate oxygens

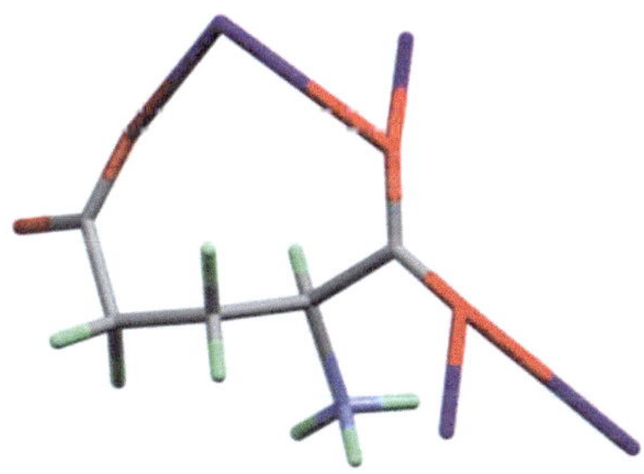

Figure 2 A view showing the plural types of metal bonding to glutamate in the $[K(\text{L-Glu})(H_2O)]_n$ complex [50], where a K^+ ion forms an intramolecular bridge between the γ-carboxylate oxygen and one of the two α-carboxylate oxygens, the second ion binds to the same α-carboxylate oxygen, and the other two metal ions simultaneously attach to the other α-carboxylate oxygen, the glutamate molecule behaving as a pentadentate ligand. Hydrogen atoms are drawn in mint color.

except for one case, $[K(\text{L-Asp})(H_2O)_2]_n$ [45], in which solely β-carboxylate oxygens are bound. As an example, Figure 2 shows the K^+ bonding in $[K(\text{L-Glu})(H_2O)]_n$ [50].

In summary, among the four groups of amino acids, the most interesting amino acids, from the viewpoint of coordination chemistry and their possible interactions with metal ions as amino acid residues of proteins, are those that belong to group *4*, followed by group *2*. Though the data are still limited, it may be concluded about the alkali metal ion binding sites at physiological pH that the polar functional groups, OH of Ser, Thr, and Tyr, SH of Cys, $CONH_2$ of Asn and Gln, which belong to the group *2* amino acids, are little significant, whereas the β- and γ-COO^- groups of Asp and Glu, respectively, which belong to group *4*, are of major importance. It is noticeable that alkali metal ion binding to the neutral amino group NH_2 is observed only once in the Li^+–Gly complex [14].

2.2 *Small Peptide Complexes*

For peptides, possible ligation sites are the amide oxygen and the nitrogen (but only when deprotonated) within the peptide backbone, polar functional groups on the side chains of the amino acids belonging to the groups *2* and *4* in Section 2.1, and the amino terminal nitrogen and the carboxy terminal oxygens. Table 3 lists the total number of alkali metal ion complexes with small peptides and Table 4 summarizes metal binding sites/modes on the peptide ligands for each complex [14, 52–56]. Antibiotic ionophore peptides are not involved here but are separately dealt with in Section 5.2.1.

Unfortunately, only eight crystal structures are available: three complexes [14, 52] for Li^+, three [53–55] for Na^+, one each for K^+ [53] and Rb^+ [56], and none for

Table 3 Total number of alkali metal ion–small peptide complexes in the solid state.[a,b]

Peptide[c]	Li^+	Na^+	K^+	Rb^+	Cs^+	Total
Glycyl[d]	3[e] (5)	2[g] (20)	1[j] (6)	1[k] (2)	0	7 (33)
Alanyl	0 (2)	0 (2)	0 (3)	0	0	0 (7)
Valyl	0 (1)	1[h] (2)	0	0	0	1 (3)
Leucyl	0	1[h] (2)	0 (2)	0	0	1 (4)
Isoleucyl	0	0	0	0	0	0
Prolyl	1[f] (2)	1[h] (5)	0	1[l] (1)	0	3 (8)
Phenylalanyl	0 (0)	0 (1)	0 (3)	0	0	0 (4)
Tryptophanyl	0	1[h] (1)	0	0	0	1 (1)
Methionyl	0	0	0	0	0	0
Seryl	0	0	0	0	0	0
Threonyl	0	0 (1)	0	0	0	0 (1)
Cysteinyl	0	0	0	0	0	0
Tyrosyl	0	1[i] (1)	0	0	0	1 (1)
Asparaginyl	0	0	0	0	0	0
Glutaminyl	0	0 (1)	0	0	0	0 (1)
Histidyl	0	0 (1)	0	0	0	0 (1)
Lysyl	0	0 (1)	0	0	0	0 (1)
Arginyl	0	0 (1)	0	0	0	0 (1)
Aspartyl	0	1[h] (1)	0	0	0	1 (1)
Glutamyl	0	0 (1)	0	0	0	0 (1)
Total	4 (10)	8 (41)	1 (14)	2 (3)	0	15 (68)

[a]Obtained from the Cambridge Structural Database (the CSD version 5.35 updated to February 2014).
[b]The number in the parenthesis denotes a total number of alkali metal ion *complexes* and alkali metal ion *salts* that have no direct metal–peptide bonding.
[c]A peptide that contains more than two kinds of amino acid residues is overlappingly counted into the item of individual amino acid residues: for example, a peptide 'GlyPro' is counted in both 'Glycyl' and 'Prolyl'.
[d]'Glycyl' denotes Gly residue containing peptides.
[e][14, 52]. [f][52]. [g][53, 54]. [h][55]. [i][54]. [j][53]. [k][56]. [l][56].

Cs^+. Above all, peptide ligands of interest that contain polar residues belonging to groups *2* and *4* are only *N*-substituted glycyltyrosine [54] and a cyclic pentapeptide, cyclo(–D-Trp-D-Asp-L-Pro-D-Val-L-Leu-) [55] (known as BQ123, a potent endothelin-1 inhibitor). BQ123 possesses a high, intrinsic affinity for Na^+ ion [57]. The Na^+ complex of BQ123 is reported [55]: there exist four independent Na^+ ions (Na1-Na4) and four independent ligand molecules (L1-L4), among which two Na^+ ions (Na1 and Na2) each are sandwiched between two peptide molecules (L1 and L2 for Na1 and L3 and L4 for Na2), forming cage-like structures, and the other two Na^+ ions (Na3 and Na4) form a cluster composed of water ligands and peptides (L1-L4).

In 'sodium-caged' structures, as shown in Figure 3, the octahedral coordination sites of Na1 and Na2 are fully occupied by D-Asp and D-Val residues from two peptide ligands (L1 and L2 for Na1, and L3 and L4 for Na2) through two amide oxy-

Table 4 Alkali metal ion binding sites/modes on small peptides in the solid state.[a,b]

Complex		Coordination geometry or number	Metal binding sites on peptide ligand and other ligands/comments, if any	CSD code or Reference code[c]	Ref.
Li^+ ion complexes					
{[Li(GlyGlyH)]·Cl}$_n$		Tet.	3O [α-CO_2^- (Mol.1, Mol.2, Mol.3): 1.930-1.963][d], O [amide[e] (Mol.4): 1.898] GlyGlyH is neutral with the N-terminal NH_2 and the C-terminal COOH.	HEFXEV	[14]
{[Li(GlyGly)]·H_2O}n		Tet.	3O [α-CO_2^- (Mol.1, Mol.2, Mol.3): 1.903-1.945], O [amide (Mol.4): 1.904] GlyGly is a monovalent anion with the N-terminal NH_2 and the C-terminal COO^-.	HEFXIZ	[14]
[Li{cyclo(Gly-L-Pro)$_3$}(SCN)]$_2$·3H_2O	Li1	Tet.	3O [amide (Mol.1, Mol.1', Mol.1"): 1.962], N [solvent: 1.899]	YUWXUJ	[52]
	Li2	Tet.	3O [amide (Mol.1, Mol.1', Mol.1"): 1.948], N [solvent: 2.012] Li1 and Li2 ride each on a threefold axis. A model for ion transport by cyclo(GlyPro)$_3$.		
Na^+ ion complexes					
[Na{N,N'-bis(O-methyl-GlyGly)-4,13-diaza-18-crown-6}]$_2$·2I·THF·H_2O	Na1	8-Coord.	2O [Gly1 amide (Mol.1, Mol.2): -[f]], 2N [diazacrown: 2.640, 2.664], 4O [diazacrown: 2.471-2.556]	SAHSEZ	[53]
	Na2	8-Coord.	2O [Gly1 amide (Mol.1, Mol.2): -], 2N [diazacrown: 2.651, 2.692], 4O [diazacrown: 2.515-2.598]		
[Na{Cu(*N*-5-bromosalicyl-aldehydeGly-L-Tyr)}$_3$Sr(H_2O)$_3$]·9H_2O		6-Coord.	3O [Tyr α-CO_2^- (Mol1, Mol.1', Mol.1"): 2.376], 3O [aqua: 2.483] Na^+ rides on a threefold axis.	BIJFOQ	[54]
[Na{cyclo(-D-Trp-D-Asp-L-Pro-D-Val-L-Leu-)}]$_4$·8H_2O·5.63(iPrOH)	Na1	6-Coord.	2O [Asp amide: 2.280 (Mol.1), 2.263 (Mol.2)], 2O [Asp β-CO_2^- 2.375 (Mol.1), 2.375 (Mol.2)], 2O [Val amide: 2.471 (Mol.1), 2.459 (Mol.4)]	he0271[c]	[55]
	Na2	6-Coord.	2O [Asp amide: 2.247 (Mol.3), 2.288 (Mol.4)], 2O [Asp β-CO_2^-: 2.387 (Mol.3), 2.338 (Mol.4)],		

(continued)

Table 4 (continued)

			2O [Val amide: 2.519 (Mol.3), 2.502 (Mol.4)]		
	Na3	5-Coord.	2O [Trp amide: 2.277 (Mol.1'), Leu amide: 2.384 (Mol.1')], 3O [aqua: 2.292-2.340]		
	Na4	6-Coord.	2O [Pro amide: 2.354 (Mol.3'), Leu amide: 2.376 (Mol.3')], 4O [aqua: 2.243-2.847] See Figure 3.		
K^+ ion complex					
[K{N,N'-bis(O-methyl-GlyGly)-4,13-diaza-18-crown-6}]·I	K1	8-Coord.	2O [Gly1 amide (Mol.1, Mol.2): -][g], 2N [diazacrown: 2.985, 2.985], 4O [diazacrown: 2.733-2.786]	FISGUJ10	[53]
	K2	8-Coord.	2O [Gly1 amide (Mol.1, Mol.2): -], 2N [diazacrown: 2.999, 2.999], 4O [diazacrown: 2.772-2.792]		
Rb^+ ion complex					
{[Rb{cyclo(L-ProGly)$_4$}(H_2O)]·-NCS·2H_2O}$_n$		Dist. oct.	4O [amide (Mol.1): 2.84-3.02], O [amide (Mol.2: 2.86)], O[aqua: 2.78]	CPRGLR	[56]
Cs^+ ion complex: None					

[a]Obtained from the Cambridge Structural Database (the CSD version 5.35 updated to February 2014).
[b]Abbreviations: Dist. oct. distorted octahedral; Dist. tet., distorted tetrahedral; iPr, isopropyl; Tet., terahedral.
[c]When the structure is not registered in the CSD version 5.35 updated to February 2014, the reference code for suplementary data is given.
[d]'3O [α-CO_2^- (Mol.1, Mol.2, Mol.3): 1.930-1.963]' denotes that the metal ion binds to three C-terminal α-carboxyl oxygens from three peptide molecules (GlyGly in this case), Mol.1, Mol.2, and Mol.3, and their bond distances (Å).
[e]'amide' denotes the amide oxygen of the peptide backbone.
[f]Not reported.
[g]'2O [Gly1 amide (Mol.1, Mol.2): –]' denotes that the metal ion binds to two amide oxygens of the first Gly residue from two peptide molecules, Mol.1 and Mol.2, and their bond distances (Å) (not reported in this case).

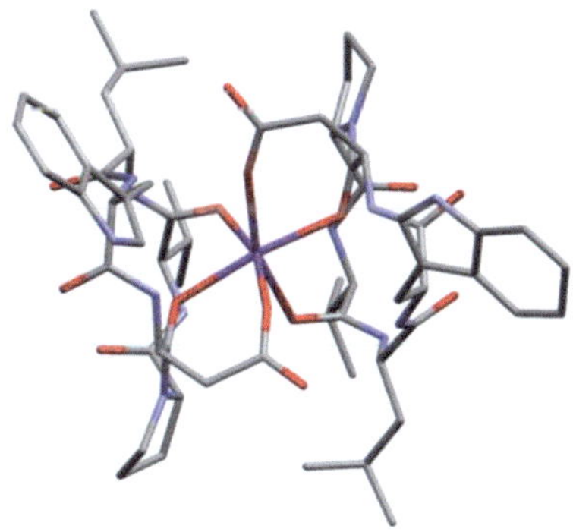

Figure 3 Structure of the Na^+ complex [55] of cyclo(–D-Trp-D-Asp-L-Pro-D-Val-L-Leu-), an opioid peptide. The Na^+ ion is sandwiched between two peptide molecules, forming a cage-like structure.

gens and two β-carboxylate oxygens of Asp and two amide oxygens of Val. Na3 and N4, which are connected to each other by a bridging water ligand, are each ligated by three or four water ligands and a peptide ligand: for Na3, through two amide oxygens of Trp and Leu residues while through two amide oxygens of Pro and Leu of L3 for Na4, and *via* these water ligands peptide ligands L2 and L4 are further connected, forming a 'two sodium-embedded' cluster.

The observed data, though they are quite limited, show that the amide carbonyl oxygen is the most preferred metal binding site, followed by the terminal carboxyl oxygens (observed in the two Li^+–GlyGly complexes [14] and a Na^+–GlyTyr complex [54]). This is also the case for antibiotic ionophore peptides, as described later in Section 5.2.1. The β-carboxylate oxygen of the Asp side-chain in an Asp-containing peptide is also involved in metal coordination in the Na^+ complex, where Asp plays a crucial role in forming the Na^+ complex [55].

3 Nucleic Acid Constituent Complexes

There are three potential metal binding groups on nucleic acids: phosphate, sugar, and base moieties. Extensive X-ray studies of metal ion interactions with nucleic acid constituents reveal [4, 75] that the phosphate groups of the mono-, di-, tri-, and oligonucleotides are good ligands for alkali and alkaline earth metal ions as well as transition and heavy metal ions. Sugar hydroxyl groups are good ligands for alkali and alkaline earth metal ions but not for transition and heavy metal ions. Ring nitrogens of bases are good targets for all types of metal ions, particularly for transition and heavy metal ions. Coordination to the ring nitrogens is favored over binding to the exocyclic amino and keto groups: the N7 site is the most preferred metal binding site for purine bases and the N3 site for cytosine. Keto substituents at O2 of cytosine and O2 and O4 of thymine and uracil are usual metal binding sites for alkali and alkaline earth metal ions. As a rule, amino substituents at N6 of adenine, N2 of guanine, and N4 of cytosine become good ligands only when they are deprotonated, with the sole exception observed in the [Ru(9-methyladenine)(trithiacyclononane)-(DMSO)]$^{2+}$ complex [76] by accompanying the N7–Ru–N6(NH_2) chelation. Nomenclature and chemical structure of the common nucleotides are presented in Figure 4. For the conformational terms of the nucleotide structures see [77].

3.1 Nucleobase Complexes

Table 5 gives the total number of alkali metal ion complexes with nucleobases and Table 6 summarizes metal binding sites/modes for each complex [58–74]. Substituted nucleobases are not involved except for N9-substituted purine bases and N1-substituted pyrimidine bases.

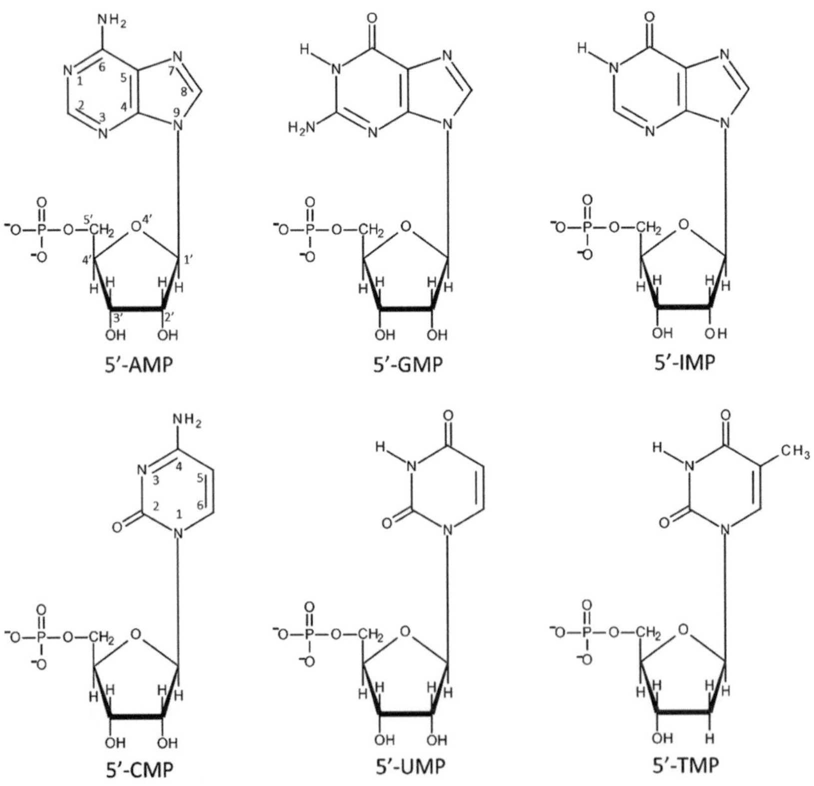

Figure 4 Nomenclature and chemical structure of the common nucleotides. The nucleotides are shown in the *syn* conformation (to save space) even though they mostly occur in the *anti* form.

Table 5 Total number of alkali metal ion–nucleobase complexes in the solid state.[a,b]

Nucleobase[c]	Li^+	Na^+	K^+	Rb^+	Cs^+	Total
Adenine	0	0	1[d] (1)	0	0	1 (1)
Guanine	0	1[e] (3)	1[f] (2)	0	0	2 (5)
Hypoxanthine	0	0	0	0	0	0
Cytosine	0	5[g] (5)	1[h] (2)	0	0	6 (7)
Thymine	0	5[i] (7)	5[j] (6)	1[k] (1)	1[l] (1)	12 (15)
Uracil	0	4[m] (4)	3[n] (3)	0	2[o] (2)	9 (9)
Total	0	15 (19)	11 (14)	1 (1)	3 (3)	30 (37)

[a]Obtained from the Cambridge Structural Database (the CSD version 5.35 updated to February 2014).
[b]The alkali metal ion–nucleobase complexes are those that contain *direct* metal bonding to nucleobases. The number in parenthesis denotes the total number of alkali metal ion-containing compounds of nucleobases, involving alkali metal ion–nucleobase complexes.
[c]Substituted nucleobases are not involved except for N(9)-substituted purine bases and N(1)-substituted pyrimidine bases.
[d][58]. [e][59]. [f][60]. [g][59, 61, 62]. [h][59]. [i][63–67]. [j][67–69]. [k][67]. [l][67]. [m][63, 70–72] [n][73, 74]. [o][59, 72].

Adenine complex: Only one complex is available, [K(9-Et-azacrown-Ade)]·PF_6 [58], where the K^+ ion is captured by the N9-tethered azacrown chelator and this enables the metal ion to bind to N3, which is the rare metal binding site for N9-substituted purine bases. On the other hand, in the corresponding Na^+ complex [58], such a trapped Na^+ ion binds *indirectly* to N3 via an O–H (aqua ligand)···N3 hydrogen bond.

Guanine complexes: In the Na^+ complex, $[Na(H_2O)_2\{Pt(9\text{-MeGua})_2(1\text{-MeCyt})(NH_3)\}]\cdot 3ClO_4\cdot 0.5H_2O$ [59], the Na^+ ion binds to the exocyclic keto substituent O6 of the guanine ligand (the Pt^{2+} ion binds to the ring nitrogen N7). On the other hand, in the K^+ complex, $[K_6(H_2O)_6\{Pt(9\text{-MeGua})(MeNH_2)_2\}\{Pt(CN)_4\}_3]_n$ [60], one of the three independent K^+ ions binds to O6 of one ligand, while the other K^+ ion binds to the same ligand through the five-membered N7–M–O6 chelation and additionally to the other ligand by forming a quite rare M–NH_2 (N2) bonding or a π interaction with the C2–N2 bond (the Pt^{2+} ion binds to the deprotonated N1).

Hypoxanthine complex: None.

Cytosine complexes: Five Na^+ complexes [59, 61–64] and a K^+ complex [59] are reported. In $[Na(H_2O)_2\{Pt(9\text{-MeGua})_2(1\text{-MeCyt})(NH_3)\}]^{3+}$ [59] (which is the same compound as that described above in the guanine complexes) and $[Na(NO_3)\{Pt(1\text{-MeCyt–H})_4Co(H_2O)\}]_2\cdot 2NO_3\cdot 4H_2O$ [61], the Na^+ ions bind to the exocyclic O2 of the cytosine ligand (the Pt^{2+} ion binds to the ring N3 in both compounds; the Co^{2+} ion binds to the deprotonated amino N4 [61]). Two analogous Na^+ complexes, $\{[Na_3(Cyt)_3(H_2O)_4(ClO_4)]\cdot 2ClO_4\}_n$ [62] and $\{[Na_3(Cyt)_3(H_2O)_3(ClO_4)_2]\cdot ClO_4\}_n$ [62] are of interest in that each O2 of three independent cytosine ligands is coordinated simultaneously to three different Na^+ ions (Na1-Na3), in a μ_3-coordination mode; Na1 is bound by four ligand molecules, Na2 by three ligands, and Na3 by two

Table 6 Alkali metal ion binding sites/modes on nucleobases in the solid state.[a]

Complex		Coordination geometry or number	Metal binding sites on nucleobases and other ligands/comments, if any[b]	CSD code	Ref.
Adenine complex					
[K(9-Et-azacrown-Ade)]·PF_6		7-Coord.	*N3* [Ade: 2.939], N [azacrown: 3.149], 5O [azacrown: 2.622-2.816]	CAFKIE	[58]
Guanine complexes					
[Na(H_2O)$_2${Pt(9-MeGua)$_2$(1-MeCyt)-(NH_3)}]·3ClO_4·0.5H_2O		5-Coord.	2*O6* [Gua: 2.261 (Mol.1), 2.379 (Mol.2)][b], *O2* [Cyt: 2.58][b], 2O[aqua: 2.308, 2.40], O [ClO_4: 2.58] The Pt^{2+} ion binds to two Gua ligands each through N7 and to a Cyt ligand through N3.	XAHZOV	[59]
[K_6(H_2O)$_6${Pt(9-MeGua)$_2$($MeNH_2$)$_2$}-{Pt(CN)$_4$}$_3$]$_n$	K1	6-Coord.	*O6* [Gua (Mol.1): 2.788], 3N [Pt(CN)$_4$: 2.774-3.142], 2O [aqua: 2.708, 2.721]	OYOVEE	[60]
	K2	4-Coord.	3N [Pt(CN)$_4$: 2.796-2.889], O [aqua: 2.737]		
	K3	8-Coord.	*N7,O6* [Gua (Mol.1): 2.987 (N7), 3.050 (O6)], *N2* [Gua (Mol.2): 3.131], 3N [Pt(CN)$_4$: 2.850-3.105], 2O [aqua: 2.893, 3.051] N7–K^+–O6 chelation. Quite rare K^+–NH_2 (N2) bonding. Pt1 binds to Gua at deprotonated N1. Pt1 and Pt3 ride on inversion centers.		
Hypoxanthine complex: None					
Cytosine complexes					
[Na(H_2O)$_2${Pt(9-MeGua)$_2$(1-MeCyt)-(NH_3)}]·3ClO_4·0.5H_2O		5-Coord.	2*O6* [Gua: 2.261 (Mol.1), 2.379 (Mol.2)], *O2* [Cyt: 2.58], 2O [aqua: 2.308, 2.40], O [ClO_4: 2.58] The Pt^{2+} ion binds to two Gua ligands each through N7 and to a Cyt ligand through N3.	XAHZOV	[59]
[Na(NO_3){Pt(1-MeCyt–H)$_4$-Co(H_2O)}]$_2$·2NO_3·4H_2O		6-Coord.	5*O2* [Cyt (Mol.1-Mol.5)], O [nitrate] Na–O distances (including nitrate) in the range of 2.230-2.644 Å. The Pt^{2+} ion binds to N3. The Co^{2+} ion binds to the deprotonated N4.	CUKMIE	[61]

$\{[Na_3(Cyt)_3(H_2O)_4(ClO_4)]\cdot 2ClO_4\}_n$	Na1	6-Coord.	4*O2* [Cyt (Mol.1-Mol.4): 2.373-2.672], 2O [aqua: 2.356, 2.373]	ACITIQ	[62]
	Na2	6-Coord.	3*O2* [Cyt (Mol.1-Mol.3): 2.422-2.521], O [perchlorate: 2.403], 2O [aqua: 2.370, 2.374]		
	Na3	6-Coord.	2*O2* [Cyt (Mol.1, Mol.2): 2.446, 2.495], O [perchlorate: 2.413], 3O [aqua: 2.343-2.356]		
$\{[Na_3(Cyt)_3(H_2O)_3(ClO_4)_2]\cdot ClO_4\}_n$	Na1	6-Coord.	4*O2* [Cyt (Mol.1-Mol.4): 2.367-2.614], 2O [aqua: 2.329, 2.370]	ACITOW	[62]
	Na2	6-Coord.	3*O2* [Cyt (Mol.1-Mol.3): 2.422-2.521], 2O [perchlorate (Mol.1, Mol.2): 2.339, 2.404], O [aqua: 2.377]		
	Na3	6-Coord.	2*O2* [Cyt (Mol.1, Mol.2): 2.446, 2.495], O [perchlorate: 2.414], 3O [aqua: 2.343-2.357]		
$\{[K\{Pt(1\text{-}MeCyt)_2(NH_3)_2\}(PF_6)_3]\cdot H_2O\}_n$		8-Coord.	2*O2* [Cyt (Mol.1, Mol.1'): 2.689], 6F [PF_6 (Mol.1-Mol.4): 2.829-3.1703] The K^+ ion rides on a twofold axis. The Pt^{2+} ion binds to two Cyt bases each at N(3).	XAHZIP	[59]
Thymine complexes					
$[Na(1\text{-}MeThy)(H_2O)_4].[AuClO_4]\cdot 1\text{-}MeThy\cdot 2H_2O$		6-Coord.	*O2* [Thy: 2.286], 5O [aqua: 2.378-2.547]	JEYVUE	[63]
$[Na(1\text{-}Pr\text{-}azacrown\text{-}Thy)_2]\cdot acetone$		6-Coord.	2*O2* [Thy: 2.468 (Mol.), 2.488 (Mol.2)], 2N [azacrown: 3.088, 3.201], 4O [azacrown: 2.416-2.713]	QIFVUW	[64]
$[Na\{MeHg(1\text{-}MeThy\text{–}H)\}_2(NO_3)]$		6-Coord.	2*O2* [Thy (Mol.1, Mol.1'): 2.436], 2*O4* [Thy (Mol.2, Mol.2'): 2.324], 2O [nitrate (Mol.1, Mol.2): 2.406] The Na^+ ion rides on an inversion center. The $MeHg^{2+}$ ion binds to the deprotonated N3 of Thy.	TYMHGB10	[65]
$\{[Na_2\{Pd(1\text{-}MeThy\text{–}H)_2(en)\}\text{-}(NO_3)_2]\cdot H_2O\}_n$	Na1	6-Coord.	2*O4* [Thy: 2.458 (Mol.1), 2.255 (Mol.2)], *O2* [Thy (Mol.3): 2.402], 3O [aqua: 2.371-2.704]	YIBWAH	[66]
	Na2	6-Coord.	2*O2* [Thy: 2.309 (Mol.1), 2.349 (Mol.2)], *O4* [Thy (Mol.4): 2.383], 3O [aqua:2.404-3.050] The Pd^{2+} ion binds to head-to-head oriented two Thy bases through		

(continued)

Table 6 (continued)

			the deprotonated N3. Intramolecular O4–Na1–O4 and O2–Na2–O2 bondings.		
[Na(1-EtThy)$_4$(H$_2$O)$_2$]·[Na(1-EtThy)$_2$-(H$_2$O)$_2$]·[AuCl$_4$)]$_2$·2(1-Et-Thy)·2H$_2$O	Na1	6-Coord.	2*O2* [Thy (Mol.1, Mol.1'): 2.368], 2*O4* [Thy (Mol.2, Mol.2'): 2.375], 2O [aqua: 2.503]	ASACOM	[67]
	Na2	Sq. pl.	2*O2* [Thy (Mol.3, Mol.3'): 2.450], 2O [aqua: 2.184] Na1 and Na2 ride on inversion centers.		
{[K(1-EtThy)$_4$]·[AuCl$_4$]}$_n$		6-Coord.	4*O2* [Thy: 2.767 (Mol.1, Mol.1'), 3.062 (Mol.2, Mol.2')], 2*O4* [Thy: 2.777 (Mol.3, Mol.3')] The K$^+$ ion rides on an inversion center.	ASABIF	[67]
{[K(1-EtThy)$_4$(AuCl$_4$)$_{0.5}$]·[AuCl$_4$]$_{0.5}$}$_n$		7-Coord.	3*O2* [Thy (Mol.1-Mol.3): 2.671-2.867], 2*O4* [Thy (Mol.4, Mol.5): 2.760, 2.791], 2Cl [AuCl$_4$: 3.240, 3.288]	ASABOL	[67]
[K(Thy–H)(H$_2$O)$_3$]		6-Coord.	*O2* [Thy: -], 5O [aqua: -] The thyminato ligand is in the N1-deprotonated form.	KTHYMT	[68]
[K(1-PrThy)$_2$(Pd$_2$Cl$_6$)$_{0.5}$]		7-Coord.	*O2* [Thy: -], *O4* [Thy: -], 5Cl [Pd$_2$Cl$_6$: -]	PRTMKP	[69]
{[K(1-MeThy)$_2${AuCl$_{1.5}$(CN)$_{2.5}$}]}$_n$		6-Coord.	2*O2* [Thy: 2.647 (Mol.1), 2.654(Mol.2), 3*O4* [Thy (Mol.3-Mol.5): 2.654-3.007], N [AuCN: 2.850]	ASABUR	[67]
{[Rb(1-EtThy)$_4$]·[AuCl$_4$]}$_n$		6-Coord.	4*O2* [Thy: 3.083 (Mol.1, Mol.1'), 2.880 (Mol.2, Mol.2')], 2*O4* [Thy: 2.856 (Mol.3, Mol.3')] The Rb$^+$ ion rides on an inversion center.	ASACAY	[67]
{[Cs(1-EtThy)$_4$]·[AuCl$_4$]}$_n$		6-Coord.	4*O2* [Thy: 3.175 (Mol.1, Mol.1'), 3.262 (Mol.2, Mol.2')], 2*O4* [Thy: 3.200 (Mol.3, Mol.3')] The Cs$^+$ ion rides on an inversion center.	ASACEC	[67]
Uracil complexes					
[Na(1-MeUra)$_4$]·[AuCl$_4$]		Sq. pl.	4*O4* [Ura: 2.301 (Mol.1, Mol.1'), 2.336 (Mol.2, Mol.2')] The Na$^+$ ion rides on an inversion center.	JEYVOY	[63]
[Na(H$_2$O)$_4${Pt(1-MeUra–H)-(NH$_3$)$_2$}$_2$Ag]·2ClO$_4$·6.5H$_2$O		6-Coord.	2*O4* [Ura (Mol.1, Mol.1'): 2.318], 4O [aqua: 2.44-2.61] The Na$^+$ ion rides on a twofold axis. The Pt^{2+} ion binds to two Ura ligands each at the deprotonated N3. The Ag$^+$ ion binds to two Ura ligands each at O4.	HIXMIK	[70]

$[Na_2(H_2O)_2$(DL-willardiine)$_4$-Pd$\{Cu(H_2O)_2\}]\cdot 9H_2O$	Na1	5-Coord.	2*O2* [Ura: 2.245 (Mol.1), 2.385 (Mol.2)], 2*O4* [Ura: 2.310 (Mol.3), 2.357 (Mol.4)], O [aqua: -]	TASHOK	[71]
	Na2	5-Coord.	2*O4* [Ura: 2.359 (Mol.1), 2.311 (Mol.2)], 2*O2* [Ura: 2.432 (Mol.3), 2.271 (Mol.4)], O [aqua: -] The Pd^{2+} ion binds to four uracil ligands each through the deprotonated N3. The Na^{+} ions are captured into the cavities constructed by the assembly of four carbonyl oxygens of the [Pd(uracilato)$_4$].		
$\{[\{Na(1\text{-}MeUra\text{–}H)_2(H_2O)_2\}_2Pd]\cdot$-$7H_2O\}_n$	Na1	8-Coord.	4*O2* [Ura (Mol.1, Mol.1', Mol.1'', Mol.1'''): 2.400], 4O [aqua: 2.445]	WIQYEA01	[72]
	Na2	8-Coord.	4*O4* [Ura (Mol.2, Mol.2', Mol.2", Mol2.'"): 2.419], 4O [aqua: 2.504] Na1 and Na2 ions and the Pd^{2+} ion ride on a fourfold axis. The Pd^{2+} ion binds to four uracil ligands each through the deprotonated N3.		
$[K_2\{PtI_2(1\text{-}MeUra\text{–}H)_2\}]\cdot 6H_2O$		8-Coord.	*O2* [Ura (Mol.1): 2.760], *O4* [Ura (Mol.2): 2.834], 6O [aqua: 2.714-3.187] The Pt^{2+} ion rides on an inversion center and binds to symmetry-related two Ura bases through the deprotonated N3.	HEFWOE	[73]
$\{[\{K(1\text{-}MeUra\text{–}H)_2(H_2O)_2\}_2Pt]\cdot$-$6H_2O\}_n$	K1	6-Coord.	4*O2* [Ura (Mol.1-Mol.4): 2.691-2.848], 2O [aqua: 2.848, 2.968]	SIYMIX	[74]
	K2	6-Coord.	4*O4* [Ura (Mol.1-Mol.4): 2.711-2.883], 2O [aqua: 2.883, 3.084] The Pt^{2+} ion binds to four Ura ligands each through the deprotonated N3.		
$\{[\{K(1\text{-}MeUra\text{–}H)_2(H_2O)_2\}_2Pd]\cdot$-$8H_2O\}_n$	K1	6-Coord.	4*O2* [Ura (Mol.1-Mol.4: 2.693-2.721], 2O [aqua: 2.84, 3.021]	LUKWOD	[72]
	K2	7-Coord.	4*O4* [Ura (Mol.1-Mol.4): 2.692-2.727], 3O [aqua: 2.85-3.15] The Pd^{2+} ion binds to four Ura ligands each through the deprotonated N3.		
$\{[Cs(H_2O)_2\{Pt(1\text{-}MeUra)(NH_3)I_2\}]\cdot$-$2H_2O\}_n$		6-Coord.	*O2* [Ura (Mol.1): 3.163], *O4* [Ura (Mol.2): 3.139], 2I [PtI_2 (Mol.1, Mol. 2): 3.936, 3.988], 2O [aqua: 3.11, 3.32] The Pt^{2+} ion binds to two Ura bases each through the deprotonated	XAHZEL	[59]

(continued)

Table 6 (continued)

			N3.		
{[{Cs(1-MeUra–H)$_2$(H$_2$O)$_2$}$_2$Pd]·-6H$_2$O}$_n$	Na1	7-Coord.	4*O2* [Ura (Mol.1-Mol.4): 2.975-3.019], 3O [aqua: 3.18-3.77]	LUKWIX	[72]
	Na2	8-Coord.	4*O4* [Ura (Mol.1-Mol.4): 2.98-3.012], 4O [aqua: 3.27-3.76]		
			The Pd^{2+} ion binds to four Ura ligands each through the deprotonated N3.		

[a]Abbreviations: en, ethylenediamine; 9-Et-azacrown-Ade, 16-(2-(9H-adenin-9-yl)ethyl)-16-aza-1,4,7,10,13-pentaoxacyclooctadecane; 1-MeThy–H, N3-deprotonated 1-methylthyminate; 1-MeUra–H, N3-deprotonated 1-methyluracilate; 1-Pr-azacrown-Thy, (3-(1-thyminyl)propyl)-4,13-diaza-18-crown-6; 1-PrThy, 1-propylthymine; Sq. pl., square planar; Willardiine, 1-(2-amino-2-carboxyethyl)uracil.

[b]For example, '2*O6* [Gua: 2.261 (Mol.1), 2.379 (Mol.2)' denotes that the metal ion binds to two O6 atoms from two guanine molecules (Mol.1 and Mol.2) with the bonding distances of 2.261 and 2.379 Å, respectively.

ligands, each through O2. In the K^+ complex, $\{[K\{Pt(1\text{-}MeCyt)_2(NH_3)_2\}(PF_6)_3]\cdot H_2O\}_n$ [59], the K^+ ion chelates to two O2 oxygens of the head-to-head oriented two cytosine ligands within the *trans*-$[Pt(1\text{-}MeCyt)_2(NH_3)_2]^{2+}$ unit (N3–Pt^{2+}–N3 coordination).

Thymine complexes: Five complexes for Na^+ [63–67] have been reported, five for K^+ [67–69], and one each for Rb^+ [67] and Cs^+ [67]. In all of the twelve complexes, the O2 site participates always and the O4 site mostly (in nine complexes) in metal coordination. There seems to be a trend that, when the metal ion is highly hydrated, it binds to O2 rather than O4, and this is also the case for $[Na(1\text{-}MeThy)(H_2O)_4]^+$ [63] and $[K(Thy\text{–}H)(H_2O)_3]$ [68], in which the metal ion binds only to O2. The failure of metal bonding to O4 but O2 in $[Na(1\text{-}Pr\text{-}azacrown\text{-}Thy)_2]$ [64] is clearly due to the metal trapping in the N1-tethered azacrown chelator. A series of complexes, $[M(1\text{-}EtThy)_4]^+$ (M: Na^+ [ASACOM], K^+ [ASABIF], Rb^+ [ASABOR], and Cs^+ [ASACEC]), which were intended to be mimics for thymine quartet structures, provide a variety of structural motifs having an exclusive metal coordination to exocyclic O2 and O4 of thymine bases. The authors show that K^+ has the ideal size to fit either in the plane of a quartet or between two quartets [67].

Uracil complexes: Four Na^+ complexes [63, 70–72], three K^+ complexes [72, 74], and two Cs^+ complexes [59, 72] are available. In these nine complexes, the O2 or O4 sites are commonly involved in metal bonding with an equal frequency of appearance. In a series of uracil complexes, $[(Na^+, K^+, Rb^+, \text{or } Cs^+)_2\{Pd(1\text{-}MeUra\text{–}H)_4\}]$ (Na^+ [WIQYEA01], K^+ [LUKWOD], and Cs^+ [LUKWIX]) [72], $[K_2\{Pt(1\text{-}MeUra\text{–}H)_4\}]$ [SIYMIX] [74], $[Na_2\{Pd(willardiine)_4\}]$ (willardiine: 1-(2-amino-2-carboxyethyl)uracil, a naturally occurring amino acid) [71], the alkali metal ions are captured in the cavities constructed by the assembly of four carbonyl O2 and O4 oxygens of the [M(1-MeUra–H or willardiine)$_4$] structure, thus, those behaving as ‘metallacryptates’.

In summary, alkali metal ion complexes of purine bases provide rare metal binding modes, M–N3 bonding for adenine [58] and five-membered N7–M–O6 chelation, and even more uncommon M–NH_2 (N2) bonding or π interaction with the C2–N2 bond for guanine [60]. For pyrimidine bases, exocyclic carbonyl oxygens, O2 for cytosine, O2 and O4 for thymine and uracil, are preferred metal binding sites for alkali metal ions. Thymine and uracil bases, which have a wealth of carbonyl oxygens, are fascinating as building blocks for constructing ‘metallaionophores’ with a variety of building architectures [67, 72].

3.2 Nucleoside Complexes

A nucleoside provides, in addition to the base moiety, the sugar part that has three hydroxyl oxygens (O2’, O3’, and O5’) as possible metal binding sites. Table 7 lists the total number of alkali metal ion complexes with nucleosides and Table 8 summarizes metal binding sites/modes for each complex [78–91].

Adenosine complex: A sole adenosine complex is $[Na(H_2O)\{Sb(Ado\text{–}2H)_2\}]_n$ [78], where the Sb^{3+} ion is coordinated by two adenosinato ligands each through the

Table 7 Total number of alkali metal ion–nucleoside complexes in the solid state.[a,b]

Nucleoside (Abbreviation)	Li^+	Na^+	K^+	Rb^+	Cs^+	Total
Adenosine (Ado)	0	1[c] (2)	0	0	0	1 (2)
Guanosine (Guo)	0	0	0 (1)	0	1[d] (1)	1 (2)
Inosine (Ino)	0	1[e] (1)	0	0	0	1 (1)
Cytidine (Cyd)	0	0	0	0	0	0
Thymidine (Thd)	0	0	0	0	0	0
Uridine (Urd)	0	1[f] (1)	0	0	0	1 (1)
Total	0	3 (4)	0 (1)	0	1 (1)	4 (6)

[a]Obtained from the Cambridge Structural Database (the CSD version 5.35 updated to February 2014).
[b]The alkali metal ion–nucleoside complexes are those that contain *direct* metal bonding to nucleosides. The number in parenthesis denotes the total number of alkali metal ion-containing compounds of nucleosides, involving alkali metal ion–nucleoside complexes.
[c][78]. [d][79]. [e][80]. [f][81].

1,2-diolato group of the sugar moiety of the nucleoside. The Na^+ ion bridges between two $[Sb(Ado–2H)_2]^-$ entities by forming two intramolecular O3'–M–O5' chelations, one chelation with one Ado–2H ligand in one $[Sb(Ado–2H)_2]^-$ and the other chelation with the other Ado–2H ligand of the other $[Sb(Ado–2H)_2]^-$.

Guanosine complex: No guanosine but an isoguanosine (isoGuo) complex of Cs^+ is reported, $[Cs(O2',O3',O5'\text{-substituted isoGuo})_{10}(CH_3CN)_2]_2^{2+}$ [79]. Isoguanosine, a guanosine isomer with transposed carbonyl and amino groups, selfassociates in the presence of cations [82]. This isoGuo derivative extracts cesium salts from water into $CHCl_3$ with an affinity and selectivity rivaling that of covalent macrocycles [83]. As shown in Figure 5, the crystal structure of the Cs^+ complex [79] reveals that the isoGuo derivatives selfassociate to give a hydrogen-bonded, planar cyclic isoGuo-pentamer and the Cs^+ ion is sandwiched between the two isoGuo-pentamers by forming ten M–O6 bondings.

Inosine complex: In the $[Na(Ino–H)(H_2O)_{2.5}]$ complex [80], the N1-deprotonated inosinate ligand adopts an uncommon *syn* conformation about the glycosidic bond, which is stabilized by an intramolecular O5'–H···N3 hydrogen bond, and the Na^+ ion binds to O3' and O5' hydroxyl oxygens, one each from two different ligands, but it does not bind to the hypoxanthine moiety.

Cytidine and Thymidine complex: None.

Uridine complex: A Na^+ complex is available, $[Na_8Cu_8(Urd–3H)_8]\cdot 5NaClO_4\cdot 48\,H_2O$ [81]. The triply deprotonated Urd ligand is three-coordinated toward three Cu^{2+} atoms by deprotonated N3, O3', and O2', this last oxygen atom bridging two Cu^{2+} atoms, which are related to each other by a twofold symmetry (space group *P*432, *Z*=3; the asymmetric unit includes one each for Na^+ and Cu^{2+} ions and the uridine ligand). The further operation by a crystallographic fourfold symmetry creates a cage-like octamer of the formula $[Cu_8(Urd–3H)_8]^{7-}$. Interestingly, this species incorporates the fully hydrated $[Na(H_2O)_6]$ ion into the center of the octamer by forming a total of eight hydrogen bonds between the water ligands and O2 oxygens of the

Table 8 Alkali metal ion binding sites/modes on nucleosides in the solid state.[a]

Complex		Coordination geometry or number	Metal binding sites on nucleosides and other ligands/comments, if any[b]	CSD code	Ref.
Adenosine complex					
$[Na(H_2O)\{(Ado–2H)_2Sb\}]_n$		Sq. pyr.	*O3',O5'* [Ado (Mol.1): 2.448 (O3'), 2.360 (O5')][b], *O3',O5'* [Ado (Mol.2): 2.390 (O3'), 2.236 (O5')], O [aqua: 2.276] The Sb^{3+} ion bridges between the deprotonated O2' and O3'oxygens. Intramolecular O3'–M–O5' chelation.	SICHES	[78]
Guanosine complex					
$[Cs(O2',O3', O5'$-substituted $isoGuo)_{10}(CH_3CN)_2]_2·2Ph_4B·-33CH_3CN·1.5H_2O$		10-Coord.	10*O6* [isoGuo (Mol.1-Mol.10): -] The Cs^+ ion ionophore: the Cs^+ ion is sandwiched between the two planar isoGuo-pentamers. See Figure 5.	LOJFIZ	[79]
Inosine complex					
$[Na(Ino–H)(H_2O)_{2.5}]$		6-Coord.	*O2'* [Ino (Mol.1): 2.385], *O5'* [Ino (Mol.2): 2.308], 4O [aqua: 2.382-2.485] Intramolecular O5'–H···N3 hydrogen bonding.	TIBWOQ	[80]
Cytidine complex: None					
Thymidine complex: None					
Uridine complex					
$[Na_8Cu_8(Urd–3H)_8]·5NaClO_4·48H_2O$	Na1	6-Coord.	2*O4* [Urd: 3.04 (Mol.1), 3.13(Mol.2)], 3O [aqua: 2.43-2.92], O [perchlorate: 3.11]	FECPIM	[81]
	Na2	6-Coord.	6O [aqua: 2.38-2.40] Cu^{2+} ions bind to the deprotonated N3, O2', and O3' atoms. A $[Na(H_2O)_6]^+$ ion is incorporated into the cavity of the $[Cu_8(Urd–3H)_8]^{8–}$ octamer.		

[a]Abbreviations: Ado, adenosine; Ado–2H, adenosine doubly deprotonated at O2'- and O3'-hydroxyl groups; Guo, guanosine; Ino–H, inosine deprotonated at N1; O2',O3',O5'-substituted isoGuo, 5'-*t*-butyl-dimethylsilyl-2',3'-di-O-isopropylidene isoguanosine; Urd–3H, uridine triply deprotonated at N3, O2', and O3'; Sq. pyr., square pyramidal.

[b]For example, '*O3',O5'* [Ado (Mol.1): 2.448 (O3'), 2.360 (O5')]' denotes that the metal ion binds to the O3' and O5' atoms of the adenosine molecule 1 (Mol.1) ($O3'–Na^+–O5'$ chelation) with the bonding distances of 2.448 Å for O3' and 2.360 Å for O5'.

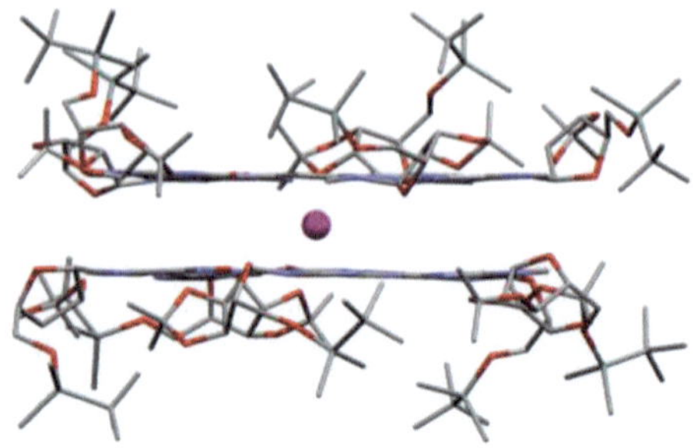

Figure 5 A side view of the structure of the Cs^+ complex with a substituted isoguanosine (isoGuo), $[Cs(O2',O3',O5'$-substituted $isoGuo)_{10}(CH_3CN)_2]^+$ [79], showing the sandwiched Cs^+ ion between the two isoGuo-pentamers.

base moiety with the water–O2 distance of 2.95 Å. One of the other Na^+ ions binds to two uridine ligands each through O4.

In summary, in the alkali metal ion complexes of nucleosides, the metal bonding to sugar hydroxyl oxygens O3', and O5', and exocyclic carbonyl oxygens, O6 of guanine and O2 and O4 of uracil is observed.

3.3 *Nucleotide Complexes*

We deal here with mono- and dinucleotides but do not cover quadruplexes (see Chapter 7 and [121]) and other oligonucleotides. X-ray crystal structures of oligonucleotide complexes involving alkali, alkaline earth, and transition metal ions before 1996 were reviewed in [122, 123] and those after 1996 and before 2009 in [4]. Because of the anionic nature of nucleotides due to the phosphates, a relatively large number of metal ion complexes of mononucleotides has been reported, most of which are Na^+ complexes. Though the crystal structures of most of these complexes were once reviewed or tabulated [4, 123, 124], in order to give original data from which some general aspects on the metal binding sites of nucleotides are derived, noted in the beginning of this section, they are briefly described here once again, together with new data.

3.3.1 Mononucleotide Complexes

Table 9 lists the total number of alkali metal ion complexes with mononucleotides and Table 10 summarizes the metal binding sites/modes for each complex [84–120].

Adenosine nucleotide complexes: Six complexes for AMP (two for 5'-AMP [86, 87], one for 3',5'-cyclicAMP [88], and three for 2'-AMP [84]) have been reported, three complexes for ADP [85, 91–93] involving the NAD complex [85], and two complexes for ATP [89, 90]. Two 5'-AMP complexes, $Na_2[(Mo_5O_{15})(5'\text{-}AMPH)_2]$

Table 9 Total number of alkali metal ion–mononucleotide complexes in the solid state.[a,b]

Nucleotide	Sugar[c]	Phosphate[d]	Li^+	Na^+	K^+	Rb^+	Cs^+	Total
AMP	R	5'	0	2[g] (2)	0	0	0	2 (2)
	R	3',5'	0	1[h] (1)	0	0	0	1 (1)
	R	2'	1[e] (1)	1[i] (1)	1[k] (1)	0	0	3 (3)
	D	5'	0	0 (1)	0	0	0	0 (1)
ADP	R	5'	1[f] (1)	0	1[l] (1)	1[m] (1)	0	3 (3)
ATP	R	5'	0	2[j] (2)	0	0	0	2 (2)
GMP	R	5'	0	3[n] (4)	0	0	0	3 (4)
	R	3',5'	0	0 (1)	0	0	0	0 (1)
	D	5'	0	1[o] (1)	0	0	0	1 (1)
IMP	R	5'	0	6[p] (6)	0	0	0	6 (6)
CMP	R	5'	0	4[q] (4)	0	0	0	4 (4)
	R	3'	0	0	0	0	0 (1)	0 (1)
	R	2',3'	0	1[r] (1)	0	0	0	1 (1)
	D	5'	0	0 (2)	0	0	0	0 (2)
CDP	R	5'	0	2[s] (3)	0	0	0	2 (3)
TMP			0	0	0	0	0	0
UMP	R	5'	0	2[t] (2)	0	0	0	2 (2)
	R	3'	0	1[u] (1)	1[x] (1)	0	0	2 (2)
	D	5'	0	1[v] (1)	0	0	0	1 (1)
UDP	R	5'	0	1[w] (1)	1[y] (2)	0	0	2 (3)
Total			2 (2)	28 (34)	4 (5)	1 (1)	0 (1)	35 (43)

[a]Obtained from the Cambridge Structural Database (the CSD version 5.35 updated to February 2014).
[b]The alkali metal ion–nucleotide *complexes* are those that contain *direct metal bonding* to nucleotides. The number in parenthesis denotes the sum of alkali metal ion-nucleotide *complexes* and alkali metal ion–nucleotide *salts*, the latter having *no direct metal bonding* to nucleotides.
[c]Abbreviations for sugars: R = ribose; D = deoxyribose.
[d]Abbreviations for phosphates: 5' = 5'-phosphate; 3',5' = 3',5'-cyclic phosphate; 2' = 2'-phosphate; 3' = 3'-phosphate; 2',3' = 2',3'-cyclic phosphate.
[e][84].
[f]A salt of nicotinamide adenine dinucleotide [85].
[g][86, 87]. [h][88]. [i][84]. [j][89, 90]. [k][84]. [l][91, 92]. [m][93]. [n][94–97]. [o][98, 99]. [p][100–105]. [q][106–109]. [r][110].
[s]Complexes of cytidine diphosphocholine [111] and cytidine diphosphoethanolamine [112].
[t][113–115], containing a complex of the phosphate group-methylated 5'-UMP [115].
[u]A complex of dihydrouridine 3'-monophosphate [116].
[v][117].
[w]A complex of uridine diphosphoglucose [118]. [x][119]. [y][120].

[86] and $Na_{8.5}[U_3O_7(5'\text{-}AMP)_3]$ [87], involve the phosphate-only metal bonding. A 3',5'-cyclicAMP complex, $Na_2(3',5'\text{-}cyclicAMP)_2 \cdot 8H_2O$ [88], is unique in that a Na^+ ion binds to the ring nitrogen N1 of one ligand (Mol.1), in addition to phosphate oxygens from two other ligands (Mol.2 and Mol.3); the second Na^+ ion binds to Mol.2 through phosphate only. In all the three 2'-AMP complexes, $M_2(2'\text{-}AMPH)_2 \cdot nH_2O$ (M: Li^+ [84], Na^+ [84], and K^+ [84]), the 2'-AMP molecule adopts

Table 10 Alkali metal ion binding sites/modes on mononucleotides in the solid state.[a]

Complex	Coordination geometry or number		Metal binding sites on nucleotides/comments, if any[b]	CSD code	Ref.
Adenosine nucleotide complexes					
$Li_2(2'\text{-}AMPH)_2 \cdot 6H_2O$	Li1	Tet.	*O3'* [Mol.1: 1.988][b], *O(P)* [Mol.2: 1.877]	–[c]	[84]
	Li2	Tet.	*O(P)* [Mol.3: 1.951] Protonation at the phosphate group.		
$Li(NAD) \cdot 2H_2O$		Tet.	*N7,O(P)* [Mol.1: 2.13 (adenine N7), 1.86 (β-phos)], 2*O(P)* [Mol.2: 1.88 (α-phos), 1.92 (β-phos)] Intramolecular N7–M–O(β-phos) chelation.	NADLIH10	[85]
$Na_2[(Mo_5O_{15})(5'\text{-}AMPH)_2] \cdot 6H_2O$		6-Coord.	*O(P)* [2.43] Protonation at N1 of the base.	ROBQII	[86]
$Na_{8.5}[U_3O_7(5'\text{-}AMP)_3]_2 \cdot 33H_2O$		6-Coord.	*O(P)* [-] Six Na^+ positions only partially (40–44%) occupied.	XAXNEQ	[87]
$Na_2(3',5'\text{-}cyclicAMP)_2 \cdot 8H_2O$	Na1	5-Coord.	*N1* [Mol.1: -], 2*O(P)* [Mol.2, Mol.3: -]	NAAMPH10	[88]
	Na2	6-Coord.	*O(P)* [Mol.2: -] Coordination sites assigned from Fig. 2 in [88].		
$Na_2(2'\text{-}AMPH)_2 \cdot 3H_2O$	Na1	6-Coord.	*O3'* [Mol.1: 2.473], *O4'* [sugar ring, Mol.2: 2.712], 2*O(P)* [2.490 (Mol.3), 2.514 (Mol.4)]	–[c]	[84]
	Na2	5-Coord.	*O3'* [Mol.2: 2.301], *O4'* [sugar ring, Mol.1: 2.451], *O(P)* [Mol.5: 2.388] Protonation at the phosphate group.		
$Na_2(ATPH2) \cdot 3H_2O$	Na1	Dist. oct.	*N7,O(P)* [Mol.1: 2.90 (N7), 2.47 (γ-phos)], 3*O(P)* [Mol.2: 2.59 (α-phos), 2.42 (β-phos), 2.53 (γ-phos)], *O(P)* [Mol.3: 2.65 (γ-phos)]	ADENTP	[89]
	Na2	Dist. oct.	*N7,O(P)* [Mol.2: 2.69 (N7), 2.32 (γ-phos)], 3*O(P)* [Mol.1: 2.45 (α-phos), 2.48 (β-phos), 2.32 (γ-phos)], *O(P)* [Mol.4: 2.54 (γ-phos)]		
	Na3	Dist. oct.	2*O(P)* [β-phos: 2.32 (Mol.5), 2.32 (Mol.6)]		
	Na4	Dist. sq. pyr.	*O2',O4'* [Mol.1: 2.80 (O2'), 2.80 (O4')], *O(P)* [Mol.5: 2.72 (α-phos)], *O2',O3'* [Mol. 7: 2.84 (O2'), 2.69 (O3')]		

			Protonation at N1 and γ-phosphate. Intramolecular N7–M–O(γ-phos) chelation.		
Na_2(ATPH2)·$2H_2O$	Na1	6-Coord.	*N7,O(P)* [Mol.1: 2.69 (N7), 2.33 (γ-phos)], 3*O(P)* [Mol.2: 2.48 (α-phos), 2.41 (β-phos), 2.41 (γ-phos)], *O(P)* [Mol.3: 2.81 (γ-phos)]	CITWUX10	[90]
	Na2	6-Coord.	*N7,O(P)* [Mol.2: 2.52 (N7), 2.39 (γ-phos)], 3*O(P)* [Mol.1: 2.32 (α-phos), 2.53 (β-phos), 2.39 (γ-phos)], *O(P)* [Mol.4: 2.89 (γ-phos)]		
	Na3	6-Coord.	2*O(P)* [β-phos: 2.34 (Mol.1), 2.30 (Mol.4)], O3' [Mol.5: 2.48]		
	Na4	6-Coord.	*O(P)* [Mol.1: 2.38 (β-phos)], *O(P)* [Mol.3: 2.34 (α-phos)] Protonation at N1 and γ-phosphate. Intramolecular N7–M–O(γ-phos) chelation. The original assignment of Na4 and OW4 in ADENTP [89] was interchanged.		
K_2(2'-AMPH)$_2$·$1.5H_2O$		6-Coord.	*O3'* [M.ol.1: 2.674], *O4'* [sugar ring, Mol.2: 3.040], 2*O(P)* [2.685 (Mol.3), 2.719 (Mol.4)] Protonation at the phosphate group.	–[c]	[84]
K(5'-ADPH2)·$2H_2O$		7-Coord.	*N3,O2'* [Mol.1: 3.15 (N3), 2.84 (O2')], 3*O(P)* [Mol.2: 2.73 (α-phos), 2.97 (β-phos, protonated), 3.26 (β-phos)], *O(P)* [Mol.3: 3.14 (α-phos)] Protonation at N1 and a β-phosphate oxygen. Intramolecular N3–M–O2' chelation.	KADPHD01	[91]
K(5'-ADPH2)·$2H_2O$		7-Coord.	*N3,O2'* [Mol.1: 3.162 (N3), 2.856 (O2')], 3*O(P)* [Mol.2: 2.767 (α-phos), 2.953 (β-phos), 3.238 (β-phos)], *O(P)* [Mol.3: 3.147 (α-phos)] The same compound as KADPHD01 [91]	KADPHD02	[92]
Rb(5'-ADPH2)·H_2O		7-Coord.	*N3,O2'* [Mol.1: 3.19 (N3), 2.91 (O2')], 3*O(P)* [Mol.2: - (α-phos), - (β-phos, protonated), - (β-phos)], *O(P)* [Mol.3: - (α-phos)] Isostructural to KADPHD01 [91].	RBADPM10	[93]
Guanosine nucleotide complexes					

(continued)

Table 10 (continued)

Na_2(5'-GMP)·7H_2O	Na1	Dist. oct.	*2N7* [2.419 (Mol.1), 2.611 (Mol.2)]	GUOPNA11	[94]
	Na2	Dist. oct.	*O2',O3'* [Mol.3: 2.471 (O2'), 2.314 (O3')] Intermolecular *cis*-N7–M–N7 bonding. Intramolecular O2'–M–O3' chelation.		
Na_2(5'-GMP)·7H_2O	Na1	Dist. oct.	*2N7* [2.415 (Mol.1), 2.613 (Mol.2)]	GUOPNA10	[95]
	Na2	Dist. oct.	*O2',O3'* [Mol.3: 2.492 (O2'), 2.305 (O3')] The same compound as GUOPNA11 [95].		
[Na(H_2O)$_4${Cu(5'-GMP)(5'-GMPH)(H_2O)$_3$}]·MeOH·6H_2O		6-Coord.	*O2',O3'* [2.506 (O2'), 2.338 (O3')] Intramolecular O2'–M–O3' chelation. The Cu^{2+} ion binds to two nucleotide ligands each through N7.	ESIWOT	[96]
Na_2[Pt(Me_4dae)(5'-GMP)$_2$]·7D_2O		7-Coord.	*O6* [Mol.1: -], *O2',O3'* [Mol.2: -], 2*O(P)* [Mol.3: -], *O(P)* [Mol.4: -] Intramolecular O2'–M–O3' chelation. The Pt^{2+} ion binds to two nucleotide ligands each through N7.	XICVIP	[97]
Na_2(5'-dGMP)·4H_2O	Na1	Oct.	*O6* [Mol.1: 2.559], O3' [Mol.2: 2.354]	SDGUNP	[98]
	Na2	Dist. sq. pyr.	*O6* [Mol.1: 2.355], 2*O(P)* [2.383 (Mol.3), 2.322 (Mol.4)]		
Na_2(5'-dGMP)·4H_2O	Na1	Oct.	*O6* [Mol.1: 2.553], *O3'* [Mol.2: 2.348]	SDGUNP01	[99]
	Na2	Dist. sq. pyr.	*O6* [Mol.1: 2.367], 2*O(P)* [2.368 (Mol.3), 2.318 (Mol.4)] The same compound as SDGUNP [98].		
Inosine nucleotide complexes					
Na(5'-IMPH)·8H_2O		Dist. oct.	*O2',O3'* [2.48 (O2'), 2.33 (O3')] Protonation at a phosphate oxygen. Intramolecular O2'–M–O3'chelation.	NAINPH10	[100]
Na_2(5'-IMPH)$_2$·20H_2O	Na1	6-Coord.	*O2',O3'* [- (O2'), - (O3')]	KIKPUP	[101]
	Na2	7-Coord.	*O2',O3'* [- (O2'), - (O3')] Protonation at a phosphate oxygen. Intramolecular O2'–M–O3' chelation.		
Na_2[Cu(5'-IMP)$_2$(imidazole)$_{0.8}$-(H_2O)$_{3.2}$]·12.4H_2O		7-Coord.	*O2',O3'* [- (O2'), - (O3')] Intramolecular O2'–M–O3' chelation. The Cu^{2+} ion binds to two nucleotide ligands each through N7.	DAGWUD	[102]
Na_2·[Pt(5'-IMP)$_2$(en)]		–[d]	–[d] The Pt^{2+} ion binds to two nucleotide ligands each through N7.	EPTIMP	[103]
Na_2[Pt(5'-IMP)$_2$(NH_3)$_2$]$_{0.86}$·(5'-		7-Coord.	*O2',O3'* [2.564 (O2'), 2.347 (O3')]	IMPPTS	[104]

$IMP)_{0.28} \cdot 16H_2O$			Intramolecular O2'–M–O3' chelation. The Pt^{2+} ion binds to two nucleotide ligands each through N7.		
$Na_2[Cu(5'\text{-}IMP)_2(dien)] \cdot 10H_2O$		6-Coord.	*O2',O3'* [2.430 (O2'), 2.280 (O3')] Intramolecular O2'–M–O3' chelation. The Cu^{2+} ion binds to two nucleotide ligands each through N7.	SINPCU	[105]
Cytidine nucleotide complexes					
$Na_4(5'\text{-}CMP)_2 \cdot 13H_2O$ (Space group $P2_1$)	Na1	5-Coord.	*N3,O2* [Mol.1: 2.400 (N3), 2.855 (O2)], *O(P)* [Mol.2: 2.361]	ACEWAG	[106]
	Na2	5-Coord.	*O(P)* [Mol.1: 2.334], *N3* [Mol.3: 2.521]		
	Na3	5-Coord.	*O2',O3'* [Mol.4: 2.413 (O2'), 2.306 (O3')] Intramolecular N3–M–O2 chelation for Na1. Intramolecular O2'–M–O3' chelation for Na3.		
$Na_4(5'\text{-}CMP)_2 \cdot 13H_2O$ (Space group $P2_12_12_1$)	Na1	6-Coord.	*O2',O3'* [Mol.1: 2.42 (O2'), 2.31 (O3')]	HAKBEB	[107]
	Na2	5-Coord.	*N3* [Mol.2: 2.52], *O(P)* [Mol.3: 2.34]		
	Na3	5-Coord.	*N3,O2* [Mol.4: 2.41 (N3), 2.83 (O2)], *O(P)* [Mol.5: 2.36] Intramolecular O2'–M–O3' chelation for Na1. Intramolecular N3–M–O2 chelation for Na3.		
$Na_4(5'\text{-}CMP)_2 \cdot 12.14H_2O \cdot 0.54MeOH$ (Space group $P2_12_12_1$)	Na1	5-Coord.	*N3,O2* [Mol.1: 2.3852 (N3), 2.8008(O2)], *O(P)* [Mol.2: 2.3621]	FOCFOT	[108]
	Na2	6-Coord.	*O2',O3'* [Mol.3: 2.3995 (O2'), 2.3093 (O3')]		
	Na3	5-Coord.	*N3* [Mol.4: 2.5201], *O(P)* [Mol.5: 2.3252] Intramolecular N3–M–O2 chelation for Na1. Intramolecular O2'–M–O3' chelation for Na2.		
$Na_8(5'\text{-}CMP)_4 \cdot 32.5H_2O$	Na1	6-Coord.	*O2',O3'* [Mol.1: 2.36 (O2'), 2.33 (O3')]	HAKBAX	[107]
	Na2	6-Cood.	*O2',O3'* [Mol.2: 2.82 (O2'), 2.28 (O3')]		
	Na3	5-Coord.	*O2,O2'* [Mol.3: 2.28 (O2), 2.53 (O2')]		
	Na4	6-Coord.	*O3'* [Mol.4: 2.38] Intramolecular O2'–M–O3' chelation for Na1 and Na2. Intramolecular O2–M–O2' chelation for Na3.		
$Na_8(5'\text{-}CMP)_4 \cdot 37H_2O$	Na1	6-Coord.	*O2',O3'* [Mol.1: 2.45 (O2'), 2.33 (O3')]	HAJZUO	[107]
	Na2	6-Coord.	*O2',O3'* [Mol.2: 2.68 (O2'), 2.28 (O3')]		
	Na3	6-Coord.	*O2',O3'* [Mol.3: 2.44 (O2'), 2.32 (O3')]		
	Na4	6-Coord.	*O2,O2'* [Mol.4: 2.40 (O2), 2.48 (O2')]		

(continued)

Table 10 (continued)

			Intramolecular O2'–M–O3' chelation for Na1–Na3. Intramolecular O2–M–O2' chelation for Na4.		
Na(5'-5-bromo-CMPH)·1.25H_2O		6-Coord.	*N3,O2* [Mol.1: - (N3), - (O2)], *O2* [Mol.2: -], 2*O*(P) [Mol.3: -] Protonation at a phosphate oxygen. M–N/O distances of 2.219-3.070 Å. Intramolecular N3–M–O2 chelation.	WERKUZ	[109]
Na_2(2',3'-cyclicCMP)$_2$·4H_2O	Na1	6-Coord.	*N3,O2* [Mol.1: 3.322 (N3), 2.508(O2)], *O2* [Mol.2: 2.527], *O(P)* [Mol.3: 2.461]	CYTCYP20	[110]
	Na2	6-Coord.	*N3*,O2 [Mol.2: 2.829 (N3), 2.630 (O2)], O5' [Mol.4: 2.292] Intramolecular N3–M–O2 chelation for Na1 and Na2.		
Na(5'-CDPcholine)·5H_2O		Dist. oct.	*N3,O2* [Mol.1: 2.39 (N3), - (O2)], *O2',O3'* [Mol.2: - (O2'), - (O3')], *O(P)* [Mol.3: 2.31 (α-phos)] Intramolecular N3–M–O2 chelation. Intramolecular O2'–M–O3' chelation.	CYPCHO	[111]
Na(5'-CDPethanolamine)·7H_2O		Dist. oct.	*N3,O2* [Mol.1: 2.455 (N3), 2.970(O2)], *O2',O3'* [Mol.2: 2.389 (O2'), 2.387 (O3')], *O(P)* [Mol.3: 2.220 (α-phos)] Intramolecular N3-M-O2 chelation. Intramolecular O2'–M–O3' chelation.	NACYTD	[112]
Thymidine nucleotide complex: None					
Uridine nucleotide complexes					
Na_2(5'-UMP)·7H_2O	Na1	8-Coord.	*O2',O3'* [Mol.1: -], *O2',O3'* [Mol.2: -], 2*O(P)* [Mol.3: -, Mol.4, -]	SURIPH10	[113]
	Na2	7-Coord.	2*O2* [Mol.2: -, Mol.5: -], *O(P)* [Mol.6: -] Intramolecular O2'–M–O3' chelation for each of Mol.1 and Mol.2 for Na1.		
Na_2(5'-UMP)·7H_2O		7-Coord.	*O2',O3'* [2.55 (O2'), 2.27 (O3')] The same compound as SURIPH10 [113]. A neutron study to distinguish between the water molecules and Na^+ ions. Intramolecular O2'–M–O3' chelation.	–[c]	[114]
Na(5'-*O*-Me-UMP)·MeOH		Dist. oct.	*O2* [Mol.1: 2.410], *O4* [Mol.2: 2.350], *O2',O3'* [Mol.2: 2.332 (O2'), 2.383 (O3')], 2*O(P)* [Mol.3: 2.306, 3.213] Intramolecular O2'–M–O3' chelation.	SUROMM	[115|]

Na_2(3'-UMP) $4H_2O$		7-Coord.	*O2',O3'* [Mol.1: 2.336 (O2'), 2.697 (O3' ester)], *O5'* [Mol.2: 2.350], *O(P)* [Mol.3: 2.867] Intramolecular O2'–M–O3' chelation.	SURIDP	[116]
Na_4(5'-dUMP)$_2$·$10H_2O$	Na1	Dist. oct.	*O4* [Mol.1: 2.257], *O3'* [Mol.2: 2.545], *O(P)* [Mol.3: 2.500]	NADOUR	[117]
	Na2	5-Coord.	*O4* [Mol.4: 2.274], *O3'* [Mol.5: 2.566], *O(P)* [Mol.6: 2.436]		
Na_2(5'-UDPglucose)·$2H_2O$	Na1	Dist. oct.	*O2* [Mol.1: 2.356], *O4* [Mol.2: 2.324], *O(P)* [β-phos Mol.3: 2.993], *O* [glucose hydroxyl Mol.4: 2.299],	CEKLUZ	[118]
	Na2	Dist. oct.	*O2* [Mol.5: 2.343], 2*O* [glucose hydroxyl Mol.4: 2.415, 2.456], *O* [glucose ring Mol.3: 2.800], *O* [glucose hydroxyl Mol.3: 2.308]		
K(3'-dihydroUMPH)·$0.5H_2O$		8-Coord.	2*O4* [Mol.1: 2.766, Mol.2: 2.791], *O2',2O(P)* [Mol.3: 2.779 (O2'), 2.910 (OP), 3.061(OP)], 2*O(P)* [Mol.4: 2.879, 3.062] Protonation at a phosphate oxygen. Intramolecular O2'–M–O(P) chelation.	KURDMP	[119]
K_4(5'-UDPH)$_2$·$6H_2O$	K1	Dist. oct.	2*O2* [Mol.1: 2.786, Mol.2: 2.841]	KURDPI	[120]
	K2	Dist. oct.	*O4* [Mol.3: 2.711], *O(P)* [α-phos Mol.4: 2.590], 2*O(P)* [β-phos: 2.739 (Mol.1), 2.778 (Mol.5)]		
	K3	Dist. oct.	*O4* [Mol.2: 2.680], *O3'* [Mol.7: 3.008], *O(P)* [α-phos Mol.1: 2.580], 2*O(P)* [β-phos: 2.692 (Mol.4), 3.005 (Mol.6)]		
	K4	Dist. oct.	2*O(P)* [α-phos: 3.145 (Mol.1), 3.030 (Mol.1)], 2*O(P)* [β-phos: 2.695 (Mol.1), 2.741 (Mol.4)] Protonation at a β-phosphate oxygen.		

[a]Abbreviations: α-phos, α-phosphate; β-phos, β-phosphate; γ-phos, γ-phosphate; dien, diethylenetriamine; M, metal ion; Me_4dae, tetramethyl-1,2-diaminoethane; NAD, 5'-nicotinamide-ribosyl-5'-adenyl-pyrophosphate; O(P), phosphate oxygen; Dist. oct, distorted octahedral; Dist. sq. pyr., distorted square pyramidal; Dist. tri. bipyr., distorted trigonal bipyramidal; Oct., octahedral; Tet., tetrahedral.

[b]For example, '*O3*'[Mol.1: 1.988]' denotes that the metal ion binds to the hydroxyl O3' of the nucleotide molecule 1 (Mol.1) with the bonding distance of 1.988 Å.

[c]Not registered in the CSD version 5.35 updated to February 2014.

[d]Not mentioned.

an uncommn *syn* conformation, which is stabilized by the formation of an intramolecular O5'–H···N3 hydrogen bond, and metal ions commonly bind to hydroxyl O3' and phosphate oxygens. In addition, in the Na^+ and K^+ complexes, the ether oxygen O4' of the sugar ring is involved in metal bonding.

The Li^+ complex of NAD, which possesses the ADP entity, Li(NAD)·$2H_2O$ [85], is of particular interest because the tetrahedral Li^+ ion forms the intramolecular N7–M–O(β-phosphate) chelation which stabilizes the *anti* conformation of the nucleotide molecule, and *vice versa*, and additionally forms an intramolecular O(α-phosphate)–M–O(β-phosphate) chelation with the second ligand. The K^+ and Rb^+ complexes of ADP, K(5'-ADPH2)·$2H_2O$ [91, 92] and Rb(5'-ADPH2)·H_2O [93], which are isostructural to each other, also exhibit an interesting metal binding mode, that is, the Rb^+ ion bridges O2' and N3 of the same ligand molecule that assumes the *anti* conformation.

Two humidity-dependent polymorphic Na^+ complexes of ATP, Na_2(ATPH2)·$3H_2O$ [89] and Na_2(ATPH2)·$2H_2O$ [90], provide an additional, valuable example in which the metal ion bridges N7 of the base and the phosphate oxygen (of γ-phosphate) of the same ATP molecule with the *anti* conformation.

Guanosine nucleotide complexes: Four Na^+ ion complexes of GMP are available, among which three are those of 5'-GMP, Na_2(5'-GMP)·$7H_2O$ [94], [$Na(H_2O)_4${Cu(5'-GMP)(5'-GMPH)$(H_2O)_3$}]·MeOH·$6H_2O$ [96], and Na_2[Pt(Me_4dae)(5'-GMP)$_2$]·$7D_2O$ [97], and one is the complex of 5'-dGMP, Na_2(5'-dGMP)·$4H_2O$ [98, 99]. The three 5'-GMP complexes commonly involve the intramolecular O2'–M–O3' chelation. Interestingly, metal bonding to N7 or O6 of the base are observed in the Na^+ complex [94] and the Na^+/Pt^{2+} complex [97], respectively. Metal bonding to the phosphate group occurs only in the Na^+/Pt^{2+} complex [97]. In the 5'-dGMP complex [98, 99], the base O6, the sugar O3', and phosphate oxygens are all involved in metal bonding.

Inosine nucleotide complexes: Six Na^+ ion complexes of 5'-IMP are reported, Na(5'-IMPH).$8H_2O$ [100], Na_2(5'-IMPH)$_2$.$20H_2O$ [101], Na_2[Cu(5'-IMP)$_2$(imidazole)$_{0.8}$$(H_2O)_{3.2}$]·$12.4H_2O$ [102], Na_2[Pt(5'-IMP)$_2$(en)]·nH_2O [103], Na_2[Pt(5'-IMP)$_2$$(NH_3)_2$]$_{0.86}$·(5'-IMP)$_{0.28}$·$16H_2O$ [104], and Na_2[Cu(5'-IMP)$_2$(dien)]·$10H_2O$ [105]. Four complexes [102–105] have the common structural unit [M(5'-IMP)$_2$] (M: Cu^{2+} or Pt^{2+}) in which the metal ion binds to two nucleotide ligands each through N7. Somewhat surprisingly, in all the 5'-IMP complexes (except for the Na^+/Pt^{2+}/en complex [103] for which a Na^+ ion binding site is not mentioned), the intramolecular O2'–M–O3' chelate formation is commonly and solely observed. Lack of a participation of phosphates in the direct metal ion interaction is, at least in part, due to extensive hydration around the phosphate oxygens.

Cytidine nucleotide complexes: Out of seven Na^+ complexes reported, four are 5'-CMP complexes [106–109], one is a 2',3'-cyclicCMP complex [110], and the other two are complexes of 5'-CDPcholine [111] and 5'-CDPethanolamine [112]. Five humidity-dependent polymorphic complexes of 5'-CMP, Na_4(5'-CMP)$_2$·$13H_2O$ (space group $P2_12_12_1$) [106], Na_4(5'-CMP)$_2$·$13H_2O$ (space group $P2_1$) [107], Na_4(5'-CMP)$_2$·$12.14H_2O$·0.54MeOH [108], Na_8(5'-CMP)$_4$·$32.5H_2O$ [107], and Na_8(5'-CMP)$_4$·$37H_2O$ [107], commonly involve the intramolecular O2'–Na^+–O3' chelation and further exhibit two types of humidity-dependent metal binding modes, that is, (i) when the extent of hydration is not so large (this is the case for the former three complexes), the phosphate O atoms provide a ligand for the coordination and the base moiety is also involved in metal coordination to form the intramolecular N3–M–O2 chelation or the single M–N3 bonding, and (ii) when the extent of hydration is large (for the latter two complexes), the Na^+ ions interact with the phosphate only indirectly through water bridges, and the N3 site of the base is also hydrated so that the metal ions bind only to O2. As an additional and extreme case, in the Na^+ salt of 5'-dCMP, $Na_2\cdot$(5'-dCMP)$\cdot 7H_2O$ [125], not only N3 and the phosphate but also O2 is hydrated, thus preventing the access of metal ions to these sites. In the two Na(5'-5-bromo-CMPH)$\cdot 1.25H_2O$ [109] and Na_2(2',3'-cyclicCMP)$_2\cdot 4H_2O$ [110] complexes, the Na^+ ions bind to the phosphates, and N3–M–O2 chelation or single M–O2 bonding are observed, but the metal binding mode to the sugar moiety is different, namely none in the former complex and O5' bonding in the latter complex. The two 5'-CDP complexes, Na(5'-CDPcholine)$\cdot 5H_2O$ [111] and Na(5'-CDPethanolamine)$\cdot 7H_2O$ [112] involve the three sets of metal binding modes, namely N3–M–O2 chelation, O2'–M–O3' chelation, and M–phosphate bonding.

Thymidine nucleotide complex: None.

Uridine nucleotide complexes: These include two Na^+ complexes of 5'-UMP, Na_2(5'-UMP)$\cdot 7H_2O$ [113, 114] and Na(5'-*O*-Me-UMP)$\cdot$MeOH [115], a Na^+ complex of 3'-UMP, Na_2(3'-UMP)$\cdot 4H_2O$ [116], a K^+ complex of 3'-dihydroUMP, K(3'-dihydroUMPH)$\cdot 0.5H_2O$ [119], a Na^+ complex of 5'-dUMP, Na_4(5'-dUMP)$_2\cdot 10H_2O$ [117] and a Na^+ complex of 5'-UDPglucose, Na_2(5'-UDPglucose)$\cdot 2H_2O$ [118], and a K^+ complex of 5'-UDP, K_4(5'-UDPH)$_2\cdot 6H_2O$ [120]. Metal bonding to O2 or O4 of the base, the hydroxyl oxygens of the sugar, and to the phosphates are a rule for uridine nucleotides with a few exceptions [113, 116, 118]. The crystal structure of Na_2(5'-UMP)$\cdot 7H_2O$ was first determined by X-ray diffraction [113] and later redetermined by neutron diffraction [114] to reveal that the O2 and O4 sites as well as the phosphate are fully hydrated and only O2'–M–O3' chelation occurs. In the case of Na_2(3'-UMP)$\cdot 4H_2O$ [116], O2 and O4 participate in the formation of hydrogen bonds between the two UMP ligands themselves in the crystal lattice. The other one which deviates from this rule is the Na^+–5'-UDPglucose complex [118] in which the metal–sugar bonding is missing.

In summary, for the five kinds of mononucleotides examined (no data are available for thymidine nucleotides), the functional groups on the base moiety, the hydroxyls of the sugar, and the phosphates all provide preferred metal binding sites for alkali metal ions. The potential sites on the bases for metal coordination are N7 and N1 of adenine, N7 and O6 of O6-oxo purines, N3 and O2 of cytosine, and O2 and O4 of uracil. Intramolecular N3–M–O2 chelation is dominant in the cytidine nucleotide complexes (of Na^+). Metal bonding to the sugar moiety by forming the O2'–M–O3' chelation is the most frequently observed metal binding mode in mononucleotide complexes (of Na^+) (three complexes of 5'-GMP, four of 5'-IMP, seven of 5'-CMP and 5'-CDP, and three of 5'-UMP). Interestingly, the M–O2' bond distance is always longer than that of the M–O3' bond, that is, the metal ion binds more strongly to the hydroxyl O3' than to O2', as observed in a total of twelve complexes for which the M–O distances are available, with the exception of the Na^+–5'-O-MeUMP complex (where the Na^+–O2' and –O3' bond distances are 2.332(5) and 2.382(5) Å, respectively) [115].

Additional interesting metal binding modes include the intramolecular N7–M–O (β-phosphate) chelation in the Li^+–NAD complex [85], the N7–M–O (γ-phosphate) chelation in the Na^+–5'-ATP complexes [89, 90], and the intramolecular N3–M–O2' chelation in the K^+ [91, 92] and Rb^+ [93] complexes of 5'-ADP. Uncommon but important metal binding sites include the ring ether oxygen O4' of the sugar [84] and the phosphate ester oxygens (of 3'-phosphate [116]). Finally, it should be noted that the identification of the sodium ion is still problematic because of the mixed occupancy of Na^+ and water molecules (the sodium ion and the water molecule carry the same number of electrons): this is the case for the Na^+–5'-ATP [89, 90, 126], and the Na^+–5'-UMP complex [113, 114].

3.3.2 Dinucleotide Complexes

Table 11 summarizes the metal binding sites/modes in alkali metal ion complexes with dinucleotides [127–132].

A total of six Na^+ ion complexes are available and none for K^+, Rb^+ and Cs^+ ions. $[Na_2(ApU)_2(H_2O)_6]\cdot 6H_2O$ [127] and $[Na(GpC)(H_2O)_{3.5}]\cdot 5.5H_2O$ [128] had received much attention because self-complementary ApU and GpC dinucleotides form Watson–Crick-type mini-double helices with structural features reminiscent of A-RNA. In the ApU complex, one of the two Na^+ ions occupies a position on the pseudodyad and is ligated by both uracil O2 atoms located in the minor groove. The other Na^+ ion binds to two phosphate groups through both the adenosine O3' ester oxygens and the non-esterified phosphate oxygens. On the other hand, in the GpC complex, the Na^+ ion binds only to the phosphate groups, indicating a discrimination of the Na^+ ion between base sequences at the dinucleotide level.

Such a direct metal ion bridge has also been observed in crystal structures of a $[d(CGCGAATTCGCG)]_2$ duplex (Dickerson–Drew dodecamer), a single monovalent metal ion is located in the minor grove at the 5'ApT3' step for Na^+ (1.40 Å resolution) [133], K^+ (1.2 Å) [134], Rb^+ (1.2 Å) [135], and Cs^+ (1.8 Å) [136], where an octahedral $[M(H_2O)_2]^+$ ion makes *inner*sphere contacts with two O2 oxygen atoms of T7 and T19 and two O4' ether oxygen atoms of T8 and T20. $[Na(CpG)_2(H_2O)_5]\cdot$(acridine orange)$\cdot 20H_2O$ [129] and $[Na\{d(CpG)d(C^+pG)\}]\cdot 7H_2O$ [130] also form Watson–Crick-type miniduplexes. The Na^+ ion binds only to the O2' hydroxyl of the cytidine nucleotide in the CpG complex [129], whereas the Na^+ ion shows multiple binding to the guanine O6 of one ligand molecule and to both the guanine N7 and the cytidine hydroxyl O5' of the other molecule in the d(CpG) complex [130].

In $Na_2(pTpT)\cdot 13.25H_2O$ [131], one of the two Na^+ ions binds to three different pTpT molecules through interactions with a non-esterified oxygen of the bridging phosphate group and with O2 on two thymine bases. The other Na^+ ion is disordered at three positions, one of which is coordinated to the terminal phosphate group and also has a weak interaction with the thymine O2 of a different molecule.

Table 11 Alkali metal ion binding sites/modes on dinucleotides in the solid state.[a,b]

Complex	Coordination geometry or number		Metal binding sites on dinucleotides/comments, if any	CSD code	Ref.
$[Na_2(ApU)_2(H_2O)_6]\cdot 6H_2O$	Na1	Dist. oct.	*O3'* [A(Mol.1): 2.50][c], *O3'* [A(Mol.2): 2.58], 2*O(P)* [Mol.1: 2.37, Mol.2: 2.32]	ADYPUR10	[127]
	Na2	Dist. oct.	*O2* [U(Mol.1): 2.37], *O2* [U(Mol.2): 2.34] Watson–Crick-type of miniduplex. Interduplex metal ion bridge for Na1. Intraduplex metal ion bridge for Na2 in the minor groove.		
$[Na(GpC)(H_2O)_{3.5}]\cdot 5.5H_2O$		Oct.	*O(P)* [Mol.1: 2.39], *O(P)* [Mol.2: 2.30] Watson–Crick-type of miniduplex with the exact twofold axis symmetry. Interduplex O(P)–M–O(P) metal ion bridge.	GUPCYT20	[128]
$[Na(CpG)_2(H_2O)_5]\cdot$(acridine orange)$\cdot$-$20H_2O$		Dist. oct.	*O2'* [Cyt: -] Watson–Crick-type of miniduplex with an acridine orange molecule intercalated between the two C-G base pairs.	GIFBAY	[129]
$[Na\{d(CpG)d(C^{+}pG)\}]\cdot 7H_2O$		5-Cood.	*O6* [G(Mol.1): 2.450], *N7* [G(Mol.2): 2.582], *O5'* [C(Mol.2): 2.704] Watson–Crick-type of miniduplex.	DINYII10	[130]
$Na_2(pTpT)\cdot 13.25H_2O$	Na1	6-Coord.	*O2* [T1(Mol.1): 2.41], *O2* [T2(Mol.2): 2.54], *O(P)* (bridging, Mol.3: 2.46)	THYTHY10	[131]
	Na2-1, Na2-2, Na2-3	3-Coord.	*O2* [T2(Mol.1): -], *O(P)* [terminal, Mol.1: -] Na2 is disordered over three sites (Na2-1, -2, and -3 with the occupancy factors of 0.5, 0.25, and 0.25).		
$[Na_4(ApppA)(H_2O)_7]\cdot 5H_2O$	Na1	6-Coord.	*O2',O3'* [A1(Mol.1): -, -]	HIFKEM	[132]
	Na2	6-Coord.	*N7,O(P2)* [A2(Mol.1): -, -], *N7,O(P3)* [A1(Mol.2): -, -],		
	Na3	6-Coord.	*O(P2),O(3P),O(4P)* [Mol.1: -, -, -]		
	Na4	6-Coord.	*O2',O3'* [A2(Mol.1): -, -] Intramolecuar O2'–M–O3' chelation for Na1 and Na4.		

[a]Obtained from the Cambridge Structural Database (the CSD version 5.35 updated to February 2014).
[b]Abbreviations: A, adenine base; ApppA, P^1,P^4-bis(5'-adenosyl)tetraphosphate; ApU, adenylyl-3',5'-uridine; CpG, cytidylyl-3',5'-guanosine; GpC, guanylyl-3',5'-cytidine; M, metal ion; pTpT, 5'-phosphoryl-thymidylyl-3',5'-thymidine; U, uracil base; Oct., octahedral; Dist. oct., distorted octahedral.
[c]For example, '*O3'* [A(Mol.1)]' denotes that the metal ion binds to the hydroxyl O3' of the adenine base of the ApU molecule 1 (Mol.1).

Finally, in $[Na_4(Ap_4A)(H_2O)_7]\cdot 5H_2O$ [132] (for biological activities of Ap_4A see [137]), the two adenosine entities in the Ap_4A molecule, which are connected by the four-phosphate chain, are folded back to form a stack between the two adenine base moieties. Four independent Na^+ ions exist, two of which bind to the sugar moieties of both adenosines of the Ap_4A molecule, each by forming the O2'–M–O3' chelation. The third Na^+ ion plays a role to stabilize the folded $A_{(1)}p_{(1)}p_{(2)}p_{(3)}p_{(4)}A_{(2)}$ structure and also connects two Ap_4A molecules, by bridging N7 of the adenosine $A_{(2)}$ and a phosphate oxygen of the second phosphate group $p_{(2)}$ in the same ligand molecule and additionally by bridging N7 of $A_{(1)}$ and a phosphate oxygen of $p_{(3)}$ in the second ligand. The forth Na^+ ion attaches to phosphate oxygens over the three phosphate groups in the same molecule.

In summary, sodium ion exhibits all of the three M–base, –sugar, and –phosphate binding modes. The observed, predominant metal binding sites on the bases are N7 of adenine, N7 and O6 of guanine, and O2 of uracil or thymine.

4 Simple-Carbohydrate Complexes

Carbohydrates are ubiquitous molecules in both living organisms and an inexhaustible source of raw material for chemical industry. As possible metal ligation sites, they possess polyhydroxyl oxygens and the ring ether oxygen for monosaccharides, and additionally the ether oxygen that bridges between the two sugar residues for polysaccharides. As simple carbohydrates, we survey here only selected saccharides that are biologically of special relevance, including monosaccharides (five-membered ribose, ribulose, and xylulose; six-membered glucose, mannose, galactose, and fructose and a few of their phosphate or sulfate esters) and disaccharides (sucrose, maltose, and lactose). Oligosaccharides (α-, β-, and γ-cyclodextrins) are excluded. Table 12 lists the total number of alkali metal ion complexes formed with these carbohydrates and Table 13 summarizes metal binding sites/modes for each complex [138–156]. Nomenclature and chemical structure of the representative saccharides are presented in Figure 6.

The CSD search provided three complexes only for glucose among the six-membered monosaccharides examined, four for glucose 1- and 6-phosphates, one for galactose 1-phosphate, one for fructose 1-phosphate, four for disaccharides, and none for the five-membered monosaccharides. The three glucose complexes, $\{[Na(\text{D-Glc})_2]\cdot Cl\}_{3n}$ [138], $\{[Na(\alpha\text{-D-Glc})_2]\cdot Cl\cdot H_2O\}_{3n}$ [139], and $\{[Na(\alpha\text{-D-Glc})_2]\cdot Cl\cdot 0.78H_2O\}_n$ [140], have a common $[Na(\text{D-Glc})_2]$ structural unit, where the Na^+ ion is coordinated by six hydroxyls originating from four separate glucose molecules. The observed interaction of Na^+ and glucose is relevant to the mechanism of action of the glucose·Na^+ cotransporter in biological systems [157].

Three complexes of glucose 1-phosphate, one for Na^+ [141, 142] and two for K^+ [118, 143–145] and a Na^+ complex of glucose 6-phosphate [146, 147] are available. No structural details are given in the first report [141] of the Na^+ complex, $Na_2(\alpha\text{-D-Glc-1-phosphate})\cdot 3.5H_2O$; later, the same authors mention that the Na^+ ion binds to

Table 12 Total number of alkali metal ion–carbohydrate complexes in the solid state.[a,b]

Carbohydrate[c]	Li^+	Na^+	K^+	Rb^+	Cs^+	Total
Ribose	0	0	0 (1)	0	0	0 (1)
Ribulose	0	0	0 (1)	0	0	0 (1)
Xylulose	0	0	0	0	0	0
Glucose	0	3[d] (3)	0	0	0	3 (3)
Mannose	0	0	0	0	0	0
Galactose	0	0	0	0	0	0
Fructose	0	0	0	0	0	0
Ribose-phosphate	0	0	0	0	0	0
Glucose 1-phosphate	0	1[e] (1)	2[f] (2)	0	0	3 (3)
Glucose 6-phosphate	0	1[g] (1)	0	0	0	1 (1)
Glucose 6-sulfate	0	0	1[h] (1)	0	0	1 (1)
Galactose 1-phosphate	0	0	1[i] (1)	0	0	1 (1)
Fructose 6-phosphate	0	1[j] (1)	0	0	0	1 (1)
Fructose 1,6-diphosphate	0	1[k] (1)	0	0	0	1 (1)
Sucrose	0	2[l] (2)	0	0	0	2 (2)
Maltose	0	1[m] (1)	1[n] (1)	0	0	2 (2)
Lactose	0	0	0	0	0	0
Total	0	10 (10)	5 (7)	0	0	15 (17)

[a]Obtained from the Cambridge Structural Database (the CSD version 5.35 updated to February 2014).
[b]The alkali metal ion–carbohydrate *complexes* are those that contain *direct metal bonding* to carbohydrates. The number in parenthesis denotes the sum of alkali metal ion-carbohydrate *complexes* and alkali metal ion–carbohydrate *salts*, the latter having *no direct metal bonding* to carbohydrates.
[c]Only selected saccharides that are biologically of special relevance are surveyed.
[d][138–140]. [e][141, 142]. [f][143–146]. [g][147, 148]. [h][149]. [i][150]. [j][151]. [k][152]. [l][153–155]. [m][156]. [n][157].

the ring ether O5 and the hydroxyl O6 of the same ligand and additionally to phosphate oxygen O(P) of a different ligand [149]. The crystal structure of the same compound was redetermined by another author [142] and the above mentioned argument is confirmed. This coordination mode is also observed in the two K^+ complexes, K(α-D-Glc-1-phosphate) [143] and K_2(α-D-Glc-1-phosphate)·$2H_2O$ [118, 144, 145]. On the other hand, the Na^+ complex of glucose-6-phosphate, Na(α-D-Glc-6-phosphate) [146, 147] is of interest in that O5, O6, and an O(P) of the same ligand are simultaneously involved in metal coordination.

Glucose 6-sulfate is a component of sulfated polysaccharides (for example, chondroitins). There is evidence that cations associated with sulfate groups play an essential role in the mechanism of polysaccharide gelation [158]. In the K^+ complex, K(β-D-Glc-6-sulfate) [148], the K^+ ion binds to four sulfate oxygens from four ligand molecules in addition to two hydroxyl oxygens from two ligands.

The K^+ complex of galactose 1-phosphate, K_2(α-D-Gal-1-phosphate)·$5H_2O$ [149], where galactose is an epimer (at C4) of glucose, performs simultaneous,

Table 13 Alkali metal ion binding sites/modes on carbohydrates in the solid state.[a]

Complex	Coordination geometry or number		Metal binding sites on carbohydrates[b]/comments, if any	CSD code	Ref.
Glucose complexes					
$\{[Na(D\text{-}Glc)_2]\cdot Cl\}_{3n}$ (Space group $P3_1$)	Na1	Dist. oct.	*O1,O2*[b] [Mol.1: 2.397 (O1), 2.380 (O2)][c], *O1,O2* [Mol.2: 2.383 (O1), 2.343 (O2)], 2*O4* [2.373 (Mol.3), 2.345 (Mol.4)]	XOZVOY	[138]
	Na2.	Dist. oct.	*O1,O2* [Mol.5: 2.388 (O1), 2.358 (O2)], *O1,O2* [Mol.6: 2.396 (O1), 2.357 (O2)], 2*O4* [2.346 (Mol.7), 2.390 (Mol.8)]		
	Na3	Dist. oct.	*O1,O2* [Mol.9: 2.349 (O1), 2.359 (O2)], *O1,O2* [Mol.10: 2.437 (O1), 2.340 (O2)], 2*O4* [2.304 (Mol.11), 2.437 (Mol.12)] Three independent $\{[Na(D\text{-}Glc)_2].Cl\}_n$ polymeric chains. Intramolecular O1–M–O2 chelation.		
$\{[Na(\alpha\text{-}D\text{-}Glc)_2]\cdot Cl\cdot H_2O\}_{3n}$	Na1	Dist. oct.	*O1,O2* [Mol.1: 2.355 (O1), 2.528 (O2)], *O1,O2* [Mol.2: 2.388 (O1), 2.430 (O2)], 2*O4* [2.341 (Mol.3), 2.394 (Mol.4)]	SOFJUS	[139]
	Na2	Dist. oct.	*O1,O2* [Mol.5: 2.373 (O1), 2.430 (O2)], *O1,O2* [Mol.6: 2.344 (O1), 2.526 (O2)], 2*O4* [2.370 (Mol.7), 2.360 (Mol.8)]		
	Na3	Dist. oct.	*O1,O2* [Mol.9: 2.356 (O1), 2.462 (O2)], *O1,O2* [Mol.10: 2.361 (O1), 2.490 (O2)], 2*O4* [2.376 (Mol.11), 2.356 (Mol.12)] Three independent $\{[Na(D\text{-}Glc)_2].Cl.H_2O\}_n$ polymeric chains.		

			Intramolecular O1–M–O2 chelation.		
{[$Na_{0.5}$(α-D-Glc)]·0.5Cl·0.39H_2O}$_n$		6-Coord.	*O1,O2* [Mol.1: 2.361 (O1), 2.480 (O2)], *O1,O2* [Mol.1': 2.361 (O1), 2.480 (O2)], 2*O4* [2.359 (Mol.2, Mol.2')] The Na^+ ion rides on a twofold axis. Intramolecular O1–M–O2 chelation.	VEGLOI	[140]
Glucose 1-phosphate, glucose 6-phosphate, and glucose 6-sulfate complexes					
Na_2(α-D-Glc-1-phosphate)·3.5H_2O	Na1	6- Coord.	*O(P)* [-]	CIMDUX	[141]
	Na2	6- Coord.	*O(P)* [-]		
	Na3	5-Coord.	Ref. [149] mentions that the Na^+ ion binds to O5 and O6 of the same sugar.		
Na_2(α-D-Glc-1-phosphate)·3.5H_2O	Na1	5-Coord.	*O2* [Mol.1: 2.420], *O(P)* [Mol.2: 2.313]	CIMDUX01	[142]
	Na2	6- Coord.	2*O(P)* [Mol.2, Mol.2': 2.527]		
	Na3	6- Coord.	*O4* [Mol.3, Mol.3': 2.410], 2*O5,O6* [Mol.2, Mol.2': 2.443 (O5), 2.389 (O6)] Na2 and Na3 ride on twofold axes.		
K(α-D-Glc-1-phosphate)		6- Coord.	*O2* [Mol.1: 2.753], *O3* [Mol.2: 2.806], *O4* [Mol.3: 2.713], *O5,O6* [Mol.4: 2.789 (O5), 2.774 (O6)], *O(P)* [Mol.5: 2.660] Protonation at phosphate. Intramolecular O5–M–O6 chelation.	JUGTAG	[143]
K_2(α-D-Glc-1-phosphate)·2H_2O	K1	7-Coord.	*O2* [Mol.1: 3.05], *O3* [Mol.2: 3.14], *O6* [Mol.3: 2.94], *O(P)* [Mol.4: 2.97]	KGLUCP	[144]
	K2	5-Coord.	*O3* [Mol.1: 2.77], *O4* [Mol.5: 2.73], *O6* [Mol.2: 2.61] , *O(P)* [Mol.6: 2.61] $R = 0.172$		
K_2(α-D-Glc-1-phosphate)·2H_2O		–	–	KGLUCP02	[118]

(continued)

Table 13 (continued)

			The refinement of KGLUCP [144]: $R = 0.029$		
K_2(α-D-Glc-1-phosphate)·$2H_2O$	K1	6-Coord.	–	KGLUCP13	[145]
	K2	8-Coord.	–		
			The same compound as KGLUCP [144]: $R = 0.075$. Ref. [149] mentions that the K^+ ion binds to O5 and O6 of the sugar (Mol.1) and a phosphate oxygen (Mol.2).		
Na(α-D-Glc-6-phosphate)		7-Coord.	*O1,O5,O(P)* [Mol.1: 2.866 (O1); 2.414 (O5), 2.875 (OP)], *O1* [Mol.2: 2.599], 3*O(P)* [Mol.1-Mol.3: 2.378-2.596] Protonation at phosphate. Intramolecular O1–M–O5 and O5–M–O(P) chelation.	DACHEU	[146]
Na(α-D-Glc-6-phosphate)		7-Coord.	*O1,O5,O(P)* [Mol.1: 2.86 (O1); 2.40 (O5), 2.87 (OP)], *O1* [Mol.2: 2.59], 3*O(P)* [Mol.1-Mol.3: 2.60-2.38] The same compound as DACHEU [146].	DACHEU01	[147]
K(β-D-Glc-6-sulfate)		6-Coord.	*O1* [Mol.1: 2.759], *O5* [Mol.2: 2.796], 4*O(S)* [sulfate: 2.715-2.793] M–O5(ring ether) bonding.	–[d]	[148]
Galactose 1-phosphate, fructose 6-phosphate, and fructose 1,6-diphosphate complexes					
K_2(α-D-Gal-1-phosphate)·$5H_2O$	K1	6-Coord.	*O2,O3* [Mol.1: 2.882 (O2), 2.861 (O3)], *O6* [Mol.2: 2.785]	JEYDAS	[149]
	K2	7-Coord.	*O5,O6,O(P)* [Mol.3: 2.826 (O5), 2.748 (O6), 2.631 (OP)], *O3* [Mol.4: 2.901]		
Na_2(α-D-Fru-6-phosphate)·$7H_2O$	Na1	6-Coord.	Two hydroxyl oxygens [-], 4O [aqua: -]	–[d]	[150]
	Na2	5-Coord.	*O2* [-], 4O [aqua: -] Na–O distances (including aqua ligands) of 2.353–2.648 Å for Na1 and 2.330–2.476 Å for Na2.		
Na_2(α-D-Fru-1,6-diphosphate)·$8H_2O$	Na1	6-Coord.	*O2,O3* [Mol.1: 2.449 (O2), 2.379 (O3)], *O4* [Mol.2: 2.428],	–[d]	[151]

			O1,O5 [Mol.3: 2.612 (phosphate ester O1), 2.452 (ring ether O5)], O [aqua: 2.322]		
	Na2	6-Coord.	6O [aqua: 2.348-2.454]		
	Na3	6-Coord.	6O [aqua: 2.325-2.526]		
Sucrose complexes					
Na(sucrose)·Br·2H_2O		6-Coord.	*O4* [Glc (Mol.1): 2.62], *O6* [Glc (Mol.2): 2.42], *O6* [Fru (Mol.3): 2.32] Crystal structure was determined by using (0*kl*), (*h*0*l*), and (*hk*0) reflection data only.	DINYOO01	[152]
Na(sucrose)·Br·2H_2O		6-Coord.	*O4* [Glc (Mol.1): 2.533], *O6* [Glc (Mol.2): 2.493], *O6* [Fru (Mol.3): 2.431] Redetermination of DINYOO01 [152]. $R = 0.024$.	DINYOO10	[153]
Na_3(sucrose)$_2$·3I·3H_2O	Na1	7-Coord.	*O5,O6* [Glc (Mol.1): 2.599 (O5), 2.419 (O6)], *O2,O6* [Fru (Mol.1): 2.483 (O2), 2.397 (O6)], *O6* [Fru (Mol.2): 2.585]	ZZZNBS02	[154]
	Na2	7-Coord.	*O5,O6* [Glc (Mol.3): 2.691 (O5), 2.439 (O6)], *O2,O6* [Fru (Mol.3): 2.464 (O2), 2.442 (O6)], *O6* [Fru (Mol.4): 2.573]		
	Na3	8-Coord.	*O1,O2* [Glc (Mol.1): 2.747 (O1), 2.437 (O2)], *O1,O3* [Fru (Mol.1): 2.510 (O1), 2.310 (O3)], *O1,O2* [Glc (Mol.3): 2.737 (O1), 2.432 (O2)], *O1,O3* [Fru (Mol.3): 2.485 (O1), 2.540 (O3)]		
Maltose complexes					
[Na_2(1,6-anhydro-β-maltose)$_2$(H_2O)$_3$]·O_2	Na1	7-Coord.	O3,O5 [1,6-anhydro-Glc (Mol.1): 2.435 (O3), 2.467 (O5)], O1',O2' [Glc' (Mol.1): 2.617 (O1'), 2.511 (O2')],	–[d]	[155]

(continued)

Table 13 (continued)

			O2',O3' [Glc' (Mol.2): 2.307 (O2'), 2.600 (O3')]		
	Na2	7-Coord.	O2',O3' [Glc' (Mol.1): 2.330 (O2'), 2.685 (O3')], O2,O5 [1,6-anhydro-Glc (Mol.2): 2.427 (O2), 2.491 (O5)], O1',O2' [Glc' (Mol.2): 2.631 (O1'), 2.508 (O2')] 'Carbohydrate ionophore'.		
$[K_2(1,6\text{-anhydro-}\beta\text{-maltose})_3(NCS)_2\text{-}(MeOH)]_n$	K1	Dist. oct.	*O2,O5* [1,6-anhydro-Glc (Mol.1): 2.98 (O2), 2.74 (O5)], *O1',O2'* [Glc' (Mol.1): 2.81 (O1'), 2.85 (O2')], *O3,O6* [1,6-dihydro-Glc (Mol.2): 2.78 (O3), 2.74 (O6)]	OWUVOS	[156]
	K2	9-Coord.	*O2,O5* [1,6-anhydro-Glc (Mol.3): 2.72 (O2), 2.83 (O5)], *O1',O2'* [Glc' (Mol.3): 2.94 (O1'), 2.77 (O2')], *O3,O6* [1,6-dihydro-Glc (Mol.4): 3.51 (O3), 2.81 (O6)], *O5',O6'* [Glc' (Mol.5): 3.30 (O5), 2.69 (O6')] K^+ inclusion: sugar ionophore. See Figure 7.		

[a]Abbreviations: D-Fur: D-fructofuranose; D-Gal, D-galactopyranose; D-Glc, D-glucopyranose; luctose, β-D-galactopyranosyl-D-gluconate; maltose, 4-*O*-α-D-glucopyranosyl-D-glucose; M, metal ion; O(P), phosphate oxygen; O(S), sulfate oxygen; *R*, reliable factor; sucrose, α-D-glucopyranosyl-(1→2)-β-D-fructofuranoside; Dist. oct, distorted octahedral.
[b]For the numbering of oxygen atoms, see Figure 6.
[c]For example, '*O1,O2* [Mol.1: 2.397 (O1), 2.380 (O2)]' denotes that the metal ion binds to the hydroxyl O1 and O2 of the sugar ligand molecule 1 (Mol.1) with the bonding distances of 2.397 and 2.380 Å, respectively.
[d]Not registered in the CSD version 5.35 updated to February 2014.

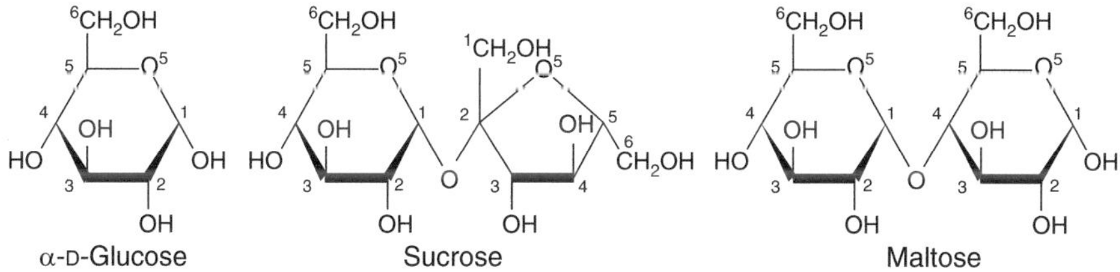

Figure 6 Nomenclature and chemical structure of representative simple carbohydrates.

intramolecular O5–, O6–, and O(P)–M bonding. On the other hand, in the Na^+ complex of fructose 6-phosphate, Na_2(α-D-Fru-6-phosphate)·$7H_2O$ [150], highly (fourfold) hydrated Na^+ ions bind to the sugar ligands through only one or two hydroxyl oxygens and besides there exists no direct interaction with phosphate oxygens, which are surrounded by one or two water molecules for each of three oxygens by forming hydrogen bonds. The Na^+ complex of fructose 1,6-diphosphate, Na_2(α-D-Fru-1,6-diphosphate)·$8H_2O$ [151], is quite unique in that it involves the intramolecular O1 (phosphate ester oxygen)–M–O5 (ring ether oxygen) chelation.

Two sucrose complexes are reported: Na(sucrose)·Br·$2H_2O$ [152, 153] and Na_3(sucrose)$_2$·3I·$3H_2O$ [154]. The former Br complex (metal:ligand = 1:1) exhibits a simple metal coordination environment: three ligand molecules are arranged around the Na^+ ion by ligating each through a hydroxyl oxygen (i.e., two hydroxyl oxygens from the Glc residue and one from the Fru residue). The latter I complex (metal:ligand = 3:2) includes three independent Na^+ ions (Na1, Na2, and Na3), each of which is trapped between the two ligand molecules by forming five Na^+–oxygen bonds for both the Na1 and Na2 and eight bonds for Na3. The coordination mode is the same for Na1 and Na2: among the four ligation sites from one ligand molecule, two each are from Glc and Fru residues (a hydroxyl and a ring ether oxygen from Glc and two hydroxyl oxygens from Fru), and the fifth ligation site is occupied by a hydroxyl oxygen of Fru of the second ligand. Na3 is fully occupied by eight oxygens, four each from the two ligand molecules with the same set of the ligation sites, that is, for each ligand, a hydroxyl oxygen and a bridging ether oxygen O1 from Glc and two hydroxyl oxygens from Fru.

Two maltose (to be exact, its derivative) complexes are reported: [Na_2(1,6-anhydro-β-maltose)$_2$(H_2O)$_3$]·O_2 [155] and [K_2(1,6-anhydro-β-maltose)$_3$(NCS)$_2$(MeOH)]$_n$ [156]. In both the Na^+ and K^+ complexes, the metal ions are trapped among the ligand molecules, thus this maltose derivative behaves as 'carbohydrate ionophore'. The Na^+ complex contains two independent Na^+ ions (Na1 and Na2), which show similar metal coordination mode/sites: four oxygens from one ligand (a hydroxyl oxygen and a ring ether oxygen of the 1,6-anhydro-Glc residue and a hydroxyl oxygen and a bridging ether oxygen O1 of the Glc residue) and two hydroxyl oxygens of the Glc residue from the other ligand. The K^+ complex

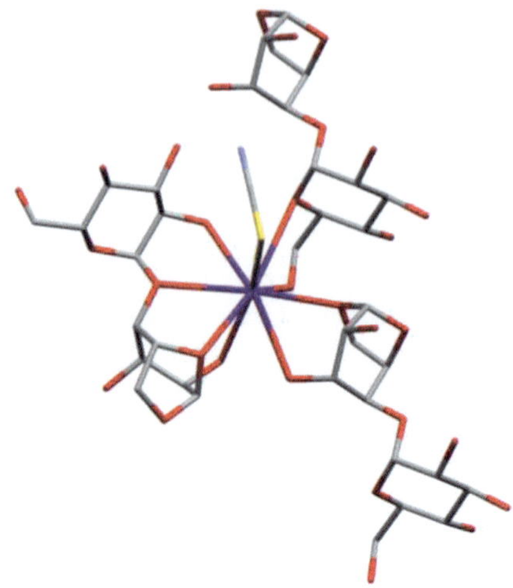

Figure 7 A view showing the trapping of a K^+ ion among the three ligand molecules in the $[K_2(1,6\text{-anhydro-}\beta\text{-maltose})_3(NCS)_2(MeOH)]_n$ complex [156]. Ligation sites are: four oxygens from one ligand (a hydroxyl oxygen and a ring ether oxygen of the 1,6-anhydro-Glc residue and a hydroxyl oxygen and a bridging ether oxygen O1 of the Glc residue), two oxygens from the second ligand (a hydroxyl oxygen and a ring ether oxygen of the 1,6-anhydro-Glc residue), and two oxygens (a hydroxyl oxygen and a ring ether oxygen of the Glc residue) from the third ligand. The ninth site is occupied by a NCS solvent (*yellow*: sulfur atom).

exhibits essentially the same coordination mode for two independent K^+ ions (see Figure 7).

In summary, it is quite certain that the three ligation sites of carbohydrates, hydroxyl oxygens and the ring ether oxygen for monosaccharides and additionally the bridging ether oxygen for disaccharides, are commonly preferred in alkali metal ion coordination in the solid state. Sugars having phosphate or sulfate groups provide the phosphate and sulfate oxygens as additional, major ligation sites, including the phosphate ester oxygens [151].

5 Naturally Occurring Antibiotic Ionophore Complexes

Ionophores can be characterized as receptors which form stable, lipophilic complexes with charged hydrophilic species such as Na^+, K^+, Ca^{2+}, etc., and thus are able to transport them into lipophilic phases, for example across natural or artificial membranes. There are two types: mobile ion carriers and channel-forming ionophores. Ion carrier ionophores are further classified into three types: depsipeptides, macrotetrolides, and polyether antibiotics. Figure 8 presents chemical structures of some representative ionophores.

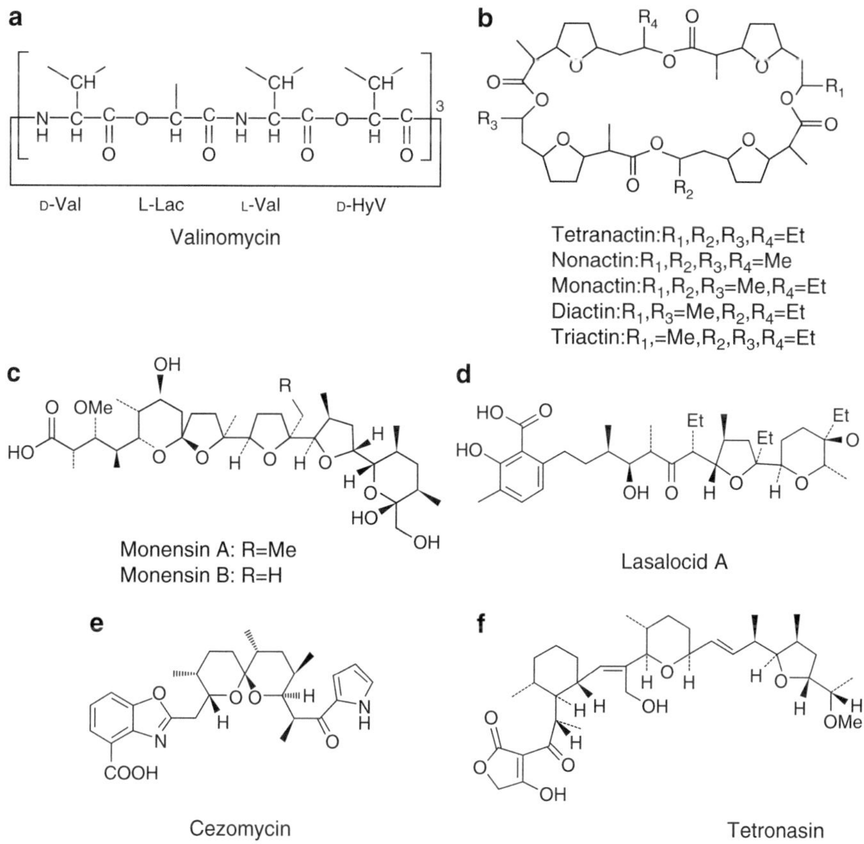

Figure 8 Nomenclature and chemical structures of representative antibiotic ionophores.

Good reviews on X-ray structures of metal complexes with depsipeptides [6], macrotetrolides [6], and polyether ionophore antibiotics [6, 7] reported prior to 1982 and, most recently, those for polyether antibiotics [8] are available. Table 14 lists the total number of crystal structures of alkali metal ion complexes with naturally occurring antibiotic ionophores, including their chemical derivatives [161–190, 193–240], containing those cited in [6–8]. Uncomplexed structures are not considered here. We also do not deal with synthetic ionophores, such as crown ethers, cryptands, or synthetic polyethers, though they could mimic natural ionophores in the sense that some of them display similar selective complexation and transfer as naturally occurring ligands (for a review on crystal structures of synthetic polyethers and their metal complexes prior to 1982, refer to [6]).

Table 14 Total number of alkali metal ion–natural ionophore complexes in the solid state.[a]

Antibiotic[b]	Li^+	Na^+	K^+	Rb^+	Cs^+	Total
1. Channel-type ionophore						
Gramicidin	0	1[c]	2[d]	1[e]	3[f]	7
2. Carrier-type ionophores						
Depsipeptides						
Antamanide	2[g]	1[h]	0	0	0	3
Enniatin B	0	1[i]	3[j]	2[k]	0	6
Prolinomycin	0	0	0	1[l]	0	1
Valinomycin	0	1[m]	4[n]	1[o]	1[p]	7
Macrotetrolides						
Nonactin	0	1[q]	1[r]	0	1[s]	3
Tetranactin	0	1[t]	2[u]	2[v]	1[w]	6
Boromycin	0	0	0	1[x]	1[y]	2
Polyether antibiotics						
Class 1a: Monovalent polyethers						
Alborixin	0	1[z]	1[aa]	0	0	2
CP-54838	0	0	0	1[ab]	0	1
Kijimicin	0	0	0	1[ac]	0	1
Monensin	3[ad]	20[ae]	2[af]	2[ag]	0	27
Mutaromycin	0	0	3[ah]	0	0	3
Nigericin	0	1[ai]	1[aj]	0	0	2
Noboritomycin	0	0	0	1[ak]	0	1
Salinomycin	0	3[al]	0	0	0	3
Class 1b: Monovalent glycoside polyethers						
A204A	0	1[am]	0	0	0	1
CP-80219	0	0	0	1[an]	0	1
CP-84657	0	0	0	1[ao]	0	1
Dianemycin	0	1[ap]	1[aq]	3[ar]	0	5
Endusamycin	0	0	0	1[as]	0	1
K-41	0	2[at]	0	0	0	2
Septamycin	0	0	0	1[au]	0	1
Class 2a: Divalent polyethers						
Lasalocid A	0	4[av]	0	0	0	4
Class 2b: Divalent glycoside ionophore						
None						
Class 3: Divalent pyrrole ether ionophore						
Cezomycin	0	2[aw]	0	0	0	2
Class 4: Acyl tetronic acid ionophore						
Tetronasin	0	1[ax]	0	0	0	1
Total	5	42	20	20	7	94

[a]Obtained from the Cambridge Structural Database (the CSD version 5.35 updated to February 2014) or Protein Data Bank (January 2015) for gramicidin.
[b]Synthetic analogs are included.
[c][161]. [d][162, 163]. [e][164]. [f][165–167]. [g][169, 170]. [h][171]. [i][172]. [j][173–175]. [k][176, 177]. [l][178]. [m][179]. [n][180–183]. [o][184]. [p][185]. [q][186]. [r][187]. [s][188]. [t][189]. [u][189]. [v][184, 189]. [w][188]. [x][190]. [y][190]. [z][193]. [aa][194]. [ab][195]. [ac][196]. [ad][197–199]. [ae][200–214]. [af][205, 215]. [ag][216, 217]. [ah][218]. [ai][219]. [aj][220]. [ak][221]. [al][222–224]. [am][225]. [an][226]. [ao][227]. [ap][228]. [aq][229]. [ar][226, 230, 231]. [as][232]. [at][233]. [au][234]. [av][235–237]. [aw][238, 239]. [ax][240].

5.1 Channel-Forming Ionophore Complexes

This class of ionophores includes gramicidins and peptaibols (alamethicin, zervamicin, suzukallin, trichotoxins, etc. [159]). Among these, no crystal structures of metal ion complexes have been reported except for those of gramicidins, which are described here.

Gramicidin is a linear pentadecapeptide antibiotic composed of hydrophobic amino acids with alternating L- and D- configurations, HCO-L-Val_1-Gly_2-L-Ala_3-D-Leu_4-L-Ala_5-D-Val_6-L-Val_7-D-Val_8-L-Trp_9-D-Leu_{10}-L-Trp_{11}-D-Leu_{12}-L-Trp_{13}-D-Leu_{14}-L-Trp_{15}-$NHCH_2CH_2OH$. Naturally occurring gramicidin D (gD) is a mixture of isoforms differing in amino acid composition at position 1, Val1/Ile1, and position 11, Trp11(gA)/ Phe11(gB)/Tyr11(gC) (gA, gB, gC: gramicidin A, B, C). It forms membrane channels that are specific for the transport of monovalent cations, with no measurable permeability to anions or polyvalent cations; ions and water move through a pore whose wall is formed by the peptide backbone; and the single-channel conductance and cation selectivity vary when the amino acid sequence is varied, even though the permeating ions make no contact with the amino acid side chains [160].

Seven alkali metal ion complexes have been reported: one Na^+ [162], two K^+ [162, 163], one Rb^+ [164], and three Cs^+ [165–167] complexes. In each of $[Na_{1.26}(\text{gramicidin D})_4]\cdot I_{1.21}\cdot 18.5H_2O\cdot 2MeOH$ [PDB code 3L8L] [161], $[K_{1.1}(\text{gramicidin D})_4]\cdot I_x\cdot nH_2O$ [2IZQ] [163], $[Rb_{2.8}(\text{gramicidin D})_4]\cdot Cl_{2.6}\cdot nH_2O\cdot EtOH$ [1W5U] [164], and $[Cs_2(\text{gramicidin A})_4]\cdot Cl_2\cdot nH_2O$ [1AV2] [165], four polypeptide

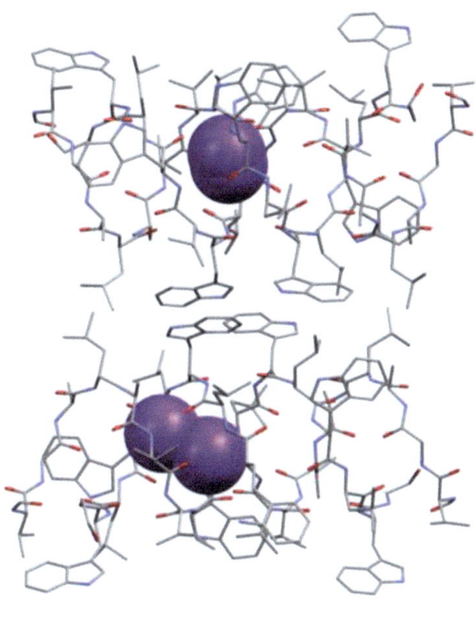

Figure 9 Structure of the Na^+ complex of gramicidin D, $[Na_{1.26}(\text{gramicidin D})_4]\cdot I_{1.21}$ [161], showing the formation of the two pairs of *right*-handed double-stranded β-sheets (channels 1 and 2) intertwined in an antiparallel fashion and the therein incorporated Na^+ ion positions. The peptide chains and the Na^+ ions are depicted as 'wires' and 'spheres', respectively.

monomers form two pairs of *right*-handed double-stranded β-sheets (the two β-helical double-strands: channels 1 and 2) intertwined in an antiparallel fashion, where the two peptide backbones form the wall of the channel with the carbonyl groups directing inward and the hydrophobic side chains outward (Figure 9 shows the Na^+ complex [161]).

The number of the metal binding sites within each channel (each gramicidin dimers) and the sum of the partial occupancies (the total number of metal ions) per two channels (two dimers) vary: two sites with 1.26 occupancies for Na^+ [161], three sites with 1.10 occupancies for K^+ [163], seven sites with 2.81 occupancies for Rb^+ [164], and three sites with 2.0 occupancies for Cs^+ [165]. Each ion makes more than two contacts to the π clouds of the carbonyl and peptide bonds: three contacts (from 2.3 to 3.3 Å) for Na^+, 2–4 contacts (from 2.6 to 3.7 Å) for K^+, 3–5 contacts (from 3.1 to 3.8 Å) for Rb^+, and six contacts (from 3.6 to 4.5 Å) for Cs^+. On the other hand, the formation of the *left*-handed, anti-parallel, double-stranded β-helix was also reported in crystal structures of [K(gramicidin A)]·NCS [1GMK] [162] and [Cs(gramicidin A)]·Cl [1C4D] [166], but these structures have been claimed to be incorrect [165, 168].

5.2 Ion Carrier Ionophore Complexes

Table 15 summarizes alkali metal ion binding sites/modes on ion carrier ionophore antibiotics in the solid state [169–240].

5.2.1 Depsipeptide Ionophore Complexes

Depsipeptide ionophores are cyclic compounds constructed by sequences of amino and hydroxy carboxylic acids (usually α-amino and α-hydroxy acids), commonly but not necessarily regularly alternating. These involve antamanide, enniatins, beauvericin, prolinomycin, and valinomycin (though, strictly speaking, antamanide is not a depsipeptide but a peptide, yet it is conventionally classified in this class of ionophores). A total number of so far reported crystal structures of alkali metal ion complexes are: three [169–171] for antamanide, six [172–177] for enniatin B, zero for beauvericin, one [178] for prolinomycin, and seven [179–185] for valinomycin (Table 15).

Commonly in these complexes, the polar moieties of the ionophore molecule are turned toward the interior while the exterior surface contains the hydrophobic side groups. Interestingly, metal binding sites/modes are different among these ionophores: For the antamanide ionophore, cyclo(−Val-Pro-Pro-Ala-Phe-Phe-Pro-Pro-Phe-Phe-), four amide oxygens are commonly involved in metal coordination in three complexes with Li^+ [169, 170] and Na^+ [171] ions. For enniatin B, cyclo(−L-MeVal-D-HyIv-)$_3$, (i) three amide and three ester oxygens are involved in a discrete

Table 15 Alkali metal ion binding sites/modes on carrier type of ionophore antibiotics in the solid state.[a]

Complex		Coordination geometry or number	Metal binding sites on ionophore antibiotics[b] and other ligands/comments, if any	CSD code or CCDC number	Ref.
Depsipeptides					
$[Li(antamanide)(CH_3CN)]\cdot Br\cdot 2CH_3CN$		Sq. pyr.	4O [amide: 2.04-2.24][c], N [solvent: 2.07] Antamanide: cyclo(-Val-Pro-Pro-Ala-Phe-Phe-Pro-Pro-Phe-Phe-); high selectivity for Na^+ over K^+.	ANTAML10	[169]
$[Li(H_2O)(Cha^5,Cha^6,Cha^9,Cha^{10}$-antamanide)$Li(H_2O)_3]\cdot 2Br\cdot 2CH_3CN$	Li1	Dist. sq. pyr.	4O [amide (Mol.1): 1.996-2.178], O [solvent: 2.019]	DOLJAP	[170]
	Li2	Dist. tet.	O [amide (Mol.1): 2.006], 3O [solvent: 1.851-2.053] The polar backbone exposed toward the exterior surface, corresponding to the complete loss of antitoxic potency despite its ability to form ion complexes in the same manner as antamanide.		
$[Na(Phe^4,Val^6$-antamanide)(EtOH)]$\cdot$Br			4O [amide: 2.25-2.36], O [solvent: 2.28]	PVANSB	[171]
$Na\cdot$(enniatin B)$\cdot Ni(NO_3)_3\cdot 1.5MeOH\cdot 3H_2O$		–[d]	–[d] Enniatin B: cyclo(-L-MeVal-D-Hyi-L-MeVal- D-Hyi-L-MeVal-D-Hyi-).	–[e]	[172]
$3\{[K(enniatin B)]_4\cdot 4NCS\}_n$	K1	6-Coord.	3O [ester (Mol.1): 2.760], 3O [ester (Mol.4): 2.790]	IHECUT	[173]
	K2	6-Coord.	3O [amide (Mol.1): 2.705], 3O [amide (Mol.2): 2.651]		
	K3	6-Coord.	3O [ester (Mol.2): 2.780], 3O [ester (Mol.3): 2.714]		
	K4	6-Coord.	3O [amide (Mol.3): 2.705], 3O [amide (Mol.4): 2.631]		
	K5	6-Coord.	3O [amide (Mol.5): 2.653], 3O [amide (Mol.8): 2.742]		
	K6	6-Coord.	3O [ester (Mol.5): 2.818], 3O [ester (Mol.6): 2.755]		
	K7	6-Coord.	3O [amide (Mol.6): 2.695], 3O [amide (Mol.7): 2.652]		
	K8	6-Coord.	3O [ester (Mol.7): 2.767], 3O [ester (Mol.8): 2.723]		
	K9	6-Coord.	3O [ester (Mol.9): 2.819], 3O [ester (Mol.12): 2.707]		

(continued)

Table 15 (continued)

	K10	6-Coord.	3O [amide (Mol.9): 2.627], 3O [amide (Mol.10): 2.689]		
	K11	6-Coord.	3O [ester (Mol.10): 2.701], 3O [ester (Mol.11): 2.792]		
	K12	6-Coord.	3O [amide (Mol.11): 2.653], 3O [amide (Mol.12): 2.753] Each enniatin molecule has a molecular threefold axis and is located on one of the three crystallographic threefold axes.		
[K(enniatin B)$_2$]·I·H_2O		6-Coord.	3O [amide (Mol.1): 2.64-2.74], 3O [amide (Mol.2): 2.62-2.82]	PEKFEQ	[174]
[K(enniatin B)]·I		Oct.	6O [3O amide (Mol.1) and 3O ester (Mol.1): 2.6-2.8]	ENNBKI	[175]
[Rb(enniatin B)(NCS)]$_n$		Dist. oct.	5O [1O amide (Mol.1), 2O ester (Mol.1), 1O amide (Mol.2), 1O ester (Mol.2): 2.88-2.98], N [solvent: 3.06] Infinite [-ionophore-Rb-]$_n$ "sandwich" spiral around a 4_1 axis.	MVHIRB10	[176]
[Rb(LDLLDL enniatin B)]·NCS		–	– A LDLLDL analog of enniatin B (LDLDLD). Infinite [-ionophore-Rb-]$_n$ "pseudosandwich" complex.	–	[177]
[Rb(prolinomycin)]$_2$·2picrate·3toluene·-2$CHCl_3$	Rb1	6-Coord.	6O [amide (Mol.1): 2.861-2.928]	PROMYC10	[178]
	Rb2	6-Coord.	6O [amide (Mol.2): 2.798-3.014] Prolinomycin: cyclo(-D-Val-L-Pro-L-Val-D-Pro-)$_3$. Each of two independent ionophore molecules has an inversion center, on which each of two independent K^+ ions ride.		
[Na(valinomycin)(picrato)(H_2O)]·-*m*-xylene		6-Coord.	3O [ester: 2.33-2.54] 2O [picrate: 2.37, 2.64], O [aqua: 2.33] Valinomycin: cyclo(-D-HyIv-D-Val-L-Lac-L-Val-)$_3$.	BINFIN	[179]
[K(valinomycin)(picrato)]·*m*-xylene		7-Coord.	6O [ester: 2.67-2.81], O [picrate: 3.86] (See Figure 10)	VALKPC10	[180]
[K(valinomycin)]·[$AuCl_4$]		6-Coord.	6O [ester: 2.7-2.8]	GIFBEC	[181]
[K(valinomycin)]·I_5		6-Coord.	6O [ester: 2.69-2.83]	VALINK	[182]
[K(*meso*-valinomycin)] [$AuCl_4$]		6-Coord.	6O [ester: 2.73] *Meso*-valinomycin: cyclo(-D-Val-L-HyIv-L-Val-D-HyIv-)$_3$	TEFBAH	[183]
[Rb(valinomycin)]·SF		6-Coord.	6O [ester: -]	BASFUW	[184]
[Cs(valinomycin)(picrato)]		7-Coord.	6O [ester: 2.85-3.07], O [picrate: 3.26]	DOWDAU	[185]

Macrotetrolides: 32-membered cyclic tetralactones				
[Na(nonactin)]·NCS	8-Coord.	4O [carbonyl: 2.40-2.44], 4O [ether: 2.74-2.79] The $Cs^+ > Na^+$ selectivity in nonactin. The Na^+ rides on a two-fold rotation axis and the nonactin molecule has a crystallographic diad axis of symmetry with an approximate S_4 (–4) symmetry.	NONACS	[186]
[K(nonactin)]·NCS	8-Coord.	4O [carbonyl: 2.73-2.81], 4O[ether: 2.81-2.88] The K^+ rides on a twofold rotation axis and the nonactin molecule has a crystallographic diad axis of symmetry with an approximate S_4 (–4) symmetry. (See Figure 11)	NONKCS	[187]
[Cs(nonactin)]·NCS	8-Coord.	4O [carbonyl: 3.133-3.181], 4O [ether: 3.069-3.163] The Cs^+ rides on a twofold rotation axis and the nonactin molecule has a crystallographic diad axis of symmetry with an approximate S_4 (–4) symmetry.	NONACU	[188]
[Na(tetranactin)]·NCS (Space group *C*2/*c*)	8-Coord.	4O [carbonyl: 2.43-2.45], 4O [ether: 2.70-2.94] The $Na^+ > Cs^+$ selectivity in tetranactin.	TRANNA10	[189]
[K(tetranactin)]·NCS (Space group $P2_1/n$)	8-Coord.	4O [carbonyl: 2.78-2.79], 4O [ether: 2.85-2.91]	TETINK11	[189]
[K(tetranactin)]·NCS (Space group *C*2/*c*)	8-Coord	4O [carbonyl: 2.75-2.81], 4O [ether: 2.83-2.92]	TETINK20	[189]
[Rb(tetranactin)]·NCS (Space group $P2_1/n$)	8-Coord.	4O [carbonyl: 2.88-2.93], 4O [ether: 2.90-2.98]	TETINR20	[189]
[Rb(tetranactin)]·SF	8-Coord.	4O [carbonyl:-], 4O [ether: -]	BASGAD	[184]
[Cs(tetranactin)]·NCS	8-Coord.	4O [carbonyl: 3.057-3.161], 4O [ether: 3.031-3.097]	TETRCS	[188]
[Rb(des-valino-boromycin)]·2MeOH	8-Coord.	[2O ether, 4O hydroxyl, 2O ester: 2.80-3.17]	RBORMY	[190]
Cs.(des-valino-boromycin)	–	–	QQQBRP01	[190]
Polyether antibiotics (Polyether carboxylic antibiotics)				
Class 1a: Monovalent polyethers				
[Na(6-demethyl-alborixin)]·0.85DMF	5-Coord.	2O [ether: 2.43, 2.54], 2O [hydroxyl: 2.41, 2.47],	BOSSEH10	[193]

(continued)

Table 15 (continued)

		O[carboxylate: 2.38] The $K^+ > Na^+$ selectivity and the $K^+ > Rb^+ > Na^+ > Cs^+ > Li^+$ cation transport property in alborixin. Refer to Figure 2 in [241].		
[K(alborixin)]	8-Coord.	3O[ether: 2.76-3.07], 4O [hydroxyl: 2.69-2.98], O [carboxylate: 2.89]	KALBOR	[194]
[Na(CP-54838)]	6-Coord	[2O ether, 2O carbonyl, 1O methoxy, 1O phenoxide: 2.32-2.64]	–	[195]
[Rb(kijimicin)]·0.5hexane	7-Coord.	4O [ether: -], O [hydroxyl: -], O [methoxy: -], O [carboxylate] Refer to Figure 20 in [8].	JIGPUK	[196]
[Li(monensin B)]·H_2O	6-Coord.	4O [ether: -], O [hydroxyl: -], O [CH_2OH: -]	DOPVEJ	[197]
[Li(monensin A 1-naphthylmethyl-ester)(H_2O)]·ClO_4	Sq. pyr.	3O [ether: 1.955-2.280], O [hydroxyl: 1.984], O [solvent: 2.027] Refer to Figure 56 in [8].	AMIDUW	[198]
[Li(monensin)]·MeCN	6-Coord.	4O [ether: 2.173-2.633], O [hydroxyl: 2.022], O [CH_2OH: 2.114] Refer to Figure 54 in [8].	MIPSIO	[199]
[Na(monensin B)]·H_2O	6-Coord.	4O [ether: 2.35-2.47], O [hydroxyl: 2.33], O[CH_2OH: 2.35]	BELDAX	[200]
[Na(monensin)]·Br	Dist. oct.	4O [ether: 2.366-2.503], O [hydroxyl: 2.349], O[CH_2OH: 2.419] Refer to Figure 3c in [241].	MONSBR	[203]
[Na(monensin)]	Dist. oct.	4O [ether: 2.408-2.529], O [hydroxyl: 2.346], O[CH_2OH: 2.381] Refer to Figure 2d in [241].	NAMNSB	[202]
[Na(monensin)]·$2H_2O$ (Space group $P2_12_12_1$)	Dist. oct.	4O [ether: 2.358-2.543], O [hydroxyl: 2.336], O[CH_2OH: 2.449]	QQQCYA01	[202]
[Na(monensin)]·$2H_2O$ (Space group $P2_1$)	6-Coord.	4O [ether: 2.33-2.60], O [hydroxyl: 2.34], O [CH_2OH: 2.38]	QQQCYA	[203]
[Na(monensin)]·$2H_2O$ (Space group $P2_12_12_1$)	Dist. oct.	4O [ether: 2.365-2.520], O [hydroxyl: 2.336], O [CH_2OH: 2.373]	QQQCYA02	[204]
[Na{7-O-(4-ethylbenzylmonensin)}].-isopropyl ether	6-Coord.	4O [ether: 2.394-2.543], O [hydroxyl: 2.585], O [CH_2OH: 2.40]	YISBAD	[205]
[Na(monensin A)]·MeCN	Dist. oct.	4O[ether: 2.328-2.603], O [hydroxyl: 2.341], O [CH_2OH: 2.383] Refer to Figure 52 in [8].	DEYGAQ	[206]
[Na(monensin A allyl urethane)]·-0.75MeCN	5-Coord.	4O [ether: 2.340-2.575], O [hydroxyl: 2.327], O [carboxylate: 2.268]	XIBKAW	[207]

[Na(monensin A phenylurethane)]		5-Coord.	3O [ether: 2.315-2.547], O [hydroxyl: 2.365], O [carboxylate: 2.340] Refer to Figure 57 in [8].	IPIWEK	[208]
[Cu{Na(monensin A)}$_2$Cl$_2$]·MeCN		6-Coord.	4O [ether: 2.339-2.474], O [hydroxyl: 2.331], O [CH$_2$OH: 2.338] The Cu^{2+} ion rides on a twofold axis; only one [Na(monensin)] molecule is independent.	LULPOY	[209]
[Na(Me-monensin A)]·-[Na(H-monensin A)]·[CuCl$_2$]·Cl		6-Coord.	4O [ether: 2.359-2.473], O [hydroxyl: 2.340], O [CH$_2$OH: 2.374]	LULPEO	[209]
[Mn{Na(monensin A)}$_2$Cl$_2$]·H$_2$O		6-Coord.	4O [ether: 2.342-2.480], O [hydroxyl: 2.333], O [CH$_2$OH: 2.364] The Mn^{2+} ion rides on a twofold axis; only one [Na(monensin)] molecule is independent.	XITDAG	[210]
[Co{Na(monensin A)}$_2$Cl$_2$]·H$_2$O		6-Coord.	4O [ether: 2.342-2.475], O [hydroxyl: 2.332], O [CH$_2$OH: 2.351] The Co^{2+} ion rides on a twofold axis; only one [Na(monensin)] molecule is independent.	XITDEK	[210]
[Na(monensin A)]$_2$·2ClO$_4$	Na1	6-Coord.	4O [ether (Mol.1): 2.378-2.512], O [hydroxyl (Mol.1): 2.346], O [CH$_2$OH (Mol.1): 2.427]	866477[f]	[211]
	Na2	6-Coord.	4O [ether (Mol.2): 2.331-2.497], O [hydroxyl (Mol.2): 2.329], O [CH$_2$OH (Mol.2): 2.396] Refer to Figure 3b in [241].		
[Na(monensin A)]·Cl·MeCN		6-Coord.	4O [ether: 2.335-2.572], O [hydroxyl: 2.332], O [CH$_2$OH: 2.434]	866478[f]	[211]
[Na(monensin *N*-phenylamide)]·-Cl·MeCN		6-Coord.	4O [ether: 2.382-2.562], O [hydroxyl: 2.409], O [CH$_2$OH: 2.410]	710817[f]	[212]
[Na(monensin 1-naphthylmethyl-ester)]·ClO$_4$		7-Coord.	4O [ether: 2.375-2.549], 2O [hydroxyl: 2.341, 2.865], O [CH$_2$OH: 2.395]	860451[f]	[213]
[Na(demethylmonensin A)]·2H$_2$O		6-Coord.	4O [ether:-], O [hydroxly: -], O [CH$_2$OH: -]	793152[f]	[214]
[Na(dehydroxylmonensin A)]·1.5H$_2$O		6-Coord.	3O [ether: -], O [hydroxyl: -], 2O [carboxylate: -]	793153[f]	[214]
[K(monensin)]·2H$_2$O		6-Coord.	4O [ether: -], O [hydroxyl: -], O [CH$_2$OH: -]	QQQCYD	[203]
[K(monensin A)]·2H$_2$O		6-Coord.	4O [ether: 2.660-2.796], O [hydroxyl: 2.650], O [CH$_2$OH: 2.786] Refer to Figure 2e in [241].	FECROU10	[215]

(continued)

Table 15 (continued)

[K{7-O-(4-ethylbenzylmonensin)}]·-acetone		6-Coord.	4O [ether: 2.627-2.798], O [hydroxyl: 2.695], O [CH_2OH: 2.712]	YIRZUU	[205]
[Rb(monensin)]·$2H_2O$		6-Coord.	4O [ether: 2.788-2.901], O [hydroxyl: 2.807], O [CH_2OH: 2.875] Refer to Figure 2f in [241].	RITLIQ01	[216]
[Rb(monensin)]·$2H_2O$		7-Coord.	4O [ether: 2.800-2.913], 2O [hydroxyl: 2.787, 3.143], O [CH_2OH: 2.918] (See Figure 12)	RITLIQ	[217]
[K(mutaromycin)]		8-Coord.	4O [ether: 2.666-2.982], 2O [hydroxyl: 2.698, 2.972], 2O [carboxylate: 2.702, 3.250]	–	[218]
[K(2-*epi*-mutaromycin)]·$2H_2O$		8-Coord.	4O [ether: 2.697-3.051], 2O [hydroxyl: 2.651, 3.201], 2O [carboxylate: 2.666, 3.079]	VATGUS	[218]
[K(28-*epi*-mutaromycin)]		7-Coord.	4O [ether: 2.665-3.084], O [hydroxyl: 2.665], 2O [carboxylate: 2.679, 2.789] Refer to Figure 15 in [8].	VATHAZ	[218]
[Na(nigericin)]		5-Coord.	3O [ether: 2.38-2.52], O [methoxy: 2.44], O [carboxylate: 2.25]	NIGERI	[219]
[K(nigericin)]		5-Coord.	3O [ether 2.73-2.81], O [methoxy: 2.67], O [carboxylate: 2.48]	–	[220]
[Rb(6-chloro-noboritomycin)]	Rb1	7-Coord.	2O [ether (Mol.1): 2.791, 3.010], 2O [hydroxyl (Mol.1): 2.972, 2.944], 2O [carbonyl (Mol.1): 2.799. 2.939], O [carboxylate (Mol.1): 2.825]	–	[221]
	Rb2	7-Coord.	2O [ether (Mol.2): 2.901, 3.150], 2O [hydroxyl (Mol.2): 3.102, 3.221], 2O [carbonyl (Mol.2): 2.925, 2.960], O [carboxylate (Mol.2): 2.878]		
[Na(salinomycin)(H_2O)]·-[Na(salinomycin)$(H_2O)_2$]·MeCN	Na1	5-Coord.	2O [ether (Mol.1): 2.415, 2.487], O [carbonyl (Mol.1): 2.436], O [carboxylate (Mol.1): 2.495], O [aqua: -]	GIZDIC	[222]
	Na2	6-Coord.	2O [ether (Mol.2): 2.402, 2.486], O [carbonyl (Mol.2): 2.356], O [carboxylate(Mol.2): 2.433], 2O [aqua: -] The relative affinity of salinomycin for complex formation in the order $K^+ > Na^+ > Cs^+ > Sr^{2+} > Ca^{2+}, Mg^{2+}$.		

[Na(20-oxosalinomycin)]		5-Coord.	2O [ether: -], O [carbonyl: -], O [hydroxyl: -], O [carboxylate: -] Refer to Figure 46 in [8].	BIDWER	[223]
[Na(20-deoxysalinomycin)]·0.69MeOH		5-Coord.	2O [ether: -], O [carbonyl: -], 2O [carboxylate: -] Refer to Figure 45 in [8].	ESUZEX	[224]
Class 1b: Monovalent glycoside polyethers					
[Na(A204A)]·acetone		6-Coord.	3O [ether: -], O [methoxy: -],2O [carboxylate: -]	KALBOR	[225]
[Rb(CP-80219)]		7-Coord.	4O [ether: -], 2O [hydroxyl: -], O [CH_2OH: -]	–	[226]
[Rb(CP-84657)]·H_2O		–	–	–	[227]
[Na(dianemycin)(MeOH)]·MeOH		7-Coord.	4O [ether: 2.327-2.793], O [hydroxyl: 2.493], O [CH_2OH: 2.327], O [solvent: 2.298]	FATBAF	[228]
[K(dianemycin)(H_2O)]		7-Coord.	4O [ether: -], O [hydroxyl: -], O [CH_2OH: -], O [solvent: -]	PDINMC01	[229]
[Rb(dianemycin*)]		7-Coord.	4O [ether: -], 2O [hydroxyl: -], O [CH_2OH: -]	JIHNUJ	[226]
[Rb(*iso*-dianemycin)]·2MeOH		–	–	SAVCEX	[230]
[Rb(19-*epi*-dianemycin)]		–	–	–	[231]
[Rb(endusamycin)]·3MeOH		7-Coord.	4O [ether: -], 2O [hydroxy: -], O [CH_2OH: -] Refer Figure 13 in [8].	SAWGIG	[232]
[Na(K-41 *p*-Br-benzoate)]·H_2O·hexane		6-Coord.	3O [ether: -], O [methoxy: -], 2O [carboxylate: -]	ANTBRN	[233]
[Na(K-41 *p*-I-benzoate)]·H_2O·hexane		6-Coord.	3O [ether: -], O [methoxy: -], 2O [carboxylate: -]	ANTINA	[233]
[Rb(27-methoxyseptamycin)]		–	–	PAFRET	[234]
Class 2a: Divalent polyethers					
[Na(5-Br-lasalocid)]$_2$ (Space group $P2_12_12_1$)	Na1	6-Coord.	2O [ether (Mol.1): -],2O [hydroxyl (Mol.1): -], O [carbonyl (Mol.1): -], O [carboxyl (Mol.2): -]	BLASAL	[235]
	Na2	6-Coord.	2O [ether (Mol.2): -], 2O [hydroxyl (Mol.2): -], O [carbonyl (Mol.2): -], O [carboxyl (Mol.1): -] A 2:2 "head-to-tail" dimer.		
[Na(5-Br-lasalocid)]$_2$ (Space group $C222_1$)	Na1	6-Coord.	2O [ether (Mol.1): -], 2O [hydroxyl (Mol.1): -], O [carbonyl (Mol.1): -], O [carbonyl (Mol.2): -]	BLASAL01	[235]

(continued)

Table 15 (continued)

	Na2	6-Coord.	2O [ether (Mol.2): -], 2O [hydroxyl (Mol.2): -], O [carbonyl (Mol.2): -], O [carbonyl (Mol.1): -] A 2:2 "head-to-head" dimer.		
[Na(lasalocid A)(H_2O)]$_2$	Na1	7-Coord.	2O [ether (Mol.1): 2.42, 2.47], 2O [hydroxyl (Mol.1): 2.56, 2.67], O [carbonyl (Mol.1): 2.67], O [carbonyl (Mol.2): 2.42], O [solvate: 2.45]	NALASC	[236]
	Na2	6-Coord.	2O [ether (Mol.2): 2.40, 2.44], 2O [hydroxyl (Mol.2): 2.47, 2.72], 2O [solvent]: 2.37, 2.40] (See Figure 13)		
[Na(lasalocid A)(MeOH)]		6-Coord.	2O [ether (Mol.1): -],2O [hydroxyl: -], O [carbonyl: -], O [solvent: -]	SLASAM	[237]
Class 3: Divalent pyrrole ether ionophores					
[Na{demethyl(C-11)cezomycin}$_2$]·-3H_2O		6-Coord.	N [benzoxazole (Mol.1):-], O [carbonyl (Mol.1): -], O[carboxylate (Mol.1): -], N [benzoxazole (Mol.2): -], O [carbonyl (Mol.2): -], O [carboxylate (Mol.2): -] (see Figure 14) Refer to Figure 9 in [8].	ARELIS	[238]
[Na(cezomycin)(H_2O)]$_4$	Na1	6-Coord.	N [benzoxazole (Mol.1):-], O [carbonyl (Mol.1): -], O [carboxylate (Mol.1): -], O [solvent (Mol.1): -], O [carboxylate (Mol.2): -], O [carboxylate (Mol.3: -)	–	[239]
	Na2	6-Coord.	N [benzoxazole (Mol.2):-], O [carbonyl (Mol.2): -], O [carboxylate (Mol.2): -], O [solvent (Mol.2): -], O [carboxylate (Mol.1): -], O [carboxylate (Mol.4): -] Due to a crystallographic twofold symmetry in the [Na(ionophore)(H_2O)]$_4$ tetramer, the [Na(ionophore)(H_2O)]$_2$ dimer unit is independent.		
Class 4: Acyl tetronic acid ionophores					
[Na(4-Br-dinitrobenzoyltetronasin)-(H_2O)]		6-Coord.	2O [ether: -], 2O [carbonyl: -] O [methoxy: -], O [solvent: -] (See Figure 15)	BAGZUE	[240]

[a]Abbreviations: Antamanide, cyclo(−Val-Pro-Pro-Ala-Phe-Phe-Pro-Pro-Phe-Phe-); Enniatin B, cyclo(−L-MeVal-D-Hyi-L-MeVal-D-Hyi-L-MeVal-D-Hyi-). Cha, cyclohexylalanyl; Dianemycin*, 19-De(5-Me-4-methoxy-pyran-2-yl)-27-(5-Me-4-methoxy-pyran-2-yl)dianemycin; DMF, *N*,*N*-dimethylformamide; Hyi, hydroxyisovaleric acid residue; SFH, 3,5-di-*t*-butyl-4-hydroxybenzylidene-malononitrile; Valinomycin, cyclo(−D-HyIv-D-Val-L-Lac-L-Val-)$_3$; *Meso*-valinomycin, cyclo(−D-Val-L-HyIv-L-Val-D-HyIv-)$_3$.

[b]'Amide' and 'ester' denote the carbonyl oxygens of the amide and the ester groups, respectively.

[c]'4O [amide: 2.04–2.24]' denotes that the metal ion binds to four amide oxygens with the distances of 2.04–2.24 Å.

[d]Not reported.

[e]Not registered in the CSD version 5.35 updated to February 2014.

[f]CCDC number.

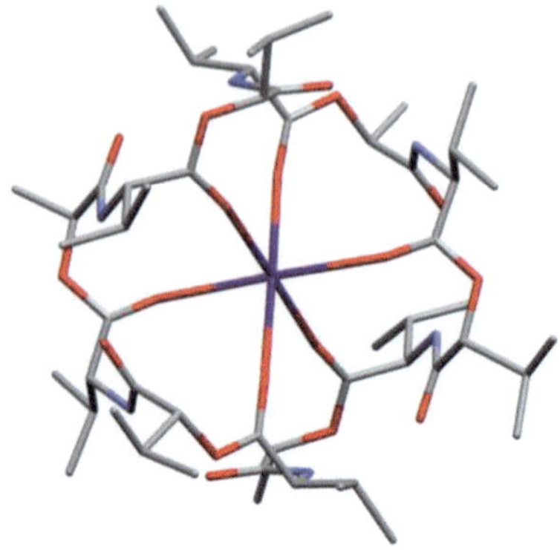

Figure 10 Structure of the K^+ complex of valinomycin (refer to Figure 8), [K(valinomycin)(picrato)-(H_2O)] [180]. Six ester carbonyl oxygens are involved in the metal bonding. The picrato and the aqua ligands are omitted for clarity.

[K(ionophore)] complex [175], (ii) six amide oxygens (three each from each of the two ionophore molecules) in a discrete [K(ionophore)$_2$] complex [174], and (iii) in infinite [–(K^+ or Rb^+)-ionophore-]$_n$ "sandwich"-type complexes [173, 176], (iii-1) six amide oxygens (three from the upper molecule and three from the lower molecule) and six ester oxygens (three each from the upper and the lower molecule) in the K^+ complex [173], or (iii-2) two amide (one each from the upper and the lower molecule) and three ester oxygens (two from the upper and one from the lower molecule) in the Rb^+ complex [176]. For valinomycin (refer to Figure 8), cyclo(–D-HyIv-D-Val-L-Lac-L-Val-)$_3$, six ester carbonyl oxygens are always involved in metal bonding in Na^+ [179], K^+ [180–183], Rb^+ [184], and Cs^+ [185] complexes, as shown in Figure 10, while six amide oxygens are involved in the Rb^+ complex with prolinomycin [178], cyclo(–D-Val-L-Pro-L-Val-D-Pro-)$_3$, a synthetic proline analog of valinomycin.

5.2.2 Macrotetrolide Ionophore Complexes

Macrotetrolide ionophores are 32-membered cyclic tetralactones which are built up by four ω-hydroxycarboxylic acid subunits of alternating enantiomerism bound to each other by esterification, including nonactin, monactin, dinactin, trinactin, and tetranactin (refer to Figure 8). Among these, X-ray structures of alkali metal ion complexes are available only for nonactin and tetranactin. Commonly in all of these complexes, four ester carbonyl oxygens and four ether oxygens from the tetrahydrofuran moieties are all directed toward the interior of the ionophore molecule to form eight coordination bonds with the metal ion (Na^+ [186], K^+ [187], and Cs^+ [188] ions for nonactin and Na^+ [189], K^+ [189], Rb^+ [184, 189], and Cs^+ [188] ions for tetranactin), with the periphery of the complex molecules being highly hydrophobic, a favorable feature for an effective alkali metal ion transport, as shown in Figure 11. Boromycin is a D-valine ester of a Böeseken complex of boric acid with a macrodiolide. Unfortunately, limited information is given on metal bonding in its Rb^+ as well as Cs^+ complexes [190].

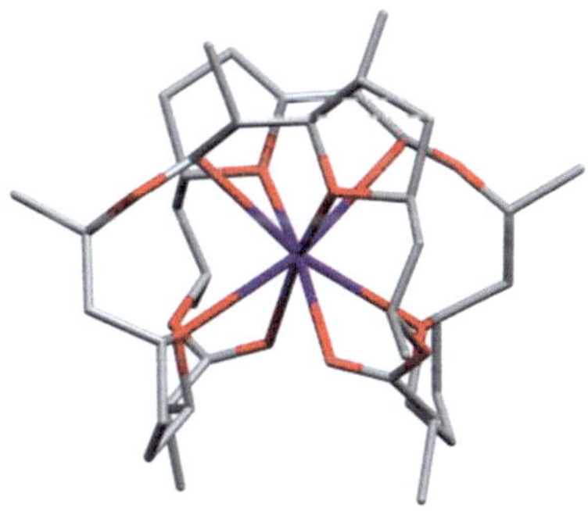

Figure 11 Structure of the K^+ complex of nonactin (refer to Figure 8), [K(nonactin)]·NCS [187]. Four ester carbonyl oxygens and four ether oxygens from the tetrahydrofuran moieties are involved in metal bonding.

5.2.3 Polyether Ionophore Complexes

Polyether ionophores, also designated as 'carboxylic acid ionophores', are linear and contain a terminal carboxyl group and one or two hydroxyl groups at the other end of tetrahydrofuran and tetrahydropyrane rings. According to Westley [191], they fall into four broad classes: monovalent (class *1a*), monovalent glycoside (class *1b*), divalent (class *2a*), and divalent pyrrole ether (class *3*). Monovalent ionophores are those that are not able to transport divalent cations, including monensin and nigericin as representatives; monovalent glycoside ionophores include dianemycin, septamycin, and the antibiotic A204A; divalent ionophores include lasalocids A-E, iso-lasalocid, lysocellin, and ionomycin; and divalent pyrrole ether ionophores contain antibiotic-A23187 and X-14547A. Structures and properties of polyether ionophores were repeatedly reviewed [5–8, 192].

A large body of crystal structures of alkali metal ion complexes with polyether ionophores are available, most of which belong to classes *1a* or *1b* ionophores (Table 14), while much fewer exist for classes *2a* and *3* ionophores. A common structural feature characteristic of these polyether ionophore complexes is that the ionophore molecule is wrapped around the metal ion and held in this conformation by one or two head-to-tail hydrogen bonds between the carboxylate group and the one or two hydroxyls or pyrrole functions of the terminal moiety. A large number of ring ether oxygens (from two to four) always participate in metal bonding (with one exception in the cezomycin–Na^+ complex [238]). Characteristic metal binding sites/modes for representative ionophores are described below.

Class 1a, monovalent polyether ionophores: For monensin, which exhibits a high metal ion selectivity for Na^+, it seems a rule that the metal ion is coordinated by four ring ether oxygens, one hydroxyl oxygen, and one CH_2OH oxygen, as observed in three Li^+ complexes [197–199], twenty Na^+ complexes [200–210], three K^+ complexes [203, 205, 215], and two Rb^+ complexes [216, 217] (see Figure 12). The involvement of the CH_2OH oxygen in metal coordination is unique for monensin among class *1a* ionophores. As exceptions, carboxylate oxygen (or oxygens) instead of CH_2OH oxygen is (are) involved in metal bonding in two Na^+ complexes with monensin substituted at the CH_2OH group by CH_2O-urethane [208] or methyl

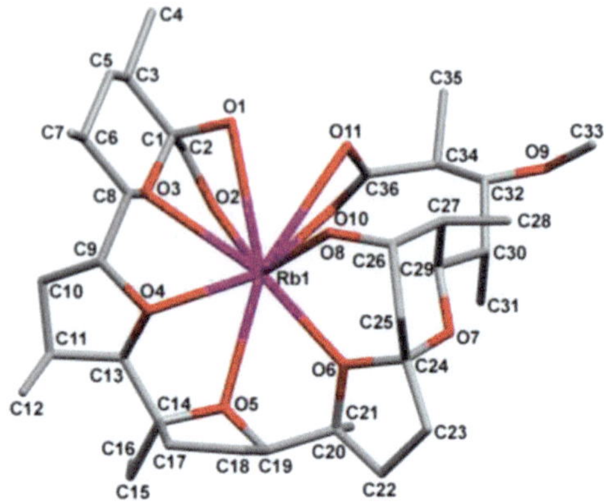

Figure 12 Structure of the Rb^+ complex of monensin (refer to Figure 8), [Rb(monensin)]·H_2O [217]. Four ring ether oxygens, one hydroxyl oxygen, and one CH_2OH oxygen are involved in metal bonding.

[214] residues. (Refer also to a recent review on structures and properties of monensin A, where many crystal structures of monensin complexes are drawn [241]). For nigericin, metal binding sites are three ether oxygens, one carboxylate oxygen, and one methoxy oxygen in Na^+ [219] and K^+ [220] complexes. Ionophores that contain the carbonyl functional group have the common metal–O=C bonding in a Na^+ complex of CP54838 [195], a Rb^+ complex of noboritomycin [221], and three Na^+ complexes of salinomycin [222–224].

Class 1b, monovalent glycoside ionophores: For structurally related dianemycin, endusamycin, and CP-80219, which have a deoxysugar substituent at different positions, four ether oxygens, one (for Na^+ and K^+ ions) or two (for Rb^+ ion) hydroxyl oxygens, and one CH_2OH oxygen are common metal binding sites, in Na^+ [228], K^+ [229], and Rb^+ [226] complexes of dianemycin, a Rb^+ complex of endusamycin [232] and a Rb^+ complex of CP-80219 [226]. On the other hand, in Na^+ complexes of A204A [225] and K-41 [233], which belong to a group of antibiotics similar to septamycin, three ether oxygens, one methoxy oxygen, and two carboxylate oxygens are common metal binding sites.

Class 2a, divalent polyether ionophores: Crystal structures of alkali metal ion complexes with class *2a* ionophores are quite limited, only four Na^+ complexes of lasalocid A or its derivatives are available [235–237]. Lasalocids are unique in that their backbones are considerably shorter than those of the other polyether ionophores, making it difficult for the ionophore molecule to fully shield the complexed metal ion from the solvent. This difficulty is overcome by forming ionophore:metal 2:2 complexes as observed in three Na^+ complexes [235, 236]. In this dimeric structure, one of the two metal ions (Na1) binds to one ligand (L1) through two ether oxygens, two hydroxyl oxygens, and a carbonyl oxygen, and the sixth position is occupied by one carbonyl oxygen of the other ligand (L2), and the second metal ion (Na2) binds to the L2 ligand through two ether oxygens, two hydroxyl oxygens in [Na(lasalocid A)(H_2O)]$_2$ [236] (see Figure 13) or additionally a carbonyl oxygen from the L1 ligand in [Na(5-Br-lasalocid)]$_2$ [235]. As a minor case, the simple

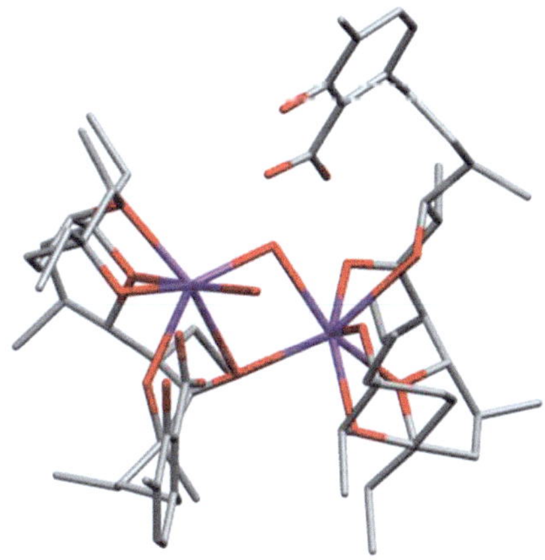

Figure 13 Structure of the Na^+ complex of lasalocid A (refer to Figure 8), $[Na(lasalocid\ A)(H_2O)]_2$ [236]. Because of the small size of the ionophore molecule, in order to overcome the difficulty for the ionophore molecule to fully shield the metal ion from the solvent, the ligand:metal 2:2 complex is formed.

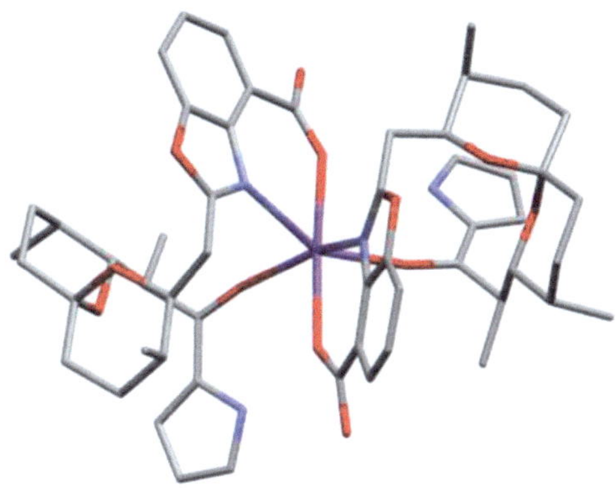

Figure 14 Structure of the Na^+ complex of cezomycin (refer to Figure 8), [Na{demethyl(C-11)-cezomycin}$_2$]·$3H_2O$ [238]. Because of the small size of the ionophore molecule, in order to fully shield the metal ion from the solvent, the ligand:metal 2:1 complex is formed.

ligand:metal 1:1 complex is also formed in [Na(lasalocid)(MeOH)] [237] in which the Na^+ ion is coordinated to the same five ligand oxygens as in the dimer structure, but it is capped by the oxygen of a methanol molecule. This monomeric structure is assumed to be a model for cation uptake and release in polar media, whereas the actual cation transport process is achieved by the lipophilic dimer [237].

Class 2b, divalent glycoside ionophore: Antibiotic 6016. No X-ray structure of any alkali metal ion complex is reported.

Class 3, divalent pyrrole ethers ionophores: These include A-23187 (calcimycin), A-83094A (16-deethylindanomycin), AC7230, cezomycin, CP-61405 (routiennocin), and X-14547A (indanomycin), each of which contains a pyrrole-2-carbonyl chromophore. Among these, only two X-ray structures of alkali metal ion complexes are known for cezomycin [238, 239]. Because of the small size of the ionophore molecule, in order to fully shield the metal ion from the solvent, ligand:metal 2:1 or 4:4 complexes are formed in [Na{demethyl(C-11)-cezomycin}$_2$] [238] (Figure 14) and $[Na(cezomycin)]_4$ [239], respectively, where the octahedral Na^+ ion binds to a carboxylate oxygen, a carbonyl oxygen, and a benzoxazole nitrogen of

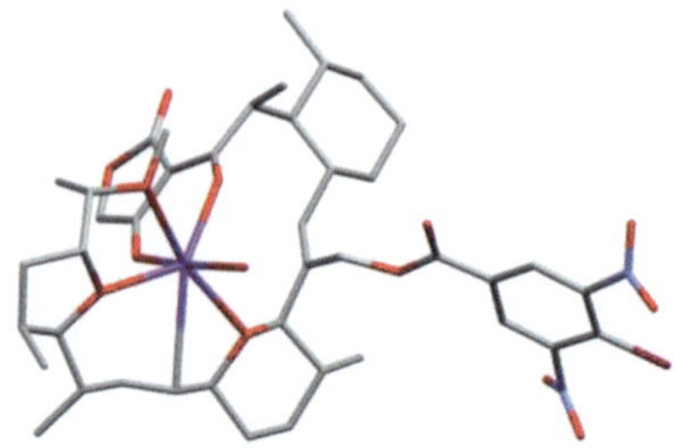

Figure 15 Structure of the Na^+ complex of tetronasin (refer to Figure 8), [Na(4-Br-dinitrobenzoyl-tetronasin)(H_2O)] [240]. Two ether oxygens, one methoxy oxygen, and two carbonyl oxygens are involved in metal bonding.

one ligand and to the same set of the three sites from the other ligand in the former 2:1 complex, while each of four Na^+ ions binds to the same set of the three sites noted above from one ligand and two carboxylate oxygens (one from the second ligand and the other from the third ligand) in the tetrameric latter complex. No ether oxygen is involved in metal coordination.

Class 4, acyl tetronic acid ionophores: Tetronacin (M139603) and tetronomycin are unique in that they contain a tetronic acid moiety instead of the essential carboxylic acid function. The octahedral Na^+ ion is coordinated by two ether oxygens, one methoxy oxygen, two carbonyl oxygens, and a water solvent in [Na(4-Br-dinitrobenzoyltetronasin)(H_2O)] [240] (Figure 15).

In summary, extensive X-ray crystallographic studies have revealed that the remarkable selectivity of some ionophores is attributed to the size of the cage. Only cations with an appropriate radius fit the cavity perfectly, larger ones have to deform it, while smaller ones find a nonoptimal coordination geometry. Over 120 naturally occurring ionophores are known [192] and still new antibiotic ionophores are reported every year. Just as the early works of antibiotic ionophores as selective alkali cation carriers originated the synthesis of so-called polyether ligands such as crown ethers and led to the rapid development into the fields of host–guest chemistry and supramolecular chemistry, it is hoped that the discovery of the new types of antibiotic ionophores will continue to play such a role.

6 Concluding Remarks and Outlook

In this chapter, by surveying hitherto reported X-ray studies, we have tried to present a summary of the state of our knowledge regarding alkali metal ion interactions with biologically important ligands, that is, amino acids and small peptides, nucleic acid constituents, simple carbohydrates, and naturally occurring antibiotic ionophores. X-ray studies have revealed some trends or certain general aspects on metal ion binding sites/modes toward these ligands and their structures or functions, especially for antibiotic ionophores. However, somewhat surprisingly, there is still quite limited knowledge of interactions of amino acids and peptides and carbohydrates with alkali metal ions, in particular from the viewpoint of structural chemistry. Because these substances possess a wealth of functional groups with 'hard' nature preferred by 'hard' alkali meal ions, it is hoped that further X-ray studies will add new observations, for example, the formation of novel 'carbohydrate ionophores' or 'metallacryptands' with novel architectures, with one of the ultimate goals being directed towards the design of more highly ion-selective hosts.

Though not treated here, it is clear that comparative studies of the complexes involving alkaline earth metal ions and transition metal ions are needed, in order to enhance our understanding of the nature and the biological relevance of the alkali metal ion complexes. Of course, such studies have to be complemented by solution studies.

Abbreviations

Regarding the abbreviations of amino acids see Table 1.

aa	amino acid
Ade	adenine
Ado–2H	adenosinate dianion
ADP	adenosine diphosphate
ADPH2	diprotonated adenosine diphosphate
Ala–H	alaninate monoanion
AMP	adenosine monophosphate
AMPH	monoprotonated adenosine monophosphate
ApppA (Ap_4A)	P^1,P^4-bis(5'-adenosyl)tetraphosphate
ApU	adenylyl-3',5'-uridine
ApT	adenylyl-3',5'-thymidine
ATP	adenosine 5'-triphosphate
ATPH2	diprotonated adenosine 5'-triphosphate
C^+	cytosinium monocation
CCDC	Cambridge Crystallographic Data Centre
CDP	cytidine diphosphate
CDPcholine	cytidine diphosphocholine
CDPethanolamine	cytidine diphosphoethanolamine
Cha	cyclohexylalanyl
CMP	cytidine monophosphate
CMPH	monoprotonated cytidine monophosphate
CpG	cytidylyl-3',5'-guanosine
CSD	Cambridge Structural Database
Cyt	cytosine
d	deoxyribose

D	dexter (optical isomer named after Latin)
dCMP	deoxycytidine monophosphate
d(CpG)	deoxycytidylyl-3',5'-deoxyguanosine
dien	diethylenetriamine
dGMP	deoxyguanosine monophosphate
DMSO	dimethylformamide
en	ethylenediamine
9-Et-azacrown-Ade	16-(2-(9H-adenin-9-yl)ethyl)-16-aza-1,4,7,10,13-pentaoxa-cyclooctadecane
1-EtThy	N1-ethylthymine
Fru	fructose
gA, gB, gC, gD	gramicidin A, B, C, D
Gal	galactose
Glc	glucose
GlyGly	glycylglycine
Gly–H	glycinate monoanion
GlyTyr	glycyltyrosine
GMP	guanosine monophosphate
GMPH	monoprotonated guanosine monophosphate
GpC	guanylyl-3',5'-cytidine
Gua	guanine
HyIv	hydroxyisovalerate
Hyp	hypoxanthine
IMP	inosine monophosphate
IMPH	monoprotonated inosine monophosphate
Ino–H	inosinate monoanion
iPr	isopropyl
isoGuo	isoguanosine
L	leavus (optical isomer named after Latin)
L	ligand
Lac	lactic acid
M	metal ion
1-Me-Cyt	N1-methylcytosine
1-MeCyt–H	N1-methylcytosinate monoanion
Me_4dae	*N,N,N',N'*-tetramethyl-1,2-diaminoethane
9-MeGua	N9-methylguanine
1-MeThy	N1-methylthymine
1-MeThy–H	N1-methylthyminate monoanion
1-MeUra–H	N1-methyluracilate monoanion
MeVal	methylvalyl
NAD	5'-nicotinamide-ribosyl-5'-adenyl-pyrophosphate
O(P)	oxygen atom of the phosphate group
1-Pr-azacrown–Thy	(3-(1-thyminyl)propyl)-4,13-diaza-18-crown-6
Pro–H	prolinate monoanion

pTpT	5'-phosphoryl-thymidylyl-3',5'-thymidine
THF	tetrahydrofuran
Thr–H	threonate monoanion
Thy	thymine
Thy–H	thyminate monoanion
TMP	thymidine monophosphate
UDP	uridine 5'-diphosphate
UDPglucose	uridine diphosphate glucose
UDPH	monoprotonated uridine diphosphate
UMP	uridine monophosphate
UMPH	monoprotonated uridine monophosphate
Ura	uracil
Urd–3H	uridinate trianion

References

1. G. Wu, J. Zhu, *Prog. Nucl. Magn. Reson. Spect.* **2012**, *61*, 1–70, and major review articles and books cited therein.
2. W. Saenger, *Principles of Nucleic Acid Structure*, Springer-Verlag, New York, 1984, pp. 202–203.
3. S. Yano, M. Otsuka, *Met. Ions Biol. Syst.* **1996**, *32*, 27–60.
4. K. Aoki, K. Murayama, *Met. Ions Life Sci.* **2012**, *10*, 43–102, and major review articles and books cited therein.
5. M. Dobler, *Ionophores and Their Structures*, New York, London, John Wiley, 1981, pp 379.
6. R. Hilgenfeld, W. Saenger, *Topics Curr. Chem.* **1982**, *101*, 1–82.
7. E. N. Duesler, I. C. Paul, in *Polyether Antibiotics*, Ed J. W. Westley, Marcel Dekker, New York, 1983, Vol. 2, pp. 87–195.
8. J. Rutkowski, B. Brzezinski, *BioMed Res. International*, **2013**, *2013*, Article ID 162513.
9. J. A. Subirana, M. Soler-López, *Ann. Rev. Biophys. Biomol. Struct.* **2003**, *32*, 27–45.
10. L. Randaccio, M. Furlan, S. Geremia, M. Slouf, I. Srnova, D. Toffoli, *Inorg. Chem.* **2000**, *39*, 3403–3413.
11. K. R. Rao, C. Aneesh, H. L. Bhat, S. Elizabeth, M. S. Pavan, T. N. G. Row, *Cryst. Growth Des.* **2013**, *13*, 97–105.
12. Y.-M. Jiang, J.-H. Cai, Z.-M. Liu, X.-H. Liu, *Acta Cryst.* **2005**, *E61*, m878–m880.
13. L. Pauling, *The Nature of the Chemical Bond*, 3rd edn, Cornell University Press, Ithaca, New York (1960).
14. G. Müller, G.-M. Maier, M. Lutz, *Inorg. Chim. Acta*, **1994**, *218*, 121–131.
15. T. Balakrishnan, K. Ramamurthi, J. Jeyakanthan, S. Thamotharan, *Acta Cryst.* **2013**, *E69*, m60–m61.
16. M. R. Hudson, D. G. Allis, W. Ouellette, P. M. Hakey, B. S. Hudson, *J. Mol. Struct.* **2009**, *934*, 138–144.
17. J. Baran, M. Drozd, A. Pietraszko, M. Trzebiatowska, H. Ratajczak, *Pol. J. Chem.* **2003**, *77*, 1561–1577.
18. J. Baran, M. Drozd, H. Ratajczak, A. Pietraszko, *J. Mol. Struct.* **2009**, *927*, 43–53.
19. M. Fleck, K. Schwendtner, A. Hensler, *Acta Cryst.* **2006**, *C62*, m122–m125.
20. D. Dutta, A. D. Jana, M. Debnath, A. Bhaumik, J. Marek, M. Ali, *Dalton Trans. (Discussion of Faraday Soc.)* **2010**, *39*, 11551–11559.
21. K. Jayalakshmi, M. A. Sridhar, J. S. Prasad, M. N. Bhat, S. M. Dharamprakash, *Mol. Cryst. Liq. Cryst. Sci. Technol.* **2003**, *A393*, 95–103.

22. Y. I. Smolin, A. E. Lapshin, G. A. Pankova, *J. Struct. Chem.* **2007**, *48,* 708–710.
23. Y. Wang, D. Xiao, Y. Qi, E. Wang, J. Liu, *J. Cluster Sci.* **2008**, *19*, 367–378.
24. G. Aromí, J. J. Novoa, J. Ribas-Ariño, S. Igarashi, Y. Yukawa, *Inorg. Chim. Acta* **2008**, *361*, 3919–3925.
25. M. Fleck, L. Bohatý, *Acta Cryst.* **2006**, *C62*, m22–m26.
26. E. Kita, H. Marai, T. Muzioł, K. Lenart, *Transition Met. Chem.* **2011**, *36*, 35–44.
27. X.-Y. Jiang, X.-Y. Wu, R.-M. Yu, D.-Q. Yuan, W.-Z. Chen, *Inorg. Chem. Commun.* **2011**, *14*, 1546–1549.
28. C. Gabriel, M. Kaliva, J. Venetis, P. Baran, I. Rodriguez-Escudero, G. Voyiatzis, M. Zervou, A. Salifoglou, *Inorg. Chem.* **2009**, *48*, 476–487.
29. J. Liu, J. Zhang, D. Xiao, E. Wang, *J. Cluster Sci.* **2007**, *18*, 909–920.
30. X.-Y. Wu, C.-Z. Lu, Q.-Z. Zhang, S.-M. Chen, X.-J. Xu, *J. Coord. Chem.* **2006**, *59*, 2047–2054.
31. S.-M. Hu, S.-C. Xiang, J.-J. Zhang, T.-L. Sheng, R.-B. Fu, X.-T. Wu, *Eur. J. Inorg. Chem.* **2008**, 1141–1146.
32. K. van Hecke, E. Cartuyvels, T. N. Parac-Vogt, C. Görller-Walrand, L.van Meervelt, *Acta Cryst.* **2007**, *E63*, m2354.
33. H. An, T. Xu, E. Wang, C. Meng, *Inorg. Chem. Commun.* **2007**, *10*, 1453–1456.
34. A. Khatib, F. Aqra, D. Deamer, A. Oliver, *J. Chem. Res.* **2009**, 98–100.
35. T. T. Ong, P. Kavuru, T. Nguyen, R. Cantwell, Ł. Wojtas, M. J. Zaworotko, *J. Am. Chem. Soc.* **2011**, *133*, 9224–9227.
36. T. U. Devi, N. Lawrence, R. R. Babu, S. Selvanayagam, H. Stoeckli-Evans, K. Ramamurthi, *Crys. Growth Des.* **2009**, *9*, 1370–1374.
37. L.-Y. Wang, S. Igarashi, Y. Yukawa, Y. Hoshino, O. Roubeau, G. Aromi, R. E. P. Winpenny, *Dalton Trans.* (*Discussion of Faraday Soc.*) **2003**, 2318–2324.
38. R. I. Yousef, M. Bette, G. N. Kaluderović, R. Paschke, C. Yiran, D. Steinborn, H. Schmidt, *Polyhedron* **2011**, *30*, 1990–1996.
39. P. J. Nichols, C. L. Raston, *Dalton Trans. (Discussion of Faraday Soc.)* **2003**, 2923–2927.
40. S.-C. Xiang, S.-M. Hu, J.-J. Zhang, X.-T. Wu, J.-Q. Li, *Eur. J. Inorg. Chem.* **2005**, 2706–2713.
41. T. Konno, T. Kawamoto, R. Kuwabara, T. Yoshimura, M. Hirotsu, *Chem. Lett.* **2002**, 304–305.
42. J. R. Knox, C. K. Prout, *Acta Cryst.* **1969**, *B25*, 1857–1866.
43. R. Yoshida, S. Ogasahara, H. Akashi, T. Shibahara, *Inorg. Chim. Acta* **2012**, *383*, 157–163.
44. H. An, Z. Han, T. Xu, C. Meng, E. Wang, *Inorg. Chem. Commun.* **2008**, *11*, 914–917.
45. H. Schmidbaur, I. Bach, D. L. Wilkinson, G. Müller, *Chem. Ber.* **1989**, *122*, 1427–1431.
46. W. S. Sheldrick, E. Hauck, S. Korn, *J. Organomet. Chem.* **1994**, *467*, 283–292.
47. R. Bergs, K. Sünkel, W. Beck, *Chem. Ber.* **1993**, *126*, 2429–2432.
48. O. Versiane, J. Felcman, J. L. de Miranda, R. A. Howie, J. M. S. Skakle, J. L. Wardell, *Acta Cryst.* **2006**, *E62*, m52–m55.
49. F. Wiesbrock, H. Schmidbaur, *CrystEngComm* **2003**, *5*, 262–264.
50. C. Sano, N. Nagashima, T. Kawakita, Y. Iitaka, *Anal. Sci.* **1989**, *5*, 121–122.
51. H. Schmidbaur, P. Mikulcik, G. Muller, *Chem. Ber.* **1990**, *123*, 1001–1004.
52. L. M. Thomas, N. Ramasubbu, K. K. Bhandary, *Biopolymers* **1994**, *34*, 1007–1013.
53. B. D. White, J. Mallen, K. A. Arnold, F. R. Fronczek, R. D. Gandour, L. M. B. Gehrig, G. W. Gokel, *J. Org. Chem.* **1989**, *54*, 937–947.
54. W. Liu, Y. Song, Y. Li, Y. Zou, D. Dang, C. Ni, Q. Meng, *Chem. Commun.* **2004**, 2348–2349.
55. M. Doi, A. Asano, T. Ishida, Y. Katsuya, Y. Mezaki, M Sasaki, A. Terashima, T. Taniguchi, H. Hasegawa, M. Shiono, *Acta Cryst.* **2001**, *D57*, 628–634.
56. Y.-Y. H. Chiu, L. D. Brown, W. N. Lipscomb, *J. Am. Chem. Soc.* **1977**, *99*, 4799–4803.
57. L. C. M. Ngoka, M. L. Gross, *Biochem. Biophys. Res. Commun.* **1999**, *257*, 713–719.
58. A. E. Gibson, C. Price, W. Clegg, A. Houlton, *J. Chem. Soc., Dalton Trans.* **2002**, 131–133.

59. E. Freisinger, A. Schneider, M. Drumm, A. Hegmans, S. Meier, B. Lippert, *J. Chem. Soc., Dalton Trans.* **2000**, 3281–3287.
60. B. Müller, W.-Z. Shen, P. J. S. Miguel, F. M. Albertí, T. van der Wijst, M. Noguera, L. Rodríguez-Santiago, M. Sodupe, B. Lippert, *Chem. Eur. J.* **2011**, *17*, 9970–9983.
61. A. Hegmans, E. Zangrando, E. Freisinger, F. Pichierri, L. Randaccio, C. Mealli, M. Gerdan, A. X. Trautwein, B. Lippert, *Chem. Eur. J.* **1999**, *5*, 3010–3018.
62. D. Armentano, G. D. Munno, R. Rossi, *New J. Chem. (Nouv. J. Chim.)*, **2006**, *30*, 13–17.
63. B. Fischer, H. Preut, B. Lippert, H. Schöllhorn, U. Thewalt, *Polyhedron* **1990**, *9*, 2199–2204.
64. S. L. D. Wall, L. J. Barbour, O. F. Schall, G. W. Gokel, *J. Chem. Cryst.* **2000**, *30*, 227–231.
65. F. Guay, A. Beauchamp, *Inorg. Chim. Acta* **1982**, *66*, 57–63.
66. W. Micklitz, B. Lippert, F. Lianza, A. Albinati, *Inorg. Chim. Acta* **1994**, *227*, 5–12.
67. E. Freisinger, A. Schimanski, B. Lippert, *J. Biol. Inorg. Chem.* **2001**, *6*, 378–389.
68. C. J. L. Lock, P. Pilon, B. Lippert, *Acta Cryst.* **1979**, *B35*, 2533–2537.
69. B. L. Kindberg, E. H. Griffith, E. L. Amma, *J. Chem. Soc., Chem. Commun.* **1975**, 195–196.
70. F. Zamora, H. Witkowski, E. Freisinger, J. Müller, B. Thormann, A. Albinati, B. Lippert, *J. Chem. Soc., Dalton Trans.* **1999**, 175–182.
71. M. Mizutani, K. Jitsukawa, H. Masuda, H. Einaga, *Chem. Commun.* **1996**, 1389–1390.
72. M. Mizutani, S. Miwa, N. Fukushima, Y. Funahashi, T. Ozawa, K. Jitsukawa, H. Masuda, *Inorg. Chim. Acta* **2002**, *339*, 543–550.
73. O. Renn, B. Lippert, I. Mutikainen, *Inorg. Chim. Acta* **1994**, *218*, 117–120.
74. L. Holland, W.-Z. Shen, W. Micklitz, B. Lippert, *Inorg. Chem.* **2007**, *46*, 11356–11365.
75. B. Lippert, *Coord. Chem. Rev.* **2000**, *200–202*, 487–516.
76. N. Shan, S. J. Vickers, H. Adams, M. D. Ward, J. A. Thomas, *Angew. Chem. Int. Ed.* **2004**, *43*, 3938–3941.
77. W. Saenger, *Principles of Nucleic Acid Structure*, Springer-Verlag, New York, 1984, pp. 17–24.
78. P. Klüfers, P. Mayer, *Z. Anorg. Allg. Chem.* **2007**, *633*, 903–907.
79. M. Cai, A. L. Marlow, J. C. Fettinger, D. Fabris, T. J. Haverlock, B. A. Moyer, J. T. Davis, *Angew. Chem., Int. Ed. Engl.* **2000**, *39*, 1283–1285.
80. P. Klüfers, P. Mayer, *Acta Cryst.* **1996**, *C52*, 2970–2972.
81. J. Galy, A. Mosset, I. Grenthe, I. Puigdoménech, B. Sjöberg, F. Hultén, *J. Am. Chem. Soc.* **1987**, *109*, 380–386.
82. T. Golas, M. Fikus, Z. Kazimierczuk, D. Shugar, *Eur. J. Biochem.* **1976**, *65*, 183–192.
83. J. T. Davis, S. K. Tirumala, A. L. Marlow, *J. Am. Chem. Soc.* **1997**, *119*, 5271–5272.
84. G. S. Padiyar, T. P. Seshadri, *J. Biomol. Struct. Dyn.* **1998**, *15*, 803–821.
85. (a) W. Saenger, B. S. Reddy, K. Mühlegger, G. Weimann, *Nature (London)* **1977**, *267*, 225–229. (b) B. S. Reddy, W. Saenger, K. Mühlegger, G. Weimann, *J. Am. Chem. Soc.* **1981**, *103*, 907–914.
86. M. Inoue, T. Yamase, *Bull. Chem. Soc. Jpn.* **1996**, *69*, 2863–2868.
87. Z. Szabó, I. Furó, I. Csöregh, *J. Am. Chem. Soc.* **2005**, *127*, 15236–15247.
88. K. I. Varughese, C. T. Lu, G. Kartha, *J. Am. Chem. Soc.* **1982**, *104*, 3398–3401.
89. O. Kennard, N. W. Isaacs, W. D. S. Motherwell, J. C. Coppola, D. L. Wampler, A. C. Larson, D. G. Watson, *Proc. R. Soc. London, Ser. A* **1971**, *325*, 401–436.
90. Y. Sugawara, N. Kamiya, H. Iwasaki, T. Ito, Y. Satow, *J. Am. Chem. Soc.* **1991**, *113*, 5440–5445.
91. D. A. Adamiak, W. Saenger, *Acta Cryst.* **1980**, *B36*, 2585–2589.
92. P. Swaminathan, M. Sundaralingam, *Acta Cryst.* **1980**, *B36*, 2590–2597.
93. M. A. Viswamitra, M. V. Hosur, Z. Shakked, O. Kennard, *Nature (London)* **1976**, *262*, 234–236.
94. (a) S. K. Katti, T. P. Seshadri, M. A. Viswamitra, *Curr. Sci.* **1980**, *49*, 533–535. (b) S. K. Katti, T. P. Seshadri, M. A. Viswamitra, *Acta Cryst.* **1981**, *B37*, 1825–1832.

95. C. L. Barnes, S. W. Hawkinson, *Acta Cryst.* **1982**, *B38*, 812–817.
96. P. Zhou, H. Li, *Dalton Trans. (Discussion of Faraday Soc.)* **2011**, *40*, 4834–4837.
97. M. Benedetti, G. Tamasi, R. Cini, L. G. Marzilli, G. Natile, *Chem. Eur. J.* **2007**, *13*, 3131–3142.
98. D. W. Young, P. Tollin, H. R. Wilson, *Acta Cryst.* **1974**, *B30*, 2012–2018.
99. T. P. Seshadri, M. A. Viswamitra, *Pramana* **1974**, *3*, 218–235.
100. S. T. Rao, M. Sundaralingam, *J. Am. Chem. Soc.* **1969**, *91*, 1210–1217.
101. M. Sriram, Y.-C. Liaw, Y.-G. Gao, A. H.-J. Wang, *Acta Cryst.* **1991**, *C47*, 507–510.
102. M. D. Poojary, H. Manohar, *Inorg. Chem.* **1985**, *24*, 1065–1069.
103. R. Bau, R. W. Gellert, S. M. Lehovec, S. Louie, *J. Clin. Hematol. Oncol.* **1977**, *7*, 51–61.
104. T. J. Kistenmacher, C. C. Chiang, P. Chalilpoyil, L. G. Marzilli, *J. Am. Chem. Soc.* **1979**, *101*, 1143–1148.
105. C. C. Chiang, T. Sorrell, T. J. Kistenmacher, L. G. Marzilli, *J. Am. Chem. Soc.* **1978**, *100*, 5102–5110.
106. G. Borodi, A. Hernanz, I. Bratu, M. Pop, R. Navarro, *Acta Cryst.* **2001**, *E57*, m514–m516.
107. Y. Sugawara, A. Nakamura, Y. Iimura, K. Kobayashi, H. Urabe, *J. Phys. Chem. B* **2002**, *106*, 10363–10368.
108. S. V. Gonzalez, K. Larsen, W. H. Nelson, E. Sagstuen, C. H. Görbitz, *Acta Cryst.* **2005**, *E61*, m554–m556.
109. S. S. Mande, T. P. Seshadri, M. A. Viswamitra, *Acta Cryst.* **1994**, *C50*, 876–879.
110. C. L. Coulter, *J. Am. Chem. Soc.* **1973**, *95*, 570–575.
111. M. A. Viswamitra, T. P. Seshadri, M. L. Post, O. Kennard, *Nature* **1975**, *258*, 497–501.
112. S. K. Katti, M. A. Viswamitra, *Acta Cryst.* **1981**, *B37*, 1058–1063.
113. T. P. Seshadri, M. A. Viswamitra, G. Kartha, *Acta Cryst.* **1980**, *B36*, 925–927.
114. R. Chitra, R. Ranjan-Choudhury, M. Ramanadham, *Appl. Phys. A* **2002**, *74*, S1576–S1578.
115. J. D. Hoogendorp, C. Romers, *Acta Cryst.* **1978**, *B34*, 2724–2728.
116. M. A. Viswamitra, B. S. Reddy, M. N. G. James, G. J. B. Williams, *Acta Cryst.* **1972**, *B28*, 1108–1116.
117. M. A. Viswamitra, T. P. Seshadri, M. L. Post, *Acta Cryst.* **1980**, *B36*, 2019–2024.
118. Y. Sugawara, H. Iwasaki, *Acta Cryst.* **1984**, *C40*, 389–393.
119. J. Emerson, M. Sundaralingam, *Acta Cryst.* **1980**, *B36*, 537–543.
120. M. A. Viswamitra, M. L. Post, O. Kennard, *Acta Cryst.* **1979**, *B35*, 1089–1094.
121. N. H. Campbell, S. Neidle, *Met. Ions Life Sci.* **2012**, *10*, 119–134.
122. K. Aoki, in *Comprehensive Supramolecular Chemistry*, Ed J.-M. Lehn, Vol. 5, Pergamon Press, Oxford, 1996, pp. 249–294.
123. K. Aoki, *Met. Ions Biol. Syst.* **1996**, *32*, 91–134.
124. K. Aoki, in *Landolt-Börnstein: Nukleinsäuren: Teilband b; Kristallographische und Structurelle Daten II*, Ed W. Saenger, Springer-Verlag, Berlin, 1989, pp. 171–246.
125. J. Pandit, T. P. Seshadri, M. A. Viswamitra, *Acta Cryst.* **1983**, *C39*, 342–345.
126. A. C. Larson, *Acta Cryst.* **1978**, *B34*, 3601–3604.
127. N. C. Seeman, J. M. Rosenberg, F. L. Suddath, J. J. P. Kim, A. Rich, *J. Mol. Biol.* **1976**, *104*, 109–144.
128. J. M. Rosenberg, N. C. Seeman, R. O. Day, A. Rich, *J. Mol. Biol.* **1976**, *104*, 145–167.
129. A. H.-J. Wang, G. J. Quigley, A. Rich, *Nucleic Acids Res.* **1979**, *6*, 3879–3890.
130. M. Coll, X. Solans, M. Font-Altaba, J. A. Subirana, *J. Biomol. Struct. Dyn.* **1987**, *4*, 797–811.
131. N. Camerman, J. K. Fawcett, A. Camerman, *J. Mol. Biol.* **1976**, *107*, 601–621.
132. D. Watanabe, M. Ishikawa, M. Yamasaki, M. Ozaki, T. Katayama, H. Nakajima, *Acta Cryst.* **1996**, *C52*, 338–340.
133. (a) X. Shui, L. McFail-Isom, G. G. Hu, L. D. Williams, *Biochemistry* **1998**, *37*, 8341–8355. (b) L. MacFail-Isom, X. Shui, L. D. Williams, *Biochemistry* **1998**, *37*, 17105–17111.
134. C. C. Sines, L. McFail-Isom, S. B. Howerton, D. van Derveer, L. D. Williams, *J. Am. Chem. Soc.* **2000**, *122*, 11048–11056.

135. V. Tereshko, G. Minasov, M. Egli, *J. Am. Chem. Soc.* **1999**, *121*, 3590–3595.
136. K. K. Woods, L. McFail-Isom, C. C. Sines, S. B. Howerton, R. K. Stephens, L. D. Williams, *J. Am. Chem. Soc.* **2000**, *122*, 1546–1547.
137. J. Lüthje, *Klin. Wochenschr.* **1989**, *67*, 317–327.
138. K. C. Wong, A. Hamid, S. Baharuddin, C. K. Quah, H.-K. Fun, *Acta Cryst.* **2009**, *E65*, m1308–m1309.
139. G. Ferguson, B. Kaitner, B. E. Connett, D. F. Rendle, *Acta Cryst.* **1991**, *B47*, 479–484.
140. Y. Cho, R. B. Honzatko, *Acta Cryst.* **1990**, *C46*, 587–590.
141. N. Narendra, T. P. Seshadri, M. A. Viswamitra, *Acta Cryst.* **1984**, *C40*, 1338–1340.
142. A. E. Kozioł, *Pol. J. Chem.* **1991**, *65*, 455–463.
143. T. Lis, *Carbohydr. Res.* **1992**, *229*, 33–39.
144. C. A. Beevers, G. H. Maconochie, *Acta Cryst.* **1965**, *18*, 232–236.
145. N. Narendra, M. A. Viswamitra, *Curr. Sci.* **1984**, *53*, 1018–1020.
146. T. Lis, *Carbohydr. Res.* **1985**, *135*, 187–194.
147. N. Narendra, M. A.Viswamitra, *Acta Cryst.* **1985**, *C41*, 1621–1624.
148. D. Lamba, W. Mackie, B. Sheldrick, P. Belton, S. Tanner, *Carbohydr. Res.* **1988**, *180*, 183–193.
149. R. Krishnan, T. P. Seshadri, *Acta Cryst.* **1990**, *C46*, 2299–2302.
150. T. Lis, *Acta Cryst.* **1986**, *C42*, 1745–1747.
151. N. Narendra, T. P. Seshadri, M. A. Viswamitra, *Acta Cryst.* **1985**, *C41*, 31–34.
152. C. A. Beevers, W. Cochran, *Proc. R. Soc. London, Ser. A* **1947**, *190*, 257–272.
153. C. A. Accorsi, F. Bellucci, V. Bertolasi, V. Ferretti, G. Gilli, *Carbohydr. Res.* **1989**, *191*, 105–116.
154. C. A. Accorsi, V. Bertolasi, V. Ferretti, G. Gilli, *Carbohydr. Res.* **1989**, *191*, 91–104.
155. T. Kato, T. Fujimoto, A. Tsutsui, M. Tashiro, Y. Mitsutsuka, T. Machinami, *Chem. Lett.* **2010**, *39*, 136–137.
156. T. Fujimoto, T. Kato, Y. Usui, O. Kamo, K. Furihata, K. Tsubono, T. Kato, T. Machinami, M. Tashiro, *Carbohydr. Res.* **2011**, *346*, 1991–1996.
157. J. K. Wright, R. Seckier, P. Overath, *Annu. Rev. Biochem.* **1986**, *55*, 225–248.
158. M. Tako, S. Nakamura, Y. Kohda, *Carbohydr. Res.* **1987**, *161*, 247–255.
159. J. F. D. S. Daniel, E. R. Filho, *Natural Product Reports* **2007**, *24*, 1128–1141.
160. D. A. Kelkar, A. Chattopadhyay, *Biochim. Biophys. Acta* **2007**, *1768*, 1103–1113.
161. A. Olczak, M. L. Glówka, M. Szczesio, J. Bojarska, Z. Wawrzak, W. L. Duax, *Acta Cryst.* **2010**, *D66*, 874–880.
162. D. A. Doyle, B. A. Wallace, *J. Mol. Biol.* **1997**, *266*, 963–977.
163. A. Olczak, M. L. Glówka, M. Szczesio, J. Bojarska, W. L. Duax, B. M. Burkhart, Z. Wawrzak, *Acta Cryst.* **2007**, *D63*, 319–327.
164. M. K. Glowka, A. Olczak, J. Bojarska, M. Szczesio, W. L. Duax, B. M. Burkhart, W. A. Pangdorn, D. A. Langs, Z. Wawrzak, *Acta Cryst.* **2005**, *D61*, 433–441.
165. B. M. Burkhart, N. Li, D. A. Langs, W. A. Pangborn, W. L. Duax, *Proc. Natl. Acad. Sci. USA* **1998**, *95*, 12950–12955
166. B. A. Wallace, K. Ravikumar, *Science* **1988**, *241*, 182–187.
167. R. E. Koeppe, K. O. Hodgson, L. Stryer, *J. Mol. Biol.* **1978**, *121*, 41–54.
168. W. L. Duax, V. Pletnev, B. M. Burkhart, *J. Mol. Struct.* **2003**, *647*, 97–111.
169. I. L. Karle, *J. Am. Chem. Soc.* **1974**, *96*, 4000–4006.
170. I. L. Karle, *Proc. Nat. Acad. Sci. USA* **1985**, *82*, 7155–7159.
171. I. L. Karle, *Biochem.* **1974**, *13*, 2155–2162.
172. N. E. Zhukhlistova, G. N. Tishchenko, *Sov. Phys. Cryst.* **1981**, *26*, 700–704.
173. N. E. Zhukhlistova, *Cryst. Reports* **2002**, *47*, 433–442.
174. N. E. Zhukhlistova, G. N. Tishchenko, L. Refaat, M. M. Woolfson, *Cryst. Reports* **1998**, *43*, 45–52.
175. M. Dobler, J. D. Dunitz, J. Krajewski, *J. Mol. Biol.* **1969**, *42*, 603–606.
176. G. N. Tishchenko, Z. Karimov, *Sov. Phys. Cryst.* **1978**, *23*, 409–416.

177. N. E. Zhukhlistova, G. N. Tishchenko, K. M. Polyakov, *Sov. Phys. Cryst.* **1982**, *27*, 176–181.
178. J. A. Hamilton, M. N. Sabesan, L. K. Steinrauf, *Acta Cryst.* **1980**, *B36*, 1052–1057.
179. L. K. Steinrauf, J. A. Hamilton, M. N. Sabesan, *J. Am. Chem. Soc.* **1982**, *104*, 4085–4091.
180. J. A. Hamilton, M. N. Sabesan, L. K. Steinrauf, *J. Am. Chem. Soc.* **1981**, *103*, 5880–5885.
181. M. Pinkerton, L. K. Steinrauf, P. Dawkins, *Biochem. Biophys. Res. Comm.* **1969**, *35*, 512–518.
182. K. Neupert-Laves, M. Dobler, *Helv. Chim. Acta* **1975**, *58*, 432–442.
183. V. Z. Pletnev, I. N. Tsygannik, Yu. D. Fonarev, I. Yu. Mikhaylova, Yu. V. Kulikov, V. T. Ivanov, D. A. Lengs, V. L. Dyueks, *Bioorg. Khim.* **1995**, *21*, 828–833.
184. Y. Nishibata, A. Itai, Y. Iitaka, Y. Nawata, *Acta Cryst.* **1981**, *A37*, C75.
185. L. K. Steinrauf, K. Folting, *Isr. J. Chem.* **1984**, *24*, 290–296.
186. M. Dobler, R. P. Phizackerley, *Helv. Chim. Acta* **1974**, *57*, 664–674.
187. (a) B. T. Kilbourn, J. D. Dunitz, L. A. R. Pioda, W. Simon, *J. Mol. Biol.* **1967**, *30*, 559–563. (b) M. Dobler, J. D. Dunitz, B. T. Kilbourn, *Helv. Chim. Acta* **1969**, *52*, 2573–2583.
188. T. Sakamaki, Y. Iitaka, Y. Nawata, *Acta Cryst.* **1977**, *B33*, 52–59.
189. T. Sakamaki, Y. Iitaka, Y. Nawata, *Acta Cryst.* **1976**, *B32*, 768–774.
190. (a) J. D. Dunitz, D. M. Hawley, D. Mikloš, D. N. J. White, Y. Berlin, R. Marušić, V. Prelog, *Helv. Chim. Acta* **1971**, *54*, 1709–1713. (b) W. Marsh, J. D. Dunitz, D. N. J. White, *Helv. Chim. Acta* **1974**, *57*, 10–17.
191. J. W. Westley, in *Polyether Antibiotics*, Ed J. W. Westley, Marcel Dekker, New York, 1982, Vol.1, pp. 1–20.
192. C. J. Dutton, B. J. Banks, C. B. Cooper, *Natural Product Reports* **1995**, *12*, 165–181.
193. P. van Roey, W. L. Duax, P. D. Strong, G. D. Smith, *Isr. J. Chem.* **1984**, *24*, 283–289.
194. M. Alléaume, B. Busetta, C. Farges, P. Gachon, A. Kergomard, T. Staron, *J. Chem. Soc., Chem. Commun.* **1975**, 411–412.
195. J. Bordner, P. C. Watts, E. B. Whipple, *J. Antibiot.* **1987**, *40*, 1496–1505.
196. Y. Takahashi, H. Nakamura, R. Ogata, N. Matsuda, M. Hamada, H. Naganawa, T. Takita, Y. Iitaka, K. Sato, T. Takeuchi, *J. Antibiot.* **1990**, *43*, 441–443.
197. D. M. Walba, M. Hermsmeier, R. C. Haltiwanger, J. H. Noordik, *J. Org. Chem.* **1986**, *51*, 245–247.
198. A. Huczyński, J. Janczak, B. Brzezinski, *J. Mol. Struct.* **2011**, *985*, 70–74.
199. A. Huczyński, M. Ratajczak-Sitarz, A. Katrusiak, B. Brzezinski, *J. Mol. Struct.* **2007**, *871*, 92–97.
200. Y. Barrans, M. Alléaume, G. Jéminet, *Acta Cryst.* **1982**, *B38*, 1144–1149.
201. D. L. Ward, K.-T. Wei, J. G. Hoogerheide, A. I. Popov, *Acta Cryst.* **1978**, *B34*, 110–115.
202. W. L. Duax, G. D. Smith, P. D. Strong, *J. Am. Chem. Soc.* **1980**, *102*, 6725–6729.
203. M. Pinkerton, L. K. Steinrauf, *J. Mol. Biol.* **1970**, *49*, 533–546.
204. F. A. A. Paz, P. J. Gates, S. Fowler, A. Gallimore, B. Harvey, N. P. Lopes, C. B. W. Stark, J. Staunton, J. Klinowski, J. B. Spencer, *Acta Cryst.* **2003**, *E59*, m1050–m1052.
205. A. Nagatsu, T. Takahashi, M. Isomura, S. Nagai, T. Ueda, N. Murakami, J. Sakakibara, K. Hatano, *Chem. Pharm. Bull.* **1994**, *42*, 2269–2275.
206. A. Huczyński, M. Ratajczak-Sitarz, A. Katrusiak, B. Brzezinski, *J. Mol. Struct.* **2007**, *832*, 84–89.
207. A. Huczyński, J. Janczak, B. Brzezinski, F. Bartl, *J. Mol. Struct.* **2013**, *1043*, 75–84.
208. A. Huczyński, M. Ratajczak-Sitarz, J. Stefańska, A. Katrusiak, B. Brzezinski, F. Bartl, *J. Antibiot.* **2011**, *64*, 249–256.
209. I. N. Pantcheva, P. Dorkov, V. N. Atanasov, M. Mitewa, B. L. Shivachev, R. P. Nikolova, H. Mayer-Figge, W. S. Sheldrick, *J. Inorg. Biochem.* **2009**, *103*, 1419–1424.
210. P. Dorkov, I. N. Pantcheva, W. S. Sheldrick, H. Mayer-Figge, R. Petrova, M. Mitewa, *J. Inorg. Biochem.* **2008**, *102*, 26–32.
211. A. Huczyński, J. Janczak, D. Łowicki, B. Brzezinski, *Biochim. Biophys. Acta* **2012**, *1818*, 2108–2119.

212. D. Łowicki, A. Huczyński, M. Ratajczak-Sitarz, A. Katrusiak, J. Stefańska, B. Brzezinski, F. Bartl, *J. Mol. Struct.* **2009**, *923*, 53–59.
213. A. Huczyński, J. Janczak, B. Brzezinski, *J. Mol. Struct.* **2012**, *1030*, 131–137.
214. W. Hüttel, J. B. Spencer, P. F. Leadlay, *Beilstein J. Org. Chem.* **2014**, *10*, 361–368.
215. W. Pangborn, W. Duax, D. Langs, *J. Am. Chem. Soc.* **1987**, *109*, 2163–2165.
216. A. Huczyński, M. Ratajczak-Sitarz, A. Katrusiak, B. Brzezinski, *J. Mol. Struct.* **2008**, *888*, 224–229.
217. S. Ö. Yildirim, V. McKee, F.-Z. Khardli, M. Mimouni, T. B. Hadda, *Acta Cryst.* **2008**, *E64*, m154–m155.
218. T. Fehr, M. Kuhn, H.-R. Loosli, M. Ponelle, J. J. Boelsterli, M. D. Walkinshaw, *J. Antibiot.* **1989**, *42*, 897–902.
219. Y. Barrans, M. Alléaume, L. David, *Acta Cryst.* **1980**, *B36*, 936–938.
220. A. J. Geddes, B. Sheldrick, W. T. J. Stevenson, L. K. Steinrauf, *Biochem. Biophys. Res. Commun.* **1974**, *60*, 1245–1251.
221. J. W. Westley, R. H. Evans, L. H. Sello, N. Troupe, C.-M. Liu, J. F. Blount, R. G. Pitcher, T. H. Williams, P. A. Miller, *J. Antibiot.* **1981**, *34*, 139–147.
222. E. F. Paulus, M. Kurz, H. Matter, L. Vértesy, *J. Am. Chem. Soc.* **1998**, *120*, 8209–8221.
223. E. F. Paulus, L. Vértesy, *Z. Kristallogr. New Cryst. Struct.* **2004**, *219*, 184–186.
224. E. F. Paulus, L. Vértesy, *Z. Kristallogr. New Cryst. Struct.* **2003**, *218*, 575–577.
225. N. D. Jones, M. O. Chaney, J. W. Chamberlin, R. L. Hamill, S. Chen, *J. Am. Chem. Soc.* **1973**, *95*, 3399–3400.
226. J. P. Dirlam, L. Presseau-Linabury, D. A. Koss, *J. Antibiot.* **1990**, *43*, 727–730.
227. J. P. Dirlam, A. M. Belton, J. Bordner, W. P. Cullen, L. H. Huang, Y. Kojima, H. Maeda, H. Nishida, S. Nishiyama, J. R. Oscarson, A. P. Picketts, T. Sakakibara, J. Tone, K. Tsukuda, *J. Antibiot.* **1990**, *43*, 668–679.
228. V. Kumpiņš, S. Belyakov, Ë. Bizdëna, M. Turks, *Z. Kristallogr. New Cryst. Struct.* **2012**, *227*, 145–148.
229. E. W. Czerwinski, L. K. Steinrauf, *Biochem. Biophys. Res. Commun.* **1971**, *45*, 1284–1287
230. J. R. Hauske, G. Kostek, *J. Org. Chem.* **1989**, *54*, 3500–3504.
231. W. P. Cullen, J. Bordner, L. H. Huang, P. M. Moshier, J. R. Oscarson, L. A. Presseau, R. S. Ware, E. B. Whipple, Y. Kojima, H. Maeda, S. Nishiyama, J. Tone, K. Tsukuda, K. S. Holdom, J. C. Ruddock, *J. Ind. Microbiol.* **1990**, *5*, 365–374.
232. J. R. Oscarson, J. Bordner, W. D. Celmer, W. P. Cullen, L. H. Huang, H. Maeda, P. M. Moshier, S. Nishiyama, L. Presseau, R. Shibakawa, J. Tone, *J. Antibiot.* **1989**, *42*, 37–48.
233. M. Shiro, H. Nakai, K. Nagashima, N. Tsuji, *J. Chem. Soc., Chem. Commun.* **1978**, 682–683.
234. J. P. Dirlam, A. M. Belton, J. Bordner, W. P. Cullen, L. H. Huang, Y. Kojima, H. Maeda, S. Nishiyama, J. R. Oscarson, A. P. Ricketts, T. Sakakibara, J. Tone, K. Tsukuda, M. Yamada, *J. Antibiot.* **1992**, *45*, 331–340.
235. P. G. Schmidt, A. H.-J. Wang, I. C. Paul, *J. Am. Chem. Soc.* **1974**, *96*, 6189–6191.
236. G. D. Smith, W. L. Duax, S. Fortier, *J. Am. Chem. Soc.* **1978**, *100*, 6725–6727.
237. C. C. Chiang, I. C. Paul, *Science* **1977**, *196*, 1441–1443.
238. K. D. Klika, J .P. Haansuu, V. V. Ovcharenko, K. K. Haahtela, P. M. Vuorela, R. Sillanpää, K. Pihlaja, *Z. Naturforsch., Teil B* **2003**, *58*, 1210–1215.
239. J. W. Westley, C.-M. Liu, J. F. Blount, L. H. Sello, N. Troupe, P. A. Miller, *J. Antibiot.* **1983**, *36*, 1275–1278.
240. D. H. Davies, E. W. Snape, P. J. Suter, T. J. King, C. P. Falshaw, *J. Chem. Soc., Chem. Commun.* **1981**, 1073–1074.
241. D. Łowicki, A. Huczyński, *BioMed Research International*, **2013**, *2013*, Article ID 742149.

Chapter 4
Discriminating Properties of Alkali Metal Ions Towards the Constituents of Proteins and Nucleic Acids. Conclusions from Gas-Phase and Theoretical Studies

Mary T. Rodgers and Peter B. Armentrout

Contents

M.T. Rodgers (✉)
Department of Chemistry, Wayne State University, Detroit, MI 48202, USA
e-mail: mrodgers@chem.wayne.edu

P.B. Armentrout (✉)
Department of Chemistry, University of Utah, Salt Lake City, UT 84112, USA
e-mail: armentrout@chem.utah.edu

A. Sigel, H. Sigel, and R.K.O. Sigel (eds.), *The Alkali Metal Ions: Their Role for Life*,
Metal Ions in Life Sciences 16, DOI 10.1007/978-3-319-21756-7_4

Abstract Quantitative insight into the structures and thermodynamics of alkali metal cations interacting with biological molecules can be obtained from studies in the gas phase combined with theoretical work. In this chapter, the fundamentals of the experimental and theoretical techniques are first summarized and results for such work on complexes of alkali metal cations with amino acids, small peptides, and nucleobases are reviewed. Periodic trends in how these interactions vary as the alkali metal cations get heavier are highlighted.

Keywords Amino acids • Collision-induced dissociation • Infrared multiple photon dissociation • Nucleobases • Peptides

Please cite as: *Met. Ions Life Sci.* 16 (2016) 103–131

1 Introduction

Alkali cations interact with biological molecules via noncovalent, predominantly electrostatic interactions that are non-directional. In real biological systems, this means that the ion can interact with many components of its surroundings simultaneously. As a consequence, it is difficult to quantitatively characterize the strengths of such interactions. One means to overcome the lack of thermodynamic data on these systems is to examine models in the gas phase, where pairwise interactions between individual components of biological systems can be quantitatively measured with accuracy and precision. This can be achieved both for simple systems and for more complex interactions that evolve with larger systems. Coupled with an understanding of the detailed structure of these model complexes, such pairwise interactions can be used to provide estimates of the energetics of actual biological systems that can include subtle thermodynamic aspects of these complex systems. We have previously discussed the development of such a thermodynamic "*vocabulary*" for metal ion interactions in biological systems [1], which leads to the "*phrases, sentences, paragraphs, and book chapters*" that comprise real biological systems.

One strength of the relatively small gas-phase systems explored in this chapter is that the quantitative experimental information can be directly compared with high level theoretical results. Such comparisons play two equally important roles. First, they elucidate subtle features in the periodic trends in the metal cation affinities and how these vary with the properties of the biological molecules. It is the trends in these features (elucidated below) that transcend the gas-phase and allow application to condensed phase systems. Second, these comparisons provide important benchmarks to which levels of theory are useful approaches to describe such interactions and with what degree of accuracy.

2 Experimental and Theoretical Methods

Because alkali cations are naturally charged, their interactions with biomolecules are readily studied using a range of mass spectrometric (MS) techniques, which provide specificity regarding the identity of the complexes being studied. More advanced methods must be combined with MS in order to elucidate structure and thermodynamic information. In this section, we introduce the gas-phase experimental and theoretical methods that have been used to quantitatively explore the interactions of alkali metal cations with biomolecules. This section is designed to provide an overview of the methods along with appropriate literature references leading to more detailed expositions for the interested reader. The focus of this chapter is two experimental techniques, namely infrared multiple photon dissociation (IRMPD) and threshold collision-induced dissociation (TCID) utilized by the authors, but these approaches legitimately provide much of the available quantitative information on the systems of interest. Other useful experimental methods are also introduced.

2.1 Infrared Multiple Photon Dissociation Experiments

One means of providing structural information regarding gas-phase ions is to measure their infrared (IR) spectrum, either in the fingerprint region of 500 – 2000 cm^{-1} or in the hydrogen stretching region from 2800 – 3600 cm^{-1}. The difficulty with acquiring the desired IR spectrum for ions is that the concentrations of ions cannot easily be raised to the level needed for absorption measurements. Unlike solution phase species, where the concentration can be easily adjusted over a very wide range, the concentration of gas-phase ionic species is limited by the Coulombic repulsion between the ions. To overcome this limitation, several "action" spectroscopies have been developed in which the absorption of a photon (or multiple photons) induces dissociation, which is easily monitored using mass spectrometers. In a "tagging" infrared photodissociation experiment, a weakly bound species (usually a rare gas atom) is attached to the ion of interest. For such a molecular ion, absorption of a single IR photon is generally sufficient to cleave the noncovalent interaction with the tag.

Generation of such tagged molecules usually requires cryogenic conditions and the tagging atom may perturb the absorption spectrum. Alternatively, the ion can be probed directly if multiple photons are absorbed in an infrared multiple photon dissociation experiment. IRMPD requires only routine ion production methods but necessitates the use of higher powered lasers to provide sufficient fluence to allow absorption of multiple photons. Thus, the use of free electron lasers (FELs) has been commonplace in such studies, with facilities in the Netherlands (the Free Electron Lasers for Infrared eXperiments, FELIX) and France (Centre Laser Infrarouge d'Orsay, CLIO) being used extensively for such work.

Furthermore, the need for the absorption of multiple photons can distort the spectrum (often to the red to compensate for anharmonicity of vibrational modes). IRMPD generally relies on efficient intramolecular vibrational redistribution (IVR) [2, 3], which allows the ion to absorb a photon into a specific vibrational mode and then transfer that energy into other vibrational modes such that absorption of additional photons at the same photon frequency can occur again and again [4, 5]. IVR works best for larger ions with higher state densities. IRMPD studies of ions as small as four atoms have been conducted [6], but these required an intracavity laser (the Free Electron Laser for IntraCavity Experiments, FELICE, at the FELIX facility), where the photon flux is orders of magnitude higher than for an externally coupled laser.

Structural information for ions is generally obtained from IRMPD results by comparing the experimental spectrum with that generated from quantum chemical calculations (see Section 2.4). A major caveat is that the latter provides a prediction for absorption of a single photon, generally in the harmonic limit (although explicit anharmonic calculations are increasingly performed) [7–9]. Because of the multiple photon nature of the IRMPD spectrum, bands can be shifted according to the anharmonicity of the vibrational mode and intensities can vary appreciably (with weak bands showing little or no intensity because insufficient photons are absorbed to induce fragmentation). Nevertheless, the agreement between IRMPD and theoretical spectra is often sufficient for unequivocal identification of the conformation of the ion involved, even when multiple low-energy conformations may be present.

2.2 Collision-Induced Dissociation Experiments

Collision-induced dissociation (CID) experiments, also known as collisional activation (CA) or collision activated dissociation (CAD), take advantage of the fact that ions are charged and therefore easily accelerated to hyperthermal kinetic energies using electric fields. If these accelerated ions collide with an unreactive species (usually one of the rare gases but other gases or even surfaces can be used), translational to internal energy transfer occurs and provides sufficient internal energy to induce fragmentation. The extent of dissociation observed depends on the amount of energy transferred, the complexity of the ion, and the time scale over which the fragmentation is detected. In analytical applications, CID can be used to sequence complicated biomolecules (peptides, proteins, DNA, and RNA) simply by measuring the masses of fragment ions. Cleavage of various peptide (or phosphate ester) bonds along the backbone leads to a sequence of fragment ions that, if complete enough, enable the original primary structure of the biomolecule to be elucidated.

CID can also be used more quantitatively to measure the thermodynamics associated with the dissociations, so-called threshold CID experiments (TCID) [10]. In this application, ions are subjected to single collisions with an inert gas with the collision energy varied from zero through the energies needed to induce fragmentation. Analysis of the resultant kinetic energy-dependent processes reveals the threshold energy for the observed fragmentation pathways, which can be related to the heights of the rate-limiting transition states. Although tandem mass spectrometers such as triple quadrupole MS (QQQ) can be used to provide such kinetic energy-dependent data, the best quantitative information is obtained using specialized instrumentation, notably a guided ion beam tandem mass spectrometer (GIBMS) [11]. This instrument is specifically designed for such quantitative studies and enables very low and well controlled collision energies and excellent collection of product ions.

Determination of the threshold energy in TCID experiments requires consideration of several experimental parameters including effects of multiple collisions [12], internal energy distributions [13], translational energy distributions [11], lifetimes for dissociation [14], competition between reaction channels [15], and sequential dissociations [16]. Methods for modeling all of these effects are detailed in the literature [17] and have proven to provide accurate thermodynamic information. Comparison of TCID results to quantities determined using equilibrium measurements (Section 2.3.1) of these quantities are generally in good agreement [18–24], although TCID has been applied to many systems where equilibrium methods are simply inaccessible. For such systems, comparison with theoretical information generally confirms the accuracy of TCID results.

It has also been realized that the thermochemistry derived from TCID measurements can be used for structural identification of fragment ions [25–30]. As for IRMPD studies, this is generally achieved by comparison of the experimental results with those derived from theory. TCID thermochemistry is insufficiently precise to distinguish between isomers or conformers that are relatively close in energy (whereas IRMPD can be very sensitive to such differences), but can be applied to fragment ions, which may not be generated in sufficient quantity to interrogate using IRMPD methods. Although different conclusions from these two approaches have been found [28, 31], they can be attributed to the harsher source conditions needed to generate sufficient ion intensities for IRMPD studies. In contrast, TCID is sensitive to the structure of the species formed at threshold, which is often but may not be the thermodynamically most favorable isomer.

2.3 *Other Experimental Approaches*

2.3.1 Equilibrium Methods

The classic means of determining thermochemistry is to measure the equilibrium concentration of reactants and products as a function of temperature and then convert to enthalpies and entropies using the van't Hoff equation. This method can provide the most rigorous thermodynamic determinations and, in the gas phase, can be achieved at both low and high pressures in the presence or absence of buffer gases, but is subject to perturbations associated with sampling the equilibrium zone and uncertainties in the absolute temperature. This method has been used to measure the relative binding energies of alkali cations to a wide range of ligands [32], albeit few biological systems partly because of their limited volatility.

2.3.2 Kinetic Method

The kinetic method was first developed by Cooks and coworkers as a simple method to estimate thermodynamic values from easily performed experiments [33, 34], namely the relative intensities of products resulting from dissociation of a proton (or cation or anion) bound dimer. It has been revised to attempt to include important entropic effects [35, 36], and has been discussed thoroughly [37–39]. Ultimately, it cannot provide true thermodynamic values because the temperature of the evaluation is ill-defined, but does provide useful relative information in the absence of more definitive data.

2.3.3 Blackbody Infrared Radiative Dissociation

Blackbody infrared radiative dissociation (BIRD) allows ions to reach equilibrium with their thermal surroundings by absorption of infrared radiation [40, 41]. Quantitative assessment of the extent of dissociation observed as a function of the temperature of the surroundings allows thermodynamic information to be extracted.

2.3.4 Ion Mobility

In ion mobility studies, ions are dragged through a viscous medium (usually He and more recently N_2) by an electric field [42–44]. The rate at which the ions move in this environment depends on their interaction cross section with the gas, such that larger ions move more slowly than compact ions. Using this technique, the size of ions can be assessed, and by comparison with theoretical structures, the structures of ions can be inferred.

2.4 Theoretical Calculations

Experimental studies of the interactions of alkali metal cations with biological molecules rely on quantum chemistry calculations to provide structural information and molecular parameters useful for interpreting the data. Although some results could be achieved in the absence of theoretical input, the synergy between experiment and theory provides much more robust and detailed results. In this chapter, we focus on theoretical results that provide such synergistic information to experiment, and do not include theoretical results for which there are no complementary experimental data.

Specific computational approaches to alkali metal cation complexes of biomolecules vary nearly as much as the number of investigators that have investigated these systems, but in the laboratories of the authors, the relatively small systems that have been studied have led to an approach that appears to provide accurate structural and thermodynamic information. For systems where all-electron calculations can be performed, structures and harmonic vibrational frequencies are typically calculated using the hybrid density functional B3LYP coupled with a double or triple zeta basis set, 6-31G(d), 6-31+G(d,p), or 6-311+G(d,p) being common. This is followed by single point energy calculations performed using a larger basis set, 6-311+G(2d,2p), needed to yield accurate energetics, again using B3LYP as well as MP2(full) and sometimes B3P86 or M06 approaches. For calculations of metal-ligand bond dissociation energies (BDEs), basis set superposition error (BSSE) corrections at the full counterpoise level are applied [45, 46]. We generally use multiple computational approaches because different levels of theory often do not agree on the relative energies of low-lying conformations and calculated metal cation-ligand binding energies generally exhibit a spread comparable to experimental uncertainties. These disparate results properly reflect the uncertainties in the computational results.

For the heavier Rb and Cs alkali metal systems, basis sets that utilize effective core potentials (ECPs) are generally used. Extensive testing has demonstrated that even when augmented by additional polarization functions [47], the commonly used LANL2DZ basis set [48] does not provide accurate thermodynamic information [49, 50]. Accurate energetics are provided by using def2-TZVP (or def2-TZVPPD) basis sets, which are balanced basis sets on all atoms at the triple-ζ level including polarization (P) and diffuse (D) functions that use ECPs on rubidium and cesium developed by Leininger et al. [51]. These basis sets are used along with B3LYP for structure and vibrational frequency calculations followed by single point energy calculations using B3LYP, MP2(full), B3P86, or M06 levels.

For the lightest alkali cation, Li^+, we have found that strongly bound ligands are sufficiently close to the 1 s core electrons of the cation that repulsive interactions play a role [52]. Larger basis sets cannot compensate for this repulsion, which requires that the core electrons be permitted to polarize away from the ligand. This can be accomplished using the correlation consistent polarized core/valence basis sets (cc-pCVnZ where n = D, T, and Q) of Woon and Dunning for lithium [53]. For consistency, cc-pVnZ or aug-cc-pVnZ basis sets are used on all other atoms. This approach has been found to yield good thermodynamic values with reasonable computational cost [52]. The def2 and cc-pCVnZ basis sets can be obtained from the Environmental Molecular Sciences Laboratory basis set exchange library [54, 55].

Theory is also used to provide the isotropic molecular polarizabilities of neutral molecules needed for thermochemical analysis of TCID data as well as examination of periodic trends as discussed below. The PBE0/6-311+G(2d,2p) level of theory has been shown to provide polarizabilities that are in good agreement with measured values [56].

As the biological molecules considered increase in size, the methods noted above become increasingly cumbersome largely because a thorough exploration of the enormous conformational space of such molecules becomes expensive. More advanced techniques such as replica exchange molecular dynamics (REMD) [57, 58] or various approximations [59–62] can then be used.

3 Alkali Metal Cations Interacting with Amino Acids

3.1 Structure

Unlike in solution, amino acids in the gas phase are generally not zwitterionic because there is no solvent to support the separated charges. Although alkali cations can yield such zwitterionic structures in a salt-bridge configuration, most alkali metal cation–amino acid complexes, $M^+(AA)$, involve charge solvated structures in which the cation is solvated by the various heteroatoms of the amino acid. For most naturally occurring amino acids and alkali metal cations, the structures of these complexes have been examined in extensive IRMPD spectroscopic studies [63–82]. In the case of proline (and several related molecules), these spectroscopic results have been augmented by structural assignments from TCID studies [83]. In both cases, structures are assigned on the basis of comparison to computational results, IR spectra or BDEs (see next section), respectively.

The results of this work are compiled in Table 1 where it can be seen that the structures evolve as the metal ion changes. For the smallest alkali metal cation, Li^+, the higher charge density leads to tridentate structures in which the metal binds to the

Table 1 Structures of $M^+(AA)$ complexes of alkali metal cations with amino acids.[a]

AA	Li^+	Na^+	K^+	Rb^+	Cs^+
Gly[b]		NO			
Pro[c]	ZW[d]	ZW[b,d]	ZW, OX[e]	ZW, OX	OX, ZW
Ser[f]	NOS	NOS	NOS, OX	NOS, OX	NOS, OX, ZW
Thr[g]	NOS	NOS	NOS, OX	NOS, OX	NOS, OX, ZW
Cys[h]	NOS	NOS	OX, NOS, ZW	OX, NOS, ZW	OX, NOS, ZW
Met[i]	NOS	NOS	NOS, ZW, OX(S)?	NOS, ZW, OX(S)?	NOS, ZW, OX(S)?
Phe[j]	NOS	NOS	NOS[j]	NOS, OS	NOS, OX, OS
Tyr[k]			NOS		
Trp[l]	NOS	NOS, OS	NOS, OS	OS, NOS	OS, NOS
Asp[m]	NOS				OXS, NOS[n]
Glu[m]	NOS				OXS, OS, NOS[n]
Asn[o]	NOS	NOS	NOS, OXS	NOS, OXS	NOS, OXS
Gln[p]	NOS	NOS, OS, OXS?	NOS, OS, OXS?		OXS, OS, NOS
Lys[q]	NOS	NOS	OS, NOS, ZW		
Arg[r]	NOS	ZW, NOS	ZW	ZW	ZW
His[s]	NOS	NOS[t]	OS, NOS, OX	OS, NOS, OX	OS, OX, NOS

[a]Isomers are given in order of approximate relative population, ?=may be present. SC=side-chain heteroatom or aromatic ring, NO=bidentate [N,CO], NOS=tridentate [N,CO,SC], OS=bidentate [CO,SC], OX=bidentate carboxylic acid [COOH], OXS=tridentate carboxylic acid [COOH,SC], ZW=bidentate zwitterion [CO_2^-], OX(S)=OX or OXS or both.
[b][63]. [c][78]. [d][83], TCID. [e][69]. [f][71]. [g][72]. [h][79]. [i][76]. [j][77]. [k][64]. [l][65]. [m][70]. [n][81]. [o][75]. [p][73]. [q][68]. [r][66,67]. [s][80]. [t][74].

amino nitrogen and carbonyl oxygen of the backbone along with the heteroatom or aromatic ring of a functionalized side-chain. This gives a tridentate metal ion coordination of [N,CO,SC] as illustrated for the sulfur side-chain of methionine in Figure 1. The single exception is proline, where the unique secondary nitrogen stabilizes a bidentate zwitterionic structure [CO_2^-] in which the metal ion binds to both oxygens of the carboxylate anion, which is hydrogen bonded to the protonated amino group forming a salt-bridge structure. For Na^+, this structural motif is largely maintained although tryptophan and glutamine exhibit bidentate structures that bind only to the backbone carbonyl and the side-chain aromatic ring or carbonyl of the amide group, [CO,SC]. The basic amino acid, arginine, also switches to having predominantly a zwitterionic structure, although the tridentate species is still observed. Glycine, where the cation binds to the amino nitrogen and carbonyl oxygen [N,CO], is also not tridentate, but only because it has no functionalized side-chain.

Because of the lower charge density on the heavier alkali cations, K^+, Rb^+, and Cs^+, these M^+(AA) complexes exhibit several different isomers, with the tridentate [N,CO,SC] structures still present but can be dominated by the bidentate [CO,SC] structures or by bidentate or tridentate structures where the ion binds to the oxygens

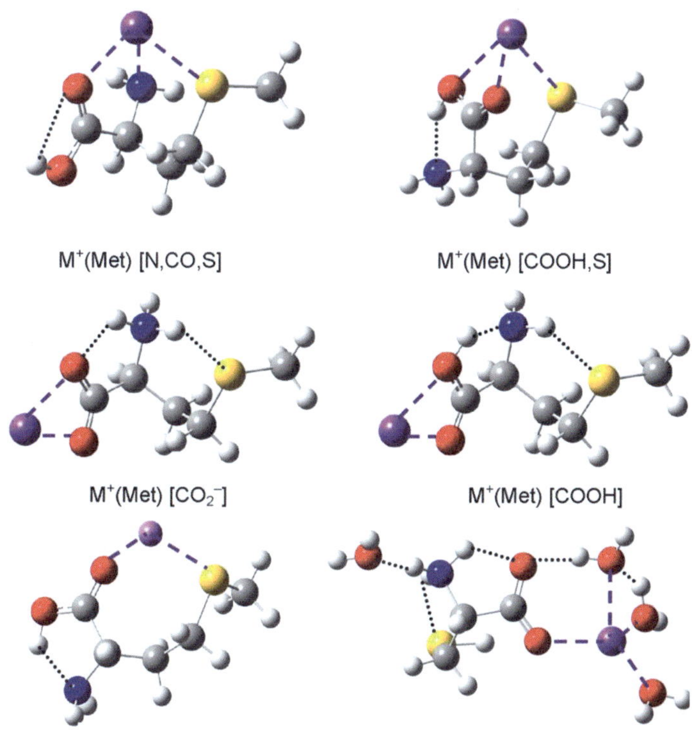

Figure 1 Structures of M^+(Met) and $Na^+(Cys)(H_2O)_4$ showing types of metal coordination (dashed lines). Dotted lines indicate hydrogen bonds.

of the carboxylic acid group, [COOH] and [COOH,SC], respectively, instead of to the amino nitrogen. Even for these metal cations, the only amino acids exhibiting dominant zwitterionic structures are proline and arginine; however, such structures are also observed for Ser, Thr, Cys, and Met. In these cases, the [CO_2^-] structures are accompanied by [COOH] and it can be realized that these two species differ only in whether the proton has transferred to the amino group [CO_2^-] or remains on the oxygen of the carboxylic acid [COOH] (see Figure 1). In both cases, this proton is shared between the backbone amino and carboxyl moieties and the metal cation binds to both carboxyl oxygens, thereby providing the driving force for moving the proton away from the oxygen. As such, these two structures actually lie in a single double-well potential that favors [CO_2^-] when the metal charge density is high (e.g., for metal dications, Li^+ and Na^+) and favors [COOH] when the metal charge density is low (Rb^+ and Cs^+) [71, 80]. Interestingly, computations indicate that the zero point energy of this proton motion generally lies above the bottom of both wells, meaning that the wavefunction associated with the ground vibrational level samples both [COOH] and [CO_2^-] potential wells. This allows simultaneous observation of the CO stretches characteristic of the carboxylate and carboxylic acid groups. Zwitterionic structures are not observed for Li^+ and Na^+ because the [N,CO,SC] structures are strongly favored, whereas the [COOH] (and accompanying [CO_2^-]) structures become competitive for Rb^+ and Cs^+ (and sometimes K^+).

In all of these cases, the structures found are generally in good agreement with computational results of the relative stabilities of the different isomers, although this requires that all accessible conformations are located. For example, assignments for the structure of Cs^+(Glu) were revised after locating a lower energy tridentate [COOH,SC] conformation [70, 81]. In addition, the computed ground isomer can vary substantially with the level of theory used. In many of our own results, we have found that the relative stabilities predicted by single point energy calculations at the MP2 level are preferred compared to B3LYP results. Finally, it has been shown that the conditions used in the source can influence the structures observed, kinetically trapping excited isomers [84].

3.2 *Thermodynamics*

The evolution of the structures of alkali metal cation-amino acid complexes in the gas-phase as elucidated by IRMPD spectroscopy is fruitfully understood by considering the strength of the interactions with the simplest amino acid, glycine, and molecules representing its various functional components, specifically, 1-propyl amine, 1-propanol, methyl ethyl ketone, propionic acid, and ethanol amine [85, 86]. For Na^+ complexes [85], the molecules having only one functional group bind in the order $OH < NH_2 < CO$. Computations indicate that Na^+ binds to both oxygens of propionic acid, which increases the BDE relative to 1-propanol (by 5 kJ/mol experimentally), but decreases the BDE compared to methyl ethyl ketone (by 13 kJ/mol experimentally). This difference can be attributed to a better orientation of the metal cation in the ketone (which prefers a ∠NaOC bond angle of 180°) coupled with an inductive effect of the hydroxyl group. Compared with the ketone, glycine binds Na^+ more strongly by 33 kJ/mol, a result of the bidentate binding to both the amino group and the carbonyl, [N,CO]. Interestingly, ethanol amine binds Na^+ more tightly than Gly by 11 kJ/mol because the sp^2 hybridization of the carboxylic acid group of Gly leads to greater conformational restrictions, whereas the sp^3 carbon in ethanol amine provides more flexibility and better overlap of the heteroatom lone pairs and the metal cation.

When a similar analysis is conducted for potassium ion complexes [86], the relative results are similar although the absolute BDEs are 74 ± 4 % of those for Na^+. The most notable difference between the metals lies in the observation that the molecules with single functional groups have BDEs to K^+ in the order $NH_2 < OH < CO$, i.e., the amino group binds less strongly (67 % compared to the Na^+ complex). Because of the poorer binding to the amino group, most levels of theory indicate that K^+(Gly) has a [COOH] ground structure, with [N,CO] (corresponding to the ground conformer for Na^+(Gly)) lying 1 – 7 kJ/mol higher in energy. One driving force for this change is that the larger potassium cation prefers to bind to the ground structure of glycine, which is stabilized by a OH•••N hydrogen bond, whereas the smaller sodium cation binds sufficiently strongly that it preferentially binds to a higher energy conformer of Gly. Results comparable to those of K^+ are expected for the larger alkali metal cations, Rb^+ and Cs^+.

Given these results, the trends in Table 1 are clearer. The alkali metal cations prefer to bind to carbonyl groups, which are always available along the backbone. The smaller cations also like to bind to the amino group and then augment these interactions by binding to the side chain. As the metal cation gets larger, the preference for binding to the amino group in the [N,CO,SC] isomers decreases compared to [COOH,SC] and [COOH] isomers. The larger metal cation also makes it more difficult to maintain strong binding to all three functional groups in the tridentate [N,CO,SC] structure such that the bidentate [CO,SC] structure is competitive.

3.3 Periodic Trends

The interactions of amino acids and alkali metal cations in the gas phase are largely dictated by electrostatics. Table 2 includes a comprehensive listing of the known M^+–AA BDEs measured by TCID [49, 50, 52, 81–83, 85–96], along with a more complete set of values for sodium complexes obtained using the kinetic method [97, 98]. For all AAs, experimental and theoretical results for M^+(AA) BDEs decrease down the periodic table, $Li^+ > Na^+ > K^+ > Rb^+ > Cs^+$. This is primarily a reflection of the increasing radius of the metal cation: 0.70, 0.98, 1.33, 1.49, and 1.69 Å, respectively [99]. The larger cation radius necessarily increases the M^+–AA bond distances, thereby reducing the electrostatic interactions. Experimentally, it is found that BDEs decrease approximately linearly with the inverse of the ionic radius, as shown in Figure 2 for AA = Gly, Pro, and His, with similar plots obtained for other AAs. (Although this inverse dependence seems to suggest a long-range Coulombic potential, the true short-range interactions must comprise combinations of

Table 2 Bond dissociation energies (kJ/mol) at 0 K of M^+(AA) complexes of alkali metal cations with amino acids and small peptides and their polarizabilities (α).[a]

AA	α (Å^3)	Li^+	Na^+	Na^{+}[b]	K^+	Rb^+	Cs^+
Gly	6.2	220.0 (8.0)[c]	164.0 (4.8)[d]	161 (8)	121.4 (4.8)[e]	108.7 (7.0)[f]	93.3 (2.5)[g]
Ala				167 (8)			
Val				173 (8)			
Leu				175 (8)			
Cys	11.2	255.8 (11.9)[h]	176.9 (5.0)[h]	175 (8)	120.7 (3.1)[h]	102.5 (2.8)[h]	96.8 (4.2)[g]
Ile				176 (8)			
Pro	10.8	254.7 (6.8)[i]	186.2 (4.8)[j]	196 (8)	143.5 (4.5)[j]	125.2 (4.5)[f]	107.9 (4.6)[g]
Asp	10.4		194.9 (5.8)[k]	201 (8)[l]	147.6 (6.8)[m]	126.1 (6.8)[n]	108.9 (6.7)[n]
Glu	12.2		198.8 (4.8)[k]	203 (8)[l]	152.4 (6.8)[m]	134.2 (7.7)[n]	112.6 (6.5)[n]
Ser	8.6	280.8 (12.5)[o]	199.7 (7.7)[o]	192 (8)	144.7 (6.8)[o]	115.7 (4.9)[f]	102.3 (4.1)[g]
Met	14.6	292.0 (12.2)[p]	201.7 (10.6)[p]		141.8 (10.6)[p]	121.0 (7.0)[q]	102.8 (6.6)[q]
Thr	10.5	284.6 (13.5)[o]	203.6 (9.6)[o]	197 (8)	148.6 (9.6)[o]	122.1 (4.6)[f]	105.4 (4.3)[g]
Phe	18.1		205.5 (6.8)[r]	198 (8)	150.5 (5.8)[r]	123.8 (7.2)[q]	112.9 (5.5)[q]
Tyr	18.8		209.4 (9.6)[r]	201 (8)	155.3 (8.7)[r]	125.8 (7.4)[q]	115.6 (6.9)[q]
Asn	11.2		209.4 (5.8)[k]	217 (8)[l]	156.3 (6.8)[m]	138.4 (7.1)[n]	115.3 (6.9)[n]
Gln	13.0		213.2 (5.8)[k]	222 (8)[l]	161.1 (7.7)[m]	144.2 (9.2)[n]	128.1 (8.5)[n]
Trp	22.0		217.1 (7.7)[r]	210 (8)	165.0 (5.8)[r]	138.1 (7.5)[q]	125.0 (6.8)[q]
Lys		<376 (10)[r]	219.0 (10.0)[s]	>213 (8)[l]	155.3 (10.0)[s]	127.4 (8.8)[s]	111.0 (10.0)[s]
His	15.2		222.5 (11.0)[t]	228 (8)[l]	163.5 (8.5)[t]	137.4 (5.7)[t]	118.2 (6.4)[t]
Arg				242 (8)[l]			
Gly_2	11.3		209 (13)[u]	203 (8)[v]	149 (7)[u]		
Gly_3	16.2		240 (17)[u]	237 (9)[v]	183 (15)[u]		
Gly_4	21.1			261 (11)[v]			

[a]Uncertainties in parentheses. [b][97] Kinetic method.
[c][52]. [d][85]. [e][86]. [f][49]. [g][50]. [h][89]. [i][94]. [j][83]. [k][88]. [l][98]. [m][87]. [n][81]. [o][96]. [p][92]. [q][91].
[r][93]. [s][82]. [t][90]. [u][116]. [v][133].

ion-dipole (r^{-2}), ion-quadrupole (r^{-3}), and ion-induced dipole (r^{-4}) forces along with complex chelation effects.) The slopes of the lines shown in Figure 2 for Gly, Pro, and His complexes are 158±6, 190±8, and 214±15 Å kJ/mol, respectively, and properly reflect the relative strength of the interactions of different AAs with the metal cations in that Gly is the weakest binding AA and His is the strongest binding AA measured to date. Direct comparison of the BDEs shows that relative to Na^+(AA) complexes, BDEs for Li^+(AA) are 144±11 %, K^+(AA) are 74±3 %, Rb^+(AA) are 56±21 %, and Cs^+(AA) are 46±21 %.

It can be seen in Figure 2 that the BDEs for Rb^+(His) and Cs^+(His) lie below the linear correlation line, whereas those for Na^+(His) and K^+(His) lie above it, i.e., the larger metal cations show a smaller enhancement in binding associated with the His side chain. Similar effects are also observed for Ser, Thr, Cys, Asp, Glu, and Asn.

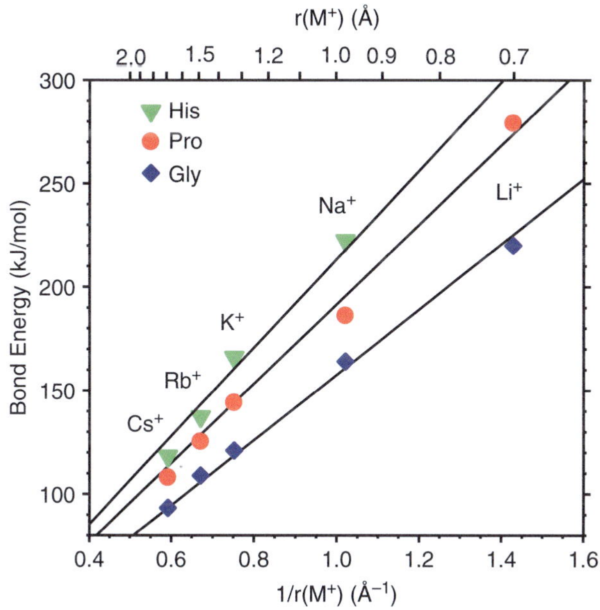

Figure 2 Experimental 0 K bond dissociation energies (in kJ/mol) for M^+(Gly) (blue diamonds), M^+(Pro) (red circles), and M^+(His) (light green triangles) for $M^+ = Li^+$, Na^+, K^+, Rb^+, and Cs^+ are plotted *versus* the inverse metal cation radius (in $Å^{-1}$). The lines are linear regression fits to the data constrained to pass through the origin. Adapted from [90].

These trends can be explored more thoroughly by examining the M^+(AA) BDEs as a function of the polarizability of the AA, as shown in Figure 3 for Na^+ and Cs^+.

Originally, it was found that the BDEs for Gly and the aromatic amino acids (Phe, Tyr, and Trp) exhibited a good correlation with polarizability [93], and this correlation was later extended to include Cys, Met, and Pro [1], as shown in Figure 3 in blue. In both cases, the lines are regression analyses of the data shown. For all alkali metal cations, Asp, Glu (in green), Ser, Thr, Asn, Gln, and His (in red) all exhibit enhanced binding compared to values predicted from the original polarizability correlation [90–92]. For Na^+ (as well as K^+), BDEs to Ser, Thr, Asn, Gln, and His form series that parallel those for the aliphatic AAs. The average enhancement for these five AAs is ~28 and 24 kJ/mol for Na^+ and K^+, respectively [90, 92]. Asp and Glu form a separate series with enhancements of ~15 kJ/mol for both Na^+ and K^+. Compared to Asn and Gln, the carboxylic acid side chains of Asp and Glu bind less tightly because the inductive effect of the hydroxyl group removes electron density from the carbonyl, as noted in Section 3.2 above. It can also be seen that Glu and Gln bind more tightly than Asp and Asn, respectively, because the longer side chain allows more flexibility in the orientation of the side-chain binding. The magnitude of these enhancements has been related to the local dipole moment of the side-chain coordinating site, small for Met, Phe, Tyr, and Trp, relatively large for Ser, Thr, Asn, and Gln, and intermediate for Asp and Glu. Thus,

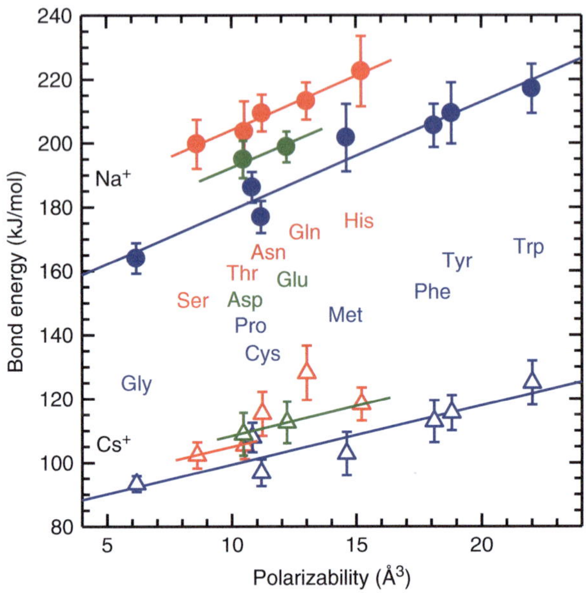

Figure 3 Experimental 0 K bond dissociation energies (in kJ/mol) for $Na^+(AA)$ (circles) and $Cs^+(AA)$ (triangles) versus the calculated polarizability (in $Å^3$) of the amino acid. The blue lines show linear regression analyses of the data, and the red and green lines are parallel to the regressions. Adapted from [81] and [90].

even though His has a polarizability (15.2 $Å^3$) comparable to Met (14.6 $Å^3$), it binds to the alkali cations much more tightly, even exceeding Trp, which has a much higher polarizability (22.0 $Å^3$). The local dipole moment can be quantified by considering molecules similar to the side chains. Specifically, acetamide (mimicking Asn and Gln) and imidazole (His) have dipole moments of 3.68 and 3.8 D, whereas the 1.70 D dipole moment of acetic acid (Asp and Glu) is about half as large [100]. Methanol (mimicking Ser and Thr) also has a dipole of 1.70 D, which suggests that the side-chain enhancements are influenced by the alignment of the dipole as well as its magnitude. This alignment is influenced by the length of the side-chain, with the two atom lengths (from backbone C_α to coordinating atom) of Ser and Thr being more restrictive than the three atom (Asn, Asp, His) and four atom (Gln, Glu) lengths.

For Cs^+ (as well as Rb^+), the polar side chains of Ser, Thr, Asp, Asn, Glu, Gln, and His also lead to enhanced binding. For Cs^+, Ser and Thr lie ~7 kJ/mol above the trend line [49, 50], Asp, Glu, and His lie ~10 kJ/mol above the trend [90], and Asn and Gln are 15 and 24 kJ/mol higher, respectively. For Rb^+, these same AAs show enhancements of 5 – 7, ~15 kJ/mol, and ~24 and ~24 kJ/mol, respectively. The changes in these side-chain enhancements relative to the smaller cations can be understood in part on the basis of the length of the side chain. For Ser and Thr, where the side chain is only two atoms long, the enhancement drops from 28 to 24 to 10 to 7 kJ/mol for Na^+ to K^+ to Rb^+ to Cs^+. This decrease is a consequence of the increasing metal cation radius and the fact that larger cations cannot take advantage of binding in a tridentate conformation as well as the smaller cations [49, 50]. Gln has a side-chain length of four atoms and an enhancement that is largely independent of the metal, 24 – 28 kJ/mol, suggesting that the longer side chain allows better tridentate binding for the strongly binding carbonyl group of the carboxamide. For Asp, Glu, and His, with three, four, and three atom side-chain lengths, the enhancements drop from 28 (Na^+) to 24 (K^+) to 15 (Rb^+) and 15 (Cs^+) kJ/mol. This is again consistent with better tridentate binding than the short side chain AAs, but less effective binding than with Gln because the side chain moieties of Asp, Glu, and His do not bind as tightly as the carboxamide. Asn also possesses this carboxamide group but a shorter side-chain than Gln and it shows a trend of 28 (Na^+), 24 (K^+), 24 (Rb^+) and 15 (Cs^+) kJ/mol, intermediate between the strong effect of Gln and the weaker effects of Asp, Glu, and His.

3.4 *Effects of Hydration*

For several of the sodiated amino acids, Gly, Pro, and Cys, TCID and theoretical studies have examined the effects that hydration exerts on the structure and binding [101–103]. In the gas-phase, hydration effects can be studied for the sequential addition of water molecules, in these cases from one to four. Experimental results (matched by theory) show that the sequential binding energies for losing one water molecule and the amino acid from $Na^+(AA)(H_2O)_x$ decrease monotonically with increasing x. Increasing ligand-ligand repulsion and decreasing effective charge on the sodium cation explain this simple trend.

In the case of $Na^+(Gly)(H_2O)_x$, $x=1–4$, theory finds that hydration changes the coordination site for Na^+ on glycine from [N,CO] for $x=0$ and 1 to [CO] coordination where one water molecule bridges to the hydroxyl group for $x=2 – 4$ [101]. Theory indicates that zwitterionic structures are destabilized by the addition of up to four water molecules because solvation of the Na^+ charge center reduces the ability of the complex to support the charge separation. Interestingly, as additional water molecules are added, zwitterionic structures can become the overall ground structures because solvation of both the Na^+ and $–NH_3^+$ charge centers can occur.

Similar conclusions have been reached in an extensive set of experiments by Williams and coworkers for $M^+(Val)(H_2O)_x$, $x=1 – 6$ and $M^+=Li^+$, Na^+, and K^+. These results include kinetic data for all systems [104, 105], hydration energies for $x=1 – 3$ for Li^+ and $x=1$ for Na^+ obtained using BIRD [106, 107], and the only spectroscopic results for hydrated alkali metal cationized amino acids [108]. The kinetic results indicate a change in metal cation coordination from [N,CO] to [COOH] at $x=3$ for Li^+ and $x=2$ for Na^+. Comparison of the hydration BDEs to related molecules were used to suggest that $Li^+(Val)(H_2O)_3$ has a zwitterionic structure, however, the more sensitive spectroscopic probe indicates that this complex retains a charge-solvated structure through $x=4$.

For $Na^+(Pro)(H_2O)_x$, $x=1-4$, the addition of a single water molecule again does not change the sodium cation coordination from zwitterionic proline [CO_2^-] [102]. Additional waters keep the zwitterionic proline but change the binding to [CO^-] coordination with one of the water molecules bridging to the second carboxylate oxygen. Theory suggests that the third water added can bind to either cationic site, Na^+ or $-NH_3^+$, with comparable energies. Four waters are sufficient to completely solvate both cationic sites. Similar conclusions were drawn for proline methylated at different positions coordinated to Li^+ and Na^+ and one and two waters in kinetic and BIRD studies [109].

For $Na^+(Cys)(H_2O)_x$, $x=0-2$, theory indicates that the sodium ion binds in the [N,CO,S] tridentate configuration with the waters attaching directly to the sodium ion [103]. Some levels of theory suggest that, similar to Gly and Pro, it is isoenergetic for Na^+ to bind to the carbonyl [CO] with one of the water molecules bridging to the hydroxyl oxygen. The third water can bind to Na^+, $-NH_3^+$, or –OH with equivalent BDEs and at $x=4$, the Na^+ and $-NH_3^+$ groups are completely solvated (Figure 1), in part because there is a strong intramolecular S•••HN hydrogen bond. This effect allows the ground structures of $Na^+(Cys)(H_2O)_4$ and $Na^+(Pro)(H_2O)_4$ to be very similar, i.e., the first hydration shells are complete with four water molecules and the AAs are both zwitterionic. Similar trends are expected for other AAs with functionalized side chains.

BIRD has also been used to examine the first and second hydration energies of Li^+(Gln) and Na^+(Gln) [110, 111] and the first hydration energies of Li^+(Lys) [112]. In all cases, hydration is not found to affect the coordination of these metal cations to the amino acid, which retain charge-solvated structures in all cases. In contrast, IRMPD spectra of $M^+(Arg)(H_2O)$ where $M^+=Li^+$ and Na^+ indicate that the coordination of Li^+ changes from charge-solvated tridentate [N,CO,N] to a zwitterionic [N,CO^-] coordination, whereas Na^+ maintains its zwitterionic [CO_2^-] coordination upon hydration [113]. Kinetic studies indicate that a second water binds to the metal cation, whereas the third likely binds to the cationic guanidinium side chain.

When comparing the hydration energies from system to system, BDEs for losing water from $Na^+(Pro)(H_2O)_x$ are 9–12 kJ/mol smaller at each x compared to $Na^+(Gly)(H_2O)_x$ [102]. This result simply reflects the stronger binding of Na^+ to proline compared to glycine. This inverse correlation has been further substantiated by thermodynamic studies of additional AAs. Using high pressure equilibrium methods, Wincel measured hydration energies for $x=1-3$ of Na^+(AA) where AA = Val, Pro, Met, Phe, and Gln [114] (with good agreement with BIRD results for Val and Gln and reasonable agreement with the TCID studies for Pro). He has also documented a similar trend for hydration of K^+(AA) where AA = Gly, Ala, Val, Met, Pro, and Phe [115].

4 Alkali Metal Cations Interacting with Peptides

4.1 Structure

4.1.1 Di- and Tripeptides

The simplest peptides, di- and triglycine, complexed with sodium cations have been examined using theory and IRMPD spectroscopy [116, 117]. The IRMPD results clearly show that the metal cation coordinates with all carbonyl groups but cannot distinguish whether the N-terminal amine group is also coordinated, in agreement with theory which shows both structures are low-lying. Similar results are obtained for complexes of Li^+ [118], Na^+ [117], K^+ [118–120], and Cs^+ [118] with di- and trialanine. Theory indicates the same is also true for potassiated di- and triglycine [116].

When the peptides are functionalized, the side chain can also participate in the metal coordination. IRMPD results for PheAla and AlaPhe demonstrate that sodium and potassium cations coordinate with the phenyl ring of the side chain (SC) in either a [CO,CO,SC] or [N,CO,SC] geometry [121]. Similar coordination is obtained for the double-π system PhePhe bound to Li^+, Na^+, K^+, and Cs^+ [122], although alkaline earth dications move to an intercalating [CO,CO,SC,SC] coordination. Potassiated HisGly is similar with an apparent [CO,CO,SC] geometry being favored [123]. Comparison of the spectra of potassiated AlaAla, AlaAlaAla, PheAla, PhePhe, and PheGlyGly also reveal extensive similarities with four major peaks near 1750, 1650, 1520, and 1150 cm^{-1} [120]. These correspond to the C-terminal carboxylic carbonyl stretch mediated by metal coordination, the amide I (CO stretch of amide linkages) and amide II (NH bend plus CN stretch, also metal mediated) bands, and the free COH hydroxyl bending mode, respectively.

IRMPD results for GlyArg and ArgGly indicate more profound differences resulting from the sequence inversion [124, 125]. For M^+(GlyArg) where $M^+=Li^+$, Na^+, and Cs^+, the metal cation coordination is [CO,CO^-] where there is a salt-bridge between the carboxylate anion and protonated guanidine side chain, a geometry favored by a nearly linear arrangement of formal charge sites. In contrast, for M^+(ArgGly) where $M^+=Li^+$, Na^+, K^+, and Cs^+, the [CO,CO^-] salt-bridge geometry is maintained for the larger cations, whereas Li^+ has a [CO,CO,SC] coordination and Na^+ shows a mixture of this and the salt-bridge structures.

4.1.2 Larger Peptides

The Ohanessian group has systematically examined the structures of Na^+ complexes of polyglycine (Gly_n) for $n=2-8$ using IRMPD [117, 126, 127]. For the smaller peptides, $n=2-5$, all peptidic carbonyls coordinate to the metal cation in globular structures. At $n=6$, the peptide is sufficiently long that hydrogen bonding interactions between the termini may occur. Larger peptides can no longer tightly chelate the metal cation with all carbonyls, but rather coordinate with six carbonyls and stabilize the complex by hydrogen bonding interactions, mainly between the termini. Although calculations suggest the zwitterionic structure in which the proton has transferred is low in energy, the experimental spectra support the presence of only the charge solvated structure.

Dunbar et al. used IRMPD to explore the effect of the metal cation identity [128] on the conformations of pentaalanine (Ala_5), finding that Na^+ could bind five carbonyls whereas the larger K^+ and Cs^+ cations prefer structures in which fewer carbonyls coordinate the metal cation allowing the C-terminal carboxyl group to hydrogen bond (as identified by very weak carbonyl bands at 1750 cm^{-1}). This occurs because the weaker binding of these metal cations allows the intramolecular hydrogen bonding to become thermodynamically competitive and influence the coordination.

Larger peptides allow more extensive hydrogen bonding, leading to the familiar helices and sheets in the secondary structures of proteins. However, the conformational space that can be explored for these large peptides is sufficiently large that theoretical calculations are challenging. Nevertheless, IRMPD studies have examined the transition from the globular structures noted above to helices in the case of $Na^+(Ala_n)$ complexes for $n = 8 - 12$ [129] (see Figure 4). The spectrum for the octapeptide indicates that both globular and helical structures contribute, the nonapeptide also has features of both types, and the larger oligopeptides switch to the helical form. Theory indicates the helical form is the lowest in energy for all of these peptides with the globular structures having accessible relative free energies. These results agree with previous ion mobility measurements that demonstrate helical forms for Ala_n where $n = 12 - 20$ complexed with all five alkali cations [130, 131]. Deviations from helicity may occur for smaller peptides (especially lithiated and

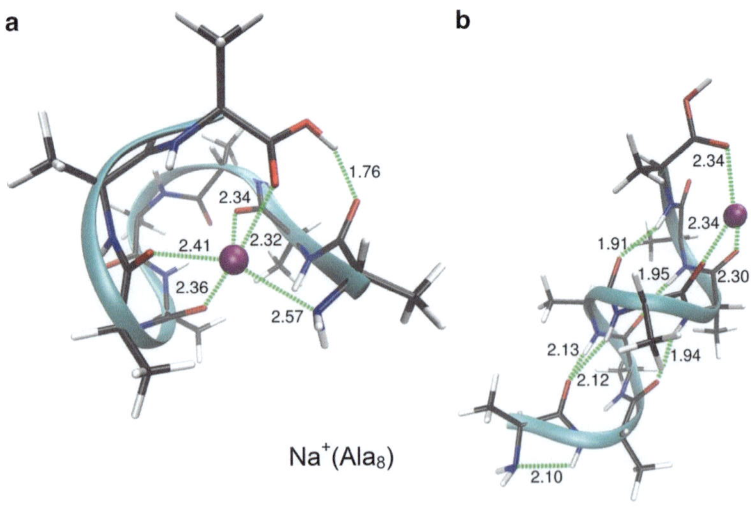

Figure 4 Structures of $Na^+(Ala_8)$ for the lowest energy globular structure (**a**) and α-helical structure (lowest energy, part **b**) as calculated at the M06/6-311+G(d,p)//M06/6-31 G(d,p) level of theory. Bond lengths (Å) for metal coordination and hydrogen bonds are indicated by green dashed lines. Reproduced from [129] with permission from the American Chemical Society; © copyright 2012.

sodiated) where experiments extend down to $n=6$, but the mobility measurements cannot definitively distinguish the two forms.

4.2 Thermodynamics

Thermodynamic studies of metal cations interacting with small peptides are less extensive than for isolated amino acids and IRMPD studies. At this writing, TCID work includes only Na^+ and K^+ bound to glycylglycine (Gly_2) and glycylglycylglycine (Gly_3) [116], which supercedes earlier studies by Klassen et al. [132]. The TCID values for the two sodiated peptides agree nicely with kinetic method results (Table 2), which also examined the sodium cation affinities of tetraglycine (Gly_4) as well as many other di- and tripeptides [133]. These experimental BDEs agree reasonably well with theoretical values [133–137] (see also [116]).

The experimental BDEs for sodiated and potassiated Gly, Gly_2, Gly_3, and Gly_4 increase with the size of the peptide, which can be related to the increasing number of carbonyl groups as well as the increasing polarizability of the peptide (Figure 5). As for any other ligand, the sodium cation affinities are greater than the potassium cation affinities. All levels of theory predict charge-solvated structures for M^+(peptide). As noted above, the metal cations prefer to bind in [CO,CO] bidentate or [N,CO,CO] tridentate configurations in $M^+(Gly_2)$, [CO,CO,CO] tridentate configuration in $M^+(Gly_3)$, and [CO,CO,CO,CO] or [N,CO,CO,CO,CO] structures in $M^+(Gly_4)$. Notably, the increasing BDE with increasing residues begins to level off at Gly_4. This is expected on the basis of decreasing charge density on Na^+ with increasing coordination number and with increased steric constraints [1].

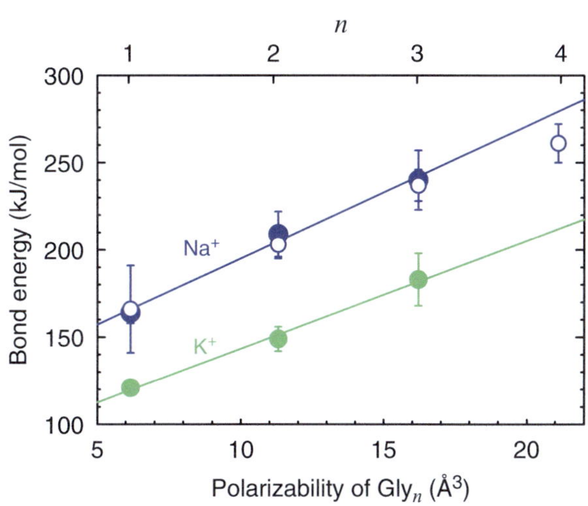

Figure 5 Experimental 0 K BDEs of $M^+(Gly_n)$ versus the theoretical molecular polarizability of Gly_n where n=1–4 and $M^+=Na^+$ and K^+ taken from [116] (solid symbols) and [133] (open symbols). Adapted from [116].

As for the amino acids (Figure 3), the BDEs correlate nicely with the polarizability of these peptides (Figure 4). Notably, this correlation differs from that shown in Figure 3, which can be appreciated by noting that the BDEs of Gly_2 and Asn (isomers of one another and having similar polarizabilities) are similar for both Na^+ and K^+. As for the side-chain enhancements in Figure 3, the local dipole moments of the coordinating carbonyl sites of the peptides must play a role (again consistent with the similarity between Gly_2 and Asn). Although theory does find that the local dipole moments of the peptides are aligned with the metal ion in these complexes [116, 137], there is no direct correlation between the BDEs and calculated dipole moments of Gly, Gly_2, and Gly_3 in their ground complexed geometries (3.2, 5.8, and 5.2 D for Na^+, respectively, and 6.2, 5.9, and 7.8 D for K^+, respectively [116]). Not surprisingly, this indicates the cation affinities of complex species like peptides must involve a superposition of electrostatic and steric effects.

Using the kinetic method, sodium cation binding energies to nineteen peptides containing 2 – 4 residues have been determined and compared favorably to theoretical values [133]. These studies conclude that sodium cation coordination involves all available carbonyls in the most stable structures and that functionalized side chains augment the binding as well. In a comparison of three isomeric pairs of dipeptides (GlyPhe and PheGly, AlaTrp and TrpAla, and GlyHis and HisGly), locating the functionalized side chain at the N-terminus augments the cation affinity more than when it is at the C-terminus by 3 – 12 kJ/mol.

5 Alkali Metal Cations Interacting with Nucleobases

5.1 *Structure*

Because of the aromaticity of the heterocyclic rings of the nucleobases, there is much less conformational flexibility compared to the amino acids. Theoretical studies indicate that metal cation-nucleobase complexes are nearly planar, with π-bound complexes lying considerably higher in energy. For adenine (A), IRMPD spectroscopy near 3500 cm^{-1} indicates that Li^+, Na^+, K^+, and Cs^+ all bind to A at the N3 and N9 positions [138] (see Figure 6 for the nomenclature used). This tautomer, which differs from the canonical form shown in Figure 6 by moving the H9 hydrogen to N7, is less relevant biologically because the sugar attaches to N9 in nucleosides and nucleotides. Addition of a single water left this binding mode unchanged. In the case of cytosine (C), IRMPD spectroscopy of complexes with all five alkali metal cations yield similar spectra, although the Rb^+ and Cs^+ complexes are richer in spectral details [139]. Although small changes in the positions and shape of the bands are observed to correlate with the size of the metal cation, all five spectra can be explained by a single isomer in which the cation binds to the O2 and N3 atoms of the canonical cytosine structure shown in Figure 6, in agreement with computations.

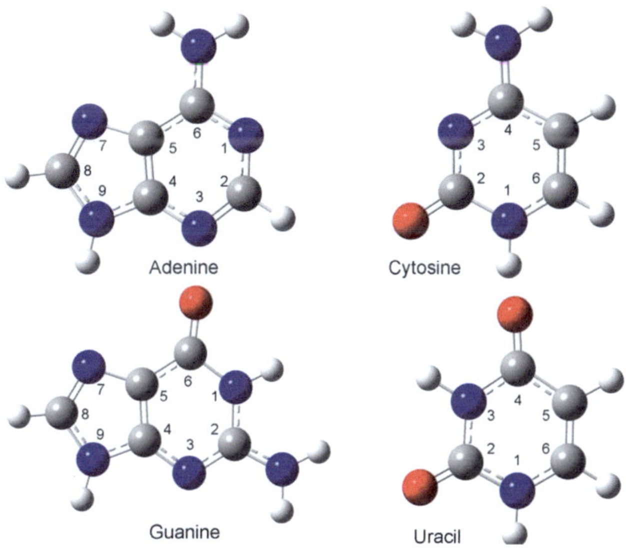

Figure 6 Structures of adenine, cytosine, guanine, and uracil.

In the case of uracil (U), IRMPD spectroscopy unambiguously shows that Li^+ and Na^+ bind at the O4 carbonyl [140, 141]. For thymine (T = 5-methyluracil = m^5U), addition of water, or addition of another base (U or T) to the lithiated complex leaves this binding motif unchanged. If sulfur is substituted at the 2 position (S^2U) without or with methyl substitutions at the 5 and 6 positions (m^5S^2U and m^6S^2U), this binding mode is also retained. In contrast, substitution at the 4 position (in both S^4U and $S^{24}{}_2U$) induces tautomization in which the N3 hydrogen moves to S4 and Na^+ binds at the O2N3 (or S2N3) positions. Likewise, the sodium cation complexes of 5-halogenated uracil retain the O4 binding site for all four halogens (F, Cl, Br, and I), with chelation to the halogen substituent providing additional stabilization [142]. In contrast, $Na^+(cl^6U)$ is found to have multiple isomers present, although the O4 binding site remains a major contributor. For both the U and C studies, various tautomeric structures were also carefully considered in these studies, however, the binding sites observed are biologically relevant because the sugar in nucleosides and nucleotides binds at N1 of uracil and cytosine.

For larger complexes, the number of studies is limited. The effect of sodiation on the conformers of all possible dinucleotides (single-strands, duplexes, and triplexes) was examined using ion mobility [143]. Three types of structural motifs were found: base stacking, perpendicular bases, or coplanar bases with small barriers (4 – 8 kJ/mol) between them. Duplexes and triplexes showed considerably less structural diversity than the single-strand complexes with only one conformer observed for all duplexes except dTG, dGT, and dTT and for all triplexes (where dTG, dGT, and dTT are the DNA dinucleotides possessing T and G = guanine nucleobase residues).

5.2 Thermodynamics

Early TCID studies of M^+(nucleobase) where $M^+ = Li^+$, Na^+, and K^+ and nucleobase = U, T, and A determined the metal cation binding energies for all nine systems [144] (Table 3), which agree reasonably well with theory and also with prior kinetic method results [145] except for Na^+(A). The latter work also includes guanine (G) (Table 3). Subsequent TCID work has extended this work to nucleobase = C [146], where the value for Na^+(C) is again in agreement with kinetic method results [147]. This study also includes the first (and as yet only) extension to Rb^+ and Cs^+ complexes, finding bond energies of 148.0 ± 3.6 and 137.2 ± 4.3 kJ/mol, respectively.

As for amino acids and peptides, the BDEs decrease as the metal cation gets larger. Comparison of experimental BDEs with theoretical results (along with the IRMPD results noted above for U and C) show that the metal cations bind preferentially at the N7 site coupled with chelation to the amino group for A, at the O2 carbonyl and N3

Table 3 Bond dissociation energies (kJ/mol) at 0 K of alkali metal cations with nucleic acid bases and their sites of attachment (SA).[a]

Nucleobase	SA	Li^+	Na^+	K^+
A[b]	N7NH$_2$	226.1 (6.1)	139.6 (4.2)	95.1 (3.2)
C	O2N3[c]		209.5 (5.0)	161.5 (4.5)
	N1O2[d]	235.2 (7.0)	177.6 (5.4)	
G[e]	O6N7	239 (8)	182 (8)	117 (8)
U[b]	O4	211.5 (6.1)	134.6 (3.4)	104.3 (2.8)
m^{1}U[f]	O4	234.0 (7.2)	150.7 (4.1)	110.9 (2.7)
m^{3}U[f]	O4	220.9 (6.7)	143.6 (3.8)	107.5 (3.3)
m^{5}U (= T)[b]	O4	210.1 (7.0)	135.3 (3.8)	104.0 (3.8)
m^{6}U[f]	O4	222.3 (6.6)	136.6 (5.8)	108.8 (5.4)
m$^{13}{}_2$U[f]	O4	239.8 (7.8)	153.6 (4.7)	118.9 (3.3)
m$^{56}{}_2$U[f]	O4	233.8 (6.5)	136.8 (5.1)	113.2 (3.2)
fl^{5}U[g]	O4F5	198.9 (4.8)	149.0 (4.3)	110.2 (3.8)
cl^{5}U[g]	O4Cl5	242.9 (8.2)	141.4 (3.4)	104.1 (2.7)
cl^{6}U[g]	O4	229.6 (7.8)	139.5 (4.4)	97.8 (2.2)
br^{5}U[g]	O4Br5	235.9 (5.3)	142.3 (4.8)	109.8 (2.3)
io^{5}U[g]	O4I5		147.5 (5.7)	117.6 (4.7)
S^{2}U[h]	O4	216.2 (5.7)	139.8 (3.3)	103.2 (2.6)
m^{5}S^{2}U[h]	O4	213.3 (5.5)	142.3 (5.5)	101.2 (2.8)
m^{6}S^{2}U[h]	O4	222.0 (6.2)	143.6 (3.8)	106.6 (3.5)
S^{4}U[h]	O2	213.1 (5.0)	125.8 (4.7)	97.3 (5.5)
S$^{24}{}_2$U[h]	S4	175.1 (4.5)	100.2 (5.8)	80.8 (2.8)[i]

[a]Uncertainties in parentheses. A = adenine, C = cytosine, G = guanine, U = uracil, m = methyl, fl = fluorine, cl = chlorine, br = bromine, io = iodine, S = sulfur. Superscripts indicate the position of the substitution.

[b][144]. [c][146]. [d][149]. [e][145] kinetic method. [f][151]. [g][152]. [h][153].

site of C, at the O6 carbonyl and N7 site of G, and at the O4 carbonyl of U and T. Note that the assignment for A differs from that from the IRMPD work [138], which can be attributed to the different ion sources used to produce the metalized nucleobase. (Interestingly, as originally suggested on the basis of theoretical work [148], earlier TCID [149] and kinetic method work [147] using different ion source methodology appears to have measured thermochemistry for excited isomers of the cytosine complexes, in which the metal cations bind at the N1O2 position, which requires tautomerization from the lowest energy isomer shown in Figure 6. Because the sugar attaches at N1, this isomer is less relevant to biological systems.)

The bidentating binding of C provides the strongest metal cation BDEs. G is also bidentate, but the geometry of the binding sites requires longer metal ligand bonds. A is also bidentate but requires rotation of the amino group in order to chelate and this is energetically costly because of the loss of π-resonance energy. The monodentate M^+(U) and M^+(T) complexes are the weakest binders for Li^+ and Na^+, whereas K^+ binds A most weakly. Notably, these interactions are sufficiently large to potentially disrupt hydrogen bonding in A::T (A::U) nucleic acid base pairs.

A number of studies have examined the effects of modification on these BDEs. Such modified nucleobases are relatively prominent in transfer RNAs, with methylation being most common [150], and can exhibit both therapeutic and harmful effects. The kinetic method has been used to show that Na^+ binds 1- and 5-methyl cytosine more tightly than unsubstituted cytosine by 10 and 5 kJ/mol, respectively [147]. A more extensive series of studies on modified uracils have been conducted using the TCID approach. Methylation at the 5, 6, and 5 and 6 positions has a relatively small effect on the BDEs with Li^+, Na^+, and K^+ (Table 3), whereas methylation at N3 increases the BDE by 5 ± 2 %, at N1 (where the sugar attaches) by 10 ± 2 %, and at both sites by 14 ± 1 % [151]. In contrast, halogenation at the 5 position increases the alkali metal cation BDEs, in part because the halogen can also chelate with the metal [152]. Halogenation at the 6 position, where it can no longer chelate, introduces a much smaller effect (actually decreasing the BDE for K^+). Notably, because halogenation is found to decrease the proton affinity of U, such substitutions should stabilize A::U base pairs, which may provide some insight into their use as antitumor and antiviral agents. Thioketo substitution at the 2 position slightly increases the metal cation BDEs, whereas 4-thioketo U decreases the BDE, especially if both positions are exchanged for sulfur [153]. The latter substitution is predicted to have a marked effect at destabilizing A::SU base pairs.

The thermodynamics of hydration of sodiated and potassiated complexes of A, C, U, and T by one and two waters have also been measured using equilibrium methods [154]. In concert with the results above, these results suggest that tautomerization of cytosine is important in determining the distribution of complexes formed by electrospray ionization. In the case of thiouracils, substitution at the 2 or 4 position has little effect on the first and second hydration energy of the sodiated complexes [155]. Results appear to be consistent with direct binding of water to the sodium cation without disruption of the alkali metal cation-nucleobase structures noted above.

6 Concluding Remarks and Future Directions

Gas-phase studies in combination with computational work are capable of providing considerable detail information regarding how alkali metal cations interact with biological systems ranging in size from simple model compounds through the constituents of proteins and nucleic acids up to peptides and nucleotides. The strength of these approaches are the quantitative structural and thermodynamic information that is afforded. Their weakness is the considerable projection that is needed to apply these data to real biological systems.

However, the exploration of such larger complexes is a very active area of research that promises to bridge the gap between the small molecule chemistry reviewed here and real systems. Such studies will not only directly yield more quantitative information about how secondary structure influences their properties and function, but should allow the lessons learned from small molecules to translate more effectively to the large complex systems.

Abbreviations and Definitions

A	adenine
AA	amino acid
Arg	arginine
Asn	asparagine
Asp	aspartic acid
BDE	bond dissociation energy
BIRD	blackbody infrared radiative dissociation
BSSE	basis set superposition error
C	cytosine
CA	collisional activation
CAD	collision activated dissociation
CID	collision-induced dissociation
CLIO	Centre Laser Infrarouge d'Orsay
Cys	cysteine
dGT	2'-deoxy-guanine-thymine dinucleotide
dTG	2'-deoxy-thymine-guanine dinucleotide
dTT	2'-deoxy-thymine dinucleotide
ECP	effective core potential
EMSL	Environmental Molecular Sciences Laboratory
FEL	free electron laser
FELICE	Free Electron Laser for IntraCavity Experiments
FELIX	Free Electron Lasers for Infrared eXperiments
G	guanine
GIBMS	guided ion beam tandem mass spectrometer

Gln	glutamine
Glu	glutamic acid
Gly	glycine
His	histidine
IR	infrared
IRMPD	infrared multiple photon dissociation
IRPD	infrared photodissociation
IVR	intramolecular vibrational redistribution
Lys	lysine
Met	methionine
MS	mass spectrometer or mass spectrometry
Phe	phenylalanine
Pro	proline
QQQ	triple quadrupole mass spectrometer
REMD	replica exchange molecular dynamics
SC	side chain
Ser	serine
T	thymine
TCID	threshold collision-induced dissociation
Thr	threonine
Trp	tryptophan
Tyr	tyrosine
U	uracil

Acknowledgment We would like to thank the many students that contributed to this work for their dedication, insight, and hard work. The research discussed in this chapter has been supported by the National Science Foundation, Grants No. CHE-1409420 (MTR), CHE-1359769 (PBA), and OISE-0730072.

References

1. M. T. Rodgers, P. B. Armentrout, *Acc. Chem. Res.* **2004**, *37*, 989–998.
2. K. K. Lehmann, G. Scoles, B. H. Pate, *Annu. Rev. Phys. Chem.* **1994**, *45*, 241–274.
3. A. Beil, D. Luckhaus, M. Quack, J. Stohner, *Ber. Bunsen Phys. Chem.* **1997**, *101*, 311–328.
4. J. G. Black, E. Yablonovitch, N. Bloembergen, S. Mukamel, *Phys. Rev. Lett.* **1977**, *38*, 1131–1134.
5. E. Grant, P. Schulz, A. S. Sudbo, Y. Shen, Y. T. Lee, *Phys. Rev. Lett.* **1978**, *40*, 115–118.
6. V. J. F. Lapoutre, B. Redlich, A. F. G. van der Meer, J. Oomens, J. M. Bakker, A. Sweeney, A. Mookherjee, P. B. Armentrout, *J. Phys Chem. A* **2013**, *117*, 4115–4126.
7. A. B. McCoy, X. Huang, S. Carter, M. Y. Landeweer, J. M. Bowman, *J. Chem. Phys.* **2005**, *122*, 061101.
8. C. M. Leavitt, A. F. DeBlase, C. J. Johnson, M. van Stipdonk, A. B. McCoy, M. A. Johnson, *J. Phys. Chem. Lett.* **2013**, *4*, 3450–3457.
9. X. Cheng, R. P. Steele, *J. Chem. Phys.* **2014**, *141*, 104–105.
10. P. B. Armentrout, *J. Am. Soc. Mass Spectrom.* **2002**, *13*, 419–434.
11. K. M. Ervin, P. B. Armentrout, *J. Chem. Phys.* **1985**, *83*, 166–189.

12. D. A. Hales, L. Lian, P. B. Armentrout, *Int. J. Mass Spectrom. Ion Processes* **1990**, *102*, 269–301.
13. R. H. Schultz, K. C. Crellin, P. B. Armentrout, *J. Am. Chem. Soc.* **1991**, *113*, 8590–8601.
14. M. T. Rodgers, K. M. Ervin, P. B. Armentrout, *J. Chem. Phys.* **1997**, *106*, 4499–4508.
15. M. T. Rodgers, P. B. Armentrout, *J. Chem. Phys.* **1998**, *109*, 1787–1800.
16. P. B. Armentrout, *J. Chem. Phys.* **2007**, *126*, 234302.
17. P. B. Armentrout, K. M. Ervin, M. T. Rodgers, *J. Phys. Chem. A* **2008**, *112*, 10071–10085.
18. N. F. Dalleska, K. Honma, P. B. Armentrout, *J. Am. Chem. Soc.* **1993**, *115*, 12125–12131.
19. N. F. Dalleska, K. Honma, L. S. Sunderlin, P. B. Armentrout, *J. Am. Chem. Soc.* **1994**, *116*, 3519–3528.
20. N. F. Dalleska, B. L. Tjelta, P. B. Armentrout, *J. Phys. Chem* **1994**, *98*, 4191–4195.
21. M. T. Rodgers, P. B. Armentrout, *J. Phys. Chem. A* **1997**, *101*, 1238–1249.
22. M. T. Rodgers, P. B. Armentrout, *J. Phys. Chem. A* **1997**, *101*, 2614–2625.
23. C. Iceman, P. B. Armentrout, *Int. J. Mass Spectrom.* **2003**, *222*, 329–349.
24. J. C. Amicangelo, P. B. Armentrout, *J. Phys. Chem. A* **2004**, *108*, 10698–10713.
25. A. L. Heaton, P. B. Armentrout, *J. Am. Chem. Soc.* **2008**, *130*, 10227–10232.
26. A. L. Heaton, S. J. Ye, P. B. Armentrout, *J. Phys. Chem. A* **2008**, *112*, 3328–3338.
27. S. J. Ye, P. B. Armentrout, *J. Phys. Chem. B* **2008**, *112*, 10303–10313.
28. P. B. Armentrout, A. L. Heaton, *J. Am. Soc. Mass Spectrom.* **2012**, *23*, 632–643.
29. P. B. Armentrout, A. A. Clark, *Int. J. Mass Spectrom.* **2012**, *316–318*, 182–191.
30. P. B. Armentrout, E. M. S. Stennett, *J. Am. Soc. Mass Spectrom.* **2014**, *25*, 512–523.
31. U. H. Verkerk, C.-K. Siu, J. D. Steill, H. El Aribi, J. Zhao, C. F. Rodriguez, J. Oomens, A. C. Hopkinson, K. W. M. Siu, *J. Phys. Chem. Lett.* **2010**, *1*, 868–872.
32. P. Kebarle, in *Encyclopedia of Mass Spectrometry*, Ed. P. B. Armentrout, Elsevier, Amsterdam, 2003, Vol. 1, pp. 319–338.
33. S. A. McLuckey, D. Cameron, R. G. Cooks, *J. Am. Chem. Soc.* **1981**, *103*, 1313–1317.
34. R. G. Cooks, P. H. Wong, *Acc. Chem. Res.* **1998**, *31*, 379–386.
35. X. Cheng, Z. Wu, C. Fenselau, *J. Am. Chem. Soc.* **1993**, *115*, 4844–4848.
36. P. B. Armentrout, *J. Am. Soc. Mass Spectrom.* **2000**, *11*, 371–379.
37. P. B. Armentrout, *J. Mass Spectrom.* **1999**, *34*, 74–78.
38. L. Drahos, K. Vekey, *J. Mass Spectrom.* **1999**, *34*, 79–84.
39. R. G. Cooks, J. T. Koskinen, P. D. Thomas, *J. Mass Spectrom.* **1999**, *34*, 85–92.
40. W. D. Price, P. D. Schnier, E. R. Williams, *Anal. Chem.* **1996**, *68*, 859–866.
41. R. C. Dunbar, T. B. McMahon, D. Thoelmann, D. S. Tonner, D. R. Salahub, D. Wei, *J. Am. Chem. Soc.* **1995**, *117*, 12819–12825.
42. S. Lee, T. Wyttenbach, M. T. Bowers, *Int. J. Mass Spectrom.* **1997**, *167/168*, 605–614.
43. C. S. Hoaglund, S. J. Valentine, D. E. Clemmer, *Anal. Chem.* **1997**, *69*, 4156–4161.
44. A. B. Kanu, P. Dwivedi, M. Tam, L. Matz, H. H. Hill, *J. Mass Spectrom.* **2008**, *43*, 1–22.
45. S. F. Boys, R. Bernardi, *Mol. Phys.* **1970**, *19*, 553–566.
46. F. B. van Duijneveldt, J. G. C. M. van Duijneveldt-van de Rijdt, J. H. van Lenthe, *Chem. Rev.* **1994**, *94*, 1873–1885.
47. E. D. Glendening, D. Feller, M. A. Thompson, *J. Am. Chem. Soc.* **1994**, *116*, 10657–10669.
48. P. J. Hay, W. R. Wadt, *J. Chem. Phys.* **1985**, *82*, 284–298.
49. V. N. Bowman, A. L. Heaton, P. B. Armentrout, *J. Phys. Chem. B* **2010**, *114*, 4107–4114.
50. P. B. Armentrout, Y. Chen, M. T. Rodgers, *J. Phys. Chem. A* **2012**, *116*, 3989–3999.
51. T. Leininger, A. Nicklass, W. Kuechle, H. Stoll, M. Dolg, A. Bergner, *Chem. Phys. Lett.* **1996**, *255*, 274–280.
52. M. T. Rodgers, P. B. Armentrout, *Int. J. Mass Spectrom.* **2007**, *267*, 167–182.
53. D. E. Woon, T. H. Dunning, Jr., *J. Chem. Phys.* **1995**, *103*, 4572–4585.
54. K. L. Schuchardt, B. T. Didier, T. Elsethagen, L. Sun, V. Gurumoorthi, J. Chase, J. Li, T. L. Windus, *J. Chem. Inf. Model.* **2007**, *47*, 1045–1052.
55. D. Feller, *J. Comput. Chem.* **1996**, *17*, 1571–1586.

56. S. M. Smith, A. N. Markevitch, D. A. Romanor, X. Li, R. J. Levis, H. B. Schlegel, *J. Phys. Chem. A* **2000**, *108*, 11063–11072.
57. D. Semrouni, C. Clavaguera, J. P. Dognon, G. Ohanessian, *Int. J. Mass Spectrom.* **2010**, *297*, 152–161.
58. N. V. Buchete, G. Hummer, *Phys. Rev. E* **2008**, *77*, 030902.
59. C. Bleiholder, T. Wyttenbach, M. T. Bowers, *Int. J. Mass Spectrom.* **2011**, *308*, 1–10.
60. T. Wyttenbach, G. v. Helden, J. J. Batka, D. Carlat, M. T. Bowers, *J. Am. Soc. Mass Spectrom.* **1997**, *8*, 275–282.
61. M. F. Mesleh, J. M. Hunter, A. A. Shvartsburg, G. C. Schatz, M. F. Jarrold, *J. Phys. Chem. A* **1996**, *100*, 16082–16086.
62. A. A. Shvartsburg, M. F. Jarrold, *Chem. Phys. Lett.* **1996**, *261*, 86–91.
63. C. Kapota, J. Lemaire, P. Maitre, G. Ohanessian, *J. Am. Chem. Soc.* **2004**, *126*, 1836–1842.
64. N. C. Polfer, B. Paizs, L. C. Snoek, I. Compagnon, S. Suhai, G. Meijer, G. von Helden, J. Oomens, *J. Am. Chem. Soc.* **2005**, *127*, 8571–8579.
65. N. C. Polfer, J. Oomens, R. C. Dunbar, *Phys. Chem. Chem. Phys.* **2006**, *8*, 2744–2751.
66. M. F. Bush, J. T. O'Brien, J. S. Prell, R. J. Saykally, E. R. Williams, *J. Am. Chem. Soc.* **2007**, *129*, 1612–1622.
67. M. W. Forbes, M. F. Bush, N. C. Polfer, J. Oomens, R. C. Dunbar, E. R. Williams, R. A. Jockusch, *J. Phys. Chem. A* **2007**, *111*, 11759–11770.
68. M. F. Bush, M. W. Forbes, R. A. Jockusch, J. Oomens, N. C. Polfer, R. J. Saykally, E. R. Williams, *J. Phys. Chem. A* **2007**, *111*, 7753–7760.
69. M. K. Drayss, D. Blunk, J. Oomens, M. Schäfer, *J. Phys Chem. A* **2008**, *112*, 11972–11974.
70. J. T. O'Brien, J. S. Prell, J. D. Steill, J. Oomens, E. R. Williams, *J. Phys. Chem. A* **2008**, *112*, 10823–10830.
71. P. B. Armentrout, M. T. Rodgers, J. Oomens, J. D. Steill, *J. Phys. Chem. A* **2008**, *112*, 2248–2257.
72. M. T. Rodgers, P. B. Armentrout, J. Oomens, J. D. Steill, *J. Phys. Chem. A* **2008**, *112*, 2258–2267.
73. M. F. Bush, J. Oomens, R. J. Saykally, E. R. Williams, *J. Phys. Chem. A* **2008**, *112*, 8578–8584.
74. R. C. Dunbar, A. C. Hopkinson, J. Oomens, C.-K. Siu, K. W. M. Siu, J. D. Steill, U. H. Verkerk, J. Zhao, *J. Phys. Chem. B* **2009**, *113*, 10403–10408.
75. A. L. Heaton, V. N. Bowman, J. Oomens, J. D. Steill, P. B. Armentrout, *J. Phys. Chem. A* **2009**, *113*, 5519–5530.
76. D. R. Carl, T. E. Cooper, J. Oomens, J. D. Steill, P. B. Armentrout, *Phys. Chem. Chem. Phys.* **2010**, *12*, 3384–3398.
77. R. C. Dunbar, J. D. Steill, J. Oomens, *Phys. Chem. Chem. Phys.* **2010**, *12*, 13383–13393.
78. M. K. Drayss, P. B. Armentrout, J. Oomens, M. Schäfer, *Int. J. Mass Spectrom.* **2010**, *297*, 18–27.
79. M. Citir, E. M. S. Stennett, J. Oomens, J. D. Steill, M. T. Rodgers, P. B. Armentrout, *Int. J. Mass Spectrom.* **2010**, *297*, 9–17.
80. M. Citir, C. S. Hinton, J. Oomens, J. D. Steill, P. B. Armentrout, *J. Phys. Chem. A* **2012**, *116*, 1532–1541.
81. P. B. Armentrout, B. Yang, M. T. Rodgers, *J. Phys. Chem. B* **2014**, *118*, 4300–4314.
82. A. A. Clark, B. Yang, R. R. Wu, M. T. Rodgers, P. B. Armentrout, work in progress.
83. R. M. Moision, P. B. Armentrout, *J. Phys. Chem. A* **2006**, *110*, 3933–3946.
84. N. C. Polfer, R. C. Dunbar, J. Oomens, *J. Am. Soc. Mass Spectrom.* **2007**, *18*, 512–516.
85. R. M. Moision, P. B. Armentrout, *J. Phys. Chem. A* **2002**, *106*, 10350–10362.
86. R. M. Moision, P. B. Armentrout, *Phys. Chem. Chem. Phys.* **2004**, *6*, 2588–2599.
87. A. L. Heaton, P. B. Armentrout, *J. Phys. Chem. B* **2008**, *112*, 12056–12065.
88. A. L. Heaton, R. M. Moision, P. B. Armentrout, *J. Phys. Chem. A* **2008**, *112*, 3319–3327.
89. P. B. Armentrout, E. I. Armentrout, A. A. Clark, T. E. Cooper, E. M. S. Stennett, D. R. Carl, *J. Phys. Chem. B* **2010**, *114*, 3927–3937.

90. P. B. Armentrout, M. Citir, Y. Chen, M. T. Rodgers, *J. Phys. Chem. A* **2012**, *116*, 11823–11832.
91. P. B. Armentrout, B. Yang, M. T. Rodgers, *J. Phys. Chem. A* **2013**, *117*, 3771–3781.
92. P. B. Armentrout, A. Gabriel, R. M. Moision, *Int. J. Mass Spectrom.* **2009**, *283*, 56–68.
93. C. Ruan, M. T. Rodgers, *J. Am. Chem. Soc.* **2004**, *126*, 14600–14610.
94. A. Mookherjee, P. B. Armentrout, *Int. J. Mass Spectrom.* **2014**, *370*, 16–28.
95. A. Mookherjee, P. B. Armentrout, *Int. J. Mass Spectrom.* **2013**, *345–347*, 109–119.
96. S. J. Ye, A. A. Clark, P. B. Armentrout, *J. Phys. Chem. B* **2008**, *112*, 10291–10302.
97. M. M. Kish, G. Ohanessian, C. Wesdemiotis, *Int. J. Mass Spectrom.* **2003**, *227*, 509–524.
98. P. Wang, G. Ohanessian, C. Wesdemiotis, *Int. J. Mass Spectrom.* **2008**, *269*, 34–45.
99. R. G. Wilson, G. R. Brewer, *Ion Beams with Applications to Ion Implantation*, Wiley, New York, 1973.
100. *CRC Handbook of Chemistry and Physics*, Ed D. R. Lide, CRC Press, Boca Raton, 2002.
101. S. J. Ye, R. M. Moision, P. B. Armentrout, *Int. J. Mass Spectrom.* **2005**, *240*, 233–248.
102. S. J. Ye, R. M. Moision, P. B. Armentrout, *Int. J. Mass Spectrom.* **2006**, *253*, 288–304.
103. S. J. Ye, P. B. Armentrout, *Phys. Chem. Chem. Phys.* **2010**, *12*, 13419–13433.
104. R. A. Jockusch, A. S. Lemoff, E. R. Williams, *J. Phys. Chem. A* **2001**, *105*, 10929–10942.
105. R. A. Jockusch, A. S. Lemoff, E. R. Williams, *J. Am. Chem. Soc.* **2001**, *123*, 12255–12265.
106. A. S. Lemoff, M. F. Bush, E. R. Williams, *J. Am. Chem. Soc.* **2003**, *125*, 13576–13584.
107. A. S. Lemoff, E. R. Williams, *J. Am. Soc. Mass Spectrom.* **2004**, *15*, 1014–1024.
108. A. Kamariotis, O. V. Boyarkin, S. R. Mercier, R. D. Beck, M. F. Bush, E. R. Williams, T. R. Rizzo, *J. Am. Chem. Soc.* **2006**, *128*, 905–916.
109. A. S. Lemoff, M. F. Bush, E. R. Williams, *J. Phys. Chem. A* **2005**, *109*, 1903–1910.
110. A. S. Lemoff, M. F. Bush, C.-C. Wu, E. R. Williams, *J. Am. Chem. Soc.* **2005**, *127*, 10276–10286.
111. A. S. Lemoff, C.-C. Wu, M. F. Bush, E. R. Williams, *J. Phys. Chem. A* **2006**, *110*, 3662–3669.
112. A. S. Lemoff, M. F. Bush, J. T. O'Brien, E. R. Williams, *J. Phys. Chem. A* **2006**, *110*, 8433–8442.
113. M. F. Bush, J. S. Prell, R. J. Saykally, E. R. Williams, *J. Am. Chem. Soc.* **2007**, *129*, 13544–13553.
114. H. Wincel, *J. Phys. Chem. A* **2007**, *111*, 5784–5791.
115. H. Wincel, *J. Am. Soc. Mass Spectrom.* **2007**, *18*, 2083–2089.
116. S. J. Ye, P. B. Armentrout, *J. Phys. Chem. A* **2008**, *112*, 3587–3596.
117. O. P. Balaj, C. Kapota, J. Lemaire, G. Ohanessian, *Int. J. Mass Spectrom.* **2008**, *269*, 196–209.
118. R. C. Dunbar, J. D. Steill, J. Oomens, *Int. J. Mass Spectrom.* **2010**, *297*, 107–115.
119. R. C. Dunbar, J. Steill, N. C. Polfer, J. Oomens, *J. Phys. Chem. B* **2009**, *113*, 10552–10554.
120. R. C. Dunbar, N. C. Polfer, G. Berden, J. Oomens, *Int. J. Mass Spectrom.* **2012**, *330–332*, 71–77.
121. N. C. Polfer, J. Oomens, R. C. Dunbar, *ChemPhysChem* **2008**, *9*, 579–589.
122. R. C. Dunbar, J. D. Steill, J. Oomens, *J. Am. Chem. Soc.* **2011**, *133*, 9376–9386.
123. R. C. Dunbar, J. Oomens, G. Berden, J. K. C. Lau, U. H. Verkerk, A. C. Hopkinson, K. W. M. Siu, *J. Phys. Chem. A* **2013**, *117*, 5335–5343.
124. J. S. Prell, M. Demireva, J. Oomens, E. R. Williams, *J. Am. Chem. Soc.* **2009**, *131*, 1232–1242.
125. J. S. Prell, T. M. Chang, J. A. Biles, G. Berden, J. Oomens, E. R. Williams, *J. Phys. Chem. A* **2011**, *115*, 2745–2751.
126. O. P. Balaj, D. Semrouni, V. Steinmetz, E. Nicol, C. Clavaguera, G. Ohanessian, *Chem.-Eur. J.* **2012**, *18*, 4583–4592.
127. D. Semrouni, O. P. Balaj, F. Calvo, C. F. Correia, C. Clavaguéra, G. Ohanessian, *J. Am. Soc. Mass Spectrom.* **2010**, *21*, 728–738.
128. R. C. Dunbar, J. D. Steill, N. C. Polfer, J. Oomens, *J. Phys. Chem. A* **2012**, *117*, 1094–1101.

129. J. K. Martens, I. Compagnon, E. Nicol, T. B. McMahon, C. Clavaguera, G. Ohanessian, *J. Phys. Chem. Lett.* **2012**, *3*, 3320–3324.
130. M. Kohtani, B. S. Kinnear, M. F. Jarrold, *J. Am. Chem. Soc.* **2000**, *122*, 12377–12378.
131. M. Kohtani, M. F. Jarrold, S. Wee, R. A. J. O'Hair, *J. Phys. Chem. B* **2004**, *108*, 6093–6097.
132. J. S. Klassen, S. G. Anderson, A. T. Blades, P. Kebarle, *J. Phys. Chem.* **1996**, *100*, 14218–14227.
133. P. Wang, C. Wesdemiotis, C. Kapota, G. Ohanessian, *J. Am. Soc. Mass Spectrom.* **2007**, *18*, 541–552.
134. M. Benzakour, M. Mcharfi, A. Cartier, A. Daoudi, *J. Mol. Struct. Theochem.* **2004**, *710*, 169–174.
135. B. A. Cerda, S. Hoyau, G. Ohanessian, C. Wesdemiotis, *J. Am. Chem. Soc.* **1998**, *120*, 2437–2448.
136. M. M. Kish, C. Wesdemiotis, G. Ohanessian, *J. Phys. Chem. B* **2004**, *108*, 3086–3091.
137. C. H. S. Wong, N. L. Ma, C. W. Tsang, *Chem. Eur. J.* **2002**, *8*, 4909–4918.
138. K. Rajabi, E. A. L. Gillis, T. D. Fridgen, *J. Phys. Chem. A* **2010**, *114*, 3449–3456.
139. B. Yang, R. R. Wu, N. C. Polfer, G. Berden, J. Oomens, M. T. Rodgers, *J. Am. Soc. Mass Spectrom.* **2013**, *24*, 1523–1533.
140. E. A. L. Gillis, K. Rajabi, T. D. Fridgen, *J. Phys. Chem. A* **2009**, *113*, 824–832.
141. Y.-w. Nei, T. E. Akinyemi, C. M. Kaczan, J. D. Steill, G. Berden, J. Oomens, M. T. Rodgers, *Int. J. Mass Spectrom.* **2011**, *308*, 191–202.
142. C. M. Kaczan, A. I. Rathur, R. R. Wu, Y. Chen, C. A. Austin, G. Berden, J. Oomens, M. T. Rodgers, *Int. J. Mass Spectrom.* **2015**, *378*, 76–85.
143. E. Shammel Baker, J. Gidden, A. Ferzoco, M. T. Bowers, *Phys. Chem. Chem. Phys.* **2004**, *6*, 2786–2795.
144. M. T. Rodgers, P. B. Armentrout, *J. Am. Chem. Soc.* **2000**, *122*, 8548–8558.
145. B. A. Cerda, C. Wesdemiotis, *J. Am. Chem. Soc.* **1996**, *118*, 11884–11892.
146. B. Yang, M. T. Rodgers, *Phys. Chem. Chem. Phys.* **2014**, *16*, 16110–16120.
147. P. Wang, M. J. Polce, G. Ohanessian, C. Wesdemiotis, *J. Mass Spectrom.* **2008**, *43*, 485–494.
148. N. Russo, M. Toscano, A. Grand, *J. Phys. Chem. B* **2001**, *105*, 4735–4741.
149. Z. Yang, M. T. Rodgers, *Phys. Chem. Chem. Phys.* **2012**, *14*, 4517–4526.
150. A. L. Lehninger, *Biochemistry, The Molecular Basis of Cell Structure and Function*, Worth Publishers, Inc., New York, 1977.
151. Z. Yang, M. T. Rodgers, *Int. J. Mass Spectrom.* **2005**, *241*, 225–242.
152. Z. Yang, M. T. Rodgers, *J. Am. Chem. Soc.* **2004**, *126*, 16217–16226.
153. Z. Yang, M. T. Rodgers, *J. Phys. Chem. A* **2006**, *110*, 1455–1468.
154. H. Wincel, *J. Am. Soc. Mass Spectrom.* **2012**, *23*, 1479–1487.
155. H. Wincel, *J. Am. Soc. Mass Spectrom.* **2014**, *25*, 2134–2142.

Chapter 5
Alkali Metal Ion Complexes with Phosphates, Nucleotides, Amino Acids, and Related Ligands of Biological Relevance. Their Properties in Solution

Francesco Crea, Concetta De Stefano, Claudia Foti, Gabriele Lando, Demetrio Milea, and Silvio Sammartano

Contents

F. Crea • C. De Stefano • C. Foti • G. Lando • D. Milea • S. Sammartano (✉)
Dipartimento di Scienze Chimiche, Università di Messina,
Viale F. Stagno d'Alcontres, 31, I-98166 Messina, Italy
e-mail: ssammartano@unime.it

A. Sigel, H. Sigel, and R.K.O. Sigel (eds.), *The Alkali Metal Ions: Their Role for Life*,
Metal Ions in Life Sciences 16, DOI 10.1007/978-3-319-21756-7_5

Abstract Alkali metal ions play very important roles in all biological systems, some of them are essential for life. Their concentration depends on several physiological factors and is very variable. For example, sodium concentrations in human fluids vary from quite low (e.g., 8.2 mmol dm^{-3} in mature maternal milk) to high values (0.14 mol dm^{-3} in blood plasma). While many data on the concentration of Na^+ and K^+ in various fluids are available, the information on other alkali metal cations is scarce. Since many vital functions depend on the network of interactions occurring in various biofluids, this chapter reviews their complex formation with phosphates, nucleotides, amino acids, and related ligands of biological relevance. Literature data on this topic are quite rare if compared to other cations. Generally, the stability of alkali metal ion complexes of organic and inorganic ligands is rather low (usually log $K<2$) and depends on the charge of the ligand, owing to the ionic nature of the interactions. At the same time, the size of the cation is an important factor that influences the stability: very often, but not always (e.g., for sulfate), it follows the trend $Li^+>Na^+>K^+>Rb^+>Cs^+$. For example, for citrate it is: log $K_{ML}=0.88$, 0.80, 0.48, 0.38, and 0.13 at 25 °C and infinite dilution. Some considerations are made on the main aspects related to the difficulties in the determination of weak complexes. The importance of the alkali metal ion complexes was also studied in the light of modelling natural fluids and in the use of these cations as probes for different processes. Some empirical relationships are proposed for the dependence of the stability constants of Na^+ complexes on the ligand charge, as well as for correlations among log K values of NaL, KL or LiL species (L = generic ligand).

Keywords Empirical relationships • Formation constants • Models • Speciation • Thermodynamic parameters

Please cite as: *Met. Ions Life Sci.* 16 (2016) 133–166

1 Introduction

As already discussed in other chapters of this book, alkali metal ions play very important roles in biological systems and some of them are essential for life. Therefore, a detailed description in this chapter of all their functions would only generate redundant and probably incomplete information. On the contrary, here it is necessary to stress that all their activity takes place in "biological" (as well as other "natural") fluids, which are, from a chemico-physical point of view, multielectrolyte aqueous solutions in which a wide number of components (metal ions, low, medium, and high MW ligands, etc.) is dissolved or dispersed (this last aspect is outside of the scope of this chapter) [1–8]. As a consequence, the control of some typical properties of the aqueous solution, i.e., the "biological fluid", becomes essential for the control of vital functions: we can just cite electroneutrality, osmotic pressure, cell potential, and ionic strength, which are mainly maintained by sodium and potassium ions (but not only) [7, 8]. Alkali metal ions other than Na^+ and K^+ are usually "less considered", resulting in a considerable lower number of studies. They are generally limited to aspects related to the medicinal use of Li^+ and Cs^+ (including radiocesium) or on the possible toxicity of Rb^+, though some works support their possible "essentiality" [7, 9].

After a brief summary on some general aspects of alkali metal complex formation in biological fluids, including concentrations in different bio-fluids, this chapter will focus on the thermodynamic aspects related to complex formation of the main classes of inorganic and organic ligands of biological relevance, paying attention to the importance of complexes formed in modelling the speciation of various biological fluids. Most relevant literature data on the stability constants of alkali metal complexes of phosphates, nucleotides, amino acids, and related ligands of biological relevance are critically reviewed together with other thermodynamic parameters (ΔG, ΔH, ΔS), where available.

2 Alkali Metal Ion Complex Formation in Biological Fluids

An aspect that is generally neglected, especially by non-chemists, is that also alkali metal ions interact with both organic and inorganic ligands in biological fluids, forming complexes of medium to low stability, which, however, may significantly affect their speciation (and therefore, their activity), since the amount of species formed depends not only on the stability constant value, but on the metal (and ligand) concentrations as well. We just remind that the chemical speciation (i.e., the distribution of an element amongst defined chemical species in a system) influences and determines many biological properties of single compounds and whole systems [10]. As a consequence, species of low stability (weak species, like those usually formed by alkali metal ions) may be formed in higher percentage than stronger ones, if the concentration of ions involved in the formation reaction is high (like in the case of Na^+ and K^+ in biological fluids) [3–7]. This aspect is sometimes expressed by means of the "uptake factor" ($K_{ML} \cdot [M]$), which controls selectivity and concentration of certain ions at certain "sites" [7]. In this light, prior to any further consideration on alkali metal complex formation in biological fluids, it appears to be useful for the reader to recall some basic concepts.

2.1 Alkali Metal Concentration in Different Biofluids

The composition of a given fluid depends on several factors like, e.g., age, sex, particular conditions and/or pathologies, kind of organism (for the sake of simplicity we will only refer in this paragraph to "human" fluids). Also the concentration of alkali metal ions and other inorganic components in most fluids is very variable and affected by different physiological conditions. For example, considering Na^+ and K^+, the most abundant ions, their concentration in human urine depends on its daily flow (from total anuria 0 mL d^{-1} to severe polyuria up to 15 L d^{-1}), renal functionality, age, sex, fast time, daily intake, pathologies [2]: their excretion via urine ranges from 80 and 40 to 560 and 100 mmol d^{-1}, respectively. However, their mean concentration in urine of a group of 2872 healthy people has been recently found to be $[Na^+]_T = 95 \pm 44$ and $[K^+] = 29 \pm 17$ (in mmol dm^{-3}, ±95 % confidence interval, unpublished data from this laboratory).

Analogously, Na^+ and K^+ concentrations in sweat and pancreatic juice depend also on the kind of secreting gland, as well as saliva, where marked differences are observed between stimulated and unstimulated saliva. Stimulation (during fast or digestion) is also important for determining their concentrations in gastric juice, while difference must be made between hepatic or cholecystic bile. During the first half of pregnancy, Na^+ and K^+ in amniotic fluid are in equilibrium with those of maternal plasma, while significant changes are observed in the second half.

Finally, regarding blood, differences must be considered between whole blood, plasma, and serum, since the presence of cells during measurements significantly alters alkali metal concentration; indeed, the difference between intracellular and extracellular concentrations of Na^+ and K^+ is very well known, with all implications for life. Further details on these aspects are available in [2], while some data on mean Na^+ and K^+ concentrations in human fluids can be found in Table 1. Concerning other alkali metal ions, a significant number of measurements has only been done in the case of human urine and blood plasma. In the former, mean excretion values are 0.12 mmol d^{-1}, 28 μmol d^{-1} and 98 nmol d^{-1}, for Li^+, Rb^+, and Cs^+, respectively; while their concentration in the latter (same order) is 4.5 μmol dm^{-3}, 1.8 μmol dm^{-3}, and 6.4 nmol dm^{-3}.

2.2 Experimental Determination of Alkali Metal Complex Formation Constants and Calculation Problems

When looking for alkali metal ion complexes in the most common stability constant databases [11–16], at least two points are evident: (i) the amount of available data is relatively low (especially if compared to complexes of other cations like, e.g., the alkaline earth ions, transition metal ions or lanthanoids); (ii) significant discrepancies are usually found between results reported by different authors. Both aspects can be attributed to some objective difficulties related to the experimental determination of the stability constants (and other related thermodynamic parameters) of weak complexes, like (usually) those of alkali metal cations (see, e.g., refs. in [17–19]). Many classical techniques may be used for this purpose, but ion-selective-H^+ and -M^+ (ISE-H^+ and ISE-M^+) potentiometries are probably the most common ones, for several reasons [17–19]: among them, the cheapness of the instrumentation and its simplicity favored their general use, though this was not always followed on by the comprehension of the key aspects of both their correct use and data interpretation, leading to the above cited discrepancies in results obtained [17–20]. For example, factors that must be taken into account are: kind of electrode calibration (in concentration or activity) [21]; changes of activity coefficients of different species

Table 1 Concentration of alkali metal cations in some selected human fluids (in mmol dm^{-3}, no particular pathologies, adults, mean values).

Biological fluid	Na^+	K^+
Urine[a]	150	70
Urine[b]	95 ± 44	29 ± 17
Sweat[c]	46.8	8.6
Saliva[d]	29.3	16.4
Gastric juice[e]	50.8	14
Pancreatic juice[e]	125	7.6
Epathic bile	146	4.8
Cholecystic bile	179	24.6
Intestinal juice	142	4.8
Synovial fluid	136	4
Cerebrospinal fluid	145	2.96
Aqueous humor	111	3.2
Tears	146	16.2
Seminal fluid	129	20.5
Amniotic fluid[f]	138.5–123.5	3.9–4.3
Maternal milk[g]	8.2	13.8
Plasma	140	4.03
Erythrocytes	7.35	99.1

[a]In mmol d^{-1}.
[b]From a group of 2872 people (unpublished data from this laboratory), ±95 % confidence interval.
[c]17–50 years.
[d]Total, flux 2.15 mol min^{-1}.
[e]Basal.
[f]Ranges from the beginning to the end of pregnancy.
[g]Mature milk.
Data compiled from [2].

(electrodes are sensitive to "activity") due to both the different composition of solutions (ionic medium, ionic strength, reagent concentrations) and dilution (during titrations); poor temperature control; and, in general, incorrect experimental design (metal to ligand ratios, pH, etc.). Moreover, pH-metric titrations (frequently used) are based on the measurement of pH changes due to proton displacement from a ligand during M^+ binding [21]. This means that, since the amount of protons displaced (i.e., the change in pH) depends on both the strength of complex formed and the concentration of species involved in the formation reaction (i.e., the ligand and the metal cation), the changes in pH may be very small and of the same order as the instrumental uncertainty, if experiments are not properly designed [19, 20].

Alternatively, a large number of formation constants of alkali metal complexes have been successfully calculated by the so-called $\Delta\log K^{H}$ or ΔpK method, i.e., by measuring the differences of the apparent protonation constants of the ligands of interest determined in "non-interacting" (better "very weakly interacting") aqueous media and those obtained using alkali metal salts as supporting electrolytes. A detailed description of the basic principles of this approach and some examples can be found, for example, in [22–25]. Briefly, for a simple monoprotic acid (HL), the lowering effect of the apparent protonation constant in an "interacting" medium ($\log K_i^{H*}$) with respect to a non-interacting one ($\log K_i^{H}$) can be interpreted in terms of complex formation between the deprotonated ligand and the cation of the supporting electrolyte:

$$\log K_i^{H*} = \log K_i^{H} - \log\left(1 + 10^{\log K^M}[M]\right)$$

For polyprotic ligands, a slightly more complicated calculus procedure should be used, though it starts from a basic assumption, i.e., that the average number of protons bound to a ligand $(\bar{p})$ is fixed in given conditions, independently of its expression. This means that it can unequivocally be calculated using only the apparent overall protonation constants (β_i^{H*}, referring to the equilibrium: i H+L=H_iL):

$$\bar{p}^* = \frac{\sum i\beta_i^{H^*}\left[H^+\right]^i}{1+\sum \beta_i^{H^*}\left[H^+\right]^i}$$

or by the effective protonation constants $(\beta_{0i}, \beta_{0i} \equiv \beta_i^{H})$ and the complex formation constants:

$$\bar{p} = \frac{\sum i\beta_{ji}\left[M^+\right]^j\left[H^+\right]^i}{1+\sum \beta_{ji}\left[M^+\right]^j\left[H^+\right]^i}$$

The equivalence of the two expressions means that the complex formation constants of (weak complexes) can be calculated by minimizing the function

$$U = \sum\left(\bar{p} - \bar{p}^*\right)^2$$

by means of suitable computer programs. To our knowledge, only one program has been specifically dedicated to the determination of weak complexes by this method, i.e., ES2WC [22]. Worth mentioning is the fact that, in this kind of calculations, the molar concentration scale (in mol dm^{-3}) must be used [22, 26, 27].

3 Inorganic Complexes

As known, inorganic ligands are very important, practically in all natural fluids [2–6]. Even though the stability of alkali metal ion–inorganic ligand complexes is generally lower than that of species with organic ligands with analogous stoichiometry

[11–16], their importance is not negligible, for the same reasons given in previous paragraphs for alkali metal cations. The concentration of inorganic ligands in natural systems is usually higher than that of organic ones.

The inorganic ligands considered in this chapter are: hydroxide (OH^-), chloride (Cl^-), sulfate (SO_4^{2-}), carbonate (CO_3^{2-}), phosphates (PO_4^{3-}, $P_2O_7^{4-}$, $P_3O_{10}^{5-}$ and phytate (Phy^{12-}). For convenience fluoride (F^-), hexacyanoferrate(II) and (III) ($Fe(CN)_6^{4-}$ and $Fe(CN)_6^{3-}$), borate ($B(OH)_4^-$), and other minor ligands (e.g., SCN^-) are treated in a separate paragraph (see Section 3.6). A comprehensive data analysis on the interaction of alkali (and alkaline earth) metal–inorganic (and organic) complexes is reported in [18]. More recently, the stabilities of many of these sodium and potassium complexes (including mixed species) at different ionic strengths and temperatures of interest for many biological fluids have been reviewed by Crea et al. [28].

3.1 Hydroxide

Daniele et al. [18] reported the formation constants for MOH ion pairs (using the ΔpK method, see Section 2.2): log K=0.53, 0.28, and 0.18 for Li^+, Na^+, and K^+ respectively, at t=37 °C and infinite dilution [29], resulting in reasonable agreement with the reference work of Baes and Mesmer [30] (log K=0.36, −0.18, and −0.46 for Li^+, Na^+, and K^+ respectively, at t=25 °C, Cs^+ interaction neglected). From the dependence of the formation constants on temperature, a ΔH value was also reported by Daniele et al. [18] for Na^+ (ΔH=1.3 kJ mol^{-1}, same conditions). There is no doubt that alkali metal cations undergo hydrolysis, but the amount of MOH in solution is too small to be estimated with acceptable error limits [18, 30]. In any case, the formation of these ion pairs is generally negligible at the pH of most biological fluids. From a pure thermodynamic point of view, worth mentioning is the work by Popov et al. [31], who determined the formation constants of the MOH species for the whole series of alkali metals (except for Rb^+), at 25 °C in $(CH_3)_4N^+/Cl^-/OH^-$ solutions at I=3.4 mol dm^{-3}, obtaining log K=−0.04, −0.93, −0.7 and −0.8 for Li^+, Na^+, K^+, and Cs^+, respectively, and reporting the last value for the first time.

3.2 Chloride

The concentration of chloride in bio-fluids is quite high and ranges between ~0.02 mol dm^{-3} in saliva and ~0.13 mol dm^{-3} in intestinal juice [2].

Chloride complexes of alkali metal cations are very weak and were studied by partial molal volume and potentiometric measurements [32, 33]. De Robertis et al. [34] calculated the formation constant (and ΔH values) of NaCl and KCl species at different temperatures and ionic strengths, by a careful analysis of literature data. At I=0 mol dm^{-3} and 25 °C, they proposed for NaCl: log K=−0.30, ΔH=−8 kJ mol^{-1}

and for KCl: log K=−0.27, ΔH=−4 kJ mol^{-1}. Other evidence for the formation of weak ion pairs of alkali metal chlorides was reported using non-empirical quantum chemical methods [35]. In particular two kinds of association models for NaCl were observed, namely, contact ion pairs and solvent separated ion pairs. Heyrovská [36] showed that the non-ideal thermodynamic properties of strong electrolytes are due to partial dissociation.

3.3 *Sulfate*

The concentration of sulfate in bio-fluids is fairly low. In urine, for example, it may reach values of ~0.01 mol dm^{-3}, whereas in saliva its concentration is ~0.001 mol dm^{-3} [2].

Many natural fluids contain sulfate or hydrogensulfate (in acidic media), and therefore its complexes with alkali metal ions were investigated by several authors (see refs in [18, 28]). As for other ligands, only a few ΔH values are reported. Generally, the stability of sulfate ion pairs with alkali metal ions is slightly higher than that of the corresponding chloride species. Daniele et al. [37] determined the formation constants of the entire series Li^+, Na^+, K^+, Rb^+, and Cs^+ sulfate complexes at 37 °C in the ionic strength range $0.03 < I$/mol $dm^{-3} < 0.5$. At infinite dilution, the values of log K=0.79 (Li^+), 0.39 (Na^+), 0.52 (K^+), 0.60 (Rb^+), and 0.69 (Cs^+) were reported. Righellato and Davies [38] found log K=0.64 (Li^+), 0.70 (Na^+), and 0.82 (K^+) at infinite dilution at 18 °C for sulfate [while it is log K=0.58 (Na^+) and 0.91 (K^+) for thiosulfate under the same conditions]. The trend of the stability constants shows a sharp minimum (for sodium) which is absent or less evident with other ligands, whereas Midgley [39] found the following trend: $Cs^+ > Rb^+ > K^+ > Na^+ > Li^+$. Evidence for sodium ion pair formation was also obtained with Raman spectroscopy [40].

3.4 *Carbonate*

Carbonate and hydrogen carbonate, the most important species at the pH of blood plasma, are very important in all biological fluids. However, in some of them (acidic urine for example) HCO_3^- is in equilibrium with $CO_{2(g)}$. The concentration of hydrogen carbonate is ~0.01 mol dm^{-3} in saliva, ~0.09 mol dm^{-3} in pancreatic juice, and ~0.03 mol dm^{-3} in aqueous humor [2].

Despite their importance, few data are reported in the literature on the interactions of carbonate with alkali metal cations. They have been reviewed in [18, 28], including the recent work by Crea et al. [41], who determined the formation constants for sodium carbonate and hydrogen carbonate species by the ΔpK method, using the experimental protonation constants determined in $(C_2H_5)_4NI$ as baseline. At infinite dilution, they found log K=1.15, 0.26, and −0.25 for $NaCO_3^-$, $Na(HCO_3)$, and Na_2CO_3(reaction : $Na^+ + NaCO_3^-$), respectively, while it is log K=0.97 and 0.11 for KCO_3^- and $K(HCO_3)$.

3.5 Phosphates

Hydrogen phosphate is, generally, the most important species of inorganic phosphorus species in biofluids. Its concentration is commonly ~1 mmol dm^{-3}; in urine it may reach ~0.03 mol dm^{-3} [2].

The chemistry of phosphate in aqueous solutions was widely studied, and data on the binding ability of inorganic phosphate towards alkali metal ions have been recently reviewed [18, 28]. The stability of dihydrogenphosphate with Na^+ and K^+ is very similar (log K=0.28 and 0.26, respectively, for the reaction $M^+ + H_2PO_4^- = M(H_2PO_4)$ [29, 42]). The stability of the alkali metal species with hydrogenphosphate (reaction $M^+ + HPO_4^{2-} = M(HPO_4)^-$, log K=1.08, 1.02 and 0.87 for Li^+, Na^+, and K^+, respectively) follows the expected trend for the charge/radius ratio [29, 42, 43]. Analogously, for phosphate (PO_4^{3-}) complexes it was found (at infinite dilution and 37 °C): log K=1.55, 1.34, 1.20 for $LiPO_4^{2-}$, $NaPO_4^{2-}$, and KPO_4^{2-}, respectively, and log K=1.17, 1.05, 0.87 for $Li_2PO_4^{2-}$, $Na_2PO_4^{2-}$, and $K_2PO_4^{2-}$ (stepwise reaction, M+ML=M_2L), with an endothermic formation reaction [29, 42].

Alkali metal complexes of pyrophosphate (diphosphate, ($P_2O_7^{4-}$) and triphosphate ($P_3O_{10}^{5-}$) are more stable than the corresponding phosphate ones, and follow the same stability trend [44]: log K=3.71, 2.25, 2.24 for $LiP_2O_7^{3-}$, $NaP_2O_7^{3-}$ and $KP_2O_7^{3-}$ and log K=3.47, 2.33, 2.33 for $LiP_3O_{10}^{4-}$, $NaP_3O_{10}^{4-}$, and $KP_3O_{10}^{4-}$. Phosphates also form mixed $NaK(H_iPO_4)$, $NaK(H_iP_2O_7)$, and $NaK(H_iP_3O_{10})$ species (i=0 and 1), whose stability is greater than that of simple Na_2H_iL and K_2H_iL species [42, 45].

Among the phosphates, a particular class of ligands of biological relevance is represented by *myo*-inositol-n-phosphates (n=1 to 6). 1,2,3,4,5,6,*myo*-inositol hexakis-phosphate (Phy^{12-}) is the most abundant among them and it is ubiquitous in eukaryotic cells, mainly as potassium, calcium and magnesium salts. Their importance is due to the fact that phytates modify the bioavailability of cations, have antioxidant and anticancer activity, reduce kidney stone formation, and are used in the treatment of many pathologies (for more information we may refer to refs. in [46]). Due to its high charge, Phy^{12-} forms several $M_iH_rL^{(12-i-r)-}$ species (M=Li^+, Na^+, K^+, Cs^+), which show a fairly high stability: for example, the log K values for the reaction: 6 M^+ + Phy^{12-}=M_6Phy^{6-} are 28.1 (Li^+), 25.9 (Na^+), 24.5 (K^+), and 24.0 (Cs^+), while, for the reaction M^+ + H_6Phy^{6-}=$M(H_6Phy)^{5-}$ it is log K=2.8 (Li^+), 2.4 (Na^+), 2.2 (K^+), and 3.7 (Cs^+) (at I=0.1 mol dm^{-3} and 25 °C) [24, 46, 47]. The formation of mixed alkali metal cation species (Li^+/Cs^+) was also observed [47].

Recently, Veiga et al. [48] reported a quantitative study of the intramolecular details of Phy^{12-} protonation and complexation equilibria also in the presence of alkali metal cations (i.e., Na^+ and K^+). The authors determined the micro-protonation constants at physiological pH, reporting that $H_4(Phy)^{8-}$ and $H_5(Phy)^{7-}$ are the major species. In the presence of Na^+ and K^+ ($[M]_T=0.15$ mol dm^{-3}), the predominant species are $M_3H_4(Phy)^{5-}$, $M_4H_3(Phy)^{5-}$, $M_5H_2(Phy)^{5-}$, $M_6(Phy)^{6-}$. Among them, those predominant in cytosol and nucleus conditions are $Na_3H_4(Phy)^{5-}$, and $Na_4H_3(Phy)^{5-}$ for Na^+ and $K_3H_4(Phy)^{5-}$ for K^+. Cigala et al. [49] reported the solubility of the $Na_{12}(Phy)$ salt using the data on the Na^+/Phy^{12-} complexes previously determined [24], observing that only 7 moles of Na^+ per mole of salt dissociate when $Na_{12}(Phy)$ is dissolved in NaCl solutions, meaning that $Na_5(Phy)^{7-}$ is the totally deprotonated species in Na^+ solution. The total solubility of $Na_{12}(Phy)$ at infinite dilution is $\log S_0^T = -0.522$, and $\log K_{S0}=-0.194$ at $[NaCl]_T=2.036$ mol kg^{-1}, relative to the reaction:

$$K_{S0} = \left[Na_5(Phy)^{7-}\right]\cdot\left[Na^+\right]^7$$

3.6 Other Inorganic Ligands

The stability data of fluoride complexes in aqueous solution were collected by Bond and Hefter [50], who reported $\log K=-0.56$ for NaF species at $I=1$ mol dm^{-3} and 25 °C. Miller and Kester [51] determined the stability of sodium-fluoride ion pairs in seawater medium. Ferrocyanide (hexacyanoferrate(II), $Fe(CN)_6^{4-}$) and its oxidized product ferricyanide (hexacyanoferrate(III), $Fe(CN)_6^{3-}$) cannot freely pass through the plasma membrane. For this reason, ferrocyanide has been used as a probe of an extracellular electron receptor in studies of redox reactions in cells [52]. Alkali metal complexes of hexacyanoferrate(II) were studied by different techniques, and the following stability trend was found: $Li^+<Na^+<K^+<Rb^+<Cs^+$ [18, 53–55]. Daniele et al. [18] reported: $\log K_{ML}=2.0$ (Li^+), 2.3 (Na^+), 2.5 (K^+), 2.7 (Rb^+), and 2.9 (Cs^+) at 25 °C and infinite dilution. In addition, thermodynamic data relative to the formation of the MHL and M_2L species are also reported in [54], together with an empirical relationship for the estimation of the stability of the M_2L (as $\log \beta_{M2L}$, referred to the overall equilibrium $2\,M+L=M_2L$) at infinite dilution and 25 °C from the simple ML species:

$$\log \beta_{M_2L} = 0.18 + 1.27 \log K_{ML}$$

ΔH (and ΔS) values were also obtained from the dependence of the stability constants on temperature. For the overall equilibria relative to the formation of ML, MHL, and M_2L species of Li^+, Na^+, and $K^+/Fe(CN)_6^{4-}$ complexes, two mean formation entropy and enthalpy changes can be proposed: $\Delta H=15$ kJ mol^{-1} and $T\Delta S=36$ kJ mol^{-1} at infinite dilution and 25 °C. The formation of mixed hetero MM'L species was also evidenced by means of experiments in Li^+/Na^+, Li^+/K^+,

Na^+/K^+ mixed chloride solutions [54]. The formation constant of the equilibrium $M_2L + M'_2L = 2\ MM'L$ showed a mean value log $K = 1.1$ for all the mixed systems, indicating a significant stabilization for the mixed complexes (the statistical value is log $K = 0.6$ [56]).

Hexacyanoferrate(III)-alkali metal ion complexes are significantly less stable (about one log unit) and follow the same trend of hexacyanoferrate(II) species [57]. For the sake of completeness, the formation of a Rb^+ complex of cobalt(II)-pentacyanide is reported, with the average value log $K = 0.57$ [58]. Boric acid, H_3BO_3, can be treated as a weak acid and in the literature some data are reported on the interaction of the anion, $B(OH)_4^-$ with alkali metal cations. In particular, the values of 0.34, −0.15, and −0.18 are given at infinite dilution for the Li^+, Na^+, and K^+ complexes, respectively [59–62]. Among other inorganic ligands, worth mentioning is also thiocyanate, which is present up to millimolar concentrations in some biological fluids (e.g., in saliva [2]). The stability constants of its NaL and KL species, together with their dependence on ionic strength and temperature, have been recently reported by Crea et al. [28] (log $K_{ML} = -0.33$ and −0.31 for the formation of Na(SCN) and K(SCN), respectively, at 25 °C and infinite dilution).

Finally, also negatively charged hydrolytic species of many cations may form fairly stable complexes with alkali metal cations, which is a very important aspect when dealing with multicomponent systems (like biological fluids). For example, the formation of a quite stable mixed hydrolytic species of dioxouranium(VI) and sodium cations is reported by Gianguzza et al. [63], with log $K = 1.2$ for the equilibrium $[(UO_2)_3(OH)_7^- + Na^+ = (UO_2)_3(OH)_7Na]$ at 25 °C and infinite dilution. This relatively high value is not surprising, since the "affinity" of Na^+ toward dioxouranium(VI) species is very well known (e.g., UO_2^{2+} salts were commonly used in the past in qualitative chemical analysis for the selective precipitation of sodium [64]). For example, the stabilization of the negatively charged dioxouranium(VI)/oxalate complexes of Na^+ is quantified by $\Delta G = -2.6\ kJ\ mol^{-1}$ [65].

4 Nucleotide Complexes

Nucleotides are a class of organic compounds in which the molecular structure comprises a nitrogen-containing unit bound to a sugar and at least one phosphate group; they are the building blocks of nucleic acids (DNA and RNA). The sugar and phosphate group make up the backbone of the DNA double helix, while the bases are located in the middle. A covalent bond between the phosphate group of one nucleotide and the sugar of a neighboring nucleotide holds the backbone together. Hydrogen bonds between the bases that are across from one another hold the two strands of the double helix together. The nitrogen-containing bases of nearly all nucleotides are derivatives of three heterocyclic compounds: pyrimidine (cytosine, thymine, and uracil), purine (adenine and guanine), and pyridine (nicotinamide). The study of modern genetics also depends on the understanding of the physical and chemical characteristics of DNA and, in turn, on its individual nucleotides. Nucleotides play a very important role in all biological systems since they are involved in most biochemical and enzymatic processes. These biochemical processes require in many cases the presence of essential metal ions that can act as catalysts. Unfortunately, the same biological functions of nucleotides are inhibited by the presence of toxic metal ions that can act as competitors with the essential ones. When studying biological and environmental processes where these substances are involved, the interaction of nucleotides with the metal ions and the corresponding binding capacity cannot be neglected [66].

The solution chemistry of nucleotides depends on the binding capacity of the phosphate groups in the molecules, where, in particular, the terminal phosphate plays a main role. However, also the nitrogen N(7) of the base participates in some cases in the complexation of the metal ions, but obviously this is strictly related to the pH value of the medium [66].

Owing to the importance of this class of biological ligands, a huge number of papers on the interaction of nucleotides with several metal ions have been published, especially regarding ATP^{4-} with alkaline earth metal ions (Ca^{2+} and Mg^{2+}). Several general [11, 12, 14–16] and specific [67] compilations containing recommended and tentative protonation and formation constants of the M^{n+}/nucleotide (AMP^{2-}, ADP^{3-}, ATP^{4-}, AQP^{5-}, etc.) systems are available, generally at low ionic strength values ($I \sim 0.1$ mol dm^{-3} and 25 °C). Nevertheless, data on the interaction of nucleotides with alkali metals are not as abundant as those of alkaline earth metals. Studies in this field were carried out by some researchers at low ionic strengths, generally up to 0.2 mol dm^{-3}, at different temperatures ($10 \leq t/°C \leq 45$) and, in some cases, in different ionic media [NaCl, KCl, $(CH_3)_4NCl$, $(C_2H_5)_4NCl$, $(C_3H_7)_4NCl$, etc.] [43, 68–77]. Investigations in tetraalkylammonium salts were done assuming initially that these salts do not form complexes with nucleotides, and with ATP^{4-} in particular. Independently of the nucleotides and of the ionic medium, the formation of simple ML and MHL ($M = Na^+$ or K^+) species was generally reported [43, 68–77]. In some cases, the formation of M_2H_2L species ($M = Na^+$ or K^+) was also observed by ISE-M^+ investigations [71]. Despite the weakness of alkali metal-nucleotide complexes (generally log $K < 2$), at pH ~ 7 the ligand is generally complexed in high yield, especially if the metal ion is in excess (like Na^+ and K^+ in biological systems). As an example, at $I = 0.1$ mol dm^{-3} and low ligand concentration, more than 70 % of ATP^{4-} is complexed by K^+ or Na^+ [76]. Further investigation on the protonation and alkali (and alkaline earth) complex formation of ATP^{4-} at different ionic strengths and in different ionic media was carried out both in alkali metal media (Li^+, Na^+, K^+, Rb^+, and Cs^+) and in different tetraalkylammonium salts [77].

For all the systems, the formation of the $M(ATP)^{3-}$ and $MH(ATP)^{2-}$ species was observed, while the formation of a binuclear $M_2(ATP)^{2-}$ species was reported only for Li^+, and in excess of cation. Despite its weakness, this species reaches about 10 % of formation at pH ~7, confirming that it cannot be neglected in a correct speciation study in biological fluids. For the simple $M(ATP)^{3-}$ ($M = Na^+$, K^+, Rb^+, and Cs^+) species, a fairly constant value of log $K = 1.23$ was calculated at $I = 0.25$ mol dm^{-3} and 25 °C, while only the $Li(ATP)^{3-}$ is more stable (log $K = 1.78$). Finally, in the same work [77], the dependence of the stability constants of various complexes on the ionic strength is similar for the same kind of species formed by different alkali metal cations. More recently, the protonation of ATP^{4-} and the ion pair formation with Na^+ and K^+ (together with the modelling of their dependence on ionic strength by different approaches) have been investigated in more detail [78] in this laboratory. At infinite dilution and 25 °C, log $K = 2.10$ and 1.95 have been proposed for the formation of $Na(ATP)^{3-}$ and $K(ATP)^{3-}$, respectively. The stability constants of alkali metal cation/ATP complexes were also reported by other authors: log $K = 2.43$ (Li^+), 1.98 (Na^+), 1.87 (K^+), 1.98 (Rb^+), and 1.94 (Cs^+) [43, 77, 79] at infinite dilution and 25 °C. In general, most authors agree that the stability of alkali metal complexes of ATP^{4-} is inversely proportional to cation size, following the trend $Li^+ > Na^+ > K^+ > Rb^+ > Cs^+$ [18].

Smith and Alberty [43] reported the formation constant of the ML species of alkali metal cations with different nucleotides in Me_4NCl at $I = 0.20$ mol dm^{-3} and 25 °C. These authors proposed log $K = 0.98$, 0.83, and 0.6 for the AMP^{2-} complexes with Li^+, Na^+, and K^+, respectively; log $K = 1.71$ (Li^+), 1.39 (Na^+) and 1.29 (K^+) for ADP^{3-}, and log $K = 2.85$ (Li^+), 2.18 (Na^+), and 2.04 (K^+) for AQP^{5-}. These stability constants for the $M(AQP)^{4-}$ complexes were recalculated by Smith et al. [67] at $I = 0.10$ mol dm^{-3} in Me_4N^+ medium to give log $K = 2.22$ for Li^+, 1.71 for Na^+, and 1.54 for K^+. The significant difference between the stability constants of AQP-alkali metal ion complexes and other nucleotides can be explained taking into account the higher ligand charge [43].

Relatively scarce is also the information on the enthalpic or entropic contribution in the formation reaction of nucleotide-alkali metal ion complexes. From calorimetric measurements and from the dependence of the formation constants on temperature it was possible to calculate the enthalpy change values for both the protonation and the complexation of ATP^{4-} [77]. Results show that alkali metal complexes of ATP^{4-} are entropically stabilized and low enthalpy changes (~1 kJ mol^{-1} for the Na^+ and K^+ complexes of ATP^{4-}) are observed: in many cases it holds $\Delta S > 160$ J mol^{-1} K^{-1} at $I = 0.25$ mol dm^{-3} and 25 °C [77].

In spite of the high number of collections containing data on the thermodynamic aqueous properties of nucleotides in simple electrolyte solutions, literature data in mixed or multicomponent media are rare. An example is the study by Crea et al. [80] on the speciation and interaction of IMP^{2-}, AMP^{2-}, ADP^{3-}, and ATP^{4-} (protonation and complex formation constants) in a multicomponent solution simulating the seawater at different salinities and containing simultaneously different cations, as well as mixed $NaCl/MCl_2$ ($M = Mg^{2+}$ or Ca^{2+}) solutions at different ionic strengths. Although Na^+ interacts weakly with nucleotides, this interaction cannot be neglected when dealing with their chemical speciation in seawater, where Na^+ is one of the main inorganic components. Some selected stability constants of ATP/alkali metal ion complexes are reported in Table 2.

Table 2 Selected log *K* values for the interaction of Li^+, Na^+, K^+, Rb^+ and Cs^+ with some ligands of biological interest ($I = 0$ mol dm^{-3}, 25 °C).[a]

Ligand	Li^+	Na^+	K^+	Rb^+	Cs^+
OH^-	0.36	−0.18	−0.46		
Cl^-		−0.30	−0.27		
SO_4^{2-}	0.79	0.39	0.52	0.60	0.69
CO_3^{2-}		1.15	0.97		
HCO_3^-		0.26	0.11		
ATP^{4-}	1.78	1.23	1.23	1.23	1.23
AMP^{2-}	0.98	0.83	0.60		
$(AA)^-$[b]		0.28	0.28		
H(AA)[b]		−0.40	−0.40		
GSH^{3-}	0.96	0.96	0.96	0.96	0.96
Cit^{3-}	0.88	0.80	0.48	0.38	0.13
$EDTA^{4-}$	2.85	1.75	0.70		
NTA^{3-}	2.50	1.30	0.70		
$H(DTPMPA)^{9-}$		3.71	3.32		
$HEDP^{4-}$		1.80	1.40		
Ox^{2-}	1.17	0.85	0.71		
Ac^-	0.20	−0.12	−0.27	−0.16	−0.12
$Lact^-$		−0.11	−0.27		

[a]Data collected from different references cited in this chapter.
[b]AA = generic simple amino acid.

5 Amino Acid and Peptide Complexes

Amino acids play a central role in many metabolic processes, as well as building blocks of proteins. In this context, it is obvious that the biological activity of proteins depends on the chemical properties of the amino acids they contain. Considering that the study of the protein structure and stability is an important research area since many years, the investigation of the structure and the properties of amino acids became necessary.

In aqueous solution (like biological fluids), the acid base properties of amino acids determine their behavior, since, at a given pH, the proton moves from the carboxylic (COOH) to the amino (NH_2) group, generating a zwitterion with a negative and a positive charge. This neutral zwitterionic form is the most stable form in the human body. The amino acids, being simultaneously "amines" and "acids", have intermediate characteristics between the two ligand classes, with the possibility to chelate cations through N- and/or O-donor interactions. As an example, Asp^{2-} and His^- have complexing characteristics between those of Suc^{2-} and Ala^-, and Ala^- and Imid, respectively [81]. On the basis of experimental evidence, another interesting aspect about the behavior of amino acids in aqueous solutions can be highlighted: since they are both O-donor and N-donor ligands, like carboxylates and amines they show interesting regularities in the stability of their metal complexes [82–85]. However, despite the importance of this class of biologically active molecules and their main role in many biochemical processes, thermodynamic data on the interaction with metal ions (and alkaline cations in particular) are relatively few [11, 12, 14–16]. Fiol et al. [86] investigated the protonation constants of Ala^- in NaCl at different ionic strengths. The authors determined the interaction parameters between Ala^- and the ions of the supporting electrolyte by means of different equations, but the possible formation of weak Na^+/Ala^- complexes was not taken into account.

Studying the acid–base properties of amines and amino acids, Casale et al. [87] observed that the dependence of the protonation constants on ionic strength is fairly constant for all the considered ligands. Some significant deviations in the ionic strength dependence parameters were explained by taking into account the interactions of the deprotonated amino acids with the cation of the background electrolyte. Analyzing the protonation constants of 18 different amino acids in different ionic media (in Na^+ or K^+ and Cl^-, NO_3^- or ClO_4^-), the authors evidenced that the stability of the alkali metal ion complexes with amino acids is independent of the type of M^+. These authors [87] estimated for a generic M(L) and MH(L) complex a mean formation constant at infinite dilution of log $K=0.28$ and -0.4, respectively. From the data reported by Casale et al. [87] it follows that the formation constants with the alkali metal ions are close to the mean value for the reaction with a monocarboxylic ligand.

De Stefano et al. [88, 89] and De Robertis et al. [81] studied the acid base properties of some amino acids (Gly^-, Ala^-, Ser^-, Glu^{2-}, Lys^-, Asp^{2-}, His^-, and IDA^{2-}) in simple and mixed electrolyte solutions. The properties of these amino acids were analyzed by grouping the ligands in three different sets on the basis of the side chain: the first group, with a neutral side chain, included Gly^-, Ala^-, and Ser^-, the second, with a basic side chain (Lys^- and His^-), and the third, with an acidic side chain (Asp^{2-}, Glu^{2-}, and IDA^{2-}). The formation of weak complexes between Na^+ and the amino acids was observed for all experimental conditions investigated and resulted to be independent of the nature of the amino acids but dependent on the side chain group. The values of the average formation constants indicate that they mostly depend on the number of carboxylate groups, in fact: log K [NaL]$=0$, 0 and 0.5 for group 1 (neutral), 2 (basic), and 3 (acidic), respectively, and log K [NaHL]$=-0.5$, -0.5, and 0 (same order). Despite the weakness of the alkali metal-amino acid complexes, their formation percentage in multicomponent solutions cannot be neglected. For example, in a synthetic seawater solution at a salinity of 35, Asp^{2-} (group 3) forms the $Na(Asp)^-$ and NaH(Asp) species, whose formation percentages reach about 10 % (at pH>10) and 20 % (pH ~ 6.5–7), respectively, in the presence of higher amounts of stronger alkaline earth metal-Asp^{2-} complexes [89]. Similar conclusions were reached for Lys^-, whose formation constants with Na^+ were determined at different ionic strengths and in different ionic media [89].

The formation of weak alkali metal species has also been reported for $Ccys^{2-}$ by Bretti et al. [90], who determined a stability of log $K = -0.40$ (at infinite dilution) for the reaction $Na^+ + H_2(Ccys) = NaH_2(Ccys)^+$.

An attempt to calculate the enthalpy changes for the complexation of amino acids with alkali and alkaline earth metal cations was also carried out by studies at different temperatures [81, 89]. Owing to the large errors associated with the enthalpy changes, these data are considered by the authors as tentative temperature gradients [81, 89]. In this light, some mean ΔH values can be given for the formation of Na(L) species at 25 °C and infinite dilution: $\Delta H \sim -3$, -7 and 11 kJ mol^{-1} for group 1 (neutral), 2 (basic), and 3 (acidic) amino acids, respectively. In the same work, the authors calculated the significance of the various species in terms of the concentration of the salt that is needed to reach a value of at least 5 % of the species formation.

By means of *ab initio* calculations, Reddy and Sastry [91] studied the interaction energies of cations (Li^+, Na^+, K^+, Mg^{2+}, Ca^{2+}, NH_4^+, Me_4N^+) with the aromatic side chains of four amino acids (Phe^-, Tyr^-, Trp^-, and His^-). They reported that metal ions interact with the aromatic moieties either through cation–π interaction or cation–heteroatom interaction. Moreover, they did not observe a situation where a metal ion is bound covalently to only one carbon of the aromatic ring. If the aromatic ring contains a heteroatom, the cation is attracted by its high electronegativity. The general trend of interaction energies of the alkali and alkaline earth cations with various aromatics is $Mg^{2+} > Ca^{2+} > Li^+ > Na^+ > K^+ \sim NH_4^+ > Me_4N^+$.

Using the data reported by Crea et al. and Cigala et al. [92, 93] for the protonation of GSH^{3-} in the alkali metal chlorides, it is possible to determine the stabilities of the alkali metal complexes using suitable computer programs [22]. The values for the $M^+ + L^{3-} = ML^{2-}$ reaction are very similar for all the alkali metals, although for Li^+ it is slightly higher. The average value at infinite dilution and 25 °C is log $K = 0.96$, comparable with the stability of a dicarboxylate ligand [for example, for $Na^+ + Ox^{2-} = Na(Ox)^-$, log $K = 0.85$]. For the reaction $M^+ + HL^{2-} = MHL^-$ under the same conditions log $K = -0.08 \pm 0.22$ was calculated, which is in turn comparable with the stability of the complex of a monocarboxylate ligand.

6 Other Ligands of Biological Relevance

6.1 Amines

The complexing properties of N-donor ligands towards alkali metal cations are very weak and were generally neglected. A few evidences about the formation of such complexes are reviewed by Daniele et al. [18]. Furthermore, due to the weakness of these interactions, salts of alkali metal cations have been generally used in the past as baseline (non-interacting) supporting electrolytes for the determination of weak complexes of amines with halides and/or tetraalkylammonium cations by the Δp*K* method [94–97]. As a consequence, few or no data are reported in the literature on alkali metal cation complexes with amines. Only the formation constants of two species between Li^+ and En were reported in the paper of Casale et al. [98]. In that paper, the authors interpreted the differences between the protonation constants determined in NaCl and LiCl media in terms of formation of two ion pairs, namely $Li(En)^+$ and $LiH(En)^{2+}$, using the Δp*K* method (Section 2.2). The stability of the two species is very low; at $I=0.1$ mol dm^{-3} it is log $K=-0.20$ and −0.65, respectively. The determination of these species may lead to the conclusion that other amines can interact with alkali metal cations as well, but at the moment no quantitative results are published, although a considerable number of papers are available with protonation constants determined in different alkali metal salts.

6.2 Carboxylates

The concentration of carboxylates in biological fluids is very variable. For example, the total concentration of organic carboxylates in urine may reach ~0.05 mol dm^{-3} [2]. They can be classified as *hard* ligands and therefore their affinity for alkali metal complexes has been extensively investigated [11, 12, 14–16]. Due to their anionic nature, carboxylate ligands are favored in the formation of complexes with alkali metal ions with respect to other basic uncharged groups (such as nitrogen-containing donors). Literature data about some important carboxylates are reported in Table 2. A quick look at the table immediately shows that the number of charges involved in the complex formation reaction is the most important factor and suggests that the species in solution are mainly formed through ionic interactions. For most of the ligands reviewed, the alkali metal cations studied are Li^+, Na^+, and K^+, and the formation constants decrease with increasing atomic number, due to the lowering of the charge/radius ratio. This trend is not confirmed when data for Rb^+ and Cs^+ are available as well: for example, for Li(Ac), Na(Ac), K(Ac), Rb(Ac), and Cs(Ac) log $K=0.20$, −0.12, −0.27, −0.16, and −0.12, respectively [99]. Other data are reviewed by Daniele et al. [18] for $Mala^{2-}$ and Cit^{3-}. Analyzing protonation data of $Lact^-$, reported in various media in the most common stability constant databases [11, 12, 14–16], the stability of the Na(Lact) and K(Lact) species can be determined and their values at infinite dilution and 25 °C are log $K=-0.11$ and −0.27 for Na^+ and K^+, respectively.

Many carboxylate ligands are multidentate and it seems reasonable that they can form bi- (or poly-) nuclear species with the alkali metal ions. Binuclear species were reported for Cit^{3-}, Tca^{3-}, $Ssal^{2-}$, Btc^{4-}, and Mlt^{6-} [18]. To give an example, the values of log K for the reaction $M^+ + M(Cit)^{2-} = M_2(Cit)^-$ are 0.88 (Li^+), 0.80 (Na^+), 0.48 (K^+), 0.38 (Rb^+), and 0.13 (Cs^+) [100]. Furthermore, both Btc^{4-} and Mlt^{6-} show the tendency to bind more than two alkali metal ions, confirming the importance of the charges involved in the complex formation reaction [101, 102]. Berto et al. [19] reported a methodological study for the determination of weak Na^+ complexes using different ion-selective electrodes, taking also into account the variation of the ionic strength during the potentiometric titrations. The weak formation constants of Mlt^{6-}, Btc^{4-}, Cit^{3-}, $Phtal^{2-}$, and $Mala^{2-}$ with Na^+, determined by Berto et al. [19] are in agreement with those reported in the literature. Midgley [39] found that for α-hydroxo-acids, the strength of the complex depends on the number of both carboxylate and α-hydroxo groups, thus the order for the most common compounds is $Cit^{3-} > Tar^{2-} > Mala^{2-} > Glyc^{-} > Lact^{-}$, while Pyr^{-} is stronger than all these ligands.

Within carboxylic acids, the polyelectrolytes may be considered separately. Katchalsky [103] found that their presence in the cells can cause strong agglutination, preceded by an adsorption of the poly-ions to the cell surface. This leads to numerous biological changes, such as inhibition of growth, lowering of toxicity, and hemolysis of erythrocytes. However, in the literature only few data are reported about these molecules. De Stefano et al. [104] reported the alkali metal complexes of polyacrylic acid (Paa) (2 and 20 kDa) and polymethacrylic acid (Pma) (5.4 kDa) with Li^+, Na^+, and K^+. At infinite dilution and 25 °C, the authors found that the stability of alkali metal complexes is similar to that of Btc^{4-} and it is fairly constant varying the MW and the alkali metal cation: for Paa, it is log K = 1.81, 1.84, and 1.83 for Li^+, Na^+, and K^+, while for Pma, it is log K = 1.82, 1.71, and 1.65 (same order). Both polyelectrolytes follow the same stability trend reported above for many other ligands. Bretti et al. [105] studied the interaction of Na^+ with two acrylic-maleic copolymers (Pca) (MW 3 and 70 kDa). The authors found that at infinite dilution the values of the NaL species is log K_{NaL} = 1.74 (close to Paa) and 2.56 for Pca 3 kDa and 70 kDa, respectively. According to Bretti et al. [105], the Pca 3 kDa can be considered in the model as a tri-anion and the 70 kDa as a tetra-anion. Comparing the data with those in Table 2 for Cit^{3-}, it is possible to affirm that data for Pca are higher than those of the simple carboxylates. It is also reported that the acid–base properties depend on the MW [106].

Some data have also been reported in the literature (see, e.g., [107–110]) on the behavior of carboxylates in mixed NaCl/KCl media (or LiCl/KCl for Paa). Among them, Bretti et al. [110] studied the formation of ternary species with Btc^{4-} in mixed media determining, at I = 1.15 mol dm^{-3} and 25 °C, log K ($K^+ + NaH_iL^{(3-i)-} = KNaH_iL^{(2-i)-}$) = 1.41, 1.29, and 0.9 for i = 0, 1, and 2, respectively. These values, compared to those of the simple NaH_iL species, indicate a significant tendency to form mixed alkali metal ion pairs. In other papers, the formation of analogous species for Cit^{3-} [107], Paa [108], and Mlt^{6-} [109] may be determined using suitable computer programs, such as ES2WC [22], comparing protonation data in mixed media (NaCl + KCl) and in pure salts (NaCl or KCl).

6.3 *Thiols*

With the exception for GSH^{3-}, whose data have already been discussed above, few data are reported in the literature about the stability of alkali metal complexes with S-donor ligands. To our knowledge only two papers [90, 111] report quantitative data. Tac^-, $Mtac^-$, Tla^{2-}, Mpa^{2-}, and Tma^{3-} were studied by Bretti et al. [111], in which the interaction with Na^+ was modelled at different ionic strengths. The main species found by the authors are NaL (for all the ligands), NaHL (for Mpa^{2-}, Tla^{2-}, and Tma^{3-}), NaH_2L, and Na_2L (only for Tma^{3-}). The stability of the NaL species increases with increasing charge (z) of the ligand. In fact, the values are log $K_{NaL}=-0.51$, 0.48, 0.74, and 1.45 for $Mtac^-$ (z=−1), Mpa^{2-} (z=−2), Tla^{2-} (z=−2), and Tma^{3-} (z=−3), respectively. In general, the Na^+ ion pairs with thiols are less stable than the corresponding carboxylate complexes. However, considering also simple carboxylates (Ac^-, Mal^{2-}, Suc^{2-}, and Tca^{3-}), the authors found the following empirical relationship, valid for various $NaH_iL^{(1+i-z)}$ species, with predictive and test purposes: $\log K_{NaL}=-1.2+0.9\cdot|z-1|$, where i is the number of protons of the species and z is the charge of the ligand.

6.4 Complexones

Polyaminocarboxylate compounds such as, $EDDA^{2-}$, $MIDA^{2-}$, NTA^{3-}, $EDTA^{4-}$, $EDDS^{4-}$, $EGTA^{4-}$, $DTPA^{5-}$, $TTHA^{6-}$, the so-called complexones, are the most used chelating agents for industrial, environmental and biomedical applications, as well as for analytical purposes. The chelation therapy is only one example of the use of complexones in the medical field, and an important review was published recently [112]. Their importance is confirmed by the large number of investigations and reviews dedicated to them [113–116]. However, the alkali metal complexes of this class of ligands are only little investigated, owing to the difficult determination of the stability of these interactions. However, to obtain thermodynamic values of protonation constants in K^+- or Na^+-containing media, these weak interactions must be considered. Anderegg [113] suggested a positive correction (0.21 log K units) for the protonation constants of $EDTA^{4-}$ in 0.1 mol dm^{-3} KNO_3 medium with respect to those in Na^+. For this reason, some accurate values of protonation constants in various media have been reported in the literature for $EDTA^{4-}$ and NTA^{3-} [79, 117–119], according to Schwarzenbach et al. [120] and Schwarzenbach and Ackerman [121]. The alkali metal complex formation constants with NTA^{3-} and $EDTA^{4-}$ were critically analyzed by Anderegg [113, 114]. The average log K for $Li(NTA)^{2-}$, $Na(NTA)^{2-}$, and $K(NTA)^{2-}$ (at $I=0.1$ mol dm^{-3} and 25 °C) are: log $K=2.50$, 1.3, and 0.7, respectively, while it holds log $K=2.85$, 1.75, and 0.7 for $Li(EDTA)^{3-}$, $Na(EDTA)^{3-}$, and $K(EDTA)^{3-}$ under the same conditions. As expected for an ion association model, the stability of the species decreases with increasing the ionic radius of the cation.

De Stefano et al. [122] studied the interaction of two highly charged complexones, $DTPA^{5-}$ and $TTHA^{6-}$, with major components of natural waters. Using $(C_2H_5)_4NI$ as baseline electrolyte the authors found the presence of many Na_jH_iL complex species. For $TTHA^{6-}$, they determined the stability of the NaL^{5-}, $NaHL^{4-}$, NaH_2L^{3-}, NaH_3L^{2-}, NaH_4L^-, NaH_5L, Na_2L^{4-}, Na_2HL^{3-}, $Na_2H_2L^{2-}$, and Na_3L^{3-} species. Owing to a lower number of carboxylic groups, for DTPA NaH_5L^+, $Na_2H_2L^-$, and Na_3L^{2-} were not found. Considering the electrostatic nature of the interaction, the stability of the Na^+ ion pairs with $TTHA^{6-}$ is higher than those with $DTPA^{5-}$: for example, at infinite dilution, it is log $K=2.8$ and 2.5 for $Na(TTHA)^{5-}$ and $Na(DTPA)^{4-}$, respectively. Analogously, Crea et al. [123] determined six Na^+ complex species with $EGTA^{4-}$, namely NaL^{3-}, $NaHL^{2-}$, NaH_2L^-, NaH_3L, Na_2L^{2-}, and Na_2HL^-, whose stability at infinite dilution is log $K=2.1$, 1.5, 0.5, −1.0, 1.2, and 0.2, respectively. Bretti et al. [124] studied the interaction of $EDDS^{4-}$ and $CDTA^{4-}$ with Na^+, determining three species, namely NaL^{3-}, $NaHL^{2-}$, and Na_2L^{2-}. The stability of these species is log $K=1.68$ and 2.26 for the formation of $Na(EDDS)^{3-}$ and $Na(CDTA)^{3-}$; log $K=0.9$ and 1.8 for the $Na(HEDDS)^{2-}$ and $Na(HCDTA)^{2-}$ species (reaction: $Na^+ + HL = NaHL$) and log $K=-0.7$ and 1.3 for the $Na_2(EDDS)^{2-}$ and $Na_2(CDTA)^{2-}$ (reaction: $Na^+ + NaL = Na_2L$). Midgley [39] reviewed the stability of alkali metal complexes with some other aminopolycarboxylate ligands, obtaining the following stability trend: $CDTA^{4-} > PDTA^{4-} > UDA^{2-} > EDTA^{4-} > NTA^{3-} > IDA^{2-}$.

6.5 *Phosphonates*

An important class of organic molecules which interacts quite strongly with alkali metal cations is represented by the molecules that contain the phosph(on)ate moiety. This class can be divided into two sub-classes depending on the kind of bond between the phosphorus and carbon atom. If there is a direct carbon-phosphorus bond we have the sub-class of the phosphonates, if there is an O-bridge between the carbon and the phosphorus atom we have the sub-class of the phosphates. Due to their importance in medical and other fields, the interaction of phosphonates with alkali metal cations was studied in many papers (see, e.g., refs. in [125] and other papers cited below). Moreover, the interaction of methylphosphonic acid (H_3C-PO_3H_2), phosphonoacetic acid (HO_2C-CH_2-PO_3H_2), and methylenediphosphonic acid (H_2O_3P-CH_2-PO_3H_2) with Na^+ was studied by Alderighi et al. [126] with log $K_{ML}=0.54$, 0.99, and 2.13, respectively. The stability of the NaHL species (reaction: $Na^+ + HL = NaHL$) is log $K=-0.05$, 0.06, and 0.95 (same order) and that for the reaction $Na^+ + H_2L = NaH_2L$ is log $K=0.42$ for methylenediphosphonic acid. In the class of biphosphonates, important ligands for the treatment of osteoporosis [127], the alkali metal complexes of the $HEDPA^{4-}$ and Ris^{4-} have been studied in the past. In the review by Popov et al. [125] the formation of the ML and MHL species of alkali metal cations with $HEDPA^{4-}$ is reported: at $I=0.5$ mol dm^{-3}, it holds log $K_{ML}=3.4$, 2.1, 1.8, and 1.6 for Li^+, Na^+, K^+, and Cs^+, respectively. At the same ionic strength, for the reaction $M + HL = MHL$, log $K=1.1$, 0.5, 0.4, and 0.2 (same order). Foti et al. [128] determined the stability of Na^+ and K^+ complexes with $HEDPA^{4-}$ at different ionic strengths and at 25 °C, proposing the formation of six species, namely MH_3L, MH_2L, MHL, ML, M_2L, and M_2HL. In the literature, some values of complex formation enthalpy changes are also reported. In $(CH_3)_4NCl$ at $I=0.5$ mol dm^{-3} Carrol and Irani [129] found the value of $\Delta H_{ML}=1.7$ kJ mol^{-1} for both Na^+ and K^+, whereas Vasil'ev et al. [130] reported a value of $\Delta H=13.3$ kJ mol^{-1}

Table 3 Values of log *K* for the interaction of Na^+ and K^+ with $HEDPA^{4-}$ and Ris^{4-} ($I = 0.1$ mol dm^{-3}, 25 °C).

	$HEDPA^{4}$ [128]		Ris^{4-} [131]
Reaction	Na^+	K^+	Na^+
$M^+ + H_3L^- = MH_3L$	0.9	0.6	−0.9
$M^+ + H_2L^{2-} = MH_2L^-$	1.7	1.21	0.04
$M^+ + HL^{3-} = MHL^{2-}$	2.07	1.56	0.48
$M^+ + L^{4-} = ML^{3-}$	1.8	1.4	1.28
$M^+ + ML^{3-} = M_2L^{2-}$	−0.4	−0.1	0.77
$M^+ + MHL^{2-} = M_2HL^-$	1.4	0.9	−0.17

for the reaction $Na + H_2L = NaH_2L$ in $[(CH_3)_4]_2SO_4$ at $I = 0.5$ mol dm^{-3}. Bretti et al. [131] recently reported the formation of Na^+ complexes with Ris^{4-} at different ionic strengths with the same speciation scheme proposed by Foti et al. [128] in the case of $HEDPA^{4-}$. For a fast comparison, the alkali metal formation constants of $HEDPA^{4-}$ and Ris^{4-} with Na^+ and K^+ are reported in Table 3 at $I = 0.1$ mol dm^{-3} and 25 °C.

Daniele et al. [18] reviewed the interaction of Na^+ with $CMDPA^{4-}$, reporting three species in $(CH_3)_4NCl$ at $I = 0.1$ mol dm^{-3}, namely NaL (log $K = 1.4$), NaHL (log $K_{Na+HL} = 0.3$), and Na_2L (log $K_{Na+NaL} = 0.8$). Another important phosphonic acid is the $DTPMPA^{10-}$, which has five phosphonate groups and therefore forms very stable complexes with many cations. Cigala et al. [25], considering the totally deprotonated ligand as $L = H(DTPMPA)^{9-}$, reported the formation of seven mononuclear and two dinuclear species, namely, ML^{8-}, MHL^{7-}, MH_2L^{6-}, MH_3L^{5-}, MH_4L^{4-}, MH_5L^{3-}, MH_6L^{2-}, M_2L^{7-}, and M_2HL^{6-} for both Na^+ and K^+, whose stability for mononuclear MH_iL species (at 25 °C and infinite dilution) is a regular function of i, according to the following equations:

$$\log K = 3.78 - 0.01\ \mathrm{i} - 0.085\ \mathrm{i}^2 \quad (Na^+)$$
$$\log K = 3.31 - 0.27\ \mathrm{i} - 0.036\ \mathrm{i}^2 \quad (K^+)$$

6.6 *Phenols*

The interaction of phenols with alkali metal cations was not thoroughly studied in the literature. Demianov et al. [132] reported the interaction with Na^+ and K^+ of several phenols, namely phenol, *o*-cresol, *p*-cresol, *o*-nitrophenol, and *p*-nitrophenol. For all of them, only one species, ML, was determined, and the stability at infinite dilution is the same and very similar to that of the Na^+/ or K^+/Ac^- systems: log $K_{NaL} = -0.05$ and log $K_{KL} = -0.1$.

Bretti et al. [133] studied the formation of Na^+ ion pairs with an important class of natural ligands, the resorcinols. In particular, two species (NaL and NaHL) were determined for the three ligands considered: Res^{2-}, $Meres^{2-}$, and $Clres^{2-}$. At infinite dilution, the stability constants for the reaction $Na^+ + L^{2-} = NaL^-$ are log $K=0.494$, 0.350, and 0.356 for Res^{2-}, $Meres^{2-}$, and $Clres^{2-}$, respectively. For the reaction $Na^+ + HL^- = NaHL$ it holds log $K=-0.48$, -1.15, and -0.88 (same order). The data reported here for resorcinols are significantly lower than the corresponding values for dicarboxylates, however, it should be remarked that Bretti et al. [133] used $(CH_3)_4NCl$ as baseline electrolyte, whereas in most of the other cases $(C_2H_5)_4NI$ was adopted.

The interaction of maltol, ethyl-maltol, and kojic acid with Na^+ and K^+ was studied by Bretti et al. [134]. All these molecules are important in many fields. In particular, kojic acid and its derivatives are often used in pharmacy in the treatment of metal overload diseases, such as Wilson's disease (see, e.g., [135] and refs therein). Bretti et al. [134] studied the ionic strength and the temperature dependence of the Na^+ and K^+ complexes with the three ligands, reporting that the alkali metal association process at infinite dilution is slightly endothermic, and entropic in nature. In the same work, Bretti et al. [134] found that the mean values for NaL and KL complexes of the three investigated maltols are log $K=0.04$ and -0.03 at infinite dilution and 25 °C, which are in agreement with the stability of a generic mono-charged anion predicted by Daniele et al. [18]. On the contrary, the values determined for maltols are higher than those of resorcinols [133], for which the mean value for the formation constant of Na^+ complexes is log $K=-0.8$ for the neutral NaHL species under the same conditions.

6.7 *Other Ligands*

Midgley [39] reported that macrocyclic compounds exhibit an unusual strong stability and selectivity towards alkali metal cations due to their hydrophilic cavity inside the molecule. Many macrocyclic molecules occur naturally (for example the antibiotics valinomycin, monoactin, nigericin, and enneatin), and they can be divided into two main classes: (i) crown polyethers and (ii) macrobicyclic diamines. Due to the nature of the interaction between the cations and the macrocycle, the stability constants are much higher than those of other organic or inorganic ligands, and generally the conditional constants for Na^+ and K^+ are similar. In addition, they generally do not depend on the counter anion of the alkali metal cation. The stability constants, whose values have been determined mainly by NMR spectroscopy [136], can range from $0.5 < \log K < 6.0$. Some recommended values relative to the thermodynamic formation parameters of alkali metal cations (not only) and crown polyethers (15C5 and 18C6) are given by IUPAC [137] and are summarized in Table 4. With respect to other alkali metal ion interactions, the enthalpy change is generally high, showing that the main contribution to the stability is enthalpic in nature.

Finally, it is worth mentioning that, besides their importance in biological systems, very few or even no literature data are available for the stability of alkali metal cation complexes with carbohydrate species [11, 12, 14–16].

Table 4 Values of log K_{ML} for the interaction of alkali metal cations with crown ethers from IUPAC [137] (infinite dilution and 25 °C).

Cation	log K	$-\Delta G$ (kJ mol^{-1})	ΔH (kJ mol^{-1})	$T\Delta S$ (kJ mol^{-1})
		15C5		
Na^+	0.80	4.6	−6.3	−1.7
K^+	0.75	4.3	−17.2	−12.9
Cs^+	0.80	4.6	−5.4	−0.8
		18C6		
Na^+	0.80	4.6	−11.0	−6.0
K^+	2.05	11.7	−25.0	−13.3
Rb^+	1.51	8.6	−16.0	−7.4
Cs^+	0.96	5.5	−17.0	−12.0

7 The Relevance of Alkali Metal Ion Complexes in Modelling Biofluids

Biological fluids, like other real systems, are very complex and different from one another, both from the chemical and physical point of view. As a consequence, their study needs, very often, the use of various "models". The importance of the alkali metal complexes in building models for biological fluids is very high. In fact, almost all natural fluids, including biological ones, contain alkali metal cations in high concentrations, mostly Na^+ and K^+. The term "complex" does not imply any particular kind of bonding in the associated species, which may be an ion pair, a solvent separated ion pair or a coordination compound [39]. The typical concentrations of these ions in several biological fluids are discussed in Section 2.1. The presence of these cations may lead to the following exchange reaction:

$$\mathrm{ML + Na = NaL + M}$$

Even if the equilibrium constant of the above reaction is generally low (log $K<1$), very often it is $[\mathrm{Na}]_T >> [\mathrm{M}]_T$, leading to the formation of non-negligible amounts of the NaL complex. For this reason some attempts of building models for the most important biological fluids were carried out in the past. Three of them were proposed by this group, regarding the speciation of particular biological fluids: SERPE (SERum sPEciation) to model the blood plasma [138], URSUS (URine Speciation and Urine Saturation) for urine [139], and SALMO (SALiva MOdel) for saliva [28]. All of them are based on a set of stability constants of species that may be formed in the fluid considered, together with their ionic strength and temperature dependent parameters. Once the composition of the fluid is fixed (e.g., pH and analytical concentrations of different components), it is possible with these models to determine the amount of species formed together with their errors, calculated from the uncertainties associated to the formation constants used. In addition, URSUS also determines the degree of saturation of urine related to the possible formation of kidney stones.

As a general consideration, two main approaches may be used in modelling (multi)electrolyte solutions: (i) the ion pairs approach, and (ii) the variation of activity coefficients [1, 3–6, 26, 27]. The first is a chemical approach based on the solution of an adequate number of mass balance equations to determine the exact concentration of all the chemical species present in the system, where the deviations from the ideality (ideal solutions) are interpreted in terms of formation of complexes and/or ion pairs, whose stability constants depend only on the effective (total) ionic strength of the solution. Some examples of computer programs that use this approach are Haltafall [140], COMICS [141], Solgaswater [142], ES4EC [143], and OLI [144]. The second is mostly a physical approach, where the deviations from the ideality are interpreted in terms of ionic interactions, which are "specific" for each couple (or more) of ions. Some models based on this approach are, e.g., SIT [5, 145], Pitzer [146, 147], Bromley [148], and the Mean Spherical Approximation [149, 150].

Whether chemical or physical approaches are used, updated databases are needed, in order to gather the necessary information about the systems under investigation. Research in this field must be supported, especially about the knowledge of reaction enthalpies, solubility and interaction parameters. At the moment, only few data are available on alkali metal ion complexes in the most common equilibrium constant databases [11–16].

8 Alkali Metal Ions as Probes in Biological Systems

The presence of alkali metal ions in biological fluids has resulted in the need to develop many analytical methods for their determination and different types of probes sensitive to their concentration. On the opposite, in some cases, ions themselves are used as probes or tracers or, indirectly, in the synthesis of recognition systems. Important examples concern the use of Na^+ in the synthesis of rotaxanes, which consist of a linear species (guest) and cyclic species (host) bound together in a threaded structure by non-covalent forces. In these syntheses, Na^+ ions template the threading of linear molecules to form pseudorotaxane. More recently (e.g., [151, 152]), it has been observed that a single urea or amide functionality in a dumbbell-shaped guest is "clipped" by a macrocycle generated from a diamine and a dialdehyde through the templating effect of a Na^+ ion.

Lithium ion was used as a probe of Na^+ channel activity [153] and, under many forms, for the determination of the oxygen concentration in tissues, by electron paramagnetic resonance (EPR). Many compounds of lithium such as unsubstituted lithium phthalocyanine, lithium naphthalocyanine or tetraphenoxy-substituted lithium phthalocyanine were proposed for this application [154–156].

Potassium (as potassium permanganate) has widely been used in genomic footprinting assays to map DNA-protein interactions [157] and to study the DNA distortion [158]. The isotopes of rubidium are widely applied in medicine, in particular in positron emission tomography (PET) to evaluate brain function, and in other similar studies of the heart and brain. In other studies, Rb^+ was used as potassium tracer considering the similarity with K^+. More recently, the non-radioactive Rb^+ efflux assay has largely replaced radioactive $^{86}Rb^+$ and found widespread application in drug discovery and development. It replaced $^{86}Rb^+$ assays for the analysis of K^+ and nonselective cation channels in the pharmaceutical industry [159].

9 General Conclusions

In this contribution, the interaction of alkali metal cations with organic and inorganic ligands of biological interest has been reviewed. In the literature, thermodynamic data on this kind of complexes have always been scarce, due to the weakness of these species (rarely log $K>2$), which led to problems in their experimental determination (mainly performed by potentiometric or conductimetric techniques): some of these data are available in the most common databases [11–16]. In Section 2.1, the concentration of Li^+, Na^+, K^+, Rb^+, and Cs^+ in various biological fluids has been reported, evidencing that, especially for Na^+ and K^+, it is high enough to lead to the formation of non-negligible amounts of alkali metal complexes. Generally, it was found that the stability of the complexes follows the trend $Li^+>Na^+>K^+>Rb^+>Cs^+$, which is inversely proportional to the size of the cations. For example, for Cit^{3-} it holds log $K_{ML}=0.88$, 0.80, 0.48, 0.38, and 0.13 (in the preceding order) at 25 °C and infinite dilution. Some exceptions are observed with $Fe(CN)_6^{4-}$, which shows an opposite trend, and sulfate, for which a minimum is observed for the stability of the $NaSO_4^-$ complexes.

Almost all classes of ligands interact with alkali metal cations. For amines, only the complexes of Li^+ with En are reported in the literature [98].

The stability of the complexes strongly depends on the number of charges involved in the formation reaction: for example, complexones show high stability constants (log $K_{ML}=2.85$ and 1.75 for $Li(EDTA)^{3-}$ and $Na(EDTA)^{3-}$, respectively).

In Table 3, a summary of the stability of alkali metal complexes of some biologically relevant ligands is presented, also to show the strength of these complexes. Furthermore, one observes how the stability of the species depends to a large part on the charge of the ligand. In fact, using the data reported in this chapter for inorganic and organic ligands, it is possible to express the dependence of the association constant [$Na^+ + H_iL^{z-} = Na(H_iL)^{(z-1)-}$] between Na^+ and the ligands as a function of its charge (z). The diagram in Figure 1 shows that the formation constant increases with increasing "z", according to a second degree polynomial function, indicating a strong electrostatic contribution to the formation of these weak species. The equation, determined by using a population of 125 data, including protonated species, is valid at infinite dilution and 25 °C:

$$\log K^{M} = -0.94 + 0.86 \cdot z - 0.036 \cdot z^2,$$

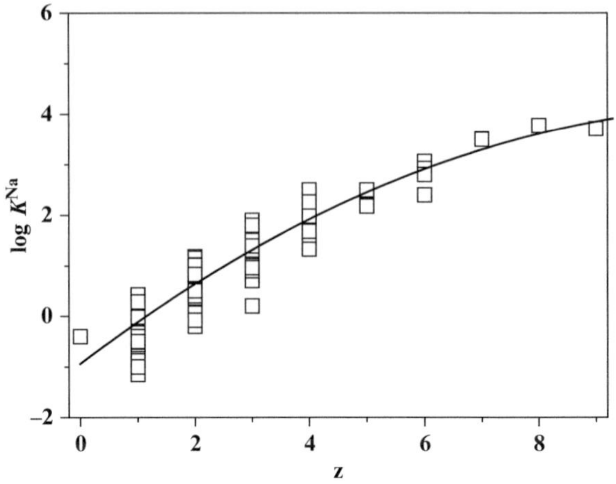

Figure 1 Dependence of the stability constant of various $Na(H_iL)^{(z-1)-}$ species [equilibrium $Na^+ + H_iL^{z-} = Na(H_iL)^{(z-1)-}$] on the anion charge "z" at infinite dilution and 25 °C.

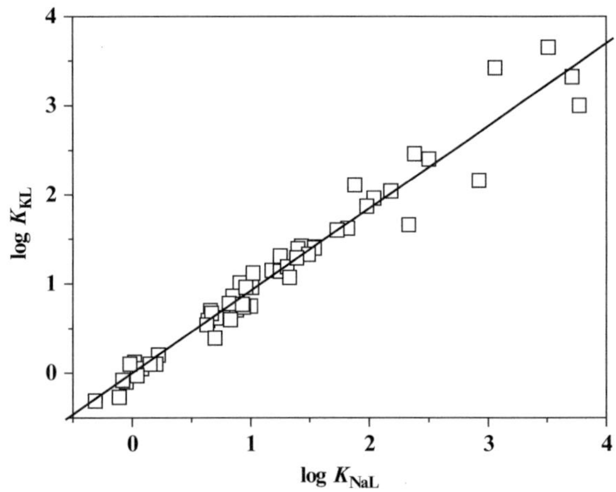

Figure 2 Correlation between the stability constants of NaL and KL species of different ligands, at infinite dilution and 25 °C.

with a standard deviation $\sigma = 0.06$ [please note that z is variable, e.g., z = 1 for $H_2(Cit)^-$, 2 for $H(Cit)^{2-}$, 3 for Cit^{3-}, etc.].

Worth mentioning is also the fact that, for a wide series of ligands of different classes (inorganic ligands, carboxylates, nucleotides, phosphonic ligands, complexones, amino acids, etc.), there is a significant linear correlation (r = 0.98 on 53 experimental data) between the formation constants of the NaL and KL species (Figure 2):

$$\log K_{KL} = 0.925 \log K_{NaL}$$

Similar considerations also hold for Li^+ and Na^+ complexes

$$\log K_{LiL} = 1.107 \log K_{NaL}$$

with r = 0.96 (on 34 experimental data).

Little knowledge exists on the thermodynamics of the alkali metal ion complexes, for example, the most reliable ΔH data are $\Delta H = 1.3$, −8, 1.0, −3, and 2.0 kJ mol^{-1} for the interaction of Na^+ with OH^-, Cl^-, nucleotides, amino acids, and $HEDPA^{4-}$, indicating that generally the main contribution to the stability is entropic in nature, except for the crown polyethers. In many cases, it was demonstrated that the formation of mixed MM'H_iL species is thermodynamically favored. For example, Na^+/K^+ species have been found with Btc^{4-}, Cit^{3-}, and phosphates, while Li^+/Cs^+ mixed species were determined with Phy^{12-}. The absence of further examples is mostly due to the lack of experiments in mixed media.

The wide use of models for the dependence on the ionic strength based on the variation of the activity coefficients, such as SIT or Pitzer, has resulted in the lack of development of specific calculation programs for the determination of weak complexes, which, in fact are predominantly developed by individual research groups. It is hoped that, in the future, the problem of the determination of the weak complexes will be taken into serious consideration, in accordance with the wide literature on this subject and the fact that the cations of the alkali metals are present in large amounts in all natural fluids.

Abbreviations and Definitions

Abbreviations of ligands in formulae refer in general to their totally deprotonated form.

ΔG	Gibbs energy
ΔH	enthalpy change
ΔS	entropy change
15C5	1,4,7,10,13-pentaoxacyclopentadecane
18C6	1,4,7,10,13,16-hexaoxacyclo-octadecane
Ac^-	acetate
ADP^{3-}	adenosine 5'-diphosphate
Ala^-	alaninate
Am	general amine
AMP^{2-}	adenosine 5'-monophosphate
AQP^{5-}	adenosine 5'-tetraphosphate
Asp^{2-}	aspartate
ATP^{4-}	adenosine 5'-triphosphate
Btc^{4-}	1,2,3,4-butanetetracarboxylate
$Ccys^{2-}$	S-carboxymethyl-L-cysteine carbocysteine
$CDTA^{4-}$	1,2-diaminocyclohexanetetraacetate
Cit^{3-}	citrate
$Clres^{2-}$	4-chloro-resorcinolate
$CMDPA^{4-}$	dichloromethylenebiphosphonate
$DTPA^{5-}$	diethylenediamine-N,N,N′,N′′,N′′-pentaacetate

$DTPMPA^{10-}$	diethylenetriamino-N,N,N′,N″,N″-pentakis(methylenephosphonate)
$EDDA^{2-}$	ethylenediamine-N,N′-diacetate
$EDDS^{4-}$	(S,S)-ethylenediamine-N,N′-disuccinic acid
$EDTA^{4-}$	ethylenediamine-N,N,N′,N′-tetraacetate
$EGTA^{4-}$	ethylene glycol-bis(2-aminoethylether)-N,N,N′,N′-tetraacetate
En	ethylenediamine
Et	ethyl group
Glu^{2-}	glutamate
Gly^{-}	glycinate
$Glyc^{-}$	glycolate
GSH^{3-}	totally deprotonated reduced glutathione
$HEDPA^{4-}$	1-hydroxyethylidenediphosphonate etidronate
His^{-}	histidinate
IDA^{2-}	iminodiacetate
Imid	imidazole
IMP^{2-}	inosine 5′-monophosphate
ISE	ion selective electrode
IUPAC	International Union of Pure and Applied Chemistry
K_{S0}	solubility product
L	general ligand
L’	general ligand different from L
$Lact^{-}$	lactate
log K or log β	equilibrium constant (as logarithm decimal base)
Lys^{-}	lysinate
M	general metal ion
M’	general metal ion different from M
Mal^{2-}	malonate
$Mala^{2-}$	malate
Me	methyl group
$Meres^{2-}$	2-methyl-resorcinolate
$MIDA^{2-}$	methyliminodiacetate
Mlt^{6-}	1,2,3,4,5,6-benzenehexacarboxylate (mellitate)
Mpa^{2-}	mercaptopropionate
$Mtac^{-}$	(methylthio)acetate
MW	molecular weight
NTA^{3-}	nitrilotriacetate
Ox^{2-}	oxalate
Paa	polyacrylate
Pca	acrylate-maleate copolymers
$PDTA^{4-}$	1,2-diaminopropanetetraacetate
Phe^{-}	phenylalaninate
$Phtal^{2-}$	*o*-phtalate
Phy^{12-}	1,2,3,4,5,6,*myo*-inositol hexakis(dihydrogen phosphate) (phytate)
Pma	polymethacrylate
Pyr^{-}	pyruvate

R	alkyl group
Res^{2-}	1,3-benzenediol resorcinolate
Ris^{4-}	(1-hydroxy-1-phosphono-2-pyridin-3-yl-ethyl)phosphonate risedronate
Ser^{-}	serinate
SIT	Specific Ion Interaction Theory
$Ssal^{2-}$	5-sulfosalicylate
Suc^{2-}	succinate
Tac^{-}	thioacetate
Tar^{2-}	tartrate
Tca^{3-}	1,2,3-propanetricarboxylate (tricarballylate)
Tla^{2-}	thiolactate
Tma^{3-}	thiomalate
Trp^{-}	tryptophanate
$TTHA^{6-}$	triethylenetetramine-N,N,N′,N″,N‴,N‴-hexaacetate
Tyr^{-}	tyrosinate
UDA^{2-}	uramil-N,N-diacetate
[X]	free concentration of the component X
$[X]_T$	analytical (total) concentration of the component X
X^{n-}	general anion
z	charge

Acknowledgment We thank the University of Messina for partial financial support.

References

1. H. S. Harned, B. B. Owen, *The Physical Chemistry of Electrolytic Solutions*, Reinhold Pub. Corp., New York, 1958, pp. 803.
2. C. Lentner, *Geigy Scientific Tables,* 8th edn., Ed C. Lentner, CIBA-Geigy, Basel, Switzerland, 1983.
3. J. Buffle, *Complexation Reactions in Aquatic Systems: An Analytical Approach*, Ellis Horwood, Chichester, 1988, pp. 692.
4. W. Stumm, J. J. Morgan, *Aquatic Chemistry. Chemical Equilibria and Rates in Natural Waters,* 3rd edn., John Wiley & Sons, Inc., New York, 1996, pp. 1040.
5. I. Grenthe, I. Puigdomenech, *Modelling in Aquatic Chemistry*, Eds I. Grenthe, I. Puigdomenech, OECD, Paris, 1997, pp. 777.
6. F. J. Millero, *Physical Chemistry of Natural Waters*, John Wiley & Sons, Inc., New York, 2001, pp. 654.
7. R. M. Roat-Malone, *Bioinorganic Chemistry: A Short Course,* 2nd edn., John Wiley & Sons, Inc., Hoboken, NJ, 2007, pp. 544.
8. R. R. Crichton, *Biological Inorganic Chemistry,* 2nd edn., *A New Introduction to Molecular Structure and Function*, Elsevier, Amsterdam, 2012, pp. 472.
9. *Elements and Their Compounds in the Environment,* 2nd edn., Eds E. Merian, M. Anke, M. Ihnat, M. Stoeppler, Wiley-VCH Verlag GmbH & Co. KGaA, Weinheim, 2004, pp. 1773.
10. D. M. Templeton, F. Ariese, R. Cornelis, L. G. Danielsson, H. Muntau, H. P. van Leeuwen, R. Lobinski, *Pure Appl. Chem.* **2000**, *72*, 1453–1470.

11. L. G. Sillén, A. E. Martell, *Stability Constants of Metal Ion Complexes, Special Publ. 17*, The Chemical Society, Wiley, London, 1964.
12. L. G. Sillén, A. E. Martell, *Stability Constants of Metal Ion Complexes. Supplement Special Publ. 25*, The Chemical Society, Wiley, London, 1964.
13. E. Hogfeldt, *Stability Constants of Metal-Ion Complexes. Part A: Inorganic Ligands, IUPAC Chemical Data Series*, Pergamon Press, Oxford, 1982, pp. 310.
14. P. M. May, K. Muray, *Talanta* **1991**, *38*, 1419–1426.
15. D. Pettit, K. Powell, *IUPAC Stability Constants Database*, Academic Software, Otley, UK, 2004.
16. A. E. Martell, R. M. Smith, R. J. Motekaitis, *NIST Standard Reference Database 46, version 8*, Gaithersburg, 2004.
17. Y. Marcus, G. Hefter, *Chem. Rev.* **2006**, *106*, 4585–4621.
18. P. G. Daniele, C. Foti, A. Gianguzza, E. Prenesti, S. Sammartano, *Coord. Chem. Rev.* **2008**, *252*, 1093–1107.
19. S. Berto, P. G. Daniele, G. Lando, E. Prenesti, S. Sammartano, *Int. J. Electrochem. Sci.* **2012**, *7*, 10976–10986.
20. A. Braibanti, G. Ostacoli, P. Paoletti, L. D. Pettit, S. Sammartano, *Pure Appl. Chem.* **1987**, *59*, 1721–1728.
21. R. P. Buck, S. Rondinini, A. K. Covington, F. G. K. Baucke, C. M. A. Brett, M. F. Camoes, M. J. T. Milton, T. Mussini, R. Naumann, K. W. Pratt, P. Spitzer, G. S. Wilson, *Pure Appl. Chem.* **2002**, *74*, 2169–2200.
22. A. De Robertis, C. De Stefano, S. Sammartano, C. Rigano, *Talanta* **1987**, *34*, 933–938.
23. P. G. Daniele, A. De Robertis, C. De Stefano, S. Sammartano, C. Rigano, *J. Chem. Soc. Dalton Trans.* **1985**, 2353–2361.
24. C. De Stefano, D. Milea, A. Pettignano, S. Sammartano, *Anal. Bioanal. Chem.* **2003**, *376*, 1030–1040.
25. R. M. Cigala, M. Cordaro, F. Crea, C. De Stefano, V. Fracassetti, M. Marchesi, D. Milea, S. Sammartano, *Ind. Eng. Chem. Res.* **2014**, *53*, 9544–9553.
26. *Activity Coefficients in Electrolyte Solutions*, Ed R. M. Pytkowicz, Vol. 1, CRC Press, Inc., 1979, pp. 288.
27. *Activity Coefficients in Electrolyte Solutions*, Ed R. M. Pytkowicz, Vol. 2, CRC Press, Inc., 1979, pp. 330.
28. F. Crea, C. De Stefano, D. Milea, A. Pettignano, S. Sammartano, *Bioinorg. Chem. Appl.* **2015**, *2015*, pp. 12, ID 267985.
29. P. G. Daniele, M. Grasso, C. Rigano, S. Sammartano, *Ann. Chim. (Rome)* **1983**, *73*, 495–515.
30. C. F. Baes, R. E. Mesmer, *Am. J. Sci.* **1981**, *281*, 935–962.
31. K. Popov, L. H. J. Lajunen, A. Popov, H. Rönkkömäki, M. Hannu-Kuure, A. Vendilo, *Inorg. Chem. Commun.* **2002**, *5*, 223–225.
32. F. J. Millero, *J. Phys. Chem.* **1970**, *74*, 356–362.
33. K. Johnson, R. M. Pytkowicz, *Am. J. Sci.* **1978**, *278*, 1428–1447.
34. A. De Robertis, C. Rigano, S. Sammartano, O. Zerbinati, *Thermochim. Acta* **1987**, *115*, 241–248.
35. A. A. Reznikov, V. A. Shaposhnik, *Russ. J. Phys. Chem. A* **2007**, *81*, 179–181.
36. R. Heyrovská, *J. Electrochem. Soc.* **1996**, *143*, 1789–1793.
37. P. G. Daniele, C. Rigano, S. Sammartano, *Inorg. Chim. Acta* **1982**, *63*, 267–272.
38. E. C. Righellato, C. W. Davies, *Trans. Faraday Soc.* **1930**, *26*, 592–600.
39. D. Midgley, *Chem. Soc. Rev.* **1975**, *4*, 549–568.
40. F. P. Daly, C. W. Brown, D. R. Kester, *J. Phys. Chem.* **1972**, *79*, 3664–3668.
41. F. Crea, C. De Stefano, A. Gianguzza, D. Piazzese, S. Sammartano, *Talanta* **2006**, *68*, 1102–1112.
42. P. G. Daniele, A. De Robertis, C. De Stefano, A. Gianguzza, S. Sammartano, *J. Solution Chem.* **1991**, *20*, 495–515.

43. R. M. Smith, R. A. Alberty, *J. Phys. Chem.* **1956**, *60*, 180–184.
44. C. De Stefano, C. Foti, A. Gianguzza, *J. Chem. Res.* **1994**, (S) 464 (M) 2639–2661.
45. C. De Stefano, C. Foti, A. Gianguzza, D. Piazzese, *Chem. Spec. Bioavail.* **1998**, *10*, 19–26.
46. F. Crea, C. De Stefano, D. Milea, S. Sammartano, *Coord. Chem. Rev.* **2008**, *252*, 1108–1120.
47. F. Crea, P. Crea, C. De Stefano, D. Milea, S. Sammartano, *J. Mol. Liq.* **2008**, *138*, 76–83.
48. N. Veiga, J. Torres, I. Macho, K. Gomez, G. Gonzalez, C. Kremer, *Dalton Trans.* **2014**, *43*, 16238–16251.
49. R. M. Cigala, F. Crea, G. Lando, D. Milea, S. Sammartano, *J. Chem. Thermodyn.* **2010**, *42*, 1393–1399.
50. A. M. Bond, G. T. Hefter, *Critical Survey of Stability Constants and Related Thermodynamic Data of Fuoride Complexes in Aqueous Solution*, Ed. IUPAC, Pergamon Press, Oxford (UK), 1980, pp. 67.
51. G. R. Miller, D. R. Kester, *Mar. Chem.* **1976**, *4*, 67–82.
52. F. J. Rawson, A. J. Downard, K. H. Baronian, *Sci. Rep.* **2014**, *4*, pp. 9, ID:5216.
53. A. De Robertis, C. Rigano, S. Sammartano, *Ann. Chim. (Rome)* **1984**, *74*, 33–39.
54. S. Capone, A. De Robertis, C. De Stefano, R. Scarcella, *J. Chem. Res.* **1986**, (S) 412.
55. S. Capone, A. De Robertis, S. Sammartano, C. Rigano, *Thermochim. Acta* **1986**, *102*, 1–14.
56. M. T. Beck, *Chemistry of Complex Equilibria*, Van Nostrand Reinhold, London, 1970, pp. 285.
57. W. A. Eaton, P. George, G. I. H. Hanania, *J. Phys. Chem.* **1967**, *71*, 2016–2021.
58. J. M. Pratt, R. J. P. Williams *J. Chem. Soc. A* **1967**, 1291–1298.
59. J. Z. Yang, P. S. Song, D. B. Wang, *J. Chem. Thermodyn.* **1997**, *29*, 1343–1351.
60. A. Bousher, *J. Coord. Chem.* **1995**, *34*, 1–11.
61. L. M. Rowe, L. B. Tran, G. Atkinson, *J. Solution Chem.* **1989**, *18*, 675–689.
62. H. R. Rogers, C. M. G. Van Den Berg, *Talanta* **1988**, *35*, 271–275.
63. A. Gianguzza, D. Milea, F. J. Millero, S. Sammartano, *Mar. Chem.* **2004**, *85*, 103–124.
64. G. Svehla, *Vogel's Textbook of Macro and Semimicro Qualitative Inorganic Analysis,* 5th edn., Longman Group Limited, London, 1979, pp. 605.
65. F. Crea, A. De Robertis, C. De Stefano, S. Sammartano, *Talanta* **2007**, *71*, 948–963.
66. *Metal Ions in Biological Systems: Volume 32: Interactions of Metal Ions with Nucleotides: Nucleic Acids, and Their Constituents*, Eds A. Sigel, H. Sigel, Marcel Dekker Inc., New York, 1996, pp. 854.
67. R. M. Smith, A. E. Martell, Y. Chen, *Pure Appl. Chem.* **1991**, *63*, 1015–1080.
68. N. C. Melchior, *J. Biol. Chem.* **1954**, *208*, 615–628.
69. M. M. Taqui Khan, A. E. Martell, *J. Am. Chem. Soc.* **1966**, *88*, 668–671.
70. J. M. Blair, *Eur. J. Biochem.* **1970**, *13*, 384–390.
71. M. S. Mohan, G. A. Rechnitz, *J. Am. Chem. Soc.* **1970**, *92*, 5839–5842.
72. G. A. Rechnitz, M. S. Mohan, *Science* **1970**, *168*, 1460.
73. N. C. Melchior, G. A. Rechnitz, M. S. Mohan, *Science* **1971**, *171*, 1267–1268.
74. E. J. Foot, G. A. Rechnitz, *Arch. Biochem. Biophys.* **1974**, *165*, 604–614.
75. R. K. Kobos, G. A. Rechnitz, *Arch. Biochem. Biophys.* **1976**, *175*, 11–20.
76. R. Calì, S. Musumeci, C. Rigano, S. Sammartano, *Inorg. Chim. Acta* **1981**, *56*, L11–L13.
77. A. De Robertis, C. De Stefano, S. Sammartano, R. Calì, R. Purrello, C. Rigano, *J. Chem. Res.* **1986**, (S) 164 (M) 1301–1347.
78. C. De Stefano, D. Milea, A. Pettignano, S. Sammartano, *Biophys. Chem.* **2006**, *121*, 121–130.
79. J. Botts, A. Chashin, H. L. Young, *Biochemistry* **1965**, *4*, 1788–1796.
80. F. Crea, C. De Stefano, A. Gianguzza, D. Piazzese, S. Sammartano, *Chem. Spec. Bioavail.* **2004**, *16*, 1–8.
81. A. De Robertis, C. De Stefano, A. Gianguzza, *Thermochim. Acta* **1991**, *177*, 39–57.
82. A. De Robertis, C. De Stefano, D. Milea, S. Sammartano, *J. Solution Chem.* **2005**, *34*, 1211–1226.

83. A. Casale, C. De Stefano, G. Manfredi, D. Milea, S. Sammartano, *Bioinorg. Chem. Appl.* **2009**, *2009*, pp. 17, ID 219818.
84. F. Crea, C. Foti, D. Milea, S. Sammartano, in *Cadmium: From Toxicity to Essentiality*, Vol. 11 of *Metal Ions in Life Sciences*, Eds A. Sigel, H. Sigel, R. K. O. Sigel, Springer Science+Business Media B.V., Dordrecht, 2013, Vol. 11, pp. 63–83.
85. D. Cucinotta, C. De Stefano, O. Giuffrè, G. Lando, D. Milea, S. Sammartano, *J. Mol. Liquids* **2014**, *200*, 329–339.
86. S. Fiol, I. Brandariz, M. S. de Vicente, *Talanta* **1995**, *42*, 797–801.
87. A. Casale, C. De Stefano, S. Sammartano, P. G. Daniele, *Talanta* **1989**, *36*, 903–907.
88. C. De Stefano, A. Gianguzza, *Ann. Chim. (Rome)* **1991**, *81*, 119–130.
89. C. De Stefano, C. Foti, A. Gianguzza, S. Sammartano, *Chem. Spec. Bioavail.* **1995**, *7*, 1–8.
90. C. Bretti, A. Giacalone, A. Gianguzza, D. Milea, S. Sammartano, *Fluid Phase Equilib.* **2007**, *252*, 119–129.
91. A. S. Reddy, G. N. Sastry, *J. Phys. Chem. A* **2005**, *109*, 8893–8903.
92. P. Crea, A. De Robertis, C. De Stefano, D. Milea, S. Sammartano, *J. Chem. Eng. Data* **2007**, 1028–1036.
93. R. M. Cigala, F. Crea, C. De Stefano, G. Lando, D. Milea, S. Sammartano, *Amino Acids* **2012**, *43*, 629–648.
94. C. Bretti, F. Crea, O. Giuffrè, S. Sammartano, *J. Solution Chem.* **2008**, *37*, 183–201.
95. C. Bretti, F. Crea, C. De Stefano, S. Sammartano, *Fluid Phase Equilib.* **2008**, *272*, 47–52.
96. S. Cascio, A. De Robertis, C. Foti, *Fluid Phase Equilib.* **2000**, *170*, 167–181.
97. C. Bretti, R. M. Cigala, C. De Stefano, G. Lando, S. Sammartano, *J. Chem. Eng. Data* **2013**, *59*, 143–156.
98. A. Casale, A. De Robertis, F. Licastro, C. Rigano, *J. Chem. Res.* **1990**, (S) 204 (M) 1601–1620.
99. A. De Robertis, C. De Stefano, C. Rigano, S. Sammartano, R. Scarcella, *J. Chem. Res.* **1985**, (S) 42, (M) 629–650.
100. P. G. Daniele, A. De Robertis, C. De Stefano, A. Gianguzza, S. Sammartano, *J. Chem. Res.* **1990**, (S) 300 (M) 2316–2350.
101. A. De Robertis, C. Foti, A. Gianguzza, *Ann. Chim. (Rome)* **1993**, *83*, 485–497.
102. A. De Robertis, C. De Stefano, C. Foti, *Ann. Chim. (Rome)* **1996**, *86*, 155–166.
103. A. Katchalsky, *Biophys. J.* **1964**, *4*, 9–41.
104. C. De Stefano, A. Gianguzza, D. Piazzese, S. Sammartano, *Mar. Chem.* **2004**, *86*, 33–44.
105. C. Bretti, F. Crea, C. Rey-Castro, S. Sammartano, *React. Funct. Polym.* **2005**, *65*, 329–342.
106. C. De Stefano, A. Gianguzza, D. Piazzese, S. Sammartano, *React. Funct. Polym.* **2003**, *55*, 9–20.
107. F. Crea, C. De Stefano, F. J. Millero, V. K. Sharma, *J. Solution Chem.* **2004**, *33*, 1349–1367.
108. R. M. Cigala, F. Crea, S. Sammartano, *J. Mol. Liq.* **2008**, *143*, 129–133.
109. R. M. Cigala, F. Crea, C. De Stefano, S. Sammartano, *J. Chem. Eng. Data* **2009**, *54*, 2137–2139.
110. C. Bretti, R. M. Cigala, F. Crea, S. Sammartano, *Talanta* **2007**, *72*, 1059–1065.
111. C. Bretti, C. De Stefano, C. Foti, O. Giuffrè, S. Sammartano, *J. Solution Chem.* **2009**, *38*, 1225–1245.
112. G. Crisponi, V. M. Nurchi, J. I. Lachowicz, M. Crespo-Alonso, M. A. Zoroddu, M. Peana, *Coord. Chem. Rev.* **2015**, *284*, 278–285.
113. G. Anderegg, *Critical Survey of Stability Constants of EDTA Complexes*, Pergamon Press, Oxford (UK), 1977, pp. 48.
114. G. Anderegg, *Pure Appl. Chem.* **1982**, *54*, 2693–2758.
115. G. Anderegg, F. Arnaud-Neu, R. Delgado, J. Felcman, K. Popov, *Pure Appl. Chem.* **2005**, *77*, 1445–1495.
116. K. Popov, H. Wanner, In *Biogeochemistry of Chelating Agents*, Eds B. Nowack, J. M. Vanbriesen, American Chemical Society, Washington DC, 2005, pp. 50.
117. G. Anderegg, *Helv. Chim. Acta* **1967**, *50*, 2333–2340.

118. G. Anderegg, *Z. Naturforsch. B* **1976**, *31*, 786–789.
119. J. I. Watters, O. E. Schupp III, *J. Inorg. Nucl. Chem.* **1968**, *30*, 3359–3362.
120. G. Schwarzenbach, F. Kampitsch, R. Steiner, *Helv. Chim. Acta* **1945**, *28*, 828–840.
121. G. Schwarzenbach, H. Ackermann, *Helv. Chim. Acta* **1947**, *30*, 1798–1804.
122. C. De Stefano, S. Sammartano, A. Gianguzza, D. Piazzese, *Anal. Bioanal. Chem.* **2003**, *375*, 956–967.
123. F. Crea, C. De Stefano, A. Gianguzza, D. Piazzese, S. Sammartano, *Chem. Spec. Bioavail.* **2003**, *15*, 75–86.
124. C. Bretti, C. De Stefano, C. Foti, S. Sammartano, *J. Solution Chem.* **2013**, *42*, 1452–1471.
125. K. Popov, H. Rönkkömäki, L. H. J. Lajunen, *Pure Appl. Chem.* **2001**, *70*, 1641–1677.
126. A. Alderighi, A. Vacca, F. Cecconi, S. Midollini, E. Chinea, S. Dominguez, A. Valle, D. Dakternieks, A. Duthie, *Inorg. Chim. Acta* **1999**, *285*, 39–48.
127. R. G. G. Russell, *Bone* **2011**, *49*, 2–19.
128. C. Foti, O. Giuffrè, S. Sammartano, *J. Chem. Thermodyn.* **2013**, *66*, 151–160.
129. R. Carrol, R. Irani, *Inorg. Chem.* **1967**, *6*, 1994–1998.
130. V. P. Vasil'ev, E. V. Kozlovskii, T. B. Marina, *Zhur. Neorg. Khim.* **1985**, *30*, 36–40.
131. C. Bretti, I. Cukrowski, C. De Stefano, G. Lando, *J. Chem. Eng. Data* **2014**, *59*, 3728–3740.
132. P. Demianov, C. De Stefano, A. Gianguzza, S. Sammartano, *Environ. Toxicol. Chem.* **1995**, *14*, 767–773.
133. C. Bretti, C. De Stefano, G. Lando, S. Sammartano, *Fluid Phase Equilib.* **2010**, *292*, 71–79.
134. C. Bretti, R. M. Cigala, C. De Stefano, G. Lando, S. Sammartano, *Int. J. Electrochem. Sci.* **2013**, *8*, 10621–10649.
135. V. M. Nurchi, G. Crisponi, M. Arca, M. Crespo-Alonso, J. I. Lachowicz, D. Mansoori, L. Toso, G. Pichiri, M. Amelia Santos, S. M. Marques, J. Niclos-Gutierrez, J. M. Gonzalez-Perez, A. Dominguez-Martin, D. Choquesillo-Lazarte, Z. Szewczuk, M. Antonietta Zoroddu, M. Peana, *J. Inorg. Biochem.* **2014**, *141*, 132–143.
136. E. Shchori, N. Nae, J. Jagur-Grodzinski, *J. Chem. Soc. Dalton Trans.* **1975**, 2381–2386.
137. F. Arnaud-Neu, R. Delgado, S. Chaves, *Pure Appl. Chem.* **2003**, *75*, 71–102.
138. M. Marangella, M. Petrarulo, C. Vitale, P. G. Daniele, S. Sammartano, D. Coseddu, F. Linari, *Clin. Sci.* **1991**, *81*, 483–490.
139. P. G. Daniele, C. De Stefano, M. Marangella, C. Rigano, S. Sammartano, *Biochim. Clin.* **1989**, *13*, 507–510.
140. N. Ingri, W. Kakołowicz, L. G. Sillén, B. Warnqvist, *Talanta* **1967**, *14*, 1261–1286.
141. D. K. Nordstrom, J. W. Ball, in *Complexation of Trace Metals in Natural Waters*, Eds C. J. M. Kramer, J. C. Duinker, Nijhoff/Junk, The Hague, 1984, pp. 149–164.
142. G. Eriksson, *Anal. Chim. Acta* **1979**, *112*, 375–383.
143. C. De Stefano, P. Princi, C. Rigano, S. Sammartano, *Comput. Chem.* **1989**, *13*, 343–359.
144. J. A. Dyer, P. Trivedi, N. C. Scrivner, D. L. Sparks, *Environ. Sci. Technol.* **2003**, *37*, 915–922.
145. G. Biederman, in *Dahlem Workshop on the Nature of Seawater*, Dahlem Konferenzen, Berlin, 1975, pp. 339–362.
146. K. S. Pitzer, *J. Phys. Chem.* **1973**, *77*, 268–277.
147. K. S. Pitzer, *Activity Coefficients in Electrolyte Solutions,* 2nd Edn., Ed. K. S. Pitzer, CRC Press, Inc., Boca Raton, FL, 1991, pp. 542.
148. L. A. Bromley, *AIChE J.* **1973**, *19*, 313–320.
149. H. Krienke, J. Barthel, *J. Mol. Liq.* **1998**, *78*, 123–138.
150. H. Krienke, J. Barthel, *Z. Phys. Chem. (Munich)* **1998**, *204*, 71–83.
151. T. H. Ho, C. C. Lai, Y. H. Liu, S. M. Peng, S. H. Chiu, *Chem. Eur. J.* **2014**, *20*, 4563–4567.
152. Y. H. Lin, C. C. Lai, Y. H. Liu, S. M. Peng, S. H. Chiu, *Angew. Chem. Int. Ed.* **2013**, *52*, 10231–10236.
153. A. N. Thompson, I. Kim, T. D. Panosian, T. M. Iverson, T. W. Allen, C. M. Nimigean, *Nat. Struct. Mol. Biol.* **2009**, *16*, 1317–1324.

154. K. J. Liu, P. Gast, M. Moussav, S. W. Norby, N. Vahidi, T. Walczak, M. Wu, H. M. Swartz, *Biophysics* **1993**, *90*, 5438–5442.
155. R. P. Pandian, S. M. Chacko, M. L. Kuppusamy, B. K. Rivera, P. Kuppusamy, *Adv. Exp. Med. Biol.* **2011**, *701*, 29–36.
156. R. P. Pandian, M. Dolgos, C. Marginean, P. M. Woodward, P. C. Hammel, P. T. Manoharan, P. Kuppusamy, *J. Mater. Chem.* **2009**, *19*, 4138–4147.
157. S. Spicuglia, S. Kumar, L. Chasson, D. Payet-Bornet, P. Ferrier, *J. Biochem. Bioph. Meth.* **2004**, *59*, 189–194.
158. S. A. Akman, J. H. Doroshow, M. Dizdaroglu, *Arch. Biochem. Biophys.* **1990**, *282*, 202–205.
159. G. C. Terstappen, *Assay Drug Dev. Technol.* **2004**, *2*, 553–559.

Chapter 6
Sodium and Potassium Interactions with Nucleic Acids

Pascal Auffinger, Luigi D'Ascenzo, and Eric Ennifar

Contents

P. Auffinger (✉) • L. D'Ascenzo • E. Ennifar
Architecture et Réactivité de l'ARN, Université de Strasbourg, IBMC, CNRS,
15 rue René Descartes, F-67084 Strasbourg, France
e-mail: p.auffinger@ibmc-cnrs.unistra.fr; l.dascenzo@ibmc-cnrs.unistra.fr;
e.ennifar@ibmc-cnrs.unistra.fr

A. Sigel, H. Sigel, and R.K.O. Sigel (eds.), *The Alkali Metal Ions: Their Role for Life*,
Metal Ions in Life Sciences 16, DOI 10.1007/978-3-319-21756-7_6

Abstract Metal ions are essential cofactors for the structure and functions of nucleic acids. Yet, the early discovery in the 70s of the crucial role of Mg^{2+} in stabilizing tRNA structures has occulted for a long time the importance of monovalent cations. Renewed interest in these ions was brought in the late 90s by the discovery of specific potassium metal ions in the core of a group I intron. Their importance in nucleic acid folding and catalytic activity is now well established. However, detection of K^+ and Na^+ ions is notoriously problematic and the question about their specificity is recurrent. Here we review the different methods that can be used to detect K^+ and Na^+ ions in nucleic acid structures such as X-ray crystallography, nuclear magnetic resonance or molecular dynamics simulations. We also discuss specific *versus* non-specific binding to different structures through various examples.

Keywords DNA • Hydration • K^+ • Monovalent ions • Metal binding • Molecular dynamics simulations • Na^+ · NMR • Potassium • RNA • Sodium • Solvation • X-ray crystallography

Please cite as: *Met. Ions Life Sci.* 16 (2016) 167–201

1 Introduction

It is common knowledge that K^+ ions are mainly intracellular while Na^+ dominates in extracellular fluids [1–6]. This imbalance was secured when the first walled cells appeared. Hence, cells required the acquisition of efficient Na^+/K^+ transporters being able to distinguish between the two cations based on small differences in their physicochemical properties and especially in their ionic radii [7–11]. Thus, since cellular processes evolved in high K^+ environments, K^+ became an absolute requirement for many cellular functions such as osmotic regulation, protein synthesis, and enzyme activation. As a result of this evolutionary process, high Na^+ or very low K^+ concentrations in the extracellular fluids are detrimental and most of the time toxic to cells, although an adaptation process exists in K^+-deficient media where increased Na^+ concentrations enhances growths [12]. Given these adaptation and other cell signaling processes such as the large increase of Na^+ (from $\approx$10 to 100 mM) occurring in dendrites during neuron synaptic peak activity [13], intracellular monovalent ions are occasionally described as second messengers [14]. Moreover, it is well appreciated that a decrease in the intracellular K^+ and Cl^- concentrations is witnessed during cell apoptosis [15].

Besides monovalent cation channel selectivity, the roles of Na^+ and K^+ in enzyme function are relatively well documented and many enzymes display increased activity in the presence of monovalent cations [16]. For most of these systems, measured selectivity for a particular cation is low. For some, however, an increase in activity is observed in the presence of the larger K^+, Rb^+ or NH_4^+ ions. Few enzymes are selective for Na^+. The thrombin enzyme, associated with blood clotting and localized in Na^+-rich extracellular domains, represents an example of a Na^+-activated protein [17–19]. These enzymes are usually inhibited in the presence of K^+, Rb^+ or NH_4^+. Such mutually exclusive ion choices suggest that cationic physicochemical properties are determinant for certain enzymatic functions [16].

Like most enzymes, nucleic acids evolved in K^+-rich environments and have most certainly fine-tuned their structures and functions to these conditions. Yet, biomolecular buffers used in *in vitro* experiments do not often reflect intracellular ionic conditions and contain extracellular Na^+ rather than intracellular K^+ ions. The origins of the choice of these experimental conditions are unclear but stem partly from the fact that such buffers might have led to the setup of stable experimental conditions that became the standards over time. Due to the handing over of lab protocols, most of these experimental conditions became difficult to change. To complicate things, apart from a few systems, exchanging Na^+ for K^+ did not lead to quantifiable differences. Such results did not encourage researchers to change their procedures and did not prompt them to use the more biologically relevant K^+ ion. For instance, nearest neighbor studies inferring thermodynamic stability of RNA base-pair motifs from melting experiments are generally carried out in 1.0 M NaCl buffers for historical consistency [20, 21]. A recent study stated that the nearest neighbor effects of Na^+ and K^+ on duplex stability are undistinguishable [22]. Despite of this, current trends remain in favor of using Na^+-containing buffers instead of switching towards the more germane K^+-containing buffers.

In a twenty-year-old study, Freund and Kalbitzer [23] suggested that, for biomolecular investigations, a key advantage of NMR compared to X-ray crystallography relies in its ability to use buffers mimicking the properties of the natural physiological solvent. However, NMR spectroscopists are still making sparse use of this advantage, favoring buffers far from physiological conditions such as buffers composed of distilled water resulting in abnormally low ionic strength, or buffers with addition of sodium chloride, which are inappropriate for intracellular proteins. Freund and Kalbitzer [23] proposed two nearly physiological buffers for NMR spectroscopy of biological samples that reflect intracellular (free K^+: 148 mM; free Na^+: 8 mM) and extracellular (free K^+: 5 mM; free Na^+: 144 mM) ionic conditions.

Since then, much progress has been made in the investigation of the roles monovalent cations play in biomolecular systems. The fact that some nucleic acids display high sensitivity to the type of monovalent cations present in buffers is currently better appreciated. For example, several ribozymes display K^+/Na^+-dependent activities [24] and NMR studies demonstrated that DNA structures have different sequence-dependent responses to physiological concentrations of K^+ or Na^+ [25]. In this respect, DNA and RNA quadruplexes represent the best-known example of

K^+/Na^+ ion selectivity in nucleic acid systems [26–29] and quadruplexes are the cornerstones of very efficient K^+ detection systems [30].

In this review, we will address issues related to the roles monovalent cations play in *in vivo, in vitro* and also *in silico* conditions for nucleic acid systems and more specifically focus on monovalent ion dependent effects on structure and function. We will discuss some solvent detection issues in biophysical methods and especially nucleic acid crystallography. Several reports describe, based on statistical surveys of the CSD and the PDB, the geometrical characteristics of ion coordination including those of monovalent cations [31–35] as well as other physico-chemical characteristics [36]. Numerous web sites and servers were also designed to analyze metal binding sites in proteins and nucleic acids [37–47]. Further, several reviews focus on nucleic acid monovalent ion binding issues [36, 48–56].

2 Identification of Na^+ and K^+ Ions in Structural Studies

2.1 *X-ray Crystallography*

2.1.1 Na^+ Prevalence in Crystal Structures – A Brief Statistical Overview

As of April 2015, the PDB embraces 95,000 X-ray structures and holds altogether significantly more structures containing Na^+ (5,116) than K^+ (1,769). For nucleic acids, the proportion remains roughly the same: 520 and 339 structures containing Na^+ and K^+ ions, respectively (note that ≈ 18 % of the latter are related to quadruplexes). Thus, ≈ 7 % of all X-ray structures and ≈ 15 % of all nucleic acid structures contain monovalent cations. These proportions are small and suggest that a certain number of technical limitations prevent the efficient detection of ions that are generally thought to be present in much higher proportions in biomolecular systems [57].

2.1.2 Monovalent Ion Detection Based on Geometrical Considerations

2.1.2.1 Coordination Distances

The most direct path to identify monovalent cations relies on their specific water coordination distances: 2.42 Å and 2.84 Å for Na^+ and K^+, respectively. These distances were derived from a careful analysis of high-resolution crystal structures deposited in CSD [31, 34, 58]. Note that coordination distances might be somewhat "author"-dependent. It has been reported that the $Na^+ \cdots O$ distance can be close to 2.30 Å from a CSD survey [32], a number that significantly contrasts with the above reported value. Coordination distances are also slightly ligand-dependent. Thus, given the larger covalent radii of N and S compared to O, the coordination distances should be augmented by ≈ 0.1 and ≈ 0.2 Å, respectively. Yet, coordination of monovalent cations to imidazole is rare [31, 34], suggesting less frequent coordination to the imine nucleobase nitrogens with a possible exception for the guanine N7 atoms that form a strong electronegative edge in association with the adjacent O6 atoms. Density peaks that are difficult to assign are also frequently observed especially in medium- to low-resolution structures. In a 2.95 Å resolution structure, K^+ was assigned to electron density peaks with coordination distances far from optimal for first and second shell coordination and need probably more insights [59].

Unfortunately, given identical ≈ 2.8 Å coordination distances, this criterion is ineffective for distinguishing K^+ ions from water molecules and one has to rely on other characteristics such as coordination patterns or density size.

2.1.2.2 Coordination Numbers and Coordination Geometry

Usually, coordination numbers are much better defined for the smaller than the larger cations. For Na^+, six is by far the commonest coordination number with well-defined octahedral coordination geometry since there is not much space around the ion to accommodate more water molecules and to deviate from this geometry (Figure 1 [62, 63]). For K^+, the coordination number is close to eight, probably never six, and the coordination geometry is fuzzier [31–35, 60]. Indeed, with the exception of quadruplex structures where the K^+ coordination is eight and nicely defined, almost no K^+ in nucleic acid structures presents a complete coordination

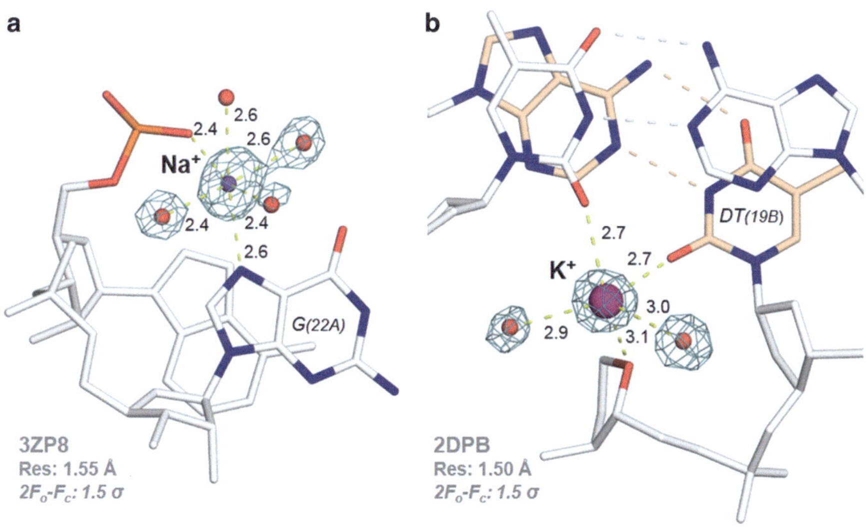

Figure 1 Na^+ and K^+ coordination spheres. (**a**) A Na^+ ion bound to a (G)N7 atom, a phosphate group, and four water molecules form a slightly distorted octahedral pattern extracted from a hammerhead ribozyme crystal structure [62]. The average coordination distances are close to 2.4 Å. (**b**) A K^+ ion in a DNA structure [63] shows an incomplete coordination pattern involving two base carbonyl groups, one O4' atom, and two water molecules. The average coordination distances are close to 2.8 Å.

sphere. Molecular dynamics (MD) simulations of this ion in solution revealed that the bound solvent molecules are also orientationally more disordered than those attached to Na^+ ions [61]. Thus, although Mg^{2+} and Na^+ share octahedral coordination geometries, the coordination geometry of Mg^{2+} is almost perfect while that of Na^+ is less optimal.

For water molecules, the coordination number is usually four. Thus, the coordination number should help distinguishing between water and K^+. Unfortunately, complete coordination patterns are rarely observed in medium- to low-resolution structures where water molecules are only occasionally detected.

2.1.3 Direct and Indirect Anomalous Signal-Based Approaches

2.1.3.1 K^+ Anomalous Signals

Anomalous diffraction techniques are of no practical use for identifying Na^+ ions for two major reasons: (i) the sodium "K edge" is far away (11.56 Å) from the attainable energy range of macromolecular crystallography synchrotron beam lines (0.7 to 1.8 Å) and (ii) the f" anomalous signal is extremely weak (0.2 to 0.1 e^-) and unusable in this range. The situation is much more favorable in the case of K^+ ions. Although the potassium "K edge" is too far for synchrotron beam lines (3.44 Å), its weak anomalous signal is appropriate to low-energy wavelength data collections (f" ~ 1.5 e^- at 1.8 Å) and several authors used these techniques to directly identify K^+ ions when high- or medium-resolution diffraction data are available [64–71] (Figure 2). Observing the K^+ anomalous signal is a valuable alternative to its replacement by another monovalent cation since no additional experiment is required. Most importantly, it does not rely on the rather **optimistic** assumption that other ions are occupying similar binding sites.

2.1.3.2 Detection Based on K^+/Tl^+ Replacement

There is no good substitute for Na^+ ions while for K^+ the toxic Tl^+ or thallium(I) ion [72] was shown to represent an interesting alternative in medium- to low-resolution structures [66, 70, 73–87]. Tl^+ has indeed a similar ionic radius and hydration enthalpy than K^+ and can also adopt a 6-octaedral or 8-coordination geometry [51]. It has also the advantage of being easily detectable either using difference electron density maps (Z_{Tl} - Z_K=61 e^-), or using its strong anomalous signal. The L-III edge of Tl is indeed well suited (0.9795 Å) for data collection on a tunable synchrotron beamline. Additionally, Tl is a rather strong anomalous scatterer (f"=5–10 e^- in the usual energy range of the multi-wavelength anomalous dispersion (MAD) beamlines) and should be detected even by using a rotating anode X-ray source. Fifteen Tl^+-containing nucleic acid structures are currently (April 2015) deposited in the PDB although Tl^+ is certainly more often used during the crystallization and data analysis processes.

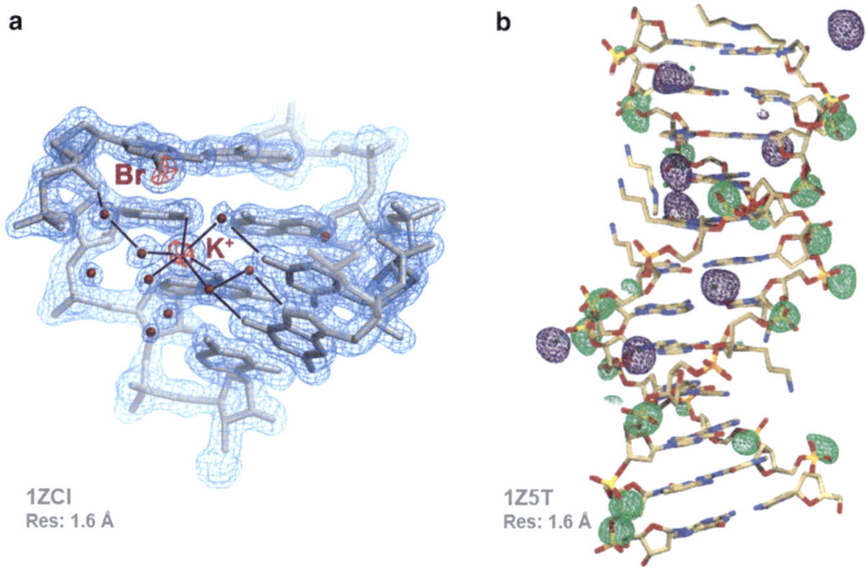

Figure 2 K^+ and Tl^+ anomalous signals. (**a**) Anomalous difference map contoured at 5 σ (in red) for K^+ and a covalently bound Br atom from a crystal structure of a DIS RNA fragment. The $2F_o$-F_c map contoured at 1.4 σ is shown in blue. K^+ ligands and some H-bonds are depicted with black lines. Reproduced by permission from [69]; © copyright 2007. (**b**) Tl^+ (violet) and phosphorus (green) anomalous densities contoured at 2.5 σ around a DNA duplex [66]. The phosphorus atoms are shown to serve as a positive control. Reproduced by permission from [66]; © copyright 2005, American Chemical Society.

Interestingly, although Tl^+ is considered a good substitute for K^+ in crystallographic and sometimes biochemical studies, in both instances one has to be aware that some induced effects are specific to Tl^+. For instance, there is a tendency of Tl^+ to react with "soft", b-type electron donors such as those belonging to "thiol" groups or the purine N7 atoms. It was noted that growing crystals in Tl^+-containing buffers rather than soaking it into crystals grown with K^+ may have led to additional Tl^+-specific binding events [88]. Besides, effects related to K^+/Tl^+ substitution in ribosomes were noted as particularly complex [89].

2.1.3.3 Detection Based on K^+/Rb^+ or K^+/Cs^+ Replacement

Besides Tl^+, crystallographers use sometimes also Cs^+ and to a lesser extend Rb^+ ions that produce large anomalous signals to infer the position of K^+ ions [77, 90–92]. In recent structures of bacterial riboswitches that control the expression of heavy metal transporters, the position of four K^+ were modeled into Cs^+ assigned densities since the authors assumed that Cs^+ typically replaces K^+ in crystals. In structures of RNA duplexes, Rb^+ was found to be unable to replace already assigned K^+ binding sites [65]. Yet, this procedure is probably not always effective given the larger 3.1 and 3.2 Å coordination distances for Rb^+ and Cs^+, respectively. Furthermore, the differences in hydration properties of these ions compared to K^+ are significant.

2.1.4 A Word of Caution

The recent tremendous increase of large X-ray RNA structures – together with advances in the identification procedures of ions has led to significant progress in the assignment of ions to solvent density peaks [57, 93, 94]. But, as stated before, replacement procedures rely on sometimes too optimistic hypotheses. In this respect, caution is certainly required with the crystallographic procedures that claim identifying Mg^{2+} ions but also K^+ and Na^+ ions in large protein or nucleic acid systems. Surely, collecting direct anomalous signals for K^+ is probably the safest route for identifying monovalent ion binding sites. In this respect, the new techniques that try to extract signals from noisy data have to make their proofs regarding solvent density assignments [95, 96].

2.2 Nuclear Magnetic Resonance

2.2.1 Nuclear Magnetic Resonance Detection of Na^+/K^+-Induced Effects

Nuclear magnetic resonance (NMR) techniques are generally unable to directly detect alkali cations in biomolecular systems but only to infer their presence by measuring induced nucleic acid relaxation property changes [97]. $^{15}NH_4^+$ was first used as a spectroscopic probe to investigate G-quadruplex features [98]. Indeed, it was shown in many instances that K^+ and NH_4^+ properties are close in nucleic acid systems [99, 100] and $^{15}NH_4^+$ remains a very effective NMR probe for inferring K^+ binding sites [101]. However, despite their clear advantage in NMR spectroscopy, these ions are non-physiological and display cell toxicity. Later, using magnetic relaxation dispersion, $^{23}Na^+$ relaxation was measured for DNA duplexes and it was suggested that Na^+ binds to these duplexes with correlation times close to 50 ns and minor groove occupancies as great as 50 % [102] although these high values are subject to debate [103]. Numerous other competition experiments were conducted that demonstrated that Na^+ and K^+ share the minor groove with water in the liquid state [97]. ^{23}Na- and ^{205}Tl-based NMR methods were also used to investigate the monovalent binding potential of DNA quadruplexes [104, 105].

^{31}P chemical shifts were measured for DNA duplexes in order to detect effects induced by the presence of Na^+ or K^+ [25]. The measured changes in chemical shifts demonstrated that each monovalent cation affects in its own way the structure of the associated duplexes. A more recent NMR study, revealed also perturbations in the NMR spectra of DNA tetraloops that are associated with the presence of one or the other cation in the experimental buffer [106]. These results demonstrate that NMR techniques are able to highlight structural effects induced by the presence of one or the other monovalent ions.

2.2.2 Prevalence of Na^+ over K^+ in Nuclear Magnetic Resonance Buffers

A rapid scan of the experimental conditions used for determining the NMR structures deposited in the PDB confirms the dominant use of Na^+ ions in buffers (≈57 % of Na^+ and ≈19 % of K^+). These numbers are in agreement with the already mentioned prevalence of Na^+ over K^+ in crystal structures and underline the distance between *in vivo* conditions and those used in biophysical experiments [23].

2.3 *Molecular Dynamics Simulations*

Molecular dynamics (MD) simulations are able, from a theoretical point of view, to provide clues about the location of high occupancy monovalent cation binding sites [107–123]. In some instances, monovalent cations were found to bind close to experimentally characterized divalent binding pockets in RNA [113, 124]. Na^+ binding sites close to the catalytic pocket of the hepatitis delta virus (HDV) ribozyme [125, 126] were confirmed by crystallographic experiments conducted in the presence of Tl^+ ions [81]. All these studies point to the fact that MD simulations are now able to shed some light on important nucleic acid ion-binding features even if significant concerns were raised regarding the reliability of the ion parameters used in early simulations. Indeed, many of these parameter sets led to the non-physical formation of NaCl or KCl aggregates [127–131].

These concerns were recently addressed through the development of improved force-field parameters for cations and anions [129, 132–137], although one study advised the use (if possible) of NaCl instead of KCl since Na^+ parameters seemed better converged and consequently less prone to generate errors [130]. These authors were putting forward technical rather than biological issues suggesting that more parameterization studies have to be undertaken. Interestingly and despite the abundance of new parameter sets, several authors including those from the Ascona B-DNA consortium (ABC) that collected microseconds of MD on DNA duplexes [138–141] turned to the use of very simple conditions, namely the SPC/E water model [142] and the Dang parameters for the ions [143]. These parameters were first tested in MD simulations of DNA duplexes and were subsequently found to avoid the formation of NaCl or KCl salt clusters [127, 128]. For some newly developed ion parameters, it was reported that the cations display an “exaggerated” tendency to aggregate with phosphates [130].

It seems clear that ion parameter issues are central to comparative studies trying to understand Na^+/K^+ binding preferences and how these ions affect the nucleic acid structures. Nevertheless, such studies are still rare. Despite a commonly shared view

[130] stating that "*in most cases a change of the force-field and salt conditions would have virtually no effect on the qualitative simulation outcome*" [144], a recent set of MD simulations of RNA duplexes demonstrated the opposite. In this study, the RNA structures seem to be significantly affected by the ion parameter choice (among other parameters such as the used force fields or water models; Figure 3 [148, 149]). If various but relatively similar K^+ ion parameters can affect the RNA structure it is doubtless that a change from K^+ to Na^+ can also significantly affect the nucleic acid structures [145]. Future force field improvements will certainly stem from the use of polarizable ion models like those described in [146, 147]. The authors of these studies claim to have characterized ion-dependent effects that were only detectable by using the more sophisticated polarizable models.

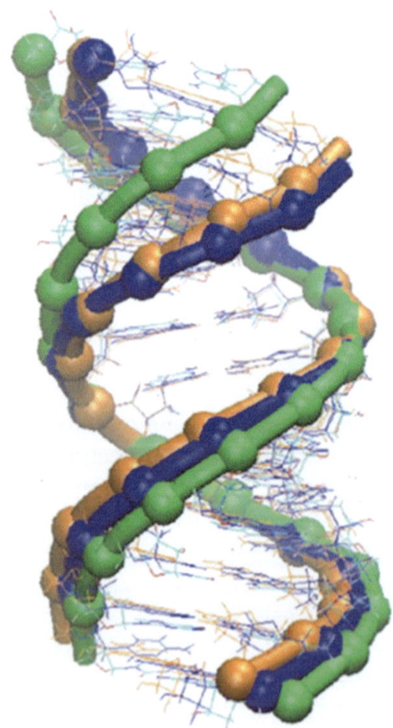

Figure 3 Differences in average structures (45–50 ns) from MD simulations of a RNA duplex with three different ion models in net-neutralized ionic (no excess salt) conditions. The displayed structures were derived from simulations using the ion parameter sets from Young and Cheatham in orange [132], Jorgensen et al. in green [148] and Roux et al. in blue [149]. Reproduced by permission from [145]; © copyright 2012, American Chemical Society.

3 Non-specific *Versus* Specific Binding of Na^+/K^+ Ions

3.1 Non-specific Binding of Monovalent Cations to Nucleic Acids

Nucleic acids are polyanions. As such, each of these molecular systems should be surrounded by an equivalent number of cations to counterbalance their charge. This shielding is provided *in vivo* by a variety of cationic systems including mono- (K^+) and divalent cations (Mg^{2+}), charged metabolites including polyamines [150] and nucleic acid binding proteins. *In vitro*, non-physiological cations such as small Li^+, Na^+ or large Cs^+, $Co(NH_3)_6^+$ and other hexammines are often used. Because the precise localization of cations was difficult to obtain in the early days of crystallography, it has for long been assumed, based on the counterion condensation theory [151, 152], that a certain percentage of monovalent cations stay in the vicinity of nucleic acids and thereby reduce their effective charge. As an outcome, it was assumed that monovalent ion binding is diffuse, highly non-specific and concentrated in the vicinity of the negatively charged phosphate groups. Consequently, the possibility of strong binding of monovalent cations to specific nucleic acid sites has been overlooked. Nowadays, theoretical calculations reformed this view and it is now recognized that sequence-specific monovalent binding sites exist, especially in major grooves of nucleic acids. These findings led to the conclusion that nucleic acid grooves should be considered as ion traps or ionophores [153].

Several studies pointed to effects related to monovalent cation size without being able to relate directly these facts to specific binding sites. It was reported that the thermal stability of a DNA hairpin increased with cation size although the effects associated with Li^+, Na^+ or K^+ are not distinguishable [154]. Melting studies on DNA duplexes suggested that the sequence affects ion-binding properties. Here again, effects associated with Na^+ and K^+ were reported to be similar [155]. On the other hand, measures of folding free energies using optical tweezers reported stabilizing effects of K^+ over Na^+ [156].

3.2 Binding of Monovalent Cations to Nucleic Acid Grooves

3.2.1 DNA Duplexes

3.2.1.1 Molecular Dynamics Perspective

Interestingly, the first clues about cation binding to DNA came from MD simulations that suggested that DNA grooves do capture monovalent cations at specific locations in the minor groove [157]. Subsequently, it was found that monovalent cations bind even more strongly to the major groove [107–110, 114, 116–118, 123]. All these findings were backed up by experimental studies using NMR and

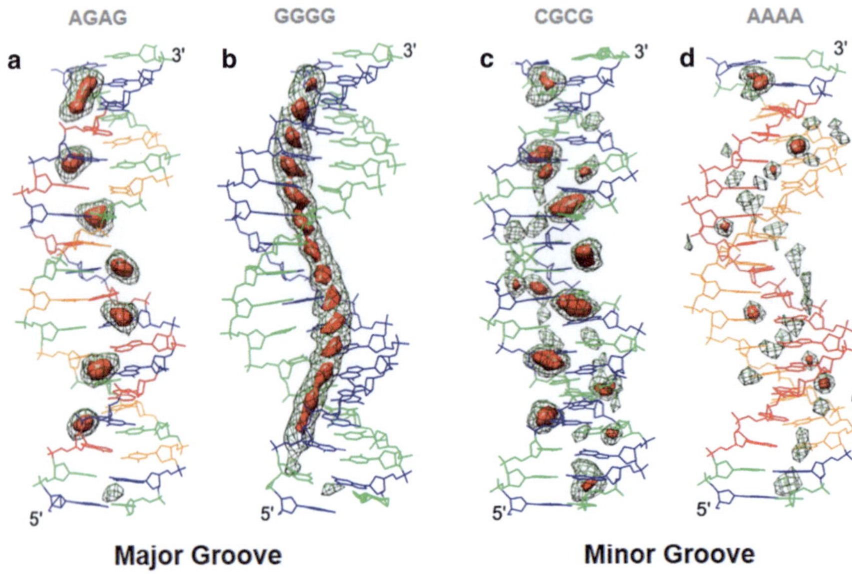

Figure 4 Examples of sequence-dependent major and minor groove K^+ densities derived from microsecond long MD simulations of B-DNA duplexes showing cartesian isomolarity surfaces at 15 M (red) and 5 M (green mesh). Adapted by permission from [141]; © copyright 2015, Oxford University Press.

crystallographic techniques [64, 66, 76, 97, 158–163]. An atomic resolution structure of the Drew-Dickerson dodecamer derived from a crystal grown in the presence of Tl^+, Mg^{2+} and spermine showed that the sites with highest occupancies (20–35 %) were located in the major groove at CpG steps and that most of them did not connect to phosphate groups [76]. Further, this structure suggested that monovalent cations do share occupancies with water molecules. It was found that monovalent cation occupancies derived from MD simulations follow the same trend although the occupancies for major groove cations (K^+) are calculated to be higher [108, 110, 141]. Similarly, Na^+ and K^+ density distributions around canonical B-DNA derived from integral equation theories reported ion-specific binding profiles [164].

Recent microsecond long MD simulations using K^+ suggested that phosphate groups are not attracting cations but that monovalent cation binding is essentially concentrated in the major groove at GpC steps (Figure 4). The flanking steps can significantly affect the ion binding. It was reported that moving from CpCpGpG to ApCpGpT stems reduces the minor groove occupancies by 90 % from 0.48 to 0.06. Overall, the maximum calculated major groove occupancy exceeds 0.8 at GpC steps and can reach 0.5 at minor groove YpG steps [141]. These results confirmed data obtained earlier on much shorter MD simulations [108, 110].

Only a small number of studies investigated the effects of Na^+ *versus* K^+ on DNA structures. Some of them reported that simulations are not sensitive to the type of ionic conditions that are used, while others mentioned significant differences. For a set of simulations of a DNA dodecamer, it was found that, while Na^+ binding in both grooves is strongly sequence-dependent with the preferred binding site in the minor groove, K^+ mainly visits the major groove and binds close to the center of the oligomer. Hence, calculations show that the DNA structure is differently affected by Na^+ and K^+ [116]. In another study, where DNA ends were constrained through the use of periodic conditions, the DNA structure was not influenced by the choice of cations, although Na^+ and K^+ binding modes were different [117]. Moreover, mobility of both water and cations in K^+ systems is faster than in Na^+ systems. Na^+ organizes and immobilizes the water structure around itself and near DNA while for K^+, water is less organized and more dynamic [117]. Other authors reported that Na^+ ions penetrate the DNA interior and condense towards the DNA exterior to a significantly larger extend than K^+ ions [118]. In the presence of K^+, the oligomer was calculated to be more flexible. Interestingly, these authors observed a significant involvement of Cl^- ions in the process. Cl^- co-ions were observed to provide more efficient screening for the K^+ ions than for the Na^+ ions contributing, thus, to the larger Na^+ condensation around DNA. Yet, this effect could also be attributed to a KCl clustering artifact observed with older AMBER force-field versions [128, 130].

Recently, a MD study on ion binding to the Drew-Dickerson-dodecamer conducted at different temperatures and with Li^+, Rb^+ or Cs^+, reported that the DNA structure becomes more flexible than with Na^+ or K^+ and described a widening of major and minor grooves at CpG and GpC steps when K^+ is present. The authors reported, at odds with the preceding study, that the smaller Li^+ and Na^+ ions prefer to interact with the free phosphate oxygen atoms while the heavier ions (Rb^+ and Cs^+) strongly interact with the base pairs [165]. These findings are consistent with MD simulations suggesting that Na^+ favors condensation around DNA compared to K^+. A study on DNA bending of A-tracts reported a slight bending reduction in NaCl and neutral (no-added salt) conditions compared with KCl conditions [166].

MD studies using sophisticated Drude polarizable models reported that the conformational properties of DNA double strands in solution are sensitive to the type of monovalent cation. More specifically, the minor groove width is smaller for the smaller cations. Interestingly, the authors state that these differences are only apparent when polarizable ion models are used and cannot be inferred with more conventional additive force fields [146, 147]. These results provide a rational explanation as to why ion-dependent conformational effects were up to now difficult to characterize theoretically and emphasize that for generating useful MD simulations, increasing the simulation length is not enough. More attention has also to be devoted to the fine adjustments that are needed to improve current force fields [167].

3.2.1.2 Experimental Views

The type of monovalent cations present in experimental buffers can alter DNA conformation of, at least, some structures. For example, crystallographic data emphasized subtle shifts of monovalent cations toward the floor of the minor groove with increasing ionic radii [159] and a solution neutron scattering study revealed a shift from a duplex to a quadruplex structure when Na^+ is replaced by K^+ at concentrations above 100 mM [168]. A higher potential in DNA compaction by Na^+ *versus* K^+ has also been reported [169] and electrophoretic measurements showed clear sequence-specific cation dependence for DNA bending [170].

Moreover, capillary electrophoresis experiments on A-tracts suggested that the binding of monovalent cations depends on the cation identity as well as on the DNA sequence. Interestingly, some A-tracts were reported not to bind K^+ ions [171]. A thorough NMR study based on measurements of phosphorus chemical shifts demonstrated that DNA duplex structures respond differently to physiological concentrations of K^+ or Na^+ by modulating the proportion of backbone conformers [25].

3.2.2 RNA Duplexes

Fewer studies were devoted to RNA double strands since RNA folds preferentially into intricate 3D structures. MD simulations conducted in parallel on the same B-DNA and A-RNA duplexes demonstrated that the RNA grooves are also excellent ionophores and that the ion-binding properties are similar for both systems although a preference for RNA can be noted [108, 110, 122, 172]. The binding to double strands is certainly dynamic and monovalent ions are expected to exchange at a relatively high frequency. MD simulations suggest binding times that can reach 1–2 ns for K^+ and are close for RNA and DNA. Moreover, MD simulations suggest shared occupancies of these sites with water molecules.

More specifically, a recent RNA duplex MD study suggested that K^+ tends to support more compact structures than Na^+ [145]. However, as already described in the MD section, the RNA structures seem to be significantly affected by ion parameter choices. If various but relatively similar K^+ ion parameters can affect the RNA structure it is doubtless that Na^+ can also lead to different average structures (Figure 3).

3.3 Specific Binding of Monovalent Cations to Quadruplex Structures

3.3.1 DNA Quadruplexes – A Perfect K^+ Coordination Geometry

Quadruplexes comprise stretches of consecutive guanines that aggregate to form four-stranded helices. They are probably the best illustration of specific binding of monovalent cations to nucleic acids. These structures that form in the presence of Na^+ or K^+ are of great biological interest as exemplified by the large guanine-rich telomere 3'-end overhang. In an *in vivo* context, it is highly probable that K^+ is the active cation and indeed DNA quadruplexes display the best affinity for this cation over the smaller Na^+. This is further reflected by the fact that Na^+ quadruplexes can

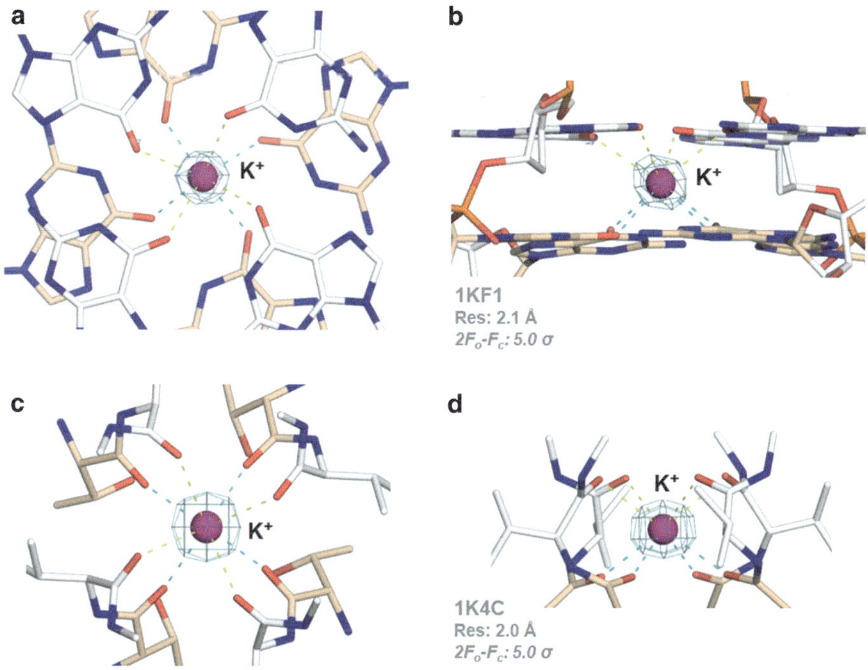

Figure 5 K^+ coordination in a DNA quadruplex [174] and in the KcsA ion channel [7]. (**a**, **b**) Top and side view of a K^+ sandwiched between two G-quartets and at ≈ 2.8 Å of the oxygen atoms of eight guanine carbonyl groups in a distorted bi-pyramidal antiprismatic geometry. (**c**, **d**) Top and side view of K^+ similarly interacting with eight backbone carbonyl groups in the ion channel, with ≈ 2.8 Å coordination distances.

adopt two forms where (i) the ion is found in a guanine quartet plane or (ii) sandwiched between two of them. Instead, K^+ was only found in the latter configuration where it interacts with eight (G)O6 atoms adopting a distorted bi-pyramidal antiprismatic geometry (Figure 5a) [71, 173]. Interestingly, the same distinctions between Na^+ (in the middle of a plane of four oxygen atoms) and K^+ binding modes (in the middle of a cage made from two planes of oxygen atoms) were gathered from K^+ channel protein MD simulations suggesting the existence of a universal recognition motif for dehydrated K^+ ions (Figure 5b) [11].

Titration experiments showed that for human telomeric structures folding requires more K^+ ions than its number of G-quartets suggesting that these ions might also interact with the loops [173]. Indeed, G-quadruplexes exhibit a remarkable and well-documented cation-induced polymorphism [175–183]. For the $d(G_4T_4G_4)$ sequence, the guanine arrangement between the K^+ and Na^+ structures is similar with, however, strikingly different loop structures that are probably related to different cation coordination events at external binding sites [173, 184]. Differences in the loop dynamics were also highlighted [177] and similar cation-induced polymorphism was observed for the *Tetrahymena* telomeric repeat $[d(T_2G_4)]_4$ and related sequences such as $d(G_2T_4G_3)$ and $d(G_4T_4G_4)$ [179, 185]. G-quadruplexes formed by human telomere repeats such as $d[A(GGGTTA)_3GGG]$ exhibit different conformations in the presence of Na^+ and K^+ [181]. It was reported that the inhibitory thrombin-binding aptamer activity is lower in the presence of Na^+ with respect to K^+ [71].

Moreover, G-rich DNA sequences can switch between two conformations: a hairpin at low K^+ concentration and a quadruplex at higher concentrations [186]. However, it was shown that not all G-rich sequences are prone to form quadruplexes. Many of the longer repeats generate homoduplexes or hairpins. The most stable tetraplexes were formed with $d(GGC)_n$ and the least stable with $d(GCG)_n$, all of them in the presence of Na^+ or K^+. The authors speculated that the formation of quadruplex structures in guanine-rich DNA strands of the fragile X chromosome repeats is unlikely [187]. Quite surprisingly a small-angle scattering study of the d(AGGGTTAGGGTTAGGGTTAGGG) sequence in the presence of Na^+ or K^+ suggested the formation of quadruplexes with the latter ion and not the former in the given experimental conditions [168].

DNA quadruplexes are still challenging structures in the field of biomolecular MD simulations, especially with respect to force-field issues associated with water and cation parameters. MD simulations on DNA quadruplexes using both Na^+ and K^+ conditions are relatively rare [131, 188] since most of them take K^+-containing buffers into account [189]. Yet, MD simulations reported, in agreement with experimental data that, while the smaller Na^+ ions can spend some time in the G-quartet plane, K^+ ions are too large and are only sandwiched between adjacent planes. The authors suggest that subtle binding differences between Na^+ and K^+ could propagate to the loops and explain their polymorphism [188]. In order to improve their models, modifications of the K^+ van der Waals parameters were suggested [190, 191]. Simulations using diverse ionic conditions were performed [131, 188, 190–192], including ones with ion parameters that led to the generation of unphysical ionic clusters [128, 131]. Through an impressive amount of work, cation-dependent penetration routes in DNA quadruplexes were proposed [131].

3.3.2 RNA Quadruplex Switches

The first RNA quadruplex structure was solved by NMR in 1992 [193] and the first crystallographic structure in 2001 [194]. A recent review stressed the numerous roles quadruplexes might play in biological systems [26]. From a structural point of view, it was shown that some guanine-rich RNA molecules could switch from a hairpin to a G-quadruplex structure in the appropriate K^+-containing ionic conditions [195]. The "Spinach" *in vitro*-selected RNA fluorophore aptamer was crystallized in the 2.2–2.4 Å resolution range and a K^+ was assigned to a density observed in a G-quartet plane based on coordination distances but not on anomalous diffraction data (Figure 6). It was found that K^+, Na^+, and NH_4^+ could enhance fluorescence although K^+ appears to be significantly more efficient [27].

A K^+-dependent RNA switch was reported to regulate human pre-miRNA maturation by possibly involving G-quadruplexes [29]. The authors found that about 16 %

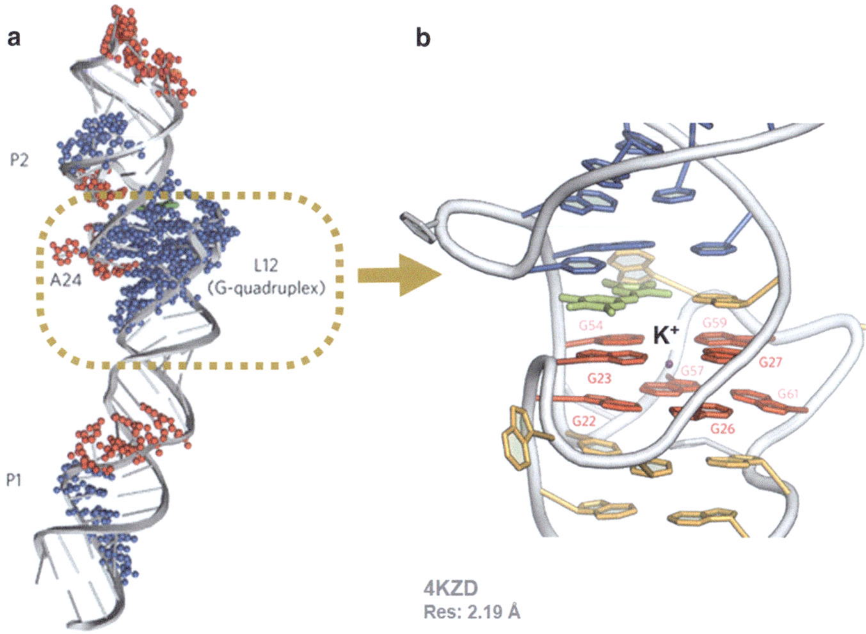

Figure 6 Crystallographic structure of the *in vitro* selected spinach RNA aptamer. (**a**) 3D structure of the aptamer. The box highlights the inferred K^+ G-quartet binding site. (**b**) Close-up view of the K^+ binding G-quartet. Adapted by permission from [27]; © copyright 2014, Nature Publishing Group.

of all pre-miRNAs contain sequences susceptible to form G-quadruplexes and that the switch from a hairpin to a quadruplex-containing structure can occur at the physiological 100 mM K^+ concentration. However, as emphasized by the authors, the *in vivo* pre-miRNA formation of G-quadruplexes needs confirmation. Furthermore, the use of RNA quadruplexes in *in vivo* conditions is questioned by the fact that, under molecular crowding conditions, RNA quadruplexes are much more stable than the corresponding DNA quadruplexes and the folded RNA structures are unaffected by the surrounding environment [182].

3.4 RNA Folding Needs Monovalent Cations

RNA folding is generally supposed to proceed in two stages. The first consists in the formation of RNA secondary structures comprising single- and double-stranded regions. Almost any ionic component that is able to screen charges and to bring polyanionic backbone segments in proximity, such as mono- and divalent cations or polyamines, stimulates this process. The second and more complex stage consists in the fine assembly of all these secondary structural elements that involves an intricate network of unusual stacking and base-pair interactions. The second stage requires finely tuned ionic conditions where Mg^{2+} is generally determinant but also where monovalent cations such as K^+ are deemed important [196]. It was found that many RNA tertiary structures fold and are stable in the presence of high concentrations of monovalent ions [24] and even that monovalent cations are needed for the folding of specific motifs [196–199].

It has been rationalized that high concentrations of monovalent cations are needed since these ions are much less effective than divalent ions in screening the negative charges carried by the polyanionic backbone. In some instances, it was reported that K^+, and often also NH_4^+, is more effective than Na^+ in stabilizing specific RNA structures [24, 100, 156, 200]. On the other hand, a mRNA pseudoknot was found to be better stabilized by the smaller alkali metal ions ($Na^+ > K^+$) [201] as well as an A-riboswitch and a tar-tar* kissing loop [24]. For the later system, data indicating that the TAR hairpin misfolded nearly twice as often in KCl than NaCl, revealed differential folding kinetic responses [156]. However, the folding kinetics of the group I *Tetrahymena* ribozyme or a human mitochondrial tRNA were reported to be similar in Na^+ and K^+ [202, 203] Further, in a recent attempt to understand thermodynamic origins of monovalent-facilitated RNA folding, it was reported that Na^+ and K^+ are strictly equivalent [204] stressing that evidencing monovalent induced effects might be method-dependent and has to be investigated further.

3.5 Specific Binding of Monovalent Cations to Complex RNA Folds

3.5.1 First Crystallographic Evidence: The P4-P6 Group I Intron Fragment

The first crystallographic evidence of specific monovalent ion binding to RNA was gathered from a crystal structure of the P4-P6 group I intron fragment through Tl^+ replacement. The authors found a monovalent binding pocket at the level of the tetraloop receptor and below an AA platform that was ascribed to K^+. The detection of this monovalent ion-binding site is the outcome of resolute investigations trying to unravel the structure of this motif and to understand its ion binding features [205, 206]. This motif specificity for a monovalent ion was confirmed by the fact that the related *Azoarcus* group I intron is six times less active in Na^+- than in K^+-containing buffers. Therefore it was suggested that this monovalent ion-binding site is important for intron splicing [73]. Hence, monovalent cations that bind specifically to nucleic acid sites were recognized, besides Mg^{2+} ions, as being important for the folding and function of this large RNA fragment [199].

Interestingly, despite these early data and a study indicating that competitive effects of monovalent ions with respect to divalent cation binding are less pronounced when K^+ replaces Na^+ [200], some studies on this important RNA motif are still favoring Na^+-containing buffers [196, 207, 208]. Unfortunately, the K^+-containing P4-P6 structure is not deposited in the PDB and the structures of the other P4-P6 fragments or the larger group I intron at resolutions ≤ 3.0 Å do not mention the presence of K^+. However, a structure at 3.1 Å for which in depth metal ion binding investigations were conducted confirms this K^+ ion-binding site [82] as well as a Na^+ ion binding to a similar motif in the large ribosomal subunit of *Haloarcula marismortui* [90]. The binding of Na^+ [209] and K^+ [199] to this motif has also received recent support from a single molecule FRET approach and a ≈ 3 kcal/mol preference of the tetraloop receptor for K^+ over other alkali metal ions was reported [24].

3.5.2 A Buried Backbone-K^+ Complex in a rRNA Fragment

A K^+ binding site was identified in a crystal structure of a 58 nucleotide 23 S rRNA domain that binds to protein L11 and that forms a knob on the ribosome surface [88]. In this case, the RNA is ≈ 3 kcal/mol more stable with K^+ than with other alkali ions [100]. A 10-fold increase in binding affinity of this rRNA fragment for ribosomal protein L11 or the antibiotic thiostrepton was measured in the presence of NH_4^+, and to a lesser extend in the presence of K^+, compared to Na^+ [210].

3.5.3 Pseudoknots Can Capture Monovalent Cations

Early studies demonstrated that, for the folding of specific ribosomal motifs, NH_4^+ or K^+ ions were required emphasizing therefore the existence of specific monovalent ion binding sites [210]. A $Li^+ > Na^+ > K^+ > Rb^+ > Cs^+ > NH_4^+$ stabilization order for a mRNA pseudoknot was also reported [201]. The crystal structure of a viral pseudoknot revealed the existence of several Na^+ and one K^+ binding site [78]. Related biochemical investigations suggested that Na^+ might not be the relevant ion in the stabilization of this structure. However, for a beet western yellow virus (BWYV) pseudoknot, no ion size-dependent stabilization was reported [24].

3.5.4 Small Ribozymes Are Active at High Monovalent Concentrations

3.5.4.1 The Hammerhead Ribozyme

It was reported that the hammerhead ribozyme is “slowly” active at molar monovalent cation concentrations in the absence of Mg^{2+} ions, NaCl being slightly more effective than KCl [211–213]. This was certainly surprising to people in the ribozyme field since the accepted dogma indicated that these ribozymes were Mg^{2+}-dependent metalloenzymes. Currently, alternative catalytic mechanisms suggest that Mg^{2+} ions are limited to a structural role and that the ribozyme catalytic mechanism is of a nucleotide-dependent acid–base type [214, 215]. A 1.55 Å resolution crystal structure of the full hammerhead ribozyme obtained by using a buffer containing 1.7 M sodium malonate and 10 mM $MgCl_2$ allowed identifying 18 well-resolved Na^+ ions but no Mg^{2+} ions [62].

In the *Schistosoma mansoni* ribozyme, it seems that a high NaCl concentration can lead to the efficient formation of 3D contacts but does not permit alone optimal catalysis to occur [216]. At temperatures above 60°, it was noted that monovalent cations protect the ribozyme against degradation [217]. This last observation extends our knowledge of the known roles of monovalent cations in biochemical systems.

3.5.4.2 The Hepatitis Delta Virus and Glms Ribozymes

The hepatitis delta virus (HDV) was also found to be "slowly" active at molar monovalent concentration with again better activities in NaCl than KCl [218]. Crystals soaked in solution containing Tl acetate revealed 15 monovalent binding sites. It was proposed that monovalent ions may stabilize the active ribozyme structure or participate actively to the catalysis since low-level activity of this ribozyme is observed in the presence of monovalent ions [81].

In this structure, it was reported that cobalt hexammine was competing for a Tl^+ binding site. Competition between Na^+ and Mg^{2+} was also observed in MD simulations [219]. MD simulations proved to be a valuable tool in exploring the HDV systems since they were also able to identify a Na^+ binding location close to the catalytic site [125, 126] before the crystal structure was known [81]. Putting all the known biochemical information together, it was suggested that the HDV ribozyme is able to utilize two distinct strategies in its cleavage reaction [220]. The leading mechanism for this ribozyme being the one where a divalent metal at the catalytic site lowers the pK_a of the active C75 [221–223].

The glms is a self-cleaving ribozyme that does not require direct participation of divalent metal ions that were found to assume only structural roles [224, 225]. Significant rate enhancements are also observed with molar concentrations of monovalent cations.

3.5.4.3 DNAzymes Are also Influenced by Monovalent Cations

DNAzymes are the DNA equivalents of ribozymes. It has been reported that in contrast to the hammerhead ribozyme, the 8–17 DNAzyme activity is not detectable in the presence of 4 M K^+, Rb^+ or Cs^+ but measurable in 4 M Li^+, NH_4^+ and, to a lesser extend, Na^+ [226]. It was suggested that this *in vitro* selected DNAzyme had developed a more stringent need for divalent metal cations than related hammerhead systems, explaining the lower observed activity in monovalent ion setups. Similarly, an RNA aptamer that was selected in the presence of high concentrations of LiCl, was found to bind ligands more strongly with the smaller alkali ions than the larger ones [24].

3.5.5 Other Systematic and Less Systematic Ion Binding Studies on RNA Systems

3.5.5.1 Thirteen Metal Ions in the HIV-1 RNA Dimerization Initiation Site

A systematic crystallographic ion binding analysis of the duplex form of two subtypes of the HIV-1 RNA dimerization initiation sites (DIS) has been conducted. This study represents an important step in investigating the variability of metal binding sites in nucleic acid systems. Although this study focused mainly on multivalent cations, K^+ binding sites could be identified for one subtype and the ineffectiveness of molecular replacement by Rb^+ or Cs^+ was noted in this particular case [65].

3.5.5.2 Signal Recognition Particle – An Example of K^+/Mg^{2+} Mixed Occupancy

The ion binding properties of the signal recognition particle (SRP), for which functional importance was suggested by their requirement for K^+ and Mg^{2+} ions during complex formation, were investigated by crystallography [77, 227, 228]. Although no Na^+ buffers were used in this study, it was reported that K^+, NH_4^+ and especially Cs^+ are more effective than Na^+ in stabilizing the SRP complex.

The crystallographic structures of SRP revealed a wealth of information on mono- and divalent ion-binding features [77]. In details, an ion bound to a low selectivity ion binding site was best modeled by assigning half-occupancy to a pentahydrated Mg^{2+} ion and half-occupancy to a K^+ ion with the monovalent ion occupying the same position as one innersphere-coordinated water molecule. The authors suggested that both ions compete at physiological concentration for the same binding site within the major groove of the RNA. Such type of solvent density modeling is rare but might better reflect the crystallographic truth than singly occupied ion binding sites.

3.5.5.3 Riboswitches and Monovalent Ions

In a lysine riboswitch, a K^+ ion is bound to a carbonyl oxygen atom of the lysine ligand. The importance of K^+ for the biological activity of this riboswitch was assessed along with a significantly decreased binding affinity when K^+ was omitted or replaced by Na^+ [70]. Similarly, for the flavin mononucleotide (FMN) riboswitch, smaller and larger cations such as Na^+ and Cs^+, potentiate FMN binding to a lesser extent than K^+ [59]. K^+ was assigned to the structures of the thiamine pyrophosphate-sensing riboswitch [229], the ydaO riboswitches [230], the aptamer domain of the FMN riboswitch [231] as well as the SAM riboswitch [232]. Less riboswitches were crystallized with Na^+. The crystal structure of the THF riboswitch contains Na^+ [233] and some structures like those of the lysine riboswitch contain both monovalent cations [70]. A recent RNA aptamer selected *in vitro* for inhibiting lysozyme catalysis was found to be dependent on the presence of monovalent cations and equimolar substitution of K^+ for Na^+ resulted in only a 1.5 fold reduction of protein binding affinity [234].

Surprisingly, it was reported that K^+ is dispensable for the tertiary structure formation of a metal-sensing regulator RNA despite the fact that four K^+ ions were found in the crystal structure. The K^+ were modeled based on the positions of Cs^+ in a derivative dataset [91]. The authors of recent structures of bacterial riboswitches that control the expression of heavy metal transporters stated that the nucleotides involved in forming contacts with K^+ are not well conserved, which supports the conclusion that K^+ ions play in these systems only supportive roles in aptamer formation [92].

3.5.5.4 Group I and Group II Introns

In order to get clues related to the catalytic mechanism of the group I and II introns, a large number of crystal structures including an almost complete palette of mono- and divalent cations were crystallized [82, 87, 235]. For the group I intron, eighteen solvent densities were assigned to metal ions (5 K^+ and 13 Mg^{2+}) and Tl^+ anomalous

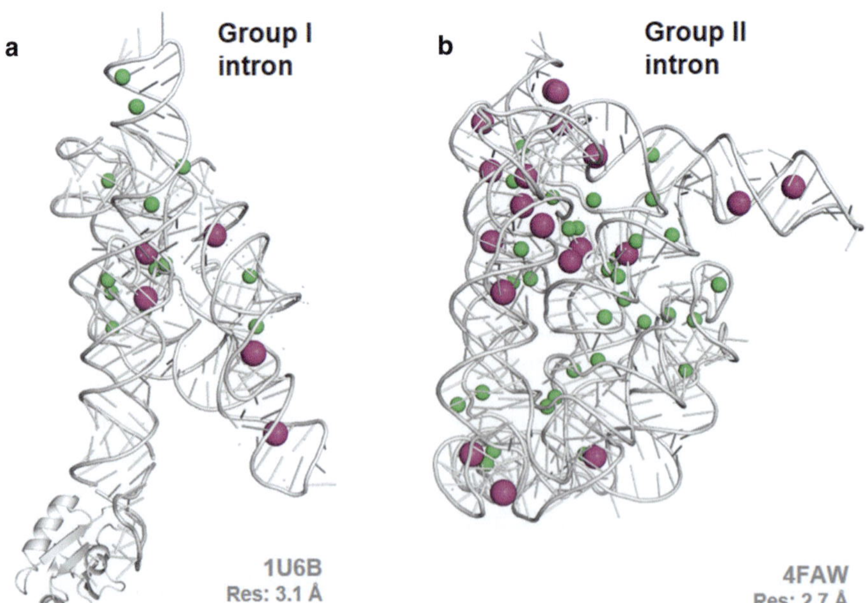

Figure 7 K^+ (magenta) and Mg^{2+} (green) ion binding sites as reported in crystal structures of the group I and II introns. (**a**) Metals in the *Azoarcus* group I intron (5 K^+ and 13 Mg^{2+}) [82]. (**b**) Metals in the *O. iheyensis* group II intron (19 K^+ and 32 Mg^{2+}) [87].

signals were obtained for the K^+ binding sites [82] (Figure 7a). For this ribozyme, it has been recognized that molar monovalent cations along with millimolar divalent cation concentrations are the most effective for folding and that monovalent salts alone are able to assist in the formation of most native tertiary contacts [197]. It was further shown that monovalent concentration changes the starting position of the RNA in its folding funnel and are essential in the formation of native tertiary contacts that promote fast and accurate folding [236]. Ion binding competitive experiments stressed the greater effectiveness of K^+ over Na^+ in stabilizing this RNA and demonstrated that in these conditions the need for Mg^{2+} ions is drastically reduced [200, 207], as it would be the case in conditions simulating crowding effects [237]. Interestingly, the *Candida* group I ribozyme was found to require much less monovalent cation concentrations to be properly folded and active than the *Tetrahymena* ribozyme [238].

For the group II intron, it is assumed that the catalytic center contains a four-metal-ion center and that the entire structure includes many other structural ions as well as delocalized ions that neutralize the high negative charge of this compact folded structure. Like for the DIS [65], SRP [77], RNase P [239] and group I intron [82], fourteen crystal structures of the group II intron were obtained in very diverse ionic conditions and in the 2.70 to 3.99 Å resolution range (Figure 7b) [87]. The catalytic metal center was reported to be heteronuclear and involving divalent (Mg^{2+}) and monovalent (K^+) ions. Based on crystallographic data, it was hypothesized that two of the detected ion-binding sites are specific to monovalent cations with strict size requirements and therefore probably occupied by K^+ under physiological conditions. The activity in Na/Mg buffers was reported to be much lower than in K/Mg, Rb/Mg or NH_4/Mg buffers [235].

3.6 *Monovalent Cation Influence on the Stability of Protein/DNA Complexes*

As described above, the use of different monovalent cations does not only affect locally the structure of the solvation shell of DNA duplexes and quadruplexes but can have important and generally not well-appreciated implications on the stability of DNA/protein complexes. For example:

(i) The thermodynamics of binding of the TATA binding protein-DNA complex is affected by the type of salt (Na^+ or K^+ acetate); i.e., for the smaller cation, the binding entropy is less favorable and the binding enthalpy mode favorable [240].
(ii) The Na^+ and K^+ ions allosterically regulate cooperative DNA binding of the human progesterone receptor [241]. Cooperative binding of the receptor to DNA is activated or repressed by Na^+ or K^+, respectively. Interestingly, the apparent binding affinities of Na^+ and K^+ are comparable to the number of intracellular ion pumps and channels specific for Na^+ or K^+. Hence, only a few

regulatory intracellular enzymes and ion transporters interact with Na^+ ions while most others seem to be functionally optimized for K^+ ions.

(iii) The affinity of the bacteriophage 434 repressor, that can distinguish between its six naturally occurring binding sites by using indirect readout, depends on the type and concentration of monovalent cations for some but not all binding sites. The stability of the repressor-O_R1 complex is cation type-dependent while that of the repressor-O_R3 complex is not [242]. Consequently, it was suggested that changing the intracellular ion concentration as it occurs in salt shock, bacteriophage infection, or apoptosis, does affect specific gene regulatory systems [243]. However, *in vivo* compensation mechanisms were reported for large changes in ionic concentrations [5]. For instance, for both *lac* repressor and RNA polymerase, it was found that formation of functional protein DNA complexes *in vivo* is only weakly dependent on ion concentration while *in vitro* both the equilibria and kinetics of binding are strongly affected [244].

(iv) Allosteric implications of K^+ ions for a poly(A)-specific ribonuclease that might be affected by different ionic conditions were also reported [245] and Na^+ was shown to possess a significantly higher potential than K^+ in DNA compaction in a crowded environment [169].

From another study, it was reported that a metal ion-binding motif in human DNA polymerase β displayed the following affinity order $Mg^{2+} < Ca^{2+} < Na^+ < K^+$, with K^+ displaying the strongest affinity. Thus, in protein/nucleic acid systems, monovalent ion binding sites in proteins (and not only in the nucleic acid components) might have a strong influence on the stability of the system [246].

Henceforth, studies suggesting equivalence between Na^+ and K^+ containing buffers should be taken with caution [247, 248] since, as described above, specific ion type dependent effects were uncovered in protein/DNA systems. More studies should certainly be devoted to these complex issues.

3.7 Ribosomal Activities Are Altered or even Inhibited by Na^+ Ions

Besides Mg^{2+}, ribosomes need K^+ for structure and function as already reported in the 70's. The importance of K^+ became apparent when significant inactivation of mammalian ribosomes was observed in K^+-free buffers [249–251]. Other studies stressed that ribosome activities are dependent on the presence of K^+ or NH_4^+ ions and that Na^+ inhibits them [252–255]. More recently, it was reported that elevated KCl concentrations stimulate picornavirus translation whereas NaCl concentrations up to 150 mM are neutral to the translation process suggesting that the association of the viral RNA with ribosomal subunits is possible in high salt conditions as witnessed during picornavirus infection [256].

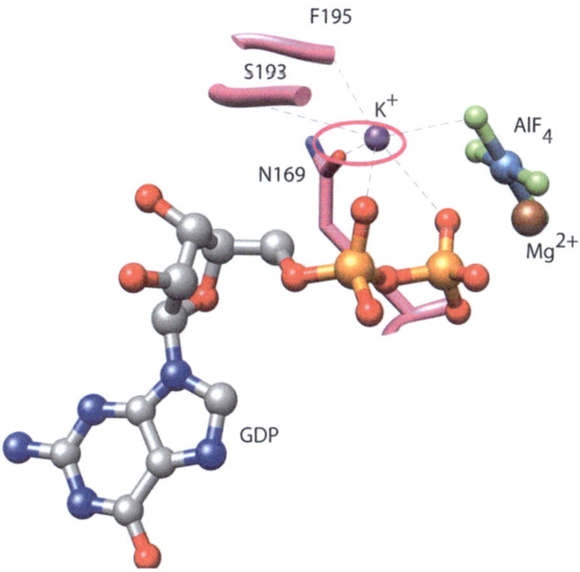

Figure 8 K^+ coordination sphere in the homology model of YqeH. This structure is exactly identical to that of GDP-bound K^+ in the MnmE GTPase [263] and shows in addition to the monovalent cation a Mg^{2+} ion and an AlF_4^- ion (PDB code: 3GJ8). Reproduced by permission from [262]; © copyright 2010.

From the crystallographic point of view, K^+ has been first identified close to the peptidyl transferase center [257]. Subsequently, 86 Na^+ and 2 K^+ were modeled into the *Haloarcula marismortui* large subunit [90]. The authors speculated that most of the identified monovalent binding sites are occupied by K^+ under physiological conditions. Na^+ and K^+ were also found in the S15-rRNA complex fragment [258] and an important K^+ binding site was located in a 58-nucleotide 23 S rRNA fragment [100]. In a recent 2.5 Å *Thermus thermophilus* crystal structure, a single K^+ was assigned [259], but K^+ was not mentioned in a 2.1 Å *Escherichia coli* structure [260] suggesting that the binding of such ions is highly dependent on crystallization condition and/or organism and/or is still difficult to characterize.

Unfortunately, no direct clues related to the mechanisms by which K^+ plays a role in ribosomal systems have been reported and, besides nucleic acid fragments, numerous RNA processing enzymes involved in the ribosomal maturation process like RNase E [261] or GTPase [262, 263], need monovalent cations. Indeed, the YqeH protein, a circularly permutated GTPase that is important for 30S maturation in *Bacillus subtilis*, was found to be K^+-dependent and inactivated in the presence of NaCl [262]. Although in the crystal structure K^+ is in contact with an AlF_4^- ion, it is also contacting two phosphate groups of a GDP molecule. Similarly, it was reported that MnmE, another GTPase that participates in tRNA modification, displays a K^+-dependent hydrolysis mechanism and a similar binding to GDP (Figure 8) [263].

Thus, K^+ is required at various stages of ribosome maturation and function and is associated with complex regulatory mechanisms in which Na^+ has an inhibitory effect. K^+ is therefore as necessary to the proper functions of *in vivo* nucleic acid based processes as the much more publicized divalent Mg^{2+} ions stressing the very subtle intra- and extracellular small metabolite ecology that are all, weak as well as strong binders, needed for the proper cellular functioning.

3.8 Are Both Na^+ and K^+ Involved in Chromatin Compaction?

We started this review by advocating that, since K^+ is the dominant intracellular ion, Na^+ should not be over-represented in *in vitro* experiments. Yet, biomolecular systems as we know are intricate and evolve in complex and crowded cellular environments that are constantly adapting to extracellular signals and intracellular modifications. As a result, we have to admit that we know few things related to the intricate regulation of ion fluxes in intracellular compartments and have only vague ideas on the metabolites that are necessary to nucleic acid systems [264].

Recent studies related to chromatin compaction revealed, in the presence of Mg^{2+}, that the addition of Na^+ promotes folding whereas in mixtures of K^+ and Mg^{2+}, folding is prevented [265]. The authors noted, as we did for other systems, that the vast majority of studies related to nucleosome arrays were carried on with buffers containing Na^+ or Mg^{2+} ions and not in more physiological environments. Although using binary salts (Na^+/Mg^{2+} and K^+/Mg^{2+}) is certainly important, these setups are nevertheless still distant from real physiological conditions. Secondary ion mass spectroscopy (SIMS) techniques revealed that all four Ca^{2+}, Mg^{2+}, Na^+, and K^+ but no other cations are present in mitotic chromosomes and stressed that cation redistribution occurs between cytosol and chromosomes during cell cycles (Figure 9). This study suggests further that these four cations are an integral part of mitotic chromatin [266] although concerns were raised regarding the possible accumulation of cations during the cell preparation process [267].

Nevertheless, these findings represent an interesting challenge for our reductionist biomolecular investigation techniques. Clearly, more refined cellular imaging techniques are needed that would allow to visualize the distribution of metals in the

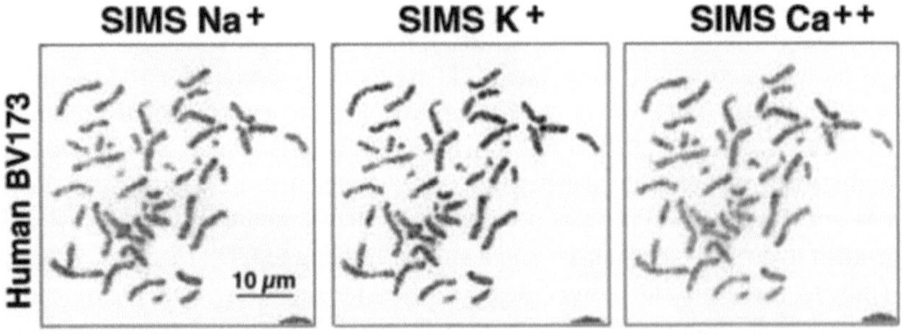

Figure 9 SIMS analytical distribution of Na^+, K^+, and Ca^{2+} in human chromosomes. Reproduced by permission from [266]; © copyright 2001, The Rockefeller University Press.

various cellular compartments during the cell life cycles [268, 269]. Moreover, the role of all ionic particles should be acknowledged, including that of anions that are also possible partners of nucleic acid systems [270, 271] and essential to cell function.

4 Conclusion

We hope that this review will help in creating awareness of the fact that Na^+ and K^+ ions are in general not neutral to nucleic acid and ribonucleoprotein particles since each of these ions, in specific structural contexts, plays different and sometimes conflicting structural and functional roles.

Consequently, much attention needs to be placed on the nature of the buffers used in biophysical experiments that much too often contain Na^+ instead of the more biologically relevant intracellular K^+ ion. Results stating that both cations are equivalent are to be taken with caution and, when comparative studies are not possible, K^+-containing conditions should be favored. In crystallographic studies for example, it might be easier to identify K^+ over Na^+, the former producing larger electron densities as well as anomalous signals that can be detected with appropriate equipment.

Yet *in vivo* systems might continue to surprise us. It has been shown that the cellular concentration of cationic species fluctuates in time and space and is associated with cell signaling [14]. Thus, partial inhibition or activation of specific cellular functions by Na^+ acting directly or indirectly on nucleic acid systems during cell life cycles can be hypothesized.

To close this review, it seems interesting to mention that the now well-documented NaCl abundance in biophysical buffers parallels that of human NaCl overconsumption associated with reduced K^+ intake that is at the origin of major health issues

Abbreviations

AMBER	Assisted Model Building and Energy Refinement
BWYV	beet western yellow virus
CSD	Cambridge Structural Database
DIS	dimerization initiation site
FMN	flavin mononucleotide
GDP	guanosine diphosphate
glms ribozyme	glucosamine-6-phosphate activated ribozyme
GTPase	guanosine triphosphate hydrolyzing enzymes
HDV	hepatitis delta virus
MAD	multi-wavelength anomalous dispersion
MD	molecular dynamics
miRNA	microRNA

NMR	nuclear magnetic resonance
PDB	Protein Data Bank
RNase	ribonuclease
rRNA	ribosomal RNA
SAM	S-adenosyl methionine
SIMS	secondary ion mass spectroscopy
SPC/E	extended simple point charge water model
SRP	signal recognition particle
THF	tetrahydrofolate
tRNA	transfer RNA

such as hypertension in industrialized populations [272, 273]. Further as noted by Di Cera, "*Abundance or absence of this simple compound has had profound effects on human health and has provided a casus belli in many important milestones in the history of man.*" [17, 274]. Indeed, these basic monovalent cations appear to play major and sometimes polemic roles in life and science.

Acknowledgments P.A. wishes to thank Prof. Eric Westhof and E.E. wishes to thank Dr. Philippe Dumas for ongoing support. This work has been published under the framework of the LABEX: ANR-10-LABX0036_NETRNA and benefits from a funding from the state managed by the French National Research Agency as part of the program "Investments for the future".

References

1. J. H. Christian, J. A. Waltho, *J. Gen. Microbiol.* **1961**, *25*, 97–102.
2. R. E. Godt, D. W. Maughan, *Am. J. Physiol.* **1988**, *254*, C591–604.
3. D. W. Maughan, R. E. Godt, *Biophys. J.* **1989**, *56*, 717–722.
4. L. N. Csonka, *Microbiol. Rev.* **1989**, *53*, 121–147.
5. S. Cayley, B. A. Lewis, H. J. Guttman, M. T. Record, Jr., *J. Mol. Biol.* **1991**, *222*, 281–300.
6. F. Lang, *J. Am. Coll. Nutr.* **2007**, *26*, 613S–623S.
7. Y. Zhou, J. H. Morais-Cabral, A. Kaufman, R. MacKinnon, *Nature* **2001**, *414*, 43–48.
8. J. Orlowski, S. Grinstein, *Curr. Opin. Cell. Biol.* **2007**, *19*, 483–492.
9. J. Arino, J. Ramos, H. Sychrova, *Microbiol. Mol. Biol. Rev.* **2010**, *74*, 95–120.
10. S. Ye, Y. Li, Y. Jiang, *Nat. Struct. Mol. Biol.* **2010**, *17*, 1019–1023.
11. I. Kim, T. W. Allen, *Proc. Natl. Acad. Sci. USA* **2011**, *108*, 17963–17968.
12. A. Rodriguez-Navarro, *Biochim. Biophys. Acta* **2000**, *1469*, 1–30.
13. C. R. Rose, A. Konnerth, *J. Neurosci.* **2001**, *21*, 4207–4214.
14. S. N. Orlov, P. Hamet, *J. Membr. Biol.* **2006**, *210*, 161–172.
15. M. I. Panayiotidis, C. D. Bortner, J. A. Cidlowski, *Acta Physiol.* **2006**, *187*, 205–215.
16. M. J. Page, E. Di Cera, *Physiol. Rev.* **2006**, *86*, 1049–1092.
17. E. Di Cera, *Mol. Aspects Med.* **2008**, *29*, 203–254.
18. J. A. Huntington, *Biol. Chem.* **2008**, *389*, 1025–1035.
19. I. Kurisaki, M. Takayanagi, M. Nagaoka, *J. Phys. Chem. B* **2015**, *119*, 3635–3642.
20. G. Chen, S. D. Kennedy, D. H. Turner, *Biochemistry* **2009**, *48*, 5738–5752.
21. D. H. Turner, D. H. Mathews, *Nucleic Acids Res.* **2010**, *38*, D280–282.
22. R. Owczarzy, B. G. Moreira, Y. You, M. A. Behlke, J. A. Walder, *Biochemistry* **2008**, *47*, 5336–5353.

23. J. Freund, H. R. Kalbitzer, *J. Biomol. NMR* **1995**, *5*, 321–322.
24. D. Lambert, D. Leipply, R. Shiman, D. E. Draper, *J. Mol. Biol.* **2009**, *390*, 791–804.
25. B. Heddi, N. Foloppe, E. Hantz, B. Hartmann, *J. Mol. Biol.* **2007**, *368*, 1403–1411.
26. K. Halder, J. S. Hartig, *Met. Ions Life Sci.* **2011**, *9*, 125–139.
27. H. Huang, N. B. Suslov, N. S. Li, S. A. Shelke, M. E. Evans, Y. Koldobskaya, P. A. Rice, J. A. Piccirilli, *Nat. Chem. Biol.* **2014**, *10*, 686–691.
28. P. Murat, J. Zhong, L. Lekieffre, N. P. Cowieson, J. L. Clancy, T. Preiss, S. Balasubramanian, R. Khanna, J. Tellam, *Nat. Chem. Biol.* **2014**, *10*, 358–364.
29. G. M. Arachchilage, A. C. Dassanayake, S. Basu, *Chem. Biol.* **2015**, *22*, 262–272.
30. X. Yang, T. Li, B. Li, E. Wang, *Analyst* **2010**, *135*, 71–75.
31. M. H. Harding, *Acta Cryst.* **2006**, *D62*, 678–682.
32. H. Zheng, M. Chruszcz, P. Lasota, L. Lebioda, W. Minor, *J. Inorg. Biochem.* **2008**, *102*, 1765–1776.
33. G. Kuppuraj, M. Dudev, C. Lim, *J. Phys. Chem. B* **2009**, *113*, 2952–2960.
34. M. J. Harding, M. W. Nowicki, M. D. Walkinshaw, *Crystallography Rev.* **2010**, *16*, 247–302.
35. I. Persson, *Pure Appl. Chem.* **2010**, *82*, 1901–1907.
36. M. Pechlaner, R. K. Sigel, *Met. Ions Life Sci.* **2012**, *10*, 1–42.
37. O. Kirillova, W. Minor, *Bioinformatics* **2006**, *22*, 1660–1661.
38. L. R. Stefan, R. Zhang, A. G. Levitan, D. K. Hendrix, S. E. Brenner, S. R. Holbrook, *Nucleic Acids Res.* **2006**, *34*, D131–134.
39. K. Y. Hsin, Y. Sheng, M. M. Harding, P. Taylor, M. D. Walkinshaw, *J. Appl. Cryst.* **2008**, *41*, 963–968.
40. K. Hemavathi, M. Kalaivani, A. Udayakumar, G. Sowmiya, J. Jeyakanthan, K. Sekar, *J. Appl. Cryst.* **2010**, *43*, 196–199.
41. C. Andreini, G. Cavallaro, S. Lorenzini, *Bioinformatics* **2012**, *28*, 1658–1660.
42. A. Philips, K. Milanowska, G. Lach, M. Boniecki, K. Rother, J. M. Bujnicki, *Bioinformatics* **2012**, *28*, 198–205.
43. J. Schnabl, P. Suter, R. K. Sigel, *Nucleic Acids Res.* **2012**, *40*, D434–438.
44. A. Tus, A. Rakipovic, G. Peretin, S. Tomic, M. Sikic, *Nucleic Acids Res.* **2012**, *40*, W352–357.
45. C. Andreini, G. Cavallaro, S. Lorenzini, A. Rosato, *Nucleic Acids Res.* **2013**, *41*, D312–319.
46. C. Andreini, G. Cavallaro, A. Rosato, Y. Valasatava, *J. Chem. Inf. Model.* **2013**, *53*, 3064–3075.
47. H. Zheng, M. D. Chordia, D. R. Cooper, M. Chruszcz, P. Muller, G. M. Sheldrick, W. Minor, *Nat. Protoc.* **2014**, *9*, 156–170.
48. M. Egli, *Chem. Biol.* **2002**, *9*, 277–286.
49. C. Hsiao, E. Tannenbaum, H. VanDeusen, E. Hershkovitz, G. Perng, A. R. Tannenbaum, L. D. Williams, in *Nucleic Acid-Metal Ion Interactions*, Ed. N. V. Hud, Royal Society of Chemistry, Cambridge, 2009, pp. 1–38.
50. J. Muller, *Metallomics* **2010**, *2*, 318–327.
51. P. Auffinger, N. Grover, E. Westhof, *Met. Ions Life Sci.* **2011**, *9*, 1–35.
52. D. Donghi, J. Schnabl, *Met. Ions Life Sci.* **2011**, *9*, 197–234.
53. M. C. Erat, R. K. Sigel, *Met. Ions Life Sci.* **2011**, *9*, 37–100.
54. A. R. Ferré-D'Amaré, W. C. Winkler, *Met. Ions Life Sci.* **2011**, *9*, 141–173.
55. Z. J. Tan, S. J. Chen, *Met. Ions Life Sci.* **2011**, *9*, 101–124.
56. K. Aoki, K. Murayama, *Met. Ions Life Sci.* **2012**, *10*, 43–102.
57. N. Echols, N. Morshed, P. V. Afonine, A. J. McCoy, M. D. Miller, R. J. Read, J. S. Richardson, T. C. Terwilliger, P. D. Adams, *Acta Cryst.* **2014**, *D70*, 1104–1114.
58. M. H. Harding, *Acta Cryst.* **2002**, *D58*, 872–874.
59. A. Serganov, L. Huang, D. J. Patel, *Nature* **2009**, *458*, 233–237.
60. S. Varma, S. B. Rempe, *Biophys. Chem.* **2006**, *124*, 192–199.
61. R. Mancinelli, A. Botti, F. Bruni, M. A. Ricci, A. K. Soper, *J. Phys. Chem. B* **2007**, *111*, 13570–13577.

62. M. Anderson, E. P. Schultz, M. Martick, W. G. Scott, *J. Mol. Biol.* **2013**, *425*, 3790–3798.
63. E. C. Juan, J. Kondo, T. Kurihara, T. Ito, Y. Ueno, A. Matsuda, A. Takenaka, *Nucleic Acids Res.* **2007**, *35*, 1969–1977.
64. V. Tereshko, C. J. Wilds, G. Minasov, T. P. Prakash, M. A. Maier, A. Howard, Z. Wawrzak, M. Manoharan, M. Egli, *Nucleic Acids Res.* **2001**, *29*, 1208–1215.
65. E. Ennifar, P. Walter, P. Dumas, *Nucleic Acids Res.* **2003**, *31*, 2671–2682.
66. T. Moulaei, T. Maehigashi, G. T. Lountos, S. Komeda, D. Watkins, M. P. Stone, L. A. Marky, J. S. Li, B. Gold, L. D. Williams, *Biochemistry* **2005**, *44*, 7458–7468.
67. E. Ennifar, J. C. Paillart, A. Bodlenner, P. Walter, J. M. Weibel, A. M. Aubertin, P. Pale, P. Dumas, R. Marquet, *Nucleic Acids Res.* **2006**, *34*, 2328–2339.
68. E. Ennifar, P. Dumas, *J. Mol. Biol.* **2006**, *356*, 771–782.
69. E. Ennifar, S. Bernacchi, P. Wolff, P. Dumas, *RNA* **2007**, *13*, 1445–1452.
70. A. Serganov, L. Huang, D. J. Patel, *Nature* **2008**, *455*, 1263–1268.
71. I. Russo Krauss, A. Merlino, A. Randazzo, E. Novellino, L. Mazzarella, F. Sica, *Nucleic Acids Res.* **2012**, *40*, 8119–8128.
72. P. Cvjetko, I. Cvjetko, M. Pavlica, *Arh. Hig. Rada. Toksikol.* **2010**, *61*, 111–119.
73. S. Basu, R. P. Rambo, J. Strauss-Soukup, J. H. Cate, A. R. Ferré-D'Amaré, S. A. Strobel, J. A. Doudna, *Nat. Struct. Biol.* **1998**, *5*, 986–992.
74. G. L. Conn, D. E. Draper, E. E. Lattman, A. G. Gittis, *Science* **1999**, *284*, 1171–1174.
75. C. C. Correll, I. G. Wool, A. Munishkin, *J. Mol. Biol.* **1999**, *292*, 275–287.
76. S. B. Howerton, C. C. Sines, D. VanDerveer, L. D. Williams, *Biochemistry* **2001**, *40*, 10023–10031.
77. R. T. Batey, J. A. Doudna, *Biochemistry* **2002**, *41*, 11703–11710.
78. M. Egli, G. Minasov, L. Su, A. Rich, *Proc. Natl. Acad. Sci. USA* **2002**, *99*, 4302–4307.
79. S. B. Howerton, A. Nagpal, L. D. Williams, *Biopolymers* **2003**, *69*, 87–99.
80. V. Kacer, S. A. Scaringe, J. N. Scarsdale, J. P. Rife, *Acta Cryst.* **2003**, *D59*, 423–432.
81. A. Ke, F. Ding, J. D. Batchelor, J. A. Doudna, *Structure* **2007**, *15*, 281–287.
82. M. R. Stahley, P. L. Adams, J. Wang, S. A. Strobel, *J. Mol. Biol.* **2007**, *372*, 89–102.
83. L. Huang, A. Serganov, D. J. Patel, *Mol. Cell.* **2010**, *40*, 774–786.
84. D. Watkins, S. Mohan, G. B. Koudelka, L. D. Williams, *J. Mol. Biol.* **2010**, *396*, 1145–1164.
85. T. Maehigashi, C. Hsiao, K. K. Woods, T. Moulaei, N. V. Hud, L. D. Williams, *Nucleic Acids Res.* **2012**, *40*, 3714–3722.
86. A. Ren, K. R. Rajashankar, D. J. Patel, *Nature* **2012**, *486*, 85–89.
87. M. Marcia, A. M. Pyle, *RNA* **2014**, *20*, 516–527.
88. G. L. Conn, A. G. Gittis, E. E. Lattman, V. K. Misra, D. E. Draper, *J. Mol. Biol.* **2002**, *318*, 963–973.
89. T. Hultin, P. H. Naslund, *Chem. Biol. Interact.* **1974**, *8*, 315–328.
90. D. J. Klein, P. B. Moore, T. A. Steitz, *RNA* **2004**, *10*, 1366–1379.
91. C. E. Dann, 3rd, C. A. Wakeman, C. L. Sieling, S. C. Baker, I. Irnov, W. C. Winkler, *Cell* **2007**, *130*, 878–892.
92. K. Furukawa, A. Ramesh, Z. Zhou, Z. Weinberg, T. Vallery, W. C. Winkler, R. R. Breaker, *Mol. Cell.* **2015**, *57*, 1088–1098.
93. C. Mueller-Dieckmann, S. Panjikar, A. Schmidt, S. Mueller, J. Kuper, A. Geerlof, M. Wilmanns, R. K. Singh, P. A. Tucker, M. S. Weiss, *Acta Cryst.* **2007**, *D63*, 366–380.
94. Z. Dauter, A. Wlodawer, W. Minor, M. Jaskolski, B. Rupp, *IUCrJ* **2014**, *1*, 179–193.
95. G. F. Schroder, M. Levitt, A. T. Brunger, *Nature* **2010**, *464*, 1218–1222.
96. J. Wang, *Acta Cryst.* **2010**, *D66*, 988–1000.
97. N. V. Hud, A. E. Engelhart, in *Nucleic Acid-Metal Ion Interactions*, Ed. N. V. Hud, Royal Society of Chemistry, Cambridge, 2009, pp. 75–117.
98. N. V. Hud, V. Sklenar, J. Feigon, *J. Mol. Biol.* **1999**, *286*, 651–660.
99. P. L. Nixon, D. P. Giedroc, *J. Mol. Biol.* **2000**, *296*, 659–671.
100. R. Shiman, D. E. Draper, *J. Mol. Biol.* **2000**, *302*, 79–91.

101. S. E. Butcher, F. H. T. Allain, J. Feigon, *Biochemistry* **2000**, *39*, 2714–2182.
102. F. C. Marincola, V. P. Denisov, B. Halle, *J. Am. Chem. Soc.* **2004**, *126*, 6739–6750.
103. F. Mocci, A. Laaksonen, A. Lyubartsev, G. Saba, *J. Phys. Chem. B* **2004**, 108, 16295–16302.
104. M. L. Gill, S. A. Strobel, J. P. Loria, *J. Am. Chem. Soc.* **2005**, *127*, 16723–16732.
105. F. C. Marincola, A. Virno, A. Randazzo, F. Miocci, G. Saba, A. Lai, *Magn. Reson. Chem.* **2009**, *47*, 1036–1042.
106. R. Fiala, N. Spackova, S. Foldynova-Trantirkova, J. Sponer, V. Sklenar, L. Trantirek, *J. Am. Chem. Soc.* **2011**, *133*, 13790–13793.
107. M. Feig, B. M. Pettitt, *Biophys. J.* **1999**, *77*, 1769–1781.
108. P. Auffinger, E. Westhof, *J. Mol. Biol.* **2000**, *300*, 1113–1131.
109. K. J. McConnell, D. L. Beveridge, *J. Mol. Biol.* **2000**, *304*, 803–820.
110. P. Auffinger, E. Westhof, *J. Mol. Biol.* **2001**, *305*, 1057–1072.
111. K. Csaszar, N. Spackova, R. Stefl, J. Sponer, N. B. Leontis, *J. Mol. Biol.* **2001**, *313*, 1073–1091.
112. K. Reblova, N. Spackova, J. E. Sponer, J. Koca, J. Sponer, *Nucleic Acids Res.* **2003**, *31*, 6942–6952.
113. P. Auffinger, L. Bielecki, E. Westhof, *J. Mol. Biol.* **2004**, *335*, 555–571.
114. N. Korolev, A. P. Lyubartsev, A. Laaksonen, L. Nordenskiold, *Biopolymers* **2004**, *73*, 542–555.
115. K. Reblova, N. Spackova, J. Koca, N. B. Leontis, J. Sponer, *Biophys. J.* **2004**, *87*, 3397–3412.
116. P. Varnai, K. Zakrzewska, *Nucleic Acids Res.* **2004**, *32*, 4269–4280.
117. Y. Cheng, N. Korolev, L. Nordenskiold, *Nucleic Acids Res.* **2006**, *34*, 686–696.
118. A. Savelyev, G. A. Papoian, *J. Am. Chem. Soc.* **2006**, *128*, 14506–14518.
119. F. Razga, M. Zacharias, K. Reblova, J. Koca, J. Sponer, *Structure* **2006**, *14*, 825–835.
120. S. E. McDowell, N. Spackova, J. Sponer, N. G. Walter, *Biopolymers* **2007**, *85*, 169–184.
121. P. Auffinger, Y. Hashem, *Curr. Op. Struct. Biol.* **2007**, *17*, 325–333.
122. P. Auffinger, in *RNA 3D Structure Analysis and Prediction*, Eds E. Westhof, N. B. Leontis, Springer Verlag, Berlin, 2012, 299–318.
123. F. Mocci, A. Laaksonen, *Soft Matter* **2012**, *8*, 9268–9284.
124. M. A. Ditzler, J. Sponer, N. G. Walter, *RNA* **2009**, *15*, 560–575.
125. M. V. Krasovska, J. Sefcikova, N. Spackova, J. Sponer, N. G. Walter, *J. Mol. Biol.* **2005**, *351*, 731–748.
126. M. V. Krasovska, J. Sefcikova, K. Reblova, B. Schneider, N. G. Walter, J. Sponer, *Biophys. J.* **2006**, *91*, 626–638.
127. A. C. Vaiana, E. Westhof, P. Auffinger, *Biochimie* **2006**, *88*, 1061–1073.
128. P. Auffinger, T. E. Cheatham, A. C. Vaiana, *J. Chem. Theory and Comput.* **2007**, *3*, 1851–1859.
129. A. A. Chen, R. V. Pappu, *J. Phys. Chem. B* **2007**, *111*, 11884–11887.
130. A. Noy, I. Soteras, F. J. Luque, M. Orozco, *Phys. Chem. Chem. Phys.* **2009**, *11*, 10596–10607.
131. R. V. Reshetnikov, J. Sponer, O. I. Rassokhina, A. M. Kopylov, P. O. Tsvetkov, A. A. Makarov, A. V. Golovin, *Nucleic Acids Res.* **2011**, *39*, 9789–9802.
132. I. S. Joung, T. E. Cheatham, 3rd, *J. Phys. Chem. B* **2008**, *112*, 9020–9041.
133. P. Banas, P. Jurecka, N. G. Walter, J. Sponer, M. Otyepka, *Methods* **2009**, *49*, 202–216.
134. J. Carlsson, J. Aqvist, *J. Phys. Chem. B* **2009**, *113*, 10255–10260.
135. A. A. Chen, M. Marucho, N. A. Baker, R. V. Pappu, *Methods Enzymol.* **2009**, *469*, 411–432.
136. H. Yu, T. W. Whitfield, E. Harder, G. Lamoureux, I. Vorobyov, V. M. Anisimov, A. D. MacKerell, B. Roux, *J. Chem. Theory Comput.* **2010**, *6*, 774–786.
137. C. Zhang, S. Raugei, B. Eisenberg, P. Carloni, *J. Chem. Theory Comput.* **2010**, *6*, 2167–2175.

138. R. Lavery, K. Zakrzewska, D. L. Beveridge, T. C. Bishop, D. A. Case, T. E. Cheatham, 3rd, S. Dixit, B. Jayaram, F. Lankas, C. Laughton, J. H. Maddocks, A. Michon, R. Osman, M. Orozco, A. Perez, T. Singh, N. Spackova, J. Sponer, *Nucleic Acids Res.* **2010**, *38*, 299–313.
139. D. L. Beveridge, T. E. Cheatham, 3rd, M. Mezei, *J. Biosci.* **2012**, *37*, 379–397.
140. R. Lavery, J. H. Maddocks, M. Pasi, K. Zakrzewska, *Nucleic Acids Res.* **2014**, *42*, 8138–8149.
141. M. Pasi, J. H. Maddocks, R. Lavery, *Nucleic Acids Res.* **2015**, *43*, 2412–2423.
142. H. J. C. Berendsen, J. R. Grigera, T. P. Straatsma, *J. Phys. Chem.* **1987**, *97*, 6269–6271.
143. L. X. Dang, *J. Am. Chem. Soc.* **1995**, *117*, 6954–6960.
144. I. Besseova, M. Otyepka, K. Reblova, J. Sponer, *Phys. Chem. Chem. Phys.* **2009**, *11*, 10701–10711.
145. I. Besseova, P. Banas, P. Kuhrova, P. Kosinova, M. Otyepka, J. Sponer, *J. Phys. Chem. B* **2012**, *116*, 9899–9916.
146. A. Savelyev, A. D. MacKerell, Jr., *J. Phys. Chem. B* **2015**, 119, 4428–4440.
147. A. Savelyev, A. D. MacKerell, Jr., *J. Phys. Chem. Lett.* **2015**, *6*, 212–216.
148. W. L. Jorgensen, D. S. Maxwell, J. Tirado-Rives, *J. Am. Chem. Soc.* **1996**, *118*, 11225–11236.
149. B. Roux, *Biophys. J.* **1996**, *71*, 3177–3185.
150. K. Igarashi, K. Kashiwagi, *Int. J. Biochem. Cell. Biol* **2010**, *42*, 39–51.
151. G. S. Manning, *Quart. Rev. Biophys.* **1978**, *11*, 179–246.
152. J. Lipfert, S. Doniach, R. Das, D. Herschlag, *Annu. Rev. Biochem.* **2014**, *83*, 813–841.
153. N. V. Hud, M. Polak, *Curr. Opin. Struct. Biol.* **2001**, *11*, 293–301.
154. E. Stellwagen, J. M. Muse, N. C. Stellwagen, *Biochemistry* **2011**, *50*, 3084–3094.
155. S. Nakano, M. Fujimoto, H. Hara, N. Sugimoto, *Nucleic Acids Res.* **1999**, *27*, 2957–2965.
156. J. Vieregg, W. Cheng, C. Bustamante, I. Tinoco, Jr., *J. Am. Chem. Soc.* **2007**, *129*, 14966–14973.
157. M. A. Young, B. Jayaram, D. L. Beveridge, *J. Am. Chem. Soc.* **1997**, *119*, 59–69.
158. B. Halle, V. P. Denisov, *Biopolymers* **1998**, *48*, 210–233.
159. X. Shui, C. C. Sines, L. McFail-Isom, D. VanDerveer, L. D. Williams, *Biochemistry* **1998**, *37*, 16877–16887.
160. X. Shui, L. McFail-Isom, G. G. Hu, L. D. Williams, *Biochemistry* **1998**, *37*, 8341–8355.
161. L. McFail-Isom, C. C. Sines, L. D. Williams, *Curr. Op. Struct. Biol.* **1999**, *9*, 298–304.
162. V. Tereshko, G. Minasov, M. Egli, *J. Am. Chem. Soc.* **1999**, *121*, 3590–3595.
163. K. Woods, L. McFail-Isom, C. C. Sines, S. B. Howerton, L. D. Williams, *J. Am. Chem. Soc.* **2000**, *122*, 1546–1547.
164. J. J. Howard, G. C. Lynch, B. M. Pettitt, *J. Phys. Chem. B* **2011**, *115*, 547–556.
165. X. Shen, B. Gu, S. A. Che, F. S. Zhang, *J. Chem. Phys.* **2011**, *135*, 034509.
166. F. Lankas, N. Spackova, M. Moakher, P. Enkhbayar, J. Sponer, *Nucleic Acids Res.* **2010**, *38*, 3414–3422.
167. S. Piana, J. L. Klepeis, D. E. Shaw, *Curr Opin Struct Biol* **2014**, *24*, 98–105.
168. R. M. Leal, S. Callow, P. Callow, M. P. Blakeley, C. J. Cardin, W. A. Denny, S. C. Teixeira, E. P. Mitchell, V. T. Forsyth, *Acta Cryst.*, *D66*, 1244–1248.
169. A. A. Zinchenko, K. Yoshikawa, *Biophys. J.* **2005**, *88*, 4118–4123.
170. R. Stefl, H. Wu, S. Ravindranathan, V. Sklenar, J. Feigon, *Proc. Natl. Acad. Sci. USA* **2004**, *101*, 1177–1182.
171. Q. Dong, E. Stellwagen, N. C. Stellwagen, *Biochemistry* **2009**, *48*, 1047–1055.
172. L. Sethaphong, Y. G. Yingling, *Molecules* **2012**, *17*.
173. D. Wei, G. N. Parkinson, A. P. Reszka, S. Neidle, *Nucleic Acids Res.* **2012**, 40, 4691–4700.
174. G. N. Parkinson, M. P. Lee, S. Neidle, *Nature* **2002**, *417*, 876–880.
175. D. Sen, W. Gilbert, *Nature* **1990**, *344*, 410–414.
176. S. Bouaziz, A. Kettani, D. J. Patel, *J. Mol. Biol.* **1998**, *282*, 637–652.
177. P. Schultze, N. V. Hud, F. W. Smith, J. Feigon, *Nucleic Acids Res.* **1999**, *27*, 3018–3028.

178. A. Risitano, K. R. Fox, *Bioorg. Med. Chem. Lett.* **2005**, *15*, 2047–2050.
179. A. E. Engelhart, J. Plavec, O. Persil, N. V. Hud, in *Nucleic Acid-Metal Ion Interactions*, Ed. N. V. Hud, Royal Society of Chemistry, Cambridge, 2009, pp. 118–153.
180. T. I. Gaynutdinov, P. Brown, R. D. Neumann, I. G. Panyutin, *Biochemistry* **2009**, *48*, 11169–11177.
181. R. D. Gray, L. Petraccone, J. O. Trent, J. B. Chaires, *Biochemistry* **2010**, *49*, 179–194.
182. D. H. Zhang, T. Fujimoto, S. Saxena, H. Q. Yu, D. Miyoshi, N. Sugimoto, *Biochemistry* **2010**, *49*, 4554–4563.
183. S. Neidle, *Therapeutic Applications of Quadruplex Nucleic Acids*, Academic Press, San Diego, USA, 2011.
184. R. Ida, G. Wu, *J. Am. Chem. Soc.* **2008**, *130*, 3590–3602.
185. G. D. Strahan, M. A. Keniry, R. H. Shafer, *Biophys. J.* **1998**, *75*, 968–981.
186. C. C. Hardin, T. Watson, M. Corregan, C. Bailey, *Biochemistry* **1992**, *31*, 833–841.
187. P. Fojtik, I. Kejnovska, M. Vorlickova, *Nucleic Acids Res.* **2004**, *32*, 298–306.
188. N. Spackova, I. Berger, J. Sponer, *J. Am. Chem. Soc.* **2001**, 123, 3925–3307.
189. S. Haider, S. Neidle, *Methods Mol. Biol.* **2010**, *608*, 17–37.
190. E. Fadrna, N. Spackova, J. Sarzynska, J. Koca, M. Orozco, T. E. Cheatham III, T. Kulinski, J. Sponer, *J. Chem. Theory. Comput.* **2009**, *5*, 2514–2530.
191. J. Sponer, X. Cang, T. E. Cheatham, 3rd, *Methods* **2012**, 57, 25–39.
192. R. Reshetnikov, A. Golovin, V. Spiridonova, A. Kopylov, J. Sponer, *J. Chem. Theory. Comput.* **2010**, *6*, 3003–3014.
193. C. Cheong, P. B. Moore, *Biochemistry* **1992**, *31*, 8406–8414.
194. J. Deng, Y. Xiong, M. Sundaralingam, *Proc. Natl. Acad. Sci. USA* **2001**, *98*, 13665–13670.
195. A. Bugaut, P. Murat, S. Balasubramanian, *J. Am. Chem. Soc.* **2012**, *134*, 19953–19956.
196. K. Takamoto, R. Das, Q. He, S. Doniach, M. Brenowitz, D. Herschlag, M. R. Chance, *J. Mol. Biol.* **2004**, *343*, 1195–1206.
197. S. A. Woodson, *Curr. Opin. Chem. Biol.* **2005**, *9*, 104–109.
198. D. Leipply, D. Lambert, D. E. Draper, *Methods Enzymol.* **2009**, *469*, 433–463.
199. N. Bisaria, D. Herschlag, *Biochem. Soc. Trans.* **2015**, *43*, 172–178.
200. T. Uchida, Q. He, C. Y. Ralston, M. Brenowitz, M. R. Chance, *Biochemistry* **2002**, *41*, 5799–5806.
201. T. C. Gluick, N. M. Wills, R. F. Gesteland, D. E. Draper, *Biochemistry* **1997**, *36*, 16173–16186.
202. I. Shcherbakova, S. Gupta, M. R. Chance, M. Brenowitz, *J. Mol. Biol.* **2004**, *342*, 1431–1442.
203. K. Dammertz, M. Hengesbach, M. Helm, G. U. Nienhaus, A. Y. Kobitski, *Biochemistry* **2011**, *50*, 3107–3115.
204. E. D. Holmstrom, J. L. Fiore, D. J. Nesbitt, *Biochemistry* **2012**, *51*, 3732–3743.
205. J. H. Cate, J. A. Doudna, *Structure* **1996**, *4*, 1221–1229.
206. J. H. Cate, A. R. Gooding, E. Podell, K. H. Zhou, B. L. Golden, C. E. Kundrot, T. R. Cech, J. A. Doudna, *Science* **1996**, *273*, 1678–1685.
207. K. J. Travers, N. Boyd, D. Herschlag, *RNA* **2007**, *13*, 1205–1213.
208. J. K. Frederiksen, N. S. Li, R. Das, D. Herschlag, J. A. Piccirilli, *RNA* **2012**, *18*, 1123–1141.
209. J. L. Fiore, J. H. Hodak, O. Piestert, C. D. Downey, D. J. Nesbitt, *Biophys J* **2008**, *95*, 3892–3905.
210. Y. X. Wang, M. Lu, D. E. Draper, *Biochemistry* **1993**, *32*, 12279–12282.
211. J. B. Murray, A. A. Seyhan, N. G. Walter, J. M. Burke, W. G. Scott, *Chem. Biol.* **1998**, *5*, 587–595.
212. E. A. Curtis, D. P. Bartel, *RNA* **2001**, *7*, 546–552.
213. J. L. O'Rear, S. Wang, A. L. Feig, L. Beigelman, O. C. Uhlenbeck, D. Herschlag, *RNA* **2001**, *7*, 537–545.
214. F. Leclerc, *Molecules* **2010**, *15*, 5389–5407.
215. W. G. Scott, L. H. Horan, M. Martick, *Prog. Mol. Biol. Transl. Sci.* **2013**, *120*, 1–23.

216. J. L. Boots, M. D. Canny, E. Azimi, A. Pardi, *RNA* **2008**, *14*, 2212–2222.
217. N. El-Murr, M. C. Maurel, M. Rihova, J. Vergne, G. Herve, M. Kato, K. Kawamura, *Naturwissenschaften* **2012**, *99*, 731–738.
218. A. T. Perrotta, M. D. Been, *Biochemistry* **2006**, *45*, 11357–11365.
219. N. Veeraraghavan, A. Ganguly, J. H. Chen, P. C. Bevilacqua, S. Hammes-Schiffer, B. L. Golden, *Biochemistry*, *50*, 2672–2682.
220. B. L. Golden, *Biochemistry* **2011**, *50*, 9424–9433.
221. A. Ganguly, P. C. Bevilacqua, S. Hammes-Schiffer, *J. Phys. Chem. Lett.* **2011**, *2*, 2906–2911.
222. A. Ganguly, P. Thaplyal, E. Rosta, P. C. Bevilacqua, S. Hammes-Schiffer, *J. Am. Chem. Soc.* **2014**, *136*, 1483–1496.
223. P. Thaplyal, A. Ganguly, S. Hammes-Schiffer, P. C. Bevilacqua, *Biochemistry* **2015**, *54*, 2160–2175.
224. A. Roth, A. Nahvi, M. Lee, I. Jona, R. R. Breaker, *RNA* **2006**, *12*, 607–619.
225. K. M. Brooks, K. J. Hampel, *Biochemistry* **2011**, *50*, 2424–2433.
226. D. Mazumdar, N. Nagraj, H. K. Kim, X. Meng, A. K. Brown, Q. Sun, W. Li, Y. Lu, *J. Am. Chem. Soc.* **2009**, *131*, 5506–5515.
227. R. T. Batey, R. P. Rambo, L. Lucast, B. Rha, J. A. Doudna, *Science* **2000**, *287*, 1232–1239.
228. R. T. Batey, M. B. Sagar, J. A. Doudna, *J. Mol. Biol.* **2001**, *307*, 229–246.
229. A. Serganov, A. Polonskaia, A. T. Phan, R. R. Breaker, D. J. Patel, *Nature* **2006**, *441*, 1167–1171.
230. A. Gao, A. Serganov, *Nat. Chem. Biol.* **2014**, *10*, 787–792.
231. Q. Vicens, E. Mondragon, R. T. Batey, *Nucleic Acids Res.* **2011**, *39*, 8586–8598.
232. C. D. Stoddard, R. K. Montange, S. P. Hennelly, R. P. Rambo, K. Y. Sanbonmatsu, R. T. Batey, *Structure* **2010**, *18*, 787–797.
233. L. Huang, S. Ishibe-Murakami, D. J. Patel, A. Serganov, *Proc. Natl. Acad. Sci. USA* **2011**, *108*, 14801–14806.
234. C. S. Padlan, V. N. Malashkevich, S. C. Almo, M. Levy, M. Brenowitz, M. E. Girvin, *RNA* **2014**, *20*, 447–461.
235. M. Marcia, A. M. Pyle, *Cell* **2012**, *151*, 497–507.
236. K. Takamoto, Q. He, S. Morris, M. R. Chance, M. Brenowitz, *Nat. Struct. Biol.* **2002**, *9*, 928–933.
237. R. Desai, D. Kilburn, H. T. Lee, S. A. Woodson, *J. Biol. Chem.* **2014**, *289*, 2972–2977.
238. Y. F. Jiang, M. Xiao, P. Yin, Y. Zhang, *RNA* **2006**, *12*, 561–566.
239. A. V. Kazantsev, A. A. Krivenko, N. R. Pace, *RNA* **2009**, *15*, 266–276.
240. R. O'Brien, B. DeDecker, K. G. Fleming, P. B. Sigler, J. E. Ladbury, *J. Mol. Biol.* **1998**, *279*, 117–125.
241. K. D. Connaghan, A. F. Heneghan, M. T. Miura, D. L. Bain, *Biochemistry* **2010**, *49*, 422–431.
242. S. A. Mauro, G. B. Koudelka, *J. Mol. Biol.* **2004**, *340*, 445–457.
243. P. Shkilnyj, G. B. Koudelka, *J. Bacteriol.* **2007**, *189*, 3115–3123.
244. B. Richey, D. S. Cayley, M. C. Mossing, C. Kolka, C. F. Anderson, T. C. Farrar, M. T. Record, Jr., *J. Biol. Chem.* **1987**, *262*, 7157–7164.
245. W. F. Liu, A. Zhang, Y. Cheng, H. M. Zhou, Y. B. Yan, *Biochem. Biophys. Res. Commun.* **2009**, *379*, 341–345.
246. H. Pelletier, M. R. Sawaya, *Biochemistry* **1996**, *35*, 12778–12787.
247. L. M. Pegram, T. Wendorff, R. Erdmann, I. Shkel, D. Bellissimo, D. J. Felitsky, M. T. Record, Jr., *Proc. Natl. Acad. Sci. USA* **2010**, *107*, 7716–7721.
248. E. Arbely, E. Natan, T. Brandt, M. D. Allen, D. B. Veprintsev, C. V. Robinson, J. W. Chin, A. C. Joerger, A. R. Fersht, *Proc. Natl. Acad. Sci. USA* **2011**, *108*, 8251–8256.
249. P. H. Naslund, T. Hultin, *Biochim. Biophys. Acta* **1970**, *204*, 237–247.
250. P. H. Naslund, T. Hultin, *Biochim. Biophys. Acta* **1971**, *254*, 104–116.
251. D. B. Willis, H. L. Ennis, *J. Bacteriol.* **1968**, *96*, 2035–2042.

252. H. L. Ennis, M. Artman, *Biochem. Biophys. Res. Commun.* **1972**, *48*, 161–168.
253. R. Miskin, A. Zamir, D. Elson, *J. Mol. Biol.* **1970**, *54*, 355–378.
254. A. Zamir, R. Miskin, D. Elson, *J. Mol. Biol.* **1971**, *60*, 347–364.
255. F. Cahn, M. Lubin, *J. Biol. Chem.* **1978**, *253*, 7798–7803.
256. M. Niepmann, *Virus. Res.* **2003**, *93*, 71–78.
257. P. Nissen, J. Hansen, N. Ban, P. B. Moore, T. A. Steitz, *Science* **2000**, *289*, 920–930.
258. A. Nikulin, A. Serganov, E. Ennifar, S. Tishchenko, N. Nevskaya, W. Shepard, C. Portier, M. Garber, B. Ehresmann, C. Ehresmann, S. Nikonov, P. Dumas, *Nat. Struct. Biol.* **2000**, *7*, 273–277.
259. Y. S. Polikanov, S. V. Melnikov, D. Soll, T. A. Steitz, *Nat. Struct. Mol. Biol.* **2015**, *22*, 342–344.
260. J. Noeske, M. R. Wasserman, D. S. Terry, R. B. Altman, S. C. Blanchard, J. H. Cate, *Nat. Struct. Mol. Biol.* **2015**, *22*, 336–341.
261. T. K. Misra, D. Apirion, *J. Biol. Chem.* **1979**, *254*, 11154–11159.
262. B. Anand, P. Surana, B. Prakash, *PLoS One* **2010**, *5*, e9944.
263. A. Scrima, A. Wittinghofer, *EMBO J.* **2006**, *25*, 2940–2951.
264. L. A. Finney, T. V. O'Halloran, *Science* **2003**, *300*, 931–936.
265. A. Allahverdi, Q. Chen, N. Korolev, L. Nordenskiold, *Sci. Rep.* **2015**, *5*, 8512.
266. R. Strick, P. L. Strissel, K. Gavrilov, R. Levi-Setti, *J. Cell Biol.* **2001**, *155*, 899–910.
267. R. Levi-Setti, K. L. Gavrilov, M. E. Neilly, *Appl. Surf. Sci.* **2006**, *252*, 6765–6769.
268. R. McRae, P. Bagchi, S. Sumalekshmy, C. J. Fahrni, *Chem. Rev.* **2009**, *109*, 4780–4827.
269. E. J. Lanni, S. S. Rubakhin, J. V. Sweedler, *J. Proteomics* **2012**, *75*, 5036–5051.
270. P. Auffinger, L. Bielecki, E. Westhof, *Structure* **2004**, *12*, 379–388.
271. L. D'Ascenzo, P. Auffinger, in *Nucleic Acid Crystallography: Methods and Protocols*, Ed. E. Ennifar, Humana Press, Berlin, 2015.
272. H. Karppanen, P. Karppanen, E. Mervaala, *J. Hum. Hypertens.* **2005**, *19*, S10–S19.
273. V. Savica, G. Bellinghieri, J. D. Kopple, *Annu. Rev. Nutr.* **2010**, *30*, 365–401.
274. M. Kurlansky, *Salt: A World History*, Random House Publishing Limited, Toronto, Canada, 2002.

Chapter 7
Role of Alkali Metal Ions in G-Quadruplex Nucleic Acid Structure and Stability

Eric Largy, Jean-Louis Mergny, and Valérie Gabelica

Contents

E. Largy • J.-L. Mergny (✉)• V. Gabelica (✉)
ARNA Laboratory, Université Bordeaux, IECB, 2, rue Robert Escarpit, F-33600 Pessac, France

ARNA Laboratory, INSERM, U869, F-33000 Bordeaux, France
e-mail: jean-louis.mergny@inserm.fr; valerie.gabelica@inserm.fr

A. Sigel, H. Sigel, and R.K.O. Sigel (eds.), *The Alkali Metal Ions: Their Role for Life*, Metal Ions in Life Sciences 16, DOI 10.1007/978-3-319-21756-7_7

Abstract G-quadruplexes are guanine-rich nucleic acids that fold by forming successive quartets of guanines (the G-tetrads), stabilized by intra-quartet hydrogen bonds, inter-quartet stacking, and cation coordination. This specific although highly polymorphic type of secondary structure deviates significantly from the classical B-DNA duplex. G-quadruplexes are detectable in human cells and are strongly suspected to be involved in a number of biological processes at the DNA and RNA levels. The vast structural polymorphism exhibited by G-quadruplexes, together with their putative biological relevance, makes them attractive therapeutic targets compared to canonical duplex DNA. This chapter focuses on the essential and specific coordination of alkali metal cations by G-quadruplex nucleic acids, and most notably on studies highlighting cation-dependent dissimilarities in their stability, structure, formation, and interconversion. Section 1 surveys G-quadruplex structures and their interactions with alkali metal ions while Section 2 presents analytical methods used to study G-quadruplexes. The influence of alkali cations on the stability, structure, and kinetics of formation of G-quadruplex structures of quadruplexes will be discussed in Sections 3 and 4. Section 5 focuses on the cation-induced interconversion of G-quadruplex structures. In Sections 3 to 5, we will particularly emphasize the comparisons between cations, most often K^+ and Na^+ because of their prevalence in the literature and in cells.

Keywords DNA • Folding • G-quadruplex • G-quartet • Interconversion • Metal ions • Methods • RNA • Stability • Structure

Please cite as: *Met. Ions Life Sci.* 16 (2016) 203–258

1 Introduction: G-Quadruplex Nucleic Acids

G-quadruplexes (G4) encompass guanine-rich nucleic acids that fold by forming successive quartets of guanines (also called G-tetrads), stabilized by intra-quartet hydrogen bonds (Figure 1A), inter-quartet stacking, and cation coordination (Figure 1B).

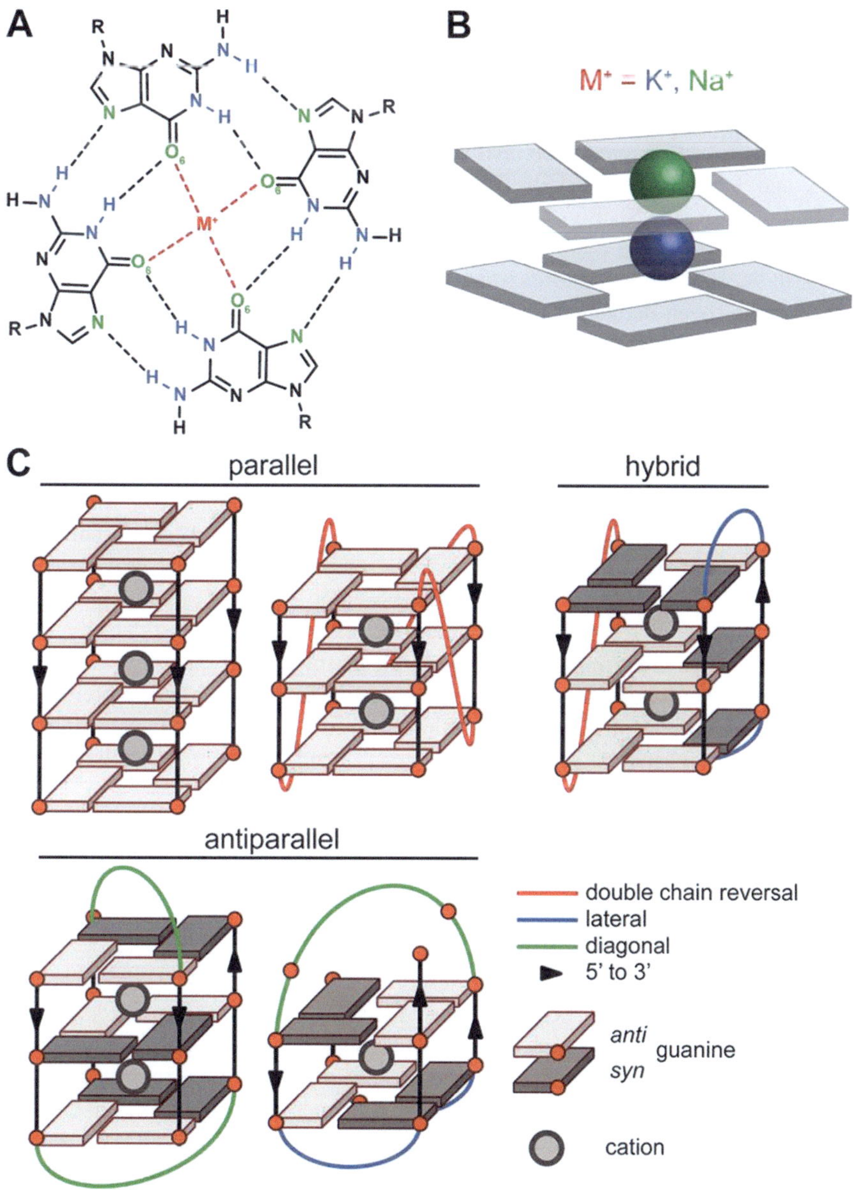

Figure 1 (**A**) Guanine tetrads contain four guanines linked by eight hydrogen bonds (donor and acceptor groups in blue and green, respectively). Guanine O6 selectively coordinate a metal cation (red). (**B**) Tetrads can stack to form G-quadruplexes. Cations of larger ionic radii are located between the tetrads (case of K^+; blue), while smaller ones can also coordinate within the plane of the tetrads, or assume an intermediate position (case of Na^+; green). (**C**) G-quadruplexes can fold into a variety of topology, which differ mainly by the relative orientation and number of strands (1 to 4), the number of tetrads (at least 2), and the geometry of the loops.

This specific although highly polymorphic type of secondary structure deviates significantly from the classical B-DNA duplex [1–3]. Such G-rich sequences are found in telomeres, and at a statistically remarkable frequency in other part of the genome, notably in promoters [4–6]. G-quadruplexes are detectable in human cells, and are strongly suspected to be involved in a number of biological processes at the DNA and RNA levels [7–15]. The vast structural polymorphism exhibited by G-quadruplexes (see Section 1.1), together with their putative biological relevance, makes them attractive therapeutic targets compared to canonical duplex DNA [16–20]. A very large – and exponentially increasing – number of studies have consequently been dedicated to G-quadruplexes involved in telomeric sequences, oncogenes, and 5′-untranslated regions (5′-UTRs) to cite the most common [21]. Most studies so far involve potassium or sodium cations because of their prevalence in human cells, but a number of other monovalent and divalent cations promote quadruplex formation (*vide infra*). However, among alkali metals, only potassium, sodium and rubidium are truly effective at stabilizing G4s.

This chapter focuses on the essential and specific coordination of alkali metal cations by G-quadruplex nucleic acids, and most notably on studies highlighting cation-dependent dissimilarities. Section 1 surveys G-quadruplex structures and their interactions with alkali metal ions while Section 2 presents analytical methods used to study G-quadruplexes. The influence of alkali cations on the stability, structure, and kinetics of formation of G-quadruplex structures of quadruplexes will be discussed in Sections 3 and 4. Section 5 focuses on the cation-induced interconversion of G-quadruplex structures. In Sections 3 to 5, we will particularly emphasize the comparisons between cations, most often K^+ and Na^+ because of their prevalence in the literature and in cells.

1.1 Overview of Structure

The core structure of all G-quadruplexes is constituted by the stacking of at least two G-tetrads (or G-quartets), each formed by the *quasi* co-planar association of four guanines linked by a network of eight hydrogen bonds (Figure 1A,B). The guanines forming the tetrads coordinate some mono- or divalent cations in their center (referred to as central stem), typically K^+ and Na^+, via their oxygen O6. The single-strand sections linking the guanine tracts are called loops, and can adopt a variety of geometries (Figure 1C): lateral (or edgewise), diagonal, and double chain reversal (or propeller). Furthermore, the four guanine tracts can be oriented (regarding their 5′ to 3′ polarity) into four different topologies: parallel (the four strands share the same polarity), antiparallel (two strands in one way, the two other in the opposite way; these pairs may correspond to two adjacent or diagonally-opposed strands, resulting in very different geometries), or hybrid (three strands sharing the same polarity, and the last one the opposite) [22, 23].

In antiparallel topologies, the strands sharing the same direction can be diagonally opposed or adjacent. This distinguishes 'chair'-type G-quadruplexes (three lateral

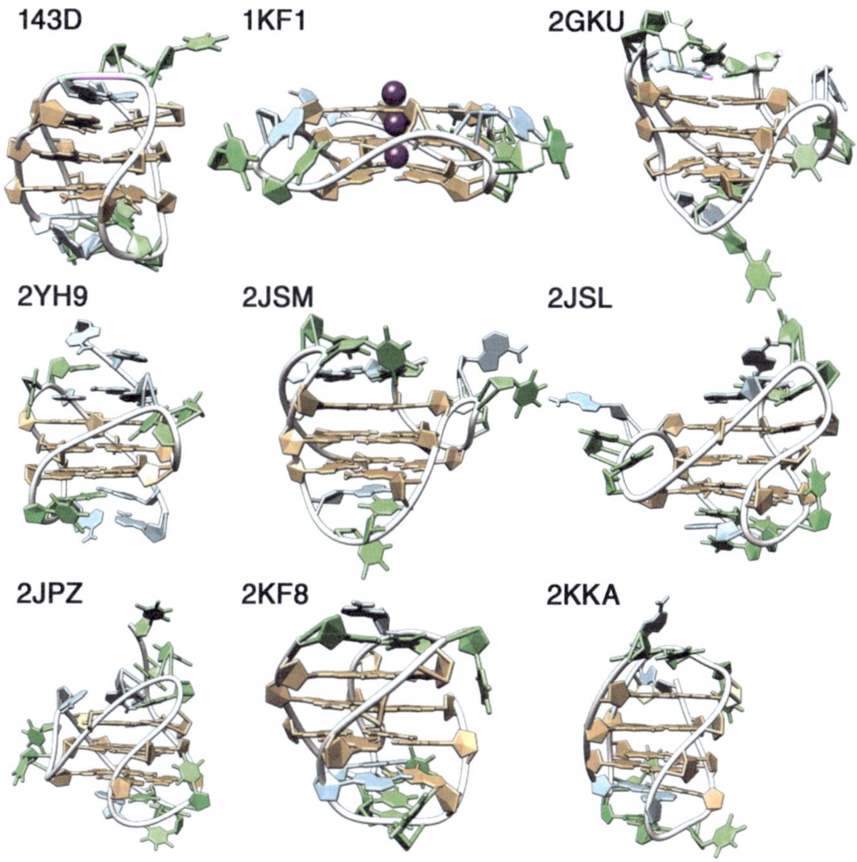

Figure 2 Examples of human telomeric G-quadruplex structures deposited in the PDB that were solved by NMR or X-ray crystallography: 143D [25], 1KF1 [26], 2GKU [27], 2YH9 [28], 2JSM and 2JSL [29], 2JPZ [30], 2KF8 [31], and 2KKA [32]. Guanosines are depicted in brown, inosine in yellow, adenosines in blue, thymidines in green, the phosphate backbones as white ribbons, and K^+ as purple spheres where available. All structures except 143D (Na^+) were solved in K^+ conditions.

loops, e.g. the thrombin binding aptamer [24]) and 'basket'-type G-quadruplexes (lateral-diagonal-lateral loops, e.g. the human telomeric sequence $d[AG_3(T_2AG_3)_3]$ in Na^+ conditions [25]; Figure 2, 143D) [25–32]. These two antiparallel conformations are fundamentally different. Intramolecular parallel structures (three double-chain reversal loops) are sometimes referred to as 'propeller' topologies (e.g., the crystal structure of $d[AG_3(T_2AG_3)_3]$ in K^+ conditions [26], Figure 2, 1KF1).

The glycosidic bond angle of guanines involved in tetrads can adopt an *anti* or *syn* geometry depending on the relative strand orientation: parallel topologies contain almost exclusively *anti* guanines (for exceptions, see [33–35]), whereas a mixture of *syn* and *anti* is observed for antiparallel and hybrid structures [36]. Consequently, parallel G-quadruplexes generally contain *anti*/*anti* stacks (*syn*/*syn* is not favored; see Section 1.2), antiparallel ones have *anti/syn* and *syn/anti*, and hybrid ones can have all of the above.

Furthermore, G-quadruplexes may be formed by the folding of a single strand or the association of two to four strands, and some sequences have a tendency to oligomerize [37–39]. One should therefore not confuse the term "quadruplex" or "G-quadruplex" (designating any structure containing stacked guanine tetrads) with the strand molecularity (for which we will adopt the nomenclature "intramolecular", "bimolecular", "trimolecular" [40], and "tetramolecular").

Finally, a number of uncommon features have been observed, such as strand bulges and snapbacks [41–43], base-pairing in loops [44], alternative tetrads [45], and other stacked planar entities (triads, pentads, hexads, heptads, and octads) [31, 46–52]. The combination of different loop geometries, strand orientations, and molecularities, implies that G-quadruplexes display an important polymorphism. It has been reported that there is a theoretical number of 26 possible G-quadruplex topologies, not taking into account unexpected folds (e.g., isolated guanines involved in tetrads [53]), but only a few of them (six) have been observed *in vitro* [22, 23], possibly meaning that some are energetically disfavored, thermodynamically or kinetically. A number of insightful review articles and books cover the G-quadruplex structures published so far [3, 12, 36, 54–59].

1.2 Stabilizing Interactions

The factors contributing to the stabilization of G-quadruplexes and yielding a particular folding topology or mixture of topologies are: stacking interactions, hydrogen bonding, solvation, and cation binding. Cation effects on G-quadruplexes differ significantly from those on duplexes (see Chapter 6 of this volume). Cation *coordination* is indeed required to form G-quadruplexes, to stabilize the G-tetrad stacks. Moreover, sufficient ionic strength is required to compensate electrostatic repulsion between the phosphate oxygens of four strands in G-quadruplexes, instead of two for duplexes (loops may be considered as unfolded single strands depending on the structure).

The total free energy can be decomposed in a number of free energy contributions (e.g., Coulombic forces, hydrogen bonding, hydration, van der Waals terms), which can themselves be decomposed in entropic and enthalpic contributions. Analysis of G-quadruplex melting and calorimetric data revealed that G-quadruplex formation is enthalpically driven [60, 61]. This results from a more negative (favorable) enthalpy of tetrad formation, only partially compensated by more negative (less favorable) entropies of tetrad formation. Below are presented some key elements to understand G-quadruplex stabilization; cation coordination *per se* is explored in Section 1.3. An excellent review discussing in details the stability of G-quadruplexes has been published by Lane et al. [62].

1.2.1 Stacking

Similarly to other nucleic acid secondary structures, G-quadruplex stabilization relies in part on $\pi-\pi$ stacking of aromatic bases, and more specifically on guanines from consecutive tetrads. Molecular mechanics simulations have predicted the relative stability order among guanine stacks to be (from 5′ to 3′): *syn/anti*>*anti/anti*> *anti/syn*>*syn/syn* [63]. This is not fully consistent with the structures solved so far as it does not explain the parallel orientation of the strands in tetramolecular assemblies, possibly because of force-field biases. Quantum mechanical (QM) dispersion-corrected density functional theory (DFT-D3) calculations on stacking of two tetrads, containing one K^+ cation gave a more consistent picture with the structures solved so far: *anti/anti*>*syn/anti*>*anti/syn*>*syn/syn* (ΔE *vs anti/anti*=1.2, 3.5, and 7.8 kcal.mol^{-1}, respectively) [64]. It was also suggested in the same study that 5′-terminal H-bonds present a stabilizing effect on 5′-terminal *syn* guanines.

Ultimately, the topology of a G-quadruplex depends on its precise sequence, which can provide additional stabilizing interactions (see below), and on buffer conditions (cations, co-solvents). Because the stacking interactions likely account for a large part of the net energetic gain, increasing the number of quartets is energetically favorable. The number of consecutive tetrads is typically equal to the length of the guanine tracts, although exceptions have been observed. For instance, the 22-mer of the human telomeric sequence d[$(G_3T_2A)_3G_3T$)] (PBD ID: 2KF8) contains four tracts of three guanines, but folds predominantly in a two-tetrad G-quadruplex (Figure 2) [31].

The formation of higher-order G-quadruplex structures (usually dimers) via stacking of external tetrads of monomer units also provides additional stabilization. Almost all published dimer structures exhibit a 5′-5′ interface [65], although 3′-3′ stacking remains possible [66, 67]. QM and 100-ns molecular dynamics (MD) simulations suggest that while 5′-5′ interface readily stack in a favorable manner (60–65° rotation), the 3′-3′ interface cannot reach the same type of geometry because of guanine-sugar clashes and therefore yields less favorable stacks (45° or 30° rotations, ΔE *versus* 5′-5′=4 and 10 kcal.mol^{-1}, respectively) [65].

Stacking of other nucleotides (non-tetrad bases) might also contribute to the overall stabilization of the G-quadruplex [31, 68]. For instance, in the aforementioned two-tetrad G-quadruplex 2KF8, loop nucleotides provide G•G•G and A•G•G triads, which stack on both tetrads [31]. Alternative planar entities have been identified, such as G•C•G•C tetrads [45], as well as the larger pentads [46, 47], hexads [48–50], heptads [51], and octads [52] that provide additional hydrogen bonds. Finally, formation of base pairs is also possible [69]. Searle et al. have described a bimolecular G-quadruplex whose two loops form mini-hairpin motifs [44]. The above-mentioned 2KF8 structure involves mismatched T•T base pairs [31]. Base-pairing can mediate the formation of stacked G-quadruplex dimers, as observed by NMR for the c-kit2 sequence d[$(CG_3)_2(CG)_2(AG_3)_2T$], which exhibits an A•A base pair at the interface [70].

1.2.2 Hydrogen Bonding

Hydrogen bonding is another stabilization element found in biomacromolecules, and in particular between bases in nucleic acids, typically interacting through two (A•T/U) or three (G•C) H-bonds. In such Watson-Crick base pairing, the stability of H-bonds is linked to donor-acceptor orbital interactions and polarization of the π system by partially neutralizing the charges in the σ-electrons system. In G-quadruplex nucleic acids, there is a network of (C2)NH_2:N7 and O6:N1H hydrogen bonds. As acceptor and donor groups are likely hydrogen-bonding to water for unfolded/unassociated strands, there is only a small enthalpy change, but the release of water molecules to the bulk medium may induce an entropy gain [62].

A combination of high-resolution variable-temperature STM and DFT calculations suggested that the stabilization of a tetrad induced by the network of eight hydrogen bonds is higher than the sum of four individual G•G pairs [71]. This cooperativity was first ascribed to π assistance, however, DFT-D calculations by Fonseca-Guerra et al. suggested that the cooperativity in guanine tetrads more likely arises from the charge separation that is associated to charge transfer between pairs of guanine, i.e., donor-acceptor orbital interactions in the σ-electron system [72]. This results in an interruption of the π-electron system, and an enhancement of both the positive charges on the H atoms of H-bond donor groups and negative charges of N and O atoms of acceptor groups. In the same study, it was evidenced that in aqueous solution, the hydrogen bond energy is significantly diminished as compared to the gas phase (−34 *versus* −80 kcal/mol). Compared to solvation, stacking of three tetrads (in water) has a weak favorable (−2 kcal/mol) effect on the cooperativity. However, introduction of two Na^+ cations in this system entirely restores the stability observed in the gas phase, although it weakens the hydrogens bonds as compared to the system without cation.

1.3 Alkali Metal Ion Coordination in G-Quartets

1.3.1 Cation Preference

Pinnavaia and co-workers' pioneering work on 5′-GMP provided the first indication of the formation of anionic cavities in G-quadruplex-type structures that can selectively complex cations small enough to fit in, but large enough to *bridge* the carbonyl oxygens [73], analogously to the binding of metal alkali to crown ethers [74]. This led the authors to propose the complexation of Na^+ in the plane of tetrads, while potassium would fit in the interplanar spacing and be classically octa-coordinated. Follow-up studies [75, 76] expanded the scope to four-stranded poly(G) strands [77], helical poly(I) structures including the binding of ammonium [78–80].

A seminal report by Blackburn et al. suggested the existence of "G•G base pairs" in the telomeric sequence of several organisms [81]. Following these findings,

Williamson, Raghuraman, and Cech proposed a model for telomeric sequences where they formally described the G-quadruplex structure for the first time [82]. The studies conducted on *Oxytricha* and *Tetrahymena* telomeric sequences provided the first hints regarding the critical importance of monovalent cations for G-quadruplex formation, stoichiometry, and stability. It was postulated that Li^+ is too small to bind in the center of G-quadruplex tetrad, whereas Na^+ fits perfectly, and larger ions such as K^+ and Cs^+ bind in the interplane cavity [82]. It is now widely accepted that cation coordination is essential for the stabilization of G-quadruplexes. By compiling a number of studies, one can estimate that G-quadruplex stabilization follows the general trend: $Sr^{2+}>\mathbf{K^+}>Ca^{2+}>NH_4^+$, $\mathbf{Na^+}$, $\mathbf{Rb^+}>Mg^{2+}>\mathbf{Li^+}\geq\mathbf{Cs^+}$ (alkali cations in bold) [83–85].

Although the bulkiness of these cations is certainly critical for binding within a G4 cavity, and is historically presented as the main explanation for the cation-dependent stability differences (*vide supra*), other factors play an important role in the stabilization of the complexes [86–89]. Cation binding to G-quadruplexes can be classified as *non-specific* (diffuse [90]), where the cations retaining their outer-sphere hydration bind the negatively charged phosphates, or *specific* (site-bound) by coordination to the guanine O6, where the hydration sphere has been lost and the binding follow the law of mass action (Figure 3). For instance, Gray and Chaires have determined that, at concentration of K^+ above 2.5 mM, $d[AG_3(T_2AG_3)_3]$ attracts up to 6–8 more cations than the predicted two specific binding sites [91]. The binding of these non-specific cations to the negatively charged phosphate backbone reduces the electrostatic repulsions and thus also promotes folding. Note that external coordination to a G-quadruplex is not necessarily diffuse: specific external coordination to loops has been suggested for Tb^{3+} on the basis of circular dichroism (CD) and luminescence data [92]. Notably, cations coordinated between the top G-quartet and the loop are often detected in the density maps provided by X-ray crystallography (see Section 1.3.3). By comparing K^+ binding to the 22-mer G-quadruplex $d[AG_3(T_2AG_3)_3]$ with the binding to the unfolded $d[T_{22}]$, it was suggested that the G-quadruplex may specifically bind up to five cations on external sites [91]. However, the use of a control oligonucleotide with such a different sequence and different structure (random coil *versus* globular) may not mimic properly the diffuse binding of the G-quadruplex.

1.3.2 Cation Binding Energetics

In specific coordination to the tetrads, the cations are involved in electrostatic and donor-acceptor orbital interactions with the lone pairs of guanine O6, yielding tight M^+–O coordination bonds [72]. Stabilization is also provided by a screening of electronic repulsion of these O6 lone pairs [93]. In solution, potassium is typically hexa-hydrated, and hence the coordination *per se* by guanine O6 instead of water molecules shall not provide a large enthalpy change. Alkali cations specifically bind within the electronegative cavity formed by the O6 of tetrad guanines in distinct fashions.

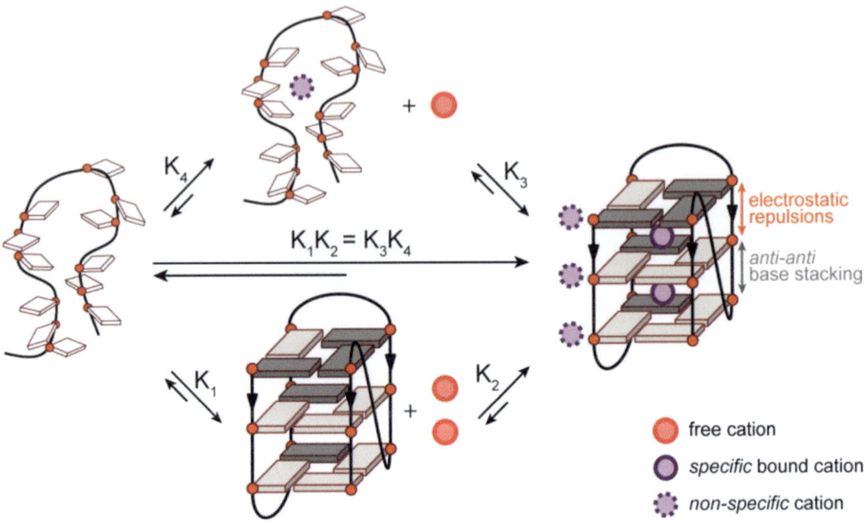

Figure 3 Multiple equilibria involved in cation coordination and G-quadruplex folding. Each cation can either bind to a preformed G-quadruplex (K_1 then K_2), or bind to the random coil that subsequently fold in a G-quadruplex (K_4 then K_3). The folded species on the right shows the differences between specific and unspecific binding as well as other stabilizing (stacking) and destabilizing (phosphate repulsions) factors.

As with crown ethers, sodium and potassium bind G-quadruplexes with different affinities, although other factors than the cation radii account for it. The difference in free energies is estimated to be around 2 ± 0.5 kcal.mol^{-1} [88, 94–97]. The thrombin-binding aptamer sequence (TBA; d($G_2T_2G_2TGTG_2T_2G_2$)), which contains only two tetrads binds a single cation, with a folding constant K_{fold} measured by monitoring the folding by spectroscopic means of 1.3×10^7 and 5.5×10^5 M^{-1} at 10 °C for K^+ and Na^+, respectively [98], or an association constant K_a of 2×10^5 M^{-1} for K^+, determined by mass spectrometry [99]. Specific ion binding has been described thermodynamically by Lane et al. by multiple equilibrium involving the folded and unfolded G-quadruplex as shown in Figure 3 [62].

Cation coordination to G-quadruplexes is typically accompanied by G-quadruplex folding that brings further stabilization, which is not experimentally distinguishable except in molecular modeling studies (see Section 2.1). Regarding the equilibrium constants defined in Figure 3, in most cases the K_1 value is large, and the apparent dissociation constant is K_1K_2 (exclusive binding mechanism) [95]. When the solution is devoid of alkali cation, the fraction of folded strand is presumably small. Cation binding to the unfolded strand is also small, and thus an important stabilization is attributed to specific cation binding (K_2/K_4 ratio) [62].

The analogy with alkali cations binding to crown ethers is noteworthy. In aqueous solutions, the crown ether 18-crown-6 binds K^+ and Na^+ with log $K \approx 2$ and 0.7, respectively, and complexation enthalpy $\Delta_r H°$ of −6 and −2 kcal.mol^{-1}, at 37 °C, including the desolvation and structural change energies [74, 100–102]. Comparatively, the net ΔG at 37 °C of three-tetrad G-quadruplex formation has been estimated to 5–10 kcal.mol^{-1}, which supports the idea that the energy of potassium binding and of subsequent conformation changes accounts for most of the stabilization [62, 91]. For instance, the 22-mer of the human telomeric sequence $d[AG_3(T_2AG_3)_3]$ has an overall −2.4 kcal.mol^{-1} folding free energy, while in 5 mM potassium, K^+ contributes to roughly −4.9 kcal.mol^{-1} [91]. This is also consistent with the unfolded state being largely populated in absence of cation (Figure 3, left). Consequently, binding of potassium or sodium is likely fast and cooperative, whereas the subsequent structural change is quasi irreversible with a net binding energy of around 5 kcal.mol^{-1}. This is the case of the human telomeric sequence for instance [95].

The apparent binding of Na^+ and K^+ to human telomeric sequences is very cooperative and lies within 5–15 and 0.5–2 mM, respectively [94, 95, 103]. Comparatively, Na^+ forms 1:1 complexes with isolated 5′-GMP with a K_a of 2.85 M^{-1} at 5 °C [104]. Feigon et al. and Gu and Leszczynski have shown that the free energy of hydration of *specific* cations can explain the stability difference observed between Na^+ and K^+ solutions, the dehydration of the former inducing a greater energetic cost [88, 105].

Indeed, unlike diffusely bound cations, tetrad-bound cations lose their whole hydration sphere [106]. The energetics of cation binding to G4s can be decomposed into a positive free energy of dehydration and a negative free energy of coordination *per se*. Na^+ gives favorable energy of coordination, as shown by NMR [88, 107], but its binding is penalized by its stronger hydration as compared to K^+. Meyer et al. have also observed by DFT calculations that solvation effects explain the favorable coordination to K^+, and shown that the cation coordination contribution accounts for 50 % of the total interaction energy of a two-tetrad construct [108]. Hydration energy of alkali cations is usually presented as being inversely proportional to their ionic radii. K^+ systematically presents the best compromise, and stabilizes G4s more than Na^+ and Rb^+ (other alkali cations provide very weak to no stabilization). However, the extent to which it does so depends hugely on the studied sequence, and in particular, on the sequence's ability to adopt different structures in the presence of different cations (see Section 3.2) [85, 89, 94, 109–119].

1.3.3 Cation Location

It is often stated that Na^+ is small enough (0.95 Å) to fit in the plane of a quartet, while any cation larger than that, such as K^+ (1.33 Å) or Rb^+ (1.52 Å), are coordinated between two planes. In reality, there is a continuum of possible binding sites resulting from (i) the abovementioned cation lone pair attraction, (ii) the presence of additional cation coordination sites (e.g., loops, dimer interface), (iii) cation-cation repulsion, and (iv) possible quartet distorsion. Indeed, when more than one cation is bound by the G4, the cations' mutual repulsion also influences their precise locations and the fine structure of the G-quadruplex. Phillips et al. obtained crystals of $[d(TG_4T)]_4$ coordinated to Na^+ and Ca^{2+} that provide a nice illustration of this phenomenon for Na^+ [120] (Figure 4A,B,C) [120, 121]. The G-quadruplex is arranged in a head-to-head dimer with eight consecutive tetrads. Out of the seven bound sodium cations, only the two outer ones lie in the G-tetrad's plane, in a four-coordinate fashion, completed by the binding of water molecules. The 5 other sodium cations are equally spaced between the two other ones, and adopt intermediate coordination positions in a continuum ending at the central cation that is sandwiched exactly halfway between two tetrads, with a bipyramidal geometry similar to the one usually adopted by K^+. As a result, the average distance between Na^+ cations (4.2 Å for outer cations, 3.6 Å for internal cations) is higher than the distance between G-tetrads. A slight distortion of the external quartets is also observed in this crystal, putatively allowing the inter-cation distance to increase. One should note however, that crystal packing forces might account for the position of the cations, which therefore would not reflect the reality of solution-based G-quadruplexes.

Creze et al. obtained a similar dimeric structure with Na^+ being coordinated increasingly more within the tetrad planes when going towards the extremities of the G-quadruplex ($d_{Na^+\text{-}Na^+} = 3.4 - 4.7$ Å), with water molecules capping at the external tetrads ($d_{Na^+-H_2O} = 2.4 - 2.5$ Å) [122]. Crystallization was carried out from a sodium solution of quadruplex using lithium sulfate as a precipitating agent. A single Li^+ ion was observed in a groove, and none of them in the central stem.

Incidentally, in the first reported crystal structure of a small molecule (daunomycin) bound to a G-quadruplex, the Na^+ cations are all coordinated in the plane of the tetrads [123]. The tetramolecular $[d(TG_4T)]_4$ G-quadruplex is also observed as a dimer, with two daunomycin molecules at the interface between the individual tetramers, which might also be linked to the difference of sodium positioning. In a more recent report on the high-resolution crystal of the dimeric $[d(TG_4T)]_4$/daunomycin complex employing *syn* glycosyl linkages, Na^+ cations are not in the plane of the tetrads, but rather sandwiched in various fashions [124]. Notably, the Na^+ closer to the 5′-interface lies almost midway between two tetrads, with a square prism geometry, due to the *syn* glycosyl orientation of the first guanine residue. Interestingly, Na^+ also occupies a site at a daunomycin-daunomycin interface.

Conversely, in the crystal structure of the intertwined dimeric quadruplex formed by the B-raf sequence $d(G_3CG_4AG_5A_2G_3A)$ (PDB ID: 4H29), the six potassium cations are all observed in-between the tetrads, in a quasi linear arrangement, and are equidistant from each other ($d_{K^+\text{-}K^+} = 3.44 \pm 0.09$ Å) (Figure 4D,E,F) [121].

The differences in location of sodium and potassium cations has been extensively investigated by NMR and X-ray crystallography for the telomeric sequence from *Oxytricha nova* $d(G_4T_4G_4)$, forming antiparallel bimolecular G-quadruplexes, where the thymines form two diagonal loops (Figure 5) [96, 107, 125–128]. The crystal structures 1JPQ and 1JRN exhibits five equidistant K^+ cations, spaced by 3.4 Å on average, three being sandwiched between the tetrads and the other two coordinated between the external quartets and within the loops [128]. The K^+-K^+ distance is the same that has been measured for the human telomeric sequence [26].

Similar results were observed with Tl^+ cations, which have a similar radius than K^+ (5 bound cations with a 3.6 Å spacing; average RMSD = 0.26 Å) [129]. Conversely, only four Na^+ cations are coordinated, the binding sites in the loops

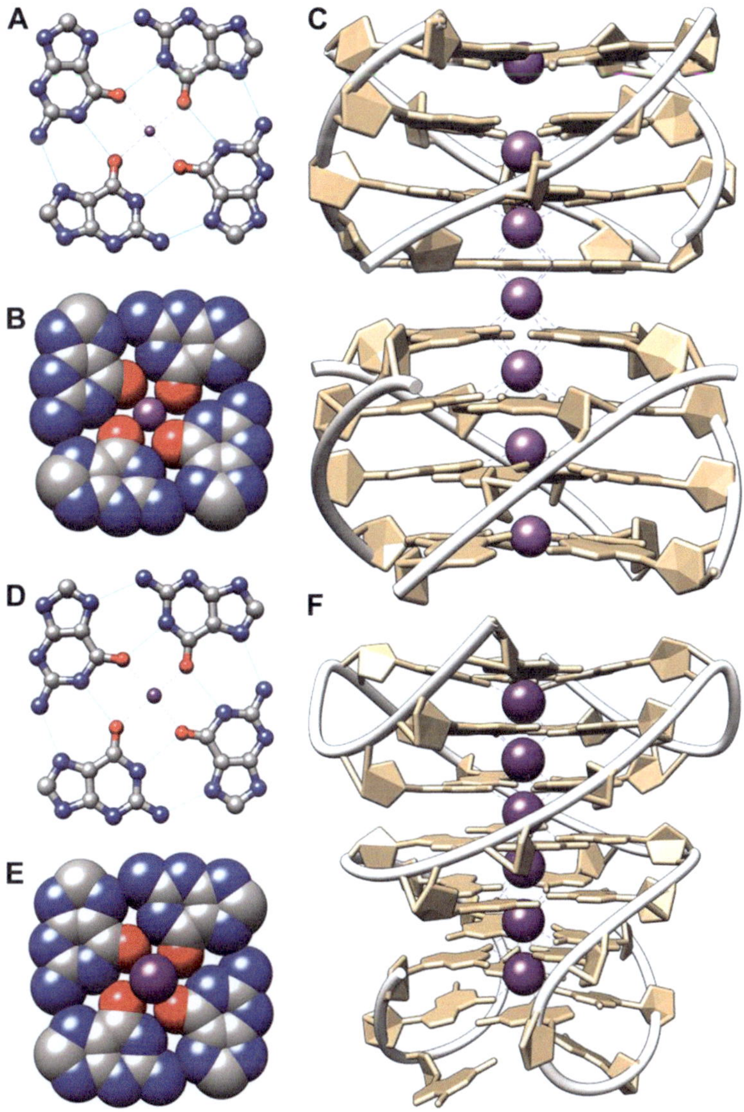

Figure 4 Crystal structure of $[d(TG_4T)]_4$ coordinating Na^+ (PDB ID 352D, [120]) (**A, B, C**), and $d(G_3CG_4AG_5A_2G_3A)$ coordinating K^+ (PDB ID 4H29, [121]) (**D, E, F**): tetrad geometry depicted as balls and sticks (**A**, **D**) and spheres (**B**, **E**), the head-to-head dimer coordinating seven Na^+ adopting positions ranging from within the tetrads (external cations) to sandwiched midway between two quartets (central cation) (**C**), and the intertwined dimer of B-raf coordinating six sandwiched K^+ cations (**F**). Hydrogen bonds are shown in blue, Na^+ and K^+ cations and coordination in purple. Carbon atoms are in grey, nitrogen in blue, and oxygen in red except in panels **C** and **F** where guanosines are colored in brown and the phosphate backbone is a white ribbon. For the sake of clarity, dT, dA, and dC residues, Ca^{2+}, H_2O, and hydrogen atoms are omitted.

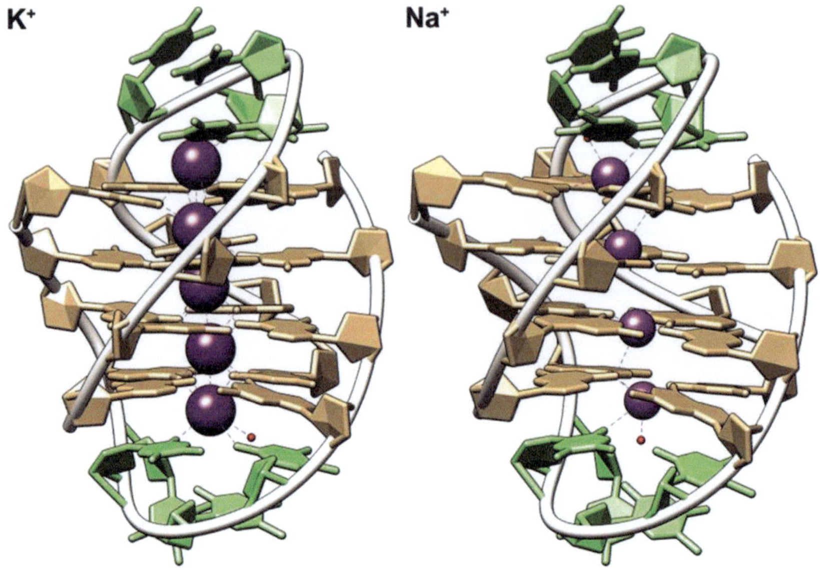

Figure 5 Crystal structure of $[d(T_4G_4T_4)]_2$ coordinating K^+ (PDB ID 1JPQ) or Na^+(PDB ID 1JB7). Guanines are shown in brown, thymines in green, the cations as purple spheres, and the oxygens from water molecules completing the spheres of coordination as red balls.

being less occupied (PDB ID 1JB7) [127]. Sodium cations are coordinated in the planes of the central tetrads while the external cations are bound slightly outside of the tetrads, towards the loops, and are coordinated by the thymines O2. This partial loop coordination leads to a distribution of Na^+ location that differs from $[d(TG_4T)]_4$, but here too the distance between sodium cations is higher than the distance between tetrads.

Overall, K^+ cations are specifically coordinated in between tetrads, in a nearly octahedral way, or alternatively within loops above tetrads, satisfying the usual hexacoordinate stereochemistry of these cations. Positioning of bases outside of the plane of the tetrad may be necessary to comply with this geometry, to an extent balanced by stacking and hydrogen bonding energies. Smaller Na^+ cations can be coordinated within the plane of tetrads, and can occupy a range of positions owing to lower steric constraints, hence reducing the electrostatic repulsions. K^+ is bound with higher affinity than Na^+ and Rb^+, while Li^+ and Cs^+ are poorly coordinated by G-quadruplexes. Although the cationic radius certainly accounts for these differences, the energy of hydration has been demonstrated to be the cause of the energetic preference for K^+ *versus* Na^+.

All these examples highlight that the stability of G-quadruplexes is massively driven by cation binding and the resulting structural (re-)organization, rather than by other weak interactions.

2 Methods to Study G-Quadruplex Nucleic Acids

2.1 *Folding Topology*

A number of high- and low-resolution analytical methods are used to determine the topology of G-quadruplexes [130], and some are presented hereafter. NMR spectroscopy, X-ray crystallography, and molecular modeling give access to atomic-scale structure information [36, 131–134], and have revealed over the last decades an impressive structural polymorphism among G-quadruplexes (see Figure 2). A number of other spectroscopic and spectrometric methods (e.g., UV absorption spectroscopy, electronic circular dichroism spectroscopy, native mass spectrometry, electrophoresis, and chromatography) are employed to determine strand orientations and stoichiometry, and molecular sizes. As each of these techniques give different types of information, and furthermore presents certain drawbacks, it is advisable to combine a number of these techniques before drawing final conclusions.

2.1.1 X-Ray Crystallography

X-Ray crystallography gives access to atomic-scale resolution structures of DNA, RNA, or LNA G-quadruplexes, including the cations (see Section 2.2.1), water molecules, and binders [132, 133]. A fair number of structures containing alkali cations (or Tl^+ as a K^+ surrogate; *vide infra*) have been solved using this method [26, 67, 120–124, 127–129, 135–161], mainly by the teams of Parkinson and Neidle, and the late Prof. Sundaralingam. Screening of sequences, as well as base modification (e.g., heavy-atom addition) and loop/flanking sequence changes are typically performed in order to obtain a sequence that crystallizes and diffracts well. These aspects, as well as crystallization protocols are discussed in detail in [133].

Almost all of the available crystal structures are tetrameric [120, 122, 123, 135–145] or dimeric, via association of distinct strands or external stacking of monomer units, often templated by binders [26, 67, 121, 124, 127–129, 145–160]. Moreover, stacked monomer units and intermolecular species [26, 158, 159] are systematically parallel-stranded within each subunit, except for the thrombin-binding aptamer bound to α-thrombin [161]. This raises questions as to the possibility to crystallize structures such as intramolecular antiparallel-stranded or hybrid structures (either because they do not crystallize or because the crystal packing forces induce conversion to parallel and multimeric folds), and this in turn casts some doubts about the complete relevance of this method for solution-based folding studies. Crystallography is nevertheless very powerful when it comes to characterize cation coordination (see Section 2.2.1), and more generally ligand binding (reviewed in [162]). It also provides better defined structures than NMR in cases where the latter is limited by internal dynamics or the presence of mixtures of conformations, with or without inter-conversion between structures.

2.1.2 Nuclear Magnetic Resonance Spectroscopy

High-field NMR is broadly used to study G-quadruplex structures and their dynamics [36, 131, 163, 164]. The full topology of many structures has now been solved, including the human telomeric DNA [25, 27–32, 45, 165–172] and RNA sequences (TERRA) [50, 173, 174], telomeres from other species [46, 68, 83, 96, 125, 126, 175–177], human oncogenes [53, 70, 178–182], minisatellites [43, 183], and aptamers against biological targets [24, 41, 184–186] or viruses [15, 69]. Besides, even when solving the full topology is not possible or required, several types of NMR experiments can provide information on particular aspects of the structure.

First, the formation of G-quadruplexes can be affirmed by the presence of guanine imino protons H1 in the 10–12 ppm range, typical of G-tetrads, whereas canonical Watson-Crick base-pairing shifts the protons to 13–14 ppm [134, 187]. Moreover, compared to Watson-Crick protons, tetrad protons – particularly the ones from central tetrads – exchange more slowly with water, which results in sharp peaks [125, 188].

Second, the number of imino proton peaks is linked to the number of tetrads (one peak per guanine, hence four peaks per tetrad, unless some protons are equivalent) [183], and a number of peaks higher than the number of guanines indicates a mixture of species. The kinetics of quadruplex formation are easily accessible given that the time-scale is compatible with NMR experiments [164].

To fully solve a structure however, a number of issues must be tackled: formation of multiple species by a single sequence, higher-order structures, and ambiguous assignments [131]. Sequences folding in multiple conformers (e.g., [167, 170]) can be characterized if the peaks are sufficiently resolved, or more frequently by screening for or favoring a given conformation by sequence modification (flanking and loop bases, modified bases [170, 172]), sample preparation, and buffer and cationic conditions (reviewed in [131]). Stoichiometry can be studied by titration [170, 176], or diffusion ordered spectroscopy (DOSY) experiments [189, 190], or alternatively by other methods described hereafter (ESI-MS, SE-HPLC, PAGE). Peak assignments were historically performed by through-space NOE experiments [125], possibly helped by base modifications (typically, guanine to inosine or bromoguanine, or thymine to uracil). Site specific ^{15}N- and ^{13}C-enrichement of defined residues allow the unambiguous elucidation of G-quadruplex structures routinely [191]. Fold determination can be typically performed by studying either the connectivity between guanine H1 and H8 from the following residue in NOE patterns, the J-couplings through tetrad hydrogen bonds [131]. Hydrogen bonds have been characterized by the H-bond scalar coupling [107].

2.1.3 Molecular Modeling

Molecular modeling is a third way to get atomic-scale information on the structure of quadruplexes. Advantageously, it gives access to individual energetic terms, dynamics, intermediates, and does not suffer from mixtures of conformations because it studies one molecule at a time [64, 65, 192–200]. As a result, the structure, energetics, and dynamics of rare conformations or of putative reaction intermediates, which would not be possible to isolate experimentally, can be explored. The quality of force fields is however an issue, notably the treatment of the loops, inter-cation repulsion, and the treatment of specific and non-specific electrostatic interactions [193–196, 201, 202], and combination with biophysical data can prove useful [203].

2.1.4 Electronic Circular Dichroism Spectroscopy

The most common and straightforward low-resolution technique for the study of G-quadruplex topology is circular dichroism [204–207]. CD is a spectroscopic method that measures the difference in absorption of left- and right-circularly polarized light by chiral compounds or by compounds in a chiral environment. G-quadruplex nucleic acids are typically characterized by their bands in the UV area (210–300 nm), arising from electronic transitions between stacked guanines. More precisely, π- π^* transitions within guanine exciton couplets occur at 279 and 248 nm, giving rise to CD bands. The sign and intensity of these bands is dictated by the relative orientation of the stacked guanines (head-to-head: *anti*/*syn*, *syn*/*anti*; head-to-tail: *anti*/*anti*, *syn*/*syn*).

As a result, the G-quadruplex topology can be inferred from the CD spectrum by analysis of the characteristic bands. Parallel topologies exhibit an intense positive band centered at 260 nm and a negative band at 240 nm, antiparallel topologies are characterized by a positive band at 290 nm, a (sometimes shallow) negative band at 260 nm, and a positive band at 245 nm, and hybrid-type structures have a positive band at 290 nm, and a shoulder at 260–270 nm. CD is relatively fast and easy to perform but has a low resolution as it only provides a global information on guanine stacks from the sample. It is not possible to easily distinguish mixtures of structures from pure species. For instance, mixtures of antiparallel and hybrid or parallel structures might lead to spectra that can be confused with a pure hybrid-type signature (see the controversy over the topology of $d[AG_3(T_2AG_3)_3]$ in Section 4.1.1).

Finally, other supramolecular features (additional base stacking, double helices, mismatches) can account for the overall spectrum and possibly lead to misinterpretations [205]. An example of additional guanine stacking modifying the CD signature is provided by the 2KF8 structure, described in Section 1.2 (Figure 2) [31]. Although this G-quadruplex has a 2-tetrad antiparallel topology where the tetrad guanines are stacked in a *syn*/*anti* fashion, its CD signature exhibits a positive band at 260 nm that could be mistakenly interpreted has a hybrid-type topology, due to additional base stacking on both tetrads of triads composed of loop nucleotides, yielding *anti*/*anti* stacks [205]. Additionally, non-classical tetrads (such as G•C•G•C as in the mutant human telomeric GGGCTA motif) may also alter the CD spectra [45]. Similarly, the presence of two loops forming a mini-hairpin motifs in an antiparallel G-quadruplex leads to the presence of a wide peak centered at 280 nm, characteristic of a B-DNA duplex, masking the expected minimum at 260 nm [37, 44].

2.1.5 UV Absorption Spectroscopy

The UV absorption spectra slightly differ between single structures, and differences in topologies are usually appreciated by *variations* in absorbance spectra, monitored by difference spectra between folded and unfolded forms. For example, thermal difference spectra (TDS) fingerprints monitor the difference between unfolded spectrum (high temperature) and the folded spectrum (low temperature) [205]. TDS fingerprints of G-quadruplexes typically exhibit local maxima at 240, 255, and 275 nm, and a minimum at 295 nm [113]. As with CD, different guanine-guanine stacks may indeed modify the intensity and wavelength of absorption bands. Hence, parallel topologies appear to be discriminated from hybrid and antiparallel topologies thanks to their greater $\Delta A_{240\text{nm}}/\Delta A_{295\text{nm}}$ ratio.

The use of TDS signatures is however not optimal because high melting temperatures leading to decreased TDS intensities, and temperature-dependent changes in molar absorption coefficient both contaminate the signature. Isothermal difference spectra (IDS), which are calculated by subtraction of the UV-spectra of a given sample acquired in the absence or presence of cation, at constant temperature. IDS are not strictly identical to TDS because they do not suffer from the drawbacks mentioned above. It is possible to study kinetics of folding and interconversion by this method [208]. In both cases the main advantages are the high throughput and ease of use that compensate their low resolution. Coordination can be indirectly monitored by either of these methods because it translates into quadruplex folding. Displacement of a cation by another can be observed by IDS, although the signature can differ significantly from case to case, and a significant change in guanine stacking is required for a signal to be observed [208].

UV-melting at 295 nm is a very common methodology to assess the stability of G-quadruplexes [209, 210]. As mentioned above, the formation of G-quadruplexes lead to a change in absorbance at 295 nm that can be followed as a function of temperature: increasing temperatures induce an unfolding of the G-quadruplexes that translates into a hypochromism at 295 nm. Alternatively, oligonucleotides tagged with two fluorophores compatible with Förster resonance energy transfer can be used for melting experiments (FRET-melting) [94, 98, 211, 212].

This is not the favored technique for structural studies since the fluorophores tend to form stacked exciton couplets that might affect both the structure and the stability of the G-quadruplex. Whatever the method used to detect the fraction folded, cation binding is related only indirectly to the fraction folded, monitored by the melting curve. Because the relationship is indirect, thermodynamic parameters extracted from such curves do not directly give access to the K_d of coordination.

2.1.6 Separative Techniques

Native gel electrophoresis can be employed to determine strand stoichiometry, although it suffers from a number of drawbacks (charge screening effects, smearing, and tedious/ambiguous quantification). It was recently combined to CD for the study of multimers [38]. Size-exclusion HPLC was shown to be a valuable alternative to get insight into oligonucleotides secondary structures, and most notably the strand stoichiometry of quadruplexes [37, 164, 213–215].

The salient advantage of this technique is the possibility to work in a variety of conditions including cation nature and concentration, and various strand concentrations (μM–mM) that are not all accessible by NMR and X-ray crystallography. Additionally, it is possible to work with a mixture of structures, which are abundant with G-quadruplex-forming sequences, whereas it is more difficult with techniques such as NMR or CD. Thermodynamics and kinetics of G-quadruplex folding can be studied with this method. It however does not provide information regarding the strand orientation or the number of quartets. Analytical ultracentrifugation (AUC) is a thermodynamically rigorous approach for the determination of absolute molecular weights and get insight into the hydrodynamic shapes of macromolecules in solution [216], and is thus used for the determination of topology and stoichiometry of quadruplexes [203, 214, 217–223].

2.1.7 Native Mass Spectrometry

Mass spectrometry (MS) in non-denaturing conditions (both in solution and in the mass spectrometer) is called native MS. Electrospray ionization (ESI) ensures non-denaturing ionization and vaporization. Native ESI-MS is another straightforward way to determine strand stoichiometry [39, 224], as well as kinetics and thermodynamics of folding of quadruplexes [225], from pure species or from mixtures, without however getting direct information on the structures. Matrix-assisted laser desorption/ionization (MALDI) has also been used to determine the strand stoichiometry [173], but this ionization method has higher risk of denaturing the complexes than ESI [226, 227].

Interestingly, the number of potassium cations bound by a given G-quadruplex can now be determined by ESI-MS (see Section 2.2.3 for more details), which can give insight into its folding topology [228, 229]. Indeed, given the preferred location of specifically bound K^+ cations, the number of tetrads of a G-quadruplex is usually equal to the number of coordinated K^+ cations plus one. In order to get additional insights into structures of macromolecules, ion-mobility spectrometry (IMS) can be coupled in-line to ESI-MS [230]. ESI-IMS-MS is a technique that, in addition to mass-to-charge ratio, further separates ions on the basis of their mobility in a drift tube filled with gas (typically, nitrogen or helium), where a static electric field is applied [231]. The travel time is related to the collision cross section of the ions with the gas: compact structures collide less with gas and travel faster as a result. Gas phase structural determination by MS- and spectroscopy-based methods have been reviewed in a recent book [232].

2.1.8 Miscellaneous

Other methods have been employed to study the topology, stability, and formation of G-quadruplexes, as well as cation binding, albeit tangentially, including single molecule FRET [233], fluorescence melting [94, 234–236], denaturing PAGE [237, 238], ^{125}I-radioprobing [239, 240], electron paramagnetic resonance (EPR) [222], surface plasmon resonance [241], fluorescence spectroscopy [242, 243], Raman spectroscopy [244, 245], calorimetry (DSC, ITC) [61, 84, 112, 246–248], temperature-jump relaxation experiments [62, 170, 236, 249, 250]. G-quadruplex characterization assays are reviewed in [62, 130, 251, 252].

2.2 Cation Coordination

2.2.1 X-Ray Crystallography

X-ray crystallography is an invaluable method for the study of cation binding in G4 nucleic acids because it directly locates cation binding sites through electron density. Crystal packing forces may however change the location of the cation as compared to solution state structures [107, 190]. Tl^+ cations can be used as probes for potassium binding sites. Tl^+ and K^+ have close atomic radii (1.44 and 1.33 Å, respectively), dehydration energies and bond lengths, and can therefore lead to similar folds, yet Tl^+ exhibits a higher X-ray scattering potential [142]. Crystals obtained in the presence of K^+ cations can be converted to Tl^+-crystals by soaking in thallium acetate [129].

2.2.2 Nuclear Magnetic Resonance Spectroscopy

NMR provides evidence not only on the location of cations within G-quadruplex structures but also on their dynamics [97, 163]. First examples of alkali cation coordination by tetrads of 5′-GMP, detected by NMR involved $^{23}Na^+$, $^{39}K^+$ and $^{87}Rb^+$ [75, 253, 254]. Notably, the resonance of $^{23}Na^+$ cations undergoes a significant line broadening and up-shifting upon binding [254]. The detection of these cations was extended to G-quadruplexes by Wu et al. [190, 255]. The coordination and dynamics of sodium cation coordination can also be directly monitored by solution- and solid-phase ^{23}Na NMR experiments [190, 256, 257]. In the same vein, ^{39}K exhibits a characteristic signature on solid-state magic-angle spinning (MAS) NMR spectra upon coordination by G-quadruplexes [258].

Indirect observation of potassium binding has been performed based on proton NMR, chemical exchange, and dynamic analysis on d($G_3T_4G_3$) and the *Oxytricha* telomeric sequence d($G_4T_4G_4$) bound to Na^+ and K^+ [88, 96]. Another indirect mean of studying K^+ coordination is the use of Tl^+ cations as surrogate, in a strategy reminiscent of the one used in X-ray crystallography studies (*vide supra*). ^{1}H-^{205}Tl scalar couplings provide important constraints for structure determination [129, 259].

$^{15}NH_4^+$ dynamics can be monitored directly with heteronuclear correlations exchange spectroscopy (^{15}N-^{1}H N_zExHSQC) [97], and serve as a reporter probe. Ammonium can indeed displace Na^+, or be displaced by K^+, allowing an indirect measurement of Na^+ and K^+ movements within G-quadruplexes [260]. Changes in topology by displacement of Na^+ by K^+ can also be used to study cation binding, albeit more indirectly [88]. Finally, Na^+ cation dynamics have been studied by a nuclear magnetic relaxation dispersion study [261].

2.2.3 Native Mass Spectrometry

Mass spectrometry seems to be an obvious choice for the study of cation binding to G-quadruplexes, since each coordinated cation increases the mass of the complex. A number of mass spectrometry studies were devoted to NH_4^+ binding determination [52, 224, 225, 231, 262–266], because ammonium acetate is a traditional volatile buffer used in biomolecule mass spectrometry. The study of alkali cations is however more tricky because (i) they tend to form clusters even at submillimolar concentrations, and (ii) they are not volatile like the classically employed ammonium acetate. The former issue leads to very noisy spectra from which no information can be inferred, while the latter means that diffuse (unspecific) coordination of alkali cation is at least partially preserved in the gas phase, which results in a higher number of bound cations than tetrad coordination sites.

These issues have recently been partially tackled by Gabelica et al. with the use of trimethyl ammonium acetate (TMAA) as a volatile co-solvent [228]. TMA is too bulky to be coordinated by the O6 of tetrad guanines and is therefore a cation of choice to fix the ionic strength and neutralize the backbone phosphates while allowing to study the specific binding of potassium (or other cations) to the tetrads. Mathematical subtraction of diffusively bound cation can be performed in a second step, by using a reference sequence that does not form a G-quadruplex [229], yielding cleaned mass spectra. Titration of G4-forming sequences with increasing potassium concentrations, and subsequent cleaning of the non-specific adducts, give access to the K_d of coordination of potassium for each tetrad binding site. This methodology is only applicable to G-quadruplexes stable at low (submillimolar) cation concentrations, which is usually not the case when working with sodium. Gross et al. have also developed a method that utilizes the gas-phase signal fractions from the bound and unbound species as input into a mathematical model that determines the binding constants [99]. This method was applied to the measure of the K_d of potassium (and strontium) to TBA.

Another bulky ammonium derivative, namely tetrabutylammonium phosphate, had already been used for the same purpose by Gray and Chaires in spectroscopic experiments involving the human telomeric sequence [91]. This allows the determination of apparent binding constants by CD. Another salient point of this study is the use of a potassium-binding benzofuran-isophthalate crown ether indicator (PBFI) to determine the concentration of free potassium by fluorescent titration, and hence deduce the concentration of coordinated cations and thermodynamic parameters [91, 267].

3 Role of Alkali Metal Ions in G-Quadruplex Stability

The stabilization of G-quadruplexes by cations has been studied for the past fifty years on increasingly complex systems. Decades after the first observation of guanosine gels [268], Gellert and coworkers have reported in 1962 the formation of gels by 5′-GMP in the presence of sodium, at acidic pH, and postulated the structuration of guanines in tetrads (or quartets), stacked upon each other to form helices [269]. Similar structures were observed a decade later using various guanosine derivatives, in the presence of potassium [270]. However, the seminal paper regarding the necessary and selective complexation of cations by such structures, at neutral pH, dates from the late seventies [73]. Pinnavaia and co-workers have demonstrated by NMR experiments that K^+, Na^+, and Rb^+ all lead to the formation of stacks of three tetrads of 5′-GMP, in a head-to-tail or alternating head-to-tail and head-to-head stacking, whereas Li^+ and Cs^+ were found to be ineffective. Additionally, it was evidenced that potassium induces a higher thermal stability than sodium. In the same manuscript, the first qualitative ranking of alkali cation-induced stability for 5′-GMP, $K^+ > Na^+, Rb^+ >> Li^+, Cs^+$ is reported, which remarkably contrasts with the ranking observed for duplex DNA where the binding constant decreases with increasing ionic radii.

Research on this topic moved to biologically relevant sequences, notably the telomeres of various species, and model G-quadruplexes such as $[d(TG_4T)]_4$. Hereafter, we will review studies highlighting the influence of alkali cations on the stability of simple tetramolecular G-quadruplexes (Section 3.1), and sets of telomeric or artificial sequences (Section 3.2). The contribution of folding and unfolding rates in the cation-dependent stability of G-quadruplexes is discussed in Section 3.3.

3.1 Case Study: $dTG_{3\text{-}5}T$ Tetramolecular G-Quadruplexes

Parallel-stranded tetramolecular G-quadruplexes formed by dTG_nT (n=3–5) sequences have been used extensively as models. These short sequences form unambiguous G-quadruplex structures devoid of loops that makes them “minimal”

Table 1 Apparent melting temperatures of tetramolecular G-quadruplexes (note: the apparent T_m depends on the temperature gradient used by the authors).

Sequence	Strand concentration (μM)	Cation concentration (mM)	T_m K^+ (°C)	T_m Na^+ (°C)	Refs
d(TG_3T)	10	110	48	16	[117]
	200	100	51	–	[273]
d(T_2AG_3)	10	110	50	17	[117]
	6400	110	≈60	–	[274]
d(T_2AG_3T)	10	110	55	24	[117]
d(TG_4T)	10	110	>94	54.5	[117]
	160	210	>94	75	[112]
d($T_2G_4T_2$)	10	110	>94	71	[117]
d(AG_4T)	10	110	>94	60	[117]
r(UG_4U)	10	110	>94	89	[117]
d(TG_5T)	10	110	>94	>94	[117]

G-quadruplexes ideal to conduct basic research. A typical measure of the stability of G-quadruplexes is the melting temperature (T_m), usually provided by UV-melting [209, 210, 271], CD-melting, and DSC [61, 84, 112, 246] experiments. Mergny et al. have reported the melting temperatures obtained by UV-melting for a number of DNA and RNA tetramolecular G-quadruplexes [117]. Typically, tetramolecular RNA G-quadruplexes are more stable than their DNA counterparts, thanks to higher association rates and lower dissociation rates. Noteworthy, RNA tetramolecular G-quadruplexes were shown as early as 1991 to be highly stable in K^+, and more than in Na^+ and Li^+ [272]. For instance, the difference in T_m is large (35 °C) between [r(UG_4U)]$_4$ and [d(TG_4T)]$_4$, in 110 mM Na^+ solutions (Table 1) [112, 117, 273, 274]. Finally, no dissociation is observed in most cases for G_5 and longer tracts in Na^+, and G_4 and longer tracts in K^+, further highlighting the cation effect on the stability of tetramolecular G-quadruplexes.

The short tetramolecular G-quadruplex [d(TG_3T)]$_4$ is characterized by an apparent melting temperature of 48 °C in the presence of 110 mM KCl and 16 °C ($\Delta T_m = 32$ °C) in the presence of 110 mM Na^+ [117]. In the case of slow association/dissociation kinetics, the melting curve is not an equilibrium curve, and therefore depends on the temperature gradient. Melting of intermolecular quadruplexes is also affected by the strand concentration. A close result was obtained in 100 mM K^+ by Balasubramanian et al. (51 °C; 200 μM strand concentration) [273]. Almost identical stability differences are observed with the addition of flanking bases mimicking the human telomeric sequence, as in [d(T_2AG_3)]$_4$ (50 *versus* 17 °C; $\Delta T_m = 33$ °C) and [d(T_2AG_3T)]$_4$ (55 *versus* 24 °C, $\Delta T_m = 31$ °C) [117]. NMR experiments on a high d(T_2AG_3) strand concentration (6.4 mM), in a 110 mM KCl buffer, produced a T_m of around 60 °C [274].

The stability of the longer $[d(TG_4T)]_4$ has been studied by different groups [89, 112, 117]. In the presence of 110 mM KCl, the melting temperature is higher than 94 °C [89, 117], indicating a very high thermal stability. An increase in Na^+ concentration did not affect significantly the stability (ΔT_m<2 °C in the 50–400 mM Na^+ range), but changed dramatically the association rate (see Section 3.3) [117]. A large difference (T_m=75 °C) can be noticed with another study by Petraccone et al. [112], performed in 210 mM NaCl but at a higher strand concentration (160 μM). Addition of thymines at both termini $[d(T_2G_4T_2)]_4$, substitution of the 5′-dT by a dA $[d(AG_4T)]_4$, or use of the RNA counterpart $[r(UG_4U)]_4$, all increase the thermal stability in 110 mM Na^+ (T_m=71, 60, and 89 °C, respectively) [117]. Further increase of the guanine tract leads to even more stable G-quadruplexes: the melting temperature is higher than 94 °C in both K^+ and Na^+ conditions for $[d(TG_5T)]_4$.

All these examples highlight the influence of the cation nature and concentration on the stability of G-quadruplexes, where little to no structural changes are expected, and thus the cation dehydration and binding *per se* accounts for most of the stabilization.

3.2 *General Trends*

3.2.1 Libraries

The Mergny group contributed several systematic UV-melting based studies on a very large number of oligonucleotides, illustrating the difference of G-quadruplex stabilization by coordination of K^+ or Na^+. In a 1998 publication illustrating the use of UV–vis spectroscopy to follow the folding of G-quadruplexes, very large differences of melting temperatures in potassium- and sodium-rich conditions were observed for the 26- and 27-mer sequences $d[T_3A_2G_3(TGTG_3)_3]$ (63 and 37 °C, respectively) and $d[G_3(TGTGTG_3)_3]$ (62 and 37 °C, respectively) [111]. In a more recent study, the melting temperature of eighty different sequences containing four tracts of three guanines with loops of variable length (between 1 and 15 bases), following the template $d(G_3L_aG_3L_bG_3L_cG_3)$, where L_a, L_b, and L_c are thymines or TTA loops, were determined in the presence of 100 mM K^+ or Na^+ [115]. Potassium stabilizes these sequences by on average 18.3 °C more than sodium, but the difference is highly variable since it can be as low as 1.2 °C, and as high as 39.2 °C. The T_m decreases when the loop length increases both in K^+ and Na^+, but to different extents – particularly for short loops – as seen from the large ΔT_m variability. For loops of 7–15 nt, the difference within a given sequence family tends to be relatively constant, from 1–2 °C (for L_a=L_c=TTA) to more than 30 °C (for L_a=L_c=T). More generally, there is a strong inverse correlation between total loop length and T_m for K^+ (each added base leads to a 2 °C drop) but the trend is less clear in Na^+. Also, the presence of adenines in the loops is favorable in the presence of sodium, when the central loop contains at least two nucleotides.

Thirty-six sequences following the general formula d($G_3T_3G_3$HNH$G_3T_3G_3$) were analyzed in similar conditions (N can be any base, H = C, T or A) [116]. The average difference of T_m is 12.7 °C in favor of potassium, but is also sequence-dependent, ranging from 9.5 °C (for ACC and TGC central loops) to 16.2 °C (for AAT). In the same study, twenty-six additional sequences that vary in length, number of quartets and loop composition and position were also investigated. The average of (measurable) potassium-sodium ΔT_ms is 14.4 °C, with a very large sequence dependence, ranging from 7.1 °C (d($G_3T_2AG_3$CGC$G_3T_2AG_3$)) to more than 40 °C (d(G_3TG_3ACTG_3TG_3)). Overall, K^+ stabilizes particularly well G-quadruplexes containing YDH loops, and poorly ACH loops, while Na^+ favors YDC loops and disfavor ACW loops (Y = C or T; H = A, C or T; W = A or T and D = A, G or T.).

Risitano and Fox have examined a randomized library of oligonucleotides based on the sequence d[T(G_3H_2)$_3G_3$], by FRET-melting, in the presence of various concentrations of K^+ and Na^+ (5–100 mM) [94]. A clear increase in T_m is observed with the concentration, ranging from 27 to 85 °C for K^+. For Na^+, lower concentrations (5 and 10 mM) do not allow sufficient folding for the T_m to be measured, and T_ms at higher concentrations (20–100 mM) range from 24 to 57 °C. The difference in stabilization, where measurable, decreases slightly with increasing cation concentration (29.4–27.6 °C).

Smargiasso et al. have investigated the melting temperature of libraries of sequences based on the sequence d($G_3W_iG_3W_jG_3W_kG_3$), where W is either a thymine or an adenine, and the loop sizes (*i*, *j*, *k*) were systematically varied between 1 and 3, yielding a total of 2,744 distinct sequences grouped as a function of their loop lengths [39]. The stability in all the groups is higher with 150 mM K^+ than with 150 mM Na^+, but again this is very sequence-dependent. The stability in K^+ is inversely dependent on loop length: shorter loops promote parallel structures with average T_m above 80 °C, whereas longer loops (at least one loop of two and one loop of three nucleotides) promote hybrid structures with average T_m in the 65—70 °C range. Conversely, longer loops yields more stable G-quadruplexes in Na^+, and the loop length has less influence. Note however that some of these sequences are suspected of forming multimeric G-quadruplexes, notably in K^+ conditions [39].

3.2.2 Human Telomeric Sequences

The human telomeric sequence is certainly the most investigated intramolecular G-quadruplex-forming sequence. Telomeres consist of thousands of tandem repeats of the sequence d(T_2AG_3), with a 3′-end overhang of 100–200 nucleotides [275, 276]. Telomere sequences are involved in the protection of the ends of chromosomes and exhibit similar repeats of guanines in numerous species such as plants (d(T_3AG_3)), *Oxytricha* (d(T_4G_4)), *Tetrahymena* (d(T_2G_4)), or *Bombyx* (d(T_2AG_2)).

Włodarczyk et al. have measured the melting temperature of the 24-mer d[(T_2AG_3)$_4$] sequence in presence of various concentration of alkali metals [85]. Table 2 summarizes the results obtained for telomeric repeats under different ionic

Table 2 Melting temperature of human telomeric G-quadruplexes.

Sequence	Cation concentration (mM)	T_m K^+ (°C)	T_m Na^+ (°C)	T_m Rb^+ (°C)	T_m Li^+ (°C)	Refs
$d[(T_2AG_3)_4]$	300	71.2	62.2	–	–	[85]
	100	59.0	50.2	40.2	32.0	[85]
	100	61	47	–	–	[271]
	70	63	49	–	–	[110]
	50	50.2	42.4	34.8	27.8	[85]
	10	44.3	–	26.4	23.0	[85]
$d[AG_3(T_2AG_3)_3)]$	110	62	55	–	–	[89]
	110	63	56	–	–	[111]
$d[(G_3T_2A)_3G_3]$	110	65	58	–	–	[111]
$d(G_3T_2AG_3)$[a]	70	42	31	–	–	[110]
$d(T_2AG_3)$[b]	110	50	17	–	–	[117]
$d(T_2AG_3T)$[b]	110	55	24	–	–	[117]
$d[(AG_2T_2)_3AG_2)]$	110	42	40	–	–	[117]
$d[T_3A_2(G_3T_2A)_3G_3]$	100	55	44	–	–	[118]
$d[T(G_3T_2A)_3G_3]$	5	45	28	–	–	[94]
	100	82	70	–	–	[94]

[a]Bimolecular structure; 20.5 μM strand concentration.
[b]Tetramolecular structure; 10 μM strand concentration.

conditions. At ~ 100 mM in K^+, Na^+, Rb^+, and Li^+, the T_m is respectively 59, 50, 40, and 32 °C, nicely illustrating the important role of alkali metal ions in G4 stability. This ranking is conserved at lower concentrations (~50 mM), ranging from 50 to 28 °C, and the G-quadruplex is still reasonably stable at 10 mM in K^+ (T_m = 44.3 °C) and to a lower extent Rb^+ (T_m = 26.4 °C). Higher Na^+ and K^+ concentrations bring further stabilization (62 and 71 °C, respectively, at ~300 mM), and partial folding is observed at room temperature with 180–230 mM Cs^+ (T_m = 27.4–27.7 °C). CD spectra were acquired at 2 °C and 20 °C with increasing concentrations of cations. Saturation was not observed for Li^+ at 20 °C and Cs^+ at both temperatures, consistent with their weak G-quadruplex stabilization properties. The presence of high Cs^+ concentrations seemingly destabilizes the G-quadruplex. Conversely, only 4 and 30 mM were enough to reach saturation for K^+ and Na^+, respectively.

Sugimoto and coworkers have found similar results in 100 mM Na^+ (~47 °C) and K^+ (~61 °C) [271], and Balagurumoorthy and Brahmachari have observed melting temperatures of 63 and 49 °C in 70 mM KCl and NaCl, respectively [110]. This is somewhat higher than found by Włodarczyk et al. [84] maybe because the values were extracted from 10-data point melting profiles. The 9-mer $d(G_3T_2AG_3)$, forming a bimolecular G-quadruplex, was also studied in the same conditions (T_m = 42 and 31 °C, respectively; 20.5 μM strand concentration). The 22-mer $d[AG_3(T_2AG_3)_3)]$ was found to be more stable in K^+ than in Na^+ by 7 °C in two distinct publications

(62/63 *versus* 55/56 °C). The difference in stability is identical for the shorter oligonucleotide 21-mer sequence d[$(G_3T_2A)_3G_3$], with both the K^+-form (T_m = 65 °C) and the Na^+-form being slightly stabilized (T_m = 58 °C) as compared to the 22-mer counterpart [111]. The short 6- and 7-mers d(T_2AG_3) and d(T_2AG_3T) forming tetramolecular G-quadruplexes are unstable in the presence of Na^+ (17 and 24 °C), and much more stable in K^+ (50 and 55 °C) [117].

Modified human telomeric sequences are also under scrutiny. An 18-mer sequence that contains repeats of only two guanines (d[$(AG_2T_2)_3AG_2$)]) is destabilized as compared to the unmodified 22-mer sequence counterpart, and has almost the same melting temperature in K^+ and Na^+ conditions (42 and 40 °C, respectively) [117]. The addition of extra nucleotides in 5′ to obtain the 26-mer d[$T_3A_2(G_3T_2A)_3G_3$] is also detrimental to the stability but yields a higher T_m difference (55 *versus* 44 °C) [118]. Risitano and Fox have examined the sequence d[$T(G_3T_2A)_3G_3$], which contains an additional dT nucleotide in 5′ as compared to the non-modified 21-mer sequence, by FRET-melting in the presence of various concentrations of K^+ and Na^+ (5–100 mM) [94]. A clear increase in T_m is observed with the concentration, ranging from 45 to 82 °C in K^+ and 28 to 70 in Na^+, with the difference of stabilization decreasing from 17 to 12 °C. The high melting temperatures could be attributed to the presence of fluorophores at both ends of the oligonucleotides that may stabilize the G-quadruplexes.

3.2.3 Other Sequences and Overview

Hardin et al. observed in 1991 that the four-repeat *Tetrahymena* telomeric sequence d[$(T_2G_4)_4$] is much more stable in K^+ than Na^+ and Li^+ conditions [215]. More recently, Tran et al. analyzed the melting temperatures of minimal (i.e., without flanking base) telomeric sequences from fourteen species [114]. All the tested sequences bearing three or more consecutive guanines (in particular $G_3T_{1\text{-}4}A$ motifs and $G_4T_{2,4}$ ciliate motifs) fold into stable G-quadruplexes both in potassium and in sodium. This is also the case of the G_2T_2A (e.g., *Bombyx mori*) and the G_2CT_2A telomeric motif sequences (e.g., *Ascaris lumbricoides*, *Caenorhabditis elegans*). Among the telomeric sequences containing a four repeated G_3 motif, coordination to K^+ induces a stabilization of on average 11 °C as compared to Na^+, at a 100 mM cation concentration. The difference is the most pronounced for *S. cerevisiae* (68 *versus* 50 °C) and the least pronounced for the human telomeres (65 *versus* 59 °C). A large difference of stability in K^+ and Na^+ conditions was equally observed in an independent study [111]. *Paramecium* and *L. esculentum* (tomato plant) have degenerated telomeric sequences. The T_m does not strongly vary among tomato variant sequences, ranging from 55 to 58 °C in NaCl and from 64 to 69 °C in KCl while *Paramecium* variant sequences spanned a broader range: from 52 to 60 °C in NaCl and from 68 to 76 °C in KCl. The ciliates *Tetrahymena* (d[$(G_4T_2)_3G_4$]) and *Oxytricha* (d[$(G_4T_4)_3G_4$]) telomeric sequences containing four repeats of four guanines form particularly stable structures in potassium-rich solutions (>80 °C), at least 16 and 14 °C more stable than in sodium solutions, respectively, which was also observed in previous studies [82, 111]. For most of these sequences, and as opposed to tetramolecular G-quadruplexes described earlier, the difference of stability as a function of the cation can be partially ascribed to the formation of distinct topologies (see Sections 1.2 and 1.3). Finally, some sequences do not form G-quadruplexes in neither K^+ nor Na^+ (yeasts *S. pombe* and *C. guillermondii*) [114].

The two-repeat *Oxytricha* telomeric sequence $d(G_4T_4G_4)$, which shares the same bimolecular, antiparallel topology in Na^+ and K^+ (*vide supra*), has been examined by various techniques. A CD-melting study from 1992 indicates that the 70 mM Na^+-stabilized G-quadruplex melts cooperatively at 55 °C (20 OD_{260} strand concentration) [109]. NMR on a 5–55 °C temperature range revealed that, in the presence of Na^+, unfolding is detected above 35 °C by the heterogeneity of the spectra [107]. Conversely, the spectra acquired with K^+ retain their homogeneity on the full temperature range. These observations were presented as being consistent with CD-melting experiments evidencing melting of $d(T_4G_4)$ in Na^+ at around 40 °C [119], but the comparison is not entirely relevant since (i) such a short sequence certainly forms a G-quadruplex of distinct topology (parallel), as seen from the CD signature, (ii) of distinct molecularity (tetramolecular) and strand concentration (2.2 *versus* 0.1 mM, respectively), and (iii) the cation concentration is different (50 *versus* 200 mM NaCl) (see Section 3.1). Incidentally, this latter study by Kallenbach et al. shows that the addition of a single 3′-dT nucleotide dramatically increases the melting temperature (+20 °C), and that the G-quadruplex formed in 200 mM KCl is not entirely melted at 100 °C [119].

Petraccone et al. have used DSC to determine the T_m of the two-repeat *Oxytricha* telomeric sequence forming a bimolecular G-quadruplex $[d(G_4T_4G_4)]_2$, in 200 mM sodium solutions [112]. Analysis of the heating curve revealed a two-step dissociation, centered around 47 and 67 °C, contrary to what has been proposed in earlier studies [109, 119]. The low-temperature transition has been attributed to an intramolecular pre-melting event. FRET-melting experiments were performed on the four-repeat $d[T(G_4T_4)_3G_4T]$ sequence, but the melting temperatures were very dependent on the heating/cooling rates, and the melting curves present a hysteresis, because of the slow dissociation rates observed in K^+ and Na^+ conditions (see Section 3.3) [249]. Overall, higher T_m were observed with K^+ than Na^+ and with increasing cation concentrations. In some cases, biphasic melting profiles were also observed.

In various studies, TBA $(d(G_2T_2G_2TGTG_2T_2G_2))$ was shown to be unfolded in Li^+, fairly unstable in 100 mM NaCl ($T_m \approx 20$ °C), and largely more stable in K^+ (48–50 °C) [89, 98, 111]. Large differences were also observed for c-myc sequences in FRET-melting experiments, notably for $d(G_4AG_4TG_4AG_4)$ ($\Delta T_m = 25$ °C at 5 mM in K^+ *versus* Na^+) [94]. The melting temperatures of the closely related sequences $d[G_3(T_2G_3)_3]$ and $d[TG_3(T_2G_3)_3T]$ in the presence of various alkali cations, span from 85 and 77 °C, respectively, in 100 mM K^+ to less than 18 °C in 100 mM Li^+ [208]. An intermediate behavior is found for Na^+ and Rb^+ that give comparable melting temperatures for both sequences, around 45 °C.

3.2.4 Summary

Overall, cation stabilization follows the order $K^+ > Na^+$, $Rb^+ > Cs^+ > Li^+$. From all these studies it can clearly be seen that the difference of G-quadruplex stability between K^+ and Na^+ conditions is always in favor of potassium, at least for natural sequences. However, one cannot predict the stability of a G-quadruplex in sodium from the result obtained in potassium or *vice versa*, since this difference spans from a couple of degrees to almost 40 °C, depending on the sequence. Differences in stability are particularly important for tetramolecular assemblies ($[TG_3T]_4$, $[TG_4T]_4$, single repeat human and *Oxytricha* telomeric sequences), where the structure is the same regardless of the cation.

Increasing cation concentrations lead to higher T_m, as expected from the law of mass action, and reduces the T_m difference between cations. This difference is classically attributed to the greater cost of Na^+ dehydration, which is only partially compensated by a more favorable free energy of binding than K^+ [88, 105]. However, the differences in T_m are also strongly sequence- and hence, sometimes structure-dependent. The influence of cations on the structure of G-quadruplexes is addressed in Section 4.

3.3 *Kinetics of Strand Association, Dissociation, Folding, and Unfolding*

Early work on tetramolecular G-quadruplex nucleic acids show that they generally display slow association and dissociation rates [277, 278]. As mentioned previously, the guanine imino proton exchange is very slow as compared to duplex DNA, and lies in the day-to-week scale in D_2O solutions for the tetramolecular RNA G-quadruplex $[r(UG_4U)]_4$ [279]. Intermolecular G-quadruplex formation is extremely slow, and the rate is very concentration-dependent (for both the strand and the cation). A key study regarding the association and dissociation of short G-rich strands in tetramolecular G-quadruplex, in Na^+ or K^+ conditions, demonstrated that this association is close to fourth order in monomer, whereas the dissociation is first order in G-quadruplex [117]. The association rate constant decreases with increasing temperatures, reflecting a negative apparent activation energy, and is increased by an increase in strand or alkali cation concentration. On the other hand, the dissociation is also temperature-dependent, but is virtually not affected by the ionic strength. The association half-time of $[d(TG_4T)]_4$ ranges from six seconds to more than a hundred years for strand concentrations decreasing from 1 mM to 1 μM, in 110 mM Na^+. Substituting Na^+ by K^+ affords a 20–50 fold increase in association constants, fairly constant in a 2–37 °C temperature range and 50–300 mM cation concentration range. Although an increase in sodium concentration from 50 to 400 mM does not affect significantly the melting temperature of $[d(TG_4T)]_4$ (see Section 3.1), it dramatically increases the association rates.

An increase of one order of magnitude of Na^+ concentration leads to a 1,000–2,000 increase in association constant. The fact that the cation-dependent increase in stability observed for many G-quadruplexes is mainly reflected in the association rate constant is similar to duplex and triplex nucleic acids (see Chapter 6 in this volume). The dissociation is not observed in most cases for G_5 and longer tracts in Na^+, and G_4 and longer tracts in K^+. Melting and temperature-jump experiments of fluorescently-tagged G-quadruplexes ($[d(TG_4T)]_4$, $[d(G_4T)]_4$, $[d(TG_4)]_4$, $[d(G_4)]_4$) at different heating/cooling rates in Na^+ solutions confirmed that the rates of annealing are very slow [236].

Another comprehensive study on the cation-dependent formation of DNA $[d(TG_nT)]_4$ (n=3–6) and RNA $[d(UG_4U)]_4$ tetramolecular G-quadruplexes has been provided by Bardin and Leroy [164]. NMR experiments demonstrated that an increase of salt concentrations favors the association rate, but that the cation nature (K^+, Na^+, Li^+) only weakly contributes. However, the lifetime of the G-quadruplex is much higher in K^+ than in Na^+ and Li^+, and increases with the guanine tract length. The authors suggest that, in 125 mM K^+ solutions, the G-quadruplex formation proceeds step-by-step via successive duplex and – kinetically limiting – triplex intermediates, and follows a reaction order of three. On the other hand, the reaction order is found to be four in Li^+, and each step is strand-concentration-dependent. The formation of mismatched G-quadruplexes, particularly at low temperature and K^+ concentrations, is believed to impede the formation of the canonical G-quadruplex by kinetic trapping. In the same vein, Kim et al. have performed PAGE experiments showing that the formation of a tetramolecular RNA G-quadruplex by the $r(GC_2GAUG_2UAGUGUG_4U)$ sequence is faster in Na^+ than K^+, despite the higher stability observed in the latter case [272]. They also suggest that the formation of unproductive intermediates impedes the formation of the G-quadruplex in potassium conditions.

The cation-dependent kinetics of intramolecular G-quadruplex formation by telomeric sequences was investigated by different groups. Raghuraman and Cech evidenced that the half-life of the *Oxytricha* telomeric sequence $[d(G_4T_4)_4]$ unfolding is shorter in 50 mM Na^+ than K^+ (4 *versus* 18 h at 37 °C) [280]. The FRET-tagged sequence $d[T(G_4T_4)_3G_4T]$ shows exceptionally slow dissociation rates in K^+ and Na^+ conditions [249]. The half-life at 37 °C is roughly 10 years in 50 mM K^+, while the corresponding association half-life is of about 0.9 s. The rate of association is 10–60 fold faster than in sodium in the 50–100 mM cation concentration range. Differences in dissociation rates are much more marked ($t_{1/2}^{37\,°C}$ = 3,900 s in 50 mM Na^+). From these results, Fox and coworkers suggested that, although there are differences in the rates of association of G-quadruplexes under different conditions, the dissociation rate constant is the most important factor that affects the relative stabilities [249]. However, as mentioned in the latter study, and others [164, 209], association rates decrease with increasing temperature and suggests a nucleation-zipper mechanism for G-quadruplex formation. Using a PNA trap, large cation-dependent differences were observed for the average unfolding time constants of the human telomeric sequence $d[G_3(T_2AG_3)_3]$, in the presence of either 100 mM K^+

(40 h at 20 °C) or Na^+ (30 min) [281]. This is qualitatively consistent with the higher melting stability observed in K^+ (see Section 3.2.2).

Time-dependent FRET monitoring as a function of complementary strand hybridization of TBA revealed a first-order kinetics, where the unfolding of the G-quadruplex is the rate-determining step [98], consistent with some previously mentioned studies [280, 281]. The unfolding rate is also strongly cation-dependent (62×10^5 and $3{,}200 \times 10^5$ s^{-1} at 10 °C, in K^+ or Na^+, respectively). If the nature of the cation affects largely the dynamics of G-quadruplex folding and unfolding, other factors are to be taken into account such as the cation concentration, and the oligonucleotide sequence. At low K^+ concentration, the human telomeric sequence displays a relatively fast dynamic behavior (<100 s) [243]. The sequence – and hence structure – that is studied also affects the dynamics of folding and unfolding. As seen above, the *Oxytricha* telomeric sequence can maintain its G-quadruplex structures for longer times than its human counterpart [249, 280], even at low K^+ concentrations [243], likely because it contains four quartets instead of three.

In the past years, several groups have explored folding intermediates and their interactions with cations (mostly K^+). G-quadruplex folding intermediates may adopt pre-organized structures inclined to fold into a G-quadruplex after cation binding. Mashimo and coworkers have investigated the folding pathways of the human telomeric sequence [282–284]. A combination of *ab initio* calculations and molecular dynamics hinted at the formation of hairpin containing Hoogsteen G•G base pairs, possibly yielding a more stable triplex intermediate. Coordination of K^+ favors each step of the folding by decreasing the electrostatic repulsions [284]. Binding to a hairpin near the second lateral TTA loop was found to be preferable, and more generally to 5′-G(*syn*)G-(*anti*)-3′/5′-G(*syn*)G(*anti*)-3′, rather than to 5′- G(*anti*)G(*anti*)-3′/5′-G(*syn*)G(*syn*)-3′.

The Chaires group also thoroughly investigated the folding and unfolding of human telomeric sequences in the presence of K^+, notably by CD, 2-aminopurine fluorescence, FRET, and molecular modeling (molecular dynamics) [285–287]. The latest results suggest a four-step pathway [285] (Figure 6A) [171, 285, 288]. The first folding intermediate I1 is a mixture of hairpin structures that is rapidly formed thanks to cation-induced collapse of the unfolded strands. Intramolecular fold over of I1 may form antiparallel G-quadruplex structures I2 upon binding to K^+ ($\tau_1 \approx 0.1$ s in 25 mM KCl, at 25 °C), the latter further converting to a triplex intermediate I3 ($\tau_2 \approx 3{,}700$ s) that finally yields the final G-quadruplex ($\tau_3 \approx 750$ s). Other groups have suggested the triplex structure as an unfolding intermediate for the human telomeric sequence [289–291], including in conformational switching context [165]. Zhang and Balasubramanian stated that, regardless of the structure of the DNA or RNA quadruplex formed, the folding pathway involves two steps initiated by the binding of a single K^+ cation, leading to a hairpin intermediate, followed by the formation of a triplex [292].

A triplex folding intermediate, stabilized by a K^+ cation, was also proposed for the folding of TBA [293, 294]. In MD calculations, K^+ was manually positioned between the two triads according to distances observed for the high-resolution crystal structure of TBA [161]. The guanine G10, despite being out of plane, seems

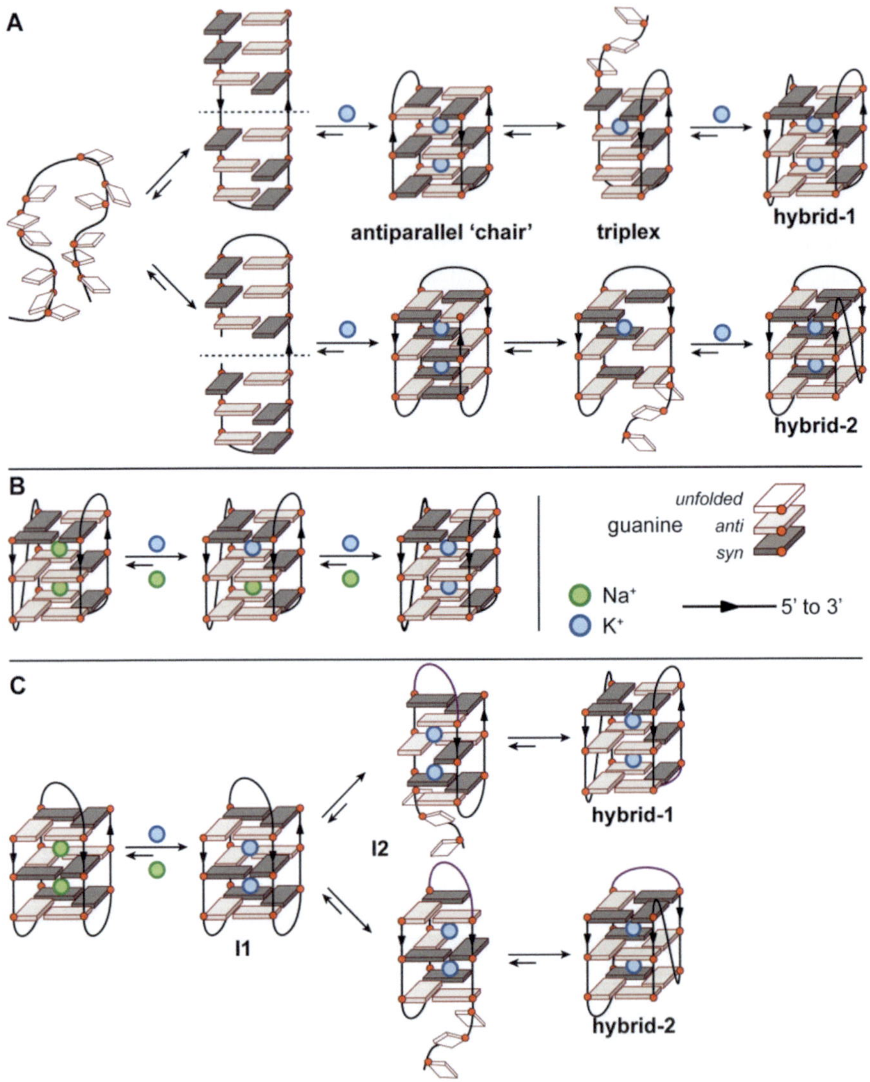

Figure 6 (**A**) Folding of the human telomeric sequence in K^+ conditions as proposed in reference [285]. (**B**) Step-wise Na^+/K^+ exchange mechanism of the human telomeric sequence proposed in reference [171]. (**C**) Na^+/K^+ exchange and structural interconversion proposed in reference [288].

well positioned to coordinate K^+ with its O6. Na^+ poorly stabilizes the triplex of TBA (folding is partially visible by NMR under 1 °C), which is not surprising as it was shown to be a poor TBA stabilizer [89, 98, 111].

Plavec et al. have recently proposed a new folding intermediate for the G-quadruplex of the *Oxytricha* sequence Oxy-1.5, namely i-Oxy-1.5, existing as two symmetric bimolecular forms containing G•G N1-carbonyl symmetric base pairs [246]. This intermediate was characterized by NMR, TDS, CD, DSC, and PAGE, nicely illustrating how the use of a combination of analytical methods allows to unearth insightful results. DOSY, PAGE and concentration-dependent T_m results demonstrated that i-Oxy-1.5 is bimolecular and has a comparable compactness as the final G-quadruplex. However, it does not contain tetrads as seen by NMR and DSC, the latter allowing to evaluate the difference in enthalpy of unfolding between i-Oxy-1.5 and Oxy-1.5 at around 48 kcal.mol^{-1}. The three-fold lower enthalpy observed for the intermediate has been attributed to twice as less H-bonds, less favorable stacking interactions, and the absence of cation coordination. Its formation is a slow kinetically governed process, while K^+ binding and subsequent restructuration consist of two fast processes.

Studies investigating folding intermediates mostly focus on K^+, and hence do not give much information about the cation nature dependence in the folding process. Comparable behaviors may be expected as compared to final G-quadruplex structures (e.g., Na^+ would stabilize the intermediates to a lower extent), as hinted by the triplex intermediate of TBA [294], but insufficient data had been gathered so far. However, they all highlight the rapidity and the importance of processes that involve cation binding as compared to strand folding and unfolding. This will be further emphasized in Sections 3.4 and 5 that focus on cation exchange and conformational switching.

3.4 Cation Exchange Mechanisms

Na^+ and K^+ have a different mobility inside G-quadruplexes. Sodium is moving faster [295], presumably because its size does not hamper passing through tetrads [97]. The mobility of Na^+ depends on the binding site, as observed by ^{23}Na NMR spectroscopy [190, 257], and $^{15}NH_4^+$ displacement NMR experiments [97]. In the bimolecular G-quadruplex formed by d($G_4T_4G_4$) (see Section 1.3 and Figure 5), Na^+ ions bound in loops have a residence lifetime of 220 µs at 15 °C (k_{off}=4.5×10^3 s^{-1}) [190, 257], whereas the lifetime in the central stem is at least two orders of magnitude higher [97]. Sodium is in a relatively dynamic exchange between coordination sites and the bulk medium, and the more affined potassium can displace sodium from these binding sites [256, 257]. Replacement of Na^+ by K^+ within the G-quadruplex central stem is completed within ~250 µs [190, 257, 296].

Hud et al. suggested that two K^+ binding events occur on the G-quadruplex [d($G_3T_4G_3$)]$_2$, where the displacement of Na^+ by increasing amount of K^+ does not impact significantly the topology [88]. Hence, each sodium cation is exchanged in a step-wise process where the intermediate is a mixed-cation species that is in fast exchange (on the NMR time-scale) with the pure species. A decade later, Plavec and coworkers have shown that the conversion of the $^{15}NH_4^+$-coordinating [d($G_3T_4G_4$)]$_2$ G-quadruplex to the K^+-form also proceeds by step-wise exchange of the cations [260]. The second binding site is not exchanged before the first one is fully occupied by potassium, yielding a mixed-cation intermediate. This can be explained by the differences between the two coordination sites; the variations in ΔG between the two cations are not identical for both sites ($\Delta\Delta G = -5.7$ and -4.3 kcal.mol^{-1}, respectively, including cation dehydration energies). ^{23}Na NMR spectroscopy also evidenced that increasing amount of K^+ displaces Na^+ cations from the abovementioned $[d(G_4T_4G_4)]_2$ G-quadruplex, without unfolding of the G-quadruplex, and that both cations may reside in the G-quadruplex simultaneously [190]. Differences in binding sites leading to mixed-cation species were also observed with quadruplexes built from lipophilic guanosine derivatives [297]. Within a four tetrad construct, the replacement of the central Na^+ by K^+ takes place before the outer cations are exchanged, whereas the wider Cs^+ cation can exclusively be coordinated by the central binding site.

The Na^+ to K^+ exchange in the human telomeric motif was also particularly intensely investigated. It was proposed on the basis of NMR and CD experiments that human telomeric sequences (26 and 22-mers) form an antiparallel 'basket-type' structure in Na^+ solutions and a hybrid-1 structure in K^+ [165], so here the cation exchange is accompanied by a dramatic change in G-quadruplex topology. Therefore the authors hypothesized that the cation exchange from the former to the latter involves a partial unfolding and restructuration steps, after the cation exchange (see Section 5). In the same vein, Gray et al. have shown that the (30 mM) Na^+ to (50 mM) K^+ exchange of the 22-mer $d[AG_3(G_3T_2A)_3]$ is also followed by a conformation change, from the Na^+-basket type to the K^+-hybrid (Figure 6C) [288]. Three distinct kinetic processes and two intermediates I1 and I2 are involved (see Section 5 for more details). The cation exchange *per se* is believed to be fast (5 ms), as compared to the ensuing conformational change (11–14 min). Chang et al. have proposed that the 23-mer sequence $d[TAG_3(T_2AG_3)_3]$ adopts the same hybrid-1 topology in both K^+ and Na^+ solutions despite very different CD spectra, and thus exclude unfolding of the G-quadruplex for the exchange to take place (Figure 6B) [171]. The exchange is complete in around 80 s, and its mechanism involves two steps, each Na^+ cation being exchanged independently via a mixed-cation intermediate as observed by aforementioned studies on bimolecular G-quadruplexes [88, 190, 260]. This step-wise exchange is proposed notably on the basis of a 22-fold difference between the binding constants of the two coordination sites.

The residence lifetimes of Na^+ cations are much shorter than tetrad breathing movements. This strongly suggests that Na^+ ions exchange through the central stem of G-quadruplexes without requiring partial tetrad opening [298]. Accordingly, all studies show that the exchange of Na^+ by K^+ does not require unfolding, although unfolding may take place after cation exchange to allow conformational switching (see Section 5). Consistent with that hypothesis is the lower residence lifetime measured for tetramolecular G-quadruplexes *versus* mono- and bimolecular G-quadruplexes, whose loops could impede cation release [189].

Solution-state NMR experiments on $^{15}NH_4^+$ evidenced that cation movement is slower at the 5′-end of G-quadruplexes, and that it is slower through an all-*syn* tetrad than through an all-*anti* one [299]. NMR also confirmed that cation movement occurs between each coordination site and the bulk medium rather than between the coordination sites. Exchange of "cation 1" by "cation 2" can thus be described as four successive steps: (i) release of cation 1, (ii) hydration of cation 1, (iii) dehydration of cation 2, and (iv) coordination of cation 2 [62]. The highest energetic cost resides in step (i) where the coordination of the cation shall be removed, and, as a direct consequence, the structure of the G-quadruplex might change.

Molecular dynamics simulations suggest that the entering or releasing of K^+ from the G-quadruplex is accompanied by significant changes in O6 distances [295]. Li^+ induces a shrinking of the structure, whereas larger cations (Rb^+, Cs^+) significantly deform the structure and hence destabilize the G-quadruplex. Re-hydration of cation 1 (step ii) is, however, favorable. The following steps (iii) and (iv) are the reverse mechanism [62]. Note, however, that the final G-quadruplex can exhibit a different structure than the starting one, which would affect the net energetic change. Therefore, this net energy change (2 ± 0.5 kcal.mol^{-1} for a three-tetrad system [91]; see Section 1.3) is a result of solvation, coordination, and structuration terms.

Reshetnikov et al. have described the cation coordination process with TBA, thanks to molecular dynamics simulation, and hybrid quantum mechanics/molecular mechanics simulations [300]. Cations enter TBA through 'gates', formed by either the G8 base at the base from the central TGT loop and the space between guanine's O6 from the upper tetrad, or by the non-canonical T–T pair and the space generated by guanine's O6 from the lower quartet. When the cation is captured by the quadruplex, it enters rapidly into the central stem, and the gates tend to close (the initial binding event lies in the ns range for both Na^+ and K^+).

Conversely, in the absence of coordinated cations, the gates are open. The TGT loop modulates the cation binding process in three ways, by (i) slowing down the cation binding by obstructing the gates, (ii) contributing to keep the coordinated cation in the quadruplex, and (iii) avoiding the quadruplex to collapse when no cation is bound. During the cation exchange between the bulk medium and other cations already coordinated, the release of the bound cation is correlated with the initial binding of the incoming cation. In short quadruplexes (small number of tetrads), the initial binding of a new cation can lead to electrostatic repulsions with the central stem-coordinated cation, eventually leading to the release of the latter.

Recently, DFT-D3 calculations have evidenced that binding of a second ion in a three-tetrad system is accompanied by a large overestimation of the cation-cation repulsion at the molecular mechanics level of experiment because of its lack of polarization term [202], leading to large energy differences with the quantum mechanics level (20 kcal.mol^{-1} in the gas phase). Indeed, the polarization effect between the central stem and the cations decreases the energy cost of coordination of the second cation, by reducing the electrostatic repulsions of the cations. Because of this bias in the force fields that are currently employed, the authors question the validity of MD simulations published so far regarding the treatment of cations (reduced coordination rate, frequent cation ejections, destabilization of the cation coordination in loops above tetrads), and future work in this direction is therefore needed.

4 Influence of Alkali Metal Ions on G-Quadruplex Structures

4.1 *Case Study: The Human Telomeric G-Quadruplex Sequence*

We have seen in Section 3.2.2 that the G4 community has a predominant interest in the human telomeric sequence. This particular attention is partly historical, with a number of seminal studies appearing in the late eighties and early nineties [81, 82, 301], partly therapeutic [302–304], but also arises from its particularly intriguing polymorphism [172]. As described below, this polymorphism depends on a number of factors, including the cation nature, the precise sequence (number of repeats, flanking sequences), the putative presence of co-solvent, and other experimental factors imposed by the chosen analytical method. Many topologies have been proposed based on various methodologies including NMR [25, 27–32, 50, 168, 170–174, 242], X-ray [26, 150], ESI-MS [231, 305], CD [85, 173, 174, 245, 306–308], single molecule FRET [233], native [271] or denaturing PAGE [237–239], ^{125}I-radioprobing [239, 240], UV-melting [271], fluorescence spectroscopy [242], Raman scattering [245], analytical ultracentrifugation and molecular modeling [203]. Available high-resolution structures of 4-repeat human telomeric G-quadruplexes are depicted in Figure 2 and schematized in Figure 7 [25–32].

4.1.1 The Intramolecular Folding of dAGGG(TTAGGG)$_3$

The 22-mer sequence d[$AG_3(T_2AG_3)_3$] containing four guanine repeats and a single flanking nucleotide (5′-dA) has been particularly studied. A first structure has been solved by NMR, in Na^+-rich conditions, as early as 1993 by Wang and Patel (PDB ID: 143D) [25]. The structure formed is an antiparallel "basket-type" G-quadruplex, characterized by three *syn•syn•anti•anti* tetrads, with each G-tract following the pattern 5′-*syn-anti-syn* or *anti-syn-anti*, connected by consecutive lateral-diagonal-lateral loops.

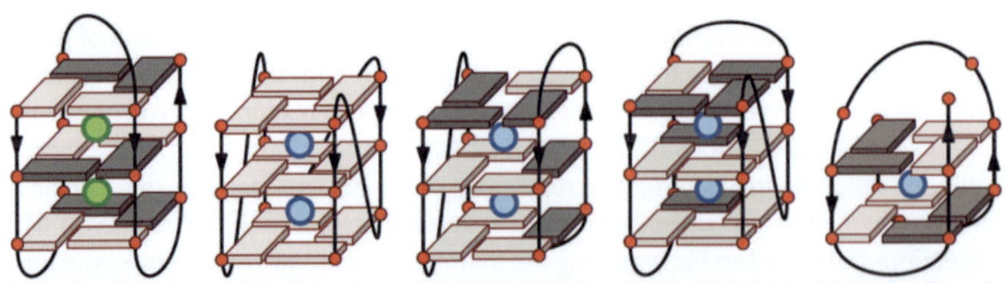

Figure 7 Topologies of the four-repeat DNA human telomeric G-quadruplex deposited in the PDB [25–32].

In contrast, in K^+ conditions, several different structures were found, highlighting the influence of the cation on G-quadruplex structures, but also, regrettably, of the structural analysis method. Parkinson et al. published the crystal structure of d[$AG_3(T_2AG_3)_3$] in 2002 (PDB ID: 1KF1) [26]. It exhibits an all parallel topology, with three tetrads formed by *anti* guanines. The K^+ cations are positioned between the tetrads at equidistance from each guanine O6 (2.7 Å), and separated from each other by ~3.4 Å (G-quartet rise: 3.1 Å). If this structure differs significantly from the solution-based Na^+ structure described above, it also differs from more recent solution-based structures obtained with K^+ samples.

d[$AG_3(T_2AG_3)_3$] in K^+ solutions actually adopts a mixture of antiparallel and hybrid structures [309, 310]. ^{125}I radioprobing suggested that an antiparallel fold is present in both Na^+ and K^+ conditions, although in potassium the topology would be chair-type (three lateral loops) rather than basket-type, and in a mixture with other topologies [239]. Vorlícková et al. have argued that the solution structure has essentially the same antiparallel topology with both cations, based on CD experiments [306], contrary to earlier reports by three other groups that support an antiparallel-to-hybrid switch upon addition of K^+ to a Na^+-solution [95, 165, 311]. The conversion from the Na^+ form to the K^+ form could be too fast to reflect a structural change as important as reported previously, and the differences in CD signatures could be assigned to changes in tetrad stacking due to specific K^+ coordination.

A similar observation (change in CD signature but no change in topology) was proposed by Chang et al. for the 23-mer d[$TAG_3(T_2AG_3)_3$] [171]. In the same vein, Bombard et al. have observed an identical platination site (namely dA13) in both Na^+ and K^+ conditions, by gel analysis of 3′-exonuclease-digested d[$AG_3(T_2AG_3)_3$], previously incubated with platinum complexes [238]. An earlier study involving *cis*- and *trans*-platinum complexes yielded similar results [237]. It is thus possible that the basket type is at least partially populated in K^+ conditions, and binding to this structure would be favored by the platinum complexes. Alternatively, different structures such as the 2-tetrad antiparallel fold could lead to identical platination sites.

It is likely that the crystal structure of Parkinson et al. is influenced by crystal packing forces [26]. Incidentally, water-depleting co-solvents such as polyethylene glycols [307, 312], ethanol [306], and acetonitrile [309] promote the parallel fold. However, other groups have stated that this fold is not the favored one in physiological conditions [312–314]. Hänsel et al. have suggested by a combination of NMR and fluorescence spectroscopy that d[$AG_3(T_2AG_3)_3$)] predominantly adopts the hybrid-1 conformation *in vivo*, *ex vivo*, and in dilute potassium-based solution, and confirm the observation of a parallel fold in water-depleted conditions [242]. Renčiuk et al. also suggested that the difference between X-ray and NMR experiments arises from the unexpected DNA concentration dependence on the human telomeric intramolecular G-quadruplex [306]. Abu-Ghazalah and coworkers have observed a mixture of antiparallel and parallel structures at high strand concentrations (2 mM), in Na^+ solutions [245]. The parallel topology was tentatively attributed to either higher-order structures formed by propeller monomer or tetramolecular assemblies.

4.1.2 Other Human Telomeric Sequences

The crystal structure of the 12-mer d[$(TAG_3T)_2$] is a parallel bimolecular G-quadruplex in K^+ [26], whereas the solution structure is a mixture of parallel and antiparallel bimolecular G-quadruplexes [170]. Bolton et al. suggested that the Na^+ fold is antiparallel [308]. The RNA human telomeric (TERRA) counterpart r[$(UAGGGU)_2$] does not exhibit the same cation-dependency since it associates into a bimolecular 'propeller' G-quadruplex in K^+ crystals [150], and in both Na^+ [173] and K^+ solutions [174], reminiscent of the DNA crystal structure albeit for changes in sugar puckering. The topology of RNA 4-repeat sequences do not exhibit any cation dependency either. CD suggests that r[$(UUAGGG)_4$] forms a parallel G-quadruplex in Na^+ [173] and K^+ solutions [174]. In the absence of 5′-flanking nucleotides, the bimolecular G-quadruplex [r$(G_3U_2AG_3U)$]$_2$ further dimerizes by 5′-5′ stacking in K^+ solutions, as seen by NMR [50]. This was also observed on the 12-mer by ESI-MS, but with ammonium ions instead of alkali cations [224].

Unlike DNA 12-mers, no structure difference was observed for the 16-mer three-repeat d[$G_3(T_2AG_3)_2T$] in Na^+ and K^+, when associating into an asymmetric, anti-parallel bimolecular G-quadruplex [168, 172]. The bimolecular G-quadruplex was also observed in Na^+ by association of two three-repeat strands, where one provides only one G-tract [168].

As stated earlier, four-repeat human telomeric sequences fold into a variety of structures. Vorlícková et al. have proposed that the 21-mer d[$G_3(T_2AG_3)_3$] folds into an antiparallel G4 in both Na^+ and K^+, alike d[$AG_3(T_2AG_3)_3$] [306]. Other groups have attributed the CD signature to a mixture of hybrid and antiparallel folds [307]. Similar conclusions were drawn from ^{125}I-radioprobing experiments, that also suggest that the presence of 5′-flanking nucleotides stabilizes an hybrid fold in K^+, while Na^+ promotes an alternative basket structure [240]. NMR experiments show that the addition of 5′ and 3′ flanking sequences stabilizes hybrid [27–30, 171] (PDB ID: 2GKU, 2HY9, 2JSM, 2JSL, 2JPZ), or 2-tetrad antiparallel topologies (2KF8, 2KKA) [31, 32], in K^+ solutions, which are all distinct from the Na^+ basket type of d[$AG_3(T_2AG_3)_3$] (143D) (Table 3) [25–32, 50, 150, 168, 170–174, 271, 306–308].

Hybrid-1 and Hybrid-2 structures are characterized by successive double chain reversal-lateral-lateral and lateral-lateral-double chain reversal loops, respectively. Chang et al. have used NMR to show that d[$TAG_3(T_2AG_3)_3$] adopts the same hybrid-1 topology in both K^+ and Na^+ solutions, despite of their different CD signatures [171]. A number of sequences starting by a guanine, including the 21-mer d[$G_3(T_2AG_3)_3$], adopt the 2-tetrad antiparallel topology (sometimes coined 'form 3'), that contains 2 tetrads, with lateral-diagonal-lateral loop bases being involved in externally stacked triplets/base pairs [31]. Gabelica et al. have evidenced by native ESI-MS experiments the binding of a single K^+ cation at low KCl concentrations, consistent with the 2-tetrad folding, while higher concentrations lead to the binding of a second K^+, either by binding between external tetrads and loop bases, or by conversion to a 3-tetrad hybrid structure [229]. Włodarczyk et al. have acquired CD spectra of d$(T_2AG_3)_4$ at 2 and 20 °C at increasing concentrations of alkali

Table 3 Structures adopted by selected DNA and RNA human telomeric sequences.

Sequence (5′ to 3′)	Cation	Topology	Assay	PDB ID	Refs
$d[(TAG_3T)_2]$	Na^+ (140 mM)	dimeric antiparallel	CD	–	[308]
	K^+ (100 mM)[a]	dimeric parallel + antiparallel	NMR	–	[170]
	K^+	dimeric parallel	X-ray	1K8P	[26]
$r[(UAGGGU)_2]$	Na^+ (215 mM)[b]	dimeric parallel	NMR	–	[173]
	K^+ (100 mM)[a]	dimeric parallel	NMR	2KBP	[174]
	K^+	dimeric parallel	X-Ray	3IBK	[150]
$r(G_3U_2AG_3U)$	K^+ (100 mM)[a]	5’-5’ stacked parallel dimer	NMR	2M18	[50]
$d[G_3(T_2AG_3)_2T]$	Na^+ (140 mM)[c]	dimeric antiparallel	NMR	2AQY	[168]
	K^+[d]	dimeric antiparallel	NMR	–	[172]
$d[G_3(T_2AG_3)_3]$	Na^+ (145 mM)	antiparallel (basket)	CD	–	[306]
	K^+ (variable)[e]	antiparallel (basket)	CD	–	[306]
	K+ (50 mM)	hybrid mixture	CD	–	[307]
	K^+ (100 mM)[a]	hybrid + (2-tetrad) antiparallel	NMR	–	[31]
$d[AG_3(T_2AG_3)_3]$	Na^+ (100 mM)	antiparallel (basket)	NMR	143D	[25]
	K^+ (diffusion)	parallel (propeller)	X-ray	1KF1	[26]
	Na^+ (145 mM)	antiparallel (basket)	CD	–	[306]
	K^+ (variable)[e]	antiparallel (basket)	CD	–	[306]
$d[T_2G_3(T_2AG_3)_3A]$	K^+ (100 mM)[a]	hybrid-1	NMR	2GKU	[27]
$d[A_3G_3(T_2AG_3)_3A_2]$	K^+ (100 mM)[a]	hybrid-1	NMR	2HY9	[28]
$d[TAG_3(T_2AG_3)_3]$	K^+ (100 mM)[a]	hybrid-1	NMR	2JSM	[29]
	K^+ (150 mM)	hybrid-1	NMR	–	[171]
	Na^+ (150 mM)	hybrid-1	NMR	–	[171]
$d[TAG_3(T_2AG_3)_3T]$		hybrid-1 + hybrid-2	NMR	–	[29]
$d[TAG_3(T_2AG_3)_3T_2]$	K^+ (100 mM)[a]	hybrid-2	NMR	2JSL	[29]
$d[(T_2AG_3)_4T_2]$	K^+ (110 mM)[f]	hybrid-2	NMR	2JPZ	[30]
$d[(T_2AG_3)_n]$[g]	Na^+ (100 mM)	antiparallel	UV-melting	–	[271]
	K^+ (100 mM)	parallel + antiparallel	UV-melting	–	[271]
$d[G_3(T_2AG_3)_3T]$[h]	K^+ (100 mM)[a]	antiparallel (2-tetrad basket)[i]	NMR	2KF8	[31]
$d[A(G_3T_2A)_2IG_2T_2AG_3T]$	K^+ (110 mM)[e]	antiparallel (2-tetrad basket)	NMR	2KKA	[32]
$r[(UUAGGG)_4]$	Na^+ (100 mM)	parallel	CD	–	[173]
	K^+ (100 mM)[a]	parallel	CD	–	[174]

[a]Calculated for 70 mM KCl + 20 mM potassium phosphate (pH 7.0).
[b]Calculated for 200 mM NaCl + 10 mM sodium phosphate (pH 6.8).
[c]Calculated for 100 mM NaCl + 10 mM sodium phosphate (pH 6.8).
[d]Unpublished.
[e]Addition of up to 100 mM in a 145 mM Na^+ solution.
[f]Calculated for 70 mM KCl + 25 mM potassium phosphate (pH 7.0).
[g]$n = 4–12$.
[h]Other sequences adopting the same fold: $G_3(T_2AG_3)_3$, $G_3(T_2AG_3)_3T_2$, and $G_3(T_2AG_3)_3T_2A$.
[i]Predominant fold (60 %). Mutation of G7 to a ^{Br}G favors this topology (90 %).

cations [85]. No conclusion was drawn regarding the relative topologies formed, but the λ_{max} were reported (290 nm for K^+, 295 nm for Na^+ and Rb^+, 300 nm for Cs^+, and 301 nm for Li^+).

Renčiuk et al. have suggested that, although flanking nucleotides can stabilize hybrid topologies, longer telomeric sequences fold in an antiparallel fashion [306]. The use of site-specifically ^{15}N-labeled G4-units in a native-like single stranded telomeric in high resolution NMR experiments revealed that the 3′-terminal and internal G4 unit predominantly coexist in 2-tetrad antiparallel basket and hybrid-2 structures, arranged in "beads-on-a-string"-like fashion [242]. Sugimoto and co-workers have suggested that long sequences (5–12 repeats) form intramolecular G-quadruplexes arranged in distinct G-quadruplex units connected by TTA linkers in a beads on a string fashion, in both K^+ and Na^+ solutions, based on UV-melting experiments [271]. Na^+ promotes the antiparallel basket topology, whereas potassium is believed to lead to a mixture of parallel and antiparallel structures, or hybrid structures.

4.1.3 Summary

Clearly, K^+ allows $d[AG_3(T_2AG_3)_3)]$ to adopt a variety of structures, whereas only the basket-type G-quadruplex has been solved in Na^+ conditions. The precise topologies adopted in potassium conditions remain controversial, the consensus being that it is likely a mixture. Other four-repeat sequences spark a similar debate, notably the minimal 21-mer sequence. A current issue is that a number of studies performed prior to the publication of the antiparallel 2-tetrad structures 2KF8 and 2KKA (in 2009–2010) often attributed antiparallel signatures to the 3-tetrad antiparallel 'basket' topology ('Na^+ fold'). In the same vein, folding/unfolding intermediates are not usually considered (see Section 3.3). Shorter two-repeat sequences also exhibit a clear cation-dependent folding topology, but not the three-repeat one. To complicate the picture, the exact sequence used clearly impacts the structure(s) formed, and the analytical methods sometimes influence those detectable.

Progress is needed to separate conformation mixtures while preserving each of them, in order to disentangle the cation effects. In contrast to DNA G-quadruplexes, the topologies formed by RNA sequences do not seem to exhibit any cation dependence. Finally, most studies on long DNA or RNA telomeric sequences (reviewed in [172]) only involve K^+, and further work is required to bring insight into the cation dependency of G-quadruplex folding.

4.2 Other Sequences

Following the work of Sen and Gilbert who first investigated the cation-dependence on the topology of various sequences [315], Hardin et al. have shown that the telomeric sequence of *Tetrahymena* $d[(T_2G_4)_4]$ folds into G-quadruplexes of different structures in relatively low K^+ and Na^+ concentrations (20 mM), by a combination of NMR, CD, gel electrophoresis, and an early use of SE-HPLC for G4 nucleic acids [215]. The latter method suggested that the K^+ form is twice as long as the Na^+ form, and this was ascribed to the formation of a tetramolecular structure in K^+. The intramolecular Na^+-structure was solved in 1994 by Wang and Patel (PDB ID 186D) [46]. Despite its tracts of four guanines, it displays only three tetrads, linked by successive lateral-lateral-double chain reversal loops, in a pattern similar to the hybrid-2 G-quadruplex formed by the human telomeric sequence [29, 30]. The 2-repeat sequence $d(TG_4T)_2$ was also examined in Na^+, and adopts two distinct, interconverting, bimolecular, antiparallel 4-tetrad structures, which differ by the arrangement of the lateral loops (head-to-head or head-to-tail) [176].

The one-repeat *Oxytricha nova* telomeric DNA sequence $d(T_4G_4)$ forms a parallel, tetramolecular G-quadruplex in Na^+ or K^+ conditions, but oligomerizes only in the latter case [119]. The two-guanine tracts counterpart $d(G_4T_4G_4)$, sometimes referred to as Oxy-1.5, retains the same fold in the presence of K^+ or Na^+ cations, whether it is studied by NMR [96, 125, 126] or X-ray crystallography [127, 128]. It is composed of four tetrads with the thymine loops connecting the strands diagonally over the external tetrads. The potassium and sodium forms differ by the position of the cations (see Figure 5). Large chemical shift differences between these species were evidenced suggesting that the geometry of the tetrads are somewhat different. Dingley et al. have evidenced by studying scalar couplings that $(C2)NH_2 \cdots N7$ H-bonds are shorter in the presence of Na^+ than K^+, in agreement with the crystal structures (2.62 Å *versus* 2.78 Å), likely because Na^+ residing in the center of the tetrads holds the guanines more tightly than K^+ [107]. The similar sequence $d(G_3T_4G_3)$ also folds into a bimolecular antiparallel structure with diagonal loops in both alkali cations [316, 317]. However, the loops are more flexible with K^+, where they adopt two possible and interconverting conformations, and the K^+-stabilized tetrads are slightly larger than with Na^+ [317]. Moreover, long incubation times after addition of K^+ on a Na^+ sample result in the likely formation of a tetramolecular G-quadruplex, while Na^+ alone does not facilitate this association.

Plavec et al. have shown that another related sequence, $d(G_4T_4G_3)$, folds into a Na^+-containing G-quadruplex similar to the above-mentioned oligonucleotide, but adopts a mixture of structures with K^+ [318], whereas $d(G_3T_4G_4)$ and $d(G_4T_4G_4)$ adopt the same bimolecular fold with both cations, evidencing that small and targeted changes can have a dramatic effect on both the G-quadruplex topology and sensitivity to cation nature [319]. Two groups have independently demonstrated by NMR that the four-guanine tracts sequence $d[G_4(T_4G_4)_3]$ ('Oxy-3.5') adopts an antiparallel topology in Na^+ (PDB ID: 201D and 230D) [175, 320]. Upon addition of K^+ cations, significant changes in the spectra were monitored, without apparent change in the global topology, as with Oxy-1.5 [320]. Single molecule FRET spectroscopy gave consistent results with Na^+ that promote an antiparallel G-quadruplex [243]. However, K^+ leads to a mixture of a parallel and an antiparallel structure, the antiparallel differing from the Na^+ fold. These different topologies were also observed with longer sequences $d(T_4G_4)_{n\geq 4}$ by native PAGE and UV-melting [119]. Potassium seems to lead to the formation of a mixture of parallel and antiparallel intra- and intermolecular G-quadruplexes, whereas Na^+ promotes the formation of antiparallel G-quadruplexes only, either intramolecular (*n*=4–7) or in a mixtures of intra- and intermolecular assemblies (*n*=8–12). In the same vein, Thomas et al. published a phase diagram for Na^+ and K^+ of the sequence $d(T_4G_4)_4$ ('Oxy-4') [244]. At low concentrations, both cations promote the formation of an intramolecular antiparallel G-quadruplex. However, at higher concentrations, K^+ was more effective at stimulating the formation of tetramolecular assemblies. The midpoints of conversion, estimated from the Raman spectra, are 65 mM and 225 mM for K^+ and Na^+, respectively.

More generally, a number of other telomeric sequences have a distinct CD signature in Na^+ *versus* K^+ solutions, including *Arabidopsis*, *L. esculentum*, *C. glabrata*, *S. cerevisiae*, and *Paramecium* [114]. These sequences have G_3 or G_4 repeats, and give antiparallel signatures in Na^+, but not in K^+ conditions, where hybrid structures or mixtures are likely. Conversely, *Bombyx* and *Ascaris* have an antiparallel signature in Na^+ and K^+. Both have repeats of two guanines only, and therefore contain most likely two tetrads, which was confirmed by NMR for *Bombyx mori* in K^+ conditions [68]. Guédin et al. examined eighty sequences containing four tracts of three guanines with loops of variable length (between 1 and 15 bases), following the template $G_3L_aG_3L_bG_3L_cG_3$, where L_a, L_b, and L_c are thymines or TTA loops, in the presence of 100 mM K^+ or Na^+ [115]. In K^+, all short-looped sequences (two one-nucleotide loops) adopt a parallel topology, while longer loops (notably with two three-nucleotide loops) seem to promote hybrid structures, which is reminiscent of the human telomeric sequence. Even in the presence of long loops (up to nine nucleotides) that allow a certain flexibility, none of the sequences display a clear antiparallel signature. Conversely, some G-quadruplexes have an antiparallel signature in the presence of Na^+ ($d(G_3T_2G_3T_3G_3T_3G_3)$, $d(G_3T_1G_3T_3G_3T_3G_3)$, $d(G_3T_4G_3T_4G_3T_4G_3)$), although short loops also promote hybrid folds.

Recently, Hartig and coworkers have shown by CD and NMR that the sequence $d[(G_4CT)_3G_4]$, from the human pathogen *Treponema pallidum*, exhibits a remarkable cation dependency [222]. Low K^+ concentrations promote a 4-tetrad intramolecular antiparallel structure, while Na^+ and Li^+ do not markedly induce G4 folding, even at high concentrations. The corresponding G_5 motif also adopts an antiparallel signature, but not the G_3 counterpart that favors a parallel conformation, as does the absence of thymine in the loops. Moreover, CD, AUC and EPR experiments demonstrate that higher K^+ concentrations (≥500 mM) lead to the formation of a parallel tetramer, which was confirmed by SE-HPLC shortly thereafter [37]. Interestingly, low concentrations of K^+ supplemented by Na^+ of Li^+ to increase the ionic strength is sufficient to trigger the formation of the tetramer.

Russo Krauss and coworkers have obtained high resolution crystal structures of TBA ($d(G_2T_2G_2TGTG_2T_2G_2)$) bound to α-thrombin in the presence of Na^+ or K^+ (PDB ID: 4DIH and 4DII) (Figure 8) [161]. The difference in G-quadruplex stability in favor of the latter cation is well known [89, 98, 111], as is the antiparallel 'chair-type' structure observed with both alkali cations.

However, this work highlighted subtle cation-induced differences in the structure that may explain differences in binding mode and potency, and as a possible

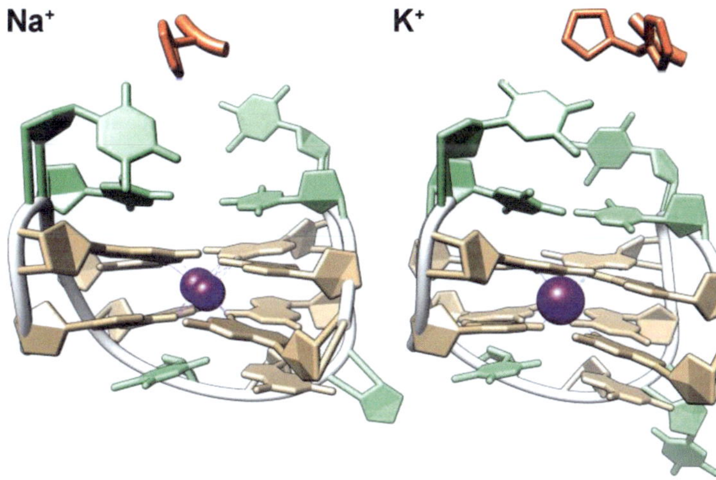

Figure 8 Crystal structure of TBA coordinating Na^+ (**left**; purple spheres: two positions; PDB ID: 4DIH) or K^+ (**right**; purple sphere: one position; PDB ID: 4DII), bound to α-thrombin (not shown except for His71, in orange) [161]. Guanosines are depicted in brown, thymidines in green, and the phosphate backbones as white ribbons.

consequence the enhanced clotting inhibitory activity of the aptamer in the presence of potassium. K^+ is coordinated at the center of the cavity between the two G-tetrads and bridges together all the eight guanine O6 atoms in a distorted anti-prism geometry ($d_{O\text{-}K}$ = 2.7–2.9 Å), thereby increasing the rigidity and the stability of the G-quadruplex. On the other hand, Na^+ can occupy two alternative coordination sites, closer to one of the two tetrads ($d_{O\text{-}K}$ = 2.2–2.7 Å), therefore conferring a higher plasticity to the aptamer that allows a better fit with the binding surface of α-thrombin. Accordingly, the ΔT_m between K^+- and Na^+-coordinating TBA decreases from 29 (free) to 12 °C (bound to the protein). Another consequence is the absence of an interaction of TBA•Na^+ with the His71 residue, crucial for the inhibition of the fibrinogen conversion to fibrin by a-thrombin, while this interaction is visible for TBA•K^+.

4.3 General Trends

Very clearly, the nature of the cation impacts the structure of DNA G-quadruplexes, in terms of topology, loop geometry, and strand stoichiometry. This does not necessarily translates into a different topology, and has sometimes more subtle effects (e.g., cation location, loop flexibility, quartet size), but even these small modifications can lead to significant outcomes [161]. Other factors influence the structure, mainly the sequence (i.e., number of guanines, of repeats, length and composition of loops), the strand concentration (oligomerization), and the presence of co-solvents. In particular, small sequence alterations can dramatically change the structure(s) formed and, more importantly within the scope of this chapter, the cation-dependency of G-quadruplexes (e.g., $d(G_4T_4G_3)$ *versus* $d(G_3T_4G_4)$ and $d(G_4T_4G_4)$ [318]).

From all the studies performed on a wide range of sequences, it seems that potassium does not promote the formation of intramolecular antiparallel structures of four-repeat G_3 motif sequences, whereas sodium can. Conversely, four-repeat 2-tetrad G-quadruplexes seem to adopt preferentially antiparallel fold in the presence of K^+ (e.g. human telomeres [31, 32], TBA [24], *Bombyx mori* and *Ascaris* telomeres [68, 114], HIV-PRO1 [69], 21CTA [45]). Sequences with G_4 and longer motifs are also able to adopt antiparallel conformations in K^+ conditions. Moreover, K^+ promotes more efficiently oligomerization and the formation of tetramolecular assemblies than Na^+ [39, 222], although (i) oligomerization in Na^+ solutions has been observed [245], and (ii) it should be noted that a number of observations have been made with high-strand-concentration samples, and may not be reflected in dilute conditions. Finally, G-rich RNA sequences fold in parallel G-quadruplexes regardless of the cation, although differences have been noticed regarding their propensity to oligomerize.

5 Cation-Dependent Conformational Switching

The first example of a cation-dependent structural switch of G-quadruplexes was reported in 1990 by Sen and Gilbert [315]. Although the authors based their conclusions on native PAGE experiments only, and discarded intramolecular G-quadruplexes as intermediate structures or by-products, it clearly shows that sodium and potassium can dramatically affect the structure(s) adopted by guanine-rich oligonucleotides, as well as their rate of formation. In 1992, Hardin et al. provided a nice example of cation-dependent hairpin-to-quadruplex conversion, which led to ranking of cations as a function of their G-quadruplex stabilization properties: $\mathbf{K^+} > Ca^{2+} > \mathbf{Na^+} > Mg^{2+} > \mathbf{Li^+}$, and $\mathbf{K^+} > \mathbf{Rb^+} > \mathbf{Cs^+}$ [321].

We have seen in Sections 3 and 4 that the nature and concentration of the cation(s) has a large effect on the stability and structure of G-quadruplex-forming sequences. It is therefore not surprising to find a number of examples of cation-triggered conformational switch from a less stable structure coordinating a cation to a more stable structure coordinating another cation. Typically, the switch is triggered by adding a more stabilizing cation, hence often by adding K^+ to a Na^+-containing solution.

The telomeric sequence is well known for its particularly pronounced polymorphism in the presence of Na^+ or K^+ (Section 4.1), so the conversion from the sodium to the potassium form(s) has been extensively studied. In 2006, Yang et al. proposed that the conversion from the Na^+-basket form to the K^+-hybrid form of human telomeric sequences involves the cation exchange, followed by a partial unfolding and restructuration [165]. Partial unfolding possibly leads to triplex intermediates, as also suggested by Sugiyama et al. and Chaires et al. [282, 285–287]. Gray et al. have shown that the (30 mM) Na^+ to (50 mM) K^+ exchange of the 22-mer d[$AG_3(G_3T_2A)_3$] is followed by a conformation change, from the Na^+-basket type to the K^+-hybrid (see Figure 6C) [288]. Three distinct kinetic processes are involved: (i) A fast cation exchange ($\tau_1 \sim 250$ μs) yields a K^+-coordinating basket G-quadruplex 'I1'; (ii) partial unfolding giving two possible triplex intermediates 'I2' (opening in 5′ or 3′; $\tau_2 \sim 50$ s); and (iii) a slower ($\tau_3 \sim 800$ s) refolding that converts a diagonal loop to a lateral loop, yielding hybrid-1 and hybrid-2 K^+-stabilized G-quadruplexes (opening in 5′ gives the hybrid-1, in 3′ the hybrid-2). The free energy barrier between the starting and final structures is relatively modest (1.4–2.4 kcal.mol^{-1}). These results were reassessed recently under the light of new folding pathway data, with $\tau_2 > \tau_3$ [285]. The presence of a folding intermediate (presumably a triplex) was also suggested in Na^+ solution for the 21-mer sequence d[$G_3(T_2AG_3)_3$] [322].

Worth mentioning, Chang et al. have argued that the Na^+ to K^+ exchange of the 23-mer sequence d[$TAG_3(T_2AG_3)_3$] does not involve a triplex intermediate, nor any unfolding, since the sequence adopts the same hybrid-1 topology in both K^+ and Na^+ solutions (see Figure 6B) [171]. They postulate that either slightly different human telomeric sequences adopt different exchange mechanism or that other studies have not satisfyingly characterized the initial and final states. Besides cation nature, a change in cation concentration can also trigger structural changes. Hartig et al. have shown that the sequence d[$(G_4CT)_3G_4$], from the human pathogen *Treponema pallidum* promotes a 4-tetrad intramolecular antiparallel structure at low K^+ concentrations, while Na^+ and Li^+ does not markedly induce G4 folding, even at high concentrations [222].

Interestingly, CD, AUC and EPR experiments demonstrated that higher K^+ concentrations (≥500 mM) lead to the formation of a parallel tetramolecular assembly, which was also observed by SE-HPLC in an independent study [37]. Furthermore, low concentrations of K^+ supplemented by Na^+ or Li^+ to increase the ionic strength is sufficient to trigger the formation of the tetramer, possibly to screen the negatively charged phosphate repulsions upon strand association although the authors did not comment on that.

Abu-Ghazalah and coworkers have observed the structural conversion of human telomeric sequences d[$G_3(T_2AG_3)_3$] ('basket'), d[$A_3G_3(T_2AG_3)_3A_2$] ('hybrid-1'), and d[$T_2AG_3(T_2AG_3)_3T_2$] ('hybrid-2') at high strand concentrations (2 mM) [245]. CD suggested the formation of parallel tetramolecular aggregates for the former sequence in 100 mM Na^+ solutions, with a relaxation time of around 10 h, which can re-dissociate rapidly upon dilution. The conversion takes place regardless of the Na^+ concentration, however, the rate of conversion increases with increasing strand and Na^+ concentrations. The two latter hybrid-type sequences can also aggregate, albeit only at high salt concentrations (1 M K^+).

6 Concluding Remarks and Future Directions

G-quadruplexes' relevance to life sciences was revealed in the late 1980s with seminal works on telomeric sequences [81, 82, 315, 323]. The role of monovalent cations was reviewed as early as 1991 [301], but an increasing number of other sequences have been investigated since. Bioinformatics studies suggest the presence of a very large number of putative G-quadruplex-forming sequences in the human genome only, ranging from 370 000 [4–6] to more than a million (Bedrat et al., unpublished results). Compared to this very large sequence space, only a few sequences have been explored in depth, but several (six) very different topologies have been observed thus far [22, 23]. Many studies have dealt with the influence of alkali cation on G-quadruplex structures, as can be seen from the large range of publications cited in this chapter. A marked difference of stability between K^+, Na^+, and Li^+ is indeed widely accepted as a good indication that G-quadruplex structures are involved.

Quadruplexes are often considered to operate as allosteric switches, implicated notably in gene regulation processes [7–15]. Hence, the interaction of intracellular potassium and sodium with G-rich sequences might play an important role in the regulation of biological processes. As a result, the importance of alkali metal coordination, most notably of sodium and potassium, on the stability and topology of structures formed by G-quadruplex-prone sequences, was intensively investigated, but there is a clear bias in favor of potassium because of their prevalence in cells. These studies were also prompted for other purposes such as artificial switch elements for DNA-based nanodevices, but to a far lower extent.

Comparison of Na^+ and K^+ in terms of G-quadruplex stabilization and structures yielded fairly consistent results. Potassium stabilizes G-quadruplexes better than sodium and the reasons are well understood. Cation coordination is the driving force towards G-quadruplex folding or interconversion, in accordance with the unfolded state of guanine-rich sequences in the absence of a suitable cation [62, 285]. However, the question as to why a sequence folds into one or several structures in particular, in the presence of a given cation, remains largely unanswered. Answering this question is difficult because the structure depends not only on the nature of the cation, but also on cation concentration, strand concentration, and temperature. Some basic trends are very well known: tetramolecular G-quadruplexes are parallel, RNA G-quadruplexes are parallel, regardless of the cation nature. However, when it comes to the folding of intramolecular DNA G-quadruplexes, which are more relevant for genomic studies, venturing an educated guess seems very risky. In fact, even the structures of some heavily studied sequences such as the human telomeric sequence in potassium solutions are still controversial. Numerous studies clearly show that minor sequence alterations lead to large structural effects, and the human telomeric sequence is the epitome of this phenomenon [172, 203].

Similarly, the pathways of folding/unfolding and interconversion of G-quadruplexes are still a matter of debate. No clear view of cation effects emerged yet, except that potassium can swiftly displace sodium, whether it is accompanied by structural alterations or not. G-quadruplexes are often presented as biological switches within the genome because of their ability to be either folded or unfolded (and in the latter case, hybridized in a canonical duplex). Moreover, even small differences in the structure of G-quadruplexes due to the binding of K^+ or Na^+ can have an effect on protein binding [161]. In this context, studies mimicking the cellular environment are needed (crowding agents, proteins, complex mixtures of cations), complemented by the use of appropriate oligonucleotides (long telomeric sequences, G-quadruplexes embedded in duplex matrices). Recent studies have started to tackle these issues [169, 242, 314, 324], but the complexity of these experiments makes it difficult to assess the influence of the cations in cell-like environments. The elucidation alkali cation effects on G-quadruplex nucleic acids structure and stability is a difficult task, yet an important milestone towards the prediction of structure from the sequence and environment, and towards the design of stimuli-responsive artificial DNA switches.

Abbreviations

AUC	analytical ultracentrifugation
CD	circular dichroism
DFT-D	dispersion-corrected density functional theory
DOSY	diffusion ordered spectroscopy
DSC	differential scanning calorimetry
EPR	electron paramagnetic resonance
ESI-MS	electrospray ionization mass spectrometry
FRET	Förster resonance energy transfer
G4	quadruplex nucleic acid
GMP	guanosine 5′-monophosphate
HSQC	heteronuclear single-quantum correlation spectroscopy
IDS	isothermal difference spectra
IMS-MS	ion-mobility spectrometry mass spectrometry
ITC	isothermal titration calorimetry
LNA	locked nucleic acid
MALDI-TOF	matrix-assisted laser desorption/ionization time-of-flight mass spectrometry
MAS	magic-angle spinning
MD	molecular dynamics
MS	mass spectrometry
NMR	nuclear magnetic resonance
NOE	nuclear Overhauser effect
nt	nucleotide
PAGE	polyacrylamide gel electrophoresis
PBFI	benzofuran-isophthalate crown ether indicator

PDB	protein data bank
PNA	peptide nucleic acid
QM	quantum mechanics
RMSD	root-mean-square deviation
SE-HPLC	size-exclusion high-performance liquid chromatography
STM	scanning tunneling microscope
TBA	thrombin binding aptamer
TDS	thermal difference spectra
TERRA	telomeric repeat-containing RNA
TMAA	trimethyl ammonium acetate
UTR	untranslated region

Acknowledgment Pictures of G-quadruplexes from the PBD entries were generated with UCSF Chimera (alpha version 1.11) [325]. Funding was provided by *Agence Nationale de la Recherche* (OligoSwitch [ANR-12-IS07–0001], 'Quarpdiem' [ANR-12-BSV8–0008–01], and 'VIBBnano' [ANR-10-NANO-04–03]).

References

1. J. T. Davis, *Angew. Chem. Int. Ed. Engl.* **2004**, *43*, 668–698.
2. *Quadruplex Nucleic Acids*, Eds S. Neidle, S. Balasubramanian, Royal Society of Chemistry, Cambridge, UK, 2006.
3. *Quadruplex Nucleic Acids*, Eds J. B. Chaires, D. Graves, Springer, Berlin, 2013.
4. J. L. Huppert, S. Balasubramanian, *Nucleic Acids Res.* **2005**, *33*, 2908–2916.
5. A. K. Todd, *Nucleic Acids Res.* **2005**, *33*, 2901–2907.
6. J. L. Huppert, S. Balasubramanian, *Nucleic Acids Res.* **2007**, *35*, 406–413.
7. G. Biffi, M. Di Antonio, D. Tannahill, S. Balasubramanian, *Nature Chem.* **2013**, *6*, 75–80.
8. G. Biffi, D. Tannahill, J. McCafferty, S. Balasubramanian, *Nature Chem.* **2013**, *5*, 182–186.
9. E. Y. N. Lam, D. Beraldi, D. Tannahill, S. Balasubramanian, *Nat. Commun.* **2013**, *4*, 1796–1796.
10. J.-L. Mergny, *Nat. Chem. Biol.* **2012**, *8*, 225–226.
11. S. Millevoi, H. Moine, S. Vagner, *WIREs RNA* **2012**, *3*, 495–507.
12. H. J. Lipps, D. Rhodes, *Trends Cell Biol.* **2009**, *19*, 414–422.
13. L. T. Gray, A. C. Vallur, J. Eddy, N. Maizels, *Nat. Chem. Biol.* **2014**, *10*, 313–318.
14. A. Baral, P. Kumar, R. Pathak, S. Chowdhury, *Mol. BioSyst.* **2013**, *9*, 1568–1575.
15. M. Métifiot, S. Amrane, S. Litvak, M. L. Andreola, *Nucleic Acids Res.* **2014**, *42*, 12352–12366.
16. S. Balasubramanian, S. Neidle, *Curr. Opin. Chem. Biol.* **2009**, *13*, 345–353.
17. T. M. Ou, Y. J. Lu, J. H. Tan, Z. S. Huang, K. Y. Wong, L. Q. Gu, *ChemMedChem* **2008**, *3*, 690–713.
18. S. A. Ohnmacht, S. Neidle, *Bioorg. Med. Chem. Lett.* **2014**, *24*, 2602–2612.
19. G. N. Parkinson, in *Guanine Quartets: Structure and Application*, Eds L. Spindler, W. Fritzsche, Royal Society of Chemistry, Cambridge, UK, 2012, pp. 237–247.
20. D. J. Patel, A. T. Phan, V. Kuryavyi, *Nucleic Acids Res.* **2007**, *35*, 7429–7455.
21. >8000 publications include the word quadruplex or tetraplex in their title, as of January 2015, according to Thomson Reuters' Web of Science, with more than 700 for 2014 only.
22. M. Webba da Silva, *Chem. Eur. J.* **2007**, *13*, 9738–9745.
23. A. I. Karsisiotis, C. O'Kane, M. Webba da Silva, *Methods* **2013**, *64*, 28–35.

24. P. Schultze, R. F. Macaya, J. Feigon, *J. Mol. Biol.* **1994**, *235*, 1532–1547.
25. Y. Wang, D. J. Patel, *Structure* **1993**, *1*, 263–282.
26. G. N. Parkinson, M. P. H. Lee, S. Neidle, *Nature* **2002**, *417*, 876–880.
27. K. N. Luu, A. T. Phan, V. Kuryavyi, L. Lacroix, D. J. Patel, *J. Am. Chem. Soc.* **2006**, *128*, 9963–9970.
28. J. Dai, C. Punchihewa, A. Ambrus, D. Chen, R. A. Jones, D. Yang, *Nucleic Acids Res.* **2007**, *35*, 2440–2450.
29. A. T. Phan, V. Kuryavyi, K. N. Luu, D. J. Patel, *Nucleic Acids Res.* **2007**, *35*, 6517–6525.
30. J. Dai, M. Carver, C. Punchihewa, R. A. Jones, D. Yang, *Nucleic Acids Res.* **2007**, *35*, 4927–4940.
31. K. W. Lim, S. Amrane, S. Bouaziz, W. Xu, Y. Mu, D. J. Patel, K. N. Luu, A. T. Phan, *J. Am. Chem. Soc.* **2009**, *131*, 4301–4309.
32. Z. Zhang, J. Dai, E. Veliath, R. A. Jones, D. Yang, *Nucleic Acids Res.* **2010**, *38*, 1009–1021.
33. J. Zhou, K. Murayama, S. Amrane, F. Rosu, H. Kashida, A. Bourdoncle, H. Asanuma, J.-L. Mergny, *Chem. Sci.* **2013**, *4*, 3693–3693.
34. V. Esposito, A. Virgilio, A. Randazzo, A. Galeone, L. Mayol, *Chem. Commun.* **2005**, 3953–3955.
35. V. Esposito, A. Virgilio, A. Pepe, G. Oliviero, L. Mayol, A. Galeone, *Bioorg. Med. Chem.* **2009**, *17*, 1997–2001.
36. M. Webba da Silva, *Methods* **2007**, *43*, 264–277.
37. E. Largy, J. L. Mergny, *Nucleic Acids Res.* **2014**, *42*, e149.
38. P. Tothova, P. Krafcikova, V. Viglasky, *Biochemistry* **2014**, *53*, 7013–7027.
39. N. Smargiasso, F. Rosu, W. Hsia, P. Colson, E. S. Baker, M. T. Bowers, E. De Pauw, V. Gabelica, *J. Am. Chem. Soc.* **2008**, *130*, 10208–10216.
40. J. Zhou, A. Bourdoncle, F. Rosu, V. Gabelica, J.-L. Mergny, *Angew. Chem. Int. Ed. Engl.* **2012**, *51*, 11002–11005.
41. V. T. Mukundan, A. T. Phan, *J. Am. Chem. Soc.* **2013**, *135*, 5017–5028.
42. H. Martadinata, A. T. Phan, *Biochemistry* **2014**, *53*, 1595–1600.
43. M. Adrian, D. J. Ang, C. J. Lech, B. Heddi, A. Nicolas, A. T. Phan, *J. Am. Chem. Soc.* **2014**, *136*, 6297–6305.
44. G. D. Balkwill, T. P. Garner, H. E. Williams, M. S. Searle, *J. Mol. Biol.* **2009**, *385*, 1600–1615.
45. K. W. Lim, P. Alberti, A. Guédin, L. Lacroix, J.-F. Riou, N. J. Royle, J.-L. Mergny, A. T. Phan, *Nucleic Acids Res.* **2009**, *37*, 6239–6248.
46. Y. Wang, D. J. Patel, *Structure* **1994**, *2*, 1141–1156.
47. N. Zhang, A. Gorin, A. Majumdar, A. Kettani, N. Chernichenko, E. Skripkin, D. J. Patel, *J. Mol. Biol.* **2001**, *311*, 1063–1079.
48. A. Kettani, A. Gorin, A. Majumdar, T. Hermann, E. Skripkin, H. Zhao, R. Jones, D. J. Patel, *J. Mol. Biol.* **2000**, *297*, 627–644.
49. T. Mashima, A. Matsugami, F. Nishikawa, S. Nishikawa, M. Katahira, *Nucleic Acids Res.* **2009**, *37*, 6249–6258.
50. H. Martadinata, A. T. Phan, *Biochemistry* **2013**, *52*, 2176–2183.
51. A. Matsugami, K. Ouhashi, M. Kanagawa, H. Liu, S. Kanagawa, S. Uesugi, M. Katahira, *J. Mol. Biol.* **2001**, *313*, 255–269.
52. N. Borbone, J. Amato, G. Oliviero, V. D'Atri, V. Gabelica, E. De Pauw, G. Piccialli, L. Mayol, *Nucleic Acids Res.* **2011**, *39*, 7848–7857.
53. A. T. Phan, V. Kuryavyi, S. Burge, S. Neidle, D. J. Patel, *J. Am. Chem. Soc.* **2007**, *129*, 4386–4392.
54. M. A. Keniry, *Biopolymers* **2001**, *56*, 123–146.
55. D. E. Gilbert, J. Feigon, *Curr. Opin. Struct. Biol.* **1999**, *9*, 305–314.
56. S. Burge, G. N. Parkinson, P. Hazel, A. K. Todd, S. Neidle, *Nucleic Acids Res.* **2006**, *34*, 5402–5415.
57. J. L. Huppert, *Philos. Trans. R. Soc., A* **2007**, *365*, 2969–2984.
58. J. L. Huppert, *Chem. Soc. Rev.* **2008**, *37*, 1375–1384.

59. *Guanine Quartets: Structure and Application*, Eds W. Fritzsche, L. Spindler, Royal Society of Chemistry, Cambridge, 2012.
60. P. Alberti, A. Bourdoncle, B. Saccà, L. Lacroix, J.-L. Mergny, *Org. Biomol. Chem.* **2006**, *4*, 3383–3391.
61. C. M. Olsen, W. H. Gmeiner, L. A. Marky, *J. Phys. Chem. B* **2006**, *110*, 6962–6969.
62. A. N. Lane, J. B. Chaires, R. D. Gray, J. O. Trent, *Nucleic Acids Res.* **2008**, *36*, 5482–5515.
63. X. Cang, J. Šponer, T. E. Cheatham, *Nucleic Acids Res.* **2011**, *39*, 4499–4512.
64. J. Šponer, A. Mládek, N. Spačková, X. Cang, T. E. Cheatham, S. Grimme, *J. Am. Chem. Soc.* **2013**, *135*, 9785–9796.
65. C. J. Lech, B. Heddi, A. T. Phan, *Nucleic Acids Res.* **2013**, *41*, 2034–2046.
66. D. Zhang, T. Huang, P. S. Lukeman, P. J. Paukstelis, *Nucleic Acids Res.* **2014**, *42*, 13422–13429.
67. G. N. Parkinson, R. Ghosh, S. Neidle, *Biochemistry* **2007**, *46*, 2390–2397.
68. S. Amrane, R. W. L. Ang, Z. M. Tan, C. Li, J. K. C. Lim, J. M. W. Lim, K. W. Lim, A. T. Phan, *Nucleic Acids Res.* **2009**, *37*, 931–938.
69. S. Amrane, A. Kerkour, A. Bedrat, B. Vialet, M.-L. Andréola, J.-L. Mergny, *J. Am. Chem. Soc.* **2014**, *136*, 5249–5252.
70. V. Kuryavyi, A. T. Phan, D. J. Patel, *Nucleic Acids Res.* **2010**, *38*, 6757–6773.
71. R. Otero, M. Schöck, L. M. Molina, E. Laegsgaard, I. Stensgaard, B. Hammer, F. Besenbacher, *Angew. Chem. Int. Ed. Engl.* **2005**, *44*, 2270–2275.
72. C. Fonseca Guerra, H. Zijlstra, G. Paragi, F. M. Bickelhaupt, *Chem. Eur. J.* **2011**, *17*, 12612–12622.
73. T. J. Pinnavaia, C. L. Marshall, C. M. Mettler, C. L. Fisk, H. T. Miles, E. D. Becker, *J. Am. Chem. Soc.* **1978**, *100*, 3625–3627.
74. H. K. Frensdorff, *J. Am. Chem. Soc.* **1971**, *93*, 600–606.
75. A. Delville, C. Detellier, P. Laszlo, *J. Magn. Reson. (1969–1992)* **1979**, *34*, 301–315.
76. E. Bouhoutsos-Brown, C. L. Marshall, T. J. Pinnavaia, *J. Am. Chem. Soc.* **1982**, *104*, 6576–6584.
77. S. B. Zimmerman, G. H. Cohen, D. R. Davies, *J. Mol. Biol.* **1975**, *92*, 181–192.
78. H. T. Miles, J. Frazier, *J. Am. Chem. Soc.* **1978**, *100*, 8037–8038.
79. F. B. Howard, H. T. Miles, *Biopolymers* **1982**, *21*, 147–157.
80. F. B. Howard, H. T. Miles, *Biochemistry* **1982**, *21*, 6736–6745.
81. E. Henderson, C. C. Hardin, S. K. Walk, I. Tinoco, E. H. Blackburn, *Cell* **1987**, *51*, 899–908.
82. J. R. Williamson, M. K. Raghuraman, T. R. Cech, *Cell* **1989**, *59*, 871–880.
83. E. A. Venczel, D. Sen, *Biochemistry* **1993**, *32*, 6220–6228.
84. B. I. Kankia, L. A. Marky, *J. Am. Chem. Soc.* **2001**, *123*, 10799–10804.
85. A. Włodarczyk, P. Grzybowski, A. Patkowski, A. Dobek, *J. Phys. Chem. B* **2005**, *109*, 3594–3605.
86. J. R. Williamson, *Annu. Rev. Biophys. Biomol. Struct.* **1994**, *23*, 703–730.
87. W. S. Ross, C. C. Hardin, *J. Am. Chem. Soc.* **1994**, *116*, 6070–6080.
88. N. V. Hud, F. W. Smith, F. A. Anet, J. Feigon, *Biochemistry* **1996**, *35*, 15383–15390.
89. B. Saccà, L. Lacroix, J. L. Mergny, *Nucleic Acids Res.* **2005**, *33*, 1182–1192.
90. J. C. Bowman, T. K. Lenz, N. V. Hud, L. D. Williams, *Curr. Opin. Struct. Biol.* **2012**, *22*, 262–272.
91. R. D. Gray, J. B. Chaires, *Biophys. Chem.* **2011**, *159*, 205–209.
92. E. Galezowska, A. Gluszynska, B. Juskowiak, *J. Inorg. Biochem.* **2007**, *101*, 678–685.
93. W. Xu, Q. Tan, M. Yu, Q. Sun, H. Kong, E. Lægsgaard, I. Stensgaard, J. Kjems, J.-G. Wang, C. Wang, F. Besenbacher, *Chem. Commun.* **2013**, *49*, 7210–7212.
94. A. Risitano, K. R. Fox, *Biochemistry* **2003**, *42*, 6507–6513.
95. R. D. Gray, J. B. Chaires, *Nucleic Acids Res.* **2008**, *36*, 4191–4203.
96. P. Schultze, N. V. Hud, F. W. Smith, J. Feigon, *Nucleic Acids Res.* **1999**, *27*, 3018–3028.
97. N. V. Hud, P. Schultze, V. Sklenár, J. Feigon, *J. Mol. Biol.* **1999**, *285*, 233–243.
98. N. Kumar, S. Maiti, *Biochem. Biophys. Res. Commun.* **2004**, *319*, 759–767.

99. J. M. Wilcox, D. L. Rempel, M. L. Gross, *Anal. Chem.* **2008**, *80*, 2365–2371.
100. T. L. Niederhauser, B. R. Brown, S. P. Ziemer, J. D. Sargent, E. M. Woolley, *J. Chem. Thermodyn.* **2004**, *36*, 1067–1077.
101. Y. Wu, M. Tabata, *J. Solution Chem.* **2004**, *33*, 777–795.
102. S. P. Ziemer, T. L. Niederhauser, E. M. Woolley, *J. Chem. Thermodyn.* **2005**, *37*, 1071–1084.
103. A. Risitano, K. R. Fox, *Bioorg. Med. Chem. Lett* **2005**, *15*, 2047–2050.
104. J. A. Walmsley, T. J. Pinnavaia, *Biophys. J.* **1982**, *38*, 315–318.
105. J. Gu, J. Leszczynski, *J. Phys. Chem. A* **2002**, *106*, 529–532.
106. V. K. Misra, D. E. Draper, *Biopolymers* **1998**, *48*, 113–135.
107. A. J. Dingley, R. D. Peterson, S. Grzesiek, J. Feigon, *J. Am. Chem. Soc.* **2005**, *127*, 14466–14472.
108. M. Meyer, A. Hocquet, J. Sühnel, *J. Comput. Chem.* **2005**, *26*, 352–364.
109. P. Balagurumoorthy, S. K. Brahmachari, D. Mohanty, M. Bansal, V. Sasisekharan, *Nucleic Acids Res.* **1992**, *20*, 4061–4067.
110. P. Balagurumoorthy, S. K. Brahmachari, *J. Biol. Chem.* **1994**, *269*, 21858–21869.
111. J. L. Mergny, A. T. Phan, L. Lacroix, *FEBS Lett.* **1998**, *435*, 74–78.
112. L. Petraccone, E. Erra, V. Esposito, A. Randazzo, L. Mayol, L. Nasti, G. Barone, C. Giancola, *Biochemistry* **2004**, *43*, 4877–4884.
113. J. L. Mergny, J. Li, L. Lacroix, S. Amrane, J. B. Chaires, *Nucleic Acids Res.* **2005**, *33*, e138.
114. P. L. T. Tran, J. L. Mergny, P. Alberti, *Nucleic Acids Res.* **2011**, *39*, 3282–3294.
115. A. Guédin, J. Gros, P. Alberti, J. L. Mergny, *Nucleic Acids Res.* **2010**, *38*, 7858–7868.
116. A. Guédin, P. Alberti, J. L. Mergny, *Nucleic Acids Res.* **2009**, *37*, 5559–5567.
117. J. L. Mergny, A. De Cian, A. Ghelab, B. Saccà, L. Lacroix, *Nucleic Acids Res.* **2005**, *33*, 81–94.
118. K. W. Lim, L. Lacroix, D. J. Yue, J. K. Lim, J. M. Lim, A. T. Phan, *J. Am. Chem. Soc.* **2010**, *132*, 12331–12342.
119. M. Lu, Q. Guo, N. R. Kallenbach, *Biochemistry* **1992**, *31*, 2455–2459.
120. K. Phillips, Z. Dauter, A. I. Murchie, D. M. Lilley, B. Luisi, *J. Mol. Biol.* **1997**, *273*, 171–182.
121. D. Wei, A. K. Todd, M. Zloh, M. Gunaratnam, G. N. Parkinson, S. Neidle, *J. Am. Chem. Soc.* **2013**, *135*, 19319–19329.
122. C. Creze, B. Rinaldi, R. Haser, P. Bouvet, P. Gouet, *Acta Crystallogr., Sect. D, Biol. Crystallogr.* **2007**, *63*, 682–688.
123. G. R. Clark, P. D. Pytel, C. J. Squire, S. Neidle, *J. Am. Chem. Soc.* **2003**, *125*, 4066–4067.
124. G. R. Clark, P. D. Pytel, C. J. Squire, *Nucleic Acids Res.* **2012**, *40*, 5731–5738.
125. F. W. Smith, J. Feigon, *Nature* **1992**, *356*, 164–168.
126. P. Schultze, F. W. Smith, J. Feigon, *Structure* **1994**, *2*, 221–233.
127. M. P. Horvath, S. C. Schultz, *J. Mol. Biol.* **2001**, *310*, 367–377.
128. S. Haider, G. N. Parkinson, S. Neidle, *J. Mol. Biol.* **2002**, *320*, 189–200.
129. M. L. Gill, S. A. Strobel, J. P. Loria, *Nucleic Acids Res.* **2006**, *34*, 4506–4514.
130. S. Neidle, *Methods* **2012**, *57*, 1–2.
131. M. Adrian, B. Heddi, A. T. Phan, *Methods* **2012**, *57*, 11–24.
132. S. Neidle, G. N. Parkinson, *Biochimie* **2008**, *90*, 1184–1196.
133. N. H. Campbell, G. N. Parkinson, *Methods* **2007**, *43*, 252–263.
134. J. Feigon, K. M. Koshlap, F. W. Smith, *Methods Enzymol.* **1995**, *261*, 225–255.
135. G. Laughlan, A. I. Murchie, D. G. Norman, M. H. Moore, P. C. Moody, D. M. Lilley, B. Luisi, *Science* **1994**, *265*, 520–524.
136. B. Pan, Y. Xiong, K. Shi, J. Deng, M. Sundaralingam, *Structure* **2003**, *11*, 815–823.
137. B. Pan, Y. Xiong, K. Shi, M. Sundaralingam, *Structure* **2003**, *11*, 1423–1430.
138. B. Pan, Y. Xiong, K. Shi, M. Sundaralingam, *Structure* **2003**, *11*, 825–831.
139. B. Pan, K. Shi, M. Sundaralingam, *J. Mol. Biol.* **2006**, *363*, 451–459.
140. B. Pan, K. Shi, M. Sundaralingam, *Proc. Natl. Acad. Sci. USA* **2006**, *103*, 3130–3134.

141. M. P. H. Lee, G. N. Parkinson, P. Hazel, S. Neidle, *J. Am. Chem. Soc.* **2007**, *129*, 10106–10107.
142. C. Cáceres, G. Wright, C. Gouyette, G. Parkinson, J. A. Subirana, *Nucleic Acids Res.* **2004**, *32*, 1097–1102.
143. I. Russo Krauss, G. N. Parkinson, A. Merlino, C. A. Mattia, A. Randazzo, E. Novellino, L. Mazzarella, F. Sica, *Acta Crystallogr., Sect. D, Biol. Crystallogr.* **2014**, *70*, 362–370.
144. J. Kondo, W. Adachi, S. Umeda, T. Sunami, A. Takenaka, *Nucleic Acids Res.* **2004**, *32*, 2541–2549.
145. P. K. Mandal, G. W. Collie, B. Kauffmann, I. Huc, *Angew. Chem. Int. Ed. Engl.* **2014**, *53*, 14424–14427.
146. S. M. Haider, G. N. Parkinson, S. Neidle, *J. Mol. Biol.* **2003**, *326*, 117–125.
147. P. Hazel, G. N. Parkinson, S. Neidle, *J. Am. Chem. Soc.* **2006**, *128*, 5480–5487.
148. N. H. Campbell, G. N. Parkinson, A. P. Reszka, S. Neidle, *J. Am. Chem. Soc.* **2008**, *130*, 6722–6724.
149. N. H. Campbell, M. Patel, A. B. Tofa, R. Ghosh, G. N. Parkinson, S. Neidle, *Biochemistry* **2009**, *48*, 1675–1680.
150. G. W. Collie, S. M. Haider, S. Neidle, G. N. Parkinson, *Nucleic Acids Res.* **2010**, *38*, 5569–5580.
151. G. W. Collie, S. Sparapani, G. N. Parkinson, S. Neidle, *J. Am. Chem. Soc.* **2011**, *133*, 2721–2728.
152. N. H. Campbell, D. L. Smith, A. P. Reszka, S. Neidle, D. O'Hagan, *Org. Biomol. Chem.* **2011**, *9*, 1328–1331.
153. N. H. Campbell, N. H. A. Karim, G. N. Parkinson, M. Gunaratnam, V. Petrucci, A. K. Todd, R. Vilar, S. Neidle, *J. Med. Chem.* **2012**, *55*, 209–222.
154. G. W. Collie, R. Promontorio, S. M. Hampel, M. Micco, S. Neidle, G. N. Parkinson, *J. Am. Chem. Soc.* **2012**, *134*, 2723–2731.
155. J. M. Nicoludis, S. T. Miller, P. D. Jeffrey, S. P. Barrett, P. R. Rablen, T. J. Lawton, L. A. Yatsunyk, *J. Am. Chem. Soc.* **2012**, *134*, 20446–20456.
156. M. Micco, G. W. Collie, A. G. Dale, S. A. Ohnmacht, I. Pazitna, M. Gunaratnam, A. P. Reszka, S. Neidle, *J. Med. Chem.* **2013**, *56*, 2959–2974.
157. C. Bazzicalupi, M. Ferraroni, A. R. Bilia, F. Scheggi, P. Gratteri, *Nucleic Acids Res.* **2013**, *41*, 632–638.
158. G. N. Parkinson, F. Cuenca, S. Neidle, *J. Mol. Biol.* **2008**, *381*, 1145–1156.
159. D. Wei, G. N. Parkinson, A. P. Reszka, S. Neidle, *Nucleic Acids Res.* **2012**, *40*, 4691–4700.
160. D. Wei, J. Husby, S. Neidle, *Nucleic Acids Res.* **2015**, *43*, 629–644.
161. I. Russo Krauss, A. Merlino, A. Randazzo, E. Novellino, L. Mazzarella, F. Sica, *Nucleic Acids Res.* **2012**, *40*, 8119–8128.
162. S. M. Haider, S. Neidle, G. N. Parkinson, *Biochimie* **2011**, *93*, 1239–1251.
163. *Nucleic Acid-Metal Ion Interactions*, Ed N. V. Hud, Royal Society of Chemistry, Cambridge, 2008, pp. 433.
164. C. Bardin, J. L. Leroy, *Nucleic Acids Res.* **2008**, *36*, 477–488.
165. A. Ambrus, D. Chen, J. Dai, T. Bialis, R. A. Jones, D. Yang, *Nucleic Acids Res.* **2006**, *34*, 2723–2735.
166. A. Matsugami, Y. Xu, Y. Noguchi, H. Sugiyama, M. Katahira, *FEBS J.* **2007**, *274*, 3545–3556.
167. A. T. Phan, K. N. Luu, D. J. Patel, *Nucleic Acids Res.* **2006**, *34*, 5715–5719.
168. N. Zhang, A. T. Phan, D. J. Patel, *J. Am. Chem. Soc.* **2005**, *127*, 17277–17285.
169. D. J. Yue, K. W. Lim, A. T. Phan, *J. Am. Chem. Soc.* **2011**, *133*, 11462–11465.
170. A. T. Phan, D. J. Patel, *J. Am. Chem. Soc.* **2003**, *125*, 15021–15027.
171. Z. F. Wang, M. H. Li, S. T. Hsu, T. C. Chang, *Nucleic Acids Res.* **2014**, *42*, 4723–4733.
172. A. T. Phan, *FEBS J.* **2010**, *277*, 1107–1117.
173. Y. Xu, K. Kaminaga, M. Komiyama, *J. Am. Chem. Soc.* **2008**, *130*, 11179–11184.
174. H. Martadinata, A. T. Phan, *J. Am. Chem. Soc.* **2009**, *131*, 2570–2578.

175. Y. Wang, D. J. Patel, *J. Mol. Biol.* **1995**, *251*, 76–94.
176. A. T. Phan, Y. S. Modi, D. J. Patel, *J. Mol. Biol.* **2004**, *338*, 93–102.
177. L. Hu, K. W. Lim, S. Bouaziz, A. T. Phan, *J. Am. Chem. Soc.* **2009**, *131*, 16824–16831.
178. S. T. Hsu, P. Varnai, A. Bugaut, A. P. Reszka, S. Neidle, S. Balasubramanian, *J. Am. Chem. Soc.* **2009**, *131*, 13399–13409.
179. A. Ambrus, D. Chen, J. Dai, R. A. Jones, D. Yang, *Biochemistry* **2005**, *44*, 2048–2058.
180. J. Dai, D. Chen, R. A. Jones, L. H. Hurley, D. Yang, *Nucleic Acids Res.* **2006**, *34*, 5133–5144.
181. A. T. Phan, Y. S. Modi, D. J. Patel, *J. Am. Chem. Soc.* **2004**, *126*, 8710–8716.
182. A. T. Phan, V. Kuryavyi, H. Y. Gaw, D. J. Patel, *Nat. Chem. Biol.* **2005**, *1*, 167–173.
183. S. Amrane, M. Adrian, B. Heddi, A. Serero, A. Nicolas, J.-L. Mergny, A. T. Phan, *J. Am. Chem. Soc.* **2012**, *134*, 5807–5816.
184. N. Q. Do, K. W. Lim, M. H. Teo, B. Heddi, A. T. Phan, *Nucleic Acids Res.* **2011**, *39*, 9448–9457.
185. A. T. Phan, V. Kuryavyi, J.-B. Ma, A. Faure, M.-L. Andréola, D. J. Patel, *Proc. Natl. Acad. Sci. USA* **2005**, *102*, 634–639.
186. N. Q. Do, A. T. Phan, *Chem. Eur. J.* **2012**, *18*, 14752–14759.
187. D. J. Patel, A. E. Tonelli, *Biopolymers* **1974**, *13*, 1943–1964.
188. Y. Wang, D. J. Patel, *Biochemistry* **1992**, *31*, 8112–8119.
189. P. Sket, J. Plavec, *J. Am. Chem. Soc.* **2010**, *132*, 12724–12732.
190. R. Ida, G. Wu, *J. Am. Chem. Soc.* **2008**, *130*, 3590–3602.
191. A. T. Phan, D. J. Patel, *J. Am. Chem. Soc.* **2002**, *124*, 1160–1161.
192. B. Islam, M. Sgobba, C. Laughton, M. Orozco, J. Šponer, S. Neidle, S. Haider, *Nucleic Acids Res.* **2013**, *41*, 2723–2735.
193. R. Stefl, T. E. Cheatham, N. Spacková, E. Fadrná, I. Berger, J. Koca, J. Šponer, *Biophys. J.* **2003**, *85*, 1787–1804.
194. E. Fadrná, N. Spacková, R. Stefl, J. Koca, T. E. Cheatham, J. Šponer, *Biophys. J.* **2004**, *87*, 227–242.
195. A. Pérez, I. Marchán, D. Svozil, J. Šponer, T. E. Cheatham, C. A. Laughton, M. Orozco, *Biophys. J.* **2007**, *92*, 3817–3829.
196. J. Šponer, N. Špačková, *Methods* **2007**, *43*, 278–290.
197. F. Fogolari, H. Haridas, A. Corazza, P. Viglino, D. Corà, M. Caselle, G. Esposito, L. E. Xodo, *BMC Struct. Biol.* **2009**, *9*, 64.
198. P. Stadlbauer, L. Trantirek, T. E. Cheatham, 3rd, J. Koca, J. Šponer, *Biochimie* **2014**, *105*, 22–35.
199. P. Stadlbauer, M. Krepl, T. E. Cheatham, 3rd, J. Koca, J. Šponer, *Nucleic Acids Res.* **2013**, *41*, 7128–7143.
200. J. Šponer, X. Cang, T. E. Cheatham, *Methods* **2012**, *57*, 25–39.
201. M. Zgarbova, F. J. Luque, J. Šponer, T. E. Cheatham, 3rd, M. Otyepka, P. Jurecka, *J. Chem. Theory Comput.* **2013**, *9*, 2339–2354.
202. K. Gkionis, H. Kruse, J. A. Platts, A. Mládek, J. Koča, J. Šponer, *J. Chem. Theory Comput.* **2014**, *10*, 1326–1340.
203. H. T. Le, W. L. Dean, R. Buscaglia, J. B. Chaires, J. O. Trent, *J. Phys. Chem. B* **2014**, *118*, 5390–5405.
204. M. Vorlíčková, I. Kejnovská, J. Sagi, D. Renčiuk, K. Bednářová, J. Motlová, J. Kypr, *Methods* **2012**, *57*, 64–75.
205. A. I. Karsisiotis, N. M. Hessari, E. Novellino, G. P. Spada, A. Randazzo, M. Webba da Silva, *Angew. Chem. Int. Ed. Engl.* **2011**, *50*, 10645–10648.
206. J. Kypr, I. Kejnovská, D. Renčiuk, M. Vorlíčková, *Nucleic Acids Res.* **2009**, *37*, 1713–1725.
207. S. Paramasivan, I. Rujan, P. H. Bolton, *Methods* **2007**, *43*, 324–331.
208. E. Largy, A. Marchand, V. Gabelica, J.-L. Mergny, *unpublished.*
209. J. L. Mergny, L. Lacroix, *Oligonucleotides* **2003**, *13*, 515–537.
210. J. L. Mergny, L. Lacroix, *Curr. Protoc. Nucleic Acid Chem.* **2009**, *Chapter 17*, Unit 17.11.

211. A. De Cian, L. Guittat, M. Kaiser, B. Saccà, S. Amrane, A. Bourdoncle, P. Alberti, M. P. Teulade-Fichou, L. Lacroix, J. L. Mergny, *Methods* **2007**, *42*, 183–195.
212. J. L. Mergny, J. C. Maurizot, *Chembiochem* **2001**, *2*, 124–132.
213. M. C. Miller, J. O. Trent, *Curr. Protoc. Nucleic Acid Chem.* **2011**, *Chapter 17*, Unit 17.13.
214. H. T. Le, M. C. Miller, R. Buscaglia, W. L. Dean, P. A. Holt, J. B. Chaires, J. O. Trent, *Org. Biomol. Chem.* **2012**, *10*, 9393–9404.
215. C. C. Hardin, E. Henderson, T. Watson, J. K. Prosser, *Biochemistry* **1991**, *30*, 4460–4472.
216. T. Laue, *Curr. Opin. Struct. Biol.* **2001**, *11*, 579–583.
217. L. Petraccone, C. Spink, J. O. Trent, N. C. Garbett, C. S. Mekmaysy, C. Giancola, J. B. Chaires, *J. Am. Chem. Soc.* **2011**, *133*, 20951–20961.
218. M. Trajkovski, M. Webba da Silva, J. Plavec, *J. Am. Chem. Soc.* **2012**, *134*, 4132–4141.
219. H. T. Le, R. Buscaglia, W. L. Dean, J. B. Chaires, J. O. Trent, *Top. Curr. Chem.* **2013**, *330*, 179–210.
220. U. Dornberger, J. Behlke, E. Birch-Hirschfeld, H. Fritzsche, *Nucleic Acids Res.* **1997**, *25*, 822–829.
221. C. S. Mekmaysy, L. Petraccone, N. C. Garbett, P. A. Ragazzon, R. Gray, J. O. Trent, J. B. Chaires, *J. Am. Chem. Soc.* **2008**, *130*, 6710–6711.
222. C. Rehm, I. T. Holder, A. Groß, F. Wojciechowski, M. Urban, M. Sinn, M. Drescher, J. S. Hartig, *Chem. Sci.* **2014**, *5*, 2809–2809.
223. M. C. Miller, H. T. Le, W. L. Dean, P. A. Holt, J. B. Chaires, J. O. Trent, *Org. Biomol. Chem.* **2011**, *9*, 7633–7637.
224. G. W. Collie, G. N. Parkinson, S. Neidle, F. Rosu, E. De Pauw, V. Gabelica, *J. Am. Chem. Soc.* **2010**, *132*, 9328–9334.
225. F. Rosu, V. Gabelica, H. Poncelet, E. De Pauw, *Nucleic Acids Res.* **2010**, *38*, 5217–5225.
226. V. Gabelica, E. Schulz, M. Karas, *J. Mass Spectrom.* **2004**, *39*, 579–593.
227. V. Gabelica, E. De Pauw, *Mass. Spectrom. Rev.* **2005**, *24*, 566–587.
228. A. Marchand, V. Gabelica, *J. Am. Soc. Mass Spectrom.* **2014**, *25*, 1146–1154.
229. A. Marchand, A. Granzhan, K. Iida, Y. Tsushima, Y. Ma, K. Nagasawa, M. P. Teulade-Fichou, V. Gabelica, *J. Am. Chem. Soc.* **2015**, *137*, 750–756.
230. F. Lanucara, S. W. Holman, C. J. Gray, C. E. Eyers, *Nature Chem.* **2014**, *6*, 281–294.
231. R. Ferreira, A. Marchand, V. Gabelica, *Methods* **2012**, *57*, 56–63.
232. *Nucleic Acids in the Gas Phase*, Ed V. Gabelica, Springer, Berlin, Heidelberg, 2014.
233. K. Okamoto, Y. Sannohe, T. Mashimo, H. Sugiyama, M. Terazima, *Bioorg. Med. Chem.* **2008**, *16*, 6873–6879.
234. R. A. Darby, M. Sollogoub, C. McKeen, L. Brown, A. Risitano, N. Brown, C. Barton, T. Brown, K. R. Fox, *Nucleic Acids Res.* **2002**, *30*, e39–e39.
235. A. Risitano, K. R. Fox, *Nucleic Acids Res.* **2004**, *32*, 2598–2606.
236. E. E. Merkina, K. R. Fox, *Biophys. J.* **2005**, *89*, 365–373.
237. S. Redon, S. Bombard, M. A. Elizondo-Riojas, J. C. Chottard, *Nucleic Acids Res.* **2003**, *31*, 1605–1613.
238. H. Bertrand, S. Bombard, D. Monchaud, E. Talbot, A. Guédin, J.-L. Mergny, R. Grünert, P. J. Bednarski, M.-P. Teulade-Fichou, *Org. Biomol. Chem.* **2009**, *7*, 2864–2871.
239. Y. He, R. D. Neumann, I. G. Panyutin, *Nucleic Acids Res.* **2004**, *32*, 5359–5367.
240. T. I. Gaynutdinov, R. D. Neumann, I. G. Panyutin, *Nucleic Acids Res.* **2008**, *36*, 4079–4087.
241. J. E. Redman, *Methods* **2007**, *43*, 302–312.
242. R. Hänsel, F. Loehr, L. Trantirek, V. Doetsch, *J. Am. Chem. Soc.* **2013**, *135*, 2816–2824.
243. J. Y. Lee, J. Yoon, H. W. Kihm, D. S. Kim, *Biochemistry* **2008**, *47*, 3389–3396.
244. T. Miura, J. M. Benevides, G. J. Thomas, *J. Mol. Biol.* **1995**, *248*, 233–238.
245. R. M. Abu-Ghazalah, S. Rutledge, L. W. Y. Lau, D. N. Dubins, R. B. MacGregor, A. S. Helmy, *Biochemistry* **2012**, *51*, 7357–7366.
246. S. Ceru, P. Sket, I. Prislan, J. Lah, J. Plavec, *Angew. Chem. Int. Ed. Engl.* **2014**, 1002–1002.
247. B. Pagano, C. A. Mattia, C. Giancola, *Int. J. Mol. Sci.* **2009**, *10*, 2935–2957.
248. B. Pagano, A. Randazzo, I. Fotticchia, E. Novellino, L. Petraccone, C. Giancola, *Methods* **2013**, *64*, 43–51.

249. N. M. Brown, P. A. Rachwal, T. Brown, K. R. Fox, *Org. Biomol. Chem.* **2005**, *3*, 4153–4157.
250. P. A. Rachwal, I. S. Findlow, J. M. Werner, T. Brown, K. R. Fox, *Nucleic Acids Res.* **2007**, *35*, 4214–4222.
251. S. Neidle, *Methods* **2007**, *43*, 245.
252. D. Sun, L. H. Hurley, *Methods Mol. Biol.* **2010**, *608*, 65–79.
253. C. Detellier, P. Laszlo, *J. Am. Chem. Soc.* **1980**, *102*, 1135–1141.
254. M. Borzo, C. Detellier, P. Laszlo, A. Paris, *J. Am. Chem. Soc.* **1980**, *102*, 1124–1134.
255. A. Wong, R. Ida, G. Wu, *Biochem. Biophys. Res. Commun.* **2005**, *337*, 363–366.
256. Q. Xu, H. Deng, W. H. Braunlin, *Biochemistry* **1993**, *32*, 13130–13137.
257. H. Deng, W. H. Braunlin, *J. Mol. Biol.* **1996**, *255*, 476–483.
258. G. Wu, A. Wong, Z. Gan, J. T. Davis, *J. Am. Chem. Soc.* **2003**, *125*, 7182–7183.
259. M. L. Gill, S. A. Strobel, J. P. Loria, *J. Am. Chem. Soc.* **2005**, *127*, 16723–16732.
260. P. Sket, M. Črnugelj, J. Plavec, *Nucleic Acids Res.* **2005**, *33*, 3691–3697.
261. K. Snoussi, B. Halle, *Biochemistry* **2008**, *47*, 12219–12229.
262. F. Rosu, V. Gabelica, C. Houssier, P. Colson, E. De Pauw, *Rapid Commun. Mass Spectrom.* **2002**, *16*, 1729–1736.
263. J. Gros, F. Rosu, S. Amrane, A. De Cian, V. Gabelica, L. Lacroix, J.-L. Mergny, *Nucleic Acids Res.* **2007**, *35*, 3064–3075.
264. L. Joly, F. Rosu, V. Gabelica, *Chem. Commun.* **2012**, *48*, 8386–8388.
265. A. Marchand, R. Ferreira, H. Tateishi-Karimata, D. Miyoshi, N. Sugimoto, V. Gabelica, *J. Phys. Chem. B* **2013**, *117*, 12391–12401.
266. F. Balthasart, J. Plavec, V. Gabelica, *J. Am. Soc. Mass Spectrom.* **2013**, *24*, 1–8.
267. R. D. Gray, J. B. Chaires, *Methods* **2012**, *57*, 47–55.
268. I. Bang, *Biochem. Zeitschrift* **1910**, *26*, 293–311.
269. M. Gellert, M. N. Lipsett, D. R. Davies, *Proc. Natl. Acad. Sci. USA* **1962**, *48*, 2013–2018.
270. P. Tougard, J. F. Chantot, W. Guschlbauer, *Biochim. Biophys. Acta* **1973**, *308*, 9–16.
271. H. Q. Yu, D. Miyoshi, N. Sugimoto, *J. Am. Chem. Soc.* **2006**, *128*, 15461–15468.
272. J. Kim, C. Cheong, P. B. Moore, *Nature* **1991**, *351*, 331–332.
273. Y. Krishnan-Ghosh, E. Stephens, S. Balasubramanian, *J. Am. Chem. Soc.* **2004**, *126*, 5944–5945.
274. E. Gavathiotis, M. S. Searle, *Org. Biomol. Chem.* **2003**, *1*, 1650–1656.
275. V. L. Makarov, Y. Hirose, J. P. Langmore, *Cell* **1997**, *88*, 657–666.
276. R. K. Moyzis, J. M. Buckingham, L. S. Cram, M. Dani, L. L. Deaven, M. D. Jones, J. Meyne, R. L. Ratliff, J. R. Wu, *Proc. Natl. Acad. Sci. USA* **1988**, *85*, 6622–6626.
277. C. C. Hardin, M. J. Corregan, D. V. Lieberman, B. A. Brown, 2nd, *Biochemistry* **1997**, *36*, 15428–15450.
278. J. R. Wyatt, P. W. Davis, S. M. Freier, *Biochemistry* **1996**, *35*, 8002–8008.
279. C. Cheong, P. B. Moore, *Biochemistry* **1992**, *31*, 8406–8414.
280. M. K. Raghuraman, T. R. Cech, *Nucleic Acids Res.* **1990**, *18*, 4543–4552.
281. J. J. Green, L. Ying, D. Klenerman, S. Balasubramanian, *J. Am. Chem. Soc.* **2003**, *125*, 3763–3767.
282. T. Mashimo, Y. Sannohe, H. Yagi, H. Sugiyama, *Nucleic Acids Symp. Ser.* **2008**, 409–410.
283. T. Mashimo, H. Sugiyama, *Nucleic Acids Symp. Ser.* **2007**, 239–240.
284. T. Mashimo, H. Yagi, Y. Sannohe, A. Rajendran, H. Sugiyama, *J. Am. Chem. Soc.* **2010**, *132*, 14910–14918.
285. R. D. Gray, J. O. Trent, J. B. Chaires, *J. Mol. Biol.* **2014**, *426*, 1629–1650.
286. R. Buscaglia, R. D. Gray, J. B. Chaires, *Biopolymers* **2013**, *99*, 1006–1018.
287. R. D. Gray, R. Buscaglia, J. B. Chaires, *J. Am. Chem. Soc.* **2012**, *134*, 16834–16844.
288. R. D. Gray, J. Li, J. B. Chaires, *J. Phys. Chem. B* **2009**, *113*, 2676–2683.
289. M. Bončina, J. Lah, I. Prislan, G. Vesnaver, *J. Am. Chem. Soc.* **2012**, *134*, 9657–9663.
290. N. An, A. M. Fleming, C. J. Burrows, *J. Am. Chem. Soc.* **2013**, *135*, 8562–8570.
291. N. An, A. M. Fleming, E. G. Middleton, C. J. Burrows, *Proc. Natl. Acad. Sci. USA* **2014**, *111*, 14325–14331.

292. A. Y. Zhang, S. Balasubramanian, *J. Am. Chem. Soc.* **2012**, *134*, 19297–19308.
293. V. Limongelli, S. De Tito, L. Cerofolini, M. Fragai, B. Pagano, R. Trotta, S. Cosconati, L. Marinelli, E. Novellino, I. Bertini, A. Randazzo, C. Luchinat, M. Parrinello, *Angew. Chem. Int. Ed. Engl.* **2013**, *52*, 2269–2273.
294. L. Cerofolini, J. Amato, A. Giachetti, V. Limongelli, E. Novellino, M. Parrinello, M. Fragai, A. Randazzo, C. Luchinat, *Nucleic Acids Res.* **2014**, *42*, 13393–13404.
295. J. Töhl, W. Eimer, *Biophys. Chem.* **1997**, *67*, 177–186.
296. N. V. Hud, J. Plavec, in *Quadruplex Nucleic Acids*, Eds S. Neidle, S. Balasubramanian, Royal Society of Chemistry, Cambridge, UK, 2006, pp. 301.
297. L. Ma, M. Iezzi, M. S. Kaucher, Y.-F. Lam, J. T. Davis, *J. Am. Chem. Soc.* **2006**, *128*, 15269–15277.
298. A. E. Engelhart, J. Plavec, Ö. Persil, N. V. Hud, in *Nucleic Acid-Metal Ion Interactions*, Eds N. V. Hud, S. Neidle, Royal Society of Chemistry, Cambridge, UK, 2009, pp. 433.
299. P. Sket, A. Virgilio, V. Esposito, A. Galeone, J. Plavec, *Nucleic Acids Res.* **2012**, *40*, 11047–11057.
300. R. V. Reshetnikov, J. Šponer, O. I. Rassokhina, A. M. Kopylov, P. O. Tsvetkov, A. A. Makarov, A. V. Golovin, *Nucleic Acids Res.* **2011**, *39*, 9789–9802.
301. D. Sen, W. Gilbert, *Curr. Opin. Struct. Biol.* **1991**, *1*, 435–438.
302. R. M. Brosh, *Aging* **2011**, *3*, 332–335.
303. M. Dong, T. Mürdter, U. Klotz, *Eur. J. Clin. Pharmacol.* **2010**, *66*, 1–3.
304. A. De Cian, L. Lacroix, C. Douarre, N. Temime-Smaali, C. Trentesaux, J.-F. Riou, J.-L. Mergny, *Biochimie* **2008**, *90*, 131–155.
305. V. Gabelica, E. S. Baker, E. D. Pauw, M.-P. Teulade-Fichou, E. De Pauw, M. T. Bowers, *J. Am. Chem. Soc.* **2007**, *129*, 895–904.
306. D. Renčiuk, I. Kejnovská, P. Skoláková, K. Bednárová, J. Motlová, M. Vorlícková, *Nucleic Acids Res.* **2009**, *37*, 6625–6634.
307. V. Viglasky, L. Bauer, K. Tluckova, P. Javorsky, *J. Nucleic Acids* **2010**, *2010*, Article ID 820356.
308. I. N. Rujan, J. C. Meleney, P. H. Bolton, *Nucleic Acids Res.* **2005**, *33*, 2022–2031.
309. M. C. Miller, R. Buscaglia, J. B. Chaires, A. N. Lane, J. O. Trent, *J. Am. Chem. Soc.* **2010**, *132*, 17105–17107.
310. R. D. Gray, L. Petraccone, J. O. Trent, J. B. Chaires, *Biochemistry* **2010**, *49*, 179–194.
311. Y. Xu, Y. Noguchi, H. Sugiyama, *Bioorg. Med. Chem.* **2006**, *14*, 5584–5591.
312. R. Buscaglia, M. C. Miller, W. L. Dean, R. D. Gray, A. N. Lane, J. O. Trent, J. B. Chaires, *Nucleic Acids Res.* **2013**, *41*, 7934–7946.
313. J. Li, J. J. Correia, L. Wang, J. O. Trent, J. B. Chaires, *Nucleic Acids Res.* **2005**, *33*, 4649–4659.
314. R. Hänsel, F. Löhr, S. Foldynová-Trantírková, E. Bamberg, L. Trantírek, V. Dötsch, *Nucleic Acids Res.* **2011**, *39*, 5768–5775.
315. D. Sen, W. Gilbert, *Nature* **1990**, *344*, 410–414.
316. M. A. Keniry, G. D. Strahan, E. A. Owen, R. H. Shafer, *Eur. J. Biochem.* **1995**, *233*, 631–643.
317. G. D. Strahan, M. A. Keniry, R. H. Shafer, *Biophys. J.* **1998**, *75*, 968–981.
318. M. Črnugelj, N. V. Hud, J. Plavec, *J. Mol. Biol.* **2002**, *320*, 911–924.
319. M. Črnugelj, P. Sket, J. Plavec, *J. Am. Chem. Soc.* **2003**, *125*, 7866–7871.
320. F. W. Smith, P. Schultze, J. Feigon, *Structure* **1995**, *3*, 997–1008.
321. C. C. Hardin, T. Watson, M. Corregan, C. Bailey, *Biochemistry* **1992**, *31*, 833–841.
322. W. Li, X. M. Hou, P. Y. Wang, X. G. Xi, M. Li, *J. Am. Chem. Soc.* **2013**, *135*, 6423–6426.
323. D. Sen, W. Gilbert, *Nature* **1988**, *334*, 364–366.
324. A. Tanaka, J. Choi, T. Majima, *RSC Adv.* **2014**, *4*, 59071–59077.
325. E. F. Pettersen, T. D. Goddard, C. C. Huang, G. S. Couch, D. M. Greenblatt, E. C. Meng, T. E. Ferrin, *J. Comput. Chem.* **2004**, *25*, 1605–1612.

Chapter 8
Sodium and Potassium Ions in Proteins and Enzyme Catalysis

Milan Vašák and Joachim Schnabl

Contents

Abstract The group I alkali metal ions Na^+ and K^+ are ubiquitous components of biological fluids that surround biological macromolecules. They play important roles other than being nonspecific ionic buffering agents or mediators of solute exchange and transport. Molecular evolution and regulated high intracellular and extracellular M^+ concentrations led to incorporation of selective Na^+ and K^+ binding sites into enzymes to stabilize catalytic intermediates or to provide optimal positioning of substrates. The mechanism of M^+ activation, as derived from kinetic studies along with structural analysis, has led to the classification of cofactor-like (type I) or allosteric effector (type II) activated enzymes. In the type I mechanism substrate anchoring to the enzyme active site is mediated by M^+, often acting in tandem with a divalent cation like Mg^{2+}, Mn^{2+} or Zn^{2+}. In the allosteric type II mechanism, M^+ binding enhances enzyme activity through conformational transitions triggered upon binding to a distant site. In this chapter, following the discussion of the coordination chemistry of Na^+ and K^+ ions and the structural features responsible for the metal binding site selectiv-

M. Vašák (✉) • J. Schnabl
Department of Chemistry B, University of Zürich,
Winterthurerstrasse 190, CH-8057 Zürich, Switzerland
e-mail: mvasak@bioc.uzh.ch; joachim.schnabl@chem.uzh.ch

A. Sigel, H. Sigel, and R.K.O. Sigel (eds.), *The Alkali Metal Ions: Their Role for Life*, Metal Ions in Life Sciences 16, DOI 10.1007/978-3-319-21756-7_8

ity in M^+-activated enzymes, well-defined examples of M^+-activated enzymes are used to illustrate the structural basis for type I and type II activation by Na^+ and K^+.

Keywords Biological fluids • Coordination chemistry • Crystal structure • Enzyme activation • Potassium • Selectivity • Sodium

Please cite as: *Met. Ions Life Sci.* 16 (2016) 259–290

1 Introduction

The sodium (Na^+) and potassium (K^+) ions are among the most abundant alkali metal ions in biology. In animals, the extracellular space contains ~140 mM Na^+ and ~5 mM K^+ concentrations whereas in the intracellular space ~15 mM Na^+ and ~100 mM K^+ concentrations are present. Thus, in spite of highly similar chemical properties of both ions, steep concentration gradients of Na^+ and K^+ across the plasma membrane exist. In living cells, the enzyme Na^+,K^+-ATPase or the sodium pump distributes these ions between the intracellular and extracellular space at the expense of chemical energy. The membrane protein Na^+,K^+-ATPase exchanges three cytoplasmic Na^+ for two extracellular K^+ for each ATP that is hydrolyzed. The differential permeability for both cations leads to a transmembrane voltage difference typically in the range of −30 mV to −70 mV (negative on the inside of the membrane) in most living mammalian cells. This represents the basis for the action potential and, hence for neuronal signaling in general. The conduction of ions across cell membranes is of central importance for a wide range of physiological processes. Regardless whether a cell is specialized for electrical excitability, absorption, secretion, or any other process requiring the electrochemical energy of the ion gradients, it appears to use basically the same Na^+,K^+-ATPase in its membrane to perform various cellular functions. Hence, the role of Na^+,K^+-ATPase in alkali cation homeostasis is crucial for key cellular functions including volume regulation, maintenance of the resting membrane potential in nerve and muscle and Na^+-coupled solute transport (see also Chapter 10 of this volume).

K^+ is the most abundant inorganic cation in plant cells where it can contribute up to 10 % of the dry mass. Unlike animals, plants lack the Na^+,K^+-ATPase. However, plant cells have developed potassium channels that play an essential role in K^+ uptake and efflux. For an optimal metabolic activity of a plant cell a K^+ concentration in the range between 50 and 250 mM has to be maintained [1]. In contrast, Na^+ is toxic at high concentrations in the cytosol. The levels at which toxicity can occur have not been investigated in detail and are likely to depend on cell types [2], but it is generally thought that the cytosolic concentration of Na^+ should be controlled below 10–30 mM [3].

Life can be found over the whole range of salt concentrations even in hypersaline environments like that found in the Dead Sea with NaCl concentrations approaching saturation where archaea, bacteria, and fungi have been described. Organisms that tolerate extreme salt concentrations are known as halophilic microorganisms [4]. Two fundamentally different strategies exist within the microbial world to cope with the osmotic stress inherent to the presence of high salt concentrations. First, a group of halophilic bacteria and halophilic archaea maintain high intracellular salt concentrations, osmotically at least equivalent to the surrounding medium. Yet, while molar concentrations of KCl are present intracellularly that of the medium usually contains NaCl. Thus the intra- and extracellular ionic compositions differ [5, 6]. In most other halophilic and halotolerant microorganisms, the osmotic balance is provided by small organic molecules, including polyols such as glycerol and arabitol, sugars and sugar derivatives, amino acids and derivatives, and quaternary amines such as glycine betaine. These molecules are either synthesized by the cells or taken up from the medium when available [4].

Although the major physiological role of Na^+ and K^+ lays in transmembrane transport and signaling, due to their high intra- and extracellular concentrations both M^+ bind nonspecifically as counter ions to proteins, nucleic acids, and various negatively charged metabolites. Nevertheless, a large group of enzymes requiring M^+ for optimal activity has been discovered in the animal and plant world [7, 8]. K^+ or Na^+ is the preferred M^+ in enzymes since enzymes take advantage of the ready availability of Na^+ outside the cell and K^+ inside the cell to optimize their catalytic function. First evidence for the involvement of monovalent alkali cations (M^+) in enzyme activation has been reported more than seventy years ago by Boyer at al. [9, 10] showing that appreciable enzymatic activity of pyruvate kinase is expressed only in the presence of K^+. For numerous systems, selectivity for a certain M^+ is low, and a weak increase in enzyme activity is accomplished by larger cations (i.e., Rb^+, NH_4^+). The weak activation of some enzymes by larger cations has been interpreted in terms of kosmotropic effects of these cations on the water structure surrounding the protein (see below).

The classification of M^+-activated enzymes has recently been introduced by Di Cera [11]. The mechanism of M^+ activation, as derived from kinetic studies along with structural analysis, has been defined as a cofactor-like (type I) or allosteric effector (type II) enzyme activation. While in the type I mechanism M^+ mediates substrate anchoring to the enzyme active site, in the type II mechanism M^+ binding enhances enzyme activity through conformational transitions triggered upon binding to a distant site. In the following the coordination chemistry of Na^+ and K^+ ions and the mechanism responsible for the selectivity in metal ion binding in M^+-activated enzymes are discussed. We use well-defined examples of M^+-activated enzymes to illustrate the structural basis for type I and II activation by Na^+ and K^+.

2 Coordination Chemistry of Sodium and Potassium Ions

Na^+ and K^+ both belong to the group I alkali metal ions (Li^+, Na^+, K^+, Rb^+, Cs^+) and have similar chemical properties and present only a single oxidation state, +1. Na^+ and K^+ are not appreciably electrophilic because of the low charge density, forming mainly electrostatic interactions. The high solubility of Na^+ and K^+ salts in water is due to large entropies of solution while heats of solution often are endothermic. This shows that the ions are weakly hydrated, the heats of hydration are small [12]. Ion hydrations and their effect on the water structure appear to play a significant role in biological processes. The large electric field around the (smaller) ions causes the dipolar water molecules to rearrange themselves in the hydration shell(s) around the ions with structures differing from that in bulk water.

To account for the way ions affect the hydrogen bonds in the aqueous bulk, the concept of structure-making (kosmotropic) and structure-breaking (chaotropic) ions has been introduced [13]. A common view of the structure-making and -breaking properties is the difference in hydrogen bonding around the ion under study [14]. For alkali metals it has been concluded that the small Li^+ ion, with its rather high charge density, is a structure maker, Na^+ is a borderline ion, and the larger K^+, Rb^+, Cs^+ ions with low charge density are structure breakers [14]. This and the relative ease of hydration and dehydration of sodium and potassium, allowing rapid association and dissociation kinetics (~10^{-9} s) play an important role in their passage through membrane channels and enzyme activation. In many instances, a weak non-specific enhancement of enzyme activity by a variety of larger monovalent cations has been observed. This effect has been attributed to structural stabilization brought about by the kosmotropic effect of larger M^+.

It is believed that a subtle difference in the hydration shell structure of Na^+ and K^+ might play a crucial role in discriminating between the two metal ions. Therefore, the structure of the solvation shell of the Na^+ and K^+ ions in aqueous solution has been investigated both experimentally and theoretically.

A variety of experimental methods has been applied to investigate the solvation structure of these two cations. These include X-ray diffraction, extended X-ray absorption fine structure (EXAFS), and neutron diffraction. However, the experimentally derived coordination numbers are considerably scattered ranging from 4 to 6 for Na^+ and from 6 to 8 for K^+. A similar range of coordination numbers has also been obtained using molecular dynamics calculations [15]. Therefore, further experimental and theoretical studies have been initiated. By using large angle X-ray scattering (LAXS) on solutions of alkali metal salts the M–O distances of 2.43 Å for Na^+ and 2.80 Å for K^+ have been obtained. Based on the strong correlation between M–O bond distances in hydrated Na^+ and K^+ ions and the coordination number, the latter distances correspond to coordination number six and seven, respectively [16]. Independently, by molecular dynamics (MD) simulations based on density functional theory the solvation structure of Na^+ and K^+ has also been explored. The reported average coordination numbers for Na^+ and K^+ were 5.3 and 6.1, respectively [17].

In proteins, the M^+ coordination is largely mediated by O atoms provided by amino acid side-chains, e.g., Asp, Asn, Glu, and Gln, and main-chain O atoms of the polypeptide backbone. M^+-π interactions involving aromatic amino acids are rare. An example is the enzyme tagatose-1,6-biphosphate dehydrogenase where the Na^+-π interaction involving the Tyr183 residue has been observed [18]. The analysis of sodium- and potassium-containing structures in the PDB with resolution ≤1.6 Å revealed target distances for $K^+\cdots O$ and $Na^+\cdots O$ of 2.42 and 2.84 Å, respectively. Coordination numbers found for Na^+ with a coordination sphere limit of 3.17 Å ranged between 4 to 7; for K^+ with a coordination sphere limit of 3.59 Å the coordination numbers ranged between 5 to 8 [19].

3 Selectivity of Sodium(I) and Potassium(I) Enzyme Activation

Many biomolecules, including ion channels, transporters, and enzymes selectively bind or transport ions. In most cases, the differentiation between ion types is critical to the function of the molecule. The discrimination between K^+ and Na^+ is particularly interesting given their identical charge, spherical nature, and similar size. In kinetic terms, the binding site selectivity of an enzyme for a given M^+ is reflected by its substantial larger activation relative to other monovalent cations. Nevertheless, several examples of widely different effects of Na^+ and K^+ on the structure and kinetics also exist.

In the case of tryptophan synthase, changes between the Na^+-bound and K^+-bound structures are significant [20], but are not matched by the differences in the kinetics of activation [21]. Similar replacement of K^+ with Na^+ in K^+-activated pyruvate kinase does not result in structural changes [22], but the enzyme is practically inactive without K^+ [9]. Despite the fact that a number of high-resolution structures of M^+-activated enzymes with either Na^+ or K^+ are available, structural features underlying metal ion selectivity are not fully understood. Studies of selectivity in small macrocyclic ligands highlighted the role of rigidity in creating a cavity that preferentially accommodates ions of a certain size [23]. By analogy, M^+ selectivity in M^+-activated enzymes has usually been attributed to a better structural fit of a M^+ into the geometrically constrained binding site. However, in view of substantial protein dynamics the generation of a rigid pre-formed binding site in M^+-activated enzymes is unlikely.

Potassium channels, tetrameric membrane-spanning proteins, represent an example of proteins in which metal selectivity has been subject of intense studies. These channels are able to distinguish between K^+ and Na^+ with up to 1000-fold preference for K^+ [24]. Selectivity is achieved in a narrow region of the channel known as the selectivity filter, which is lined with carbonyl ligands that coordinate permeating ions, creating a thermodynamic preference for binding K^+ relative to Na^+ [25]. A widely held view attributed channel selectivity to a better structural fit of K^+ into the selectivity filter binding sites than could occur for the smaller Na^+. This picture came from the structural and thermodynamic studies of the potassium channel from *Streptomyces lividans* in which a size-matched selectivity, also seen with natural and synthetic macrocyclic ligands, has been proposed for the preferred binding of K^+ [26]. However, subsequent MD simulations for the potassium channel KscA argued against this explanation of selectivity. Firstly, in such a mechanism the protein should rigidly retain a precise (sub-Ångstrom) geometry to discriminate between the ionic radius of K^+ and Na^+, which differ by only 0.38 Å, and secondly, channels are highly flexible, undergoing rapid thermal fluctuations larger than the small difference in ionic radius between K^+ and Na^+ [27].

To account for the underlying channel selectivity the following concept has been developed. The preference is at least partly due to the channel compensating better the energy cost of dehydrating K^+ than Na^+ [28, 29]. Furthermore, the channel does not select for K^+ ions by providing binding sites of the appropriate cavity size but the selectivity arises directly from the intrinsic local physical properties of the ligands coordinating the cation in the binding site. The interplay between the attractive ion–ligand (favoring smaller cation) and repulsive ligand–ligand interactions (favoring larger cations) is the basic element that governs size selectivity in flexible protein binding sites. Consequently, altering the composition of a binding site appears to provide a potent molecular mechanism to achieve and maintain a high selectivity in protein structures despite their significant conformational flexibility [27, 30]. Therefore, it would appear that a similar mechanism might also account for the metal binding site selectivity in M^+-activated enzymes.

4 Classification of Sodium(I)- and Potassium(I)-Activated Enzymes

In many metalloproteins metal cofactors located at the active sites of enzymes can be involved in intriguing biochemical reactions, in others, they appear to play a purely structural role stabilizing a protein fold. Based on crystal structures and enzyme kinetics the mechanism of M^+-activated enzymes has been classified as a cofactor-like (type I) or allosteric (type II) activation [11].

In the type I mechanism, substrate anchoring to the enzyme active site is mediated by M^+, often acting in tandem with a divalent cation like Mg^{2+}, Mn^{2+} or Zn^{2+}. In such a mechanism, M^+ coordination is required for catalysis or substrate recognition.

In the allosteric type II mechanism, M^+ binding enhances enzyme activity through conformational transitions triggered upon binding to a site where the cation makes no direct contact with the substrate. In general, allostery is defined as the binding of a ligand at one site causing a change in the affinity or catalytic efficiency of a distant site [31]. As proteins in solution are dynamic entities, the allosteric effect reveals selective stabilization of one conformation of the enzyme through M^+ complexation that may produce local and potentially long-range effects on the enzyme structure.

4.1 Cofactor-Like Potassium(I)-Activated Enzymes (Type I)

Pyruvate kinase was the first enzyme for which a dependence of the activity on monovalent cations was documented [9]. An intriguing requirement for activation of the enzyme by two divalent cations (2 Mg^{2+} or 2 Mn^{2+}) was documented much later [32–34]. Pyruvate kinase catalyzes the final reaction of glycolysis in

Table 1 Selected examples of a cofactor-like (type I) and allosteric effector (type II) activation of K^+-dependent enzymes.[a]

Enzyme	PDB entries[b]	Structure examples	Ligands[c]	References[d]
K^+-activated type I				
Branched-chain α-ketoacid dehydrogenase kinase	15	1GJV, 1GKZ; *Free:* 1GKX	5 (4-0-1)	[38]
Diol dehydratase	37	1DIO	7 (5-0-2)	[48]
Fructose 1,6-bisphosphatase	132	1FPI	4 (3-0-1)	[70]
Glycerol dehydratase	247	1IWP; *Free:* 1MMF	7 (5-0-2)	[46]
GroEL	102	1KP8	7 (2-4-1)	[56]
Hsc70	89	1HPM, 3HSC	8 (2-3-3)	[54, 55]
*Mv*RadA	2	1XU4	4 (2-1-1)	[61]
Pyridoxal kinase	35	1LHR; *Free:* 1LHP	6 (4-1-1)	[64]
Pyruvate dehydrogenase kinase	27	1Y8N, 1Y8O, 1Y8P, 3CRK	5 (4-0-1)	[41]
Pyruvate kinase	125	1A49; *Na^+-bound:* 1A5U	6 (4-1-1)	[22]
K^+-activated type II				
Branched-chain α-ketoacid dehydrogenase	51	1DTW	5 (5-0-0)	[77]
Dialkylglycine decarboxylase	17	1DKA; *Na^+-bound:* 2DKB	6 (5-1-0)	[81]
Histone deacetylase	35	1W22	6 (6-0-0)	[99]
MutL	17	1NHI; *Na^+-bound:* 1NHJ	5 (4-1-0)	[93]
IMP dehydrogenase	43	1PVN	6 (6-0-0)	[126]
Ribokinase	69	1GQT	6 (6-0-0)	[96]
Ser dehydratase	1025	1PWH; *Free:* 1PWE	6 (6-0-0)	[84]
Tryptophanase	8	1AX4	7 (4-3-0)	[85]
Tyrosine phenol-lyase	17	2EZ1	7 (4-3-0)	[88]

[a]Table adapted from [11].
[b]Number of enzyme structures deposited in the PDB as of 2014-12-31.
[c]The format is *N* (p-w-s), where *N* is the sum of ligands from the protein (p), water (w), and substrate (s).
[d]Only the most relevant references are listed.

which phosphoenolpyruvate (PEP) and ADP are converted into pyruvate and ATP, respectively. The enzyme is allosterically regulated and features a tetrameric structure composed of identical monomers of 50–60 kDa, depending on species. The active site is located in a large cleft formed by the interface between two larger domains, the $\alpha_8\beta_8$ barrel or A domain and the α-barrel of the B domain [22]. The structure of the yeast enzyme solved in the presence of the substrate analog phosphoglycolate [35] (Table 1) reveals a close interaction between K^+ and Mn^{2+} with amino acid groups and the substrate in the active site (Figure 1). The divalent cation anchors the substrate via its carboxylate and phosphoester O atoms to the side-chain O atoms of Glu242 and Asp266. Conversely, K^+ increases electrostatic coupling of the phosphate group by screening the carboxylate O of Asp84, which is engaged in

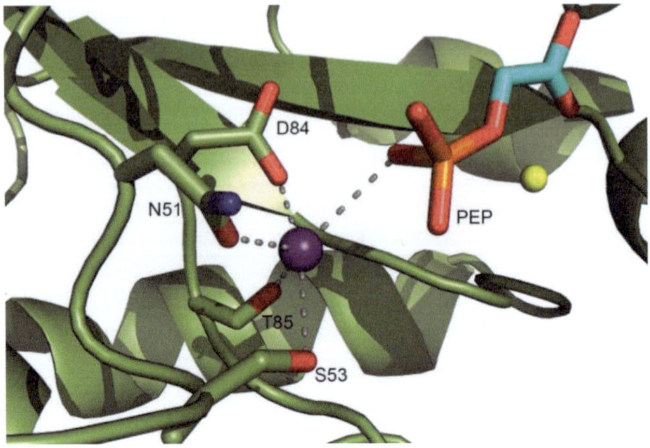

Figure 1 Structural model of the type I potassium-binding site in pyruvate kinase. Structural elements contributing to the K^+-binding site are shown in ribbons along with ball-and-stick models of the substrate, amino acids and metal ions. Directly bound to phosphoenolpyruvate (PEP) (CPK, C in cyan) is Mn^{2+} (yellow). K^+ (purple) is coordinated by the O atoms of main-chain and side-chain residues in the substrate binding pocket (PDB code 1A3W; [22]).

the coordination shell along with the main-chain O atoms of Thr85, the Oγ of Ser53, and the side-chain O atom of Asn51. Consequently, K^+ acts to electrostatically optimize substrate binding by participating in direct interaction and influencing the environment of the active site in pyruvate kinase.

Mitochondrial pyruvate dehydrogenase complex (PDC) is a large macromolecular machine catalyzing an oxidative decarboxylation of pyruvate with concomitant formation of acetyl-CoA and NADH, thereby linking glycolysis to the citric acid cycle [36]. PDC is composed of multiple copies of four components: pyruvate dehydrogenase (E1), dihydrolipoamide acetyltransferase (E2), dihydrolipoamide dehydrogenase (E3) and E3-binding protein (E3BP) [37]. The activity of this complex is regulated by mitochondrial protein kinases, molecular switches that down-regulate the oxidation of branched-chain α-ketoacids and pyruvate. Branched-chain α-ketoacid dehydrogenase kinases (BCKD kinases), together with pyruvate dehydrogenase kinases, define this novel family of serine protein kinases [36]. These mitochondrial serine protein kinases share sequence motifs essential for ATP binding and require Mg^{2+} and K^+ for optimal catalytic activity. The mammalian α-ketoacid dehydrogenase kinase complex catalyzes the oxidative decarboxylation of branched-chain α-ketoacids derived from leucine, isoleucine, and valine. The crystal structure of rat BCKD kinase in its apo-form, complexed with ADP and with a non-hydrolyzable ATP analog reveals a homodimeric arrangement. Binding of adenosine nucleotides is uniquely mediated by both magnesium and potassium (Figure 2). The Mg^{2+} is bound to the conserved Asn249 and the phosphate moieties in the nucleotide. The K^+ is coordinated by the α-phosphate of the nucleotide and the main-chain O atoms of Val298, Asp300, Phe303, and Gly337 [38] (Table 1). This type I potassium-binding site of BCKD kinase appears to

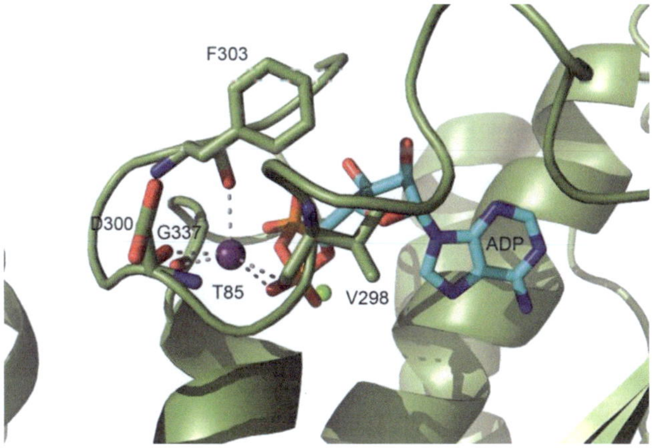

Figure 2 Structural model of the type I potassium-binding site in α-ketoacid dehydrogenase kinase. Structural elements contributing to the K^+-binding site are shown in ribbons along with ball-and-stick models of the substrate, amino acids, and metal ions. Mg^{2+} (green) is directly bound to ADP (C in cyan) in the active site. K^+ (purple) is coordinated by the α-phosphate of ADP and further ligated to the four O atoms of amino acid side-chains in the substrate binding pocket (PDB code 1GKZ; [38]).

be similar to that of pyruvate kinase, chaperonin GroEL, chaperone Hsc70, fructose 1,6-bisphosphatase, and a few other K^+-activated enzymes involved in phosphoryl transfer reactions (see Table 1).

The other protein kinase involved in the regulation of the pyruvate dehydrogenase complex and containing a type I site is pyruvate dehydrogenase kinase (PDHK). The enzyme binds to the lipoyl domain of the pyruvate dehydrogenase complex, where it monitors changes in the reduced and acetylated states of the lipoic acid ligand. In mammalian species, PDHK is represented by at least four closely related protein kinases, which are remarkably different with respect to their activities, tissue distribution, and regulation [39, 40]. The crystal structure of PDHK3 bound to the inner lipoyl-bearing domain of dihydrolipoamide transacetylase (L2), determined with or without bound adenylyl imidodiphosphate, revealed a PDHK3 dimer complexed with two L2 domains [41]. The distal portion of the PDHK3-ATP lid makes a tight turn around the polyphosphate moiety of ATP, thereby assisting in positioning of the γ-phosphate in the active site. This closed conformation creates a type I potassium-binding site, closely similar to that reported for BCKD kinase (see Figure 2). In PDHK3, the main-chain O atoms of Leu300, Asn302, Tyr305, and Gly325 serve as protein ligands for the bound potassium ion. The fifth ligand of bound K^+ comes from the Pγ O atom of ATP. This arrangement is thought to facilitate the catalysis of phosphotransfer reaction in PDHK [41, 42]. However, new insights into the molecular mechanisms responsible for the recognition of L2 and kinase activity have been provided by recent structural studies on PDHK2 in a complex with L2. The latter structure revealed, in addition to the type I potassium-binding site, a novel type II potassium-binding site located on the PDHK2 interface

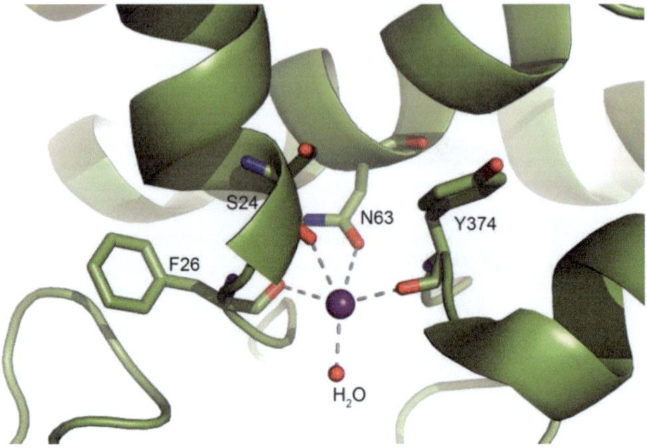

Figure 3 Structural model of the type II potassium-binding site in PDHK2. In the enzyme, both the type I and the type II potassium-binding sites are present. Structural elements contributing to the K^+-binding site are shown in ribbons along with ball-and-stick models of the substrate, amino acids, and metal ions. Bound K^+ (purple) is coordinated by the four O atoms of amino acid residues and one water molecule (red) (PDB code 3CRK; [43]).

with the L2 domain [43] (Table 1). The type II potassium-binding site plays a structural role, rigidifying the interface between PDHK2 and L2. K^+ in this binding site is coordinated by the three main-chain O atoms of Ser24, Phe26, and Tyr374. The other oxygen atoms coordinating the bound K^+ come from the side-chain oxygen of Asn63 and a water molecule (Figure 3). This structural feature dramatically improves the L2 binding, thereby facilitating targeting of PDHK2 toward its protein substrate and providing yet another route to control PDHK2 activity. In the structure of human PDHK2, the presence of similar type I and type II potassium-binding sites has also been reported [44].

Glycerol dehydratase and diol dehydratase are highly homologous isofunctional enzymes that catalyze the elimination of water from glycerol and 1,2-propanediol to the corresponding aldehyde via a coenzyme B_{12}-dependent mechanism. Both enzymes have an absolute requirement for a potassium ion for catalytic activity. The adenosylcobalamin-dependent reactions are initiated by homolysis of the Co–C bond of the enzyme-bound cofactor and catalyzed by a radical mechanism [45]. The crystal structure of substrate-free glycerol dehydratase [46] and in complex with cobalamin and propane-1,2-diol has been solved [47] (Table 1). The latter structure features a dimer of the αβγ heterotrimer [47]. The active site is located in a $(\beta/\alpha)_8$ barrel, the so-called TIM (triosephosphate isomerase) barrel, that is formed by a central region of the α subunit. The substrate propane-1,2-diol and essential cofactor K^+ are bound inside the $(\beta/\alpha)_8$ barrel above the corrin ring of cobalamin. Here, K^+ is heptacoordinated by the two hydroxyls of the substrate and five O atoms of the active-site residues (Figure 4). In the substrate-free structure, K^+ is hexacoordinated with a water molecule in place of the substrate ligands.

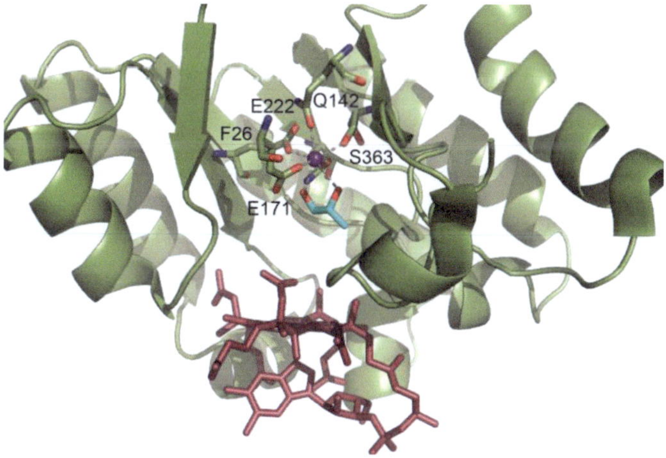

Figure 4 Structural model of the type I potassium-binding site in glycerol dehydratase. Structural elements contributing to the K^+-binding site are shown in ribbons along with ball-and-stick models of the substrate, amino acids, and metal ions. The active-site cavity is viewed from the direction parallel to the plane of the corrin ring of cobalamin (red). Active site residues interacting with the substrate (C in cyan) and K^+ (purple) are shown (PDB code 1IWP; [47]).

Although the overall polypeptide fold in glycerol dehydratase is quite similar to that of diol dehydratase, the nature and the number of the metal ions required for the enzyme activity differ. In the crystal structure of diol dehydratase two metal ions in the active site have been recognized. One metal ion has been found directly coordinated by substrate [48] and the other located near the adenine ring of the coenzyme adenosyl group [49]. In these crystallographic models of diol dehydratase both metals have been assigned to K^+. However, recent biochemical studies established that diol dehydratase contains tightly bound Ca^{2+} and is almost completely inhibited by the preincubation with EDTA or EGTA [50]. This along with enzyme kinetics and quantum mechanical/molecular mechanical (QM/MM) calculations indicated that the presence of Ca^{2+} is essential for the enzyme catalysis [50, 51]. Reevaluation of the crystal structure led to the conclusion that the substrate-coordinated metal ion is not K^+ but Ca^{2+} and that K^+, bound near the adenine ring, represents the essential potassium for the diol dehydratase catalysis. In this type II potassium-binding site K^+ is six-coordinated by four main-chain O atoms of Gly261, Ser264, Glu265, and Glu280 and two water molecules, one of which is hydrogen-bonded to N1 of the coenzyme adenine moiety (figure not shown) [50]. Thus it appears that the coenzyme B_{12}-dependent diol dehydratase is a potassium-requiring calcium metalloenzyme. Since the inhibition of highly homologous glycerol dehydratase by EDTA was also reported [52], a similar requirement of both K^+ and Ca^{2+} for its activity remains to be elucidated.

Most proteins must fold into defined three-dimensional structures to gain functional activity. However, in the cellular environment, newly synthesized proteins are at great risk of aberrant folding and aggregation, potentially forming toxic species. Therefore, cells developed a complex network of molecular chaperones, which use ingenious mechanisms to prevent aggregation and promote efficient folding. These protein families are also known as stress proteins or heat-shock proteins, as they are upregulated under conditions of stress in which the concentrations of aggregation-prone folding intermediates increase. They function via the repetitive transient association with exposed hydrophobic patches on misfolded proteins in an ATP-dependent manner [53]. Hsc70 is a well-characterized constitutively expressed chaperone that has intrinsic ATPase activity, hydrolyzing ATP into ADP. In this instance, both K^+ and Mg^{2+} are required for the activity [54]. The Hsc70s are composed of two intimately related but functionally distinct domains; the 40 kDa N-terminal nucleotide-binding domain, which binds and hydrolyzes ATP, and the 30 kDa C-terminal substrate-binding domain. In the structural studies of the active site domain of bovine Hsc70 two K^+ and one Mg^{2+} ion have been revealed. Here, K^+ provides optimal electrostatic coupling for the phosphate moiety of the substrate and optimizes the register for docking in the enzyme active site and formation of the transition state [55].

Similarly to Hsc70, GroEL, a member of the ATP-dependent chaperonin family that, along with its binding partner GroES, promotes the proper folding of many cytosolic bacterial proteins [53]. The GroEL complex consists of 14 identical subunits arranged into two heptameric rings that associate with each other in a back-to-back manner. Each subunit can be divided into three functional domains termed apical, intermediate, and equatorial. The apical domain captures unfolded protein substrates and binds GroES, an event that leads to the encapsulation of the substrate protein. The equatorial domain contains the ATP-binding site and forms contacts between the two heptameric rings. A crystal structure of the GroEL complex with ATP, Mg^{2+}, and K^+ bound to all 14 subunits revealed that potassium ions are involved in binding the triphosphate moiety of ATP [56] (Table 1). K^+ is heptacoordinated by the Pα O atom, the main-chain O atoms of Thr30 and Lys51, and four water molecules (Figure 5). Comparison with the GroEL structure obtained in the absence of potassium ions [57] allowed evaluation of the structural changes that may occur in response to cognate K^+ binding. Both structures were found to be similar, the only exception being some differences as to the temperature factor (*B*-factor) distribution [58].

Homologous recombination, in which DNA strands are exchanged between a pair of homologous sequences, is crucial for both the repair of damaged DNA and the maintenance of genomic diversity. The DNA strand exchange reaction in homologous recombination is facilitated by several M^+-dependent ATPases, known as recombinases [59], whose activity depends on ATP and Mg^{2+} or Ca^{2+} and in the case of human and yeast, but not *E. coli*, on the presence of K^+ or NH_4^+. The recombinase superfamily consists of Rad51 for eukaryotes, RadA for archaea and RecA for bacteria. The strand exchange reaction is initiated by their cooperative assembling around single-stranded DNA in the presence of ATP, forming a nucleoprotein filament in which the DNA is stretched compared with its canonical B form. This nucleoprotein filament engages a double-stranded DNA with homologous sequence thereby promoting strand exchange between the two DNA molecules [59]. While Rad51 exhibits a more efficient strand exchange in the presence of Ca^{2+} compared with Mg^{2+}, the strand exchange activity of RecA requires a high concentration of

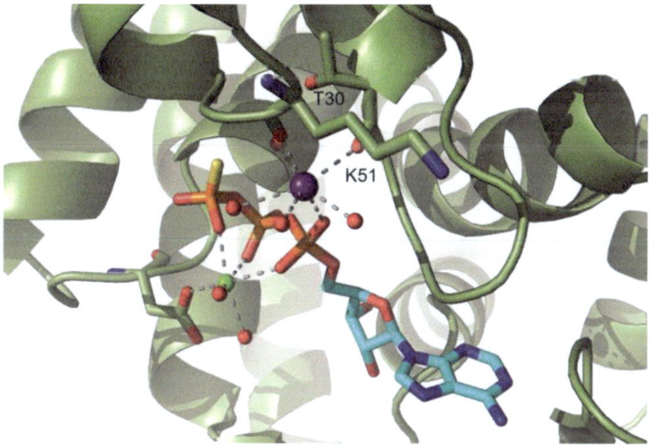

Figure 5 Structural model of the type I potassium-binding site in the GroEL chaperonin. Structural elements contributing to the K^+-binding site are shown in ribbons along with ball-and-stick models of the substrate, amino acids, and metal ions. Mg^{2+} (green) and K^+ (purple) are directly bound to ATP-γ-S (C in cyan). K^+ is further coordinated by the two O atoms of amino acid residues and four water molecules (red) (PDB code 1KP8; [56]).

Mg^{2+} [60]. The crystal structure of Rad51 homolog from *Methanococcus voltae* (*Mv*RadA) has been solved in the presence of a non-hydrolyzable ATP analog and Mg^{2+}, in the absence and presence of K^+ [61] (Table 1). The *Mv*RadA structure containing potassium revealed a typical arrangement of Mg^{2+} and two K^+ in the active site that polarize the Pγ of ATP. Each K^+ is ligated by an O atom from the Pγ and the side-chain O atoms of Asp302 or Glu151 (Figure 6). To learn more about the molecular mechanism and the role of cations in catalysis MD simulations were carried out using a dimer model of Rad51 as a minimal unit of an active form of the Rad51 filament. The results of MD simulations suggested that potassium ions are indispensable for the stabilization of the active dimer [62].

Pyridoxal kinase (PLK) catalyzes the transfer of a phosphate group from ATP to the 5′ alcohol of pyridoxine, pyridoxamine, and pyridoxal to provide pyridoxal-5′-phosphate (PLP), the form of vitamin B_6, a widely used coenzyme. The enzyme requires both monovalent and divalent cations for its function. In general, the Mg^{2+} and K^+ ions are commonly used in reactions catalyzed by kinases. While the metal ion tandem Mg^{2+} and K^+ is required for the activity of PLK from *E. coli* [63], Zn^{2+} and K^+ have been proposed to be the physiological metals needed for mammalian PLK activity [64]. This apparent discrepancy has been reconciled by further studies showing that the observed enzyme activation by Zn^{2+} has been derived from enzyme kinetics carried out under non-physiological conditions regarding substrate concentrations and/or pH (pH 6) and that under physiological conditions Mg^{2+} is the required divalent metal ion [63]. Interestingly, in a more recent kinetic study using the human enzyme Na^+ was found to elicit more than a two-fold increase in enzyme activity than K^+ [65]. Nevertheless, the fact that PLK is indeed the K^+-requiring enzyme has been concluded on the basis of its substantially increased affinity for

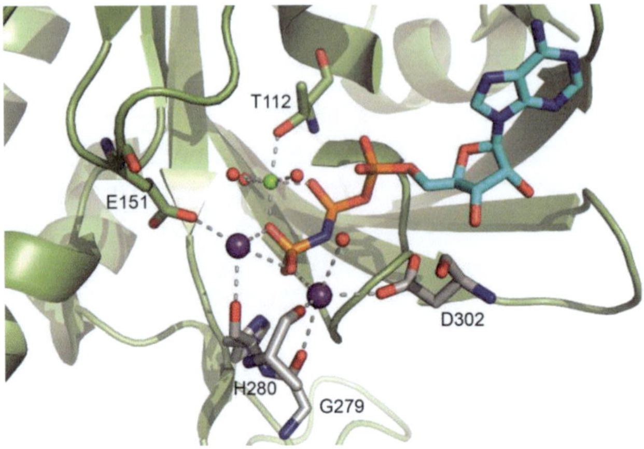

Figure 6 Structural model of the type I potassium-binding site in *Mv*RadA. Structural elements contributing to the K^+-binding site are shown in ribbons along with ball-and-stick models of the substrate, amino acids, and metal ions. The ATP analog AMP-PNP (C in cyan) in the active site is ligated to one Mg^{2+} (green) and two K^+ ions (purple). Ligands shown in grey are amino acids from the symmetry generated crystal interface (PDB code 1XU4; [61]).

the ATP and PLK substrates in the presence of K^+ as compared to Na^+ and its occurrence in intracellular space, where K^+ concentration is several-fold higher than the one of Na^+. Moreover, bound Na^+ in PLK has a five-coordination shell [64] whereas K^+ is octahedrally coordinated by the O atoms of Thr186, Asp113, Glu153, and Thr148, as well as the ATP β-phosphate group and a water molecule (Figure 7) [65]. Differences between the Na^+ and K^+ coordination spheres along with the ability of protein ligands to strip off water molecules much easier from the hydration sphere of the larger K^+ (1.33 Å) than from the smaller Na^+ (0.9 Å) [66] have been proposed to account for the observed effect of M^+ on PLK activity *in vitro* [65] (Table 1).

Another important enzyme requiring K^+ for its activity is fructose 1,6-bisphosphatase, a key enzyme in gluconeogenesis. The enzyme hydrolyzes fructose 1,6-bisphosphate to fructose 6-phosphate and inorganic phosphate [67]. The enzyme requires two divalent metal ions (Mg^{2+}, Mn^{2+} and/or Zn^{2}) for activity, and the activity is further enhanced by K^+ and inhibited by Li^+ [68, 69]. The mammalian 1,6-bisphosphatase is inhibited allosterically by AMP and competitively by fructose 2,6-bisphosphate. The crystal structure of the enzyme bound to a substrate analog and several combinations of monovalent and divalent metal ions has been determined [70, 71]. Each subunit of the tetrameric enzyme shows, besides binding sites for one substrate and one AMP molecule, also two binding sites for divalent and one for monovalent metal ions. Binding of the substrate is accomplished with the participation of a loop comprised of residues 52–72, termed the 'dynamic loop'. Of the three metal sites, loop 52–72 interacts directly with the cation at site 3 only, which has been postulated to represent the K^+ activation site. This site together with

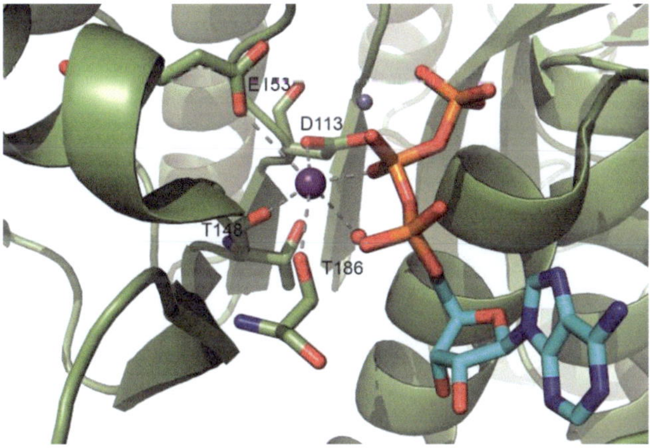

Figure 7 Structural model of the type I potassium-binding site in PLK. Structural elements contributing to the K^+-binding site are shown in ribbons along with ball-and-stick models of the substrate, amino acids, and metal ions. ATP (C in cyan) in the active site is ligated to Zn^{2+} (grey) and K^+ (purple). K^+ is octahedrally coordinated by the four O atoms of amino acid residues, ATP and one water molecule (red) (PDB code 1LHR; [64]).

Arg276, activates the enzyme by deshielding the 1-phosphoryl group of the substrate and aid in the nucleophilic attack at the phosphorus atom by OH^- [70, 72].

Similar to fructose 1,6-bisphosphatase also methionine adenosyltransferase, also known as *S*-adenosylmethionine synthase (MAT), is activated by K^+ in tandem with two Mg^{2+} ions [73, 74]. MATs represent a ubiquitous family of enzymes that utilizes ATP and methionine to produce *S*-adenosylmethionine, the most crucial methyl donor in the biological methylation of biomolecules [75]. Most of the structural studies have been carried out using the *E. coli* and rat liver enzymes in complex with various substrates, cofactors, and inhibitors. In the presence of a non-hydrolyzable ATP analog and the substrate Met, the *E. coli* structure reveals two Mg^{2+} and one K^+ in the active site anchoring the phosphate moiety of the cofactor [76]. The active site architecture is similar to that of pyruvate kinase (see Figure 1, Table 1).

4.2 Allosteric Potassium(I)-Activated Enzymes (Type II)

In type II allosteric activation the K^+ ion, located outside of the active center without direct contact with the substrate, acts as an allosteric effector that promotes substrate binding and catalysis through conformational transitions triggered upon its binding [11]. The human branched-chain α-ketoacid dehydrogenase (BCKD) complex is a member of the mitochondrial α-ketoacid dehydrogenase complex family comprising pyruvate dehydrogenase, α-ketoglutarate dehydrogenase, and BCKD complexes [37]. The BCKD complex catalyzes the oxidative decarboxylation of

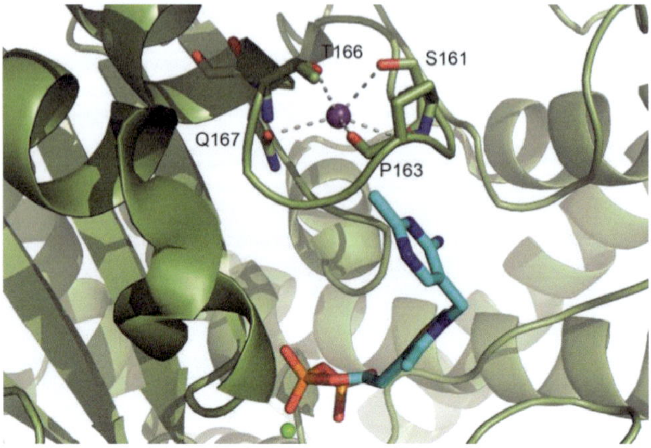

Figure 8 Structural model of the type II potassium-binding site in BCKD. Structural elements contributing to the K^+-binding site are shown in ribbons along with ball-and-stick models of the substrate, amino acids, and metal ions. Shown are the substrate thiamine diphosphate (ThDP) (C in cyan) and Mg^{2+} (green). K^+ (purple) is coordinated by the O atoms of five amino acid residues (PDB code 1DTW; [77]).

branched-chain α-ketoacids derived from the branched-chain amino acids leucine, isoleucine and valine. The activity of the BCKD complex is completely abolished by phosphorylation of Ser292 by BCKD kinase. BCKD is a thiamine diphosphate-dependent enzyme and is arranged as a $\alpha_2\beta_2$-heterotetramer. In this enzyme two important K^+-binding sites have been identified [77] (Table 1). The K^+-binding site located close to the cofactor is stabilized by a loop formed by residues 161–166 of the α subunit. The residues of the loop are directly involved in K^+ coordination through the side-chain O atoms of Ser161, Thr166, and Gln167 and the main-chain O atoms of Ser161 and Pro163 (Figure 8). The K^+ bound to this site controls the thiamine diphosphate binding and thus enzyme activity.

The second K^+-binding site, located in in the β subunit at the interface with the small C-terminal domain of the α subunit, is important for the stability of the tetrameric structure of BCKD. The coordination of K^+ in this site is octahedral and is almost exclusively maintained by the protein main-chain O atoms and a water molecule [77].

Pyridoxal 5′-phosphate, the bioactive form of vitamin B_6, acts as a cofactor for amino acid metabolizing enzymes where it is invariably bound to an active site lysine through a Schiff base linkage forming the 'internal' aldimine. In dialkylglycine decarboxylase, serine dehydratase, tryptophan synthase and tryptophanase, besides PLP also K^+ is required for their activity [78]. Dialkylglycine decarboxylase is an unusual enzyme in that a single cleavage site catalyzes both the decarboxylation and transamination reaction [79, 80]. The enzyme contains two binding sites for alkali metal ions. The crystal structure of the enzyme is an α_4 tetramer, built up as a dimer of dimers. The structure solved in the presence of K^+ with PLP bound to the active site, revealed an

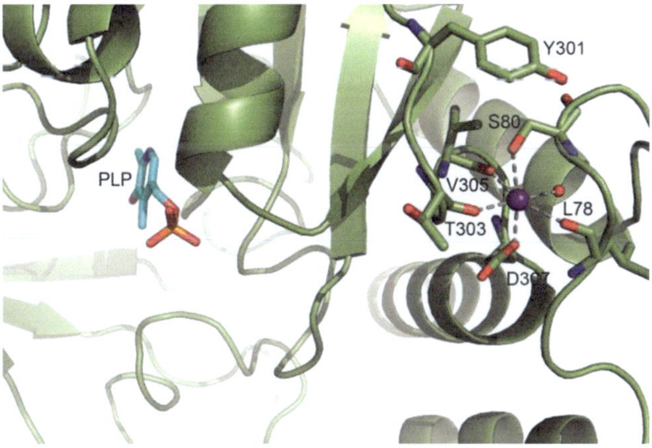

Figure 9 Structural model of the type II potassium-binding site in dialkylglycine decarboxylase. Structural elements contributing to the K^+-binding site are shown in ribbons along with ball-and-stick models of the substrate, amino acids, and metal ions. Shown is the cofactor pyridoxal 5′-phosphate (PLP) (C in cyan) bound to the active site in the presence of K^+. K^+ (purple) is hexacoordinated by the O atoms of five amino acid residues and a water molecule (red) (PDB code 1DKA; [82]).

important structural role for K^+ located near the active site (Figure 9). This potassium ion is coordinated by the side-chain O atom of Asp307, the main-chain O atoms of Leu78, Thr303, and Val305, the hydroxyl O atom of Ser80, and a water molecule. The other metal binding site is located at the C-terminus of an α helix [81, 82] (Table 1).

Serine dehydratase, the other PLP- and K^+-dependent enzyme, catalyzes dehydration of L-serine to yield pyruvate and ammonia [83]. The crystal structure of the apo- and the holoenzyme crystallized with *O*-methylserine revealed two protein domains with the active site located in the cleft between these two domains. Four O atoms of the main-chain of Gly168, Ala198, Leu223, Val225, and two side-chain O atoms of Glu194 and Ser200 coordinate the K^+ bound near the active site. The major function of this ion is structure stabilizing [84].

Similar K^+-binding sites were found in the structures of β-elimination enzymes, tryptophan synthase [20] and tryptophanase [85] (see Tables 1 and 2). The tryptophan synthase heterotetrameric $\alpha_2\beta_2$ complex catalyzes the last two reactions in the biosynthesis of L-tryptophan [86]. In the crystal structure of Trp synthase the K^+ binding site is located in the C-terminal domain of the β subunit and near the phosphate group of PLP, thereby stabilizing a conformational state of the internal aldimine. The main-chain O atoms of Gly232, Phe306, and Ser308 and one water molecule form the coordination shell of K^+ (Figure 10) [20].

The enzyme tryptophanase or tryptophan indole-lyase is a bacterial PLP-dependent lyase that catalyzes *in vivo* degradation of L-tryptophan to yield indole, pyruvate, and ammonia. The active form of tryptophanase is an α_4 tetramer and each subunit binds one molecule of PLP. The enzyme requires monovalent cations such

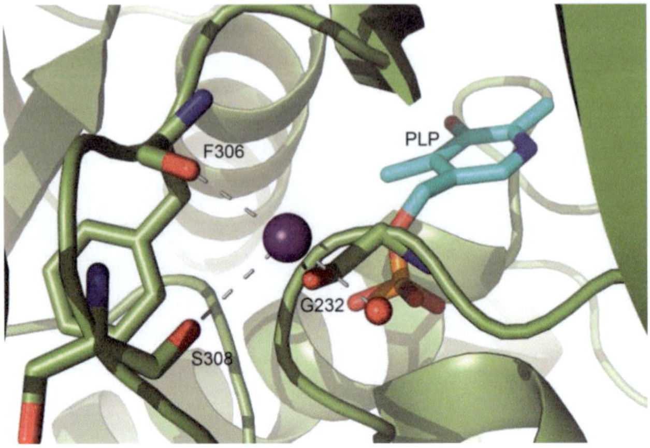

Figure 10 Structural model of the type II potassium-binding site in tryptophan synthase. Structural elements contributing to the K^+-binding site are shown in ribbons along with ball-and-stick models of the substrate, amino acids, and metal ions. Shown is the cofactor pyridoxal 5′-phosphate (PLP) (C in cyan) bound to the active site in the presence of K^+. K^+ (purple) is coordinated by the three O atoms of amino acid residues and a water molecule (red) (PDB code 1TTQ; [20]).

as NH_4^+, K^+ or Rb^+ for its activity [87]. In the crystal structure of tryptophanase the active site is located near the subunit interface with PLP bound at the bottom of the catalytic cleft. The K^+ binding site is located at the intersubunit interface in the catalytic dimer, at a large distance from the PLP moiety excluding its direct involvement in catalysis. The coordination sphere of the K^+ includes the main-chain O atoms of Gly53 and Asn271 from one subunit, the side-chain O atom of Glu70, and the main-chain O atom of Glu70 from the second subunit and three water molecules (Figure 11). Based on comparison with the crystal structure of the apoenzyme a structure stabilizing role for K^+ has been suggested [85].

Structurally, the described tryptophanase is very similar to other bacterial PLP-dependent enzyme tyrosine phenol-lyase, whose activity is also influenced by monovalent cations. The enzyme tyrosine phenol-lyase is a bacterial PLP-dependent lyase that catalyzes *in vivo* the reversible β-elimination of L-tyrosine to produce phenol and ammonium. A tetrameric molecule of tyrosine phenol-lyase binds four K^+, one K^+ per monomer. Each K^+ is bound at the interface between two related subunits stabilizing the structure of the catalytic dimer. It is coordinated by the main-chain O atoms of Gly52 and Asn262 from one subunit and by the main-chain O atoms and side-chain O atoms of Glu69 from the adjacent subunit (Table 1) [88]. All of these residues belong to the large domains of protein subunits. Additionally, there are three water molecules completing the coordination sphere of K^+ (Figure 12) [88].

DNA mismatch repair (MMR) maintains genomic stability by correcting errors that have escaped polymerase proofreading [89]. Initiation of MMR depends on the action of the MutS protein, which recognizes a mismatched base pair and recruits MutL in an ATP-dependent manner. MutL plays an essential role in MMR by

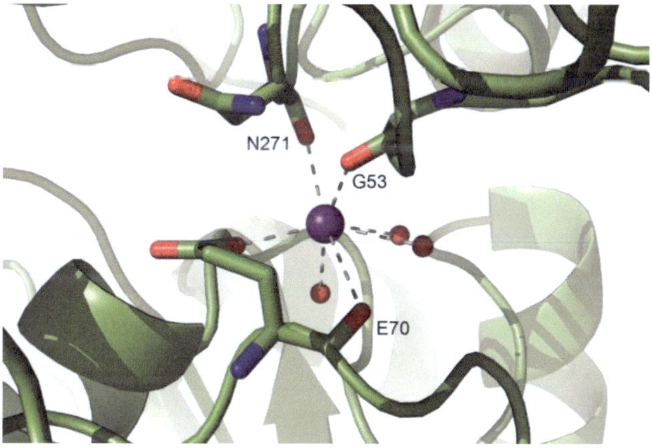

Figure 11 Structural model of the type II potassium-binding site in tryptophanase. Structural elements contributing to the K^+-binding site are shown in ribbons along with ball-and-stick models of the substrate, amino acids, and metal ions. K^+ (purple) is coordinated by the O atoms of four amino acid residues and three water molecules (red) (PDB code1AX4; [85]).

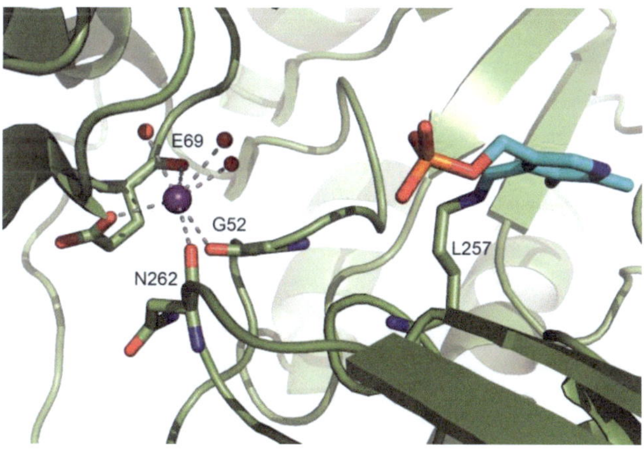

Figure 12 Structural model of the type II potassium-binding site in tyrosine phenol-lyase. Structural elements contributing to the K^+-binding site are shown in ribbons along with ball-and-stick models of the substrate, amino acids, and metal ions. K^+ (purple) is bound at the interface between two subunits. K^+ is bound to the O atoms of three amino acid residues from both subunits and three water molecules (red) (PDB ID: 2EZ1; [88])

coordinating various protein–protein interactions. The protein is composed of two structurally conserved domains connected by a variable flexible linker [90]. The N-terminal region encompasses an ATPase domain of the GHL (gyrase, Hsp90, MutL) ATPase/kinase superfamily that is conserved from bacteria to humans [91, 92]. The crystallographic and biochemical studies revealed that, in addition to a magnesium ion, a monovalent cation-binding site is present in the ATP binding

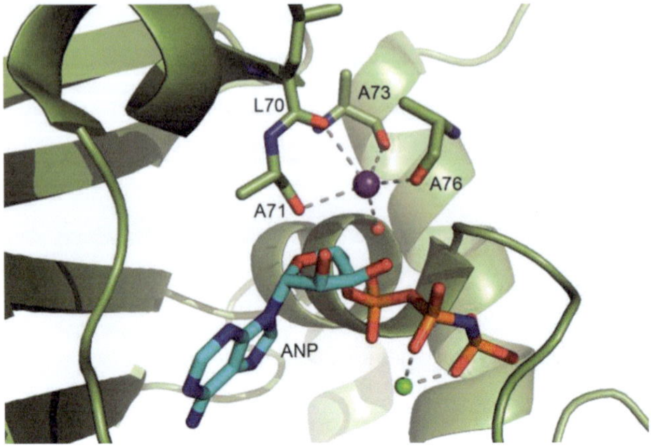

Figure 13 Structural model of the type II potassium-binding site in the ATP binding domain of MutL. Structural elements contributing to the K^+-binding site are shown in ribbons along with ball-and-stick models of the substrate, amino acids, and metal ions. One Mg^{2+} (green) is bound to the ATP analog ANP (C in cyan) in the active site. K^+ (purple) is ligated to the four main-chain O atoms and a water molecule (red) (PDB code 1NHI; [93]).

domain of MutL. ATPase activity of MutL was observed with Na^+, K^+, Rb^+ or Cs^+ ions, but not with Li^+. The crystal structure of the N-terminal 40 kDa fragment (NL40) of wild type MutL in complex with a non-hydrolytic ATP analog revealed K^+ coordination by four O atoms of the main-chain residues Leu70, Ala71, Ala73, and Ala76, originating from a tight turn following an α-helix at the N-terminal hinge of the ATP lid, and a water molecule (Figure 13). The K^+ ion in MutL does not directly contact the non-hydrolyzable ATP analog. The binding of the monovalent cation apparently stabilizes the ATP lid (Table 1) [93].

D-Ribose is widely used in biology as an energy supply and in biosynthesis of nucleotides, amino acids, and other cofactors. However, before use, D-ribose must be phosphorylated to produce ribose-5-phosphate by ribokinase. This carbohydrate kinase in a reaction requiring ATP and magnesium transfers the γ-phosphate group from ATP to ribose [94, 95]. Monovalent cations such as K^+, Cs^+, and NH_4^+, but neither Na^+ nor Li^+ are required for the ribokinase activity. Since *E. coli* ribokinase is activated by potassium with an apparent K_d of 5 mM, the enzyme should be fully active under physiological conditions. An apparent K_d of 17 mM found for cesium implies that Cs^+ can be used as an alternative ion [96]. The crystal structure of *E. coli* ribokinase bound to Cs^+, used instead of K^+ because of a larger number of electrons and thus easier detection in electron density maps, reveals a homodimer with each subunit composed of two domains. A larger domain of the enzyme provides most of the binding interactions for substrate and a smaller domain provides the dimer interface and a lid for the ribose [96]. The bound Cs^+ is heptacoordinated by the five main-chain O atoms of residues Asp249, Ile251, Ala285, Arg288, and Gly290 and the side-chain O atoms of Asp249 and Ser294. In the absence of the apo structure the proposed structural features responsible for the enzyme activation by K^+ could be only assumed [96]. Recently, to shed light into the details of the ribokinase activation mechanism by monovalent metal ions, the crystal structure of *Sa*239, a ribokinase from *Staphylococcus aureus*, in the absence of monovalent ions has been elucidated [97]. In this work, based on the comparison of the crystal structures of *Sa*239 and *E. coli* ribokinase, it has been concluded that significant conformational changes of the large ATP binding loop by M^+ binding shape the nucleotide binding pocket thereby enhancing the substrate binding affinity.

Modulation of the acetylation state of lysine residues found in the accessible N termini of core histones is one of the posttranslational chromatin modifications that affect gene expression. Acetylation and deacetylation of histones are controlled by the enzymatic activity of histone acetyltransferases and histone deacetylases (HDACs). Two related classes of eukaryotic HDACs exist based on sequence similarity. Class I HDACs are ubiquitously expressed and are primarily located in the nucleus, while the class II enzymes can shuttle between the nucleus and cytoplasm [98]. Both classes are zinc-dependent hydrolases. The crystal structure of HDAC8, a member of the class I family, comprises a single α/β domain which binds one Zn^{2+} ion and two K^+ ions *in vivo* [99–101]. The HDAC8 active site reveals a narrow pocket, likely to accommodate the acetylated lysine during the catalytic reaction, and Zn^{2+} at the bottom of this cavity. The two K^+ binding sites, designated as site 1 and site 2, are located 7 Å and 21 Å from the Zn^{2+} binding site, respectively (Table 1) [99]. K^+ in site 1 is coordinated by the side-chain O atoms of Asp176 and Ser199 and the main-chain O atoms of Asp176, Asp178, His180, and Leu200. Whereas K^+ in site 2 is ligated by two water molecules and the main-chain O atoms of Phe189, Thr192, Val195, and Tyr225 (Figure 14). In recent kinetic studies of HDAC8 and its mutant,

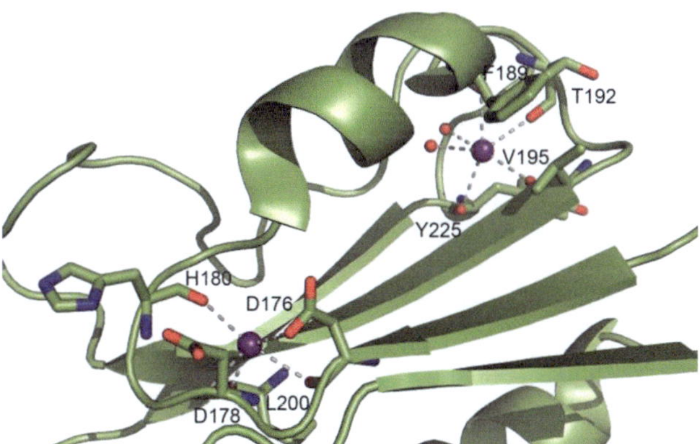

Figure 14 Structural model of the type II potassium-binding site in HDAC8. Structural elements contributing to the K^+-binding site are shown in ribbons along with ball-and-stick models of the substrate, amino acids, and metal ions. The enzyme HDAC8 contains one active site Zn^{2+} and two type II K^+ ions located in separate binding sites designated as 1 (lower part, left) and 2 (upper part, right) (K^+ in purple). K^+ (purple) in the more distant site 2 acts as an allosteric effector and is coordinated by the main-chain O atoms of Phe189, Thr192, Val195, and Tyr225 and two water molecules (red) (PDB code 1 W22; [99]).

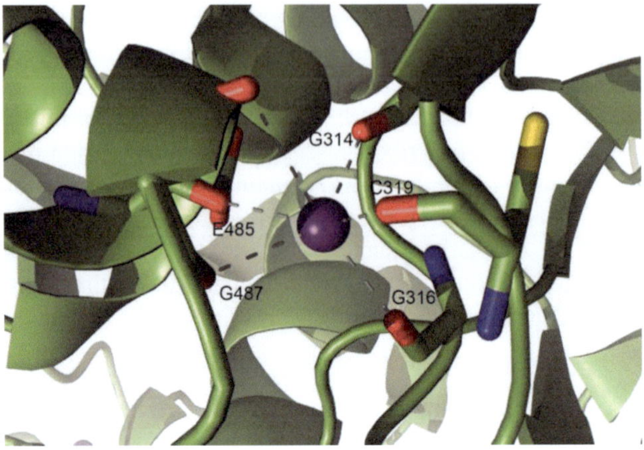

Figure 15 Structural model of the type II potassium-binding site in IMPDH. Structural elements contributing to the K^+-binding site are shown in ribbons along with ball-and-stick models of the substrate, amino acids, and metal ions. K^+ (purple) is coordinated by the six O atoms of amino acid residues (PDB code 1PVN; [126]).

the roles of the two K^+ ions have been addressed. The activating K^+ binds to the more distant site 2, consistent with its role as an allosteric effector. In contrast, the more weakly bound K^+ in site 1 decreases the enzyme activity (11-fold), likely by altering the pK_a of active site His142 present in the second coordination sphere [102].

Inosine 5′-monophosphate dehydrogenase (IMPDH) is an essential rate-limiting enzyme in the *de novo* guanine nucleotide synthetic pathway that catalyzes the conversion of IMP to XMP with the concomitant reduction of NAD to NADH. IMPDH is activated by monovalent cations such as K^+, which increases IMPDH activity by ~ 100-fold [103]. The crystal structure of IMPDH features a homotetramer with the catalytic $(\beta/\alpha)_8$ barrels arranged in square planar geometry. In the crystal structure of *Tritrichomonas foetus* IMPDH (*Tf* IMPDH) a conserved K^+ binding site has been identified (see Figure 15 and Table 1). K^+ is ligated to six main-chain O atoms, in which three are from the loop that contains the catalytic Cys319 and three from an α-helix in the C-terminal segment of the adjacent subunit (Gly314, Gly316, the catalytic Cys319, Glu485′, Gly486′, and Gly487′, where the prime denotes residue from the neighboring subunit). Hence, K^+ seems ideally positioned to promote substrate binding and/or the rate of the chemical transformations [103]. However, several observations suggested that the usual allosteric mechanism might not apply to the activation of IMPDH by monovalent cations. In a more recent study, aiming at the understanding of the IMPDH reaction, enzyme kinetics and MD simulations have been carried out to examine the effect of K^+ on individual steps in the IMPDH reaction [104]. The results revealed that chemical transformation is separated by a conformational change involving a mobile protein flap closure (residues 412–431) and that reaction specificity is controlled by differences in dynamics. It has been concluded, moreover, that IMPDH represent an example of a new paradigm of allosteric regulation via the kinetic control of protein conformation [104].

4.3 Cofactor-Like Sodium(I)-Activated Enzymes (Type I)

Fructose 1,6-bisphosphate aldolase catalyzes the cleavage of fructose 1,6-bisphosphate into dihydroxyacetone phosphate and glyceraldehyde 3-phosphate (G3P) in glycolysis. For fructose 1,6-bisphosphate aldolase two evolutionarily and mechanistically unrelated classes have been identified. Class I utilizes an active site lysine in Schiff base formation during catalysis and is mainly found in higher order organisms. Class II, found in yeast, bacteria, fungi, and blue-green algae, utilizes Zn^{2+} and Na^{+} in catalysis [105]. Both class II aldolases, the fructose 1,6-bisphosphate aldolase and the tagatose-1,6-bisphosphate aldolase, employ Zn^{2+} and Na^{+} tandem in catalysis. The active site of the *E. coli* fructose 1,6-bisphosphate aldolase is located at the C-terminal end of the $(\alpha/\beta)_8$-barrel and contains the catalytic Zn^{2+} [106]. The Na^{+} binding site is located near the catalytic zinc ion. A similar well-defined Na^{+} binding site is also present in the crystal structure of *E. coli* tagatose-1,6-bisphosphate aldolase (Table 2) [18]. Here the Na^{+} site is located about 8.5 Å from the catalytic Zn^{2+}. The Na^{+} ion is coordinated by the four main-chain O atoms

Table 2 Selected examples of a cofactor-like (type I) and allosteric effector (type II) activation of Na^{+}-dependent enzymes.[a]

Enzyme	PDB entries[b]	Structure examples	Ligands[c]	References[d]
Na^{+}-activated type I				
β-Galactosidase	181	1DP0, 1F4A, 1F4H, 4V41	5 (3-1-1)	[111]
Fructose 1,6-bisphosphate aldolase	84	1B57	6 (4-1-1)	[106]
Tagatose 1,6-bisphosphate aldolase	13	1GVF	6 (5-0-1)	[18]
Na^{+}-activated type II				
Factor VIIa	120	2A2Q, 2AER, 2FIR	4 (2-0-2)	[121]
Activated protein C	16	3F6U	7 (5-0-2)	[122]
Factor Xa	294	2BOK	6 (4-2-0)	[120]
Thrombin	899	1SFQ	6 (2-0-4)	[127]
Trp synthase	110	1BKS; *K^{+}-bound:* 1TTQ	5 (3-2-0)	[20]

[a]Table adapted from [11].
[b]Number of enzyme structures deposited in the PDB as of 2014-12-31.
[c]The format is *N* (p-w-s), where *N* is the sum of ligands from the protein (p), water (w), and substrate (s).
[d]Only the most relevant references are listed.

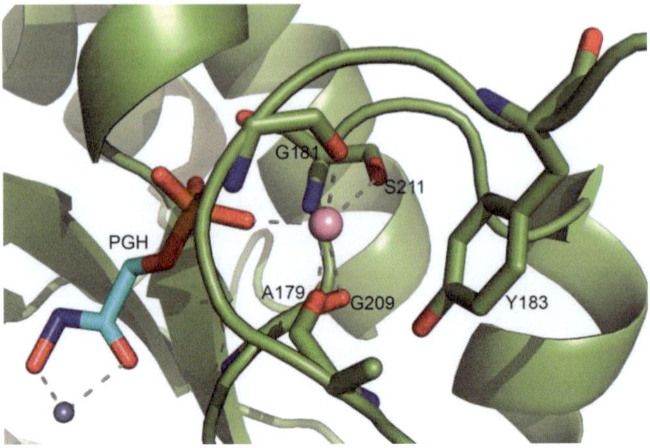

Figure 16 Structural model of the type I sodium-binding site in tagatose 1,6-bisphosphate aldolase. Structural elements contributing to the Na^+-binding site are shown in ribbons along with ball-and-stick models of the substrate, amino acids, and metal ions. The octahedral Na^+ (pink) coordination sphere is formed by four main-chain O atoms, a phosphate oxygen, and an uncommon Na^+-π interaction involving Tyr183. The blue-grey sphere represents the catalytic Zn^{2+} (PDB code 1GVF; [18]).

of residues Ala179, Gly181, Gly209, and Ser211 together with a phosphate oxygen of the transition state analogue PGH (phosphoglycolohydroxamate). In fructose 1,6-bisphosphate aldolase the octahedral Na^+ coordination sphere is completed by a water, but in tagatose-1,6-bisphosphate an uncommon cation-π interaction involving the side-chain of Tyr183 is observed (Figure 16). The Na^+ binding in class II aldolases assists the formation of the correct phosphate binding site.

The enzyme β-galactosidase catalyzes the hydrolysis of lactose, a disaccharide consisting of galactose and glucose, and other β-galactosides into monosaccharides [107]. The enzyme requires Mg^{2+} and Na^+ for full catalytic activity. The latter ion directly ligates the galactosyl O_6 hydroxyl during catalysis [108, 109]. The study of the appearance of β-galactosidase as the product of the Z gene of the *lac* operon of *E. coli* was used by Jacob and Monod in the formulation of their operon model for gene expression [110]. The crystal structure of *E. coli* β-galactosidase is a tetramer of four identical subunits each containing four catalytic sites that show no cooperativity or allosteric effectors [111]. The two metal binding sites are situated a few Å apart in the active site, both very near to an interface between two domains of the protein. In the absence of bound lactose the Na^+ site has a distorted square pyramidal geometry, including the two main-chain O atoms of residues Asn604 and Phe601 and one side-chain O atom of Asp201 together with two water molecules, or one water and the 6-OH group from the sugar upon lactose binding [111, 112]. The sodium through structural stabilization of the transition state affects the activity of the enzyme.

4.4 Allosteric Sodium(I)-Activated Enzymes (Type II)

The other pyridoxal-5′-phosphate-requiring bienzyme is tryptophan synthase. Trp synthase is a $\alpha_2\beta_2$-tetramer that catalyzes the last two steps in the biosynthesis of L-Trp. In the first reaction, the α-subunit catalyzes the cleavage of 3-indole-D-glycerol 3′-phosphate (IGP) to indole and D-glyceraldehyde 3-phosphate. In the second step, indole is channeled via the 25 Å tunnel to the β-subunit where, in a PLP-dependent reaction, it reacts with L-Ser to give L-Trp [113]. Thus, the β-subunit of the enzyme represents a member of the class of PLP enzymes. Similar to a large group of PLP-dependent enzymes, Trp synthase has an essential requirement for the binding of a monovalent cation in enzyme catalysis. This cofactor binds to a site in the C-terminal domain of the β-subunit [114, 115]. Enzyme kinetics showed that a variety of cations would satisfy this requirement, including Na^+, K^+, NH_4^+, Cs^+, and Rb^+. Crystal structures of the Trp synthase $\alpha_2\beta_2$ complex from *Salmonella typhimurium* in the presence of Na^+, K^+, and Cs^+ showed that the metal binding site is buried in the β-subunit about 8 Å from the catalytic site (Table 2) [20]. While the Na^+ ion is coordinated to the main-chain O atoms of residues Phe306, Ser308, and Gly232 and to two water molecules (Figure 17), the K^+ ion is coordinated to the same main-chain O atoms and to one water molecule (see Figure 10). The K^+-containing structure of Trp synthase differs from the Na^+-containing structure in the movement of the side-chains of Phe280 and Tyr279 from a position partially blocking the tunnel in the Na^+ structure to a position lining the surface of the tunnel in the K^+ structure [20]. These local changes near the cation site affect the interaction site between the α and β subunits. Taken together, these studies along with kinetic analyses revealed that the binding of a monovalent cation to the β-subunit is essential for both the selective

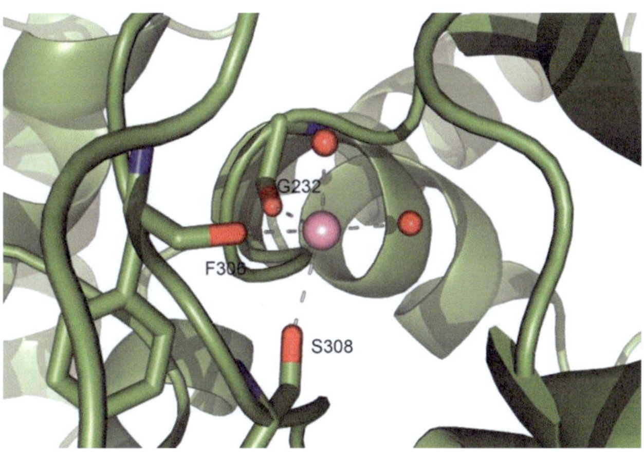

Figure 17 Structural model of the type II sodium-binding site in tryptophan synthase. Structural elements contributing to the Na^+-binding site are shown in ribbons along with ball-and-stick models of the substrate, amino acids, and metal ions. Na^+ (pink) is octahedrally coordinated by the three main-chain O atoms and two water molecules (PDB code 1BKS; [20]).

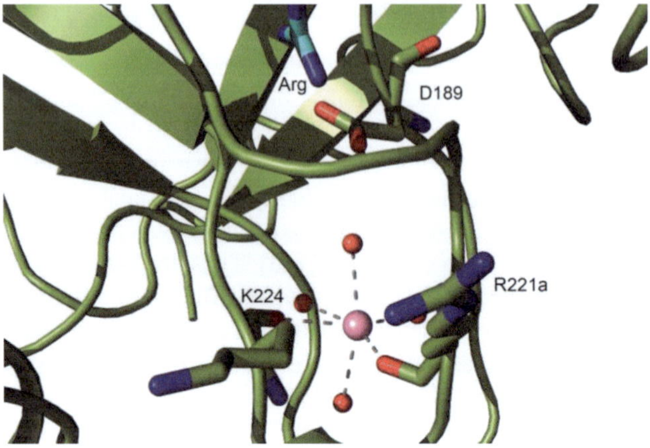

Figure 18 Structural model of the type II sodium-binding site in thrombin. Structural elements contributing to the Na^+-binding site are shown in ribbons along with ball-and-stick models of the substrate, amino acids and metal ions. The two main-chain O atoms and four buried water molecules (red) octahedrally coordinate Na^+ (pink). The Na^+-binding site is located more than 15 Å away from the residues of the catalytic triad (PDB code 1SFQ; [127]).

stabilization of active conformations and also for the allosteric communication between the subunits [113].

The clotting enzyme thrombin is a serine protease playing a key role in the vertebrate blood coagulation cascade [116]. In the final process of blood clotting the soluble protein fibrinogen is converted to the insoluble fibrin by the proteolytic action of thrombin, which in turn is produced from the inactive precursor prothrombin. The thrombin activation is accomplished by the action of a number of clotting proteases called clotting factors. Thrombin activity toward substrates is enhanced allosterically by the binding of Na^+ that converts thrombin from a "slow" sodium-free to a "fast" sodium-containing form of a higher catalytic efficiency [117]. Na^+ is located more than 15 Å away from residues of the catalytic triad His57, Asp102, and Ser195 and within 5 Å from Asp189 in the substrate binding pocket. The slow/fast thrombin transition upon Na^+ binding results in optimal orientation of Asp189 and the active site Ser195 for substrate binding. This subtle conformational alteration is mediated by a network of water molecules that connect the bound Na^+ to Ser195. In the crystal structure of thrombin the Na^+ binding site displays octahedral coordination involving the main-chain O atoms of Arg221a (where a denotes an inserted residue), Lys224, and four buried water molecules (Figure 18) [118].

The extensive structural and kinetic studies of thrombin facilitated the subsequent structural identification of the analogous Na^+ binding sites in clotting proteases such as factor Xa [119, 120], factor VIIa [121], and activated protein C [122] (see Table 2). Furthermore, in factor Xa and activated protein C the binding of Na^+, which influences the specificity of the substrate-binding pocket, is linked to the binding of Ca^{2+}.

5 Concluding Remarks

The group I alkali metal ions are ubiquitous components of biological fluids that surround biological macromolecules. They play important roles other than being nonspecific ionic buffering agents or mediators of solute exchange and transport. Molecular evolution and regulated high intracellular and extracellular M^+ concentrations led to incorporation of selective Na^+ and K^+ binding sites into enzymes to stabilize catalytic intermediates or to provide optimal positioning of substrates.

Based on the specific binding of Na^+ or K^+ into the enzyme structure and their participation in enzyme activation a cofactor-like (type I) or allosteric effector (type II) mechanism has been defined [11].

Many other enzymes have as yet undiscovered monovalent cation binding sites. However, crystallographic assignment of Na^+ and K^+ is nontrivial because Na^+ ions, with a small ionic radius (0.97 Å), can easily be misidentified as water molecules since both possess the same number of electrons. K^+ on the other hand has a higher electron density but an ionic radius (1.33 Å) almost identical to that of a water molecule.

Recently, $^{15}NH_4^+$, as a K^+ mimic, was successfully used to characterize and map the K^+ binding site in human histone deacetylase 8 and the bacterial Hsp70 homologue DnaK by NMR spectroscopy [123]. The support for the utilization of $^{15}NH_4^+$ came from the fact that (i) ammonium ions can often activate potassium-dependent enzymes *in vitro*, because of their similar ionic radii (K^+ 1.33 Å; NH_4^+ 1.44 Å), and (ii) the reports that a potassium-independent Hsc70 and pyruvate kinase can be generated by engineering a lysine residue into the structure in such a way that its terminal $R\text{-}NH_3^+$ occupies the K^+ binding site [124, 125]. The application of this NMR method will ultimately lead to a better understanding of the role of monovalent cations in proteins and enzymes.

Abbreviations

ADP	adenosine 5′-diphosphate
Ala, A	alanine
AMP	adenosine 5′-monophosphate
AMP-PNP	adenylyl-imidodiphosphate
ANP	phosphoaminophosphonic acid-adenylate ester
Asn, N	asparagine
Asp, D	aspartic acid
ATP	adenosine 5′-triphosphate
BCKD	branched-chain α-ketoacid dehydrogenase
EDTA	ethylenediamine-*N*,*N*,*N*′,*N*′-tetraacetic acid
EGTA	ethylene glycol-bis(2-aminoethylether)-*N*,*N*,*N*′,*N*′-tetraacetic acid
EXAFS	extended X-ray absorption fine structure

G3P	D-glyceraldehyde 3-phosphate
Glu, E	glutamic acid
Gly, G	glycine
HDAC	histone deacetylase
His, H	histdine
IGP	3-indole-D-glycerol 3′-phosphate
IMP	inosine 5′-monophosphate
IMPDH	inosine 5′-monophosphate dehydrogenase
LAXS	large angle X-ray scattering
Leu, L	leucine
Lys, K	lysine
MAT	*S*-adenosylmethionine synthase
MD	molecular dynamics
MMR	mismatch repair
NADH	nicotinamide adenine dinucleotide
PDC	pyruvate dehydrogenase complex
PDHK	pyruvate dehydrogenase kinase
PEP	phosphoenolpyruvate
PGH	phosphoglycolohydroxamate
Phe, F	phenylalanine
PLK	pyridoxal kinase
PLP	pyridoxal 5′-phosphate
Pro, P	proline
QM/MM	quantum mechanical/molecular mechanical
Ser, S	serine
Thr, T	threonine
TIM	triosephosphate isomerase
Trp, T	tryptophan
Tyr, Y	tyrosine
Val, V	valine
XMP	xanthosine 5′-monophosphate

Acknowledgment Financial support from the University of Zurich (J.S.) is gratefully acknowledged. M.V. would like to thank my friend and colleague Professor Roland K.O. Sigel for his hospitality and the office space at the Department of Chemistry, University of Zurich.

References

1. A. Rodríguez-Navarro, *Biochim. Biophys. Acta* **2000**, *1469*, 1–30.
2. R. Munns, M. Tester, *Annu. Rev. Plant Biol.* **2008**, *59*, 651–681.
3. D. E. Carden, D. J. Walker, T. J. Flowers, A. J. Miller, *Plant Physiol.* **2003**, *131*, 676–683.
4. A. Oren, *Microbiol. Mol. Biol. Rev.* **1999**, *63*, 334–348.
5. J. K. Lanyi, *Bacteriol. Rev.* **1974**, *38*, 272–290.
6. A. Oren, M. Heldal, S. Norland, E. A. Galinski, *Extremophiles* **2002**, *6*, 491–498.

7. C. H. Suelter, *Science* **1970**, *168*, 789–795.
8. H. J. Evans, G. J. Sorger, *Annu. Rev. Plant Physiol.* **1966**, *17*, 47–76.
9. P. D. Boyer, H. A. Lardy, P. H. Phillips, *J. Biol. Chem.* **1942**, *146*, 673–681.
10. J. F. Kachmar, P. D. Boyer, *J. Biol. Chem.* **1953**, *200*, 669–682.
11. E. Di Cera, *J. Biol. Chem.* **2006**, *281*, 1305–1308.
12. D. W. Smith, *J. Chem. Educ.* **1977**, *54*, 540.
13. R.W. Gurney, *Ionic Processes in Solution,* Chapter 10, McGraw-Hill, London, 1953.
14. Y. Marcus, *Chem. Rev.* **2009**, *109*, 1346–1370.
15. P. R. Smirnov, V. N. Trostin, *Russ. J. Gen. Chem.* **2007**, *77*, 2101–2107.
16. J. Mähler, I. Persson, *Inorg. Chem.* **2012**, *51*, 425–438.
17. A. Bankura, V. Carnevale, M. L. Klein, *J. Chem. Phys.* **2013**, *138*, 014501.
18. D. R. Hall, C. S. Bond, G. A. Leonard, C. I. Watt, A. Berry, W. N. Hunter, *J. Biol. Chem.* **2002**, *277*, 22018–22024.
19. M. M. Harding, *Acta Crystallogr. D, Biol. Crystallogr.* **2002**, *58*, 872–874.
20. S. Rhee, K. D. Parris, S. A. Ahmed, E. W. Miles, D. R. Davies, *Biochemistry* **1996**, *35*, 4211–4221.
21. E. U. Woehl, M. F. Dunn, *Coord. Chem. Rev.* **1995**, *144*, 147–197.
22. T. M. Larsen, M. M. Benning, I. Rayment, G. H. Reed, *Biochemistry* **1998**, *37*, 6247–6255.
23. D. J. Cram, *Science* **1988**, *240*, 760–767.
24. L. Heginbotham, R. MacKinnon, *Biophys. J.* **1993**, *65*, 2089–2096.
25. Y. Zhou, J. H. Morais-Cabral, A. Kaufman, R. MacKinnon, *Nature* **2001**, *414*, 43–48.
26. S. W. Lockless, M. Zhou, R. MacKinnon, *PLoS Biol.* **2007**, *5*, e121.
27. S. Y. Noskov, S. Bernèche, B. Roux, *Nature* **2004**, *431*, 830–834.
28. L. J. Mullins, *J. Gen. Physiol.* **1959**, *42*, 817–829.
29. F. Bezanilla, C. M. Armstrong, *J. Gen. Physiol.* **1972**, *60*, 588–608.
30. S. Y. Noskov, B. Roux, *Biophys. Chem.* **2006**, *124*, 279–291.
31. J. Monod, J. P. Changeux, F. Jacob, *J. Mol. Biol.* **1963**, *6*, 306–329.
32. R. K. Gupta, A. S. Mildvan, *J. Biol. Chem.* **1977**, *252*, 5967–5976.
33. A. D. Mesecar, T. Nowak, *Biochemistry* **1997**, *36*, 6803–6813.
34. A. D. Mesecar, T. Nowak, *Biochemistry* **1997**, *36*, 6792–6802.
35. M. S. Jurica, A. Mesecar, P. J. Heath, W. Shi, T. Nowak, B. L. Stoddard, *Structure* **1998**, *6*, 195–210.
36. T. E. Roche, Y. Hiromasa, *Cell. Mol. Life Sci.* **2007**, *64*, 830–849.
37. J. M. Elliott, P. J. Tayler, J. M. Young, *J. Pharm. Pharmacol.* **1978**, *30*, 27–35.
38. M. Machius, J. L. Chuang, R. M. Wynn, D. R. Tomchick, D. T. Chuang, *Proc. Natl. Acad. Sci. USA* **2001**, *98*, 11218–11223.
39. R. Gudi, M. M. Bowker-Kinley, N. Y. Kedishvili, Y. Zhao, K. M. Popov, *J. Biol. Chem.* **1995**, *270*, 28989–28994.
40. M. M. Bowker-Kinley, W. I. Davis, P. Wu, R. A. Harris, K. M. Popov, *Biochem. J.* **1998**, *329 (Pt 1)*, 191–196.
41. M. Kato, J. L. Chuang, S.-C. Tso, R. M. Wynn, D. T. Chuang, *EMBO J.* **2005**, *24*, 1763–1774.
42. M. J. Page, E. Di Cera, *Physiol. Rev.* **2006**, *86*, 1049–1092.
43. T. Green, A. Grigorian, A. Klyuyeva, A. Tuganova, M. Luo, K. M. Popov, *J. Biol. Chem.* **2008**, *283*, 15789–15798.
44. Y. Hiromasa, X. Yan, T. E. Roche, *Biochemistry* **2008**, *47*, 2312–2324.
45. C. C. Lawrence, J. Stubbe, *Curr. Opin. Chem. Biol.* **1998**, *2*, 650–655.
46. D.-I. Liao, G. Dotson, I. Turner, L. Reiss, M. Emptage, *J. Inorg. Biochem.* **2003**, *93*, 84–91.
47. M. Yamanishi, M. Yunoki, T. Tobimatsu, H. Sato, J. Matsui, A. Dokiya, Y. Iuchi, K. Oe, K. Suto, N. Shibata, Y. Morimoto, N. Yasuoka, T. Toraya, *Eur. J. Biochem.* **2002**, *269*, 4484–4494.
48. N. Shibata, J. Masuda, T. Tobimatsu, T. Toraya, K. Suto, Y. Morimoto, N. Yasuoka, *Structure* **1999**, *7*, 997–1008.

49. J. Masuda, N. Shibata, Y. Morimoto, T. Toraya, N. Yasuoka, *Structure* **2000**, *8*, 775–788.
50. T. Toraya, S. Honda, K. Mori, *Biochemistry* **2010**, *49*, 7210–7217.
51. T. Kamachi, K. Doitomi, M. Takahata, T. Toraya, K. Yoshizawa, *Inorg. Chem.* **2011**, *50*, 2944–2952.
52. Z. Schneider, E. G. Larsen, G. Jacobson, B. C. Johnson, J. Pawelkiewicz, *J. Biol. Chem.* **1970**, *245*, 3388–3396.
53. F. U. Hartl, A. Bracher, M. Hayer-Hartl, *Nature* **2011**, *475*, 324–332.
54. S. M. Wilbanks, D. B. McKay, *J. Biol. Chem.* **1995**, *270*, 2251–2257.
55. K. M. Flaherty, C. DeLuca-Flaherty, D. B. McKay, *Nature* **1990**, *346*, 623–628.
56. J. Wang, D. C. Boisvert, *J. Mol. Biol.* **2003**, *327*, 843–855.
57. K. Ishikura, M. Hasegawa, K. Nomura, T. Okamoto, S. Tanji, T. Abe, T. Fujioka, T. Ohhori, T. Kubo, *Hinyokika Kiyo* **1991**, *37*, 1229–1234.
58. P. D. Kiser, D. T. Lodowski, K. Palczewski, *Acta Crystallogr. Sect. F, Struct. Biol. Cryst. Commun.* **2007**, *63*, 457–461.
59. J. T. Holthausen, C. Wyman, R. Kanaar, *DNA Repair* **2010**, *9*, 1264–1272.
60. L. H. Fornander, K. Frykholm, A. Reymer, A. Renodon-Cornière, M. Takahashi, B. Nordén, *Nucleic Acids Res.* **2012**, *40*, 4904–4913.
61. Y. Wu, X. Qian, Y. He, I. A. Moya, Y. Luo, *J. Biol. Chem.* **2005**, *280*, 722–728.
62. Y. Kokabu, M. Ikeguchi, *Biophys. J.* **2013**, *104*, 1556–1565.
63. di Salvo, Martino L, S. Hunt, V. Schirch, *Protein Expr. Purif.* **2004**, *36*, 300–306.
64. M.-H. Li, F. Kwok, W.-R. Chang, C.-K. Lau, J.-P. Zhang, Lo, Samuel C L, T. Jiang, D.-C. Liang, *J. Biol. Chem.* **2002**, *277*, 46385–46390.
65. F. N. Musayev, M. L. di Salvo, T.-P. Ko, A. K. Gandhi, A. Goswami, V. Schirch, M. K. Safo, *Protein Sci.* **2007**, *16*, 2184–2194.
66. K. D. Collins, *Biophys. J.* **1997**, *72*, 65–76.
67. S. J. Pilkis, M. R. el-Maghrabi, T. H. Claus, *Annu. Rev. Biochem.* **1988**, *57*, 755–783.
68. F. Marcus, M. M. Hosey, *J. Biol. Chem.* **1980**, *255*, 2481–2486.
69. Y. Xue, S. Huang, J. Y. Liang, Y. Zhang, W. N. Lipscomb, *Proc. Natl. Acad. Sci. U.S.A.* **1994**, *91*, 12482–12486.
70. V. Villeret, S. Huang, H. J. Fromm, W. N. Lipscomb, *Proc. Natl. Acad. Sci. U.S.A.* **1995**, *92*, 8916–8920.
71. J. Y. Choe, H. J. Fromm, R. B. Honzatko, *Biochemistry* **2000**, *39*, 8565–8574.
72. R. Zhang, V. Villeret, W. N. Lipscomb, H. J. Fromm, *Biochemistry* **1996**, *35*, 3038–3043.
73. G. D. Markham, E. W. Hafner, C. W. Tabor, H. Tabor, *J. Biol. Chem.* **1980**, *255*, 9082–9092.
74. G. L. Cantoni, *Annu. Rev. Biochem.* **1975**, *44*, 435–451.
75. G. D. Markham, M. A. Pajares, *Cell. Mol. Life Sci.* **2009**, *66*, 636–648.
76. J. Komoto, T. Yamada, Y. Takata, G. D. Markham, F. Takusagawa, *Biochemistry* **2004**, *43*, 1821–1831.
77. A. A. Evarsson, J. L. Chuang, R. M. Wynn, S. Turley, D. T. Chuang, W. G. Hol, *Structure* **2000**, *8*, 277–291.
78. J. N. Jansonius, *Curr. Opin. Struct. Biol.* **1998**, *8*, 759–769.
79. H. G. Aaslestad, A. D. Larson, *J. Bacteriol.* **1964**, *88*, 1296–1303.
80. J. W. Keller, K. B. Baurick, G. C. Rutt, M. V. O'Malley, N. L. Sonafrank, R. A. Reynolds, L. O. Ebbesson, F. F. Vajdos, *J. Biol. Chem.* **1990**, *265*, 5531–5539.
81. M. D. Toney, E. Hohenester, S. W. Cowan, J. N. Jansonius, *Science* **1993**, *261*, 756–759.
82. M. D. Toney, E. Hohenester, J. W. Keller, J. N. Jansonius, *J. Mol. Biol.* **1995**, *245*, 151–179.
83. H. Holzer, C. Cennamo, M. Boll, *Biochem. Biophys. Res. Commun.* **1964**, *14*, 487–492.
84. T. Yamada, J. Komoto, Y. Takata, H. Ogawa, H. C. Pitot, F. Takusagawa, *Biochemistry* **2003**, *42*, 12854–12865.
85. M. N. Isupov, A. A. Antson, E. J. Dodson, G. G. Dodson, I. S. Dementieva, L. N. Zakomirdina, K. S. Wilson, Z. Dauter, A. A. Lebedev, E. H. Harutyunyan, *J. Mol. Biol.* **1998**, *276*, 603–623.

86. S. B. Ruvinov, S. A. Ahmed, P. McPhie, E. W. Miles, *J. Biol. Chem.* **1995**, *270*, 17333–17338.
87. E. E. Snell, *Adv. Enzymol. Relat. Areas Mol. Biol.* **1975**, *42*, 287–333.
88. D. Milić, D. Matković-Calogović, T. V. Demidkina, V. V. Kulikova, N. I. Sinitzina, A. A. Antson, *Biochemistry* **2006**, *45*, 7544–7552.
89. T. A. Kunkel, D. A. Erie, *Annu. Rev. Biochem.* **2005**, *74*, 681–710.
90. A. Guarné, S. Ramon-Maiques, E. M. Wolff, R. Ghirlando, X. Hu, J. H. Miller, W. Yang, *EMBO J.* **2004**, *23*, 4134–4145.
91. C. Ban, M. Junop, W. Yang, *Cell* **1999**, *97*, 85–97.
92. A. Guarné, M. S. Junop, W. Yang, *EMBO J.* **2001**, *20*, 5521–5531.
93. X. Hu, M. Machius, W. Yang, *FEBS Lett.* **2003**, *544*, 268–273.
94. J. Park, R. S. Gupta, *Cell. Mol. Life Sci.* **2008**, *65*, 2875–2896.
95. A. Anderson, R. A. Cooper, *Biochim. Biophys. Acta* **1969**, *177*, 163–165.
96. C. E. Andersson, S. L. Mowbray, *J. Mol. Biol.* **2002**, *315*, 409–419.
97. J. Li, C. Wang, Y. Wu, M. Wu, L. Wang, Y. Wang, J. Zang, *J. Struct. Biol.* **2012**, *177*, 578–582.
98. S. K. Kurdistani, M. Grunstein, *Nat. Rev. Mol. Cell Biol.* **2003**, *4*, 276–284.
99. A. Vannini, C. Volpari, G. Filocamo, E. C. Casavola, M. Brunetti, D. Renzoni, P. Chakravarty, C. Paolini, R. de Francesco, P. Gallinari, C. Steinkühler, S. Di Marco, *Proc. Natl. Acad. Sci. U.S.A.* **2004**, *101*, 15064–15069.
100. J. R. Somoza, R. J. Skene, B. A. Katz, C. Mol, J. D. Ho, A. J. Jennings, C. Luong, A. Arvai, J. J. Buggy, E. Chi, J. Tang, B.-C. Sang, E. Verner, R. Wynands, E. M. Leahy, D. R. Dougan, G. Snell, M. Navre, M. W. Knuth, R. V. Swanson, D. E. McRee, L. W. Tari, *Structure* **2004**, *12*, 1325–1334.
101. N. A. Wolfson, C. A. Pitcairn, C. A. Fierke, *Biopolymers* **2013**, *99*, 112–126.
102. S. L. Gantt, C. G. Joseph, C. A. Fierke, *J. Biol. Chem.* **2010**, *285*, 6036–6043.
103. L. Hedstrom, *Chem. Rev.* **2009**, *109*, 2903–2928.
104. T. V. Riera, L. Zheng, H. R. Josephine, D. Min, W. Yang, L. Hedstrom, *Biochemistry* **2011**, *50*, 8508–8518.
105. R. D. Kobes, R. T. Simpson, R. L. Vallee, W. J. Rutter, *Biochemistry* **1969**, *8*, 585–588.
106. D. R. Hall, G. A. Leonard, C. D. Reed, C. I. Watt, A. Berry, W. N. Hunter, *J. Mol. Biol.* **1999**, *287*, 383–394.
107. S. H. Kim, K. P. Lim, H. S. Kim, *J. Dairy Sci.* **1997**, *80*, 2264–2269.
108. D. H. Juers, T. D. Heightman, A. Vasella, J. D. McCarter, L. Mackenzie, S. G. Withers, B. W. Matthews, *Biochemistry* **2001**, *40*, 14781–14794.
109. J. Xu, McRae, Mary A A, S. Harron, B. Rob, R. E. Huber, *Biochem. Cell Biol.* **2004**, *82*, 275–284.
110. F. Jacob, J. Monod, *J. Mol. Biol.* **1961**, *3*, 318–356.
111. D. H. Juers, R. H. Jacobson, D. Wigley, X. J. Zhang, R. E. Huber, D. E. Tronrud, B. W. Matthews, *Protein Sci.* **2000**, *9*, 1685–1699.
112. D. H. Juers, B. Rob, M. L. Dugdale, N. Rahimzadeh, C. Giang, M. Lee, B. W. Matthews, R. E. Huber, *Protein Sci.* **2009**, *18*, 1281–1292.
113. M. F. Dunn, *Arch. Biochem. Biophys.* **2012**, *519*, 154–166.
114. A. Peracchi, A. Mozzarelli, G. L. Rossi, *Biochemistry* **1995**, *34*, 9459–9465.
115. E. U. Woehl, M. F. Dunn, *Biochemistry* **1995**, *34*, 9466–9476.
116. J. A. Huntington, *Biol. Chem.* **2008**, *389*, 1025–1035.
117. C. M. Wells, E. Di Cera, *Biochemistry* **1992**, *31*, 11721–11730.
118. E. Di Cera, E. R. Guinto, A. Vindigni, Q. D. Dang, Y. M. Ayala, M. Wuyi, A. Tulinsky, *J. Biol. Chem.* **1995**, *270*, 22089–22092.
119. E. Zhang, A. Tulinsky, *Biophys. Chem.* **1997**, *63*, 185–200.
120. K. Schärer, M. Morgenthaler, R. Paulini, U. Obst-Sander, D. W. Banner, D. Schlatter, J. Benz, M. Stihle, F. Diederich, *Angew. Chem. Int. Ed.* **2005**, *44*, 4400–4404.

121. S. P. Bajaj, A. E. Schmidt, S. Agah, M. S. Bajaj, K. Padmanabhan, *J. Biol. Chem.* **2006**, *281*, 24873–24888.
122. A. E. Schmidt, K. Padmanabhan, M. C. Underwood, W. Bode, T. Mather, S. P. Bajaj, *J. Biol. Chem.* **2002**, *277*, 28987–28995.
123. N. D. Werbeck, J. Kirkpatrick, J. Reinstein, D. F. Hansen, *ChemBiochem* **2014**, *15*, 543–548.
124. S. M. Wilbanks, D. B. McKay, *Biochemistry* **1998**, 37, 7456–7462.
125. L. T. Laughlin, G. H. Reed, *Arch. Biochem. Biophys.* **1997**, *348*, 262–267.
126. L. Gan, M. R. Seyedsayamdost, S. Shuto, A. Matsuda, G. A. Petsko, L. Hedstrom, *Biochemistry* **2003**, *42*, 857–863.
127. A. O. Pineda, C. J. Carrell, L. A. Bush, S. Prasad, S. Caccia, Z.-W. Chen, F. S. Mathews, E. Di Cera, *J. Biol. Chem.* **2004**, *279*, 31842–31853.

Chapter 9
Roles and Transport of Sodium and Potassium in Plants

Manuel Nieves-Cordones, Fouad Razzaq Al Shiblawi, and Hervé Sentenac

Contents

M. Nieves-Cordones • F.R. Al Shiblawi • H. Sentenac (✉)
Laboratory of Plant Biochemistry and Molecular Physiology, UMR BPMP CNRS/INRA/MontpellierSupAgro, University of Montpellier, INRA,
Place Viala, F-34060 Montpellier cedex 1, France
e-mail: sentenac@supagro.inra.fr

A. Sigel, H. Sigel, and R.K.O. Sigel (eds.), *The Alkali Metal Ions: Their Role for Life*,
Metal Ions in Life Sciences 16, DOI 10.1007/978-3-319-21756-7_9

Abstract The two alkali cations Na^+ and K^+ have similar relative abundances in the earth crust but display very different distributions in the biosphere. In all living organisms, K^+ is the major inorganic cation in the cytoplasm, where its concentration (ca. 0.1 M) is usually several times higher than that of Na^+. Accumulation of Na^+ at high concentrations in the cytoplasm results in deleterious effects on cell metabolism, e.g., on photosynthetic activity in plants. Thus, Na^+ is compartmentalized outside the cytoplasm. In plants, it can be accumulated at high concentrations in vacuoles, where it is used as osmoticum. Na^+ is not an essential element in most plants, except in some halophytes. On the other hand, it can be a beneficial element, by replacing K^+ as vacuolar osmoticum for instance. In contrast, K^+ is an essential element. It is involved in electrical neutralization of inorganic and organic anions and macromolecules, pH homeostasis, control of membrane electrical potential, and the regulation of cell osmotic pressure. Through the latter function in plants, it plays a role in turgor-driven cell and organ movements. It is also involved in the activation of enzymes, protein synthesis, cell metabolism, and photosynthesis. Thus, plant growth requires large quantities of K^+ ions that are taken up by roots from the soil solution, and then distributed throughout the plant. The availability of K^+ ions in the soil solution, slowly released by soil particles and clays, is often limiting for optimal growth in most natural ecosystems. In contrast, due to natural salinity or irrigation with poor quality water, detrimental Na^+ concentrations, toxic for all crop species, are present in many soils, representing 6 % to 10 % of the earth's land area. Three families of ion channels (Shaker, TPK/KCO, and TPC) and 3 families of transporters (HAK, HKT, and CPA) have been identified so far as contributing to K^+ and Na^+ transport across the plasmalemma and internal membranes, with high or low ionic selectivity. In the model plant *Arabidopsis thaliana*, these families gather at least 70 members. Coordination of the activities of these systems, at the cell and whole plant levels, ensures plant K^+ nutrition, use of Na^+ as a beneficial element, and adaptation to saline conditions.

Keywords Channel • Enzyme • Membrane transport • Plant • Potassium • Sodium • Transporter • Turgor

Please cite as: *Met. Ions Life Sci.* 16 (2016) 291–324

1 Introduction

K^+ is the most abundant cation in the cytosol, where the order of magnitude of its activity is 0.1 M, both in animals [1–3] and in plants [4–6]. The activity of Na^+ in the cytosol can display large variations, depending on the cell type, but is thought to usually remain lower than that of K^+ [3, 7, 8]. In plant cells, the cytosolic concentration of Na^+ seems to be controlled below 20–30 mM [9–11].

Reasons why K^+ has been selected as the major cytosolic cation during evolution and not Na^+, whereas the relative abundances of these two cations in the earth crust are quite similar, have been tentatively discussed [12, 13]. Life probably appeared in a seawater that possessed an ionic composition similar to that of the present oceans: a high concentration of Na^+, of a few hundreds of mM, and a much lower concentration of K^+, of a few tens of mM. It is thus possible that, for the first living cells, accumulation of the less abundant cation, K^+, and exclusion of the most abundant one, Na^+, was the simplest process to energize the cell membrane [13]. From a biophysical point of view, K^+ might have been selected during evolution because its hydration shell displays specific features, when compared to that of Na^+, in terms of hydration energy and hydration shell features. Weaker interactions with the first hydration shell in K^+ than in Na^+ results in larger structural flexibility and reduced disruption of the bulk water network and water arrangement close to proteins [14–16]. This might be the reason why Na^+ cannot totally replace K^+ as a coordinating ion in certain enzymatic reactions.

Molecular simulations aiming at investigating the physical and dynamical nature of the cytosol indicate that K^+ is essentially present as an unbound highly mobile osmolyte, populating the solution fairly uniformly and contributing thereby to control of the osmotic potential [17]. Na^+ can be used as osmoticum in plants, but essentially in the vacuole, where its concentration can become higher than 100 mM [18–20], and not in the cytosol [21].

With respect to the roles of Na^+ in the living world, the mechanisms underlying cell membrane energization and solute transport are fundamentally different between animals and plants. In animals, Na^+/K^+ ATPases energize the cell membrane by building up the Na^+ transmembrane electrochemical gradient that fuels Na^+-cotransporters and solute transport activity. In plants, the cell membrane is energized by H^+-ATPases (the so-called proton pumps), and active transport systems are H^+-cotransporters [22, 23]. There is so far no indication that the plasma membrane is equipped with transport systems using the Na^+ transmembrane electrochemical gradient to mediate active transport of solutes in higher plant cells. The only example of a Na^+-driven cotransport system identified in plants so far is a pyruvate transporter localized at the chloroplast envelope membrane [24, 25].

Another point that is worth to be noted in this introduction section is that most K^+ ions enter the trophic chain precisely at the moment when they are taken up by transport systems active at the plasmalemma of root cells. These transport systems are thus of major importance not only for plant K^+ nutrition and adaptation to low K^+ availability and but also for animal diet quality [26]. A major part of the present review is devoted to the presentation of the current knowledge on plant K^+ and Na^+ transport systems.

2 Potassium and Sodium Ion Concentrations in Soils

2.1 K^+ Availability in Soils

Potassium is amongst the 5 most abundant elements present in the upper continental crust, where its relative content, expressed in ppm, is close to 29×10^3 (2.9 %), a little bit above that of Na^+ (2.6 %). Therefore, soil K^+ reserves are generally large. However, important agricultural areas of the World are reported to be deficient in K^+ availability. This is the case for instance of 3/4 of the paddy soils of China, and 2/3 of the wheat belt of Southern Australia [27–29].

Three types of K^+ pools can be operationally distinguished in the soil [26]. Most of the K^+ ions in soil are in the so-called "structural form" (Figure 1), mainly comprised of K^+-bearing primary minerals such as muscovite, biotite, and feldspars. This first pool of K^+ ions is considered as slowly- or non-available to plants. However, it contributes to the plant supply in the long term [30, 31]. The other two pools contribute directly to plant nutrition [32–34]. The second pool, in terms of pool size, is constituted by potassium ions fixed in 2:1 interlayer clay minerals and is called the "non-exchangeable'" potassium pool (Figure 1) [35]. The size of this "non-exchangeable" pool depends on the soil content in clays. Illite and illite-like clays are the major sources of K^+ at least in temperate region soils [36]. The third pool is usually operationally defined as potassium ions readily "extractable" by water and/or aqueous solutions (Figure 1). Simply speaking, it comprises the pool of K^+ ions either dissolved in the soil solution or weakly bound to soil minerals. Upon K^+ concentration decreases in the soil solution, due to, e.g., root K^+ uptake, chemical equilibria result in desorption of K^+ ions from the pool of weakly bound exchangeable ions as well as in K^+ release from the "non-exchangeable" pool and even from K-feldspars [28].

The concentration of K^+ in soil solution is highly variable. Typical concentrations lie in the 10^{-5} to 10^{-3} M range [37–39]. It should be noted that, when compared with root K^+ uptake capacity, K^+ diffusion in the soil can be rate-limiting. In other words, roots are able to take up K^+ at a higher rate than this cation can diffuse from the bulk soil solution to the root surface, at least in some soils and environmental conditions. This results in K^+ depletion at the root surface, down to very low concentrations, of a few μM [40–42].

2.2 Plant K^+ Demand and K^+ Deficiency

Plant K^+ contents for optimal growth are in the range of 2–5 % of the dry weight of vegetative parts [35, 38]. In the case of maize or wheat plants grown in fertilized field conditions, K^+ uptakes of about 200 kg per hectare (ha) have been reported [43, 44]. The amount of K^+ actually removed from the field depends of course upon the plant parts that are harvested. For instance, more K^+ is removed when almost all

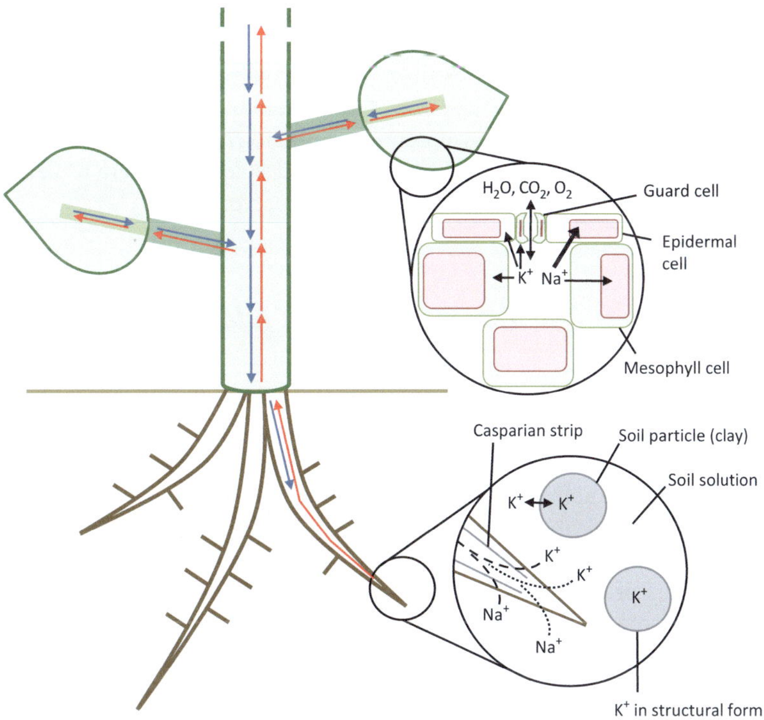

Figure 1 Overview of K^+ and Na^+ uptake and long distance transport between roots and shoots. Three K^+ pools can be operationally identified in the soil: (i) K^+ in "structural form" (slowly- or non-available to plants), (ii) "non-exchangeable" K^+, weakly bound at the surface of clay minerals, and (iii) extractable K^+ in aqueous soil solution. In root differentiated zones, due to the presence of the endodermis Casparian strip (solid gray line) that acts as a barrier preventing free diffusion within the so-called free space (cell wall continuum) towards the center of the root, K^+ and Na^+ ions (as well as other nutrient ions) have first to be selectively taken up by plasma membrane transport systems of root peripheral cells (epidermis, cortex, and endodermis) to enter the root symplasm. Then, diffusion within the root symplasm allows ions to pass the endodermis barrier (dashed line pathway) and to reach the root stele and xylem vasculature, located at the center of the root, where the flow of crude sap carries nutrient ions towards the shoots. The apical (meristematic) region of the growing root does not possess such an endodermis barrier. In these zones, ions can freely diffuse within the apoplasm and directly reach the root stele and xylem vessels (dotted line pathway), without any control by membrane transport systems. Once in the xylem, ions are transported (mass flow) towards and throughout the shoots by the xylem sap flow (red arrows). Arrived in leaves and unloaded from leaf vascular tissues, K^+ and Na^+ reach the leaf apoplasm, from which they are taken up by mesophyll, epidermal and guard cells. Na^+ ions are preferentially accumulated in leaf cell vacuoles and selectively compartmentalized in epidermal cells (wider arrow). Rapid K^+ transport into and out of guard cells controls the turgor of these cells, and thereby the diameter of the stomatal pore and the gas exchange rates between the inner leaf tissues and the atmosphere. In mature photosynthesizing (source) leaves, K^+ and Na^+ ions can also be loaded into the phloem sap (blue arrows) to be transported towards sink organs (roots, young leaves, fruits, etc.).

aboveground biomass is taken off at harvest (like in the case of sugar cane) than when only the seeds (cereals) are harvested [45]. In intensive agricultural production systems, lack of sufficient K^+ fertilizer application leads to significant depletion of available soil K^+ reserves [46]. A considerable area of farmland has become K^+-deficient [47–49]. Early visible symptoms of K^+ deficiency are brown spots at leaf surface or leaf edge and chlorosis at the tip of the oldest leaves. Severe K^+ deficiency results in further symptoms, including wilting and necrosis [26].

2.3 Saline Soils, Na^+ Toxicity, and Plant Adaptation to Salt Stress

Soil salinity results in both reduced soil water availability (due to the decrease in water potential) and ionic toxicity. Most crop plants are sensitive to soil salinity, and evidence is available at the physiological and molecular levels that Na^+ is the major cause of the toxicity in most cases. Indeed, for instance, genetic analyses and searches for QTLs of plant tolerance to salt stress have identified genes encoding Na^+ transporters or channels involved in Na^+ entry into the root or control of Na^+ transport to shoots in the model plant *Arabidopsis thaliana* [50] as well as in rice and wheat [51–53]. It should however be noted that, at high concentrations, the anion Cl^- can display its own toxicity [54].

Soils are classified as saline when their salt content affects the growth of most crop species. This may occur when the conductance of the soil solution (at soil saturation water content) displays an electrical conductance corresponding to that of 50 mM NaCl solution [38]. However, the impact of a given amount of salts on plant growth depends on the plant species, climatic conditions, soil features such as the water capacity and texture, and the nature of the salts, e.g., NaCl or Na_2SO_4, the former salt being more toxic than the latter in most species [55]. Estimates of the area of salt-affected soils vary widely, ranging from 6 % to 10 % of the earth's land area [52, 56]. Amongst "natural" saline areas are salt marshes of the temperate zones, mangrove swamps of the subtropics and tropics, and lands with saline underground water in arid and semi-arid regions where evapotranspiration is high. Besides such natural contexts, soil salinity can result from intensive agriculture and irrigation with low quality water [57]. This salinization process, which already affects 20 % of the irrigated lands, would result in loss of 3 ha of arable land from conventional crop farming every minute [58]. Enhanced demand for irrigation due to both population increase and climate change, are predicted to dramatically speed up such a loss of arable land [57, 59].

Plant adaptation to salt stress (resulting from high external Na^+ concentrations) is one of the most widely investigated domains in plant biology [52, 58, 60]. Large differences in tolerance level have been observed amongst species and even cultivars or ecotypes. At one extreme are plants named glycophytes, like the model plant *Arabidopsis thaliana*, rice or bean, which are sensitive to salt stress, being strongly affected by external NaCl concentrations higher than about 50 mM. This is the case of most plant species and crops [52]. At the other extreme are plants named halophytes, like *Aeluropus littoralis* or the saltbush Atriplex, which can thrive in environments where the salt concentration is higher than 300 mM [52, 61]. Such species constitute about 1 % of the world's flora [62]. In between are species such as barley or alfalfa, which can cope with moderate salt concentrations [52, 63].

The mechanisms involved in tolerance to salt stress are highly complex and variable amongst species. They include, for instance, biochemical responses leading to synthesis of large amounts of organic osmoprotectant molecules, such as proline or glycine betaine, or anatomical adaptations such as the presence at the leaf surface of salt glands that excrete NaCl, or of very large cells, named salt bladders, expanding from the leaf epidermis and sequestering excessive Na^+ away from internal (photosynthesizing) leaf tissues [64] (see also below, Section 3.2). Despite this large diversity of mechanisms, in every species and every soil conditions, glycophytes in the presence of low external salt concentrations or halophytes in the presence of high salt concentrations, adaptation to external Na^+ involves tight regulation of K^+ and Na^+ membrane transport activity, especially allowing selective K^+ uptake by roots, control of Na^+ uptake and translocation towards the shoots by the xylem sap, and efficient Na^+ compartmentalization into vacuoles. It is also worth noting that part of the toxic effects of Na^+ is likely to lie in disruption of the K^+ membrane transport [65, 66] and homeostasis [10].

3 Potassium and Sodium Ion Fluxes and Distribution Within the Plant

3.1 K^+ and Na^+ Uptake and Long Distance Transport Between Roots and Shoots

Plant roots exhibit a polarized anatomy in which outer layers (epidermal and cortical cells) are involved in K^+ (and other nutrient ions) selective uptake and inner layers (endodermis and central vasculature) in secretion of the nutrient ions into the xylem sap towards shoot tissues [67]. In plants, an external skeleton named cell wall, essentially made of cellulose, hemicellulose, and polygalacturonic acids, forms a matrix freely accessible to water and soluble ions that surrounds the plasma membrane of each cell. The cell wall continuum is named "free space" or apoplasm. In the root, nutrient ions from the soil can enter the apoplasm and freely diffuse towards inner cell layers. However, this pathway is interrupted by a cell layer, named endodermis, whose radial cell walls are impregnated with hydrophobic compounds, forming the so-called Casparian strip, which is impermeable to water and ions [38, 68, 69]. Thus, the endodermis is the dead-end of the apoplastic pathway [70, 71]. To go beyond this impermeable barrier towards the center of the root and the plant vasculature, nutrient ions have to enter the cytosol of a root peripheral cell (from the epidermis, cortex or endodermis). They can then diffuse from cell to cell through pores, named plasmodesmata, which connect the cytosolic milieu of two neighboring cells [72]. Diffusion within the symplasm beyond the endodermis barrier allows nutrient ions to reach the central part of the root, named stele. Once in the stele, nutrient ions have to be secreted into the stelar apoplasm, where they diffuse towards the xylem vessels, which contain the so-called crude (or xylem) sap. The sap flows upward towards the shoots (mass flow pulled by the transpiration stream or pushed by the so-called root pressure), providing the aerial parts of the plant with nutrient ions.

The Casparian strip is always absent in the apical (meristematic) region of the root and temporarily absent in the differentiated part of the root at points of secondary root emergence (Figure 1) [73]. At the level of such defects in the endodermis barrier, ions can freely diffuse within the apoplasm and reach the root stele and xylem vessels. However, this uncontrolled flow of nutrient ions towards the xylem sap remains relatively low. Most of the ions that reach the root xylem vessels have been taken up across the plasma membrane of a root peripheral cell at least one time [69]. In other words, their entry into the root symplasm has been "catalyzed", and thus controlled, by membrane transport systems, channels, transporters or co-transporters. Such a control of ion translocation from the soil to the shoots within the roots concerns all nutrient ions, including of course K^+ and Na^+, under standard physiological conditions [66, 74–76]. However, in saline soils, where the concentration of Na^+ is high, an important amount of this cation can reach the stele and xylem vessels directly through the defects in the endodermis barrier, without any control by any cell membrane. Such an uncontrolled flow of Na^+ ions contributes to the stressing effects of the external saline conditions [73, 77].

Once in the root xylem vasculature, during their mass flow-mediated transport towards the shoots (Figure 1), K^+ and Na^+ ions can be reabsorbed from the xylem sap into adjacent cells (parenchyma cells). This reabsorption is likely to result in increased root K^+ contents and thereby to favor root growth (at the expense of shoot growth) upon abiotic stresses such as drought or K^+ deficiency [78–81]. Regarding Na^+, retrieval of this cation from the xylem sap (the so-called sap-desalinization process) results in reduced Na^+ translocation towards the shoots and photosynthesizing tissues, which contributes to plant adaptation to a salinity constraint [50, 64, 82–84].

When arrived in leaves via the xylem vasculature, K^+ and Na^+ are used by various cell types. The photosynthesizing mesophyll cells, which constitute the largest part of leaf internal tissues, accumulate K^+ and Na^+ from the apoplastic and the symplastic pathways (Figure 1). Symplastic connections between mesophyll cells and epidermal cells seem to be transitory [85–87]. In the epidermis, the guard cells (osmocontractile cells that control the apertures of the stomatal pores; see below Section 4.2) are symplastically isolated. Thus, ion fluxes underlying their osmocontractility (K^+ playing the major role in this process) occur through specialized membrane proteins (essentially Shaker channels for K^+ [88, 89]; see Sections 4.2 and 5). It is worth noting that the leaf apoplasm volume is relatively small, so that rapid arrival of Na^+ ions upon saline conditions, if these ions are not rapidly taken up by leaf cells (and compartmented in the vacuoles, see below), can result in progressive increase in Na^+ concentration in the cell walls. This leads to a decrease in external water potential down to values lower than the water potential in the cell, resulting in water efflux from the cells, loss of turgor and eventually cell death.

Evidence is available that Na^+ ions are preferentially accumulated in epidermal cells (Figure 1). This compartmentalization process, at the tissue level, protects the inner photosynthesizing mesophyll cells against detrimental effects of high Na^+ concentrations. At the whole plant level, Na^+ can be preferentially accumulated in old leaves to prevent toxic accumulation in the younger ones [90]. Na^+ ions can also be re-circulated towards the shoots by the elaborated sap flow [50]. This sap, which is produced in leaves, contains photosynthates (mainly sucrose). It flows (osmotically driven mass flow) within the phloem vasculature towards sink organs (roots, young developing leaves, fruits, etc.) that it feeds. The phloem tubes, which conduct this elaborated sap (more often named phloem sap) are living cells that do not display the large central vacuole typical of many plant cell types. As living cells, they display the classical ionic composition of the cytosol, i.e., low Na^+ concentration and high K^+ concentration. Probably related to that, the rate of Na^+ recirculation from shoots to roots by the phloem sap [50] is relatively low when compared to that of K^+ [73, 91, 92]. Regarding the latter ion, it has been reported that K^+ recirculation from shoots to roots by the phloem sap can provide, at the steady state, more than 60 % of the amounts of K^+ ions transported from roots to shoots by the xylem sap, the difference corresponding to the actual uptake of K^+ from the soil solution (Figure 1) [93, 94]. This large and continuous cycling of K^+ ions from roots to shoots via the xylem vasculature and from shoots to roots via the phloem vasculature would play a role in K^+ demand signaling and regulation of K^+ uptake and distribution [79, 91, 95, 96].

3.2 Cellular Compartmentalization of K^+ and Na^+

As indicated above, with the exception of some cell types such as phloem tube cells and meristematic cells, most plant cells display a large central vacuole, which represents approximately 90 % of the cellular volume and can thereby be the main deposit for K^+ and Na^+ ions within the cell [97, 98]. K^+ accumulation in the vacuole leads to turgor build-up that importantly contributes to cellular expansion and/or osmocontractility and cell movements (see Section 4.1). The concentration of K^+ in the vacuole can also substantially vary depending on availability of this cation in the nutritive solution [6, 9, 38, 99]. Upon K^+ starvation, vacuolar K^+ ions are released into the cytoplasm, allowing a relative homeostatic control of the cytosolic K^+ concentration, close to 0.1 M [4]. Once the vacuolar K^+ pool is exhausted, the cytosolic concentration of K^+ gradually decreases [6, 100], resulting in detrimental effects on cell physiology and eventually death.

Both the plasma membrane and the vacuolar membrane (also named tonoplast) are energized by transmembrane electrochemical H^+ gradients built up by proton pumps, which are P-type H^+-ATPases in the case of the former membrane, and V-type ATPases and H^+-secreting membrane pyrophosphatases in the case of the latter one. The electrical component of the gradient across the plasma membrane can be very negative (–250 mV; negative inside) and variable, depending on H^+ secretion activity but also on K^+ external concentration (see Section 4.3) [101–103]. The magnitude of the transmembrane electrical gradient across the tonoplast is much less important, being at most of a few tens of mV, the vacuolar lumen being positive with respect to the cytosol [6, 91, 104]. K^+ and Na^+ fluxes are mediated by channels and uniporters, and active fluxes mediated by H^+-cotransporters (symporters or antiporters) [23].

Regarding K^+, strong membrane hyperpolarization of the plasma membrane allows efficient uptake of this cation through K^+ channels even from diluted external solutions, displaying K^+ concentrations as low as about 10 μM [23, 102, 105, 106]. H^+:K^+ symporters are responsible for active K^+ uptake from still more diluted solutions [23, 103, 107]. In some physiological situations and cell types (e.g., in xylem parenchyma cells bordering the xylem vessels and responsible for nutrient ion secretion into the xylem sap (see above), or in guard cells during stomatal closure (see below, Section 4.1)), the plasma membrane can become poorly polarized, allowing K^+ efflux through K^+ channels [78, 108–110].

Regarding Na^+, vacuolar accumulation of this cation decreases the cellular water potential and thus contributes to cell turgor (see below Section 4.1). It can thereby have beneficial effects on plant growth, especially when K^+ availability is low. However, since Na^+ is toxic when largely accumulated in the cytosol, plant cells try to minimize the cytosolic pool of this cation when its external concentration becomes high [110]. Such a control is critical for preventing toxicity symptoms, which mainly stem from Na^+ effects on cellular metabolism [111] (see Section 4.4). Plants prevent large accumulation of Na^+ in the cytosol by compartmentalizing this cation into the vacuole [73] or extruding it outside the cell, e.g., in soil in the case of root peripheral cells, or at the leaf surface, and not into the leaf apoplast where increased concentrations of Na^+ would have detrimental effects (see above). Some halophytes (salt-adapted species) like *Distichlis spicata* or mangroves produce exudates that are highly concentrated in Na^+, from leaf specialized structures called salt glands, which result from the modification of hydathodes [73, 112, 113]. Other species deposit significant amounts of Na^+ in either swollen vacuoles in succulent tissues or in specialized external structures called epidermal bladder cells [58, 62]. Such salt bladders seem to originate from trichomes in which the outmost cell grows until it takes a balloon-like form with a huge central vacuole [114, 115].

4 Roles of Potassium and Sodium Ions in Plants

4.1 Roles of K^+ in Cell Turgor Building

Accumulation of solutes like K^+ or Na^+ inside the cell lowers the water potential, leading to water entry into the cell. This can have beneficial effects under adverse conditions like drought in which K^+ accumulation within plant cells gives rise to an improved osmotic adjustment. In case of Na^+ excess, keeping high internal K^+ concentrations has been proven to be a key determinant of salt tolerance by helping retaining water and reducing Na^+ uptake [110, 116].

Since the cell wall restricts changes in cell volume, water uptake due to solute accumulation results in hydrostatic pressure (turgor). Such pressure is the primary force driving cell growth through cell expansion, and thereby growth at the tissue and whole plant levels. Fruit growth provides a typical example of turgor-driven cell expansion where flesh cells accumulate K^+, solutes and water to increase in size. On the other hand, cell expansion also requires changes in the cell wall architecture in order to be extensible. This involves loosening and the continued cutting and pasting of new material into the texture of the wall [117].

4.2 *Role of K^+ in Turgor-Driven Movements in Plants*

4.2.1 Regulation of Stomatal Aperture at the Leaf Surface and Control of Gas Exchanges with the Atmosphere

In terrestrial plants, a protecting waxy cuticle covers the epidermis of the aerial organs, preventing water loss and desiccation. Simultaneously, this hydrophobic barrier impedes diffusion of atmospheric CO_2 towards the inner photosynthesizing tissues. Gas exchanges between these tissues and the atmosphere mainly take place through microscopic pores, named stomata (Figure 1). Two osmocontractile cells surrounding the pore, named guard cells, control stomatal aperture. Such a control allows the plant to cope with the conflicting needs of maintaining a sufficient internal CO_2 concentration for photosynthesis and of preventing excessive transpirational water loss under diverse environmental conditions [118, 119]. Guard cells regulate the aperture of stomatal pores in response to many physiological stimuli such as light, soil water availability and leaf water status, CO_2 internal concentration and hormones [120, 121]. These cells are not connected to the symplasm of their neighboring cells (via plasmodesmata) and their cellular movements are rapid and driven by osmotic changes. Stomatal pore size will determine gas exchange rates between photosynthetic cells and the atmosphere. An increase in guard cell turgor promotes pore opening, whereas a decrease in turgor leads to stomatal closure [122].

The available information indicates that the main solutes involved in the guard cell osmocontractility are K^+, accompanying anions (malate and chloride) and sucrose, depending on the environmental conditions. During stomatal opening, guard cell volume significantly increases because of the activation of plasma membrane H^+-ATPases and inwardly-rectifying K^+ channels from the Shaker family (see Section 5.1) and organic acid production and uptake of inorganic anions (mainly Cl^- and NO_3^-) [121, 123]. Osmolyte accumulation leads to a lower water potential in guard cells, which, in turn, induces osmotic water influx into these cells. Moreover, multiple smaller vacuoles fuse to form a large central vacuole that leads to a remarkable increase in guard cell volume [124]. K^+ uptake into vacuoles during stomatal opening is dependent on H^+/K^+ antiporter activity [125, 126]. During stomatal closure, guard cell volume decreases owing to net cellular efflux of solutes. Anion efflux through anion channels induces membrane depolarization that activates outwardly-rectifying K^+ channels, leading to K^+ efflux [127, 128]. The resulting cellular export of K^+, Cl^- and organic ions results in water efflux from guard cells. At the vacuole membrane, Ca^{2+}-activated K^+ channels are involved in K^+ release into the cytoplasm [118, 129].

4.2.2 Leaf Movements and Other Organ Movements

Plants also display organ movements, which are classically sorted into two main types, tropic and nastic movements. Tropisms are oriented by the direction of the stimulus that induce the movement. In contrast, the direction of a nastic movement is independent of the stimulus's direction or position. Another difference is that the movement is irreversible in the case of tropisms, while it is generally reversible and repeatable in the case of nastic movements. The distinction between these two types of movement is however sometimes unclear.

A tropism (from Greek, tropos, to turn) relies on asymmetric growth between two opposite regions of the organ due to redistribution of growth regulators, coordinated cell division and turgor-dependent cell expansion. It also involves K^+ transport and accumulation. The difference in growth results in a curvature of the growing organ towards or away from the stimulus (e.g., light or gravity). A classical example of such a movement is root gravitropism [130, 131]. Although most nastic movements do not involve growth responses, it is worth to note that the terms epinasty and hyponasty are used to qualify bending of an organ which does involve differential growth, e.g. greater growth of the upper side than of the lower side of the leaves in plants displaying leaf epinasty [132, 133].

Other nastic movements, which are reversible, result from a gradient of turgor between two opposite regions of the organ. The direction of movement is thus determined by the anatomy of the organ rather than by the stimulus. For example, the pulvinus, a joint-like group of differentiated motor cells at the base of the petiole, is responsible for nyctinastic leaf movements in plants that can orientate their leaves to a vertical position during the dark period. A difference in turgor between cells on one side of the pulvinus, behaving as extensors, and cells on the opposite side, behaving as flexors, controls leaf angle. In a sensitive plant, seismonastic movements, by which leaves respond to mechanical stimuli, also involve gradients of cell turgor in the pulvinus. In such movements, turgor changes are triggered when K^+ ions (and accompanying anions) move into or out of the cells, and water follows by osmosis (like in guard cells during the opening and closing of stomata; see above). Voltage-gated inwardly- or outwardly-rectifying K^+ channels belonging to the Shaker family (see below, Section 5) mediate the K^+ influxes and effluxes, respectively, that underlie the changes in turgor [134, 135].

4.3 Role of K^+ in Control of the Cell Membrane Potential

K^+ concentrations at both sides of the plasma membrane have a great impact on the polarization of cell membrane potential. Electrophysiological analyses indicate that the resting plasma membrane potentials significantly vary in response to changes in K^+ external concentrations [101, 103]. K^+ channels from the Shaker family have been shown to be responsible for a large part of these variations in many cell types and physiological conditions [102, 136]. The dependency of the cell membrane potential on external and internal K^+ concentrations indirectly affect the transport of other solutes that rely on cell polarization to enter or exit the plant cell.

K^+ is also involved in electrical signals (action potentials) that plants are able to generate in response to different stimuli (cold, wounding, etc.). Unlike in neurons and other animal cells, in which Na^+ is a crucial player in action potentials (an inward flow of Na^+ ions triggering a depolarization of the cell membrane, which is thereafter rapidly repolarized by an outward flow of K^+ ions), excitation in plant cells mainly involves an efflux of anions (Cl^- or NO_3^-) for the initial depolarizing event. Then, the membrane depolarization gives rise to a repolarizing efflux of K^+ ions, like in animal cells [137]. For instance, the electrical signal allowing rhizobacteria recognition by leguminous roots, after perception of the bacterial Nod factors, would require K^+ efflux to repolarize the plasma membrane after a Cl^--induced membrane depolarization [138].

4.4 Effects of K^+ on Enzyme Activities and Roles in Metabolism

Since the initial reports that pyruvate kinase (PK) activity is strongly stimulated by K^+ [139, 140], many enzymes have been identified as being activated by monovalent cations in animals, bacteria and plants [141–143]. Activation involves selective binding of the monovalent cation to the enzyme or enzyme-substrate complex, stabilizing catalytic intermediates and enzyme structure, or providing optimal positioning of substrate. Molecular determinants of the binding of the monovalent cation to the protein have been identified in several enzymes [142], including from plants [144, 145]. The activation process involves selective interactions between the monovalent cation and the enzyme. The activating cation is most often K^+. This is the case, for instance, in various enzymes catalyzing phosphorylation of a carboxyl group or enolate anion and of molecular chaperones [142, 143]. Fewer enzymes, such as galactosidase and clotting proteases, are selective for Na^+ [142]. When activated by K^+, the enzymes are usually also significantly activated by Rb^+ and NH_4^+, while they are weakly activated by Na^+ and often not at all by Li^+ [141–143]. For instance, in plants, the starch synthetase from sweet corn displays an absolute requirement for potassium, with the optimum activation occurring at 50 mM KCI. Rb^+, Cs^+, and NH_4^+ are 80 % as effective as K^+, while Na^+ and Li^+ are respectively 21 % and 8 % as effective [146]. Conversely, enzymes displaying selectivity for Na^+ are also sensitive to Li^+, and much less to K^+, Rb^+ or NH_4^+ [141]. The mechanisms underlying the ionic selectivity are not always fully understood. For instance, in pyruvate kinase, replacement of K^+ with Na^+ does not seem to result in any apparent structural change [147], although the enzyme is practically inactive without K^+.

In plants, key enzymes (including membrane transport systems), like glutamine synthetase, phosphoenolpyruvate carboxylase (PEPC), phosphofructokinase, ADP-glucose starch synthase and some vacuolar PPases (involved in H^+ secretion across the vacuolar membrane into the vacuole) are strongly activated by K^+ [38, 141, 148, 149]. More generally, protein synthesis requires high concentrations of K^+ [4, 150]. It is worth noting that this biochemical requirement of protein synthesis for K^+ has often been taken as indirect evidence for strong homeostatic K^+ control in the cytosol [151]. Plant K^+ status can also affect plant metabolism through transcriptional and post-transcriptional regulation of metabolic enzymes. For instance, transcriptome analyses in K^+-starved *A. thaliana* plants have provided evidence for upregulation of malic enzyme and the GS/GOGAT cycle and downregulation of nitrate uptake and reduction [152, 153]. Other widely reported consequences of K^+-deficiency are accumulation of reducing sugars and depletion of organic acids and negatively charged amino acids [150].

Hence, a large set of data provides evidence that K^+ availability strongly affect plant contents of primary and secondary metabolites [150, 154]. As discussed in [153], it seems likely that at least part of such changes in metabolite contents reflects direct responses of enzyme activities and metabolism to the internal concentration of K^+, although this concentration is homeostatically controlled *in vivo* and thus poorly sensitive to large variations in the availability of K^+ in the external medium.

4.5 *Roles of Na^+ in Plants*

4.5.1 Replacement of K^+ by Na^+ as Vacuolar Osmoticum

K^+ is an essential macronutrient for plants because there are specific cellular functions that only K^+ can meet (see above). On the other hand, the role of K^+ as osmoticum in the vacuole is non-specific. In this function, this cation can be replaced by other solutes, and in particular by Na^+ [38]. Quantitatively, the K^+ cytosolic pool can represent a small fraction of the total amount of K^+ ions within a plant cell when the availability of this cation in the external medium (nutritive or soil solution) is not limiting. In such conditions, most of the cellular K^+ (around 90 %) is stored in the vacuole for osmotic purposes [155, 156]. However, when the availability of external K^+ is low, Na^+ can be substituted for K^+ as osmoticum in the vacuole, so that a relatively small amount of K^+ is sufficient to maintain K^+-specific functions in the cytosol. Indeed, it has been shown that Arabidopsis plants exhibited higher growth rates if 10 mM Na^+ was added to a 10 μM K^+ growth solution [107]. In another example, Na^+ uptake in rice plants proved to be crucial for biomass production when K^+ was limiting [157].

4.5.2 Na^+ Is a Beneficial Nutrient

Na^+ does not seem to be an essential nutrient in most higher plants (in all plants displaying C3 photosynthetic pathway, like the model plant *Arabidopsis thaliana*, and part of the plants displaying the C4 pathway, like maize and sorghum) [10]. However, because Na^+ stimulates growth in many plant species and can partly replace K^+ in some functions [39, 158–160] like osmotic adjustment of the large central vacuole, cell turgor regulation allowing cell enlargement or long-distance transport of anions (by playing the role of accompanying cation), it has been qualified as a "functional" nutrient in these species [39, 161]. Furthermore, in some plant species, partial replacement of K^+ by Na^+ can have beneficial effects even under adequate K^+ supply. In sugar beet, when 2.5 mM K^+ + 2.5 mM Na^+ replaced 5 mM K^+ in the nutrient solution, an increase in plant dry weight and sucrose concentration in the storage root was observed [38]. Thus, while most of the research programs on Na^+ in plants have been oriented towards the investigation of plant adaptation to salinity, it is clear that considering Na^+ only as a toxic ion whose uptake, translocation, and accumulation have to be tightly controlled by the plant to prevent stressing effects would be a simplistic analysis.

In halophytes, the effect of Na^+ on growth varies amongst species. Many, but not all, dicotyledonous halophytes need moderate NaCl concentrations (100–200 mM NaCl) to show optimal growth, while many monocotyledonous halophytes can display normal growth in the absence of salt or are stimulated by low NaCl concentrations, in the mM range [62, 63].

4.5.3 Essential Roles of Na^+ in Some Plant Species

Na^+ is an essential element in some plants displaying C_4 photosynthetic activity, like *Atriplex vesicaria* [162], *Echinochloa utilis* (Japanese millet), and *Portulaca grandiflotra* (rose moss) [163]. Low external concentrations of Na^+ (ca. 0.1 mM) are needed by such plants to avoid chlorosis, necrosis, and failure to set flowers even in presence of high external K^+ concentrations (>5 mM) [163]. This requirement for Na^+ is thought to reflect the involvement of this cation in at least two processes [10].

First, Na^+ has been shown to facilitate the conversion of pyruvate into phosphoenolpyruvate (PEP) (an important substrate in carbon fixation in C_4 plants), which occurs in leaf mesophyll cells prior to the Calvin cycle [164]. The molecular mechanisms underlying the effect of Na^+ in this process are however still poorly understood.

Second, it facilitates the translocation of pyruvate (which is central to the CO_2-concentrating mechanism in C_4 species) into chloroplasts [165, 166]. A transport system localized at the chloroplast envelope membrane and endowed with

Na^+-dependent pyruvate transport activity has been identified at the molecular level [24]. The identified gene (*BASS2*) is present in all plants but the encoded protein is especially abundant in plants of the sodium-dependent C_4 type [24, 25]. To our knowledge, the BASS2 protein is the only transport system identified in higher plants so far as displaying Na^+ co-transport activity.

5 Channels and Transporters Involved in Potassium and Sodium Transport in Plants

Molecular and functional analyses indicate that at least 3 families of channels, named Shaker, TPK/KCO and TPC, and 3 families of transporters, named HAK, HKT and CPA, contribute to K^+ and/or Na^+ membrane transport (uptake, distribution and compartmentalization) in plants (Figure 2).

In total, these families gather at least 70 members in the model plant *Arabidopsis thaliana*. The dichotomic classification of membrane transport systems into channels and transporters is based on the mechanisms underlying ion permeation [167, 168]. When open, channels can be regarded as selective pores through which ions move without inducing any change in the general conformation of the protein. In contrast, transporters undergo a cycle of conformational changes for each solute they transport. The maximum velocity of action of channels (up to 10^6 to 10^7 ions per second and protein) can thus be much higher than that of transporters (in the range 10–10^3 transport events per second and protein).

Furthermore, ions move always down their electrochemical gradient (passive transport) in channels, while they can move against their gradient (active transport) in transporters. The latter movement can be coupled to that of another substrate down its own electrochemical gradient, for instance H^+ [168]. It should be noted that the ionic selectivity of the transport protein does not *per se* constitute a criterion for discriminating transporters from channels since each of these categories of proteins comprise members endowed with a high or low ionic selectivity. It should also be noted that rigid dichotomization between channels and transporters is increasingly proving to be too simplistic to describe the functional diversity of these proteins [169–171].

5.1 Families of K^+-Selective Channels Identified at the Molecular Level

The two protein families, named Shaker and TPK/KCO, that have been described in plants as forming K^+-selective channels have counterparts in the animal kingdom: animal Shaker (Kv) channels for the former family [172] and K2P for the latter one [173]. K^+ channels have evolved to fulfill different functions in plants, some of which can be considered as strictly plant specific, as described below.

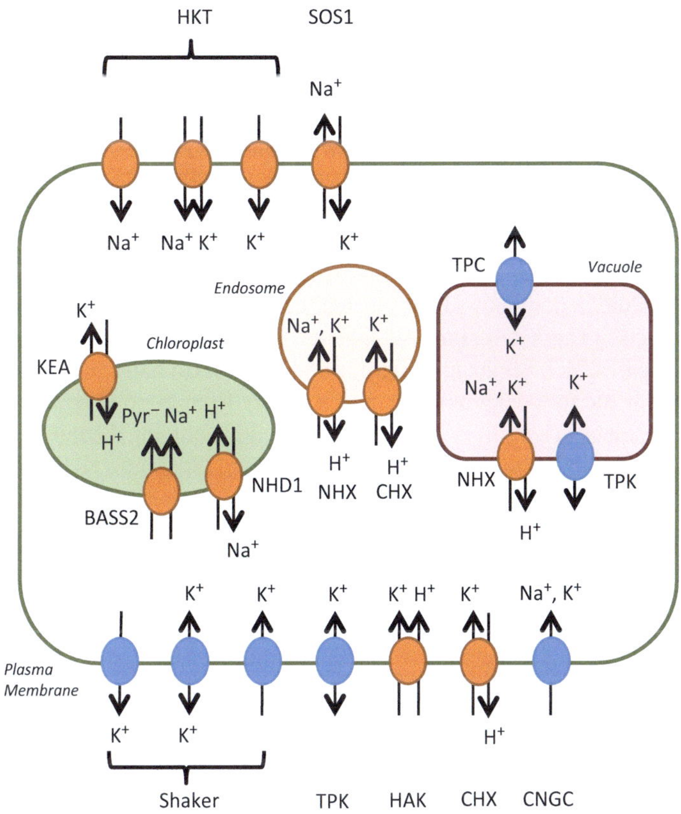

Figure 2 Schematic representation of the current knowledge of subcellular localization of members of the different families of channels and transporters permeable to Na^+ and/or K^+ in plants. Channels and transporters are displayed in blue and orange, respectively. Abbreviations: BASS2 = bile acid/Na^+ symporter family protein 2; CHX, cation/H^+ exchanger; CNGC, cyclic nucleotide-gated channels; HAK, high-affinity K^+ transporter; HKT, high-affinity K^+ transporter; KEA, K^+ efflux antiporter; NHD1, Na^+/H^+ antiporter; NHX, Na^+ and/or K^+-H^+ exchanger; SOS1, salt-overly sensitive 1 protein (NHX family); TPC, two-pore channel; TPK, tandem-pore K^+ channel. Some families are specific of a single type of membrane. This is the case of Shaker channels and HKT transporters, which are active at the plasma membrane, and of the TPC family, which display activity at the vacuolar membrane. Other families contribute to ion transport through different types of membrane. Of course, not all these proteins are expressed in the same cell type.

5.1.1 Shaker Channels

Plant Shaker channels form the main K^+ conductance at the plasma membrane in most cell types in plants (Figure 2). These channels are regulated by voltage like their animal counterparts. Functional channels result from the assembly of four subunits (alpha-subunits) that form a permeation pathway for K^+ in the center of the structure. It is worth to note that the four subunits can be the product of a single

Shaker gene (homomeric channels) or different Shaker genes (heteromeric channels) [172]. A Shaker alpha-subunit consists of a hydrophobic core displaying six transmembrane segments with both N- and C-terminal regions located on the cytosolic side of the membrane.

Two modules can be distinguished in the transmembrane core: a voltage-sensing module comprising the first four transmembrane segments (S1-S4) and a pore-forming module (S5-P-S6). In the former module, the fourth transmembrane segment (S4), which is enriched in positively charged residues, constitutes the voltage sensor. Between the fifth (S5) and sixth (S6) transmembrane segments of the pore-forming module, a pore loop (P) is located, harboring the canonical K^+ selectivity filter "TxGYG" (Thr-X-Gly-Tyr-Gly) like in animal Shaker channels. The cytosolic C-terminal part which begins just after the end of the sixth transmembrane segment (S6), contains the following domains (successively from N- to C-term): a C-linker (about 80 residues in length) [174], a cyclic-nucleotide binding domain (CNBD), an ankyrin domain (absent in some alpha-subunits) [175, 176], and a KHA domain rich in hydrophobic and acidic residues [177].

So far, there are four functional types of plant Shaker channels, which have been well documented in the model plant *Arabidopsis thaliana*. These four functional types fall into five phylogenetic groups in this species [178]. Groups 1 and 2 include five inwardly-rectifying channels: AKT1, SPIK, and AKT6 in group 1, and KAT1 and KAT2 in group 2 (Figure 2). All these channels are voltage-gated and open at hyperpolarized membrane potentials. Group 3 has AKT2 as only member, which behaves as a weakly-rectifying channel. AKT2 weakly-rectifying currents can be decomposed into two components: a voltage-independent ("Ohmic" like) component mediated by channels that are open at all membrane potentials, and an inwardly-rectifying component mediated by channels that only open at hyperpolarized potentials (like channels from groups 1 and 2) [179].

Both components correspond to two channel states and it is expected that changes between these two states are phosphorylation-dependent [180]. Like group 3, the Arabidopsis Shaker group 4 comprises a single member, named AtKC1, which seems unable to form functional channels by itself but is able to interact with alpha-subunits from groups 1, 2, and 3 to form functional heteromeric channels with distinctive functional features. It is thus regarded as a regulatory subunit. The last group, group 5, comprises two outwardly-rectifying channels, named SKOR and GORK. These two channels are voltage-gated (like the channels formed by members from group 1 to 4) but they open at depolarized membrane potentials. Thus, in Arabidopsis, 6 Shaker genes give rise to voltage-gated inwardly rectifying channel activity, 2 genes give rise to voltage-gated outwardly-rectifying channels and one gene to weakly-rectifying channel activity (mediated by homotetrameric AKT2 channels). It is worth to mention that the same basic subunit topology is found among all these channels with the exception of the ankyrin domain, which is absent in groups 1, 2, and 4.

Arabidopsis has also served as a model for determining the physiological role of many of the aforementioned Shaker functional types. Heteromeric AKT1/AtKC1 and/or homomeric AKT1 channels contribute to a large extent to K^+ uptake in root cells [102, 181–183]. Due to the hyperpolarized membrane potentials recorded in root cells, AKT1 channels can take up K^+ from solutions containing K^+ concentrations as low as 10 μM [102, 184, 185]. KAT1 and KAT2 are strongly expressed in guard cells where they mediate K^+ uptake, which leads to guard cell swelling and stomatal opening [186]. GORK is expressed in guard cells and in root periphery cells where it mediates K^+ efflux. In guard cells, this efflux gives rise to decreased turgor and stomatal closure [108]. In root periphery cells and root hairs, it could play a role in osmoregulation and possibly in signal transduction [187]. SKOR and AKT2 are preferentially expressed in vascular tissues where they take part in long-distance K^+ transport.

SKOR is expressed in root stele cells (pericycle and xylem parenchyma cells) where it mediates K^+ secretion into the xylem sap [78]. AKT2 is mainly expressed in the phloem where it contributes to K^+ load and/or unload in source and sink tissues [188, 189]. It is noteworthy that AKT2-mediated control of membrane potential (by modulating K^+ fluxes) can constitute an energy source for sucrose loading into the phloem [190]. Besides, AKT2, together with AKT1, contributes to K^+ uptake in mesophyll cells [191]. Finally, AKT6 and SPIK are expressed in reproductive tissues [189]. SPIK channels form a large K^+ inward conductance in pollen that permits germination and turgor-dependent growth of pollen tubes [192].

5.1.2 TPK/KCO Channels

The tandem-pore K^+ (TPK) channel family comprises 6 members (TPK1-TPK5 and KCO3) in the model plant *Arabidopsis thaliana*. TPK channels fall into two phylogenetic groups: TPK1 on one side and TPK2, TPK3, TPK4, and TPK5 on the other [193]. Each subunit consists of four transmembrane segments (TM) and two pore loops (P), containing the canonical K^+ selectivity filter "TxGYG", that are arranged as TM-P-TM-TM-P-TM. KCO3 is a special case since it only possesses two transmembrane segments that are separated by a pore domain (TM-P-TM topology), a structure reminiscent of that of animal Kir and bacterial KcsA channels [194–196]. The encoding *KCO3* gene is thought to have originated by gene duplication of the *TPK2* gene, followed by a partial deletion that resulted in the loss of one pore domain (a TM-P-TM module) [173, 197].

It is worth to note that plant TM-P-TM subunits have been only found in the Arabidopsis genus so far [193]. Like Shaker alpha-subunits, TPKs subunits have both their N- and C-terminal ends located in the cytosol. Binding sites for 14-3-3 proteins are found in the N-terminus while one or two Ca^{2+}-binding EF hands are present in the C-terminus. Functional TPK channels arise from the assembly of two subunits that leads to the formation of a central pore with four pore domains. Unlike plant Shaker channels, TPK channels do not have a voltage sensor domain and are thus not regulated by voltage [197]. Instead, they are regulated by Ca^{2+}, via the EF hands, and by pH. Expression analyses of *TPKs* genes have shown that the highest transcript levels in root, leaves and flowers is displayed by *TPK1*, followed by *TPK3* and *TPK5* [173]. Expression of *TPK2* and *TPK3* was essentially detected in flowers, but with low transcript levels. *TPK4* expression seems to be restricted to pollen as observed by *promoter-GUS* (β-glucuronidase) fusion analysis [198]. Albeit some TPK subunits are co-expressed in the same cell types, TPK channels seem to only result from homodimer assembly [199].

At the subcellular level, TPK1, TPK2, TPK3 and TPK5 are localized to the tonoplast (Figure 2) as shown by fusions to the fluorescent marker GFP [199]. In contrast, TPK4 seemed to be partly targeted to the plasma membrane, while an important fraction remained in the ER [198, 200]. Although general targeting sequences have not been identified in Arabidopsis TPKs, localization of rice TPKs to lytic vacuole tonoplast or to protein storage vacuoles relied on a reduced number of residues located in the C-terminus of the channel subunits [201]. Information about the physiological role of TPK channels in plants is limited at the present time. They are expected to participate in intracellular K^+ transport that involves vacuoles and organelles. For instance, TPK1 participates in vacuolar K^+ release necessary for stomatal closure and seed germination [202]. Interestingly, several vacuolar TPKs, including AtTPK1, seem to respond to mechanical stimulation, suggesting a link between osmoregulation and TPK-mediated K^+ transport [203].

5.2 K^+-Permeable Transporters from the HAK/KUP/KT Family

Plant HAK/KUP/KT transporters were first identified from their homology to bacterial KUP (K^+ Uptake) and fungal HAK (high-affinity K^+) transporters [204, 205]. The first plant transporters identified in this family were cloned from barley (HAK1 [206]) and Arabidopsis (KUP1/KT1, KUP2/KT2; K^+ transporter [207–209]). Due to the different acronyms used in these early reports, the composite name of HAK/KUP/KT is widely used to refer to the whole family in plants. Functional characterization in yeast and/or bacteria mutants devoid of endogenous K^+ uptake systems has clearly evidenced permeability to K^+ in various members of this family. Data obtained from Arabidopsis confirm that some members of the family do indeed play a role in K^+ homeostasis in roots [185, 210–212] and shoots [213]. The HAK/KUP/KT transporters characterized so far in plants are mainly permeable to K^+. Their capacity to discriminate between this cation, Rb^+ and Cs^+ is however low [172], while it is high between K^+ and Na^+, with a 10^3 difference in their corresponding apparent affinity constant (K_M close to 10 μM for K^+ and 10 mM for Na^+) [206, 214].

Based on hydropathy profiles, HAK/KUP/KT transporters would possess from 10 to 14 transmembrane segments [210]. In contrast to the Shaker and TPK families, the HAK/KUP/KT one displays a high and rather variable number of members in the different plant species genomes that have been sequenced so far. For instance, the genome of the dicot model Arabidopsis and that of the monocot model rice displays 13 and 27 genes, respectively. The physiological significance of such differences is still poorly understood [172].

HAK/KUP/KT transporters are generally classified according to their sequence homology into four clusters (I–IV [215]). Cluster I comprises well characterized transporters like AtHAK5 or OsHAK1, which have been shown to mediate K^+ uptake in roots from low external concentrations, probably by mediating H^+:K^+ co-transport (Figure 2) [214, 215]. Such capacity allows plants to thrive under low-K^+ conditions [211, 212]. So far, characterization of members belonging to cluster II has revealed a striking diversity in terms of transport properties and physiological roles. In Arabidopsis, they seem to be involved in developmental processes dependent on, or resulting in, cell expansion [213, 216, 217]. Little information is available on cluster III and cluster IV transporters.

5.3 Na^+-Selective Transporters

5.3.1 High-Affinity K^+ Transporters

High-affinity K^+ transporters (HKT) transporters are related to fungal and bacterial K^+ transporters from the Trk/Ktr families [194]. In plants however, HKT transporters display varying Na^+/K^+ permeabilities. When characterized in heterologous systems, 3 main types can be distinguished: K^+- or Na^+-selective transporters and Na^+-K^+ symporters (Figure 2) [218–221]. Initial sequence analyses and *in silico* modelling suggested that HKT transporters contain a hydrophobic core with four "MPM" domains (a MPM domain being formed by a pore loop surrounded by one transmembrane domain at each side), the N- and C-terminus being located in the cytosol. The four MPM domains assemble in such a way that the four pore loops are located at the center of the hydrophobic structure to form part of the permeation pathway. Recent crystallization of bacterial Trk/Ktr homologues has provided strong support to this configuration [222, 223].

Phylogenetic and functional analyses have led to sort HKT transporters into 2 subfamilies [224]: subfamily I, present in both monocotyledonous and dicotyledonous species, and subfamily II, identified only in monocotyledonous species so far [224]. In Arabidopsis, a single *HKT* gene, *AtHKT1,* has been identified [221] while in rice, eight or nine *HKT* genes exist depending on cultivars [225, 226]. Subfamily II HKT transporters are expected to be all K^+-permeable and can operate as Na^+-K^+ symporters [218, 226] or K^+-selective uniporters [220, 227] when heterologously expressed in yeast and/or *Xenopus* oocytes. Transporter permeability to K^+ relies on a conserved glycine residue in the middle of the selectivity filter of HKT/Trk/Ktr transporters [222, 223, 228]. In subfamily I HKT transporters, such a glycine is absent [229]. Subfamily I HKT transporters are Na^+-selective in Arabidopsis and rice [50, 53, 218, 221]. In other species, they are expected to be Na^+-selective as well because of the absence of conserved glycine in their selectivity filter [224], although indication of K^+-permeable HKT transporters within the dicotyledonous subfamily I has been reported [230–232].

Concerning their physiological role, HKT transporters have been widely associated to Na^+ transport and tolerance to Na^+ stress [172]. The AtHKT1 transporter from Arabidopsis strongly contributes to Na^+ recirculation from shoots to roots, contributing to reduce the plant sensitivity to Na^+ [50, 233]. AtHKT1 is expressed in root xylem parenchyma and root and shoot phloem and it prevents shoot Na^+ over-accumulation both by limiting the amount of Na^+ delivered to the shoots, through xylem sap desalinization, and by recirculating shoot Na^+ to the roots via the phloem sap [50, 234]. Genetic analyses have shown that several subfamily I HKT transporters, in both dicotyledonous and monocotyledonous species, are associated to QTLs of salt tolerance by limiting leaf Na^+ accumulation upon salt stress [51, 53, 235]. Less information about the role of subfamily II HKT transporters has been reported. In rice, OsHKT2;1 provides a major pathway for root high-affinity Na^+ uptake that supports plant growth under limiting K^+ supply [157]. Its possible involvement in root K^+ uptake has not been evidenced yet.

5.3.2 Na^+:Pyruvate Cotransporters

Recently, a plastidial protein named BASS2 (for "bile acid sodium symporter family" protein 2) has been characterized in several C_3 and C_4 species, including *A. thaliana,* and shown to mediate pyruvate:Na^+ co-transport into chloroplasts (Figure 2) [24]. Orthologues of BASS2 can be detected in all the land plant genomes that have been sequenced so far. Interestingly, such pyruvate:Na^+ co-transport mechanism was well established in C_4 species but not well characterized in the C_3 ones [25, 236]. In C_3 species, BASS2 function would supply pyruvate to the MEP pathway [24, 237]. To mediate pyruvate:Na^+ co-transport, it has been proposed that BASS2 requires a Na^+ gradient established by NHD1 Na^+/H^+ antiport activity (member from the NhaD family; see below).

5.4 Monovalent Cation/H^+ Antiporters from the CPA Superfamily

This superfamily of cation/H^+ antiporters (CPA) comprises three major families designated as CPA1, CPA2, and NhaD [238–240]. In plants, CPA1 includes the well-studied Na^+-K^+/H^+ exchanger (NHX) family, CPA2 includes the K^+ efflux antiporter (KEA) and cation-H^+ exchanger (CHX) families and NhaD includes Na^+/H^+ antiporters.

CPA1 transporters, which are predicted to have 10 to 12 transmembrane segments, mediate electroneutral Cation/H^+ exchange [241]. CPA1-type transporters are found in all kingdoms, including archaea, bacteria, fungi, plants and metazoa [238]. In plants, CPA1 transporters behave as Na^+/H^+ and/or K^+/H^+ antiporters (Figure 2). They are involved in salt tolerance and K^+ homeostasis by contributing either to cation compartmentalization in cellular organelles, as shown for instance for NHX transporters, or to Na^+ extrusion from the cell, as shown for the SOS1 Na^+:H^+ antiporters [242]. Arabidopsis contains eight isoforms belonging to three classes: two divergent members localized to the plasma membrane (SOS1/AtNHX7 and AtNHX8), and six intracellular isoforms that are targeted to the vacuolar (AtNHX1 to AtNHX4) or endosomal membranes (AtNHX5, AtNHX6). AtNHX1 and AtNHX2 contribute to vacuolar K^+ accumulation and pH homeostasis that is required for plant growth, stomatal functioning and flower development [125, 243, 244]. AtNHX5 and AtNHX6 seem to be involved in the establishment of pH gradients between organelles that are important for vacuolar trafficking. Interestingly, cell expansion and plant growth is greatly reduced in plants lacking both AtNHX5 and AtNHX6, which highlights the relevance of endosomal pH and ion homeostasis in these physiological processes [245, 246]. Concerning salt tolerance, these two types of NHX transporters, vacuolar or endosomal, display contrasting responses to sodium excess. Disruption of *AtNHX5* and *AtNHX6* renders plants salt-sensitive. In contrast, moderate salt can complement growth and flower defects displayed by mutants lacking AtNHX1 and AtNHX2 [244, 245]. A consistent observation is that improved Na^+ compartmentalization into the vacuole by overexpression of vacuolar NHX proteins results in increased tolerance of the plant to a salinity constraint [242]. In addition to the contribution of NHX transporters to vacuolar Na^+ accumulation, selective Na^+ excretion from the cell by plasma membrane SOS1-like transporters has been widely shown to limit Na^+ levels in the cytosol and thereby to improve plant performance under salt stress [76]. Furthermore, SOS1 activity appears to protect root plasma membrane K^+ uptake capacity from external media displaying high Na^+ concentrations [247].

Six *KEA* genes (AtKEA1 to 6) are present in the Arabidopsis genome. AtKEA1 and AtKEA2 are targeted to the chloroplast inner envelope membrane whereas AtKEA3 is targeted to the thylakoid membrane. By using Arabidopsis plants mutated in AtKEA1 to 3 transporters (single, double, and triple mutants), it has been shown that these transporters play an essential role in chloroplast osmoregulation, integrity, and ion and pH homeostasis (Figure 2) [248].

The Arabidopsis CHX family, with twenty-eight members in *A. thaliana*, is much larger than the KEA one. It comprises transporters targeted to the plasma membrane, prevacuolar membrane or the endoplasmic reticulum, where they exchange K^+ against H^+ [249–251]. Association of AtCHX proteins with endomembranes and their roles in pH and cation homeostasis suggest that these proteins play important roles in membrane trafficking, similarly to endosomal NHX transporters. AtCHX20 antiport activity in endosomes contributes to guard cell swelling and stomatal opening [126]. In Arabidopsis pollen, AtCHX21 and AtCHX23 are involved in either the reception or the transduction of female signals that target pollen tube to the ovule [250]. Intriguingly, multiple CHX genes are expressed in Arabidopsis pollen, but this is not well understood yet.

NhaD-type carriers have been identified in all vascular and non-vascular plants, including mosses and algae [252]. In *A. thaliana*, the Na^+/H^+ antiporter AtNHD1 has been shown to participate in Na^+ export from chloroplasts, a process that contributes to salt tolerance, efficient photosynthesis and plant performance (Figure 2) [253]. AtNHD1 may work in parallel to KEA exchangers at the chloroplast envelope with overlapping substrate specificity [248]. Evidence has also been obtained that NHD1 energize BASS2 pyruvate/Na^+ co-transport into the chloroplast (see Section 5.3.2) by establishing an inwardly-directed sodium gradient across the envelope [24].

5.5 Poorly Selective Transport Systems Permeable to K^+ and Na^+

5.5.1 Cyclic Nucleotide-Gated Channels

Cyclic nucleotide-gated channels (CNGCs) share structural homology with Shaker channels as they have six transmembrane segments and a long cytosolic C-terminal domain harboring a CNBD. In contrast, they lack the canonical motif TxGYG, hallmark of K^+-selective channels [66, 254]. Unlike Shaker channels, they seem to be regulated by cGMP and/or cAMP and to poorly discriminate among monovalent cations (Figure 2) [255, 256]. A notable exception is AtCNGC2, which exhibited a high degree of K^+ selectivity as opposed to Na^+ [257], a feature that is unknown in animal CNGCs [258].

Despite the fact that CNGCs seem to play a prominent role in plant immunity by probably mediating Ca^{+2} fluxes [259, 260], some of them have shown features related to K^+ and/or Na^+ transport. For instance, AtCNGC10 was shown to rescue K^+ transport mutant strains of *Escherichia coli* (LB650) and yeast (CY162), and the Arabidopsis *akt1* mutant [261]. In heterologous systems, AtCNGC3 can mediate Na^+ and K^+ uptake [129]. Moreover, promoter-driven GUS activity data has shown that AtCNGC3 is mainly expressed in epidermal and cortical root tissues in seedlings, a feature consistent with a role in K^+ and/or Na^+ transport.

5.5.2 Tandem-Pore Channels

Tandem-pore channels (TPC) proteins share structural similarities with voltage-gated cation channels, since one polypeptide consists of two repeats of the basic S1-S6 structure present in Shaker K^+ channel. Unlike their animal homologues, plant TPC channels are targeted to the tonoplast and are responsible for the so-called slow vacuolar (SV) currents (Figure 2) [262, 263]. They are activated by a rise in cytosolic calcium concentration and do not discriminate among cations, either monovalent or divalent [264–269]. In the presence of low Na^+ concentrations and of a K^+ gradient directed into the cytosol, a TPC channel is able to transport K^+ across the tonoplast in either direction, depending on the electrochemical driving forces [262]. In Arabidopsis, phenotype analyses of plants displaying a loss-of-function mutation in the single TPC family member has revealed that this channel contributes to seed germination and stomatal movements [263], but the physiological significance of these observations remains unclear.

6 General Conclusions

Important concerns and objectives, both at the biological and agricultural levels, are underlying research on roles and transport of K^+ and Na^+ in plants. K^+ is an essential element, which is required in large quantities while its availability in the soil solution is often low, in the μM range, and limiting for optimal plant growth. It is involved in a large number of crucial functions, among which are osmocontractility, for instance in guard cells which regulate the aperture of stomatal pores and gas exchanges at the leaf surface. In contrast, Na^+ is not an essential element in most plants. The major issue regarding this ion is that its concentration in the soil can be high and thus toxic, preventing plant growth and agriculture in a large proportion of the arable lands.

In the context of such concerns, efforts have been particularly made to identify and characterize channels or transporters involved in K^+ or Na^+ transport. Major advances have been made during the last 2 decades in this domain, using DNA-based strategies, cell biology, (electro)physiology, genetics and reverse genetics, and whole plant biology. For instance, this has provided valuable knowledge on transport mechanisms responsible for K^+ uptake from the soil solution [75], K^+ fluxes in guard cells [121], and Na^+ exclusion [73] or compartmentalization into vacuoles [91].

However, we are very far from having a holistic view of the functional properties and roles of the tens of Na^+ and K^+ transport systems that have been identified in the genome of the model plant *Arabidopsis thaliana* [270], and still farther from understanding the physiological significance and consequences of the differences that are revealed by phylogenetic comparison of the different families of transport systems between plant species [172]. Further progress in this direction is clearly required. It seems very reasonable to expect that this will provide new tools and strategies to improve, for instance, plant K^+ use efficiency or tolerance to soil salinity.

Abbreviations

ADP	adenosine diphosphate
AKT	Arabidopsis K^+ transport system
AtKC1	*Arabidopsis thaliana* K^+ channel 1
ATP	adenosine 5′-triphosphate
BASS2	bile acid:Na^+ symporter family protein 2
cAMP	cyclic adenosine monophosphate
cGMP	cyclic guanidine monophosphate
CHX	cation/H^+ exchanger
CNBD	cyclic nucleotide binding domain
CPA	cation/H^+ antiporter
ER	endoplasmatic reticulum

GFP	green fluorescent protein
GORK	gated outwardly-rectifying K^+ channel
GS/GOGAT	glutamine synthetase/glutamine oxoglutarate aminotransferase
GUS	β-glucuronidase
HAK	high-affinity K^+ transporter
KAT	K^+ channel in *Arabidopsis thaliana*
KEA	K^+ efflux antiporter
KT	K^+ transporter
KUP	K^+ uptake transporter
MEP	methyl erythritol phosphate
NHX	Na^+-K^+/H^+ exchanger
P	pore loop
PEP	phosphoenolpyruvate
PEPC	phosphoenolpyruvate carboxylase
PK	pyruvate kinase
QTL	quantitative trait locus
SKOR	stellar K^+ outward rectifier
SPIK	Shaker pollen inward K^+ channel
TM	transmembrane segment

Acknowledgments Manuel Nieves-Cordones is especially indebted to the Alfonso Martin Escudero Foundation for providing financial support. Fouad Razzaq Al-Shiblawi is very grateful to the University of Al-Muthanna and the Ministry of Higher Education and Scientific Research of Irak for funding. Our work on cation transport in plants has been supported by a Marie Curie Intra-European Fellowship (FP7-PEOPLE- 2010-IEF No. 272390 – KinPlants) to MN-C, a Grand Federative Project (Rhizopolis) of the Agropolis Fondation (Montpellier, France) to HS, and a grant from the French Agency for Research (Investissement d'Avenir program, grant RSNR-Demeterres) to HS.

References

1. D. A. T. Dick, S. G. Mclaughlin, *J. Physiol-London* **1969**, *205*, 61–78.
2. S. B. Horowitz, P. L. Paine, L. Tluczek, J. K. Reynhout, *Biophys J.* **1979**, *25*, 33–44.
3. A. A. Lev, *Nature* **1964**, *201*, 1132–1134.
4. R. A. Leigh, R. G. Wyn Jones, *New Phytol.* **1984**, *97*, 1–13.
5. L. N. Vorobiev, *Nature* **1967**, *216*, 1325–1327.
6. D. J. Walker, R. A. Leigh, A. J. Miller, *Proc. Natl. Acad. Sci. USA* **1996**, *93*, 10510–10514.
7. T. J. Century, I. R. Fenichel, S. B. Horowitz, *J. Cell Sci.* **1970**, *7*, 5–13.
8. S. D. Lidofsky, M. H. Xie, A. Sostman, B. F. Scharschmidt, J. G. Fitz, *J. Biol. Chem.* **1993**, *268*, 14632–14636.
9. D. E. Carden, D. J. Walker, T. J. Flowers, A. J. Miller, *Plant Physiol.* **2003**, *131*, 676–683.
10. H. J. Kronzucker, D. Coskun, L. M. Schulze, J. R. Wong, T. D. Britto, *Plant Soil* **2013**, *369*, 1–23.
11. A. Rodriguez-Navarro, F. Rubio, *J. Exp. Bot.* **2006**, *57*, 1149–1160.
12. B. Benito, R. Haro, A. Amtmann, T. A. Cuin, I. Dreyer, *J. Plant Physiol.* **2014**, *171*, 723–731.

13. B. P. Rosen, *Annu. Rev. Microbiol.* **1986**, *40*, 263–286.
14. M. Carrillo-Tripp, H. Saint-Martin, I. Ortega-Blake, *J. Chem. Phys.* **2003**, *118*, 7062–7073.
15. M. Carrillo-Tripp, M. Luisa San-Roman, J. Hernandez-Cobos, H. Saint-Martin, I. Ortega-Blake, *Biophys. Chem.* **2006**, *124*, 243–250.
16. R. Mancinelli, A. Botti, F. Bruni, M. A. Ricci, A. K. Soper, *J. Phys. Chem. B.* **2007**, *111*, 13570–13577.
17. P. B. Cossins, M. P. Jacobson, V. Guallar, *Plos Comput. Biol.* **2011**, *7*, E1002066.
18. T. J. Flowers, A. R. Yeo, *Aus. J. Plant Physiol.* **1986**, *13*, 75–91.
19. T. Matoh, J. Watanabe, E. Takahashi, *Plant Physiol.* **1987**, *84*, 173–177.
20. A. R. Yeo, T. J. Flowers, *Aus. J. Plant Physiol.* **1986**, *13*, 161–173.
21. V. S. Anil, H. Krishnamurthy, M. K. Mathew, *Physiol. Plant* **2007**, *129*, 607–621.
22. R. A. Gaxiola, M. G. Palmgren, K. Schumacher, *FEBS Lett.* **2007**, *581*, 2204–2214.
23. A. Rodriguez-Navarro, *Biochim. Biophys. Acta.* **2000**, *1469*, 1–30.
24. T. Furumoto, T. Yamaguchi, Y. Ohshima-Ichie, M. Nakamura, Y. Tsuchida-Iwata, M. Shimamura, J. Ohnishi, S. Hata, U. Gowik, P. Westhoff, A. Braeutigam, A. P. M. Weber, K. Izui, *Nature* **2011**, *476*, 472–475.
25. A. P. M. Weber, S. Von Caemmerer, *Curr. Opin. Plant Biol.* **2010**, *13*, 257–265.
26. C. Zoerb, M. Senbayram, E. Peiter, *J. Plant Physiol.* **2014**, *171*, 656–669.
27. K. Mengel, E. A. Kirkby, *Principles of Plant Nutrition*, 5th edn., Springer, New York, 2001, 479–480.
28. P. W. Moody, M. J. Bell, *Aus. J. Soil Res.* **2006**, *44*, 265–275.
29. V. Römheld, E. Kirkby, *Plant Soil* **2010**, *335*, 155–180.
30. Y. Pal, R. J. Gilkes, M. T. F. Wong, *Aus. J. Soil Res.* **2001**, *39*, 611–625.
31. Y. Pal, R. J. Gilkes, M. T. F. Wong, *Aus. J. Soil Res.* **2001**, *39*, 813–822.
32. M. Badraoui, P. R. Bloom, A. Delmaki, *Plant Soil* **1992**, *140*, 55–63.
33. N. Moritsuka, J. Yanai, M. Umeda, T. Kosaki, *Soil Sci. Plant Nutr.* **2004**, *50*, 565–573.
34. A. F. Ogaard, T. Krogstad, *J. Plant Nutr. Soil Sci.* **2005**, *168*, 80-88.
35. I. Oborn, Y. Andrist-Rangel, M. Askekaard, C. A. Grant, C. A. Watson, A. C. Edwards, *Soil Use Management* **2005**, *21*, 102–112.
36. P. Barre, B. Velde, L. Abbadie, *Biogeochemistry* **2007**, *82*, 77–88.
37. P. B. Barraclough, *Plant Soil* **1989**, *119*, 59–70.
38. P. Marschner, *Marschner's Mineral Nutrition of Higher Plants*, 3rd edn., Academic Press, San Diego, 2012.
39. G. V. Subbarao, O. Ito, W. L. Berry, R. M. Wheeler, *Crit. Rev. Plant Sci.* **2003**, *22*, 391–416.
40. J. P. Baldwin, P. H. Nye, P. B. Tinker, *Plant Soil* **1973**, *38*, 621–635.
41. N. Claassen, A. Jungk, *Zeitschrift für Pflanzenernährung und Bodenkunde* **1982**, *145*, 513–525.
42. P. R. Darrah, S. Staunton, *Eur. J. Soil Sci.* **2000**, *51*, 643–653.
43. S. A. Barber, *Soil Nutrient Bioavailability: A Mechanistic Approach*, 2nd edn., Wiley-Blackwell, Hoboken, 1995.
44. O. Strebel, W. H. M. Duynisveld, *Zeitschrift für Pflanzenernährung und Bodenkunde* **1989**, *152*, 135–141.
45. W. T. Pettigrew, *Physiol Plant.* **2008**, *133*, 670–681.
46. V. Smil, *Biosci.* **1999**, *49*, 299–308.
47. Y. Andrist-Rangel, A. C. Edwards, S. Hillier, I. Oborn, *Agr. Ecosys. Environ.* **2007**, *122*, 413–426.
48. A. Dobermann, K. G. Cassman, C. P. Mamaril, J. E. Sheehy, *Field Crop Res.* **1998**, *56*, 113–138.
49. N. M. Hoa, B. H. Janssen, O. Oenema, A. Dobermann, *Agr. Ecosys. Environ.* **2006**, *116*, 121–131.
50. P. Berthomieu, G. Conejero, A. Nublat, W. J. Brackenbury, C. Lambert, C. Savio, N. Uozumi, S. Oiki, K. Yamada, F. Cellier, F. Gosti, T. Simonneau, P. A. Essah, M. Tester, A. A. Very, H. Sentenac, F. Casse, *EMBO J.* **2003**, *22*, 2004–2014.

51. C. S. Byrt, J. D. Platten, W. Spielmeyer, R. A. James, E. S. Lagudah, E. S. Dennis, M. Tester, R. Munns, *Plant Physiol.* **2007**, *143*, 1918–1928.
52. R. Munns, M. Tester, *Annu. Rev. Plant Biol.* **2008**, *59*, 651–681.
53. Z. H. Ren, J. P. Gao, L. G. Li, X. L. Cai, W. Huang, D. Y. Chao, M. Z. Zhu, Z. Y. Wang, S. Luan, H. X. Lin, *Nat. Genet.* **2005**, *37*, 1141–1146.
54. E. V. Maas, *Tree Physiol.* **1993**, *12*, 195–216.
55. P. D. Chavan, B. A. Karadge, *Plant Soil* **1980**, *56*, 201–207.
56. A. Eynard, R. Lal, K. Wiebe, *J. Sustain. Agr.* **2005**, *27*, 5–50.
57. E. P. Eckholm, *Environment* **1975**, *17*, 9–15.
58. S. Shabala, J. Bose, R. Hedrich, *Trends Plant Sci.* **2014**, *19*, 687–691.
59. P. Rengasamy, *J. Exp. Bot.* **2006**, *57*, 1017–1023.
60. A. Hairmansis, B. Berger, M. Tester, S. J. Roy, *Rice* **2014**, *7*, 16–16.
61. Z. Barhomi, W. Djebali, A. Smaoui, W. Chaibi, C. Abdelly, *J. Plant Physiol.* **2007**, *164*, 842–850.
62. T. J. Flowers, T. D. Colmer, *New Phytol.* **2008**, *179*, 945–963.
63. E. P. Glenn, J. J. Brown, E. Blumwald, *Crit. Rev. Plant Sci.* **1999**, *18*, 227–255.
64. S. Shabala, *Ann. Bot.* **2013**, *112*, 1209–1221.
65. H. J. Kronzucker, D. T. Britto, *New Phytol.* **2011**, *189*, 54–81.
66. M. W. Szczerba, D. T. Britto, H. J. Kronzucker, *J. Plant Physiol.* **2009**, *166*, 447–466.
67. D. T. Clarkson, *Philos. Trans. R. Soc. London Series B-Biol. Sci.* **1993**, *341*, 5–17.
68. M. Barberon, N. Geldner, *Plant Physiol.* **2014**, *166*, 528–537.
69. M. Tester, R. A. Leigh, *J. Exp. Bot.* **2001**, *52*, 445–457.
70. N. Geldner, *Annu. Rev. Plant Biol.* **2013**, *64*, 531–558.
71. L. Schreiber, K. Hartmann, M. Skrabs, J. Zeier, *J. Exp. Bot.* **1999**, *50*, 1267–1280.
72. T. M. Burch-Smith, P. C. Zambryski, *Annu. Rev. Plant Biol.* **2012**, *63*, 239–260.
73. F. J. M. Maathuis, I. Ahmad, J. Patishtan, *Front Plant Sci.* **2014**, *467*, 1–9.
74. M. P. Apse, E. Blumwald, *FEBS Lett.* **2007**, *581*, 2247–2254.
75. M. Nieves-Cordones, F. Aleman, V. Martinez, F. Rubio, *J. Plant Physiol.* **2014**, *171*, 688–695.
76. T. Yamaguchi, S. Hamamoto, N. Uozumi, *Front. Plant Sci.* **2013**, *410*, 1–7.
77. A. R. Yeo, M. E. Yeo, T. J. Flowers, *J. Exp. Bot.* **1987**, *38*, 1141–1153.
78. F. Gaymard, G. Pilot, B. Lacombe, D. Bouchez, D. Bruneau, J. Boucherez, N. Michaux-Ferriere, J. B. Thibaud, H. Sentenac, *Cell* **1998**, *94*, 647–655.
79. H. Marschner, E. A. Kirkby, C. Engels, *Bot. Acta.* **1997**, *110*, 265–273.
80. J. Pritchard, R. G. W. Jones, A. D. Tomos, *J. Exp. Bot.* **1991**, *42*, 1043–1049.
81. R. E. Sharp, T. C. Hsiao, W. K. Silk, *Plant Physiol.* **1990**, *93*, 1337–1346.
82. R. J. Davenport, A. Munoz-Mayor, D. Jha, P. A. Essah, A. Rus, M. Tester, *Plant Cell Environ.* **2007**, *30*, 497–507.
83. U. Deinlein, A. B. Stephan, T. Horie, W. Luo, G. Xu, J. I. Schroeder, *Trends Plant Sci.* **2014**, *19*, 371–379.
84. H. Lessani, H. Marschner, *Funct. Plant Biol.* **1978**, *5*, 27–37.
85. P. W. Becraft, *Curr. Topics Developm. Biol.* **1999**, *45*, 1–40.
86. C. M. Duckett, K. J. Oparka, D. A. M. Prior, L. Dolan, K. Roberts, *Development* **1994**, *120*, 3247–3255.
87. M. G. Erwee, P. B. Goodwin, A. J. E. Vanbel, *Plant Cell Environ.* **1985**, *8*, 173–178.
88. T.-H. Kim, M. Boehmer, H. Hu, N. Nishimura, J. I. Schroeder, *Annu. Rev. Plant Biol.* **2010**, *61*, 561–591.
89. C. Sirichandra, A. Wasilewska, F. Vlad, C. Valon, J. Leung, *J. Exp. Bot.* **2009**, *60*, 1439–1463.
90. A. R. Yeo, T. J. Flowers, *Physiol. Plant* **1982**, *56*, 343–348.
91. I. Ahmad, F. J. Maathuis, *J. Plant Physiol.* **2014**, *171*, 708–714.
92. A. Amtmann, P. Armengaud, V. Volkov, *Membrane Transp. Plant.* **2004**, *15*, 293–339.
93. F. Jiang, C. J. Li, W. D. Jeschke, F. S. Zhang, *J. Exp. Bot.* **2001**, *52*, 2143–2150.

94. N. Wigoda, M. Moshelion, N. Moran, *J. Plant Physiol.* **2014**, *171*, 715–722.
95. M. C. Drew, L. R. Saker, *Planta* **1984**, *160*, 500–507.
96. P. J. White, *J. Exp. Bot.* **1997**, *48*, 2063–2073.
97. A. V. Barker, D. J. Pilbeam, *Handbook of Plant Nutrition*, CRC Press, Boca Raton, 2014.
98. S. Isayenkov, J. C. Isner, F. J. Maathuis, *FEBS Lett.* **2010**, *584*, 1982–1988.
99. T. A. Cuin, A. J. Miller, S. A. Laurie, R. A. Leigh, *J. Exp. Bot.* **2003**, *54*, 657–661.
100. D. J. Walker, C. R. Black, A. J. Miller, *Plant Physiol.* **1998**, *118*, 957–964.
101. J. M. Cheeseman, J. B. Hanson, *Plant Physiol.* **1979**, *63*, 61–61.
102. R. E. Hirsch, B. D. Lewis, E. P. Spalding, M. R. Sussman, *Science* **1998**, *280*, 918–921.
103. F. J. M. Maathuis, D. Sanders, *Proc. Natl. Acad. Sci. USA* **1994**, *91*, 9272–9276.
104. M. Nieves-Cordones, A. J. Miller, F. Aleman, V. Martinez, F. Rubio, *Plant Mol. Biol.* **2008**, *68*, 521–532.
105. F. Aleman, M. Nieves-Cordones, V. Martinez, F. Rubio, *Plant Cell Physiol.* **2011**, *52*, 1603–1612.
106. R. Haro, M. A. Banuelos, A. Rodriguez-Navarro, *Plant Cell Physiol.* **2010**, *51*, 68–79.
107. F. J. M. Maathuis, D. Sanders, *Planta* **1993**, *191*, 302–307.
108. E. Hosy, A. Vavasseur, K. Mouline, I. Dreyer, F. Gaymard, F. Poree, J. Boucherez, A. Lebaudy, D. Bouchez, A. A. Very, T. Simonneau, J. B. Thibaud, H. Sentenac, *Proc. Natl. Acad. Sci. USA* **2003**, *100*, 5549–5554.
109. F. F. Nocito, G. A. Sacchi, M. Cocucci, *New Phytol.* **2002**, *154*, 45–51.
110. S. Shabala, T. A. Cuin, *Physiol. Plant* **2008**, *133*, 651–669.
111. F. J. M. Maathuis, A. Amtmann, *Ann. Bot.-London* **1999**, *84*, 123–133.
112. M. C. Ball, *Funct. Plant Biol.* **1988**, *15*, 447–464.
113. N. Suarez, E. Medina, *Braz. J. Plant Physiol.* **2008**, *20*, 131–140.
114. P. Adams, D. E. Nelson, S. Yamada, W. Chmara, R. G. Jensen, H. J. Bohnert, H. Griffiths, *New Phytol.* **1998**, *138*, 171–190.
115. A. Smaoui, Z. Barhoumi, M. Rabhi, C. Abdelly, *Protoplasma* **2011**, *248*, 363–372.
116. R. Munns, *New Phytol.* **2005**, *167*, 645–663.
117. S. Kalve, D. De Vos, G. T. Beemster, *Front. Plant Sci.* **2014**, *362*, 1–25.
118. E. A. Macrobbie, *Philos. Trans. R. Soc. Lond. B Biol. Sci.* **1998**, *353*, 1475–1488.
119. M. R. Roelfsema, R. Hedrich, *New Phytol.* **2005**, *167*, 665–691.
120. J. I. Schroeder, J. M. Kwak, G. J. Allen, *Nature* **2001**, *410*, 327–330.
121. J. M. Ward, P. Maser, J. I. Schroeder, *Annu. Rev. Physiol.* **2009**, *71*, 59–82.
122. R. Hedrich, *Physiol. Rev.* **2012**, *92*, 1777–1811.
123. J. C. Shope, D. B. Dewald, K. A. Mott, *Plant Physiol.* **2003**, *133*, 1314–1321.
124. X. Q. Gao, C. G. Li, P. C. Wei, X. Y. Zhang, J. Chen, X. C. Wang, *Plant Physiol.* **2005**, *139*, 1207–1216.
125. Z. Andres, J. Perez-Hormaeche, E. O. Leidi, K. Schluecking, L. Steinhorst, D. H. Mclachlan, K. Schumacher, A. M. Hetherington, J. Kudla, B. Cubero, J. M. Pardo, *Proc. Natl. Acad. Sci. USA* **2014**, *111*, E1806–E1814.
126. S. Padmanaban, S. Chanroj, J. M. Kwak, X. Li, J. M. Ward, H. Sze, *Plant Physiol.* **2007**, *144*, 82–93.
127. B. U. Keller, R. Hedrich, K. Raschke, *Nature* **1989**, *341*, 450–453.
128. J. I. Schroeder, K. Raschke, E. Neher, *Proc. Natl. Acad. Sci. USA* **1987**, *84*, 4108–4112.
129. A. Gobert, G. Park, A. Amtmann, D. Sanders, F. J. M. Maathuis, *J. Exp. Bot.* **2006**, *57*, 791–800.
130. S. Vanneste, J. Friml, *Cell* **2009**, *136*, 1005–1016.
131. A. Vieten, M. Sauer, P. B. Brewer, J. Friml, *Trends Plant Sci.* **2007**, *12*, 160–168.
132. J. K. Polko, L. A. Voesenek, A. J. Peeters, R. Pierik, *AoB Plants* **2011**, Plr031.
133. V. M. Ursin, K. J. Bradford, *Plant Physiol.* **1989**, *90*, 1341–1346.
134. N. Moran, *FEBS Lett.* **2007**, *581*, 2337–2347.
135. M. Moshelion, D. Becker, K. Czempinski, B. Mueller-Roeber, B. Attali, R. Hedrich, N. Moran, *Plant Physiol.* **2002**, *128*, 634–642.

136. A. A. Very, H. Sentenac, *Annu. Rev. Plant Biol.* **2003**, *54*, 575–603.
137. J. Fromm, S. Lautner, *Plant Cell Environ.* **2007**, *30*, 249–257.
138. H. H. Felle, E. Kondorosi, A. Kondorosi, M. Schultze, *Plant J.* **1998**, *13*, 455–463.
139. P. D. Boyer, H. A. Lardy, P. H. Phillips, *J. Biol. Chem.* **1943**, *149*, 529–541.
140. J. F. Kachmar, P. D. Boyer, *J. Biol. Chem.* **1953**, *200*, 669–682.
141. H. J. Evans, G. J. Sorger, *Annu. Rev. Plant Physiol.* **1966**, *17*, 47–76.
142. M. J. Page, E. Di Cera, *Physiol Rev.* **2006**, *86*, 1049–1092.
143. C. H. Suelter, *Science* **1970**, *168*, 789–795.
144. A. Credali, A. Diaz-Quintana, M. Garcia-Calderon, M. A. De La Rosa, A. J. Marquez, J. M. Vega, *Planta* **2011**, *234*, 109–122.
145. S. Green, C. J. Squire, N. J. Nieuwenhuizen, E. N. Baker, W. Laing, *J. Biol. Chem.* **2009**, *284*, 8652–8660.
146. R. E. Nitsos, H. J. Evans, *Plant Physiol.* **1969**, *44*, 1260–1266.
147. T. M. Larsen, M. M. Benning, I. Rayment, G. H. Reed, *Biochemistry* **1998**, *37*, 6247–6255.
148. E. J. Kim, R. G. Zhen, P. A. Rea, *Proc. Natl. Acad. Sci. USA* **1994**, *91*, 6128–6132.
149. Y. Nakanishi, T. Saijo, Y. Wada, M. Maeshima, *J. Biol. Chem.* **2001**, *276*, 7654–7660.
150. A. Amtmann, S. Troufflard, P. Armengaud, *Physiol. Plant.* **2008**, *133*, 682–691.
151. F. J. M. Maathuis, *Curr. Opin. Plant Biol.* **2009**, *12*, 250–258.
152. A. Amtmann, P. Armengaud, *Curr. Opin. Plant Biol.* **2009**, *12*, 275–283.
153. P. Armengaud, R. Sulpice, A. Miller, J., M. Stitt, A. Amtmann, Y. Gibon, *Plant Physiol.* **2009**, *150*, 772–785.
154. S. Troufflard, W. Mullen, T. R. Larson, I. A. Graham, A. Crozier, A. Amtmann, P. Armengaud, *BMC Plant Biol.* **2010**, *10*, 172.
155. G. V. Subbarao, R. M. Wheeler, G. W. Stutte, *Life Support Biosph. Sci.* **2000**, *7*, 225–232.
156. A. Wakeel, A. R. Asif, B. Pitann, S. Schubert, *J. Plant Physiol.* **2011**, *168*, 519–526.
157. T. Horie, A. Costa, T. H. Kim, M. J. Han, R. Horie, H.-Y. Leung, A. Miyao, H. Hirochika, G. An, J. I. Schroeder, *EMBO J.* **2007**, *26*, 3003–3014.
158. P. Battie-Laclau, J.-P. Laclau, M. D. C. Piccolo, B. C. Arenque, C. Beri, L. Mietton, M. Almeida Muniz, R., L. Jordan-Meille, M. S. Buckeridge, Y. Nouvellon, J. Ranger, J.-P. Bouillet, *Plant Soil.* **2013**, *371*, 19–35.
159. R. Erel, A. Ben-Gal, A. Dag, A. Schwartz, U. Yermiyahu, *Tree Physiol.* **2014**, *34*, 1102–1117.
160. J. N. Gattward, A.-A. F. Almeida, J. O. Jr. Souza, F. P. Gomes, H. J. Kronzucker, *Physiol. Plant.* **2012**, *146*, 350–362.
161. E. A. H. Pilon-Smits, C. F. Quinn, W. Tapken, M. Malagoli, M. Schiavon, *Curr. Opin. Plant Biol.* **2009**, *12*, 267–274.
162. P. F. Brownell, J. G. Wood, *Nature* **1957**, *179*, 635–636.
163. P. F. Brownell, C. J. Crossland, *Plant Physiol.* **1972**, *49*, 794–797.
164. M. Johnston, C. P. L. Grof, P. F. Brownell, *Aus. J. Plant Physiol.* **1988**, *15*, 749–760.
165. J. Ohnishi, U. I. Flugge, H. W. Heldt, R. Kanai, *Plant Physiol.* **1990**, *94*, 950–959.
166. J. I. Ohnishi, R. Kanai, *FEBS Lett.* **1987**, *219*, 347–350.
167. H. Logan, M. Basset, A. A. Very, H. Sentenac, *Physiol. Plant.* **1997**, *100*, 1–15.
168. W. D. Stein, *Channels, Carriers and Pumps. An Introduction to Membrane Transport*, Academic Press, San Diego, 1990.
169. L. J. Defelice, T. Goswami, *Annu. Rev. Physiol.* **2007**, *69*, 87–112.
170. W. A. Fairman, R. J. Vandenberg, J. L. Arriza, M. P. Kavanaugh, S. G. Amara, *Nature* **1995**, *375*, 599–603.
171. D. C. Gadsby, *Nature* **2004**, *427*, 795–797.
172. A. A. Very, M. Nieves-Cordones, M. Daly, I. Khan, C. Fizames, H. Sentenac, *J. Plant Physiol.* **2014**, *171*, 748–769.
173. C. Voelker, J. L. Gomez-Porras, D. Becker, S. Hamamoto, N. Uozumi, F. Gambale, B. Mueller-Roeber, K. Czempinski, I. Dreyer, *Plant Biol. (Stuttg).* **2010**, *12 Suppl. 1*, 56–63.
174. M. Nieves-Cordones, A. Chavanieu, L. Jeanguenin, C. Alcon, W. Szponarski, S. Estaran, I. Cherel, S. Zimmermann, H. Sentenac, I. Gaillard, *Plant Physiol.* **2014**, *164*, 1415–1429.

175. J. A. Anderson, S. S. Huprikar, L. V. Kochian, W. J. Lucas, R. F. Gaber, *Proc. Natl. Acad. Sci. USA* **1992**, *89*, 3736–3740.
176. H. Sentenac, N. Bonneaud, M. Minet, F. Lacroute, J. M. Salmon, F. Gaymard, C. Grignon, *Science* **1992**, *256*, 663–665.
177. T. Ehrhardt, S. Zimmermann, B. Muller-Rober, *FEBS Lett.* **1997**, *409*, 166–170.
178. G. Pilot, R. Pratelli, F. Gaymard, Y. Meyer, H. Sentenac, *J. Mol. Evol.* **2003**, *56*, 418–434.
179. I. Dreyer, E. Michard, B. Lacombe, J. B. Thibaud, *FEBS Lett.* **2001**, *505*, 233–239.
180. E. Michard, I. Dreyer, B. Lacombe, H. Sentenac, J. B. Thibaud, *Plant J.* **2005**, *44*, 783–797.
181. B. Reintanz, A. Szyroki, N. Ivashikina, P. Ache, M. Godde, D. Becker, K. Palme, R. Hedrich, *Proc. Natl. Acad. Sci. USA* **2002**, *99*, 4079–4084.
182. Y. Wang, L. He, H. D. Li, J. Xu, W. H. Wu, *Cell Res.* **2010**, *20*, 826–837.
183. J. Xu, H. D. Li, L. Q. Chen, Y. Wang, L. L. Liu, L. He, W. H. Wu, *Cell* **2006**, *125*, 1347–1360.
184. Y. J. Pyo, M. Gierth, J. I. Schroeder, M. H. Cho, *Plant Physiol.* **2010**, *153*, 863–875.
185. F. Rubio, M. Nieves-Cordones, F. Aleman, V. Martinez, *Physiol. Plant* **2008**, *134*, 598–608.
186. A. Lebaudy, A. Vavasseur, E. Hosy, I. Dreyer, N. Leonhardt, J. B. Thibaud, A. A. Very, T. Simonneau, H. Sentenac, *Proc. Natl. Acad. Sci. USA* **2008**, *105*, 5271–5276.
187. N. Ivashikina, D. Becker, P. Ache, O. Meyerhoff, H. H. Felle, R. Hedrich, *FEBS Lett.* **2001**, *508*, 463–469.
188. R. Deeken, D. Geiger, J. Fromm, O. Koroleva, P. Ache, R. Langenfeld-Heyser, N. Sauer, S. T. May, R. Hedrich, *Planta* **2002**, *216*, 334–344.
189. B. Lacombe, G. Pilot, E. Michard, F. Gaymard, H. Sentenac, J. B. Thibaud, *Plant Cell.* **2000**, *12*, 837–851.
190. P. Gajdanowicz, E. Michard, M. Sandmann, M. Rocha, L. G. Correa, S. J. Ramirez-Aguilar, J. L. Gomez-Porras, W. Gonzalez, J. B. Thibaud, J. T. Van Dongen, I. Dreyer, *Proc. Natl. Acad. Sci. USA* **2011**, *108*, 864–869.
191. K. L. Dennison, W. R. Robertson, B. D. Lewis, R. E. Hirsch, M. R. Sussman, E. P. Spalding, *Plant Physiol.* **2001**, *127*, 1012–1019.
192. K. Mouline, A. A. Very, F. Gaymard, J. Boucherez, G. Pilot, M. Devic, D. Bouchez, J. B. Thibaud, H. Sentenac, *Genes Dev.* **2002**, *16*, 339–350.
193. J. L. Gomez-Porras, D. M. Riano-Pachon, B. Benito, R. Haro, K. Sklodowski, A. Rodriguez-Navarro, I. Dreyer, *Front Plant Sci.* **2012**, *167*, 1–13.
194. C. Corratge-Faillie, M. Jabnoune, S. Zimmermann, A. A. Very, C. Fizames, H. Sentenac, *Cell Mol. Life Sci.* **2010**, *67*, 2511–2532.
195. K. Czempinski, S. Zimmermann, T. Ehrhardt, B. Muller-Rober, *EMBO J.* **1997**, *16*, 2565–2575.
196. D. A. Doyle, J. Morais Cabral, R. A. Pfuetzner, A. Kuo, J. M. Gulbis, S. L. Cohen, B. T. Chait, R. Mackinnon, *Science* **1998**, *280*, 69–77.
197. T. Sharma, I. Dreyer, J. Riedelsberger, *Front Plant Sci.* **2013**, *224*, 1–16.
198. D. Becker, D. Geiger, M. Dunkel, A. Roller, A. Bertl, A. Latz, A. Carpaneto, P. Dietrich, M. R. G. Roelfsema, C. Voelker, D. Schmidt, B. Mueller-Roeber, K. Czempinski, R. Hedrich, *Proc. Natl. Acad. Sci. USA* **2004**, *101*, 15621–15626.
199. C. Voelker, D. Schmidt, B. Mueller-Roeber, K. Czempinski, *Plant J.* **2006**, *48*, 296–306.
200. M. Dunkel, A. Latz, K. Schumacher, T. Muller, D. Becker, R. Hedrich, *Mol. Plant.* **2008**, *1*, 938–949.
201. S. Isayenkov, J. C. Isner, F. J. Maathuis, *Plant Cell* **2011**, *23*, 756–768.
202. A. Gobert, S. Isayenkov, C. Voelker, K. Czempinski, F. J. M. Maathuis, *Proc. Natl. Acad. Sci. USA* **2007**, *104*, 10726–10731.
203. F. J. Maathuis, *New Phytol.* **2011**, *191*, 84–91.
204. M. A. Banuelos, R. D. Klein, S. J. Alexander-Bowman, A. Rodriguez-Navarro, *EMBO J.* **1995**, *14*, 3021–3027.
205. M. Schleyer, E. P. Bakker, *J. Bacteriol.* **1993**, *175*, 6925–6931.
206. G. E. Santa-Maria, F. Rubio, J. Dubcovsky, A. Rodriguez-Navarro, *Plant Cell.* **1997**, *9*, 2281–2289.

207. H. H. Fu, S. Luan, *Plant Cell* **1998**, *10*, 63–73.
208. E. J. Kim, J. M. Kwak, N. Uozumi, J. I. Schroeder, *Plant Cell* **1998**, *10*, 51–62.
209. F. J. Quintero, M. R. Blatt, *FEBS Lett.* **1997**, *415*, 206–211.
210. M. Gierth, P. Maser, *FEBS Lett.* **2007**, *581*, 2348–2356.
211. M. Nieves-Cordones, F. Aleman, V. Martinez, F. Rubio, *Mol. Plant.* **2010**, *3*, 326–333.
212. Z. Qi, C. R. Hampton, R. Shin, B. J. Barkla, P. J. White, D. P. Schachtman, *J. Exp. Bot.* **2008**, *59*, 595–607.
213. R. P. Elumalai, P. Nagpal, J. W. Reed, *Plant Cell* **2002**, *14*, 119–131.
214. M. A. Banuelos, B. Garciadeblas, B. Cubero, A. Rodriguez-Navarro, *Plant Physiol.* **2002**, *130*, 784–795.
215. F. Rubio, G. E. Santa-Maria, A. Rodriguez-Navarro, *Physiol. Plant* **2000**, *109*, 34–43.
216. Y. Osakabe, N. Arinaga, T. Umezawa, S. Katsura, K. Nagamachi, H. Tanaka, H. Ohiraki, K. Yamada, S.-U. Seo, M. Abo, E. Yoshimura, K. Shinozaki, K. Yamaguchi-Shinozaki, *Plant Cell* **2013**, *25*, 609–624.
217. S. Rigas, G. Debrosses, K. Haralampidis, F. Vicente-Agullo, K. A. Feldmann, A. Grabov, L. Dolan, P. Hatzopoulos, *Plant Cell.* **2001**, *13*, 139–151.
218. M. Jabnoune, S. Espeout, D. Mieulet, C. Fizames, J.-L. Verdeil, G. Conejero, A. Rodriguez-Navarro, H. Sentenac, E. Guiderdoni, C. Abdelly, A.-A. Very, *Plant Physiol.* **2009**, *150*, 1955–1971.
219. F. Rubio, W. Gassmann, J. I. Schroeder, *Science* **1995**, *270*, 1660–1663.
220. A. Sassi, D. Mieulet, I. Khan, B. Moreau, I. Gaillard, H. Sentenac, A. A. Very, *Plant Physiol.* **2012**, *160*, 498–510.
221. N. Uozumi, E. J. Kim, F. Rubio, T. Yamaguchi, S. Muto, A. Tsuboi, E. P. Bakker, T. Nakamura, J. I. Schroeder, *Plant Physiol.* **2000**, *122*, 1249–1259.
222. Y. Cao, X. Jin, H. Huang, M. G. Derebe, E. J. Levin, V. Kabaleeswaran, Y. Pan, M. Punta, J. Love, J. Weng, M. Quick, S. Ye, B. Kloss, R. Bruni, E. Martinez-Hackert, W. A. Hendrickson, B. Rost, J. A. Javitch, K. R. Rajashankar, Y. Jiang, M. Zhou, *Nature* **2011**, *471*, 336–340.
223. R. S. Vieira-Pires, A. Szollosi, J. H. Morais-Cabral, *Nature* **2013**, *496*, 323–328.
224. J. D. Platten, O. Cotsaftis, P. Berthomieu, H. Bohnert, R. J. Davenport, D. J. Fairbairn, T. Horie, R. A. Leigh, H.-X. Lin, S. Luan, P. Maeser, O. Pantoja, A. Rodriguez-Navarro, D. P. Schachtman, J. I. Schroeder, H. Sentenac, N. Uozumi, A.-A. Very, J.-K. Zhu, E. S. Dennis, M. Tester, *Trends Plant Sci.* **2006**, *11*, 372–374.
225. B. Garciadeblas, M. E. Senn, M. A. Banuelos, A. Rodriguez-Navarro, *Plant J.* **2003**, *34*, 788–801.
226. R. Oomen, J. F. J., B. Benito, H. Sentenac, A. Rodriguez-Navarro, M. Talon, A.-A. Very, C. Domingo, *Plant J.* **2012**, *71*, 750–762.
227. X. Yao, T. Horie, S. Xue, H. Y. Leung, M. Katsuhara, D. E. Brodsky, Y. Wu, J. I. Schroeder, *Plant Physiol.* **2010**, *152*, 341–355.
228. P. Maser, Y. Hosoo, S. Goshima, T. Horie, B. Eckelman, K. Yamada, K. Yoshida, E. P. Bakker, A. Shinmyo, S. Oiki, J. I. Schroeder, N. Uozumi, *Proc. Natl. Acad. Sci. USA* **2002**, *99*, 6428–6433.
229. F. Hauser, T. Horie, *Plant Cell Environ..* **2010**, *33*, 552–565.
230. Z. Ali, H. C. Park, A. Ali, D. H. Oh, R. Aman, A. Kropornicka, H. Hong, W. Choi, W. S. Chung, W. Y. Kim, R. A. Bressan, H. J. Bohnert, S. Y. Lee, D. J. Yun, *Plant Physiol.* **2012**, *158*, 1463–1474.
231. D. J. Fairbairn, W. Liu, D. P. Schachtman, S. Gomez-Gallego, S. R. Day, R. D. Teasdale, *Plant Mol. Biol.* **2000**, *43*, 515–525.
232. H. Su, E. Balderas, R. Vera-Estrella, D. Golldack, F. Quigley, C. Zhao, O. Pantoja, H. J. Bohnert, *Plant Mol. Biol.* **2003**, *52*, 967–980.
233. P. Maser, B. Eckelman, R. Vaidyanathan, T. Horie, D. J. Fairbairn, M. Kubo, M. Yamagami, K. Yamaguchi, M. Nishimura, N. Uozumi, W. Robertson, M. R. Sussman, J. I. Schroeder, *FEBS Lett.* **2002**, *531*, 157–161.

234. H. Sunarpi, T. Horie, J. Motoda, M. Kubo, H. Yang, K. Yoda, R. Horie, W. Y. Chan, H. Y. Leung, K. Hattori, M. Konomi, M. Osumi, M. Yamagami, J. I. Schroeder, N. Uozumi, *Plant J.* **2005**, *44*, 928–938.
235. M. J. Asins, I. Villalta, M. M. Aly, R. Olias, D. E. M. P. Alvarez, R. Huertas, J. Li, N. Jaime-Perez, R. Haro, V. Raga, E. A. Carbonell, A. Belver, *Plant Cell Environ.* **2013**, *36*, 1171–1191.
236. N. Aoki, J. Ohnishi, R. Kanai, *Plant Cell Physiol.* **1992**, *33*, 805–809.
237. A. Hemmerlin, J. F. Hoeffler, O. Meyer, D. Tritsch, I. A. Kagan, C. Grosdemange-Billiard, M. Rohmer, T. J. Bach, *J. Biol. Chem.* **2003**, *278*, 26666–26676.
238. C. L. Brett, M. Donowitz, R. Rao, *Am. J. Physiol. Cell Physiol.* **2005**, *288*, C223–39.
239. F. Cellier, G. Conejero, L. Ricaud, D. T. Luu, M. Lepetit, F. Gosti, F. Casse, *Plant J.* **2004**, *39*, 834–846.
240. M. H. J. Saier, *Microbiol. Mol. Biol. Rev.* **2000**, *64*, 354–411.
241. S. Chanroj, G. Wang, K. Venema, M. W. Zhang, C. F. Delwiche, H. Sze, *Front Plant Sci.* **2012**, *25*, 1–18.
242. E. Bassil, E. Blumwald, *Curr. Opin. Plant Biol.* **2014**, *22*, 1–6.
243. V. Barragan, E. O. Leidi, Z. Andres, L. Rubio, A. De Luca, J. A. Fernandez, B. Cubero, J. M. Pardo, *Plant Cell* **2012**, *24*, 1127–1142.
244. E. Bassil, H. Tajima, Y. C. Liang, M. A. Ohto, K. Ushijima, R. Nakano, T. Esumi, A. Coku, M. Belmonte, E. Blumwald, *Plant Cell* **2011**, *23*, 3482–3497.
245. E. Bassil, M. A. Ohto, T. Esumi, H. Tajima, Z. Zhu, O. Cagnac, M. Belmonte, Z. Peleg, T. Yamaguchi, E. Blumwald, *Plant Cell* **2011**, *23*, 224–239.
246. A. Martiniere, E. Bassil, E. Jublanc, C. Alcon, M. Reguera, H. Sentenac, E. Blumwald, N. Paris, *Plant Cell* **2013**, *25*, 4028–4043.
247. Z. Qi, E. P. Spalding, *Plant Physiol.* **2004**, *136*, 2548–2555.
248. H. H. Kunz, M. Gierth, A. Herdean, M. Satoh-Cruz, D. M. Kramer, C. Spetea, J. I. Schroeder, *Proc. Natl. Acad. Sci. USA* **2014**, *111*, 7480–7485.
249. S. Chanroj, Y. Lu, S. Padmanaban, K. Nanatani, N. Uozumi, R. Rao, H. Sze, *J. Biol. Chem.* **2011**, *286*, 33931–33941.
250. Y. Lu, S. Chanroj, L. Zulkifli, M. A. Johnson, N. Uozumi, A. Cheung, H. Sze, *Plant Cell* **2011**, *23*, 81–93.
251. J. Zhao, N. H. Cheng, C. M. Motes, E. B. Blancaflor, M. Moore, N. Gonzales, S. Padmanaban, H. Sze, J. M. Ward, K. D. Hirschi, *Plant Physiol.* **2008**, *148*, 796–807.
252. J. Barrero-Gil, A. Rodriguez-Navarro, B. Benito, *J. Exp. Bot.* **2007**, *58*, 2839–2849.
253. M. Muller, H. H. Kunz, J. I. Schroeder, G. Kemp, H. S. Young, H. E. Neuhaus, *Plant J.* **2014**, *78*, 646–658.
254. I. N. Talke, D. Blaudez, F. J. Maathuis, D. Sanders, *Trends Plant Sci.* **2003**, *8*, 286–293.
255. V. Demidchik, F. J. M. Maathuis, *New Phytologist* **2007**, *175*, 387–404.
256. Q. F. Gao, C. F. Fei, J. Y. Dong, L. L. Gu, Y. F. Wang, *Mol. Plant.* **2014**, *7*, 739–743.
257. B. G. Hua, R. W. Mercier, Q. Leng, G. A. Berkowitz, *Plant Physiol.* **2003**, *132*, 1353–1361.
258. K. B. Craven, W. N. Zagotta, *Annu. Rev. Physiol.* **2006**, *68*, 375–401.
259. Y. Guan, J. Guo, H. Li, Z. Yang, *Mol. Plant.* **2013**, *6*, 1053–1064.
260. W. Moeder, W. Urquhart, H. Ung, K. Yoshioka, *Mol. Plant.* **2011**, *4*, 442–452.
261. B. Kaplan, T. Sherman, H. Fromm, *FEBS Lett.* **2007**, *581*, 2237–2246.
262. R. Hedrich, I. Marten, *Mol. Plant.* **2011**, *4*, 428–441.
263. E. Peiter, F. J. Maathuis, L. N. Mills, H. Knight, J. Pelloux, A. M. Hetherington, D. Sanders, *Nature* **2005**, *434*, 404–408.
264. G. J. Allen, D. Sanders, *Plant J.* **1996**, *10*, 1055–1069.
265. L. Coyaud, A. Kurkdjian, R. Kado, R. Hedrich, *Biochim. Biophys. Acta* **1987**, *902*, 263–268.
266. S. Ranf, P. Wunnenberg, J. Lee, D. Becker, M. Dunkel, R. Hedrich, D. Scheel, P. Dietrich, *Plant J.* **2008**, *53*, 287–299.

267. J. Scholz-Starke, A. De Angeli, C. Ferraretto, S. Paluzzi, F. Gambale, A. Carpaneto, *FEBS Lett.* **2004**, *576*, 449–454.
268. B. Schulzlessdorf, R. Hedrich, *Planta* **1995**, *197*, 655–671.
269. J. M. Ward, J. I. Schroeder, *Plant Cell* **1994**, *6*, 669–683.
270. P. Maser, S. Thomine, J. I. Schroeder, J. M. Ward, K. Hirschi, H. Sze, I. N. Talke, A. Amtmann, F. J. Maathuis, D. Sanders, J. F. Harper, J. Tchieu, M. Gribskov, M. W. Persans, D. E. Salt, S. A. Kim, M. L. Guerinot, *Plant Physiol.* **2001**, *126*, 1646–1667.

Chapter 10
Potassium *Versus* Sodium Selectivity in Monovalent Ion Channel Selectivity Filters

Carmay Lim and Todor Dudev

Contents

C. Lim (✉)
Institute of Biomedical Sciences, Academia Sinica, Taipei 11529, Taiwan

Department of Chemistry, National Tsing Hua University, Hsinchu 300, Taiwan
e-mail: carmay@gate.sinica.edu.tw

T. Dudev (✉)
Faculty of Chemistry and Pharmacy, Sofia University,
1 James Bourchier Blvd., BG-1164 Sofia, Bulgaria
e-mail: t.dudev@chem.uni-sofia.bg

A. Sigel, H. Sigel, and R.K.O. Sigel (eds.), *The Alkali Metal Ions: Their Role for Life*,
Metal Ions in Life Sciences 16, DOI 10.1007/978-3-319-21756-7_10

Abstract Transport of Na^+ and K^+ ions across the cell membrane is carried out by specialized pore-forming ion channel proteins, which exert tight control on electrical signals in cells by regulating the inward/outward flow of the respective cation. As Na^+ and K^+ ions are both present in the body fluids, their respective ion channels should discriminate with high fidelity between the two competing metal ions, conducting the native cation while rejecting its monovalent contender (and other ions present in the cellular/extracellular milieu). Indeed, monovalent ion channels are characterized by remarkable metal selectivity. This striking ion selectivity of monovalent ion channels is astonishing in view of the close similarity between Na^+ and K^+: both are spherical alkali cations with the same charge, analogous chemical and physical properties, and similar ionic radii. The monovalent ion channel selectivity filters (SFs), which dictate the selectivity of the channel, differ in oligomericity, composition, overall charge, pore size, and solvent accessibility. This diversity of SFs raises the following intriguing questions: (1) What factors govern the metal competition in these SFs? (2) Which of these factors are exploited in achieving K^+ or Na^+ selectivity in the different types of monovalent channel SFs? These questions are addressed herein by summarizing results from recent studies. The results show that over billions of years of evolution, the SFs of potassium and sodium ion channels have adapted to the specific physicochemical properties of the cognate ion, using various strategies to enable them to efficiently select the native ion among its contenders.

Keywords Acid-sensing ion channels • Continuum dielectrics • Coordination number • Density functional theory • Epithelial sodium channels • Ion selectivity • Potassium channels • Voltage-gated sodium channels

Please cite as: *Met. Ions Life Sci.* 16 (2016) 325–347

1 Introduction

Monovalent Na^+ and K^+ ions are indispensable players in several processes that ensure normal functioning of living organisms. For instance, they are involved in regulating homeostasis of blood and body fluids, cardiac, skeletal, and smooth muscle contraction, taste and pain sensation, hormone secretion, and signal transduction in animals [1–4]. These ions coexist in various biological compartments with different relative concentrations: K^+ is the principal cation in the cytoplasm with a $[K^+]:[Na^+]$ concentration ratio equal to 139:12 mM, while Na^+ dominates the extracellular space with $[Na^+]:[K^+] = 145:4$ mM [5].

Na^+ and K^+ transport across the cell membrane is carried out by specialized pore-forming membrane-spanning ion channel proteins, which, by regulating the inward/outward flow of the respective cation, exert tight control on electrical signals in cells [2]. Due to the different intracellular and extracellular K^+ and Na^+ concentrations, potassium and sodium channels open to allow K^+ to exit and Na^+ to enter the cell, respectively, under physiological conditions. Na^+ influx promotes cell excitability, whereas K^+ efflux decreases it.

Altered ion selectivity of potassium or sodium channels due to gene mutations, pathological and physiological conditions (e.g., acidic pH) [6] often results in various channelopathies of the heart, brain, skeletal muscles, and lung [7]. Thus potassium and sodium channels are key drug targets for various disorders including cardiac arrhythmias, heart attack, stroke, migraine, epilepsy, pain, cancer, and autoimmune disorders [8].

1.1 From Potassium Channels to Sodium Channels

The earliest channels appear to be the linear leak and voltage-gated potassium (K_v) channels, which appeared about three billion years ago in bacteria and occur in all organisms [9, 10]. Linear leak potassium channels have a pore domain consisting of two membrane-spanning helices and a pore-forming loop (Figure 1).

K_v channels are homotetramers, each monomer consisting of a pore domain as well as a voltage-sensing domain formed by four membrane-spanning helices. Homotetrameric K_v channels are probably the founding members of the family of tetrameric channels, where each monomer consists of six transmembrane segments [10]. Early in eukaryote evolution, the gene for a six-transmembrane-helix channel duplicated yielding a protein with two domains, whose gene in turn duplicated to create a protein with four domains that can form an ion channel on its own without the need for oligomerization (Figure 1).

Such a four-domain channel evolved permeability to Ca^{2+}, which conveniently became an intracellular signaling messenger. The simplest and oldest animals and their eukaryote relatives, which lack nervous systems, possess only Ca^{2+}-selective channels [11–13]. As more complex nervous systems in eukaryotes evolved, separation between Ca^{2+}- and Na^+-dependent signaling in the cell was required: Na^+ currents are better suited to generate membrane excitability in complex nervous systems than Ca^{2+} currents, as they enable fast and accurate signaling and do not interfere with intracellular Ca^{2+} signaling and exert cytotoxicity [2]. Thus, eukaryotic voltage-gated sodium (Na_v) channels are thought to have evolved from voltage-gated calcium (Ca_v) channels [2, 11], as evidenced by the fact that their four homologous domains are more similar to those of Ca_v channels than to each other. Their relationship to prokaryotic Na_v channels, which are homotetramers of subunits containing six transmembrane segments like K_v channels, is unclear [10].

1.2 Ion Selectivity of Monovalent Ion Channels

As Na^+ and K^+ ions are both present in the body fluids, their respective ion channels should discriminate with high fidelity between the two competing metal ions, conducting the native ion while rejecting its monovalent contender and other ions in the

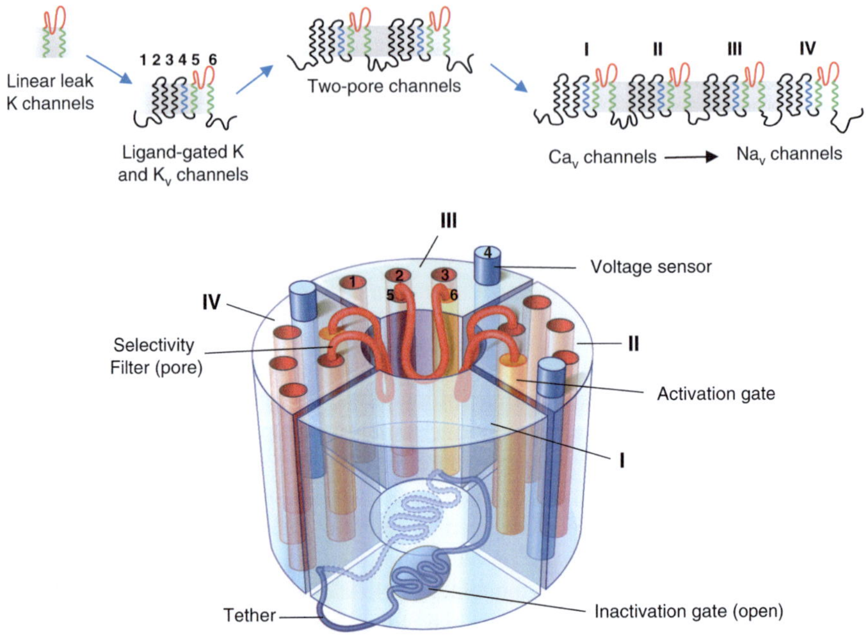

Figure 1 From potassium channels to sodium channels. The top panel from left to right shows (i) a linear leak potassium channel composed of two membrane-spanning helices (green) and a pore loop containing conserved amino acid residue(s) that determines ion selectivity (red); (ii) ligand-gated and voltage-gated potassium (K_v) channels where each monomer consists of six transmembrane helices (TMα, labeled 1–6) with voltage sensor helix 4 in blue, (iii) two-pore channels with two 6TMα domains, and (iv) Ca_v and Na_v channels with four 6TMα domains. Adapted from [10]. The bottom panel illustrates the structural organization of channels with four homologous domains (I–IV), each of which has six transmembrane helices. Adapted from [9].

cellular/extracellular milieu. Indeed, monovalent ion channels do exhibit remarkable metal selectivity, as determined by equilibrium binding experiments that characterize the free energy minima of the metal-binding site or by non-equilibrium flux measurements that estimate permeability of the native ion relative to that of its rival. Potassium channels select K^+ over Na^+ by a ratio of 100–1000:1 based on binding affinities [2, 14, 15], or a K^+:Na^+ permeability ratio of 1,000 [2] to 10,000 [16], but are inaccessible to anions and exclude divalent ions [2]. This striking ion selectivity is astonishing considering that both Na^+ and K^+ have the same charge, analogous physicochemical properties, and ionic radii that differ by only 0.36 Å [17], less than the crystallographic thermal B-factors [18]. Yet, the channel is able to "sense" the cognate ion and selectively let it pass through.

Compared to potassium channels, sodium channels are generally less selective: the epithelial Na^+ channel (ENaC) is the most selective for Na^+ over K^+ with a Na^+/K^+ selectivity ratio of 100 [19] to 500 [20]. Although the acid-sensing ion channel (ASIC) belongs to the epithelial/degenerin superfamily of ion channels, their Na^+/K^+ selectivity, which ranges from 3 to 13 [4], is an order of magnitude lower than ENaC. Compared to ASIC, vertebrate Na_v1 channels appear to be slightly more Na^+/K^+ selective with permeability ratios of 12–21 [2], while their bacterial Na_v channels exhibit varying Na^+/K^+ permeability ratios from 5 to 171 [21–24].

1.3 Selectivity Filter Characteristics of Monovalent Ion Channels

The selectivity filter (SF) is defined as a constricted region of the open pore in the conductive (as opposed to a non-conductive) conformation of the channel that dictates ion selectivity. It is a mono- or multilayered ring-like structure, lined with several conserved metal-binding residues from each of the channel pore-forming domains. The metal-ligating groups protrude toward the pore lumen and interact with the passing metal ion (Figure 2).

X-ray structures of the tetrameric potassium channel [16, 18, 25–33] show that the SF has four contiguous metal-binding sites, each formed by eight (mostly backbone carbonyl) oxygen atoms at the vertices of a square antiprism (Figure 2a). The SF pore size matches bare (dehydrated) K^+ [18]. In contrast, X-ray structures of the tetrameric bacterial Na_v channels [34–37] show a wide SF pore lined with four conserved glutamates (EEEE) that can accommodate partially hydrated Na^+. Crystal structures of the homotrimeric Na^+-selective ASIC in an open conformation [38] also show a wide SF pore lined with three Gly carbonyl groups (GGG) that fits fully hexahydrated Na^+.

Although open-state X-ray structures of other Na^+-selective channels with bound cognate ions remain unsolved, site-directed mutagenesis and channel-blocker binding experiments have revealed their SF oligomeric structure, composition, and pore size. Both ENaC [19] and Na_v1 [39] channels have narrow SF pores that fit dehydrated ions, but their SFs differ in oligomeric structure and composition. The ENaCs have heterotrimeric SFs lined with serine side chain and/or backbone carbonyl groups from the conserved (G/S)XS tract [40–42]. Eukaryotic Na_v channels have asymmetric SFs lined by conserved Asp, Glu, Lys, and Ala donated by each of the four homologous domains [43, 44], unlike prokaryotic Na_v channels with four identical domains. Surprisingly, a conserved Lys is crucial for selectivity, as its mutation to another residue drastically reduces (or even reverses) the channel's Na^+/K^+ selectivity [39, 43, 45], whereas mutation to an aspartate or glutamate rendered the channel Ca^{2+}-selective [45, 46]. The different Na_v channels differ in whether this conserved Lys comes from the second or third domain: Na_v1 channels of higher (bilaterian) animals (e.g., vertebrates, cephalochordates, urochordates, mollusks, annelids, and arthropods) have DEKA SFs (Figure 2b), whereas $Na_v2.5$ channels in cnidarian (sea anemones, corals, hydras, and jellyfish) possess a less Na^+/K^+-selective DKEA motif [12, 13, 45].

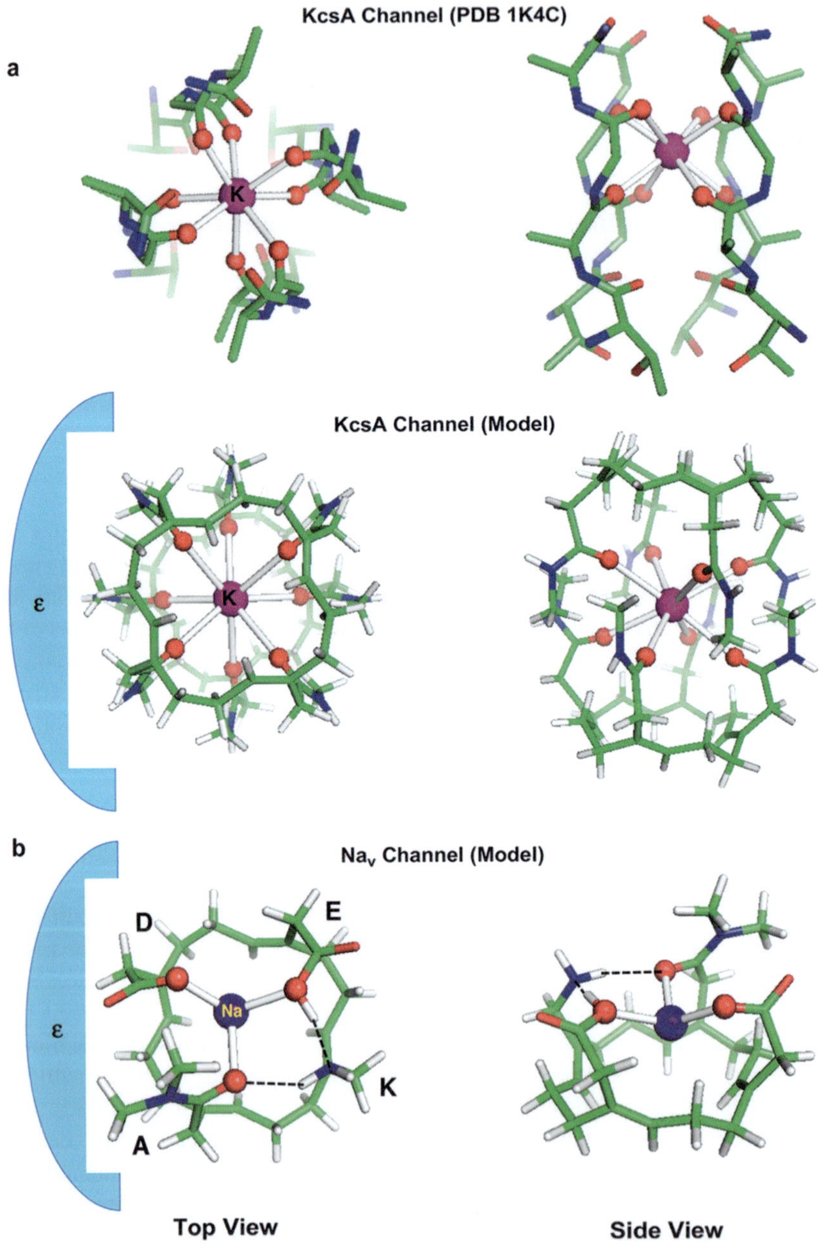

Figure 2 (**a**) The PDB structure (1K4C) of K^+ bound to eight backbone carbonyl groups in the homotetrameric KcsA channel SF (top) is modeled by K^+ bound to 8 $-CONHCH_3$ residues attached to a carbon-hydrogen ring scaffold via methylene spacers with the rest of the filter represented by an effective continuum dielectric constant, ε; the fully optimized B3LYP/6-31+G(3d,p) structure is shown (bottom). Adapted from [60]. (**b**) Fully optimized B3LYP/6-31+G(3d,p) structure of Na^+ bound to the model DEKA SF. Adapted from [88].

1.4 Questions Addressed

The above short survey shows that different types of monovalent ion channels possess different SFs that vary in the overall symmetry, number, arrangement, charge, and chemical type of the metal-ligating groups as well as pore size (see Table 1).

The diversity of SFs found in monovalent channels raises intriguing questions; e.g.,

1. How does the competition between Na^+ and K^+ in a given SF depend on factors such as (a) the inherent cation properties, (b) the ligating strength of the metal-ligating residues, and (c) the protein matrix (protein architecture and coupled protein-solvent interactions) controlling the SF properties?
2. Do the various SFs exhibit higher affinity for the cognate ion than the noncognate one; i.e., is the binding free energy for the native ion to a given SF site more favorable than that for its rival (thermodynamic selectivity)? If so, what special properties of the SF allow this channel region to discriminate the native ion from other competing ions in the ambient solution?
3. How does the SF enable its cognate ion to efficiently permeate through its pore (kinetic selectivity)?
4. Which factors are exploited in achieving K^+ or Na^+ selectivity in the different types of monovalent channel SFs?

Here, we attempt to address these questions by summarizing results from recent studies.

Table 1 Characteristic features in monovalent ion channel selectivity filters.

Channel type	P_{Na}/P_K	SF motif	Charge[a]	Pore radius (Å)
KcsA	0.001–0.0001 [2, 16]	TxGxG	0	2.3[b] (1K4D, 2.3 Å) [28]
ENaC	100–500 [19, 20]	(G/S)xS	0	1–1.4[c] [19]
ASIC	3–13 [4]	GGG	0	3.7[b] (4NTW, 2.1 Å) [38]
Bacterial Na_v	5–171 [21–24]	EEEE	−4e	4.3[b] (4F4L, 3.5 Å) [37]
Eukaryotic Na_v1	12–21 [2]	DEKA	−1e	1.8[d]
Eukaryotic $Na_v2.5$	4 [12]	DKEA	−1e	–[e]

[a]Net charge of the conserved residues comprising the SF motif.
[b]Defined by the mean distance from the pore center (center of the plane formed by the metal-ligating atoms) to the metal-ligating atoms in the open-state crystal structure whose PDB code and resolution are given in parentheses.
[c]Estimated by the largest permeant ion (Na^+) and the smallest nonpermeant (K^+) ion.
[d]Established by the size of the only permeable ammonium cation in the series of ammonium ions in [39].
[e]Pore radius is not known.

1.5 Aims and Scope

Since only the open-state metal-bound structures of potassium channels [32, 33, 47], ASIC [38], and bacterial Na_v channels [37, 48] are available, but those of the other sodium channels have yet to be solved, we focus on the contributions of the SF (rather than other segments of the pore or bulk properties [49–52]) to ion selectivity when the channel is in its open, ion-conducting conformation. We aim to elucidate how monovalent channel SFs varying in oligomeric structure, ligand composition, local charge, and pore size can achieve the same goal of selecting the native ion from competing ions.

In the following we will first briefly outline the key approaches to study ion selectivity using available metal-bound open-pore structures of potassium and sodium channels as well as methods based on SF models in cases where such structures are not available. Next, we summarize the various factors governing the outcome of Na^+ *versus* K^+ competition in the SFs of potassium and sodium channels. We then discuss which of these factors are exploited to achieve high selectivity for the native ion in the SFs of potassium channels, bacterial Na_v channels, and ASIC (for which crystal structures are available) followed by ENaC and eukaryotic Na_v1 channels. For channels where multi-ion free energy profiles for ion permeation based on the respective crystal structures have been computed, we discuss both thermodynamic and kinetic selectivity based on free energy minima in the SF and their connecting barriers. For channels lacking crystal structure and associated free energy profiles for ion permeation, we limit discussion to thermodynamic selectivity. We conclude by summarizing the different strategies used by the various SFs of potassium and sodium channels to achieve their ion selectivity.

2 Methodology

We briefly outline the two main computational approaches that have been used to study ion selectivity and refer the reader to the original publications to provide details about the models and computational methods. If a high-resolution crystal structure of the ion channel in an open conductive conformation with the cognate ion bound is available, the free energy surface (free energy minima and barriers) underlying ion permeation can be obtained using equilibrium/nonequilibrium molecular dynamics (MD) simulations of ions moving through the SF pore [53–57]. In addition, the free energy differences between equilibrium states for different ions bound by the channel can be computed from free energy perturbation and umbrella sampling methods [58, 59].

In the absence of open metal-bound X-ray structures of the ion channel, insight into the factors governing ion selectivity can be obtained from reduced SF models, which treat the ion and its ligands explicitly and the rest of protein and solvent implicitly using combined quantum mechanical and continuum dielectric calculations [60, 61]. Assuming equilibrium binding to a well-defined site, the outcome of the competition between the bulk solvent and the protein ligands for the native cation in a model SF characterized by an effective dielectric constant $\varepsilon = x$ can be assessed by computing the free energy ΔG^x for replacing K^+ bound in the SF, [K^+-filter], with Na^+:

$$[Na(H_2O)_6]^+ + [K^+\text{-filter}] + nH_2O \rightarrow [Na^+\text{-filter}] + [K(H_2O)_{6+n}]^+ \qquad (1)$$

where n = 0, 1 or 2. A positive ΔG^x implies a K^+-selective filter, whereas a negative ΔG^x implies a Na^+-selective one.

One way to compute the free energy for equation (1) is *via* a thermodynamic cycle (2):

$$\begin{array}{llcl} & & \Delta G^1 & \\ \text{Gas-phase } (x=1) & \text{Reactants} & \rightarrow & \text{Products} \\ & \Delta G^x_{\text{solv}} \downarrow & & \downarrow \Delta G^x_{\text{solv}} \\ \text{SF } (x=4-30) & \text{Reactants} & \rightarrow & \text{Products} \\ & & \Delta G^x & \end{array} \qquad (2)$$

The free energy for equation (1) can be computed as a sum of (i) the gas-phase free energy (electronic effects) and (ii) the solvation free energy difference between the products and reactants (solvation effects). Thus, the outcome of the competition between the native ion and its rival depends on the interplay between electronic and solvation effects.

Electronic effects reflect interactions between Na^+ or K^+ and its ligands, which in turn depend on the different charge-accepting ability of the two monocations and their coordination numbers (CNs) as well as the ligating strength of the metal ligands (see below). On the other hand, solvation effects reflect the solvation free energy difference between the metal aqua complexes, the SF pore size and solvent accessibility and thus the effective dielectric constant in the SF: A narrow pore that fits dehydrated ions would have a lower effective dielectric constant than a wide one that accommodates a partially/fully hydrated ion. Note that if the SF were quite flexible, it cannot maintain its pore size, as its ligands could adapt to the CN or geometry preference of a rival cation; in this way, protein dynamics/flexibility also affects ion selectivity. An upper limit of rigidity effects can be obtained by computing the free energies for equation (1) in a fully rigid SF pore that is optimized for its native ion and prohibited from relaxing upon binding a noncognate ion.

3 Factors Governing the Competition Between K^+ and Na^+ in Monovalent Ion Channel Selectivity Filters

Both experimental and theoretical studies have revealed the factors that help potassium [6, 16, 18, 25–31, 33, 53–60, 62–84] and sodium [12, 20, 23, 41, 43, 50–52, 85–99] channel SFs achieve the desired metal selectivity [61, 100–107]. Below, we discuss these factors, which encompass the inherent properties of the metal ions, the metal-ligating residues, the protein matrix and coupled metal binding and protein conformational changes.

3.1 Metal Coordination Number

This captures properties of both the metal cation and its ligands and reflects the strength of the metal–ligand interactions [108]. Compared to Na^+, K^+ forms longer and weaker bonds with a given ligand [109], reducing ligand–ligand repulsion and allowing more ligands to be bound, thus it prefers a larger CN in proteins. Hence, increasing the number of metal-ligating residues; i.e., the metal CN, increases the competitiveness of K^+ over Na^+. Conversely, decreasing the metal CN enhances selectivity for Na^+ over K^+, as it reduces the steric repulsion among the bulky protein ligands around Na^+ more than that around the larger K^+.

3.2 Ligand Ligating Strength

The various metal-ligating residues differ in their net charge and/or charge-donating ability and thus ligating strength. Negatively charged Asp/Glu side chains possess stronger ligating strength than neutral metal-ligating groups; among the latter, the Asn/Gln/backbone carbonyl group has a larger dipole moment and stronger charge-donating ability than the Ser/Thr hydroxyl group [93]. Increasing the ligating strength of a protein ligand favors the cation with better electron-accepting ability; e.g., Na^+ compared to K^+ [108]. Hence, high-field ligands such as negatively charged carboxylates prefer Na^+ to K^+ [110], as their interactions with Na^+ are more favorable than those with K^+ and suffice to offset the Na^+ dehydration penalty (−98 kcal/mol) [111]. On the other hand, medium to low field ligands such as neutral carbonyl or hydroxyl groups favor K^+ over Na^+ [110], as their interactions with K^+ may outweigh the K^+ dehydration penalty (−81 kcal/mol) [111] but may not suffice to compensate for the Na^+ dehydration penalty.

3.3 Protein Matrix

This can control the SF pore size, solvent accessibility, rigidity, stability, metal CN, and the number of metal-binding sites. A narrow, dry SF pore that fits a dehydrated Na^+, but not the bulkier K^+, is selective for Na^+ over K^+ [88]. A wide, solvent-accessible SF pore that fits a partially/fully hydrated ion can also be selective for Na^+ over K^+ because Na^+, being a stronger Lewis acid, polarizes the first-shell water molecules more than K^+, resulting in stronger Na^+–water ··· ligand interactions than the respective K^+–water ··· ligand interactions [99]. The SF pore should possess some degree of stiffness to maintain an optimal size that snugly fits the native ion, and not be so flexible that it can adapt to the coordination geometry requirements of non-native ions. The protein matrix can rigidify and stabilize the SF through hydrogen bonds and long-range electrostatic interactions with the SF residues. It can also modulate the free energy well depths of the SF metal-binding sites and thus affect the number of sites.

3.4 *Coupled Metal Binding and Selectivity Filter Conformational Changes*

The native ion could contribute to its own selectivity by inducing conformational changes that lead to a "conductive" SF, whereas the noncognate ion yields a non-conducting conformation. This is reminiscent of the well-known coupled metal binding and conformational change of Zn-finger peptides [112], where the native Zn^{2+} induces conformational changes to the native structure but other transition metal ions such as Ni^{2+} or Cu^{2+} lead to a non-native fold [113, 114]. Binding of the native ion to the SF could induce the SF ligands to re-orient their side chains to interact with the cation, thus inducing a metal-binding site, and/or lead to charge–charge repulsion with nearby bound ions, thus destabilizing the ions and lowering the barriers for ion permeation through the SF. The non-native ion, however, cannot inflict the same conformational changes induced by the native ion, resulting in reduced affinity for the SF and/or increased barriers for ion permeation through the SF.

4 K^+ *Versus* Na^+ Competition in K^+-Selective Channel Selectivity Filters

The ion-selectivity mechanisms of potassium channels are the most extensively studied because several high-resolution crystal structures of these channels in both closed [16, 18, 25–28, 30, 31, 33, 77, 115, 116] and open [32, 33, 47] conformations are available. These structures show in common a SF consisting of four contiguous metal-binding sites (labeled S_1, S_2, S_3, and S_4), each lined with eight oxygen atoms that bind directly to dehydrated K^+ (Figure 2a). All four SF sites seem necessary for K^+/Na^+ selectivity: High-resolution crystal structures of wild-type NaK and mutant NaK2CNG and NaK2K show SFs with two, three, or four contiguous ion-binding sites, respectively, that are superimposable onto equivalent sites in the K^+-selective KcsA potassium channel, but only NaK2K is K^+-selective [106]. NaK2K has two K^+/Na^+ -selective SF sites, whereas the nonselective NaK2CNG has only one, suggesting that a single high-affinity K^+ site cannot yield high K^+ selectivity during ion permeation [18, 82]. Under physiological conditions, all four metal-binding sites

(S_1–S_4) are likely occupied, resulting in repulsive ion–ion interactions, which drive efficient permeation through the SF [83].

Previous studies have examined how the selectivity for K^+ over Na^+ in potassium ion channels depends on

(i) the dehydration penalty [28, 101, 104] and hydration number [60] of the permeating cations,
(ii) the number and ligating strength of the metal-coordinating groups lining the pore [56, 60, 62, 65–69, 71, 75, 76, 101, 102],
(iii) the architecture of the metal-binding site influencing the flexibility and solvent exposure (effective dielectric constant) of the SF pore [28, 60, 69, 70, 72–75, 78, 79, 81, 94, 100, 104, 117, 118],
(iv) the number and occupancy of SF binding sites [32, 57, 58, 80, 82, 83, 106, 107], and
(v) kinetic barriers for an ion to enter the SF from the intra or extracellular side [53–56, 79, 83, 119].

These studies show that a well-balanced combination of various factors involving the native ion, the metal-coordinating ligands, and the protein matrix favors K^+ over Na^+ binding to a SF site in the potassium channel. For some potassium channels such as the KcsA channel, the native K^+ could contribute to its own selectivity, as it can stabilize the SF in a conductive conformation, whereas the rival Na^+ yields a nonconducting one [18, 82, 106]. The number and ligating strength of the SF ligands also contribute to the preference for K^+ over Na^+: Each SF site provides eight (mostly backbone) oxygen atoms, which increases the ligand repulsion around Na^+ more than that around the larger K^+, whereas the medium-field strength backbone carbonyl groups provide insufficient favorable interactions with Na^+ to offset its dehydration penalty. The potassium channel matrix could further enhance K^+/Na^+ selectivity in the SF by various means: It could constrain the orientation of the SF ligands to bind optimally to K^+ and rigidify the SF pore, forcing Na^+ to adopt the CN of K^+ rather than its preferred smaller CN [74]. It could decrease the SF pore solvent exposure, which correlates with a larger hydration number of K^+ relative to that of Na^+ [74], thus more water molecules are released upon binding of K^+ in the pore compared to binding of Na^+ [60].

The above discussion underscores the "thermodynamics" factors that enhance the affinity of the potassium channel SF site for K^+ over Na^+. However, potassium channels exhibit not only strong K^+ selectivity, but also near-diffusion-limited permeation efficiency [2]. Hence, the SF site should *not* bind K^+ so strongly that K^+ gets stuck and cannot permeate rapidly through the pore. With no other ions in the SF, a single K^+ ion would not easily escape from the deep well of a high-affinity K^+ site (see above). However, when two K^+ ions occupy adjacent sites; e.g., S_2 and S_3, they would be destabilized when a third K^+ ion binds at S_4 due to ion–ion electrostatic repulsion, enabling the K^+ ions at S_2 and S_3 to escape from the shallower wells and advance to S_1 and S_2, respectively [83]. This shows how metal binding to multiple sites could modulate the free energy wells and thus barriers for ion permeation, creating a "conductive" conformation. Hence, one way to reduce the high affinity of the SF site for K^+ is via coupled metal binding and conformational change of the SF from a nonconductive conformation to a conductive one (see above). Nonequilibrium MD simulations [57] show that binding of K^+ and Na^+ to the SF induces different structural rearrangements that facilitate K^+ conduction, but blocks Na^+: Binding of intracellular K^+ to the SF (S_4 site) induces formation of other K^+-binding sites, reducing the free energy for subsequent K^+ ions to translocate, whereas binding of Na^+ to the SF induces disorder and destabilization of Na^+-binding sites, raising the energy cost for the entry of subsequent Na^+ ions.

5 Na^+ *Versus* K^+ Competition in Na^+-Selective Channel Selectivity Filters

5.1 Prokaryotic Na_v Channel Selectivity Filters

Whereas eight backbone carbonyl groups line the SF of potassium channels, four Glu carboxylate side chains line the SF of bacterial Na_v channel (EEEE SF) [34–37, 48]. The X-ray structures of prokaryotic Na_v channels from *Arcobacter butzleri* (Na_vAb) [34] and *Rickettsiales* sp. HIMB114 (Na_vRh) [36] have closed SF pores, whereas the crystal structure of bacterial Na_v channel from *Magnetococcus* sp. (Na_vMs) containing only the pore domain has an open pore [37]. Nevertheless, both closed-pore Na_vAb [34] and open-pore Na_vMs [37] structures show a relatively wide pore that can accommodate Na^+ bound indirectly to two glutamates via water molecules. This is supported by MD simulations [23, 24, 91, 95, 96, 120–123] of *bacterial* Na_v channels, which show that the ion inside the EEEE pore is indeed partially hydrated and is coordinated to one or two Glu side chains and indirectly to two other Glu carboxylates via bridging water molecules. The EEEE binding site, which is located near the extracellular end of the channel, is followed by two other metal-binding sites formed by backbone carbonyl groups that fit fully hexahydrated ions. Thus, the outer, high field-strength EEEE binding site partly dehydrates the permeating ion, while the inner, lower field-strength sites rehydrate and conduct Na^+.

MD [23, 24, 91, 96, 120, 121, 123] simulations as well as combined density functional theory and continuum dielectric calculations of a reduced SF model [92, 93] show that the Na^+/K^+ selectivity in the bacterial Na_v channels can be attributed to the following factors:

Balanced SF charge density: If the EEEE pore were so constricted that three or four glutamates coordinate the ion directly, then it would favor Ca^{2+} over Na^+ [92, 93], as in the case of Ca_v channels [124, 125]. Hence, a fine balance between *direct* ion–carboxylate and *indirect* ion–water ··· carboxylate interactions in the bacterial Na_v channel SF is crucial for Na^+/K^+ selectivity. *Direct* ion–carboxylate interactions favor Na^+ over the weaker electron-acceptor K^+. This is supported by the finding that the Na^+/K^+ selectivity of the *bacterial* Na_v channel is decreased or abolished if the SF glutamates were replaced by uncharged residues [91] or protonated by lowering the pH [23]. *Indirect* ion–water ··· carboxylate interactions also favor Na^+ over K^+ because Na^+ polarizes its first-shell water molecules better than K^+, resulting in more favorable interactions with the SF walls than its rival K^+ (see Section 3).

Solvent-accessible and wide SF pore: The aperture of the wide EEEE pore fits nicely a partially hydrated Na^+, ensuring optimal interaction between Na^+ and the filter [91]. However, it may be too small for the bulkier hydrated K^+ to fit in the plane of the glutamates with water molecules bridging to the carboxylates, as seen in MD simulations of the Na_vAb channel [91, 121], so K^+ binding inside the EEEE filter would be less favorable. First-shell to second-shell hydrogen bonds between the metal-free Glu carboxylate oxygen atoms and backbone nitrogen atoms from neighboring residues, found in the Na_vAb crystal structures [34], help to rigidify the SF and maintain its pore size: Loss of these supporting interactions from the SF glutamates with second-shell residues was found to increase the pore helix mobility, thus destabilizing the SF [35].

Low kinetic barrier: MD simulations show that Na^+ encounters a much smaller barrier between the outer EEEE binding site and an inner site than the noncognate K^+, resulting in faster conduction [23, 24, 119, 121].

5.2 *Epithelial/Degenerin Ion Channel Selectivity Filters*

5.2.1 Acid-Sensing Ion Channel

ASICs belong to the ENaC or degenerin superfamily of ion channels [126–128], but their Na^+/K^+ permeability ratio of 3–13 [4] is 1–2 orders of magnitude lower than that of the ENaC (100–500) [19, 20]. The open-state X-ray structure of the ASIC1a/snake toxin complex [38] shows a homotrimeric SF lined with backbone carbonyl ligands from the Gly residue of the (G/S)XS motif (GGG SF) with a flexible, wide pore of radius ~3.7 Å that fits hydrated Na^+, whose radius is 3.6 Å [129]. This wide GGG SF favors Na^+ over K^+ because the hydrated Na^+ fits snugly in the pore and its shorter and more polar $Na^+–O^{water}$ bonds compared to $K^+–O^{water}$ bonds enable more hydrogen-bond contacts and stronger electrostatic interactions with the SF wall (Figure 3a, negative ΔG^x, $x = 1–30$).

5.2.2 Epithelial Na^+ Channel

Unlike the ASIC channel which has a flexible, wide SF pore, the ENaC has a less flexible, narrow SF pore of radius <1.4 Å that fits dehydrated metal ions [19]: cations such as $^+HNH_2OH$ and guanidinium can readily pass the ASIC1 channel [98], but not ENaC [20]. The ENaC consists of α, β, and γ subunits that form heterotrimers with SFs containing a conserved (G/S)xS motif [40, 42, 130, 131]. In this asymmetric SF, the conserved Ser residues from the α and γ subunits and the conserved Gly from the β subunit play a key role in conferring Na^+/K^+ selectivity [42]. Notably, substitution of the C-terminal Ser of the conserved (G/S)xS motif with larger residues significantly reduced the high selectivity of ENaC for Na^+ [41].

Since no crystal structures for ENaC have yet been solved, it is unclear whether backbone or side chain oxygen atoms or a combination of both coordinate the permeating ions. Hence, two "limits" of the ENaC SF, i.e., lined by three backbone groups (BBB SF, Figure 3b) or three Ser hydroxyl groups (SSS SF, Figure 3c), have been modeled. The computed free energies ΔG^x for replacing K^+ in these model SFs with Na^+ [99] indicate that the high Na^+/K^+ selectivity can be attributed to the following factors:

"Undercoordination" of rival K^+: Undercoordination of three SF ligands to K^+, which prefers a larger coordination sphere than Na^+, contributes to Na^+/K^+ selectivity: increasing the CN to four in a SF lined with four serines resulted in a K^+-selective SF (see [88]).

Medium ligand-ligating strength: Metal coordination to backbone oxygen atoms rather than Ser side chains enhances Na^+/K^+ selectivity (more negative ΔG^1 in the BBB SF than in the SSS SF).

Relatively rigid, constricted SF pore: A rigid ENaC SF strongly enhances Na^+/K^+-selectivity: the ΔG^x in a rigid pore prohibited from relaxing upon K^+ binding (Figure 3, numbers in parentheses) are much more negative than those in a flexible one (Figure 3, numbers without parentheses), regardless of whether side chain or backbone oxygen atoms coordinate the permeating ions.

In summary, the ENaC SF appears to select the cognate Na^+ over its key rival, K^+, via a narrow, rigid pore [40, 42, 130] that does not fit the bulkier K^+, whereas its three relatively weak ligating residues undercoordinate the bulkier K^+, which prefers a larger CN than Na^+. Departing from these design principles in the case of the ASIC SF with a wide, less rigid pore that accommodates *hexa*hydrated ions makes the ASIC SF less selective for Na^+ over K^+ than the ENaC SF, as observed experimentally.

5.3 Eukaryotic Na_v Channel Selectivity Filters

Whereas prokaryotic Na_v channels have symmetric EEEE SFs, eukaryotic Na_v channels have asymmetric DEKA/DKEA SFs. Interestingly, like prokaryotic Na_v channels, the high voltage-activated Ca_v channel also has the same EEEE motif but is Ca^{2+}-selective because its narrower pore fits monohydrated Ca^{2+} [132], whose higher charge density secures more favorable interactions with four negatively charged SF glutamates compared to Na^+ [92, 93]. Mutation of the third domain Glu in the EEEE SF of human cardiac Ca_v channel to Lys suffice to generate a Na^+-selective channel, as the channel became more permeable to Na^+ than to Ba^{2+} [133]. Thus, the presence of a Lys in the SF has played a key role in converting Ca^{2+}-selective EEEE/DEEA SFs to Na^+-selective DEKA/DKEA ones [12, 13, 134] during the course of evolution.

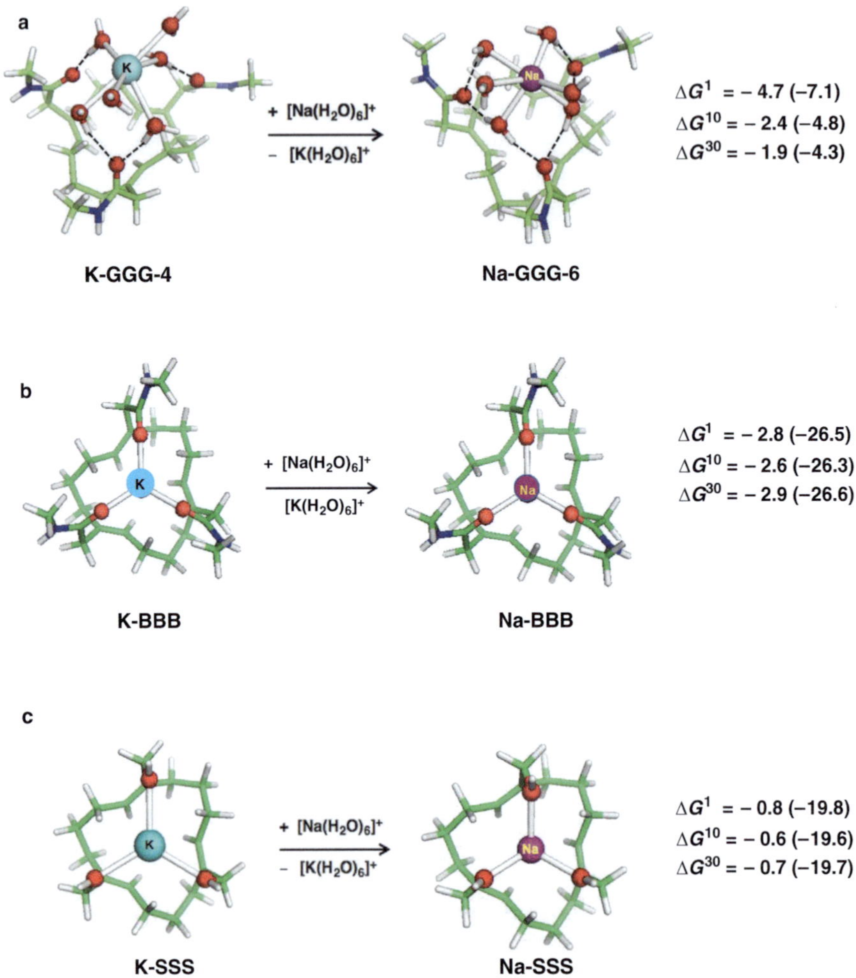

Figure 3 The free energies, ΔG^x (in kcal/mol), for replacing K^+ bound to (**a**) three $-CONHCH_3$ ligating groups (representing backbone peptide groups from glycines denoted by G) in the ASIC GGG filter with Na^+, (**b**) three $-CONHCH_3$ ligating groups (representing backbone peptide groups denoted by B) in the ENaC BBB filter to differentiate them from the backbone peptide groups of the conserved glycines lining the ASIC, and (**c**) three OH-ligating groups (representing Ser side chains) in the ENaC SSS filter. ΔG^1 refers to the metal exchange free energy in the gas phase, whereas ΔG^{10} and ΔG^{30} refer to the metal exchange free energies in an environment, characterized by an effective dielectric constant of 10 and 30, respectively. The free energies for metal exchange in a rigid Na^+-optimized GGG, BBB, and SSS filters prohibited from relaxing upon K^+ binding are in parentheses. The metal-ligating groups are coordinated to the permeating ion and attached to a carbon–hydrogen ring scaffold via methylene spacers. Shown are B3-LYP/6-31+G(3d,p) fully optimized structures of K^+ and Na^+ bound to the model SFs.

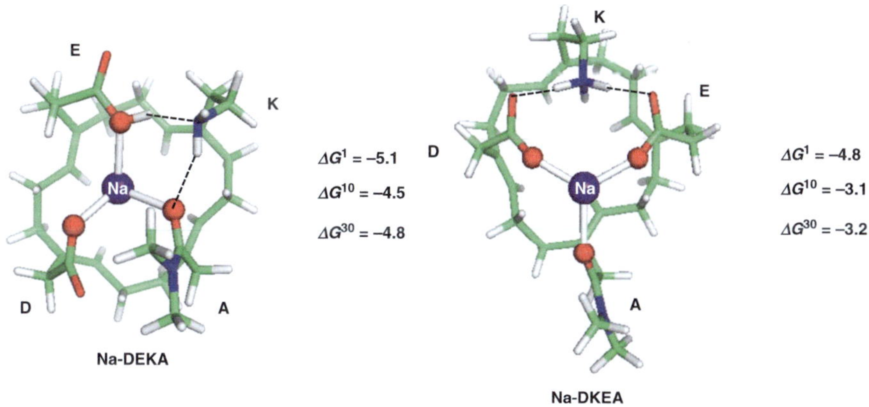

Figure 4 The free energies, ΔG^x (in kcal/mol), for replacing K^+ in model DEKA (**top**) and DKEA (**bottom**) SFs with Na^+. ΔG^1 refers to the metal exchange free energy in the gas phase, whereas ΔG^{10} and ΔG^{30} refer to the metal exchange free energies in an environment characterized by an effective dielectric constant of 10 and 30, respectively. Shown are B3-LYP/6-31+G(3d,p) fully optimized structures of Na^+-bound to model SFs with $-CH_2-COO-$, $-CH_2-CH_2-COO-$, and $-CH_2-CH_2-NH_3^+$ modeling the Asp, Glu, and Lys side chains, respectively, and $-CON(CH_3)_2$ representing the Ala backbone peptide group. The dashed lines denote hydrogen bonds.

In accord with experimental findings, the DEKA or DKEA SFs are both predicted to be selective for Na^+ over K^+ from combined quantum mechanical and continuum dielectric calculations on model SFs (Figure 4): The ΔG^x for replacing K^+ with Na^+ in the model SFs are all negative (−3 and −5 kcal/mol). Also in agreement with experimental observations [2, 12, 13, 43–45], the DEKA SF is predicted to be more Na^+-selective than the DKEA SF (more negative $\Delta G^{10/30}$ for DEKA than for DKEA). Na_v channels with DEKA SFs are highly Na^+/K^+-selective due to the favorable combination of several factors:

"Undercoordination" of rival K^+: As Lys does not bind the metal directly, the ion is coordinated to three rather than four protein ligands in the SF (see Figure 4). A small CN of three favors Na^+ over K^+ binding (see above).

Balanced SF charge density: Interactions with two high-field strength residues (D or E) favor Na^+, which has better electron-accepting ability than K^+, thus enhancing Na^+/K^+ selectivity. Increasing further the number of anionic ligands in the SF would favor Na^+ in its competition with K^+, but not in its competition with Ca^{2+}: SFs with three or four acidic residues binding directly to the metal cation were predicted to decrease or reverse Na^+/Ca^{2+} selectivity [92, 93]. Indeed, mutation of the Lys in the DEKA SF to an aspartate/glutamate rendered the channel Ca^{2+}-selective [45, 46].

Constricted SF pore: The DEKA SF is permeable only to ammonium (of radius 1.8 Å), but not to larger mono-, di-, tri-, or tetramethylammonium ions [39], indicating that its pore is relatively rigid and constricted. This is due in part to the Lys in the DEKA SF, which constricts and rigidifies the SF pore via hydrogen bonding interactions with its neighbors (Figure 4). Thus, rigidifying the pore to optimally fit the native Na^+ enhances Na^+ selectivity.

Coupled metal binding and conformational change: Simulations of mutant Na_vRh with a DEKA SF show that binding of a second metal cation to the SF induces a change in the SF conformation that stabilizes Na^+ more than K^+ [95].

5.3.1 Why the DEKA Selectivity Filter Is More Na^+-Selective than the DKEA Selectivity Filter

Interestingly the position of the residues in the SF ring is important: swopping the Glu and Lys residues in the wild-type DEKA SF to yield the DKEA SF reduces the Na^+:K^+ permeability from 10 to 2.7 [45]. In accord with the experimental finding, the free energies for the DEKA SF are more negative than those for the DKEA SF. Since the different Lys position in the DEKA and DKEA SFs does not change the metal CN or the SF net charge, how does it affect ion selectivity? The Lys in the DEKA SF rigidifies and constricts the pore more than the Lys in the DKEA one, making the DEKA SF more Na^+-selective than the DKEA one: The Lys interacts with the Asp and Glu metal-*free* O atoms in the Na-DKEA SF, but with the Glu and Ala metal-*bound* O atoms in the Na-DEKA SF (Figure 4), making the latter more rigid than the former – rigidifying the DEKA pore increases Na^+/K^+ selectivity more than rigidifying the DKEA pore (by ~1.5 kcal/mol). The Lys also makes the DEKA pore narrower than the DKEA one, as the three metal-bound O atoms are closer to each other in the DEKA SF than those in the DKEA SF [134]. Hence, in the DKEA SF, Na^+ is nearly in the plane formed by the three metal-ligating O atoms but in the DEKA SF, Na^+ has sunk below this plane.

6 Concluding Remarks and Future Directions

Over billions of years of evolution, the SFs of potassium and sodium ion channels have adapted to the specific physicochemical properties of the cognate ion, using various strategies to enable them to efficiently select the native ion among its contenders. They all exhibit higher affinity for the native ion than for its rival cation using a well-balanced combination of various factors encompassing the inherent properties of the metal ions and the metal-ligating residues as well as the protein matrix, as summarized below.

The potassium channel SF provides eight medium-field strength carbonyl ligands that bind optimally to dehydrated K^+, but "overcoordinate" Na^+, which prefers a smaller CN. Like the potassium channel SF, the SFs of ENaC and eukaryotic Na_v channels also bind dehydrated ions in a narrow pore. They help to select Na^+ over K^+ by providing three metal-ligating residues that "undercoordinate" K^+. Furthermore, the three metal-ligating residues possess a well-balanced ligating strength that is strong enough to favor Na^+ over K^+ but not too strong to prefer dicationic Ca^{2+}. Na^+/K^+ selectivity would be enhanced if the pore were sufficiently rigid and constricted to geometrically fit the cognate Na^+ but not rival cations, which cannot fit in the narrow Na^+-optimized pore without an energy penalty. The DEKA SFs of eukaryotic Na_v channels seem to exemplify this strategy: the non metal-ligating Lys reduces the metal CN, the net ligand charge, and the pore size and flexibility via a tight network of hydrogen bonds to its neighbors.

In contrast to the SFs of ENaCs and eukaryotic Na_v channels, the SFs of ASICs and bacterial Na_v channels have wider, less rigid pores that bind a fully or partially hydrated cation. Such a SF pore preferentially binds the cognate Na^+ over K^+, because Na^+ polarizes its first-shell water molecule(s), resulting in more favorable interactions with the SF walls than the rival K^+ ion. Thus, both a relatively flexible, wide pore (GGG/EEEE SF) and a less flexible, narrow pore (DEKA/DKEA SF) can achieve Na^+/K^+ selectivity.

Although the different types of channels have different SFs that are selective for the cognate ion, channels belonging to the same family with the same SF motif could exhibit varying relative permeability ratios (see Table 1). This may be due to the different measurement conditions; it may also reflect the different interactions between each protein matrix and its SF, which could affect the SF pore size, solvent accessibility, flexibility, stability or the number of metal-binding sites in the SF. Efficient permeation through the different channel SFs likely requires coupled metal binding and conformational changes that modulate the kinetic barriers controlling the ion conduction rates. Such modulations differ depending on the type and number of cations. Binding of a native ion to a metal-bound SF could induce conformational changes that facilitate its conduction, but binding of a noncognate ion could result in "nonconductive" conformational changes.

Clearly, high-resolution X-ray structures of various epithelial and eukaryotic sodium channels bound to their cognate ions in open, conductive conformations are needed. When they become available, they would enable quantum mechanical/molecular mechanical calculations or umbrella sampling MD simulations, which could help validate the key factors found to influence thermodynamic selectivity from reduced SF models. However, the reliability of the simulation results depends on (i) the use of appropriate force fields that can accurately account for electronic effects such as the differential amounts of charge transfer from the ligands to the different metal cations, polarization effects, and proton transfer [135, 136] as well as (ii) sufficient sampling of all relevant conformations [54].

The results from accurate nonequilibrium MD simulations using appropriate force fields will help to map out the free energy surface (free energy minima and barriers) for ion translocation through the SF and determine to what extent the conserved residues lining the SF, the protein matrix, the coupling between ions and kinetic barriers contribute to K^+ or Na^+ selectivity in the channel. They could also help to elucidate the intricate and complex coupling between selective ion binding and changes in the SF conformation/stability; e.g., how high K^+ concentrations trigger a change in the SF conformation from a nonconductive to a conductive structure. Ultimately, an in-depth understanding of ion selectivity mechanisms in monovalent channels could aid to design drugs that modulate ion selectivity for drug target channel proteins.

Abbreviations

Ala, A	alanine
ASIC	acid-sensing ion channel
Asp, D	aspartic acid
B	backbone carbonyl group
Ca_v	voltage-gated calcium
CN	coordination number
ENaC	epithelial Na^+ channel
Glu, E	glutamic acid
Gly, G	glycine
K_v	voltage-gated potassium channel
Lys, K	lysine
MD	molecular dynamics
Na_v	voltage-gated sodium channel
Ser, S	serine
SF	selectivity filter

Acknowledgments We thank Prof. James Wang, Prof. Thomas E. deCoursey and Prof. John Straub for helpful comments and Ms. Karine Mazmanian for technical help. This work was supported by Academia Sinica, MOST, Taiwan (Grant NSC-98- 2113-M-001-011). T.D. is supported by the Institute of Biomedical Sciences at Academia Sinica and EU Grant "Beyond Everest", FP7-REGPOT-2011-1.

References

1. I. Mano, M. Driscoll, *Bioessays* **1999**, *21*, 568–578.
2. B. Hille, *Ionic Channels of Excitable Membranes*, Sinauer Associates, Sunderland, MA, 2001, p. 814.
3. P. M. Snyder, *Endocr. Rev.* **2002**, *23*, 258–275.
4. S. Kellenberger, L. Schild, *Pharmocol. Rev.* **2015**, *67*, 1–35.
5. D. Voet, J. G. Voet, *Biochemistry*, John Wiley & Sons, New York, 1990.
6. H. Chen, F. C. Chatelain, F. Lesage, *Trends Pharm. Sci.* **2014**, *35*, 461–469.
7. R. S. Kass, *J. Clin. Invest.* **2005**, *115*, 1986–1989.
8. (a) S. England, M. J. de Groot, *Br. J. Pharmacol.* **2009**, *158*, 1413–1425. (b) B. S. Zhorov, D. B. Tikhonov, *Trends Pharm. Sci.* **2013**, *34*, 154–161.
9. D. Nelson, M. Cox, Lehninger Principles of Biochemistry, W. H. Freeman and Company, New York, 2005, p. 1216.
10. H. H. Zakon, *Proc. Natl. Acad. Sci. USA* **2012**, *109*, 10619–10625.
11. J. D. Spafford, A. N. Spencer, W. J. Gallin, *Recept. Channels* **1999**, *6*, 493–506.
12. M. Gur Barzilai, A. M. Reitzel, J. E. M. Kraus, D. Gordon, U. Technau, M. Gurevitz, Y. Moran, *Cell Rep.* **2012**, *2*, 1–7.
13. A. Senatore, A. Monteil, J. van Minnen, A. B. Smit, J. D. Spafford, *PLoS One* **2013**, *8*, e55088.
14. J. Neyton, C. Miller, *J. Gen. Physiol.* **1988**, *92*, 569–596.
15. M. LeMasurier, L. Heginbotham, C. Miller, *J. Gen. Physiol.* **2001**, *118*, 303–314.

16. D. A. Doyle, J. Morais Cabral, R. A. Pfuetzner, A. Kuo, J. M. Gulbis, S. L. Cohen, B. T. Chait, R. MacKinnon, *Science* **1998**, *280*, 69–77.
17. R. D. Shannon, *Acta Crystallogr. A* **1976**, *32*, 751–767.
18. Y. Zhou, J. H. Morais-Cabral, A. Kaufman, R. MacKinnon, *Nature* **2001**, *414*, 43–48.
19. S. Kellenberger, M. Auberson, I. Gautschi, E. Schneeberger, L. Schild, *J. Gen. Physiol.* **2001**, *118*, 679–692.
20. L. G. Palmer, *J. Membr. Biol.* **1982**, *67*, 91–98.
21. L. Yue, B. Navarro, D. Ren, A. Ramos, D. E. Clapham, *J. Gen. Physiol.* **2002**, *120*, 845–853.
22. D. Shaya, M. Kreir, R. A. Robbins, S. Wong, J. Hammon, A. Brüggemann, D. L. J. Minor, *Proc. Natl. Acad. Sci. USA* **2011**, *108*, 12313–12318.
23. R. K. Finol-Urdaneta, Y. Wang, A. Al-Sabi, C. Zhao, S. Y. Noskov, R. J. French, *J. Gen. Physiol.* **2014**, *157–171*, 804–818.
24. M. B. Ulmschneider, C. Bagnéris, E. C. McCusker, P. G. DeCaen, M. Delling, D. E. Clapham, J. P. Ulmschneider, B. A. Wallace, *Proc. Natl. Acad. Sci. USA* **2013**, *110*, 6364–6369.
25. Y. Jiang, A. Lee, J. Chen, M. Cadene, B. T. Chait, R. MacKinnon, *Nature* **2002**, *417*, 523–526.
26. A. Kuo, J. M. Gulbis, J. F. Antcliff, T. Rahman, E. D. Lowe, J. Zimmer, J. Cuthbertson, F. M. Ashcroft, T. Ezaki, D. A. Doyle, *Science* **2003**, *300*, 1922–1926.
27. Y. Jiang, A. Lee, J. Chen, V. Ruta, M. Cadene, B. T. Chait, R. MacKinnon, *Nature* **2003**, *423*, 33–41.
28. E. Gouaux, R. MacKinnon, *Science* **2005**, *310*, 1461–1465.
29. N. Shi, S. Ye, A. Alam, L. Chen, Y. Jiang, *Nature* **2006**, *440*, 570–574.
30. S. B. Long, X. Tao, E. B. Campbell, R. MacKinnon, *Nature* **2007**, *450*, 376–382.
31. M. Nishida, M. Cadene, B. T. Chait, R. MacKinnon, *EMBO J.* **2007**, *26*, 4005–4015.
32. A. Alam, Y. Jiang, *Nat. Mol. Biol.* **2009**, *16*, 30–34.
33. A. N. Miller, S. B. Long, *Science* **2012**, *335*, 432–436.
34. J. Payandeh, T. Scheuer, N. Zheng, W. A. Catterall, *Nature* **2011**, *475*, 353–359.
35. J. Payandeh, T. M. G. El-Din, T. Scheuer, N. Zheng, W. A. Cattarall, *Nature* **2012**, *486*, 135–140.
36. X. Zhang, W. Ren, P. DeCaen, C. Yan, X. Tao, L. Tang, J. Wang, K. Hasegawa, T. Kumasaka, J. He, J. Wang, D. E. Clapham, N. Yan, *Nature* **2012**, *486*, 130–134.
37. E. C. McCusker, C. Bagnéris, C. E. Naylor, A. R. Cole, N. D'Avanzo, C. G. Nichols, B. A. Wallace, *Nat. Commun.* **2012**, *2*, 1102.
38. I. Baconguis, C. J. Bohlen, A. Goehring, D. Julius, E. Gouaux, *Cell* **2014**, *156*, 717–729.
39. Y. M. Sun, I. Favre, L. Schild, E. Moczydlowski, *J. Gen. Physiol.* **1997**, *118*, 693–715.
40. P. M. Snyder, D. R. Olson, D. B. Bucher, *J. Biol. Chem.* **1999**, *274*, 28484–28490.
41. S. Kellenberger, I. Gautshi, L. Schild, *Proc. Natl. Acad. Sci. USA* **1999**, *96*, 4170–4175.
42. S. Sheng, C. J. Perry, O. B. Kashlan, T. R. Kleyman, *J. Biol. Chem.* **2005**, *280*, 8513–8522.
43. I. Favre, E. Moczydlowski, L. Schild, *Biophys. J.* **1996**, *71*, 3110–3125.
44. P. H. Backx, D. T. Yue, J. H. Lawrence, E. Marban, G. F. Tomaselli, *Science* **1992**, *257*, 248–251.
45. T. Schlief, R. Schonherr, K. Imoto, S. H. Heinemann, *Eur. Biophys. J.* **1996**, *25*, 75–91.
46. S. H. Heinemann, H. Terlau, W. Stuhmer, K. Imoto, S. Numa, *Nature* **1992**, *356*, 441–443.
47. L. G. Cuello, V. Jogini, D. M. Cortes, E. Perozo, *Nature* **2010**, *466*, 203–208.
48. C. Bagnéris, P. G. DeCaen, C. E. Naylor, D. C. Pryde, I. Nobeli, D. E. Clapham, B. A. Wallace, *Proc. Natl. Acad. Sci. USA* **2014**, *111*, 8428–8433.
49. B. Eisenberg, *Biophys. Chem.* **2003**, *100*, 507–517.
50. D. Boda, W. Nonner, M. Valisko, D. Henderson, B. Eisenberg, D. Gillespie, *Biophys. J.* **2007**, *93*, 1960–1980.
51. E. Csanyi, D. Boda, D. Gillespie, T. Kristof, *Biochim. Biophys. Acta* **2012**, *1818*, 592–600.
52. I. Kaufman, D. G. Luchinsky, R. Tindjong, P. V. E. McClintock, R. S. Eisenberg, *Phys. Rev. E* **2013**, *88*, 052712.

53. A. N. Thompson, I. Kim, T. D. Panosian, T. M. Iverson, T. W. Allen, C. M. Nimigean, *Nat. Struct. Mol. Biol.* **2009**, *16*, 1317–1324.
54. B. Egwolf, B. Roux, *J. Mol. Biol.* **2010**, *401*, 831–842.
55. C. M. Nimigean, T. W. Allen, *J. Gen. Physiol.* **2011**, *137*, 405–413.
56. B. Roux, S. Berneche, B. Egwolf, B. Lev, S. Y. Noskov, C. N. Rowley, H. Yu, *J. Gen. Physiol.* **2011**, *137*, 415–426.
57. V. N. Ngo, D. Stefanovski, S. Haas, R. A. Farley, *PLoS One* **2014**, *9*, e86079.
58. V. B. Luzhkov, J. Aqvist, *Biochim. Biophys. Acta* **2001**, *1548*, 194–202.
59. C. Maffeo, S. Bhattacharya, J. Yoo, D. Wells, A. Aksimentiev, *Chem. Rev.* **2012**, *112*, 6250–6284.
60. T. Dudev, C. Lim, *J. Am. Chem. Soc.* **2009**, *131*, 8092–8101.
61. B. Roux, *J. Phys. Chem. B* **2012**, *116*, 6966–6979.
62. B. Hille, *J. Gen. Physiol.* **1973**, *61*, 669–686.
63. M. S. Shapiro, T. E. DeCoursey, *J. Gen. Physiol.* **1991**, *97*, 1227–1250.
64. T. W. Allen, A. Bliznyuk, A. P. Rendell, S. Kuyucak, S.-H. Chung, *J. Chem. Phys.* **2000**, *112*, 8191–8204.
65. I. H. Shrivastava, P. D. Tieleman, P. C. Biggin, M. S. P. Sansom, *Biophys. J.* **2002**, *83*, 633–645.
66. S. Y. Noskov, S. Berneche, B. Roux, *Nature* **2004**, *431*, 830–834.
67. D. Asthagiri, L. R. Pratt, M. E. Paulaitis, *J. Chem. Phys.* **2006**, *125*, 24701–24706.
68. S. Y. Noskov, B. Roux, *Biophys. Chem.* **2006**, *124*, 279–291.
69. D. L. Bostick, C. L. Brooks III, *Proc. Natl. Acad. Sci. USA* **2007**, *104*, 9260–9265.
70. S. Y. Noskov, B. Roux, *J. Gen. Physiol.* **2007**, *129*, 135–143.
71. M. Thomas, D. Jayatilaka, B. Corry, *Biophys. J.* **2007**, *93*, 2635–2643.
72. S. Varma, S. B. Rempe, *Biophys. J.* **2007**, *93*, 1093–1099.
73. P. W. Fowler, K. Tai, M. S. P. Sansom, *Biophys. J.* **2008**, *95*, 5062–5072.
74. S. Varma, D. Sabo, S. B. Rempe, *J. Mol. Biol.* **2008**, *376*, 13–22.
75. D. L. Bostick, K. Arora, C. L. Brooks III, *Biophys. J.* **2009**, *96*, 3887–3896.
76. D. L. Bostick, C. L. Brooks III, *J. Am. Chem. Soc.* **2010**, *132*, 13185–13187.
77. S. Ye, Y. Li, Y. Jiang, *Nat. Struct. Mol. Biol.* **2010**, *17*, 1019–1023.
78. P. D. Dixit, D. Asthagiri, *J. Gen. Physiol.* **2011**, *137*, 427–433.
79. S. Furini, C. Domene, *J. Mol. Biol.* **2011**, *409*, 867–878.
80. I. Kim, T. W. Allen, *Proc. Natl. Acad. Sci. USA* **2011**, *108*, 17963–17968.
81. S. Varma, D. M. Rogers, L. R. Pratt, S. B. Rempe, *J. Gen. Physiol.* **2011**, *137*, 479–488.
82. D. B. Sauer, W. Zeng, J. Canty, Y. Lam, Y. Jiang, *Nat. Commun.* **2013**, *4*, 2721.
83. D. A. Köpfer, C. Song, T. Gruene, G. M. Sheldrick, U. Zachariae, B. L. de Groot, *Science* **2014**, *346*, 352–355.
84. R. Horn, B. Roux, J. Aqvist, *Biophys. J.* **2014**, *106*, 1859–1863.
85. B. Hille, *J. Gen. Physiol.* **1972**, *59*, 637–658.
86. D. T. Campbell, *J. Gen. Physiol.* **1976**, *67*, 295–307.
87. G. M. Lipkind, H. A. Fozzard, *J. Gen. Physiol.* **2008**, *131*, 523–529.
88. T. Dudev, C. Lim, *J. Am. Chem. Soc.* **2010**, *132*, 2321–2332.
89. I. Baconguis, E. Gouaux, *Nature* **2012**, *489*, 400–406.
90. M. D. Carattino, M. C. Della Vecchia, *J. Biol. Chem.* **2012**, *287*, 12927–12934.
91. B. Corry, M. Thomas, *J. Am. Chem. Soc.* **2012**, *134*, 1840–1846.
92. T. Dudev, C. Lim, *Phys. Chem. Chem. Phys.* **2012**, *14*, 12451–12456.
93. T. Dudev, C. Lim, *J. Phys. Chem. B* **2012**, *116*, 10703–10714.
94. S. Furini, C. Domene, *Biophys. J.* **2012**, *103*, 2106–2114.
95. M. D. Xia, H. H. Liu, Y. Li, N. Yan, H. P. Gong, *Biophys. J.* **2013**, *11*, 2401–2409.
96. C. l. Boiteux, I. Vorobyov, T. W. Allen, *Proc. Natl. Acad. Sci. USA* **2014**, *111*, 3454–3459.
97. T. Dudev, C. Lim, *Acc. Chem. Res.* **2014**, *47*, 3580–3587.
98. L. Yang, L. G. Palmer, *J. Gen. Physiol.* **2014**, *144*, 245–255.
99. T. Dudev, C. Lim, *Sci. Rep.* **2015**, *5*, 7864.
100. L. J. Mullins, *J. Gen. Physiol.* **1960**, *43*, 105–117.

101. G. Eisenman, in *Symposium on Membrane Transport and Metabolism*, Eds A. Kleinzeller, A. Kotyk, Academic Press, New York, 1961, pp. 163–179.
102. F. Bezanilla, C. M. Armstrong, *J. Gen. Physiol.* **1972**, *53*, 342–347.
103. G. Eisenman, R. Horn, *J. Membrane Biol.* **1983**, *76*, 197–225.
104. A. Laio, V. Torre, *Biophys. J.* **1999**, *76*, 129–148.
105. B. Corry, S.-H. Chung, *Cell. Mol. Life Sci.* **2006**, *63*, 301–315.
106. A. Alam, Y. Jiang, *J. Gen. Physiol.* **2011**, *137*, 397–403.
107. S. Durgadi, S. Y. Noskov, *Channels (Austin)* **2011**, *5*, 198–200.
108. T. Dudev, C. Lim, *Chem. Rev.* **2014**, *114*, 538–556.
109. G. Kuppuraj, M. Dudev, C. Lim, *J. Phys. Chem. B* **2009**, *113*, 2952–2960.
110. G. Eisenman, *Biophys. J.* **1962**, *2*, 259–323.
111. H. L. Friedman, C. V. Krishnan, in *Water: A Comprehensive Treatise*, Ed F. Franks, Plenum Press, New York, 1973, Vol. 3, pp. 1–118.
112. J. S. Hanas, D. J. Hazuda, D. F. Bogenhagen, F.-H. Wu, C.-W. Wu, *J. Biol. Chem.* **1983**, *258*, 14120–14125.
113. J. M. Berg, H. A. Godwin, *Annu. Rev. Biophys. Biomol. Struct.* **1997**, *26*, 357–371.
114. T. Dudev, C. Lim, *J. Am. Chem. Soc.* **2007**, *129*, 12497–12504.
115. S. B. Long, E. B. Campbell, R. MacKinnon, *Science* **2005**, *309*, 897–903.
116. S. G. Brohawn, J. D. Marmol, R. MacKinnon, *Science* **2012**, *335*, 436–441.
117. A. Grottesi, C. Domene, S. Haider, M. S. P. Sansom, *IEEE Trans Nanobioscience* **2005**, *4*, 112–120.
118. P. D. Dixit, S. Merchant, D. Asthagiri, *Biophys. J.* **2009**, *96*, 2138–2145.
119. S. Furini, C. Domene, *Biophys. J.* **2013**, *105*, 1737–1745.
120. V. Carnevale, W. Treptow, M. L. Klein, *J. Phys. Chem. Lett.* **2011**, *2*, 2504–2508.
121. S. Furini, C. Domene, *PLoS Comp. Biol.* **2012**, *8*, e1002476.
122. H. Qui, R. Shen, W. Guo, *Biochim. Biophys. Acta* **2012**, *1818*, 2529–2535.
123. N. Chakrabarti, C. Ing, J. Payandeh, N. Zheng, W. A. Catterall, R. Pomes, *Proc. Natl. Acad. Sci. USA* **2013**, *110*, 11331–11336.
124. B. Corry, T. W. Allen, S. Kuyucak, S.-H. Chung, *Biophys. J.* **2001**, *80*, 195–214.
125. D. Boda, M. Valisko, D. Henderson, B. Eisenberg, D. Gillespie, W. Nonner, *J. Gen. Physiol.* **2009**, *133*, 497–509.
126. S. Kellenberger, L. Schild, *Physiol. Rev.* **2002**, *82*, 735–767.
127. S. Grunder, X. Chen, *Int. J. Physiol. Pathophysiol. Pharmacol.* **2010**, *2*, 73–94.
128. O. B. Kashlan, T. R. Kleyman, *Am. J. Physiol. Renal Physiol.* **2011**, *301*, F684–F696.
129. E. R. Nightingale Jr., *J. Phys. Chem.* **1959**, *63*, 1381–1387.
130. S. Kellenberger, N. Hoffmann-Pochon, I. Gautschi, E. Schneeberger, L. Schild, *J. Gen. Physiol.* **1999**, *114*, 13–30.
131. J. Jasti, H. Furukawa, E. B. Gonzales, E. Gouaux, *Nature* **2007**, *449*, 316–324.
132. E. W. McCleskey, W. Almers, *Proc. Natl. Acad. Sci. USA* **1985**, *82*, 7149–7153.
133. S. Tang, G. Mikala, A. Bahinski, A. Yatani, G. Varadi, A. Schwartz, *J. Biol. Chem.* **1993**, *268*, 13026–13029.
134. T. Dudev, C. Lim, *J. Am. Chem. Soc.* **2014**, *136*, 3553–3559.
135. D. Sakharov, C. Lim, *J. Comp. Chem.* **2009**, *30*, 191–202.
136. D. Bucher, U. Rothlisberger, *J. Gen. Physiol.* **2010**, *135*, 549–554.

Chapter 11
Sodium as Coupling Cation in Respiratory Energy Conversion

Günter Fritz and Julia Steuber

Contents

Dedicated to Peter Dimroth on the occasion of his 75th birthday.

G. Fritz (✉)
Department of Neuropathology, University of Freiburg,
Breisacherstrasse 64, D-79106 Freiburg, Germany
e-mail: guenter.fritz@uniklinik-freiburg.de

J. Steuber (✉)
Institute of Microbiology, University of Hohenheim (Stuttgart),
Garbenstrasse 30, D-70599 Stuttgart, Germany
e-mail: julia.steuber@uni-hohenheim.de

A. Sigel, H. Sigel, and R.K.O. Sigel (eds.), *The Alkali Metal Ions: Their Role for Life*,
Metal Ions in Life Sciences 16, DOI 10.1007/978-3-319-21756-7_11

Abstract Among the alkali cations, Na^+ has an extraordinary role in living cells since it is used to charge the battery of life. To this end, sophisticated protein complexes in biological membranes convert chemical energy obtained from oxidation of NADH, or hydrolysis of ATP, into an electrochemical gradient of sodium ions. Cells use this so-called sodium-motive force stored in energy-converting membranes for important processes like uptake of nutrients, motility, or expulsion of toxic compounds. The Na^+ pumps act in concert with other enzymes embedded in the lipid membrane, and together they form the respiratory chain which achieves the oxidation of NADH derived from nutrients under formation of an electrochemical sodium (or proton) gradient. We explain why Na^+ pumps are important model systems for the homologous, proton-translocating complexes, and hope to convince the reader that studying the Na^+-translocating ATP synthase from the unimpressive bacterium *Ilyobacter tartaricus* had a big impact on our understanding of energy conversion by human ATP synthase. The Na^+-translocating systems described here are either driven by the oxidation of NADH, the carrier of redox equivalents of cells, or by the hydrolysis of adenosine 5′-triphosphate, the universal high-energy compound of cells. The electrochemical energy provided by these respiratory Na^+ pumps, the NADH dehydrogenase or the ATPase, drives other Na^+ transport systems like the bacterial flagellum discussed in the last part of this chapter. The flagellar motor does not represent a Na^+ pump, but like ATPase, it operates by a rotational mechanism. By comparing these two Na^+ -translocating, rotary machines, we obtain new insight into the possible mechanisms of Na^+ transport through the stator proteins of the flagellar motor. Na^+ pumps are widespread in pathogenic bacteria where they play an important role in metabolism, making them novel targets for antibiotics.

Keywords ATP synthase • Electrochemical Na^+ potential • Membrane protein • NADH dehydrogenase • Respiration • Sodium binding site • Transport mechanism

Please cite as: *Met. Ions Life Sci.* 16 (2016) 349–390

1 Sodium Bioenergetics: An Overview

1.1 Two Currencies of Energy: Proton- Versus Sodium-Motive Force

Respiration is a sophisticated reaction chain catalyzed by membrane-bound redox enzymes and electron carriers which convert the energy of nutrients to the high-energy compound adenosine 5′-triphosphate (ATP). ATP may also be formed during fermentation, for example during ethanol production by yeast, but the yield of ATP per converted substrate is much lower when compared to respiration [1]. The employment of an electron transport chain to deliver electrons to an exogenous acceptor with a highly positive redox potential – such as molecular oxygen, O_2 – is called oxidative phosphorylation (OXPHOS) which drastically increases the energetic yield to up to about 36 molecules of ATP per molecule of glucose, compared to only two molecules of ATP per glucose obtained by glycolysis [1, 2].

The mitochondrial respiratory chain comprises several redox enzyme complexes located in the mitochondrial inner membrane. The electrons enter at either the NADH:ubiquinone oxidoreductase (complex I) or the succinate dehydrogenase (complex II) which reduce ubiquinone (Q) to ubiquinol (QH_2). Ubiquinones are hydrophobic, low-molecular weight electron carriers in the lipid bilayer. Ubiquinol formed by complex I or complex II is oxidized by the ubiquinol:cytochrome *c* oxidoreductase (also called cytochrome-*bc₁*, or complex III) under reduction of cytochrome *c*, a soluble heme protein in the intermembrane space between the inner and outer mitochondrial membrane.

This compartment (P-phase) is equivalent to the periplasm of a bacterial cell; the N-phase represents the mitochondrial matrix or bacterial cytoplasm. Reduced cytochrome *c* acts as electron donor for cytochrome *c* oxidase (complex IV) which catalyzes the reduction of O_2 to H_2O [2]. In most textbooks, these respiratory chain complexes are depicted as separate entities floating in the mitochondrial membrane. Yet, it is now generally accepted that respiratory chain complexes form supercomplexes, also called respirasomes, in the inner mitochondrial membrane (reviewed in [3]) and in energy-converting membranes of bacteria [4]. There is an ongoing debate if this supercomplex formation affects the kinetics of reduction and re-oxidation of the ubiquinone pool in the membrane [5, 6].

What is the physicochemical principle behind biological energy conservation? In the respiratory chain, the flow of electrons is determined by the midpoint potential E_m (unit, Volt) of the cofactors and electron carriers involved in respiration, from redox centers with more negative to redox centers with more positive midpoint potential. As first predicted by Mitchell in his famous chemiosmotic theory [7, 8], the energy which is released as the electrons are transferred through the respiratory chain is coupled to the pumping of protons out of the mitochondrial matrix across the inner membrane, resulting in an electrochemical proton gradient, $\Delta\mu H^+$ (equation 1):

$$\Delta\mu H^+ = -F\Delta\psi + 2.3RT\ \Delta pH \tag{1}$$

The energy stored in the membrane thus comprises an electrical term, $\Delta\psi$, and the chemical gradient of protons, ΔpH. The unit is kJ per mol of translocated protons. To strengthen the fact that the electrochemical proton potential which can be established by a respiratory electron transfer chain is determined by the difference in midpoint potentials of an electron donor, like the NADH/NAD couple, and an electron acceptor, for example the QH_2/Q couple, Mitchell introduced the term "proton motive force" (PMF), or Δp, which is defined as follows (equation 2):

$$\mathrm{PMF} = \frac{\Delta\mu\mathrm{H}^+}{F} = \Delta\psi = \frac{2.3RT}{F}\Delta\mathrm{pH} \quad (2)$$

The unit of PMF is voltage. In respiring mitochondria, the PMF amounts to approximately 180–220 mV, of which 150–180 mV are provided by the membrane potential, $\Delta\Psi$ [8]. F is the Faraday constant. In Section 3.1, we will discuss how the ATP synthase (complex V of the respiratory chain) uses the PMF (or the "sodium motive force", SMF, see below) to catalyze the endergonic synthesis of ATP from ADP and inorganic phosphate.

Many respiratory complexes of mitochondria, chloroplasts, bacteria, and archaea share common evolutionary ancestors [2]. In the bacterial and archaeal kingdoms of life, there is a plethora of respiratory pathways utilizing many more electron donors and electron acceptors than NADH and O_2, the prominent substrates of respiration in the eukaryotic kingdom of life. As long as the free energy liberated by the overall redox reaction is sufficient to energize the respiratory membrane, chemiosmotic ATP synthesis by bacteria or archaea may utilize an impressive variety of organic or inorganic electron donor and acceptor pairs [9]. Besides this flexibility with respect to respiratory substrates, some bacteria and archaea benefit from an additional "battery of life", namely an electrochemical Na^+ gradient established by electrogenic Na^+ pumps. This Na^+ gradient does not depend on so-called secondary Na^+/H^+ antiporters which may generate an electrochemical Na^+ gradient at the expense of the PMF.

The properties and significance of Na^+/H^+ antiporters are discussed in Chapter 12 of this volume. The Na^+ battery of life is charged by the action of sophisticated membrane protein complexes using chemical energy to drive the uphill transport of Na^+ across respiratory membranes. When we look at those Na^+ pumps, we find that, despite their altered cation selectivity (Na^+ rather than H^+), they often (but not always) are evolutionary related to their proton-translocating counterparts. In Section 3.1. we compare the highly related H^+- and Na^+-translocating ATPases (complex V of the respiratory chain). We will ask what makes H^+ or Na^+ the preferred cation in these otherwise very similar cation pumps, and discuss the structural determinants of cation selectivity. Using the information of primary sequences of pumps in the many sequenced genomes, and knowing which critical, conserved amino acid residues act as ligands for Na^+ or H^+, one would expect that it is possible to deduce if the Last Universal Common Ancestor of cells (LUCA) ran on a proton- or on a sodium-battery of life. This problem is of special interest if we assume that chemiosmosis, not fermentation, was the first mode of energy conservation in Life [10]. It is, however, still under debate if the PMF [10] or the SMF [11] was first in storing energy during oxidative phosphorylation.

The chemiosmotic theory was first applied to H^+-pumping respiratory complexes [7]. At that time, the importance of Na^+ gradients for living cells was recognized, but only for non-respiring membranes, like the outer membrane of eukaryotic cells. Here, an imbalance of external and internal Na^+ and K^+ concentrations ($[Na^+]_{out} > [Na^+]_{in}$, $[K^+]_{out} < [K^+]_{in}$) is established by the Na^+/K^+ ATPase, generating the so-called resting potential. The important role of an electrochemical Na^+ potential for every higher cell was undisputed, but it relied on the electrochemical proton potential formed in mitochondria. This is because Na^+ expulsion by the Na^+/K^+ ATPase depends on ATP provided by the mitochondrial ATP synthase, which is driven by the electrochemical proton potential. But the primacy of protons in bioenergetics was soon lost.

In 1977, Unemoto and colleagues showed that oxidation of NADH with O_2 by the marine bacterium *Vibrio alginolyticus* was specifically stimulated by sodium ions [12], and concluded that "Na^+ probably acts as an essential cofactor both for the active transport of metabolites and for respiratory activity in marine bacteria". In 1979, the same group reported that the Na^+-stimulated respiratory complex represented a NADH:quinone oxidoreductase (later identified as the Na^+-translocating NADH:quinone oxidoreductase, or Na^+-NQR, described below), but the build-up of an electrochemical Na^+ gradient by this enzyme was not envisaged [13]. When Dimroth, in 1980, reported that the oxaloacetate decarboxylase of an enterobacterium acted as a primary Na^+ pump [14], he paved the way for an extension of Mitchell's chemiosmotic theory, now including Na^+ as a possible coupling ion in energized membranes capable of ATP synthesis. Shortly thereafter, Tokuda and Unemoto showed that the Na^+-stimulated NADH:quinone oxidoreductase from *V. alginolyticus* actually translocates Na^+ ions [15, 16]. This explained why the enzyme, strictly coupling the electron transfer reaction to the transport of Na^+ across the membrane, was stimulated by Na^+ in its NADH:quinone oxidoreduction activity. In bioenergetics, the latter is called the "chemical" reaction of the pump, whereas Na^+ transport represents the "vectorial" reaction of the pump.

In the absence of Na^+, the Na^+-NQR should not be capable of NADH oxidation. This is because electron transfer comes to a halt when, in the absence of the coupling cation, the vectorial reaction cannot proceed. This strict interdependence of the chemical and the vectorial reactions, often called "coupling", can also be observed with Na^+-NQR in detergent micelles, e.g., when the pump is not embedded in membranes [17]. This hallmark feature of Na^+ pumps, namely stimulation of the chemical reaction by the coupling cation (Na^+), helped to identify the Na^+-NQR [17] and other energy-conserving Na^+ pumps [18, 19] in cell extracts and during purification. It also represents a measure for the integrity of a Na^+ pump which typically is a rather labile membrane protein complex. In Section 2.1 we describe the high-resolution structure of the Na^+-NQR from *Vibrio cholerae*, the pathogenic bacterium causing Cholera disease, and propose a mechanism for NADH-driven Na^+ translocation. With the help of the Na^+-NQR, *V. cholerae*, like many other bacteria operating primary Na^+ pumps, generates an electrochemical Na^+ gradient, $\Delta\mu Na^+$ (equation 3):

$$\Delta\mu Na^+ = -F\Delta\psi + 2.3RT\ \Delta pNa^+ \tag{3}$$

This energy which ultimately is gained during the exergonic oxidation of NADH with Q by the Na^+-NQR thus comprises an electrical term, $\Delta\psi$, and a chemical gradient of Na^+, ΔpNa^+. The unit is kJ per mol of translocated sodium ions. This energy is utilized, very much like the PMF, for uptake of substrates, motility, and even synthesis of ATP, if the organism possesses a Na^+-dependent ATP synthase. Indeed, some microorganisms like *Propionigenium modestum* live on the Na^+ battery only, with a primary Na^+ pump (in the case of *P. modestum*, it is a Na^+-translocating decarboxylase), and a F_1F_O-type ATP synthase driven by the electrochemical Na^+ potential, establishing a cycling of Na^+ ions across the inner membrane of this obligate anaerobic bacterium [20]. But most microorganisms, including *V. cholerae*, operate a respiratory Na^+ pump in addition to H^+-coupled respiratory complexes, utilizing $\Delta\psi$ (formed by expulsion of a positive charge, H^+ or Na^+), ΔpH (with $[H^+]_{out} > [H^+]_{in}$), or ΔpNa^+ (with $[Na^+]_{out} > [Na^+]_{in}$) (Figure 1) [21, 22].

It is obvious that microorganisms living in alkaline habitats, where $[H^+]_{out} < [H^+]_{in}$, will benefit from the action of primary Na^+ pumps for the generation of $\Delta\psi$, since unlike the electrochemical proton potential, the electrochemical Na^+ potential is not affected by the inverted pH gradient. Under alkaline conditions, secondary, $\Delta\psi$-driven systems like electrogenic Na^+/H^+ antiporters are required for the import of H^+ and export of Na^+, maintaining the cytoplasm around neutral pH [23] (see Chapter 12). In analogy to the PMF, we could define the SMF by dividing the electrochemical Na^+ gradient by F, the Faraday constant (equation 2). This is helpful for *in vitro* systems, studying, for example, a Na^+-driven, bacterial ATP synthase reconstituted in liposomes [24], but care should be taken when studying the dependency of energy-consuming processes of bacterial cells on the electrochemical Na^+ or H^+ gradient. On the energy input side, it is difficult to experimentally rule out that both batteries (Na^+ and H^+) are continuously recharged by Na^+ and H^+ pumps. On the side of energy consumers, it is not easy to discriminate which driving force ($\Delta\mu Na^+$ or $\Delta\mu H^+$) is actually required. This is especially problematic if the proteins cannot be isolated and reconstituted in proteoliposomes under retention of biological function, as is the case with the H^+ (or Na^+, or perhaps also K^+) gradient-driven flagellum from different bacterial species (see Section 3.2).

In this chapter, some primary, Na^+ -translocating complexes which are important players in energy conservation of methanogens are not discussed. These are the Na^+-translocating N5-methyltetrahydromethanopterin:coenzyme M methyltransferase, the Na^+ (or H^+)-dependent A_1A_O ATP synthases, and the energy-converting Na^+ (or H^+)-dependent hydrogenases [25–27].

1.2 Na^+ Pumps as Model Systems in Membrane Energetics

1.2.1 Methods to Study Na^+ Transport

In respiring bacterial cells, ^{23}Na NMR spectroscopy is a powerful tool to quantify the amount of intracellular Na^+, and to follow Na^+ extrusion upon addition of a substrate which is metabolized by the cells [28]. ^{23}Na NMR spectroscopy makes use of the chemical shift of Na^+ in the presence of a shift reagent. Differentiation of intracellular and extracellular sodium ion signals in ^{23}Na NMR spectra is possible with

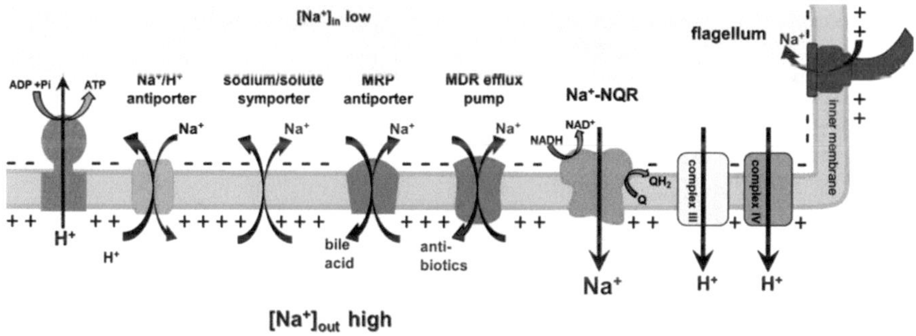

Figure 1 Respiration-driven cycling of Na^+ and H^+ across the inner cell membrane from *Vibrio cholerae*. The inner membrane separates the bacterial cytoplasm from the external environment which may change rapidly in osmolality and pH. The cytoplasmic aspect of the inner membrane is negatively charged, the outer aspect is positively charged. The buildup and maintenance of this transmembrane voltage ($\Delta\psi$) is achieved by primary H^+ pumps (complex III, complex IV) and the Na^+ pump (Na^+-translocating NADH:quinone oxidoreductase, Na^+-NQR) of the respiratory chain. Overall respiration results in the oxidation of NADH with O_2 as electron acceptor. In this process, the Na^+-NQR catalyzes the first step, namely oxidation of NADH with quinone (Q) under formation of NAD and quinol (QH_2). This exergonic reaction is coupled to the endergonic translocation of Na^+ across the cytoplasmic membrane under formation of an electrochemical Na^+ potential. Oxidation of QH_2 with cytochrome *c* by complex III, and oxidation of reduced cytochrome *c* with O_2 by complex IV, is coupled to the formation of an electrochemical proton potential. Conversion of electrochemical Na^+ in H^+ gradients, and *vice versa*, is achieved with the help of Na^+/H^+ antiporters [21]. The electrochemical Na^+ gradient drives other important processes in *V. cholerae*, such as the rotation of the flagellum [127], the efflux of toxic compounds by the multidrug resistance (MDR) complex [132], the expulsion of bile acids by the MRP system [133], and the uptake of nutrients by transporters belonging to the sodium/solute symporter family [134].

the help of a nontoxic shift reagent that does not penetrate the cell membrane [28, 29]. To quantify changes in intracellular Na^+ concentration, a reference capillary with known Na^+ concentration in the presence of another shift reagent is inserted into the NMR tube containing the (respiring) cell suspension. As a result, three distinct ^{23}Na signals $\left(Na^+_{in}, Na^+_{out}, Na^+_{reference}\right)$ are observed simultaneously which can be quantified by NMR. Substrates like glucose are taken up and metabolized by the cells to yield NADH in the cytoplasm, which is a substrate for respiration-dependent Na^+ transport. ^{23}Na NMR spectroscopy has also been applied to follow Na^+ extrusion by so-called inside-out vesicles from bacterial membranes where the site of NADH oxidation by the respiratory chain is exposed to the external buffer. This is the opposite orientation when compared to cells. With inside-out vesicles, respiratory substrates like NADH can be added directly to study respiratory Na^+ transport [29].

There is a disadvantage when using ^{23}Na NMR to study Na^+ transport since it represents a discontinuous measurement with limited temporal resolution. The acquisition of a single spectrum takes several minutes, and this also determines the dead time between addition of substrate and the recording of the first data point. In an alternative but again discontinuous approach with improved temporal resolution,

Na^+ entrapped in vesicles or liposomes which is transported by the membrane-embedded Na^+ pump is quantified. This requires removal of external Na^+ on a cation exchange column, followed by elution of vesicles or liposomes from the column with H_2O. The Na^+ in the eluate, after correction for the endogenous Na^+ content of the vesicle lumen at the start of the reaction, represents the Na^+ which was transported by the pump. It is quantified by atomic absorption spectroscopy, or by γ-counting of $^{22}Na^+$ added as tracer. To study the coupling of vectorial and chemical reactions, the same batch of liposomes or vesicles used for the transport (vectorial) reaction is used for determination of the electron transfer (chemical) reaction in a parallel manner. For example, the complex I-type NADH:quinone oxidoreductase of the enterobacterium *Klebsiella pneumonia* (see Section 2.3) exhibited specific activities of 2.4 $\mu mol \cdot min^{-1} \cdot mg^{-1}$ for quinol formation and 4.7 $\mu mol \cdot min^{-1} \cdot mg^{-1}$ for Na^+ translocation. Since two electrons are required to generate quinol from quinone, this corresponds to a Na^+/electron ratio of 1 [30]. Compared to proton pumps, determination of the transport stoichiometry of a Na^+ pump is more robust since the permeability coefficient of Na^+ in lipid bilayers is only 1.2×10^{-14} cm s^{-1} [31], compared to 10^{-7} to 10^{-3} cm s^{-1} for H^+ determined in various types of phospholipid vesicles [32]. Continuous measurement of Na^+ transport is possible using the sodium-responsive fluorophore sodium green incorporated into proteoliposomes during reconstitution of the Na^+ pump. The assay was first described for the Na^+-translocating F_1F_O-ATPase [33]. In case of the Na^+-NQR, enrichment of Na^+ within the proteoliposomes sufficient for significant fluorescence required the presence of a regenerative system for NADH consisting of lactate dehydrogenase and pyruvate [34].

1.2.2 Sodium Pumps: Pitfalls and Advantages

As exemplified by the F_1F_O ATP synthase (see Section 3.1.), many H^+-translocating pumps have Na^+ -translocating homologs. Studying the latter gives valuable insights into the mechanism of the former, but to unequivocally demonstrate that Na^+ is transported by the action of a primary pump is not a trivial task. When we ask if respiratory NADH oxidation is coupled to Na^+ transport, we must consider the contribution of other, secondary Na^+-extruding systems, like Na^+/H^+ antiporters. In organisms like *V. cholerae* operating both a Na^+ and a H^+ cycle (Figure 1), oxidation of quinol with O_2 by downstream respiratory complexes III and IV generates a PMF which might drive secondary Na^+ extrusion. To discriminate between respiratory, primary Na^+ transport, and Na^+ transport due to secondary, PMF-driven systems, the transport experiment is performed in the presence of a protonophore. Protonophores, or proton translocators, are synthetic chemicals which may undergo protonation and deprotonation under retention of their hydrophobic characters in both states. Due to their extensive π-orbital systems, they are able to delocalize the negative charge of their anionic form which, like the neutral, protonated form, is highly soluble in lipids and lipid bilayers [2].

As a consequence, protonophores dissipate an electrochemical proton potential formed by respiring membranes by allowing equilibration of protons across the lipid bilayer. If Na^+ transport is observed despite the presence of (sufficient) amounts of a protonophore, it is caused by a primary Na^+ pump. Na^+ transport rates may even increase in the presence of a protonophore [17], because protonophores may dissipate both the chemical proton gradient, ΔpH, and the transmembrane voltage, $\Delta\psi$ [35]. As a consequence, the accumulation of Na^+ in liposomes or vesicles by the Na^+ pump is facilitated. This is because the $\Delta\psi$ (inside positive) generated during turnover, which would prevent further accumulation of positively charged sodium ions in the lumen of the vesicle, is continuously degraded due to the efflux of protons bound to the protonophore. Working with proton pumps, the proton concentration is usually restricted to a very narrow pH range of 6 to 8, depending on the proton pump being studied.

In contrast, variation of the coupling ion concentration is very easy in the case of Na^+ pumps. Therefore, Na^+ pumps offer the possibility to identify Na^+-dependent step(s) during the catalytic cycle. With the Na^+-NQR, studies on the Na^+ dependence of intraelectron transfer reactions led to the hypothesis that Na^+ uptake is coupled to the electron transfer from the [2Fe-2S] cluster on subunit NqrF to the covalently attached flavin mononucleotide (FMN) on subunit NqrC, while electron transfer from the covalently bound FMN on subunit NqrB to the riboflavin also located on NqrB triggers the translocation of Na^+ across the membrane [36]. Together with the information on the Na^+ binding site obtained by crystallographic approaches (see Section 3.1.3), and by studying variants of the Na^+ pump obtained by mutagenesis of predicted Na^+ ligands or cofactors, functional and structural information is obtained which, in combination, might unravel the coupling mechanism of the Na^+ pump.

2 Redox-Driven Na^+ Pumps

2.1 Na^+-Translocating NADH:Quinone Oxidoreductase

2.1.1 Respiratory NADH:Quinone Oxidoreductases

In bacteria, three different types of respiratory NADH:quinone oxidoreductases (NQR) have been described: the electrogenic complex I [37–39], also called NDH I in bacteria [40], the non-electrogenic NADH:quinone oxidoreductases (NDH II) [41], and the Na^+-translocating NADH:quinone oxidoreductases (Na^+-NQR) described in this section. The common function of these membrane-bound enzymes in respiration is to oxidize NADH using ubiquinone (Q) as electron acceptor. The net reaction thus yields ubiquinol (QH_2), the reducing substrate of enzyme complexes further along the respiratory chain (Figure 1), and NAD^+, which is used as oxidizing agent in numerous cellular processes. NADH:quinone oxidoreductases of the NDH II type do not contribute to the generation of electrochemical potentials, i.e., are non-electrogenic. Complex I and Na^+-NQR, though not related to each other in terms of primary sequences or cofactor compositions, both use the exergonic energy released by the redox reaction to pump H^+ or Na^+ from the cytosol to the periplasm. There is an ongoing debate if the proton-translocating complex I also transports Na^+, and if so, by what type of mechanism (see Section 2.3).

2.1.2 Structure of the NADH:Quinone Oxidoreductases

2.1.2.1 Subunits and Cofactors of the NADH:Quinone Oxidoreductases

We already discussed how the discovery of a respiratory chain driving Na^+ extrusion in *V. alginolyticus* by Tokuda, Unemoto, and Hayashi led to the identification of the Na^+-NQR as the first redox-driven Na^+ pump ever described [12, 13, 15, 16]. The notion that energy conservation exclusively depended on the proton motive force had to be revised, and soon, other primary Na^+ pumps were discovered, including a F-type ATP synthase driven by the sodium motive force (see Section 3.1).

Unemoto and colleagues went on to purify and characterize the Na^+-NQR from *V. alginolyticus* which then was thought to be composed of three subunits [42]. Sequencing of the *nqr* operon revealed six structural genes encoding for subunits NqrABCDEF [43, 44], and the four largest subunits (NqrAFCB, see Table 1) were identified in the *V. alginolyticus* Na^+ -NQR by Dimroth and coworkers [17, 45]. The small, hydrophobic NqrD and NqrE subunits were also present in this highly active NQR preparation which exhibited Na^+ transport activity when reconstituted into liposomes [17]. In 1998, Unemoto and colleagues identified all six Nqr subunits in the NQR complex from *V. alginolyticus* [46]. For the related NQR from *V. cholerae* which went into structure determination, we demonstrated the presence of one copy of each subunit in the complex and determined an apparent molecular mass of 221 kDa determined by sedimentation velocity [47]. This is in accord with the calculated mass of the complex (213 kDa) assuming one copy of each subunit.

Initially, the NQR was considered to contain non-covalently bound flavin adenine dinucleotide (FAD) and non-covalently bound FMN [42], but it was later shown that upon denaturation of the enzyme, FAD but not FMN was released into the supernatant [45]. It came as a surprise when the membrane-bound NqrB and NqrC subunits were shown to contain covalently linked FMN attached in a novel mode of flavin linkage [48, 49]. The phosphate of the ribityl side chain of the FMN is bound to a threonine residue of NqrB or NqrC in a phosphodiester linkage, maintaining the attachment of the phosphate group at the 5′-position of the ribityl as in authentic FMN [50]. The flavinylation of NqrB and NqrC requires a specific enzyme described recently by Bogachev and coworkers [51]. Besides FAD (on subunit NqrF), and two covalently bound FMNs (one on subunit NqrB and NqrC, respectively), there is a fourth flavin cofactor in the NQR. Barquera and coworkers discovered this functionally active riboflavin in the NQR from *V. cholerae* [52, 53], a finding which was later confirmed for the enzyme from *V. harveyi* [54]. Riboflavin binding to the *V. cholerae* NQR required the presence of the NqrB subunit [47], and

Table 1 Properties of the subunits from *V. cholerae* Na^+-NQR.

Subunit	NqrA	NqrB	NqrC	NqrD	NqrE	NqrF
No. residues	446	415	256	209	198	408
Calculated M_w (Da)	48 624	45 357	27 488	22 706	21 498	45 095
Apparent M_w on SDS-PAGE (kDa)	52	27	32	14	13	47
TM helices (from 3D structure)	0	10	1	6	6	1
Cofactors	none	Covalent FMN, riboflavin	Covalent FMN	Ligands for Fe center	Ligands for Fe center	FAD, [2Fe-2S]
Proposed function	Quinone binding	Electron transfer, Na^+ translocation	Electron transfer	Electron transfer	Electron transfer	NADH oxidation
Homologous proteins based on primary sequence[a]	NfoC/RnfC	NfoD/RnfD	NfoG/RnfG	NfoE/RnfE	NfoA/RnfA	AMOr
Structurally related proteins[b]	Nqo1	urea transporter, ammonium transporter	none	none	none	2Fe ferredoxin, ferredoxin:$NADP^+$ reductase

[a]From a recent similarity search performed by Reyes-Prieto and coworkers [69]. AMOr, aromatic (phenol, toluene and benzene) monooxygenases comprising a ferredoxin-like and a ferredoxin:$NADP^+$ reductase-like domain [131].

[b]By comparison with the 3D structure of the NQR complex [55].

riboflavin on NqrB, FAD on NqrF and the FMNs on NqrB and NqrC, were identified in the 3D structure of the NQR [55]. The cytoplasmic NqrF subunit also harbors a [2Fe-2S] cluster, acting as electron acceptor for FAD which is reduced by NADH in the initial reaction step of the enzyme [17, 56].

Table 1 summarizes the subunit and cofactor composition of the Na^+-NQR from *V. cholerae*, including the nomenclature for the homologous subunits of the RNF (or NFO) complex (see Section 2.2). For crystallization, NQR from *V. cholerae* with a polyhistidine tag fused to the N-terminus of the peripheral NqrA subunit was used, and the enzyme was produced in a *V. cholerae* host lacking the chromosomal *nqr* operon [57]. First crystals and diffraction patterns were obtained with native protein [58], but structure determination heavily relied on the positioning of selenomethionine residues in the electron density maps of crystals from Se-Met labeled NQR [55].

The last reaction step of the NQR is the reduction of ubiquinone-8 under formation of quinol. Ubiquinone-8 (Q8) was co-purified with the NQR (ca. 0.6 mol per mol NQR which was solubilized with *n*-dodecyl-D-maltoside), but the amount depended on the detergent used for solubilization. If NQR was solubilized in lauryldimethylamine *N*-oxide, Q8 was lost from the enzyme [59, 60]. The electron density map of crystals of the NQR purified in *n*-dodecyl-D-maltoside did not reveal a region which could be assigned to a bound ubiquinone-8 molecule [55], but a binding site interacting with Q8 or its short-chain derivative, ubiquinone-1 (Q1), was identified on the peripheral NqrA subunit [60] located adjacent to the membrane-bound NqrB subunit [55] (Figure 2).

With eight isoprenoid units, Q8 is a very hydrophobic molecule, but its redox-reactive quinone head group is rather hydrophilic. The 3D structure of the respiratory complex I (H^+-translocating NADH:Q oxidoreductase, Section 2.3) revealed a binding site for ubiquinone in a subunit located well above the membrane-embedded part of the complex [38]. The situation is similar in the NQR where ubiquinone binds to the peripheral NqrA subunit exposed to the cytoplasm. It seems to be a common aspect of both complex I and the NQR that their substrates, NADH and ubiquinone, bind to peripheral subunits oriented towards the cytoplasm (in bacteria) or the matrix (in mitochondrial complex I). We find homologous structures for the conversion of substrates at the inner side of the membrane both in complex I and in the NQR. Subunit NqrA of the NQR (catalyzing Q reduction) and subunit Nqo1 of complex I (catalyzing NADH oxidation) exhibit most similar 3D structures [55], although they are not related to each other with respect to primary sequences. NqrA comprises a deep solvent accessible cavity that is formed by residues of a Rossmann-fold domain and an ubiquitin-like domain. This cavity is large enough to accommodate ubiquinone-8. In case of Nqo1 of complex I the corresponding cavity harbors the isoalloxazine moiety of the FMN cofactor, which accepts a hydride from NADH in the initial reaction step catalyzed by complex I.

It seems likely that ubiquinone-8 which was co-purified with the NQR [59, 60] represents an intrinsic cofactor which does not exchange with the ubiquinones present in the inner membrane. We recently showed that the ubiquinone binding site of the membrane-associated NqrA subunit may accommodate two ubiquinone molecules in close proximity [61]. This finding has important implications for the mechanism of ubiquinone reduction by the NQR: The enzyme-bound Q8 might act as terminal redox cofactor of the electron transfer chain within the NQR, donating electrons to a neighboring substrate ubiquinone. Upon full reduction (and protonation) of this substrate ubiquinone, it is released and enters the pool of ubiquinones present in the membrane.

There is evidence that subunit NqrB influences the binding of Q to the NQR, or may house a second Q binding site. Korormicin, an antibiotic, kills *V. alginolyticus* wild-type cells but is inactive against a mutant strain carrying a Gly140 to Val mutation in subunit NqrB of the NQR. Korormicin also acts as a noncompetitive inhibitor against NQR *in vitro*. From this, it was concluded that NqrB comprises a ubiquinone binding site [62]. Mutations replacing the corresponding Gly141 in NqrB from *V. cholerae* NQR, and the adjacent Gly140, resulted in NQR variants with decreased affinity towards Q1 [63]. In their follow-up study, Barquera and colleagues reported that Gly140 and Gly141 on NqrB exert their effect on ubiquinone binding in an indirect manner, and confirmed binding of Q to NqrA [64]. We searched for a bound Q8 in the proximity of $NqrB_{G140}$ and $NqrB_{G141}$ in the electron density map of NQR crystals but found no evidence for the presence of ubiquinone.

Very unexpectedly, the crystal structure of the NQR revealed an additional Fe center coordinated by four conserved cysteine residues in the midst of the membrane parts of subunits NqrD and NqrE. These are Cys26 and Cys120 from subunit NqrD, and Cys29 and Cys112 from subunit NqrE [55]. Subunits NqrD and NqrE are evolutionary related to each other and assume an inverted topology with respect to the orientation of their transmembrane helices. Notably, the conserved cysteines required for Fe ligation are positioned at the tip of hydrophobic helices spanning half of the membrane [55]. The positions of the redox cofactors within the NQR are shown in Figure 2, panel c. NqrF catalyzes NADH oxidation, upon which electrons are transferred from the FAD_{NqrF} to the [2Fe-2S] cluster residing at an edge-to-edge distance of 9.8 Å to the isoalloxazine ring moiety of the FAD. The edge-to-edge distance between the [2Fe–2S] cluster in NqrF and the Fe in NqrDE is 33.4 Å.

This is far too large for efficient electron transfer [65], and one must assume that NqrF undergoes a large conformational change during turnover, moving the [2Fe-2S] cluster in its ferredoxin domain closer to the membrane. Once the electron has reached the Fe site in NqrDE, transfer to FMN_{NqrC} will be very rapid since the edge-to-edge distance is only 7.9 Å. The edge-to-edge distance between FMN_{NqrC} and FMN_{NqrB} is 21.4 Å. Assuming some movement of FMN_{NqrC} away from the Fe center in NqrDE upon reduction, this distance lies within the range of the 15 Å determined by electron paramagnetic resonance (EPR) analysis of spin-spin interactions between the radical state of both FMN cofactors [66]. Continuing the transfer route within the NQR complex, and in accord with kinetic data [67], electrons on FMN_{NqrB} are delivered to the riboflavin in the hydrophobic/hydrophilic transition zone at the cytoplasmic side of NqrB. With 29.3 Å, the edge-to-edge distance between FMN_{NqrB} and riboflavin is too large for fast electron transfer, and again, we must assume some

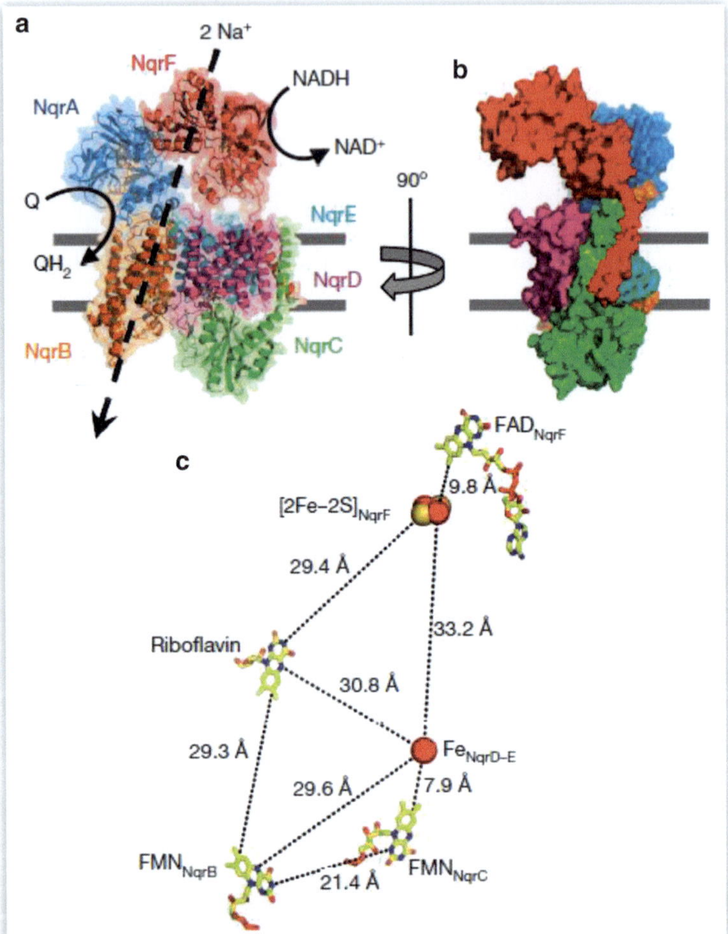

Figure 2 Structure of Na^+-NQR and edge-to-edge distances of redox cofactors of the complex. Panels **a**, **b**: Na^+-NQR is composed of six subunits, NqrA–NqrF. NqrB (orange), NqrD (magenta), and NqrE (cyan) are integral membrane proteins, whereas NqrF (red) and NqrC (green) are each anchored to the membranous subunits by single transmembrane helices. NqrA (blue) does not contain a transmembrane helix. The membrane plane is indicated by grey lines. NADH oxidation occurs at NqrF and ubiquinone (Q) has been shown to bind to NqrA [61]. Electron transfer from NADH at NqrF to quinone at NqrA drives the translocation of sodium ions. Panel **b**: Solvent accessible surface of Na^+-NQR, rotated by 90 %. Panel **c**: The arrangement of the redox cofactors in Na^+-NQR. Edge-to-edge distances of the cofactors are indicated by broken lines. The figure is taken from [55], with kind permission of the publisher; © copyright 2014.

movement of FMN_{NqrB} or riboflavin due to a conformational change during turnover of the pump.

2.1.2.2 Functional Modules of the NADH:Quinone Oxidoreductases

The Na^+-NQR shares three integral membrane subunits (NqrBDE) and one periplasmic subunit (NqrC) with the RNF/NFO redox pumps (Section 2.2), and it is safe to assume that these four subunits represent a minimum catalytic unit which is necessary (but not sufficient) for redox-driven Na^+ transport. Viewed from the membrane plane, with the transmembrane helix of NqrC positioned at the outer edge of the NqrDE unit and opposite to NqrB (Figure 2, panel a), this functional core contains an electron-accepting module, NqrDE with the novel Fe site, shielded by the large periplasmic domain of NqrC with its covalently linked FMN directed towards the Fe center. This electron-accepting module is fused to a transport module, NqrB, which offers a passage for Na^+ identified with the help of the program CAVER [68] (Figure 3). A superposition of the ammonium transporter with NqrB revealed an almost perfect overlay of the presumed pathways for NH_4^+ or Na^+. A notable exception in NqrB is the short helix VIII which narrows the channel with the help of the aromatic side chains of Phe338 and Phe342. Above the constriction, and opening towards the cytoplasm, is a vestibule with the conserved D346 which can coordinate Na^+ [55] (Figure 3).

Helix VIII of NqrB is not found in the ammonium transporters, and its redox-dependent movement to open the channel for the passage of Na^+ is a key event in the presumed transport mechanism (see below). NqrB also harbors the covalently attached FMN_{NqrB} to accept electrons from FMN_{NqrC}, and a riboflavin for subsequent electron transfer to the quinone on subunit NqrA (Figure 2, panel c). The cytoplasmic NqrA and NqrF subunits are attached to the cytoplasmic side of subunits NqrBCDE. Rotation around an axis perpendicular to the membrane reveals the rather open arrangement of NqrF, with its C-terminal, remote FAD-binding domain, its ferredoxin domain carrying the [2Fe-2S] center closer to the membrane part, and the N-terminal, single transmembrane helix tightly packed to the single transmembrane helix of NqrC extending from the periplasmic side of the NQR complex (Figure 2, panel b). NqrA contacts the membrane-bound NqrB subunit with its N- and C-termini. In addition, the N-terminal residues 38–51 of NqrB reside in a deep groove of NqrA, hereby stabilizing NqrA which lacks any transmembrane helices.

As already mentioned, NqrA shares its overall architecture with the NADH-oxidizing Nqo1 subunit from the respiratory complex I [55], although this is not reflected in a similarity of the primary sequences. The alignment of NqrA sequences from 90 species revealed that its sequence is highly variable [69], and the authors from this study concluded that this variability resulted from the lack of functional restraints on NqrA due to the lack of catalytic function of this subunit [69]. The specific interactions of NqrB with NqrA, together with the conserved domains which are catalytically important in the Nqo1 subunit of complex I, give strong

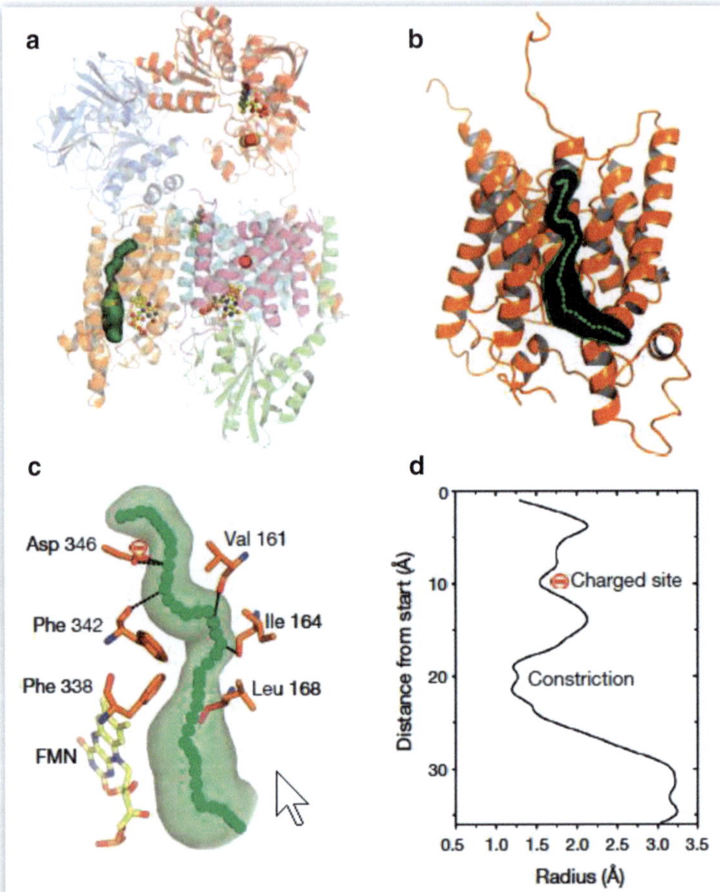

Figure 3 A Na^+ channel in subunit NqrB of the NQR. Panel **a**: Analysis of the transmembrane subunits with CAVER [68] revealed a hydrophilic channel (green) in the membrane-embedded NqrB subunit. Panel **b**: Shown in black is the close-up view of the channel in NqrB. Transmembrane helix I is removed for clarity. The green spheres indicate the proposed path of the Na^+ through NqrB. Panel **c**: Key residues of the putative Na^+ channel. The negatively charged side chain of Asp346 can coordinate Na^+ at the entry of the channel. The backbone carbonyl residues of Val161, Ile164, and Leu168 located on transmembrane helix III can coordinate the Na^+ on the predicted path. Panel **d**: Ile164, Leu168 and on the opposite site Phe338 and Phe342 located on transmembrane helix VIII narrow the channel, forming a constriction. The figure is taken from [55], with kind permission of the publisher; © copyright 2014.

evidence of a catalytic role of NqrA despite its weak degree of conservation among the NQR complexes from different bacteria.

2.1.3 Function of the NADH:Quinone Oxidoreductases

2.1.3.1 Interaction of NADH:Quinone Oxidoreductases with Alkali Cations

To understand how Na^+ transport is achieved by the Na^+-NQR, we must describe the mode of interaction of the enzyme in its various redox states with its coupling cation. The Na^+-NQR from *V. cholerae* was reported to bind three sodium ions per enzyme, both in the oxidized or in the reduced state [70]. These values were determined in a dialysis experiment after equilibrium had been reached. Using ^{23}Na NMR spectroscopy, Bogachev, Verkhovsky and coworkers determined the affinities of the NQR for Na^+ in the oxidized (K_D ca. 24 mM) *versus* reduced state (K_D ca. 30 μM) [71], indicating that the pump possesses at least one Na^+ binding site which, directly or indirectly, senses the redox state of one (or several) redox cofactors. There is evidence that changes in Na^+ affinity upon reduction of the NQR are conformational-driven, and therefore indirect.

Using a combined electrochemical and Fourier transform infrared (FTIR) spectroscopic approach, Hellwig, Barquera and coworkers [72] showed that the Na^+-NQR exhibits large conformational changes which depend both on the redox state of the enzyme, and of the presence of alkali cations. Specifically, Asp397 of the membrane-bound NqrB subunit was proposed to directly interact with Na^+ or Li^+. This carboxylic residue was shown to be of functional importance since replacement by alanine resulted in the loss of the Na^+ -stimulated quinone reductase activity [73]. The 3D structure of the Na^+-NQR [55] revealed that Asp397 of subunit NqrB, located at the interface between subunit NqrB and NqrA, is solvent-accessible. $Asp397_{NqrB}$ was proposed to capture Na^+ in a reaction step which precedes the transport of Na^+ through the membrane-embedded part of NqrB [72, 74]. The 3D structure of the Na^+-NQR does not support this assumption since there is no vestibule followed by a channel in proximity to $Asp397_{NqrB}$ [55]. $Asp397_{NqrB}$ is located at the cytoplasmic end of the C-terminal helix which contacts subunit NqrA located on top of subunit NqrB. The N-terminal helix of subunit NqrB also extends towards the cytoplasm, and together with the C-terminal helix ensures tight binding of NqrA to the membrane-bound NqrB subunit. A likely explanation for the observed effects of mutations at position $Asp397_{NqrB}$ with respect to cation-dependent conformational changes [72] and uncoupling [75] is an overall change in the mode of interaction between NqrB and NqrA.

A direct mechanism leading to the observed, redox-dependent change in affinity of the NQR towards Na^+ [71] should also be considered. In this case, one would expect a dependency of the redox potential of cofactors on Na^+ [67, 76]. Indeed, there is evidence for a modulation of the redox properties of flavin cofactors in the NQR by Na^+. The midpoint potentials of the flavoquinone/anionic flavosemiquinone and the anionic flavosemiquinone/flavohydroquinone redox couples were shifted towards more negative values when Na^+ was replaced by K^+ [76]. These flavin redox transitions were assigned to the FMNs on subunit NqrB and/or NqrC [67, 76], and their shift in redox potential towards positive values in the presence of Na^+ might lead to the acceleration of at least one electron transfer reaction between

$FADH_2$ in the cytoplasmic NqrF and the FMN_{NqrC} located in the periplasm. Barquera and colleagues showed that the electron transfer from the [2Fe-2S] center on subunit NqrF to FMN on subunit NqrC is accelerated in the presence of Na^+ [36]. The 3D structure revealed the existence of an additional Fe center in the electron transfer pathway from $[2Fe\text{-}2S]_{NqrF}$ to FMN_{NqrC} [55] (Figure 2). Additional studies will now be required to determine if Na^+ stimulates the electron transfer from [2Fe-2S] to the novel Fe center, or from the Fe center to the FMN_{NqrC}. It is noteworthy that the midpoint potential of $[2Fe\text{-}2S]_{NqrF}$ is not influenced by Na^+ [72].

There is a strong effect of cations on the NADH:Q oxidoreduction activity of the NQR. Quinone reduction was stimulated by Na^+ and Li^+, two cations which are also translocated by the NQR. In contrast, neither K^+ nor Rb^+ were transported by the enzyme, though K^+ enhanced electron transfer when tested in combination with Na^+ [70]. With the highly related Na^+-NQR from *V. alginolyticus*, Dimroth and coworkers demonstrated that KCl or $MgCl_2$ led to an increase of the affinity of the enzyme towards Na^+, as is evident from the decrease in the apparent K_m [17]. Barquera and coworkers reported that Rb^+ was inhibitory for electron transfer by the NQR [70]. From their results, the authors proposed that in addition to Na^+ binding site(s) relevant for transport, the NQR comprises at least one additional, regulatory cation binding site which may bind K^+, leading to activation, or Rb^+, leading to inactivation of the enzyme.

2.1.3.2 Mechanism of Redox-Driven Na^+ Transport by the NADH:Quinone Oxidoreductases

A comparison of the subunit composition of NQR with its homolog, the RNF complex (Section 2.2), reveals that these redox pumps are built from similar functional modules coupling electron flux between the substrates (NADH and Q in case of NQR, ferredoxin and NAD(H) in case of RNF) to Na^+ (or H^+) translocation. The central core includes all three membrane-bound plus one periplasmic subunit with a membrane anchor and comprises the following elements which are crucial for redox-driven Na^+ transport: Flavin cofactors on both sides of the membranes, the novel Fe site in the membrane part, a channel (in NqrB), one FMN (of NqrB) which is located in close proximity to the predicted channel, and another FMN (of NqrC) acting as electron wire between the Fe center and FMN_{NqrB}. The hypothetical mechanism of electron transfer-driven Na^+ transport by the NQR (Figure 4) outlined below critically depends on the widening of the Na^+ channel in NqrB caused by the redox-dependent movement of helix VIII from NqrB.

It is assumed that the overall transport stoichiometry of the Na^+-NQR is 2 electrons transferred per 2 Na^+ transported, or 1 NADH oxidized (or Q reduced) per 2 Na^+ translocated [77]. The position of the redox centers (Figure 2), and available kinetic studies (summarized in [67, 78]), are in accord with the following sequence of electron transfer events in the Na^+ -NQR:

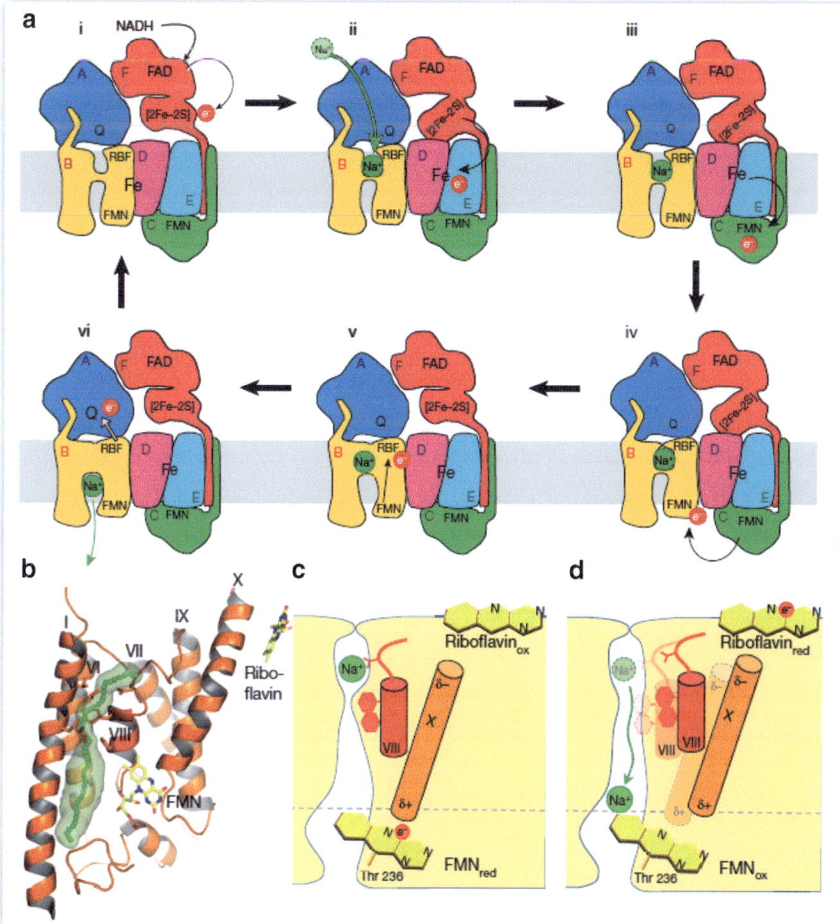

Figure 4 Mechanism of redox-driven Na^+ translocation by the Na^+-NQR. The membrane is indicated by a grey bar. **Top**, cytoplasm; **bottom**, periplasm. Panel **a**: Catalytic half cycle showing the transport of $1Na^+$ per 1 electron transferred. Panel **b** shows the Na^+ channel through NqrB (shaded in green) close to the short helix VIII which constricts the channel. Panels **c** and **d**: The redox state of FMN_{NqrB} controls the position of helices X and VIII. The figure is taken from [55], with kind permission of the publisher; © copyright 2014.

$$\mathrm{NADH} \rightarrow \mathrm{FAD}_{\mathrm{NqrF}} \rightarrow [\mathrm{2Fe-2S}]_{\mathrm{NqrF}} \rightarrow \mathrm{Fe}_{\mathrm{NqrDNqrE}} \rightarrow \mathrm{FMN}_{\mathrm{NqrC}} \rightarrow$$
$$\mathrm{FMN}_{\mathrm{NqrB}} \rightarrow \mathrm{Riboflavin}_{\mathrm{NqrB}} \rightarrow \mathrm{Q}_{\mathrm{NqrA}}$$

The structure of the NQR [55] revealed a channel in NqrB flanked by two redox-reactive flavins located at the periplasmic and cytoplasmic side, respectively, suggesting that passage of Na^+ through the channel is controlled by the redox state of these two flavins. We combine all available information and propose the following mechanism of redox-driven Na^+ transport by the NQR (Figure 4, panel a):

(i) The reaction starts with NADH donating a hydride to FAD_{NqrF}.
(ii) From there, one electron is delivered to the [2Fe-2S] cluster, eliciting a movement of the ferredoxin domain towards NqrD, and facilitating electron transfer to the Fe center located between NqrD and NqrE. At the same time, Na^+ approaches through the cytoplasmic vestibule in NqrB and is coordinated by Asp346 above the constriction of the channel.
(iii) The electron is delivered from the Fe center to FMN_{NqrC} positioned in the periplasmic domain of NqrC).
(iv) Electron transfer from FMN_{NqrC} to FMN_{NqrB}, and occlusion of Na^+ in NqrB.
(v) Electron transfer from FMN_{NqrB} to the riboflavin (RBF) positioned at the cytoplasmic side of NqrB triggers the shift of helix VIII of NqrB (also shown in panels b, c and d), followed by entry of Na^+ into the channel, and re-orientation of the ferredoxin domain of NqrF towards FAD_{NqrF}.
(vi) Upon electron transfer from the $riboflavin_{NqrB}$ to ubiquinone, the escape of Na^+ back to the cytoplasm is prevented, and Na^+ is forced to leave NqrB towards the periplasm.

With steps (i) to (vi), the first half cycle leading to the transport of one Na^+ per one electron transferred is finished. The second half cycle is initiated by transfer of the second electron from FAD to the [2Fe-2S] cluster, and the reaction steps (Figure 4, panel a) proceed again to couple transfer of the second electron to transport of the second Na^+.

The question arises how a redox reaction might trigger conformational changes in the NQR which promote uphill transport of Na^+ against the electrochemical Na^+ gradient. We note that the properties of the Na^+ channel in NqrB (shaded in green) (Figure 4, panel b) are largely determined by the short helix VIII which constricts the channel, preventing passage of Na^+. Helix VIII is in close contact to the tilted helix X which points with its positive dipole end towards FMN_{NqrB} positioned below. The riboflavin binds near the negative end of the helix X dipole. It is proposed that the redox state of FMN_{NqrB} controls the position of helix X pointing with the positive end of the helix dipole directly towards the N5 of the isoalloxazine ring of FMN_{NqrB}. (Figure 4, c and d). A net negative charge on FMN_{NqrB} keeps helix X in its downward position, maintaining the position of helix VIII and blocking the pathway for Na^+.

Electron transfer from FMN_{NqrB} to $riboflavin_{NqrB}$ removes the negative charge from FMN_{NqrB}, hereby weakening the interaction with the positive dipole end of helix X. Helix X is released upward, moving helix VIII with its Phe338 and Phe342, and opening the pathway for the translocation of the Na^+ to the periplasmic half-channel (Figure 4, panels c and d). Here, Na^+ immediately becomes hydrated by water molecules entering from the periplasmic side.

The NQR stabilizes anionic flavosemiquinone(s) in the "as isolated state", that is in the absence of electron donors, and despite the presence of oxygen. These anionic flavosemiquinones have been assigned to the covalently attached FMNs on NqrB and/or NqrC ([78], and references cited therein). The mechanism proposed here offers a rationale for this unusual property of a redox enzyme: In starving cells, when the electron donor NADH is limited, FMN_{NqrB} must be stabilized in its one-electron reduced, anionic state to keep the Na^+ pathway closed, preventing unspecific leakage of Na^+ across the cytoplasmic membrane.

2.2 NAD:Ferredoxin Oxidoreductase

NqrB, NqrC, NqrD, and NqrE are homologous to subunits RnfD, RnfG, RnfE, and RnfA of the RNF (*Rhodobacter* nitrogen fixation) complex [69] (Table 1). The genes encoding for RNF were first discovered by studying *Rhodobacter capsulatus* mutants deficient in nitrogen fixation, suggesting a role of the complex in electron transfer to nitrogenase [79]. The RNF complex from *Acetobacterium woodii* was shown to catalyze redox-driven Na^+ transport [80, 81]. Because the RNF complex uses ferredoxin (ox/red) and NAD(H) as substrates, it is also termed NFO (NAD^+:ferredoxin oxidoreductase) [82, 83]. Both Na^+-NQR and the RNF/NFO complexes are composed of six subunits, respectively, but only the central core of the complexes is conserved with regard to primary sequences and cofactor composition. This core is made up by the three integral membrane subunits of the RNF complex, RnfD (NqrB), RnfE (NqrD), and RnfA (NqrE), and the membrane-anchored RnfG (NqrC) (homologous Nqr subunits in brackets). These four subunits are considered to represent the minimal, functional unit of the NQR/RNF family of redox-driven cation pumps. NqrC is anchored via a single transmembrane helix to the membrane core of the NQR complex [55], and a similar arrangement is to be expected for RnfG in the RNF complex.

NqrC/RnfG, like NqrB/RnfD, carries a covalently attached FMN cofactor [82, 84]. As outlined above, the covalently attached FMN of NqrB might control the passage of Na^+ through the presumed channel in NqrB in a redox-dependent manner. The related RnfD subunit also carries a covalently attached FMN, further supporting the notion that both NQR and RNF complexes operate a very similar mechanism of redox-driven Na^+ (or H^+) transport. This may be true although the electron-accepting and electron-donating subunits of the complexes are only weakly related (NqrA/RnfC) [69], or, like RnfB, have no homolog in the NQR complex [69]. RnfB comprises a site for interaction with exogenous ferredoxin used as electron donor or acceptor, depending on the direction of cation transport through the NFO/RNF complex [83]. NqrA comprises a quinone binding site [61]. In the RNF complex, RnfC which shows some weak relation to NqrA is expected to interact with NADH or NAD, to either accept or donate a hydride [83]. Other redox centers in the RNF complex which are not found in the NQR are two [4Fe-4S] centres and one FMN on subunit RnfC which interacts with NAD(H). RnfB which interacts with ferredoxin) contains motifs for four [4Fe-4S] clusters and one [3Fe-4S] cluster [83]. It was also suggested that RnfB may ligate six [4Fe-4S] clusters [69]. These assignments were based on sequence analyses and await biochemical verification.

In summary, RNF and NQR complexes differ in cofactor composition in their electron input and output modules (NqrF and NqrA, versus NfoC and NfoB), but utilize the same set of cofactors in their highly conserved, membrane-embedded RnfAEGD or NqrEDCB subunits which catalyze redox-dependent cation transport. There is recent evidence for proton transport by the RNF complex from *Clostridium ljungdahlii* [85], while all NQR complexes studied so far exclusively acted as Na^+ pumps. This makes the RNF complex an interesting model system: Comparing the molecular properties of H^+- with Na^+-translocating RNF complexes will give insight into the general mechanism of cation transport by the NQR/RNF family of redox pumps, as exemplified for the H^+ or Na^+ -translocating F_1F_O ATPases described in Section 3.1.

2.3 Primary Versus Secondary Na^+ Transport by Complex I

Every other respiratory system for redox-driven voltage generation is dwarfed by complex I, the respiratory NADH:quinone oxidoreductase of eukaryotes and bacteria. There is no other machine like this. The mitochondrial complex I from eukaryotes is composed of 44 subunits, including those 14 core subunits which are homologs of the bacterial complex I. There is a chain of redox cofactors in the peripheral arm of this L-shaped complex for electron transfer, starting with the FMN at the tip which accepts a hydride from NADH, via seven FeS centers, to ubiquinone in the connecting part between the peripheral and membrane arms.

The membrane arm is composed of subunits highly related to cation/H^+ antiporters. There is no evidence for the presence of redox cofactors in the membrane arm of complex I [40, 86]. This strongly supports the notion of conformational changes elicited by NADH:quinone oxidoreduction in the peripheral arm which promote H^+ (or Na^+) translocation through the three antiporter-like subunits in the membrane arm. The stunning 3D structures of complex I from the bacterium *Thermus thermophilus* [87], of bovine complex I [37], and of complex I from the yeast *Yarrowia lipolytica* [38] support this assumption, since they revealed that the large, antiporter-related subunits (termed NuoL, NuoM, and NuoN in the bacterial complex I) share characteristic structural features with secondary, cation/H^+ antiporters like NhaA from *E. coli* discussed in Chapter 12. This Na^+/H^+ antiporter, like the complex I subunits NuoL, M and N, comprises charged or polar amino acid residues within the membrane region which are situated in close proximity to broken or partially unwound transmembrane helices. In NhaA, two adjacent, interrupted transmembrane helices form a site for Na^+ binding, and it is assumed that this critical region of the transporter undergoes a conformational change during exchange of Na^+ against H^+ [88].

Complex I is generally considered to translocate protons from the N- to the P-side in the vectorial reaction of the pump driven by NADH:quinone oxidoreduction, its chemical reaction (Figure 5).

In bacteria, the positive charged (P-) side represents the periplasmic aspect of the energy-conserving membrane. In mitochondria, the P-side of the inner membrane faces the space between the inner and outer mitochondrial membrane. The negatively charged (N-) side represents the bacterial cytoplasm, or the mitochondrial matrix. A strict selectivity for protons as coupling cation was reported for mitochondrial [89] and bacterial complex I [40]. This concept was put into question when complex I from an enterobacterium was reported to exclusively transport Na^+ from the N- to the P-side coupled to NADH:Q oxidoreduction [39].

In another study, the enterobacterial complex I was shown to pump protons from the N- to the P-side driven by oxidoreduction, together with a proposed Na^+/H^+ antiporter catalytic activity with protons re-entering the N-compartment, promoting Na^+ transport to the P-compartment [90]. In another bacterial complex I, redox-driven H^+ transport from the N- to the P-side under concomitant flux of Na^+ from the P- to the N-side, probably caused by H^+/Na^+ antiport, was reported [29, 91]. In mitochondrial complex I from the yeast *Yarrowia lipolytica*, we observed NADH oxidation-driven Na^+ transport (from N- to P-side), with no evidence for concomitant transport of protons [92].

A recent study on bovine mitochondrial complex I showed that the enzyme acts as a Na^+/H^+ antiporter in its deactive (D-) state, with H^+ transport from the P- to the N-side, and Na^+ transport from the N- to the P-side. This antiporter activity of complex I was not observed with enzyme in its active (A-) state [93]. For a detailed description of the H^+ and Na^+ transport experiments performed with complex I, and a critical discussion of its H^+ or Na^+ selectivity, the reader is referred to [94, 95]. Figure 5 summarizes the transport activities reported for complex I from different

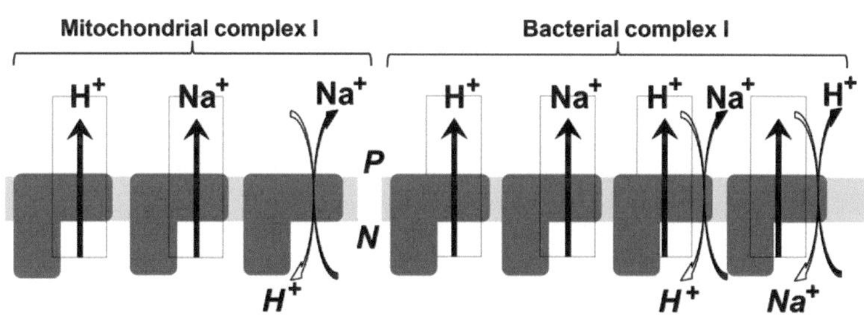

Figure 5 Directionality and cation selectivity of complex I. Complex I, the electrogenic NADH:quinone oxidoreductase, is an L-shaped molecular machine found in mitochondria and many bacteria. Its peripheral part catalyzes NADH oxidation, the membrane-embedded part acts as transporter for cations. Observed cation transport activities of mitochondrial complex I, from left to right: A proton pump in the A-state, a Na^+ pump, a Na^+/H^+ antiporter in the D-state. Observed cation transport activities of bacterial complex I, from left to right: A proton pump, a Na^+ pump, a proton pump with Na^+/H^+ antiport activity (Na^+ flux from N- to P-side), a proton pump with H^+/Na^+ antiport activity (Na^+ flux from P- to N-side). In mitochondria, the positive (P-) side of the inner membrane faces the inner membrane space which exchanges small solutes with the cytoplasm of the eukaryotic cell, the negative (N-) side faces the mitochondrial matrix. In bacteria, the P- side faces the periplasm which exchanges small solutes with the surrounding medium, the N-side faces the cytoplasm.

organisms, and illustrates that both primary, redox-driven cation transport events (indicated by straight arrows) and secondary, gradient-driven H^+ or Na^+ translocation events (indicated by bent arrows) must be considered. If we assume that complexes I from bacteria and mitochondria, highly related machines with respect to their core subunits, operate with an identical mechanism of redox-driven cation transport, these reported transport activities seem to be contradictory. But we should also consider that complexes I from different organisms might exhibit different cation selectivities (H^+ or Na^+), as exemplified by F-ATPases. Moreover, the antiporter-related NuoL, M, and N subunits of complex I (bacterial nomenclature) may act in a "coupled" manner, transporting cations only during oxidoreduction, or in an "uncoupled" mode, where cation translocation is driven by existing electrochemical gradients across the membrane. So obviously, much work remains to be done by all who want to understand the transport mechanism of this fascinating, huge redox pump.

3 Na^+ Translocation by a Rotational Mechanism

3.1 F_1F_O ATP Synthase

3.1.1 Overall Architecture and Function

F_1F_O ATP synthases are ubiquitous enzymes present in energy-converting membranes from bacteria, archaea, or eukaryotes, where they use the proton motive force or sodium motive force to synthesize adenosine-5′-triphosphate [96, 97]. The structure of ATP synthases has been first described as a 'lollipop,' as the soluble F_1 catalytic unit is attached to the electrochemically energized membranes via the F_O part of the complex (Figure 6). Here we consider the bacterial F-type ATPases which are true bidirectional enzymes operating either in the ATP hydrolysis or synthesis mode, depending on the energetic demands of the bacterium. For a discussion of the related A-type ATPases from archaea, or related V-type ATPases from eukaryotes or bacteria, the reader is referred to [26, 98, 99] and references therein.

Whereas the F_1 part is responsible for the conversion of ADP and inorganic phosphate (P_i) to ATP (in the synthesis mode), or hydrolysis of ATP to ADP and P_i (in the hydrolysis mode), the membrane-bound F_O part (O standing for oligomycin, an inhibitor of ATP synthase specifically interacting with F_O) [100] converts the flux of coupling cations (H^+ or Na^+) to rotation. This rotational energy is not wasted but generates torque, forcing a central shaft (composed of the ε- and γ-subunits, and connected to the c-ring) to alter the nucleotide-binding β-subunits in F_1 [96]. These conformational changes, driven by the flux of the coupling cation along its electrochemical gradient, promote the formation of ATP from ADP and P_i under release of H_2O. *Vice versa*, the hydrolysis of ATP in F_1 will rotate the shaft together with the c-ring against subunit a, pushing the coupling cations uphill, and creating an electrochemical gradient in an endergonic reaction driven by the exergonic hydrolysis of ATP (Figure 7).

For a discussion of the mechanism of ATP synthesis in F_1, and a description of techniques used to visualize the rotation of the shaft (and the c-ring) with respect to F_1, the reader is referred to [96, 97]. The overall architecture of F_1F_O is shown in Figure 6 which combines the current knowledge about high-resolution structures of individual parts of the F_1F_O complex. The a-subunit acts as a brushing against the rotating c-ring, and probably keeps F_1 in position with the help of the b-subunits extending from the a-subunit to the tip of F_1. The interface between subunit a and the c-subunits of the c-ring is of particular interest since it offers ligands for the coupling cation (H^+ or Na^+), promoting access or release of the cation through distinct pathways leading from and to the interface. A recent 3D structure of the holo-

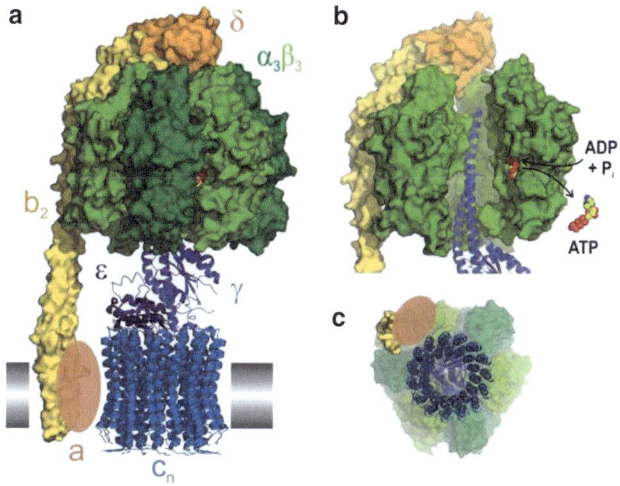

Figure 6 Structural model of the F_1F_O ATP synthase based on 3D structures of subcomplexes from different organisms. Panel **a**: Holo-complex in a horizontal view with respect to the membrane plane, with the $\alpha_3\beta_3$ subunits (dark and light green) of the F_1 part located in the cytoplasm of the bacterium, and the c-ring composed of n c-subunits (light blue) of the F_O part embedded in the inner bacterial membrane (grey). Rotation of the c-ring is transmitted to F_1 via the shaft composed of the ε-subunit (dark blue) and γ-subunit (magenta). The outer stalk formed by two b-subunits (light yellow) and subunit δ (dark yellow), together with the membrane-bound a-subunit (brown ellipsoid) serve as stator, tethering the β-subunits in F_1. Each β-subunit harbors one nucleotide binding site. Subunit a, together with its interacting c-subunits, transports the coupling cation (H^+ or Na^+) through the F_O part via a binding site located at the interface of subunit a and the c-ring. Panel **b**: α-subunits are omitted to visualize the central shaft formed by the asymmetric α-subunit, which protrudes into the nucleotide binding pocket of one of the β-subunits with bound ADP (shown in sphere representation). Panel **c**: view from the periplasmic side of a bacterial cell. The model shows the structure of bovine (*Bos taurus*) F_1 subcomplex (PDB entry 1E79 [135] together with different bacterial subcomplexes: The c-ring rotor composed of n = 15 c-subunits (pdb entry 2WIE [111]), the δ-subunit (PDB entry 2A7U [136]), the membrane domain of subunit b (PDB entry 1B9U [137]), and subunit b devoid of its membrane domain (PDB entry 3K5B [138]). Reprinted from [108], with kind permission from Springer; © copyright 2013.

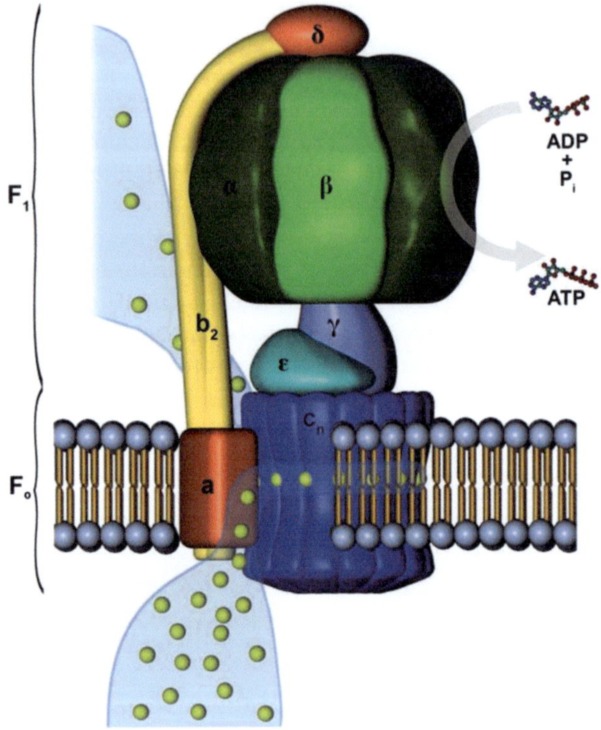

Figure 7 Coupling of cation flux to ATP synthesis by F_1F_O ATP synthase. The ATP synthase consists of the water-soluble F_1 and the membrane-embedded F_O subcomplexes. Rotating elements are the membrane-embedded ring of n c-subunits (shown in blue, with n = 8–15) and subunits γ and ε of the shaft which projects into the F_1 part. In the ATP synthesis mode illustrated here, the coupling cations (Na^+ or H^+; yellow spheres) in the periplasm (bottom of Figure) approach the binding site in the interface of subunits a and c. Compensation of the negative charge of a conserved carboxylate (Glu65) of subunit c by Na^+ (in the *I. tartaricus* ATP synthase) promotes the vertical movement of the Na^+ loaded c-subunit out of the a-c interface, and into the hydrophobic environment formed by the aliphatic chains of membrane lipids (golden bars). Taking up one Na^+ after the other, the c-subunits move from left to right, inducing step-wise rotation of the c_n-ring against subunit a. Each Na^+ bound to subunits c travels almost full circle, to be released as soon as it comes into contact with the a-c interface again, and only then to leave F_O via an exit pathway provided by subunit a. This release event is obscured by subunit a. Rotation of the c-ring results in rotation of the inherently asymmetric γ-subunit which causes conformational changes in the nucleotide binding sites of the β-subunits of F_1. The F_1 headpiece contains three β subunits and three α subunits which are homologs of the β subunits but do not possess functional nucleotide binding sites. The catalytically competent binding sites on the β subunits undergo three defined catalytic states, depending on the position of the inner shaft, which are designated open (O), loose (L) or tight (T). In the O state, affinity to ADP and P_i is poor. In the L state, ADP and P_i are loosely bound. The T state represents the catalytically active state harboring the product, ATP. In the binding change mechanism of ATP synthesis [139], each binding site on one of the three β-subunit undergoes the $O \rightarrow L \rightarrow T$ transition. Therefore, a 360° rotation of the shaft with respect to the $\alpha_3\beta_3$ subcomplex results in net synthesis of three ATP molecules (see also Figure 6). The $\alpha_3\beta_3$ subcomplex is held in place by subunit δ and two subunits b, connected to subunit a embedded in the membrane. Together, the non-rotating subunits $ab_2\alpha_3\beta_3\delta$ (green and orange/yellow) form the stator of this rotary machine. The figure was created by Paolo Lastrico, Max-Planck-Institute of Biophysics, Frankfurt. Reprinted from [97], with kind permission from Cambridge University Press; © copyright 2011.

ATP synthase at 6.2 Å showed that in the a-c interface, four long, hydrophobic helices of subunit a are positioned almost perpendicular to the transmembrane helices of the proteolipids forming the c-ring [101]. Mutational, biochemical, and modelling studies led to a sophisticated model, which describes how the interaction of the coupling cation with its binding site pushes the c-ring (linked to the central shaft) against subunit a in the synthesis or hydrolysis mode, respectively.

3.1.2 "Pumped" Protons *Versus* "Chemical" Protons: A Controversy Solved with the Help of the Na^+-Dependent F_1F_O ATPase

A look back in time reveals the importance of Na^+ -dependent ATPase to solve controversial issues regarding the general mechanism of F_1F_O ATP synthases, including the H^+-dependent ones. Laubinger and Dimroth unequivocally demonstrated the existence of a Na^+-dependent F_1F_O ATP synthase in the anaerobic bacterium *P. modestum* [19, 102]. Together with Kaim, Dimroth later showed that a hybrid ATP synthase formed from the F_1 part of the H^+-translocating ATP synthase fused to the F_O part of the Na^+-translocating enzyme retained its preferred selectivity towards Na^+ [103]. At that time, an important open question concerned the identity of "chemical" and "pumped" protons during the catalytic cycle. Citing Mitchell [104], this hypothesis assumed "the participation of protons, or their hydroxide or oxide ion equivalents, as shared intermediaries between the chemical and osmotic actions of the F_OF_1 ATPases". Another hypothesis, citing again Mitchell, assumed a "conformationally coupled binding-change mechanism" relying on "an essential energy-transmitting device, interposed between supposedly separate proton-pumping and hydrolytic domains in F_OF_1 ATPases" [104].

It was obvious that, due to the high degree of conservation with H^+-dependent ATPases, the general mechanism of Na^+- and H^+-dependent F_1F_O ATP synthases would be identical. Yet, Na^+ clearly does not represent an intermediary in the formation or cleavage of a phosphoanhydride bond, therefore the first hypothesis was refuted. Today we know that the energy-transmitting device described in the second hypothesis [104] represents the rotating shaft plus c-ring of the ATP synthase complex. This is one of many examples illustrating how studies on the Na^+-dependent respiratory complexes greatly contributed to our understanding of the mechanism of their H^+-dependent homologs which are more widespread in nature.

3.1.3 Structure of the Membrane-Embedded F_O Part of ATP Synthase

In the same April 2005 issue of Science, two groups reported, independently, the X-ray-crystallographic, high-resolution structures of the membrane-embedded rotors of bacterial Na^+-ATPases [99, 105]. This was a major step forward in our understanding of the mechanism of cation translocation by ATPases. Both structures revealed the ligands of the coupling cation, Na^+, in a site formed by two neighboring α-helices of the so-called ATPase proteolipids. These highly hydrophobic subunits form the rotor ring of ATPases. They can be enriched by an extraction of membranes with organic solvents, very much like lipids, hence their name. In the F_1F_O-type ATPases, the proteolipid (termed c-subunit) comprises two membrane-spanning helices, and together with one helix of the adjacent c-subunit, they offer the ligands for Na^+ in a site positioned in the approximate middle of the membrane [105] (Figure 8).

Coordination of Na^+ by the proteolipid from V-type ATPases follows the same structural principles, but here, the proteolipid comprises four helices, suggesting a multiplication event in the evolution of genes coding for the rotor subunits of ATPases [106, 107]. The number of binding sites for the coupling cation in a given F-rotor is determined by the number of c-subunits (n) which is constant in a given organism. Known to date are c-rings composed of eight to fifteen c-subunits [98, 108], with a corresponding increase in the diameters of the ring, and an increase in the $Na^+(H^+)$/ATP ratio. Since a 360° rotation of the c-rotor is required for the net formation of 3 ATP molecules, the calculated coupling cation/ATP ratio for an F-ATPase with eight c-subunits is 2.7, and 5.0 for an enzyme with a rotor consisting of fifteen c-subunits. One would assume that ATPases operating mainly, *in vivo*, as ATP-hydrolyzing cation pumps, benefit from a high number of Na^+ (or H^+) binding sites (due to a large number of c-subunits), whereas enzymes operating in the ATP synthesis mode have a c-ring with a small number of c-subunits, keeping the number of cations passing through the complex per 360° rotation as low as possible. Indeed, if the chemical concentration gradient (ΔpH in proton-dependent ATP synthases, or ΔpNa^+ in Na^+-dependent ATP synthases) represents the sole driving force for ATP synthesis, this argument would hold. Yet, the transmembrane voltage ($\Delta\psi$) also stimulates ATP synthesis, and in coupled, respiring membranes, $\Delta\psi$ never collapses to 0. *In vitro* studies with purified, reconstituted H^+- or Na^+-dependent ATP synthases indicate that a $\Delta\psi$ component always is required to drive ATP synthesis, even in the chloroplast F-ATP synthase which operates at very high ΔpH [24, 109].

In the following, we will only consider the Na^+ binding site in the rotor from F-type ATPases [105, 110], but it is important to note that both F- and V-ATPases operate with a very similar set and arrangement of ligands, including the conserved glutamate and glutamine residues which directly coordinate the sodium ion. These are, among others, Glu65 and Gln32 in the F-ATPase rotor (Figure 8) [105, 110], and Glu139 and Gln110 in the V-ATPase [97]. In rotors from H^+-translocating F-ATPases, the conserved glutamate residue is replaced by an aspartate which accepts and releases a proton during rotation of the c-ring [96]. The type of coordinating residues, as well their arrangement in the binding site, determines the (preferred) selectivity (Na^+ or H^+) of the ATPase [111, 112]. It should be mentioned that Na^+-dependent F-ATPases may use H^+ as coupling ion at very low Na^+ concentrations and more acidic pH [113], in further support of a common coupling mechanism of sodium- and proton-dependent F-ATPases.

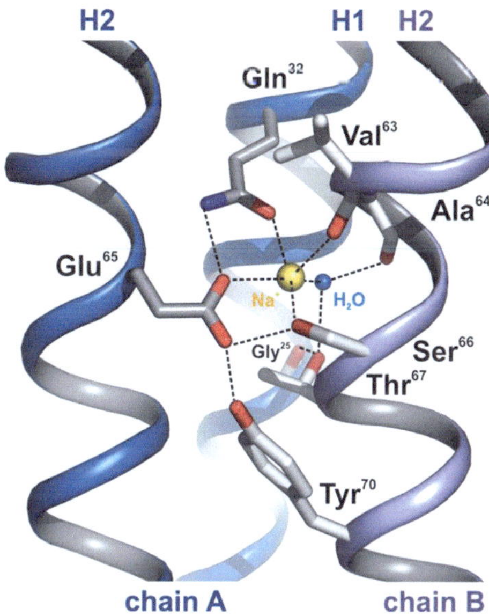

Figure 8 Ion-coordination structure in the rotor ring of the Na^+-dependent F_1F_O ATP synthase from *Ilyobacter tartaricus*. The ring is composed of 11 c-subunits, each comprising two transmembrane helices (H1 and H2), connected by a short loop. One out of eleven Na^+ binding sites is shown, with Na^+ (yellow sphere) located between adjacent subunits. The amino acids involved in ion coordination are depicted in stick representation. The network of hydrogen bonds in the site is indicated by broken lines. Thr67 and Ala64 (chain B) interact with the water molecule within the binding site. Gly25 (chain A) and Tyr70 (chain B) do not directly coordinate Na^+ but are part of the hydrogen bonding network. Reprinted from [110], with permission from Elsevier; © copyright 2009.

3.1.4 Interaction of F_O with Na^+

A major problem arising in X-ray structures of Na^+ pumps is to decide whether an electron density peak results from Na^+ (10 electrons), H_2O (8 electrons) or HO^- (9 electrons), since the number of electrons differs by only 1 or 2. Therefore, an electron density resulting from a sodium ion bound to a site on the c-ring of the F-ATPase, could be erroneously assigned to a water molecule during the refinement procedure (or *vice versa*). Figure 8 depicts the Na^+ binding site of the F-ATPase rotor ring from *Ilyobacter tartaricus* [110]. Glu65, Gln32, Ser66, and the carbonyl backbone of Val63 directly coordinate Na^+, identified in the 2005 structure of the rotor ring [105] with the help of valence screening [114]. Careful re-evaluation of the original diffraction data [105] in combination with molecular dynamics, free energy calculations and analysis of coordination geometry [110] led to the identification of a structurally well-defined H_2O molecule in the binding site which acts as the fifth ligand for Na^+. Therefore, each F- rotor with its 11 c-subunits binds a total of 11 sodium ions, each coordinated by four oxygen atoms from the protein and one oxygen from water. It is important to point out that ligation of Na^+ requires ligands from two distinct, but adjacent c-subunits, indicated by chain A and chain B in Figure 8: Val63 is part of chain B, while Glu65 and Gln32 belong to chain A. But a single transmembrane α-helix also provides ligands for two Na^+ binding sites: Ser66 (chain B) and the preceding Glu65 on chain B interact with different sodium ions from adjacent binding sites. This ligand arrangement contributes to the high stability of the Na^+-loaded c-ring and prevents its dissociation under denaturing conditions on SDS-PAGE [115, 116]. In Figure 8, only one Na^+ coordinated by Glu65 of chain A is shown.

The structure of the c-ring with its bound Na^+ [105, 110] is considered to represent a snapshot of the catalytic cycle of F-ATPases. Yet, the structure was determined in the absence of subunit a which could alter the conformation of Na^+ ligands in the a-c interface. Therefore, the relevance of the determined structure for the mechanism of the F-ATPase was experimentally tested using many other techniques. An essential and universally conserved acidic residue of the rotor, like Glu65 in subunit c of *I. tartaricus* F_O, is deprotonated when facing a conserved, positively charged arginine residue on subunit a in the a-c interface, but must bind Na^+ (or, a proton), when exposed to the hydrophobic environment of the inner lipid bilayer [117]. This arginine residue plays a critical role in Na^+ translocation through F_O, as shown by *in vitro* Na^+ transport analyses of mutated variants of F_O [118]. The conserved arginine residue of subunit a is considered to fulfil an important function: Since subunit a contains two pathways, leading to and from the Na^+ binding site in the a-c interface [119], a short-circuit leading to Na^+ transport without intermediate loading of the c-subunit binding site must be prevented. The positive charge of the conserved arginine residue separates the two pathways to prevent uncoupled Na^+ flux down the electrochemical gradient which does not drive rotation. In the direction of ATP synthesis, and viewed from the periplasm, Na^+ enters the a-c interface via a pathway through subunit a (Figure 7). Binding of Na^+ compensates the negative charge of Glu65, allowing movement of the Na^+-loaded c-subunit in clockwise direction. Travelling almost full circle, Na^+ enters the a-c interface, where it faces a more hydrophilic environment, allowing hydratization and release of Na^+ via the exit pathway [117].

For the bifunctional F-ATP (synth)ases, this mechanism is fully reversible. It works in the clockwise and counter-clockwise rotation of the rotor ring, depending on the operational mode of the enzyme, generating torque in either direction. This model of torque generation by microscopic rotation of the c-subunits of rotor is widely accepted, but the thermodynamic forces leading to torque generation at the a-c interface have not yet been identified unambiguously. The answer to this problem probably will require a high-resolution structure of the holo-ATPase, giving a detailed view of the a-c interface in different catalytic states of the enzyme. The recent cryo-electron microscopic structure of the holo-ATP synthase dimer revealed an unexpected, almost horizontal arrangement of four transmembrane helices of subunit a adjacent to the c-subunits rotor ring [101]. The 3D structure unequivocally confirmed that the critical arginine residue of subunit a is in close proximity to the proton (or Na^+-) binding carboxylic residue (glutamate in Na^+-dependent, and aspartate in proton-dependent ATP synthases), in accord with the mechanism outlined above.

3.2 *The Flagellar Motor*

3.2.1 Composition of the Flagellar Stator Complex

Motility in bacteria is achieved via extracellular helical filaments, the flagella, which rotate clockwise or counterclockwise. The bacterial flagellum consists of three distinct parts: the flagellum, a hollow filament with a length of 15–20 μm composed of flagellin proteins; the hook, which connects the filament to the motor complex; and the membrane embedded motor complex consisting of a rotor and a stator (Figure 1) [120]. The flagellum is driven either by a proton or a sodium gradient, but the mechanism converting ion flux into torque still remains unclear. The motor is surrounded by 11–12 stator units which dynamically associate with, or dissociate from, the motor without disrupting the flagellar rotation [121]. The highest rotational speeds (1700 Hz) were observed with the Na^+-dependent, polar flagellum from *Vibrio alginolyticus* [122]. The stator complex defines the specificity to either proton or sodium ions, with four MotA and two MotB subunits forming a H^+-dependent $MotA_4MotB_2$ stator, and the homologous PomA and PomB subunits forming the Na^+-dependent $PomA_4PomB_2$ stator (Figure 9). The flux of the coupling cation along the electrochemical gradient through the stator complexes is proposed to alter the conformation of the FliG component of the C ring which is part of the rotor (Figure 9).

This creates torque and forces the rotor to move with respect to the $PomA_4B_2$ (or $MotA_4B_2$) stator. Despite considerable effort, critical amino acid residue(s) conveying H^+ or Na^+ specificity in the stator could not be identified [123]. Some flagellar motors might even use both Na^+ and K^+ as coupling ions [124]. PomA possesses four transmembrane helices and a large cytoplasmic loop between helix II and III. PomB consists of one transmembrane helix and a large periplasmic domain which harbors a peptidoglycan-binding motif at the C-terminus [125]. This C-terminal part of subunit B which is essential for flagellar function extends into the periplasmic space to attach PomB to the peptidoglycan layer (Figure 9). It was proposed that helices III and IV of the A subunit and the single transmembrane helix of subunit B are in close proximity to form an ion channel [126].

3.2.2 Interaction of the Flagellar Stator with Na^+

Conserved, charged amino acid residues within the transmembrane regions of the stator were studied with respect to their putative role as a ligand for Na^+ in the access channel between PomA and PomB. It was shown that Asp24 in PomB of the flagellum of *Vibrio alginolyticus* [122] or the homologous Asp23 in the highly

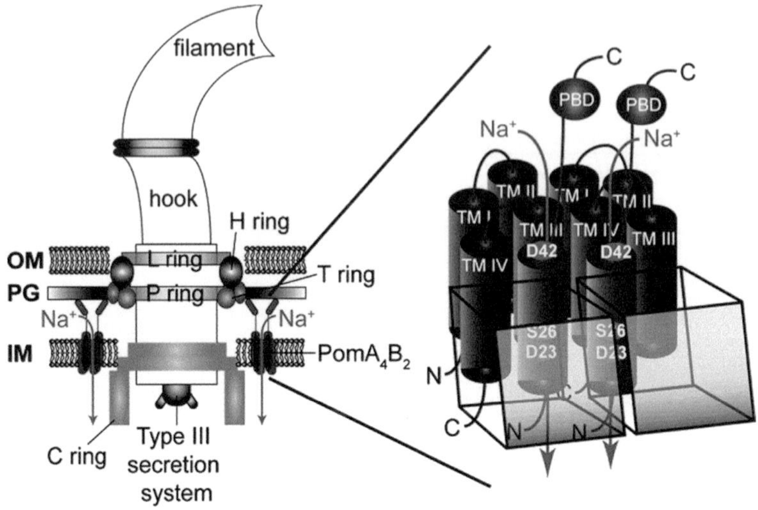

Figure 9 Rotation by the single polar flagellum from *Vibrio cholerae* is driven by Na^+ flux through the $PomA_4PomB_2$ stator complex. **Left**: The bacterial flagellum consists of rotating and static parts. The flagellum, the hook, which connects the flagellum to the motor complex, the membrane embedded L and P rings in the outer membrane (OM), and the C ring in the inner membrane (IM) rotate against several stator complexes, each composed of four PomA and two PomB proteins. The H ring functions as a brushing to enable rotation against the peptidoglycan layer (PG). The T-ring connects the stator complexes with the peptidoglycan layer. Export of protein components of the flagellum is catalyzed by a type III secretion system located at the base of the flagellum. **Right**: Four PomA and two PomB subunits form the flagellar stator complex. The flux of Na^+ through the stator complex (arrows) drives the rotation of the C-ring against the stator elements. It is assumed that Na^+ ions pass through a channel built by helices III and IV of PomA and the single transmembrane helix of PomB. The peptidoglycan binding domain (PGB) at the C-terminus of PomB stabilizes the stator complexes within the peptidoglycan layer. The conserved amino acid residues Asp42 (D42), Ser26 (S26), and Asp23 (D23) of PomB promote the transport of Na^+ through the stator. Reproduced with permission from [128].

conserved PomB from *V. cholerae* [127] is required for flagellar function. This conserved aspartate was proposed to interact with Na^+ [122]. A helical wheel analysis suggests that Ser26 is located one helix turn above Asp23 (viewed from the periplasmic side of the membrane, Figure 9), so the carboxylate of Asp23, and the OH-group of Ser26, could both line a putative channel within the PomA/PomB interface. In the Na^+-driven flagellar motor from *V. cholerae*, Ser26 was shown to be required for the hypermotile phenotype of swimming cells [128]. Asp42 is not expected to be buried in the membrane, but this residue is of interest since it is conserved in PomB from Na^+-dependent stator complexes [122] (Figures 9 and 10).

The high-resolution structure of the c- ring of the ATP synthase (Figure 8) revealed that the coordination of a sodium ion in the hydrophobic region of a membrane protein requires a carboxylate as a ligand for Na^+, so the question arose if the $PomA_4PomB_2$ stator represents a Na^+ transporter with a distinct binding site for Na^+, exemplified by the F_O part of the Na^+ -translocating F_1F_O ATP synthase. Another, more channel-like interaction of the stator with Na^+ should also be considered; examples are the K^+ channel [129] or the Na^+ channel [130]. A comparison of the positions of conserved, charged amino acid residues in the c-ring of the F_O part of the ATP synthase, and in the PomA and PomB transmembrane segments, is shown in Figure 10.

The glycine-rich stretch of the inner helix of subunit c (Figure 8) is a prerequisite for the tight packing of the c-ring, since the glycine residues result in a smaller diameter of the inner ring of helices. The membrane-spanning part of helix III of PomA also contains three glycine residues which might allow for tight packing of helices in the inner part of the $PomA_4PomB_2$ stators.

The mechanisms of these two Na^+-driven rotational machines, the flagellum and the F_1F_O ATP synthase, are clearly different. H^+- and Na^+-dependent systems are found with both types of machines, yet the Na^+-driven ATPase may operate with H^+ in the absence of Na^+ and at low pH [97], whereas the Na^+-dependent, polar flagellum obligatorily requires Na^+ for its function [122]. It seems likely that this functional difference is the result of a different interaction pattern of Na^+ with the c ring or the flagellar stator complex. In the Na^+-F_1F_O ATP synthase from *I. tartaricus*, Na^+ engages in direct coordination with a glutamate carboxyl group (Glu65) in a cation binding site constructed by several amino acid residues from one inner and two outer helices of two adjacent c subunits (Figure 8). A carboxylic acid like Glu65 reacts with *N,N′*-dicyclohexylcarbodiimide (DCCD) to form an intermediate

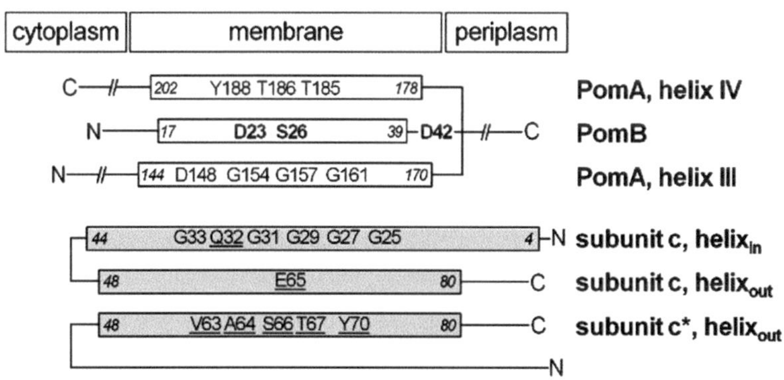

Figure 10 The transmembrane segments of the $PomA_4PomB_2$ stator: a comparison with subunits c of the Na^+-dependent F_1F_O ATPase. Transmembrane helices III and IV of PomA, together with the single transmembrane helix of PomB, likely provide interaction sites for Na^+ in the $PomA_4PomB_2$ stator complex of the flagellar motor. The carboxylate of Asp23 (D23) (PomB) is strictly required for flagellar function. Residues Ser26 and Asp42 affect flagellar performance in a pH- and salt-dependent manner [127]. In the membrane-embedded c ring of the Na^+-translocating F_1F_O ATPase from *Ilyobacter tartaricus*, seven residues (E65, Q32, Y70, T67, S66, V63, and A64, which are underlined) of three transmembrane helices from two neighboring c subunits (c and c*) are required to form a Na^+ binding site. Due to conserved glycine residues in subunit c, the N-terminal ring of helices ($helix_{in}$) exhibits a smaller diameter than the C-terminal ring of helices ($helix_{out}$) (see also Figure 8). Na^+ engages in direct coordination with the carboxylate of Glu65 of the C-terminal ring of helices. (Figure modified from [127], with permission of the publisher; © copyright 2013).

(*O*-acylisocarbamide) which is unstable in the presence of water. In hydrophobic environments, such as membrane-spanning regions of proteins, a rearrangement occurs, and the stable *N*-acylcarbamide is formed. If DCCD is labeled with ^{14}C, this results in a radioactive modification which can be detected after separation of the membrane proteins on SDS-PAGE.

In case of the Na^+-translocating F_1F_O ATPase, Na^+ prevented modification of the carboxylate of Glu65 by DCCD, demonstrating direct coordination of Na^+ to this critical ligand within the Na^+ binding site [97]. In contrast, we found no evidence for a protection from DCCD modification by Na^+ with PomB from *V. cholerae* [127]. Another argument against a direct competition between Na^+ and H^+ for binding to the critical Asp23 within the transmembrane part of PomB (Figure 11) comes from the pH profile of flagellar rotation. In case of the Na^+-dependent ATP synthase, shifting the pH towards the more alkaline range results in the activation of the enzyme (at a fixed Na^+ concentration) [97].

If H^+ and Na^+ compete for binding to the carboxylate of Asp23, lowering the H^+ concentration would promote flagellar rotation. Yet, the contrary was observed: with the Na^+-dependent flagellum from *V. cholerae*, rotation was enhanced by a raise in external proton concentration. Ser26 within the single transmembrane helix of PomB was critical for fast rotation of the flagellum under a broad range of physiological pH and salt concentrations [127]. Ser26 resides on the same side as the critical Asp23 and faces the inside of a transmembrane channel built by $PomA_4PomB_2$. Removal of the OH group (S26A), or even maintaining the OH group, but at a slightly different position (S26T), strongly impaired fast swimming of cells, but in a pH- and salt-dependent manner [127]. This strong effect is difficult to reconcile with different protonation states of the OH groups in serine or threonine residues at a given pH. Instead, the observed effects of salt and H^+ concentrations on stator function could be due to a rearrangement of water molecules or hydrogen bonds within the water-filled access channel of the stator complex (Figure 11). Asp23 is positioned below Ser26 when viewed from the periplasmic side of the inner membrane (Figure 9). It is proposed that Ser26, together with Asp23, are important for fast transport through the stator channel by perturbing the hydration shell of Na^+, preparing it for passage through a selectivity filter located below Asp23. This constriction in the passageway for the Na^+ will not depend on the presence of charged amino acid side chains as long as the critical diameter is maintained which is selective for dehydrated Na^+ (Figure 11). This model would be consistent with the finding that studies with chimeric stators made up of domains from subunits derived from proton- or Na^+-dependent stators so far failed to identify critical amino acids conferring cation selectivity [123].

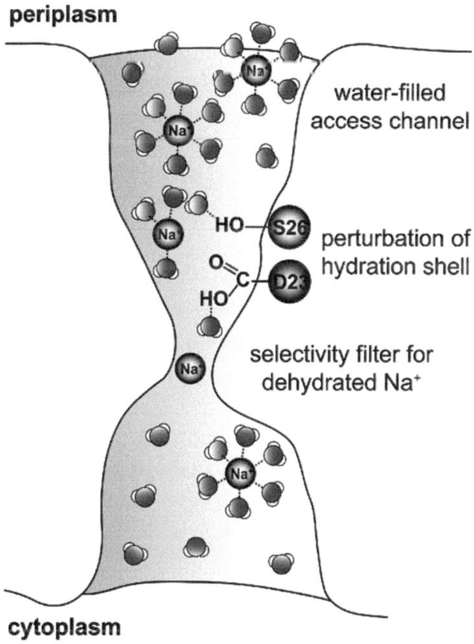

Figure 11 Na^+ flux through the flagellar $PomA_4PomB_2$ complex by a channel-like mechanism. Na^+ enters the $PomA_4PomB_2$ stator complex in its hydrated state from the periplasm via a water-filled access channel. The conserved amino acid residues Ser26 and Asp23 within the membrane-bound region of PomB interact with water molecules surrounding the central Na^+, hereby perturbing its hydration shell. The sodium ion passes the narrow constriction site, or selectivity filter, in its dehydrated state. After passage through the filter, Na^+ is hydrated and released towards the cytoplasmic side of the stator. The downhill flux of Na^+ from the periplasm to the cytoplasm is driven by the electrochemical Na^+ gradient across the inner membrane.

4 The Na^+-Translocating NADH:Quinone Oxidoreductase: An Outlook

4.1 Conformational Coupling of Electron Transfer and Na^+ Transport

Though the structure of the Na^+-NQR supports the idea that the membrane-bound NqrB subunit comprises a Na^+ channel, this awaits independent proof. Techniques to verify or falsify this hypothesis include determination of the structure of the Na^+-NQR with bound Rb^+. Unlike Na^+ which is difficult to discriminate from H_2O or OH^- in the electron density map, the larger Rb^+ can be detected.

We need to unravel the ligands, the electronic structure, and the redox properties of the novel Fe center located between subunits NqrD and NqrE in the membrane-bound part of the Na^+-NQR. The knowledge thus gained will be important to evaluate the contribution of this unprecedented, novel Fe site to intramolecular electron transfer in the Na^+-NQR.

Ultimately, we want to understand the mechanism of coupling of electron transfer to Na^+ transport. The 3D structure of the Na^+-NQR suggests that electron transfer from the [2Fe-2S] center on cytoplasmic NqrF to the Fe center within the membrane, and then to periplasmic FMN on NqrC, cannot proceed unless there is a large conformational rearrangement, bringing NqrF closer to the cytoplasmic side of the membrane. This facilitates electron transfer from cytoplasmic NqrF to the Fe site formed by the membrane-bound NqrD and NqrE subunits, and from there to the FMN on NqrC.

As a working model, we assume that this conformational change is triggered by the reduction of the complex by NADH, in analogy to the movement of the Rieske domain of the bc_1 complex upon its reduction by quinol. The presumed, large conformational rearrangement of the Na^+-NQR is thought to be a prerequisite for subsequent uphill transport of Na^+ across the membrane. Single molecule analysis of specifically labeled variants of the Na^+-NQR will be required to verify or falsify the hypothesis of large domain movements of the flexible part of NqrF towards the membrane-bound NqrD and NqrE subunits. In this respect, the Na^+-NQR offers the unique opportunity to unravel general principles of the mechanism of redox-driven cation transport.

4.2 *Na^+ Pump-Dependent Production of Virulence Factors*

Several lines of evidence indicate that the sodium cycle plays an important role in the pathogenicity of *V. cholerae.* The dissipation of the sodium motive force by ionophores, by mutations in the *nqr* operon, or by NQR inhibitors stimulate the expression of the cholera toxin and the toxin coregulated pili (reviewed in [22]). The Na^+-NQR is operative in many other pathogenic bacteria, suggesting that this primary pump is beneficial for survival within the host. It is not yet understood how the presence of the Na^+-NQR influences the regulation of the production of virulence factors. *V. cholerae* is an excellent model organism to describe the cycling of sodium ions in a pathogen, from the cytoplasm to the periplasm and back, in a qualitative and quantitative manner. This is supported by the genome information of many *V. cholerae* strains.

Which Na^+ efflux systems are active under a given environmental condition?
How fast are these systems?
Which uptake systems are driven by the electrochemical Na^+ gradient?
How does the steady-state concentration of Na^+ in the cytoplasm vary in response to changes in external pH and osmolarity?
Does the Na^+-NQR affect this steady-state concentration of Na^+?

The answers to these questions are fundamental when we want to decide if the expression of a given virulence factor, for example the Cholera toxin, is directly under control of the Na^+-NQR, or is only indirectly affected by changes in cytoplasmic Na^+ concentration. Proteome analyses will be crucial to identify important key transporters produced under a given environmental condition. They should be combined with an analysis of critical metabolites likely to be affected by the action of the Na^+-NQR, like NAD/NADH, or acetyl intermediates, which also may modulate virulence gene expression.

Abbreviations and Definitions

ADP	adenosine 5′-diphosphate
Ala, A	alanine
Asp, D	aspartic acid
ATP	adenosine 5′-trihosphate
ComplexI	electrogenic NADH:ubiquinone oxidoreductase from mitochondria
Cys, C	cysteine
DCCD	*N,N′*-dicyclohexylcarbodiimide
EPR	electron paramagnetic resonance
F	Faraday constant
F_1F_O synthase	H^+ (or Na^+) -transporting two-sector ATPase (E.C. 3.6.3.14)
FAD	flavin adenine dinucleotide
FMN	flavin mononucleotide
FTIR	Fourier transform infrared
Gln, Q	glutamine
Glu, E	glutamic acid
Gly, G	glycine
Ile, I	isoleucine
Leu, L	leucine
LUCA	Last Universal Common Ancestor
MDR	multidrug resistance
Met, M	methionine
MRP	multidrug resistance protein
Na^+-NQR	Na^+-translocating NADH:quinone oxidoreductase
NAD	nicotinamide adenine dinucleotide
NADH	nicotinamide adenine dinucleotide (reduced)
NDH I	NADH dehydrogenase I (bacterial homolog of Complex I)
NDH II	NADH dehydrogenase II (non-electrogenic)
NFO	NAD:ferredoxin oxidoreductase
NMR	nuclear magnetic resonance
OXPHOS	oxidative phosphorylation

PG	peptidoglycan
PGB	peptidoglycan binding domain
Phe, F	phenylalanine
P_i	inorganic phosphate
PMF	proton motive force
PMF, Δp	proton motive force
Q	ubiquinone
Q8	ubiquinone with eight isoprenoid units
QH_2	ubiquinol
R	universal gas constant, 8.3144621 J K^{-1} mol^{-1}
RBF	riboflavin
RNF complex	*Rhodobacter* nitrogen fixation complex
SDS-PAGE	sodium dodecylsulfate polyacrylamine gel electrophoresis
Ser, S	serine
SMF	sodium motive force
T	temperature [K]
Thr, T	threonine
Tyr, Y	tyrosine
Val, V	valine
$\Delta\psi$	transmembrane voltage

Acknowledgments We thank Thomas Meier, Max-Planck Institute of Biophysics in Frankfurt, for critically reading the manuscript. This work was supported by contract research 'Methoden in den Lebenswissenschaften' of the Baden-Württemberg Stiftung P-LS-Meth/4 (to J.S. and G.F.), and by the Deutsche Forschungsgemeinschaft grant FR 1321/3-1 (to J.S.) and grant FR 1488/3-2 (to G.F.).

References

1. D. Voet, J. G. Voet, *Biochemistry,* 3rd edn., John Wiley & Sons, Inc., New York, 2004.
2. D. G. Nicholls, S. J. Ferguson, Academic Press Limited, London, 2013
3. Y. Chaban, E. J. Boekema, N. V. Dudkina, *Biochim. Biophys. Acta* **2014**, *1837*, 418–426.
4. P. M. Sousa, M. A. Videira, A. M. Melo, *FEBS Lett.* **2013**, *587*, 2559–2564.
5. J. N. Blaza, R. Serreli, A. J. Jones, K. Mohammed, J. Hirst, *Proc. Natl. Acad. Sci. USA* **2014**, *111*, 15735–15740.
6. E. Lapuente-Brun, R. Moreno-Loshuertos, R. Acin-Perez, A. Latorre-Pellicer, C. Colas, E. Balsa, E. Perales-Clemente, P. M. Quiros, E. Calvo, M. A. Rodriguez-Hernandez, P. Navas, R. Cruz, A. Carracedo, C. Lopez-Otin, A. Perez-Martos, P. Fernandez-Silva, E. Fernandez-Vizarra, J. A. Enriquez, *Science* **2013**, *340*, 1567–1570.
7. P. Mitchell, *Nature* **1961**, *191*, 144–148.
8. P. Mitchell, J. Moyle, *Eur. J. Biochem.* **1969**, *7*, 471–484.
9. R. K. Thauer, K. Jungermann, K. Decker, *Bacteriol. Rev.* **1977**, *41*, 100–180.
10. N. Lane, J. F. Allen, W. Martin, *Bioessays* **2010**, *32*, 271–280.
11. A. Y. Mulkidjanian, P. Dibrov, M. Y. Galperin, *Biochim. Biophys. Acta* **2008**, *1777*, 985–992.

12. T. Unemoto, M. Hayashi, *J. Biochem.* **1977**, *82*, 1389–1395.
13. T. Unemoto, M. Hayashi, *J. Biochem.* **1979**, *85*, 1461–1467.
14. P. Dimroth, *FEBS Lett.* **1980**, *122*, 234–236.
15. H. Tokuda, T. Unemoto, *Biochem. Biophys. Res. Commun.* **1981**, *102*, 265–271.
16. H. Tokuda, T. Unemoto, *J. Biol. Chem.* **1982**, *257*, 10007–10014.
17. X. D. Pfenninger-Li, S. P. Albracht, R. van Belzen, P. Dimroth, *Biochemistry* **1996**, *35*, 6233–6242.
18. W. Krebs, J. Steuber, A. C. Gemperli, P. Dimroth, *Mol. Microbiol.* **1999**, *33*, 590–598.
19. W. Laubinger, P. Dimroth, *Biochemistry* **1988**, *27*, 7531–7537.
20. W. Hilpert, B. Schink, P. Dimroth, *EMBO J.* **1984**, *3*, 1665–1670.
21. C. C. Häse, N. D. Fedorova, M. Y. Galperin, P. A. Dibrov, *Microbiol. Mol. Biol. Rev.* **2001**, *65*, 353–370.
22. J. Steuber, P. Halang, T. Vorburger, W. Steffen, G. Vohl, G. Fritz, *Biol. Chem.* **2014**, *395*, 1389–1399.
23. T. A. Krulwich, G. Sachs, E. Padan, *Nat. Rev. Microbiol.* **2011**, *9*, 330–343.
24. G. Kaim, P. Dimroth, *EMBO J.* **1999**, *18*, 4118–4127.
25. T. Lienard, B. Becher, M. Marschall, S. Bowien, G. Gottschalk, *Eur. J. Biochem.* **1996**, *239*, 857–864.
26. G. Grüber, M. S. Manimekalai, F. Mayer, V. Müller, *Biochim. Biophys. Acta* **2014**, *1837*, 940–952.
27. R. K. Thauer, A. K. Kaster, H. Seedorf, W. Buckel, R. Hedderich, *Nat. Rev. Microbiol.* **2008**, *6*, 579–591.
28. A. M. Delort, G. Gaudet, E. Forano, *Anal. Biochem.* **2002**, *306*, 171–180.
29. A. P. Batista, B. C. Marreiros, R. O. Louro, M. M. Pereira, *Biochim. Biophys. Acta* **2012**, *1817*, 1810–1816.
30. A. C. Gemperli, P. Dimroth, J. Steuber, *J. Biol. Chem.* **2002**, *277*, 33811–33817.
31. H. Hauser, M. C. Phillips, M. Stubbs, *Nature* **1972**, *239*, 342–344.
32. J. Gutknecht, *Proc. Natl. Acad. Sci. USA* **1987**, *84*, 6443–6446.
33. C. von Ballmoos, P. Dimroth, *Anal. Biochem.* **2004**, *335*, 334–337.
34. V. Muras, B. Claussen, M. Karuppasamy, C. Schaffitzel, J. Steuber, *Anal. Biochem.* **2014**, *459*, 53–55.
35. E. Padan, D. Zilberstein, H. Rottenberg, *Eur. J. Biochem.* **1976**, *63*, 533–541.
36. O. Juárez, J. E. Morgan, M. J. Nilges, B. Barquera, *Proc. Natl. Acad. Sci. USA* **2010**, *107*, 12505–12510.
37. K. R. Vinothkumar, J. Zhu, J. Hirst, *Nature* **2014**, *515*, 80–84.
38. V. Zickermann, C. Wirth, H. Nasiri, K. Siegmund, H. Schwalbe, C. Hunte, U. Brandt, *Science* **2015**, *347*, 44–49.
39. A. C. Gemperli, P. Dimroth, J. Steuber, *Proc. Natl. Acad. Sci. USA* **2003**, *100*, 839–844.
40. M. Sato, J. Torres-Bacete, P. K. Sinha, A. Matsuno-Yagi, T. Yagi, *J. Bioenerg. Biomembr.* **2014**, *46*, 279–287.
41. A. M. Melo, T. M. Bandeiras, M. Teixeira, *Microbiol. Mol. Biol. Rev.* **2004**, *68*, 603–616.
42. M. Hayashi, T. Unemoto, *Biochim. Biophys. Acta* **1987**, *890*, 47–54.
43. P. Beattie, K. Tan, R. M. Bourne, D. Leach, P. R. Rich, F. B. Ward, *FEBS Lett.* **1994**, *356*, 333–338.
44. M. Hayashi, K. Hirai, T. Unemoto, *FEBS Lett.* **1995**, *363*, 75–77.
45. X. D. Pfenninger-Li, P. Dimroth, *FEBS Lett.* **1995**, *369*, 173–176.
46. Y. Nakayama, M. Hayashi, T. Unemoto, *FEBS Lett.* **1998**, *422*, 240–242.
47. M. S. Casutt, T. Huber, R. Brunisholz, M. Tao, G. Fritz, J. Steuber, *J. Biol. Chem.* **2010**, *285*, 27088–27099.
48. Y. Nakayama, M. Yasui, K. Sugahara, M. Hayashi, T. Unemoto, *FEBS Lett.* **2000**, *474*, 165–168.
49. W. Zhou, Y. V. Bertsova, B. Feng, P. Tsatsos, M. L. Verkhovskaya, R. B. Gennis, A. V. Bogachev, B. Barquera, *Biochemistry* **1999**, *38*, 16246–16252.

50. M. S. Casutt, A. Schlosser, W. Buckel, J. Steuber, *Biochim. Biophys. Acta* **2012**, *1817*, 1817–1822.
51. Y. V. Bertsova, M. S. Fadeeva, V. A. Kostyrko, M. V. Serebryakova, A. A. Baykov, A. V. Bogachev, *J. Biol. Chem.* **2013**, *288*, 14276–14286.
52. B. Barquera, W. Zhou, J. E. Morgan, R. B. Gennis, *Proc. Natl. Acad. Sci. USA* **2002**, *99*, 10322–10324.
53. O. Juárez, M. J. Nilges, P. Gillespie, J. Cotton, B. Barquera, *J. Biol. Chem.* **2008**, *283*, 33162–33167.
54. A. V. Bogachev, D. A. Bloch, Y. V. Bertsova, M. I. Verkhovsky, *Biochemistry* **2009**, *48*, 6299–6304.
55. J. Steuber, G. Vohl, M. S. Casutt, T. Vorburger, K. Diederichs, G. Fritz, *Nature* **2014**, *516*, 62–67.
56. K. Türk, A. Puhar, F. Neese, E. Bill, G. Fritz, J. Steuber, *J. Biol. Chem.* **2004**, *279*, 21349–21355.
57. M. Tao, M. S. Casutt, G. Fritz, J. Steuber, *Biochim. Biophys. Acta* **2008**, *1777*, 696–702.
58. M. S. Casutt, S. Wendelspiess, J. Steuber, G. Fritz, *Acta Crystallogr. Sect. F Struct. Biol. Cryst. Commun.* **2010**, *66*, 1677–1679.
59. B. Barquera, P. Hellwig, W. Zhou, J. E. Morgan, C. C. Häse, K. K. Gosink, M. Nilges, P. J. Bruesehoff, A. Roth, C. R. Lancaster, R. B. Gennis, *Biochemistry* **2002**, *41*, 3781–3789.
60. M. S. Casutt, R. Nedielkov, S. Wendelspiess, S. Vossler, U. Gerken, M. Murai, H. Miyoshi, H. M. Möller, J. Steuber, *J. Biol. Chem.* **2011**, *286*, 40075–40082.
61. R. Nedielkov, W. Steffen, J. Steuber, H. M. Möller, *J. Biol. Chem.* **2013**, *288*, 30597–30606.
62. M. Hayashi, N. Shibata, Y. Nakayama, K. Yoshikawa, T. Unemoto, *Arch. Biochem. Biophys.* **2002**, *401*, 173–177.
63. O. Juárez, Y. Neehaul, E. Turk, N. Chahboun, J. M. DeMicco, P. Hellwig, B. Barquera, *J. Biol. Chem.* **2012**, *287*, 25678–25685.
64. M. Strickland, O. Juárez, Y. Neehaul, D. A. Cook, B. Barquera, P. Hellwig, *J. Biol. Chem.* **2014**, *289*, 23723–23733.
65. C. C. Page, C. C. Moser, X. Chen, P. L. Dutton, *Nature* **1999**, *402*, 47–52.
66. M. I. Verkhovsky, A. V. Bogachev, A. V. Pivtsov, Y. V. Bertsova, M. V. Fedin, D. A. Bloch, L. V. Kulik, *Biochemistry* **2012**, *51*, 5414–5421.
67. B. Barquera, *J. Bioenerg. Biomembr.* **2014**, *46*, 289–298.
68. M. Petrek, M. Otyepka, P. Banas, P. Kosinova, J. Koca, J. Damborsky, *BMC Bioinformatics* **2006**, *7*, 316.
69. A. Reyes-Prieto, B. Barquera, O. Juárez, *PLoS One* **2014**, *9*, e96696.
70. O. Juárez, M. E. Shea, G. I. Makhatadze, B. Barquera, *J. Biol. Chem.* **2011**, *286*, 26383–26390.
71. A. V. Bogachev, Y. V. Bertsova, O. Aitio, P. Permi, M. I. Verkhovsky, *Biochemistry* **2007**, *46*, 10186–10191.
72. Y. Neehaul, O. Juárez, B. Barquera, P. Hellwig, *Biochemistry* **2013**, *52*, 3085–3093.
73. O. Juárez, K. Athearn, P. Gillespie, B. Barquera, *Biochemistry* **2009**, *48*, 9516–9524.
74. M. E. Shea, O. Juárez, J. Cho, B. Barquera, *J. Biol. Chem.* **2013**, *288*, 31241–31249.
75. M. E. Shea, K. G. Mezic, O. Juárez, B. Barquera, *Biochemistry* **2015**, *54*, 490–496.
76. Y. Neehaul, O. Juárez, B. Barquera, P. Hellwig, *Biochemistry* **2012**, *51*, 4072–4077.
77. A. V. Bogachev, R. A. Murtazina, V. P. Skulachev, *FEBS Lett.* **1997**, *409*, 475–477.
78. M. I. Verkhovsky, A. V. Bogachev, *Biochim. Biophys. Acta* **2010**, *1797*, 738–746.
79. M. Schmehl, A. Jahn, A. Meyer zu Vilsendorf, S. Hennecke, B. Masepohl, M. Schuppler, M. Marxer, J. Oelze, W. Klipp, *Mol. Gen. Genet.* **1993**, *241*, 602–615.
80. E. Biegel, V. Müller, *Proc. Natl. Acad. Sci. USA* **2010**, *107*, 18138–18142.
81. V. Hess, K. Schuchmann, V. Müller, *J. Biol. Chem.* **2013**, *288*, 31496–31502.
82. C. D. Boiangiu, E. Jayamani, D. Brugel, G. Herrmann, J. Kim, L. Forzi, R. Hedderich, I. Vgenopoulou, A. J. Pierik, J. Steuber, W. Buckel, *J. Mol. Microbiol. Biotechnol.* **2005**, *10*, 105–119.

83. W. Buckel, R. K. Thauer, *Biochim. Biophys. Acta* **2013**, *1827*, 94–113.
84. J. Backiel, O. Juárez, D. V. Zagorevski, Z. Wang, M. J. Nilges, B. Barquera, *Biochemistry* **2008**, *47*, 11273–11284.
85. P.-L. Tremblay, T. Zhang, S. A. Dar, C. Leang, D. R. Lovley, *mBIO* **2013**, *4* e00406–00412.
86. J. Hirst, *Annu. Rev. Biochem.* **2013**, *82*, 551–575.
87. R. Baradaran, J. M. Berrisford, G. S. Minhas, L. A. Sazanov, *Nature* **2013**, *494*, 443–448.
88. L. Kozachkov, E. Padan, *Mol. Membr. Biol.* **2013**, *30*, 90–100.
89. A. Galkin, S. Dröse, U. Brandt, *Biochim. Biophys. Acta* **2006**, *1757*, 1575–1581.
90. S. Stolpe, T. Friedrich, *J. Biol. Chem.* **2004**, *279*, 18377–18383.
91. A. P. Batista, B. C. Marreiros, M. M. Pereira, *ACS Chem. Biol.* **2011**, *6*, 477–483.
92. P. C. Lin, A. Puhar, J. Steuber, *Arch. Microbiol.* **2008**, *190*, 471–480.
93. P. G. Roberts, J. Hirst, *J. Biol. Chem.* **2012**, *287*, 34743–34751.
94. W. Steffen, J. Steuber, *Biochem. Soc. Trans.* **2013**, *41*, 1280–1287.
95. A. P. Batista, B. C. Marreiros, M. M. Pereira, *IUBMB Life* **2012**, *64*, 492–498.
96. J. E. Walker, *Biochem. Soc. Trans.* **2013**, *41*, 1–16.
97. T. Meier, J. Faraldo-Gómez, M. Börsch, *Molecular Machines in Biology* **2011**, 208–238.
98. A. G. Stewart, E. M. Laming, M. Sobti, D. Stock, *Curr. Opin. Struct. Biol.* **2014**, *25*, 40–48.
99. T. Murata, I. Yamato, Y. Kakinuma, A. G. Leslie, J. E. Walker, *Science* **2005**, *308*, 654–659.
100. J. Symersky, D. Osowski, D. E. Walters, D. M. Mueller, *Proc. Natl. Acad. Sci. USA* **2012**, *109*, 13961–13965.
101. M. Allegretti, N. Klusch, D. J. Mills, J. Vonck, W. Kühlbrandt, K. M. Davies, *Nature* **2015**.
102. W. Laubinger, P. Dimroth, *Eur. J. Biochem.* **1987**, *168*, 475–480.
103. G. Kaim, P. Dimroth, *Eur. J. Biochem.* **1993**, *218*, 937–944.
104. P. Mitchell, *FEBS Lett.* **1985**, *182*, 1–7.
105. T. Meier, P. Polzer, K. Diederichs, W. Welte, P. Dimroth, *Science* **2005**, *308*, 659–662.
106. V. Müller, A. Lingl, K. Lewalter, M. Fritz, *J. Bioenerg. Biomembr.* **2005**, *37*, 455–460.
107. R. L. Cross, V. Müller, *FEBS Lett.* **2004**, *576*, 1–4.
108. T. Meier, D. Pogoryelov, *Encyclopedia of Biophysics* **2013**, 129–134.
109. J. Petersen, K. Forster, P. Turina, P. Gräber, *Proc. Natl. Acad. Sci. USA* **2012**, *109*, 11150–11155.
110. T. Meier, A. Krah, P. J. Bond, D. Pogoryelov, K. Diederichs, J. D. Faraldo-Gómez, *J. Mol. Biol.* **2009**, *391*, 498–507.
111. D. Pogoryelov, O. Yildiz, J. D. Faraldo-Gómez, T. Meier, *Nat. Struct. Mol. Biol.* **2009**, *16*, 1068–1073.
112. A. Krah, D. Pogoryelov, J. D. Langer, P. J. Bond, T. Meier, J. D. Faraldo-Gómez, *Biochim. Biophys. Acta* **2010**, *1797*, 763–772.
113. C. Kluge, P. Dimroth, *Biochemistry* **1993**, *32*, 10378–10386.
114. M. Nayal, E. Di Cera, *J. Mol. Biol.* **1996**, *256*, 228–234.
115. T. Meier, U. Matthey, C. von Ballmoos, J. Vonck, T. Krug von Nidda, W. Kühlbrandt, P. Dimroth, *J. Mol. Biol.* **2003**, *325*, 389–397.
116. T. Meier, P. Dimroth, *EMBO Rep.* **2002**, *3*, 1094–1098.
117. D. Pogoryelov, A. Krah, J. D. Langer, O. Yildiz, J. D. Faraldo-Gómez, T. Meier, *Nat. Chem. Biol.* **2010**, *6*, 891–899.
118. F. Wehrle, G. Kaim, P. Dimroth, *J. Mol. Biol.* **2002**, *322*, 369–381.
119. W. Junge, H. Sielaff, S. Engelbrecht, *Nature* **2009**, *459*, 364–370.
120. D. Stock, K. Namba, L. K. Lee, *Curr. Opin. Biotechnol.* **2012**, *23*, 545–554.
121. M. C. Leake, J. H. Chandler, G. H. Wadhams, F. Bai, R. M. Berry, J. P. Armitage, *Nature* **2006**, *443*, 355–358.
122. S. Zhu, S. Kojima, M. Homma, *Front. Microbiol.* **2013**, *4*, 410.
123. Y. Sowa, R. M. Berry, *Q. Rev. Biophys.* **2008**, *41*, 103–132.
124. N. Terahara, M. Sano, M. Ito, *PLoS One* **2012**, *7*, e46248.
125. S. Zhu, M. Takao, N. Li, M. Sakuma, Y. Nishino, M. Homma, S. Kojima, K. Imada, *Proc. Natl. Acad. Sci. USA* **2014**, *111*, 13523–13528.

126. E. A. Kim, M. Price-Carter, W. C. Carlquist, D. F. Blair, *Biochemistry* **2008**, *47*, 11332–11339.
127. P. Halang, S. Leptihn, T. Meier, T. Vorburger, J. Steuber, *J. Bacteriol.* **2013**, *195*, 4888–4899.
128. P. Halang, T. Vorburger, J. Steuber, *Plos One* **2015**, 10(4): e0123518. doi:10.1371/journal.pone.0123518.
129. E. Gouaux, R. Mackinnon, *Science* **2005**, *310*, 1461–1465.
130. J. Payandeh, T. Scheuer, N. Zheng, W. A. Catterall, *Nature* **2011**, *475*, 353–358.
131. P. A. Karplus, H. R. Faber, *Photosynth. Res.* **2004**, *81*, 303–315.
132. J. Song, C. Ji, J. Z. Zhang, *Proteins* **2014**, *82*, 240–249.
133. J. Dzioba-Winogrodzki, O. Winogrodzki, T. A. Krulwich, M. A. Boin, C. C. Häse, P. Dibrov, *J. Mol. Microbiol. Biotechnol.* **2009**, *16*, 176–186.
134. J. Reizer, A. Reizer, M. H. Saier, Jr., *Biochim. Biophys. Acta* **1994**, *1197*, 133–166.
135. C. Gibbons, M. G. Montgomery, A. G. Leslie, J. E. Walker, *Nat Struct Biol* **2000**, *7*, 1055–1061.
136. S. Wilkens, D. Borchardt, J. Weber, A. E. Senior, *Biochemistry* **2005**, *44*, 11786–11794.
137. O. Dmitriev, P. C. Jones, W. Jiang, R. H. Fillingame, *J. Biol. Chem.* **1999**, *274*, 15598–15604.
138. L. K. Lee, A. G. Stewart, M. Donohoe, R. A. Bernal, D. Stock, *Nat. Struct. Mol. Biol.* **2010**, *17*, 373–378.
139. P. D. Boyer, *Biochim. Biophys. Acta* **1993**, *1140*, 215–250.

Chapter 12
Sodium-Proton (Na^+/H^+) Antiporters: Properties and Roles in Health and Disease

Etana Padan and Meytal Landau

Contents

E. Padan (✉)
Alexander Silberman Institute of Life Sciences, Hebrew University of Jerusalem, 91904 Jerusalem, Israel
e-mail: etana@vms.huji.ac.il

M. Landau
Department of Biology, Technion-Israel Institute of Technology, Haifa, Israel
e-mail: mlandau@technion.ac.il

A. Sigel, H. Sigel, and R.K.O. Sigel (eds.), *The Alkali Metal Ions: Their Role for Life*, Metal Ions in Life Sciences 16, DOI 10.1007/978-3-319-21756-7_12

Abstract The transmembranal Na^+/H^+ antiporters transport sodium (or several other monovalent cations) in exchange for H^+ across lipid bilayers in all kingdoms of life. They are critical in pH homeostasis of the cytoplasm and/or organelles. A particularly notable example is the SLC9 gene family, which encodes Na^+/H^+ exchangers (NHEs) in many species from prokaryotes to eukaryotes. In humans, these proteins are associated with the pathophysiology of various diseases. Yet, the most extensively studied Na^+/H^+ antiporter is Ec-NhaA, the main Na^+/H^+ antiporter of *Escherichia coli*.

The crystal structure of down-regulated Ec-NhaA, determined at acidic pH, has provided the first structural insights into the antiport mechanism and pH regulation of an Na^+/H^+ antiporter. It reveals a unique structural fold (called the NhaA fold) in which transmembrane segments (TMs) are organized in inverted-topology repeats, including two antiparallel unfolded regions that cross each other, forming a delicate electrostatic balance in the middle of the membrane. This unique structural fold (The NhaA fold) contributes to the cation binding site and facilitates the rapid conformational changes expected for Ec-NhaA. The NhaA fold has now been recognized to be shared by four Na^+/H^+ antiporters (bacterial and archaeal) and a Na^+ symporter. Remarkably, no crystal structure of any of the human Na^+/H^+ antiporters exists. Nevertheless, the Ec-NhaA crystal structure has enabled the structural modeling of NHE1, NHE9, and NHA2, three human plasmalemmal proteins that are members of the SLC9 family that are involved in human pathophysiology. Moreover, as outlined in this review, developments in the field, including cellular and biophysical methods that enable ion levels and fluxes to be measured in intact cells as well as in knockout mice, have led to striking advances in the identification and characterization of plasma membrane NHEs and NHA.

Very little is known about the endomembrane isoforms of NHE. These intracellular exchangers may serve a function in cation homeostasis and/or osmoregulation, and not in pH regulation as is the case for the plasmalemmal isoforms. This intriguing possibility should be borne in mind when designing future studies

Future progress towards gaining an understanding of the SLC9 gene family, including its structure–function relationships and regulatory mechanisms in health and in disease, is likely to include insights into the pathophysiology of multiple diseases.

Keywords Cation proton antiporter (CPA) superfamily • Membrane protein • Na^+/H^+ antiporter • NHE • NHA • NhaA • NhaA structural fold

Please cite as: *Met. Ions Life Sci.* 16 (2016) 391–458

1 Introduction

Na^+ and H^+ are among the most prevalent ions in living cells and are essential in cell bioenergetics. An appropriate concentration of these ions within the cell is crucial for protein function, whereas an overly high or low concentration of either ion is a powerful stressor to cell viability [1]. Thus, all living cells are critically dependent on homeostatic mechanisms that regulate intracellular pH, Na^+ content, and, as a result, cell volume [2, 3].

In 1974, Peter Mitchell and colleagues [4] discovered sodium proton antiport activity in bacterial cells and suggested that Na^+/H^+ antiporter proteins have primary roles in the homeostasis of these cations. Over the 40 years since, sodium proton antiporters have been identified in the cytoplasmic and organelle membranes of almost all cells, including those of plants, animals, and microorganisms [5, 6]. Furthermore, increasing numbers of these antiporters are being identified as human drug targets [7, 8].

The genome project has yielded a multiplicity of genes that encode putative Na^+/H^+ antiporters; these antiporters have been classified into families on the basis of evolutionary origin [9]. According to the transporter classification database (http://www.tcdb.org) [10], Na^+/H^+ antiporters are members of the monovalent cation proton antiporter (CPA) superfamily, which contains homologues ranging from bacteria to human, and these antiporters may share a similar structural fold [9, 11–13]. The CPA1 family includes the ubiquitous and pharmacologically important NHEs, NHA, and Nhx-2 from animals, NHXs from yeast and plants, Nha1 and Nha2 from yeasts, NhaP1 from archaea, NhaPs and YvgP from bacteria, and others. The generalized transport reaction catalyzed by functionally characterized members of the CPA1 family is [10, 14]:

$$Na^+(out) + H^+(in) \rightleftharpoons Na^+(in) + H^+(out)$$

The CPA2 family includes K^+/H^+ antiporters, but also specific bacterial $Na^+/Li^+/H^+$ antiporters. The generalized transport reaction catalyzed by members of the CPA2 family is:

$$M^+(in) + nH^+(out) \rightleftharpoons M^+(out) + nH^+(in)$$

The CPA2 family also contains some members showing a channel-mediated mode with the transport reaction:

$$M^+(in) \rightleftharpoons M^+(out)$$

NhaA from *Escherichia coli* and related bacteria belong to a separate, small family called the NhaA $Na^+:H^+$ antiporter family. The generalized transport reaction catalyzed by NhaA is [15]:

$$Na^+(in) + 2H^+(out) \rightleftharpoons Na^+(out) + 2H^+(in)$$

In this review we will focus on the physiological and biochemical properties of prokaryotic antiporters whose crystal structures have been determined. As the crystal structures of eukaryotic antiporters have not yet been determined, we will describe them on the basis of recent data.

2 Prokaryotic Na^+/H^+ Antiporters

2.1 The Proteins and Their Properties

With the exception of one species [16], Na^+/H^+ antiporters have been identified in all prokaryotes, and some species harbor multiple antiporters. For example, *E. coli* encodes more than five Na^+/H^+ antiporters, a fact that points to the high importance of these antiporters to cellular function. Here, we will describe in detail the physiological and biochemical properties of prokaryotic antiporters whose crystal structures have been determined (see Section 2.2 for details on the crystal structures).

2.1.1 Ec-NhaA, the Na^+/H^+ Antiporter of *Escherichia coli*, a Prototype for Na^+/H^+ Antiporters

Ec-NhaA is the principal Na^+/H^+ antiporter in *E. coli* and is indispensable for cell adaptation to high salinity, for challenging Li^+ toxicity, and for growth at alkaline pH (in the presence of Na^+ [17]). Ec-NhaA homologues are highly prevalent in enterobacteria [1] and have orthologues throughout the biological kingdoms, including humans [9]; Ec-NhaA displays low sequence homology with human NHE1 (12 % identity) [11] and with NHA2 (~15 % identity) [13]. Recently, a homologue of Ec-NhaA has been shown to be essential for *Yersinia pestis* virulence (associated with the bubonic plague) and is therefore a novel drug target [18]. Furthermore, a population genetics approach has revealed the *nhaA* operon as a landmark of the B2 pathogenic strain of *E. coli*, pointing to NhaA as a novel drug target [19] (see Section 6).

Several biochemical characteristics of Ec-NhaA underpin its physiological roles (Table 1):

(a) very high turnover [20]; (b) strong pH dependence [20], a property Ec-NhaA shares with other prokaryotic antiporters [17] as well as eukaryotic Na^+/H^+ antiporters [6, 21–23]. Ec-NhaA is active only above pH 6.5 and reaches maximal activity at pH 8.5 [17]. (c) Ec-NhaA is an electrogenic transporter; its directly measured stoichiometry is $2H^+/Na^+$ [15]. The electrogenicity of Ec-NhaA can also be assessed indirectly: In the absence of permeant ions, the turnover of Ec-NhaA produces a net

Table 1 Functional characterization of transporters sharing the NhaA structural fold.

Protein	Specificity and Transport Mode[a]	Apparent K_m (mM)[b] Na$^+$	Li$^+$	Turnover[c]	Stoichiometry Electrogenicity[d]	pH Dependence[e]	Oligomer[f]
Ec-NhaA	Na^+, Li^+/H^+ antiport	0.2	0.02	10^5/min RT	$2H^+/1Na^+$	6.5–8.5	dimer
Tt-NapA	Na^+, Li^+/H^+ antiport	4.0	0.41	100/min	electrogenic	6–8	dimer
Nm-ASBT	Na^+/taurocholate symport				$2Na^+$/taurocholate		monomer ?
Yf-ASBT	Na^+/taurocholate symport				$2Na^+$/taurocholate		monomer ?
Mj-NhaP1	Na^+, Li^+/H^+ antiport	0.84		100/min RT	electroneutral	5–9	dimer
Pa-NhaP	Na^+, $Li^+/Tl^+/H^+$ antiport	0.5	0.024	~4/min RT	electroneutral	5–7	dimer

The parameters were collected from the following references referred to in brackets according to the six parameters selected:

Ec-NhaA: [a][20, 393, 394]; [b][20, 393, 394]; [c][20]; [d][15]; [e][20, 83]; [f][25, 27, 28].

Tt-NapA: [a][35]; [b][35]; [c][33]; [d][35]; [e][33]; [f][35].

Nm-ASBT: [a][36].

Yf-ASBT: [a][37].

Mj-NhaP1: [a][43, 49, 57, 395]; [b][43]; [b][43]; [c][43]; [d][43]; [e][43, 57].

Pa-NhaP: [a-f][38].

RT, room temperature.

one positive charge across the membrane, which makes the rate very sensitive to membrane potential.

When measured in the absence of a co-ion, the antiporter's rate of activity is very slow because of the difference in potential that it produces across the membrane [2]. In the presence of a co-ion (valinomycin is often used in the presence of K^+), collapse of the membrane potential results in at least a 5-fold increase in the rate of activity of Ec-NhaA. An electroneutral antiporter, in contrast, is indifferent to changes in membrane potential. Importantly, we have recently shown that Ec-NhaA mutants showing a change in a rate-limiting step of the turnover cycle can also yield an electroneutral phenotype in an electrogenic transporter [24]. This suggests that, when assessing the stoichiometry of Ec-NhaA variants, direct measurement is preferred.

Like many secondary transporters, Ec-NhaA is a dimer in the native membrane and in DDM (n-dodecyl-β-D-maltoside) micelles, as shown by genetic complementation [25], biochemical pull-down experiments [25], intermolecular cross-linking [25], electron spin resonance (ESR) studies [26, 27], cryo-electron microscopy of 2D crystals [28–30] and the crystal structures of the Ec-NhaA monomer [31] (Figure 1a; PDB 1ZCD) and dimer [32] (Figure 1b; PDB 4AU5) (see Section 2.2.1).

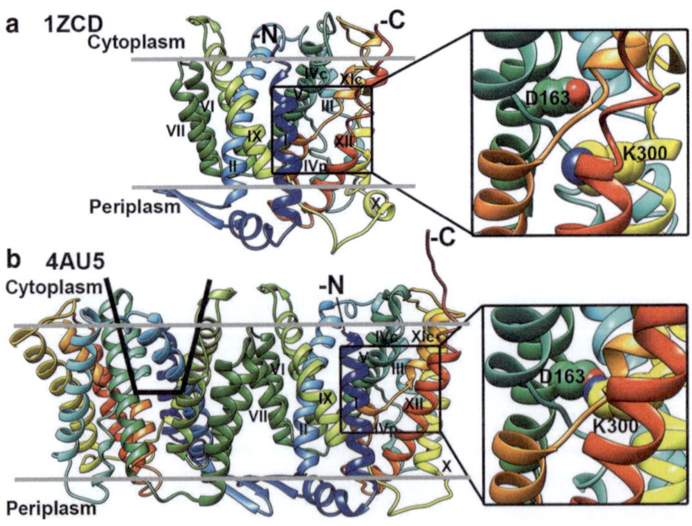

Figure 1 Crystal structures of Ec-NhaA. Two crystal structures of Ec-NhaA are displayed: (**a**) monomeric (PDB code 1ZCD, [31]) and (**b**) dimeric (PDB code 4AU5, [32]). The structures are colored in rainbow colors from the N- to the C-termini. The numbers of the TM helices are indicated in roman numerals. The molecules are viewed at the membrane plane with the intracellular side facing upward. The cytoplasmic funnel is marked on the left monomer of the 4AU5 structure. The tentative boundaries of the membrane are indicated. The insets show a zoom into the extended chains region and the putative cation binding site. The main difference between the two structures is in TM X (see Section 2.2.1). In the 4AU5 structure, residues D163 of TM V (*green*) and K300 of TM X (*yellow*) form a hydrogen bond/salt bridge (2.5 Å), whereas in the 1ZCD structure, TM X is displaced by one helix turn and the distance between D163 and K300 (11.7 Å) does not allow the formation of a salt bridge. All molecular graphics were drawn with the UCSF Chimera package [396].

2.1.2 Tt-NapA, the Na^+/H^+ Antiporter of *Thermus thermophilus*

Tt-NapA, the Na^+/H^+ antiporter of *Thermus thermophilus*, displays 21 % sequence identity to human NHA2 and <15 % sequence identity to Ec-NhaA [33].

Expression of Tt-NapA in the Na^+/Li^+ sensitive *E. coli* strain EP432 [34] has been shown to confer Na^+/Li^+ resistance. Na^+/Li^+ antiport activity was measured in everted membrane vesicles isolated from this strain using a fluorescent probe for ΔpH (Table 1). Very similar results were obtained in proteoliposomes reconstituted with pure Tt-NapA and F_1F_0 ATP synthase (for energization of the proteoliposomes by ATP) [35]. Although its Na^+/H^+ stoichiometry has not been determined, Tt-NapA is suggested to be electrogenic because imposed membrane potential can drive its antiport activity [35].

2.1.3 Nm-ASBT from *Neisseria meningitidis* and Yf-ASBT from *Yersinia frederiksenii*, Bacterial Homologues of the Human Apical Sodium-Dependent Bile Acid Transporter

Two bacterial homologues of the human bile acid symporter (SLC10, sodium bile acid co-transporter family) have been cloned: The first is Nm-ASBT from *Neisseria meningitidis*, which shares 26 % sequence identity and 54 % similarity to human ASBT [36]. Yf-ASBT from *Yersinia frederiksenii* shares 40 % sequence identity to Nm-ASBT and 22 % sequence identity and 59 % similarity to human ASBT [37]. Whole cells expressing Nm-ASBT show Na^+-dependent uptake of taurocholate (Table 1) (apparent K_m of around 50 μM, a value similar to that of rat and human ASBT) [36], which is sensitive to human ASBT inhibitors [36]. Purified Yf-ASBT reconstituted into proteoliposomes shows similar Na^+-dependent symport of taurocholate [37] (Table 1). Accordingly, both bacterial transporters can serve as valid models of the mammalian bile acid symporters, which are useful drug targets (see Section 5).

2.1.4 Pa-NhaP from *Pyrococcus abyssi and* Mj-NhaP1 from *Methanocaldococcus jannaschii*, Homologues of Human NHE1

The gene of Pa-NhaP has been cloned from *Pyrococcus abyssi* [38], and that of Mj-NhaP1 has been cloned from *Methanocaldococcus jannaschii* [39, 40]. The latter are thermophilic archaea (optimum growth temperature in marine hot vents for *Methanocaldococcus jannaschii* is 85 ° C with salinity of 10–15 mg/L and a pH of 4–6 [41]). Mj-NhaP1 uses Na^+ to excrete H^+ from the cell [42]. The Pa-NhaP and Mj-NhaP1 proteins resemble the human NHE1 and the plant (*Arabidopsis thaliana*) SOS1 (18–21 % sequence identity). Mj-NhaP1 is very similar to Mj-NhaP2 (45 % sequence identity), which does not show Na^+/H^+ exchange activity, as well as to the antiporters of *Enterococcus hirae* and *Clostridium acetobutylicum* (27–31 % sequence identity). It is less similar to Ec-NhaA (10–16 % sequence identity), or to the cation/proton transporter Ec-YcgO (20–22 % sequence identity), which plays a role in osmoregulation [39].

Na^+/H^+ antiport activity has been measured for both Mj-NhaP1 [43] and Pa-NhaP [38] (Table 1). At room temperature, the turnover rate for Na^+ of Pa-NhaP is very low, but extrapolation to 100 °C, the physiological temperature of the bacterium, suggests a turnover rate of 5000 ions $\cdot$ sec^{-1} [38]. The apparent K_m was not affected by temperature. The transport is electroneutral. Remarkably, Pa-NhaP transports Tl^+ in addition to Na^+ and Li^+ but does not transport K^+ [38].

Both Mj-NhaP1 and Pa-NhaP are dependent on pH (Table 1). At pH 6 but not at pH 5, the transport of Pa-NhaP is cooperative [38]. Unlike Pa-NhaP transport, the antiport of Mj-NhaP1 does not show cooperativity at any pH [43]. Mj-NhaP1 can be expressed in *E. coli*, but because the intracellular pH of *E. coli* is pH 7.5 where Mj-NhaP1 is not active [44], it does not complement growth of the Na^+/Li^+-sensitive EP432 strain [39].

In summary, out of the six secondary transporters that share the NhaA fold (see Section 2.2 and Table 1), four are Na^+/H^+ antiporters and two are Na^+/symporters of taurocholate and other cholate derivatives. Three of the antiporters are Na^+/Li^+-specific, but Pa-NhaP1 is less specific and transports Tl^+ in addition to Na^+ and Li^+. All antiporters are pH-dependent. Ec-NhaA has the highest measured turnover rate.

2.2 *Crystal Structures of Prokaryotic Na^+/H^+ Antiporters and the Structural Insights They Yield*

Figure 2 (and Figure 3 in Section 2.2.2) show a projection of calculated evolutionary conservation scores on the crystal structures of proteins [31, 35, 397–399] that share the NhaA fold, emphasizing the functionally important regions at the core of each protein.

2.2.1 The Crystal Structure of Ec-NhaA

The crystal structure of the monomeric, down-regulated Ec-NhaA was obtained at pH 4 [31] (Figure 1a). It provided the first structural insights into an antiporter's transport mechanism and pH regulation [45]. Ec-NhaA consists of 12 transmembrane helices (TMs) with N- and C-termini at the cytoplasmic side of the membrane (N^{in}-C^{in} topology, Figures 1 and 2a). The helices of the monomeric Ec-NhaA are organized in a unique fold of two densely-packed domains [31] (Figures 1 and 2a). A core domain, highly conserved (Figure 2a), is composed of two structurally-related bundles that are topologically inverted to each other despite very weak sequence homology between them (TMs III, IV, V and TMs X, XI, XII). The dimer interface domain (Figure 1b) comprises a linear bundle that also contains inverted-topology repeats (TMs I, II and VIII, XI), leaving TMs VI and VII as "outsiders" [46].

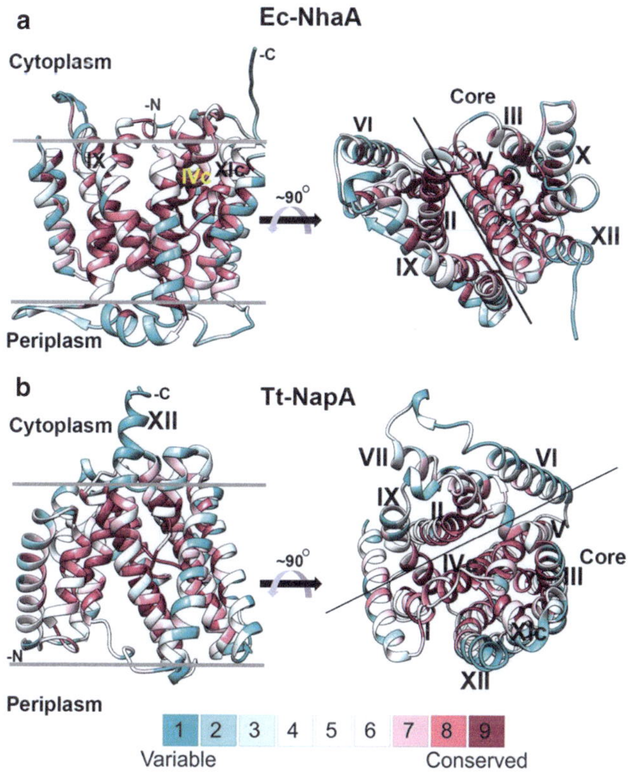

Figure 2 Evolutionary conservation scores projected on the crystal structures of the Ec-NhaA and Tt-NapA antiporters of the NhaA fold. The crystal structures of the (**a**) Ec-NhaA (PDB code 4AU5, [31]) and (**b**) Tt-NapA (PDB code 4BWZ, [35]) antiporters of the NhaA fold are displayed in a ribbon representation The ribbons are colored according to evolutionary conservation scores calculated with the ConSurf server with turquoise-through-maroon indicating variable-through-conserved [397–399]. The images on the *left side* are viewed in the membrane plane, with the intracellular side facing upward. Tentative boundaries of the membrane are shown. In the images on the *right side* the molecule is rotated about 90°, and viewed from the cytoplasmic side into the membrane. The NhaA fold is composed of two main domains; one is the core domain, containing the extended chains and the putative binding sites. The other is called the panel domain, or the interface domain that mediates dimerization (See Section 2.2).

Interestingly, one TM in each Ec-NhaA bundle (IV and XI) is interrupted by an extended chain in the middle of the membrane, leaving two short helices oriented to the cytoplasm (c) or periplasm (p) (IVc, IVp, and XIc, XIp, respectively) [31] (Figures 1, 2a, and 4a in Section 2.2.3). The interrupted TMs cross each other at the extended chains in the middle of the membrane. This crossing, in which the positive dipole ends of TM IVc and TM XIp face each other, and the negative dipole ends of TM IVp and TM XIc face each other, is referred to as the TM IV/XI assembly (Figures 1, 2a, 4a). The dipoles are compensated by residues Asp133 and Lys300, respectively [31]. This non-canonical TM IV/XI assembly creates a delicately-balanced electrostatic environment in the middle of the membrane at the ion binding site(s). Most likely, this assembly plays a critical role in the cation exchange activity of the antiporter [31].

The crystal structure of Ec-NhaA reveals an inward-open conformation: Between the core and the interface domain, a major, negatively-charged funnel opens to the cytoplasm (Figure 1b). It is composed of helices II, V, IX, and IVc and protrudes deeply towards the crossing of the extended chains. Residue D164 is located deeply at the funnel bottom [31]. An additional, shallow funnel, comprising TMs II, VIII, and XIp, opens to the periplasm (Figure 1). This funnel is separated from the cytoplasmic funnel by a barrier of hydrophobic residues [31].

In 2013, the Ec-NhaA dimer was crystallized at pH 3.5 [32] (PDB IDs, 4AU5 for wild-type (3.7 Å) and 4ATV for triple mutant A109T_Q277G_L296M (3.5 Å)) (Figure 1b). The monomers in the original and the new structures are identical, aside from several loops and the position of TM X (Figure 1). In the new dimeric structure, Lys300 (TM X) and Asp163 (TM V) form a salt bridge, while in the original structure they are about one turn of a helix apart. In the original structure, TM X had the highest B factor, implying instability [31]. This, together with the medium resolution of the original structure, may have led to an incorrect assignment of TM X in the original structure. However, a pH-induced movement of helix X has been observed experimentally [47], which is in line with the high B factor of TM X. It is therefore possible that the two Ec-NhaA structures represent two different Ec-NhaA conformations.

Interestingly, in addition to the two NhaA crystal structures obtained at low pH, 2D crystals of Ec-NhaA dimers were also obtained at acidic pH [28, 29, 48]. These results suggest that acidic pH is energetically favorable for Ec-NhaA crystallization.

The dimeric structure of Ec-NhaA [32] (Figure 1b) supports and extends previous data obtained by cryo-electron microscopy of 2D crystals [28–30, 48] and biochemical [25] and biophysical assays performed *in situ* [26, 27]. The crystal structure reveals the precise interactions between the NhaA monomers forming the dimer: At the periplasmic side, a β-sheet is formed from the β-hair pin located between TMs I and II of each monomer; at the cytoplasmic side, there are few contacts between TM IX of one monomer and TM VII of the other monomer (Figure 1b). It is likely that lipids fill the remaining space in the dimer interface.

Since the determination of the structure of Ec-NhaA [31], additional transporters have been crystallized and shown to share the NhaA structural fold (Figures 2, 3, 4, and 5): Tt-NapA [35], Nm-ASBT [36], Yf-ASBT [37], Pa-NhaP and Mj-NhaP1 [38, 43]. Cryo-electron microscopy of 2D crystals also indicates that Mj-NhaP1 has the NhaA structural fold [49]. In fact, it has been suggested that the fold is shared by all members of the superfamily of cation proton antiporters (CPA) [49]. Nevertheless, the Tt-NapA, Nm- and Yf-ASBT, Pa-NhaP and Mj-NhaP1 structures also differ from Ec-NhaA: while Ec-NhaA has 12 TMs (Figure 4a), the ASBTs are composed of 10 TMs (see Figure 5 in Section 2.2.3), and Tt-NapA (Figure 4b) and Pa- NhaP [38] and Mj-NhaP1 [43] are each composed of 13 TMs. In addition, the determined structural conformation of Tt-NapA resides in a periplasm-, rather than

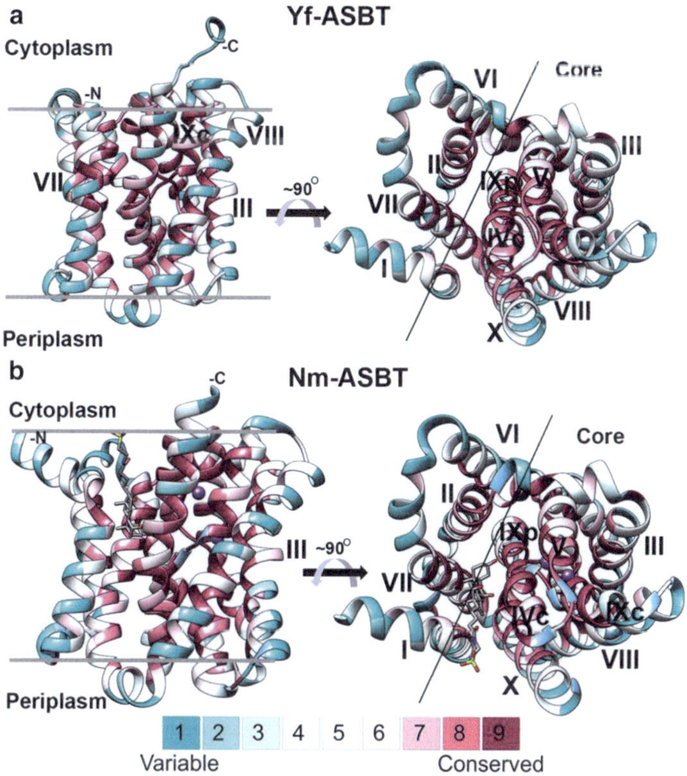

Figure 3 Evolutionary conservation scores projected on the crystal structures of the Yf-ASBT and Nm-ASBT antiporters of the NhaA fold. The crystal structures of the (**a**) Yf-ASBT (PDB code 4N7W, [37]) and (**b**) Nm-ASBT (PDB code 3ZUY, [36]) antiporters of the NhaA fold are displayed in a ribbon representation. The ribbons are colored according to evolutionary conservation scores calculated with the ConSurf server, with turquoise-through-maroon indicating variable-through-conserved [397–399]. The images on the *left side* are viewed in the membrane plane, with the intracellular side facing upward. Tentative boundaries of the membrane are shown. In the images on the *right side* the molecule is rotated about 90°, and viewed from the cytoplasmic side into the membrane. The NhaA fold is composed of two main domains; one is the core domain, containing the extended chains and the putative binding sites. The other is called the panel domain, or the interface domain that mediates dimerization (see Section 2.2). In the Nm-ASBT structure, taurocholate, bound in an intracellular cavity, is displayed in sticks representation colored by atom type. Two sodium cations, represented by *purple* atom spheres, are bound in the extended-chain core area.

a cytoplasm-facing conformation (Figure 2b and 4b), and the dimer interface of Tt-NapA resembles that of Mj-NhaP1 [43] more closely than it resembles that of Ec-NhaA [35].

Several secondary transporter structures, which share neither structural fold nor sequence homology with Ec-NhaA, have been determined. Remarkably, similarly to Ec-NhaA, these structures contain topologically-inverted repeats with unwound helices [50] having similar functional implications (reviewed in [51–54]). Together, these structures have led to the conclusion that secondary transporters are structured in three main structural folds: MFS, LeuT and NhaA [50].

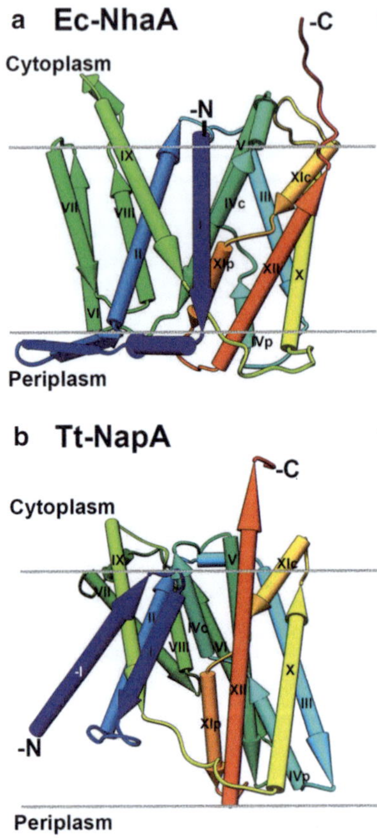

TM	Range
I	10-31
II	59-86
III	94-114
IV	133-142 122-128
V	149-175
VI	180-200
VII	204-219
VIII	222-237
IX	246-272
X	286-314
XI	338-350 326-335
XII	356-379

TM	Range
-I	3-26
I	30-43
II	53-74
III	83-110
IV	112-125,127-138
V	143-173
VI	177-200
VII	214-234
VIII	236-248
IX	252-279
X	288-313
XI	316-327,331-345
XII	349-384

Figure 4 Topologies of the Ec-NhaA and Tt-NapA crystal structures of the NhaA fold. The topologies of the (**a**) Ec-NhaA (PDB code 4AU5, [31]) and (**b**) Tt-NapA (PDB code 4BWZ, [35]) crystal structures of the NhaA fold are presented. The structures are colored in rainbow colors from the N- to the C-termini. TM helices are presented as cylindrical arrows indicating the direction of the helix (N- to C-termini). The numbers of the TM helices are indicated in roman numerals. The sequence range of each TM helix is indicated in the adjacent table. Tt-NapA has an additional helix at the N-terminus (numbered -I for convenience), which crosses the membrane from the periplasm to cytoplasm. The tentative boundaries of the membrane are indicated.

2.2.2 The Crystal Structure of Tt-NapA

The dimeric crystal structure of Tt-NapA was obtained at pH 9 (3.7 Å resolution), and the structure of a triple Cys mutant (M20C,V166C,V326C), showing similar behavior to the wild-type Tt-NapA, was obtained at pH 7.8 (3 Å resolution) [35]. Tt-NapA is composed of 13 TMs with an N^{out}-C^{in} topology (Figures 2b and 4b). It has one additional helix compared to Ec-NhaA at its N-terminus; to facilitate comparison to Ec-NhaA, this additional helix is referred to as TM(-I). The crystal structure of Tt-NapA is very similar to that of Ec-NhaA (Figure 2). The set TM 1 to TM 5 and the set TM 7 to TM 12 are topologically similar to each other but

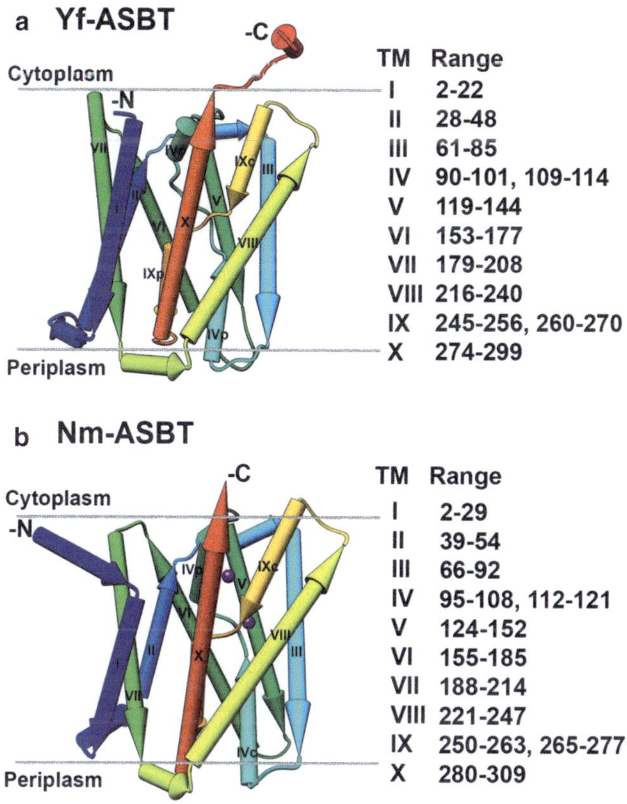

Figure 5 Topologies of the Yf-ASBT and Nm-ASBT crystal structures of the NhaA fold. The topologies of the (**a**) Yf-ASBT (PDB code 4N7W, [37]) and (**b**) Nm-ASBT (PDB code 3ZUY, [36]) crystal structures of the NhaA fold are presented. The structures are colored in rainbow colors from the N- to the C-termini. TM helices are presented as cylindrical arrows indicating the direction of the helix (N- to C-termini). The numbers of the TM helices are indicated in roman numerals. The sequence range of each TM helix is indicated in the adjacent table. The tentative boundaries of the membrane are indicated. In the Nm-ASBT structure, two sodium cations, represented by purple atom spheres, are bound in the extended chains core area.

oppositely oriented in the plane of the membrane (Figure 4b). These inverted-topology six-TM repeats intertwine to form two domains: (i) a core, which is highly conserved (Figure 2) and where translocation occurs, and (ii) a dimerization (interface) domain. The repeats are linked together via TM 6 (Figure 2). Tt-NapA also shares the Ec-NhaA fold in which two interrupted helices cross each other (TM IV and TM XI of Ec-NhaA and TM 4 and TM 11 in Tt-NapA; Figures 2 and 4a and b respectively). In both the Tt-NapA and Ec-NhaA structures, the putative binding site is located at the crossing. As discussed above (see Section 2.2.1), as a result of the crossing, the positive and negative dipole ends of TM 4 and TM 11 face each other. In Ec-NhaA, Asp133 and Lys300 compensate the dipoles. In Tt-NapA, Lys305 might fulfill a role similar to that of Lys300 in Ec-NhaA [35]. However, Asp133 in

Ec-NhaA is replaced with a serine residue in Tt-NapA, and Glu333 may compensate the positive dipoles [35].

An interesting structural characteristic of Tt-NapA is the large difference between its structure and that of Ec-NhaA in terms of the position of the core with respect to the dimer interface [35]. Specifically, the Ec-NhaA core domain is rotated by 21° relative to the Tt-NapA core domain (Figure 2). In Tt-NapA, at the contact between the core and the dimer interface, a large, negatively-charged cavity is open to the outside, and the interaction of TM 2, TM 4 and TM 5 tightly closes the cytoplasmic side of the cavity (Figure 4b). Models of Tt-NapA's inward-facing conformation, based on the structure of Ec-NhaA, have led to the suggestion that Tt-NapA's antiport function relies on a two-domain elevator mechanism ([35] and Section 2.5).

The Tt-NapA dimeric interface buries a surface area of 1800 $Å^2$ formed by tight hydrophobic helix-helix packing between TM 1 on one monomer and TM 7 on the other [35]. Contacts also exist between the ends of TM 2 and TM 9. Whereas a β-sheet comprises most of the contacts between Ec-NhaA monomers, Tt-NapA lacks such a β-sheet (see Section 2.2.1 and Figure 1b). The dimer interface of Tt-NapA more closely resembles the dimeric interfaces of Mj-NhaP1 [43, 49] and of Pa-NhaP [38] than that of Ec-NhaA (see Section 2.2.4).

2.2.3 The Crystal Structure of Nm- and Yf-ASBT

The crystal structure of Nm-ASBT was obtained at 2.2 Å resolution at pH 4.5 [36]. It is composed of ten TMs with an N^{in}-C^{in} topology (Figures 3 and 5). The sets TMs 1–5 and TMs 6–10 are topologically inverted repeats. Each repeat is made up of an N-terminal V-motif (TMs 1,2 and 6,7) and a core motif (TMs 3–5 and 8–10). The core motifs from the two repeats form the core domain, whereas the two V motifs create a panel-like domain (Figure 3). In each repeat, one TM (TMs 4 and 9, respectively, in the two repeats) is interrupted by an extended chain that crosses the other in the middle of the membrane, similarly to the case of Ec-NhaA, Tt-NapA (Figures 2 and 4) and Pa-NhaP [38] and Mj-NhaP1 [43]. From about this crossing point, a funnel opens to the cytoplasm, and separates the core from the panel-like domain. The extracellular side is tightly sealed by TMs 1, 2, 4b, 7, 9b, and 10. The sealed cavity is hydrophobic, but near the bottom there are charged residues and water molecules. The structure was solved with a bound bile-acid molecule, taurocholate, which is situated between the core and panel domains in this large hydrophobic cavity. The binding site of taurocholate has been identified and verified by mutagenesis [36]. The cavity is larger than taurocholate, perhaps reflecting the large variety of compounds recognized by Nm-ASBT.

The first structural insight obtained from the crystal structure of Nm-ASBT relates to the fact that, despite being a symporter that transports taurocholate, Nm-ASBT shows remarkable structural similarity to Ec-NhaA and to the other Na^+/H^+ antiporters that share the NhaA fold [36] (Figures 2, 3, 4, and 5). This similarity emphasizes the remarkable plasticity of transporters, which enables them to use a common fold to translocate different substrates.

Another important insight obtained from the Nm-ASBT structure is the identification of the Na^+ binding site, which has not yet been observed in any of the other similar structures [36]. Specifically, two Na^+ ions were observed at the crossing of the interrupted TMs (Figures 3b and 5b).

A model of the outward-facing conformation of Nm-ASBT was generated by using the symmetry of the inverted-topology repeats according to [55]. The largest difference between the two conformations led the authors [36] to assume that transport relies on a two-domain elevator mechanism (see Section 2.5). Notably, TM 10 of Nm-ASBT and TM IX of Ec-NhaA line the transport pathway. However, the conformation of the Ec-NhaA dimer-interface domain is more stable than that of the equivalent panel domain of Nm-ASBT. This difference may be due to the smaller Ec-NhaA ligand or it could indicate that the model of Nm-ASBT represents another conformation (occluded state) [36].

Yf-ASBT was crystallized in lipidic cubic phase in the unliganded inward-facing conformation (at 1.95 Å resolution) [37]. Similarly to Nm-ASBT, it has ten TMs folded into a panel domain (TMs 1, 2, 6, and 7) and a core domain (TMs 3–5 and 8–10) with inverted-topology repeats and interrupted TMs (TMs 4 and 9) crossing each other (Figures 3a and 5a), such that at the point of crossing an inward-facing funnel opens to the cytoplasm. In addition to the TMs, Yf-ASBT contains four amphipathic helices, which are probably located at the interface between the membrane and bulk solution. Two Na^+ binding sites were identified by mutagenesis and by comparison to Nm-ASBT. To obtain Nm-ASBT in an alternative conformation, one of the sites deduced to bind Na^+ (denoted Na^+1) was mutated (E254A), leading to a crystal structure (2.5 Å resolution) of an outward, open, unliganded conformation [37]. On the basis of these conformations, the two-domain elevator mechanism was adopted to describe transport (see Section 2.5).

2.2.4 The Crystal Structures of Pa-NhaP and of Mj-NhaP1

Recently, crystal structures of dimeric Pa-NhaP in two different conformations were obtained at pH 4 and pH 8 (3.15 Å resolution) [38]. Each protomer has 13 TMs. The structural fold of Pa-NhaP is very similar to the NhaA fold; TMs 4–6 and TMs 11–13 form the core, a six-helix bundle. TMs 1–3 and 7–10 form the dimer interface. TMs 1–6 are an inverted repeat of TMs 8–13, and the two repeats are connected by TM 7. TMs 5 and 12 are interrupted by antiparallel extended chains that cross each other in the center of the membrane.

The conformation of the structure obtained at pH 8 is inward-facing. Between the core and the dimer interface, a deep, negatively-charged funnel opens to the cytoplasm. Interestingly, a second polar funnel extends from the cytoplasm to almost the tip of the cytoplasmic funnel. On the periplasmic side, a deep cavity starts between the protomers and protrudes into the protein. It contains lipids. A Tl^+ binding site was revealed at the deepest point of the cytoplasmic funnel by soaking the crystals at pH 8 with thallium acetate. The binding site is accessible from the cytoplasm but not from the periplasm.

In the crystal structure obtained at pH 4 (3.5 Å resolution), as in the pH 8 crystal structure, the active site is accessible from the cytoplasm via the cytoplasmic funnel, but not from the periplasmic side. Both structures, therefore, show an inward-open state. In the pH 4 structure, in contrast to the pH 8 crystal structure, the second narrow polar pathway is blocked.

Cryo-electron microscopy of 2D crystals of Mj-NhaP1, obtained at pH 4, yielded projection maps (8 Å [56], 7 Å [57], and 6 Å [49]) showing that Mj-NhaP1 (47 kDa) is a dimer. The protomer consists of 13 TMs packed into two domains: a six-helix bundle core domain and a dimer interface domain. A model structure of Mj-NhaP1 was obtained on the basis of a detailed sequence comparison with the Ec-NhaA crystal structure, and the 7 Å projection map [57].

Remarkably, the crystal structure (3.5 Å resolution) of dimeric Mj-NhaP1 has recently been determined at pH 8; the structure indicates that the 13 helices are characterized by N^{out}-C^{in} topology [43]. Like Pa-NhaP, Mj-NhaP1 is characterized by the NhaA fold, with a core comprising a six-TM bundle, and a dimer interface domain with a row of seven helices. In the core, TMs 5 and 12 are each interrupted by an extended chain, and the two extended chains cross each other. A negatively-charged cytoplasmic funnel protrudes into the monomer.

The dimer interface of Mj-NhaP1 is somewhat less interwoven than that of Pa-NhaP. A deep hydrophobic cavity spans nearly the entire thickness of the membrane between the monomers, and is probably filled with lipids [43].

Attempts to determine the crystal structure of Mj-NhaP1 without Na^+ or at pH 4 have thus far been unsuccessful. To obtain insights into these structures, cryo-electron microscopy (EM) of 2D crystals grown at pH 4 was used, and a 3D map was obtained [49] (in-plane resolution of 6 Å) (and see above in this chapter). The Mj-NhaP1 crystal structure was fitted manually to the EM map to obtain an atomic model, and a 3D difference map was calculated. Remarkably, clear differences in orientation were observed between the EM map and the crystal structure obtained at pH 8, primarily in the core; this led to identification of an outward-open conformation.

The effect of pH on Mj-NhaP1 was tested *in situ* on the electron-microscope grid by incubating the 2D crystals at various pH levels [56]. At pH 6 and above, a major change in density distribution occurred within the helix bundle in addition to a 2-Å shift in the position of the helix bundle relative to the dimer interface. Similar changes were seen in each monomer, suggesting that, as in Ec-NhaA [58], the monomer is the functional unit in Mj-NhaP1. In another recent, almost identical experiment, in which 2D crystals were first obtained at different pH levels without Na^+ [49], differences in density between pH 4 to 8 were in the background level, suggesting either that the changes occur in side chains that cannot be seen at this resolution (6 Å) or that pH alone does not cause changes in the Mj-NhaP1 conformation.

To examine conformational changes induced purely by Na^+, 2D crystals grown without Na^+ at pH 4 were incubated with 20–500 mM NaCl at constant pH values. Addition of Na^+ caused a change in the helix bundle, giving rise to three sets of major peaks in difference maps. The same three sets of positive/negative difference peaks were observed at pH 8. The peaks increased when the Na^+ concentration increased between 20 and 500 mM. Addition of Li^+ produced the same results, whereas addition of Mg^{2+} did not induce any change. Addition of K^+ did have an effect, which differed from the effects of exposure to Na^+/Li^+, suggesting that the changes induced by Na^+/Li^+ are specific to these molecules. K_d values for Na^+ were calculated on the basis of the Na^+ titration curve: 30 mM at pH 8, and 280 mM at pH 4.

Taken together, the structures described above highlight the interesting phenomenon whereby the NhaA fold is uniquely characterized by the antiparallel crossing of inverted TM bundles, interrupted by extended chains [31, 50, 59]. In Ec-NhaA, for example, the positive dipole ends of TM IVc and TM XIp face each other, and the negative dipole ends of TM IVp and XIc face each other (Figures 1, 2, and 4). In Ec-NhaA the dipoles are proposed to be neutralized by the side-chains of Asp133 and Lys300, respectively [31].

Extended chains are highly suitable for cation binding for the following reasons [31, 60]: (i) Unlike intact helices, extended chains are not saturated by hydrogen bonds, so they are flexible and can provide space for ions in an otherwise densely-packed environment typical to α-helical membrane proteins. (ii) The free peptide bonds of extended chains can provide main chain carbonyl oxygen and amide for ion coordination. (iii) The dipole moment of the helix termini connected to the extended chains can also contribute to ion coordination. (iv) Transport proteins change conformation during turnover, and extended chains can confer flexibility to these changes at low energetic cost compared with α-helices. Indeed, in many transporters that depend on ion transport, extended chains are involved in ion binding [36, 52, 53, 60, 61]. Moreover, the cation-binding sites in the NhaA fold have been shown or predicted to be located at the crossing of the extended chains interrupting the TMs [62].

2.3 *Functional Organization Insights Provided by the Crystal Structures*

The mechanism of the vast majority of secondary transporters, including Na^+/H^+ antiporters, can best be described using the alternate-access model [63]. In its simplest form, the model is based on a shift between two conformations, one in which the ligand is accessible to the binding site from the cytoplasm, and another in which the ligand is accessible from the periplasm. Hence, these transporters are 'nano machines'. Revealing the molecular details of the entire transport mechanism requires, at the very least, determination of the crystal structures of the active site and all involved conformations. On top of this structural information, additional essential functional knowledge is required: (i) the antiporter's functional organization, i.e., the location on the structure of functionally important residues outside the active site and the dynamic domains; (ii) the functional dynamics, i.e., the kinetics of the conformational changes: Which rate constants are involved? Are the rates in the pure protein in detergent equal to those measured under physiological conditions? (iii) Which signals elicit the dynamics? To answer these questions, it is necessary to carry out *in situ* interdisciplinary studies employing molecular genetics, biochemistry, biophysics, and computations, which will shed light on the functional organization of the antiporter and its functional dynamics in the membrane. The remainder of this section reviews what is currently known about the functional organization of Na^+/H^+ antiporters.

2.3.1 The Active Site of the Na^+/H^+ Antiporters

The active site is best revealed via determination of a crystal structure of the ligand-bound protein. However such structures are difficult to obtain, especially with Na^+, for which X-ray diffraction is similar to that of a water molecule.

The crystal structure of Nm-ASBT reveals Na^+ binding sites in the NhaA fold (2.2 Å) [36] (Figures 3b and 5b). Specifically, in the highly conserved core domain of Nm-ASBT, two bound Na^+ ions were identified on the basis of coordination and bond distances. The first binding site (Na^+1) is located approximately 10 Å from the cytoplasmic surface between TM 4b and TM 5 and also interacts with Glu260 (TM 9a). The second binding site (Na^+2) is located 8 Å from Na^+1 near the membrane center at the crossover point of the TM 4/TM 9 assembly. Four backbone carbonyl oxygen atoms coordinate Na^+2 (including highly conserved Glu260, Gln264, Gln77). Glu260 is an essential residue in human ASBT. In Nm-ASBT, Ala replacement of Glu260 and Gln77 abrogates transport [36]. Hence, both Na^+ ions are needed for transport, as indicated by the stoichiometry. Na^+2 neutralizes the partial negative dipole of TM 9a and by doing so stabilizes the interaction with TM 4a. Neutralization of the helix dipoles by Na^+/Li^+ seems to be a conserved feature for this fold [36].

Although bound Na^+ is not observed in the Yf-ASBT crystal structure, an assay of Na^+ binding to purified Yf-ASBT showed the existence of two binding sites (EC_{50} of 5.37 mM and a Hill coefficient of 1.56, suggesting positive cooperativity). Accordingly, the mutations E254A and Q258A in the putative binding sites Na^+1 and Na^+2, respectively, reduced binding and reduced the Hill coefficient to 1. The mutation E254A enabled the outward-open conformation to be crystallized (see Section 2.2.3). Ligand-binding assays in which taurocholate and Na^+ were bound to purified Yf-ASBT [37] showed that Na^+ binds first.

Neither Na^+ nor Li^+ have been observed in the crystal structures of Ec-NhaA and Tt-NapA. In these antiporters, the cation-binding sites have been experimentally and computationally deduced. In Ec-NhaA, Asp163 and Asp164 have been suggested to form the binding site; this proposition is based on the residues' position at the bottom of the cytoplasmic funnel in close proximity to the crossing of the Ec-NhaA TM IV/XI assembly [31], evolutionary conservation [46], mutants [45, 64], isothermal calorimetry experiments [62], and molecular dynamics (MD) simulations [65]. In the Tt-NapA structure, Asp156 and Asp157 (homologues of Ec-NhaA Asp163 and Ap164) have been suggested to coordinate the ions on the basis of their position in the structure, evolutionary conservation, mutations (D156A, D157A [33] D156N, D147N [35]) and Ec-NhaA data. Since Tt-NapA was crystallized at pH 7.8, the Asp residues are likely to be deprotonated. Indeed, unlike Asp164 of Ec-NhaA, Asp157 is oriented towards the center of the cavity rather than hydrogen-bonded with the backbone of TM 4. In addition, MD simulations predicted that both Asp157 in Tt-NapA and Asp164 in Ec-NhaA bind Na^+ [35, 65].

Interestingly, in Pa-NhaP, a binding site for Tl^+ was identified by soaking crystals grown at pH 8 in Tl^+. A trigonal bipyramidal geometry of bound Tl^+ was observed and suggested to characterize Na^+ as well [38], although the ionic radius of monovalent Tl^+ (1.5 Å) is similar to that of K^+ (1.44 Å) and larger than that of Na^+ (1.12 Å). Accordingly, Tl^+ is able to replace Na^+ in Pa-NhaP. Similar cases were found [66–68]. The striking selectivity of Na^+ over K^+ observed in Pa-NhaP is thought to be related to ion solvation [38] as in Na^+ channels [69].

The Tl^+ binding site of Pa-NhaP is composed of the acidic side chains Glu73 (TM 3) and Asp159 (TM 6), which coordinate Tl^+ directly. Asp130 in the extended chain of TM 5 binds the cation via a water molecule, and the main chain carbonyl of Thr129 and the hydroxyl side chain of Ser155 of TM6 provide additional ligands. The ion-binding site at the end of the cytoplasmic funnel is accessible from a polar cavity near the narrow polar pathway via Thr129, Ser155, and Asn158. To further support these observations, the residues involved in Tl^+ coordination were mutated and their activity measured. Replacement of both ion-coordinating aspartates (D130S and D159S) eliminated Na^+ transport. E73A increased activity, S155A had little effect, but T129V inhibited activity. H292C shows intermolecular cross-linking and the effect of cross-linking on transport [38].

The substrate-binding site of Mj-NhaP1 was not resolved in the crystal structure. Structural homology to Pa-NhaP suggests that the cation-binding site involves Ser157, Asp161, Asp132, and the backbone of Thr131 in the extended chain of TM 5. Asn160, part of the ND characteristic motif of the CPA superfamily, interacts indirectly with the ligand. The mutation N160A inactivated transport; the mutation N160D reduced activity, but transport remained electro-neutral [43].

It should be stressed that the location of the binding site of the antiported ligand H^+ is still unknown in all members of the NhaA fold.

2.3.2 Where on the Ec-NhaA Crystal Structure, Outside of the Active Site, Are Functionally Important Residues Located?

A major functional insight obtained from the Ec-NhaA crystal structure was the elucidation of the functional organization of Ec-NhaA [45, 59]; the structure has provided molecular interpretations for the results of previous mutagenesis studies and inspired many biochemical, biophysical and computational studies as follows:

Since the cloning of the Ec-NhaA gene [70], Padan and colleagues have advanced selection methods to obtain various types of Ec-NhaA mutants out of randomly mutagenized *nha*A [24]. All mutants were characterized biochemically in isolated membrane vesicles or in proteoliposomes; in the former, a fluorescent probe was used to monitor Na^+/Li^+-induced changes in ΔpH across the membrane; in the latter, ΔpH-driven $^{22}Na^+$ uptake was measured. These classical biochemical transport assays have since been combined with interdisciplinary approaches: isothermal titration calorimetry (ITC) [62] and electrophysiology [71] have provided critical insights; the residues in the active site were identified by ITC [62] combined with mutagenesis [62]; the study of the thermodynamics of the antiport, a previously neglected research area was initiated. Electrophysiology based on solid support membrane (SSM) was also applied [71, 72]. The latter experiments showed that Ec-NhaA is a symmetric transporter [72] and advanced a kinetic model of the Ec-NhaA transport cycle and pH regulation (see Section 2.4.2).

As discussed in [59], there are differences between the results obtained by electrophysiology and those obtained by classical transport assays (ΔpH-driven $^{22}Na^+$ uptake in proteoliposomes; respiration-driven Na^+ uptake in everted membrane vesicles using acridine orange or similar fluorescent probes of ΔpH). These differences probably stem from the different time resolutions of the biochemical assays (seconds to minutes) and electrophysiology (milliseconds) and the different driving forces (chemical potential gradient in electrophysiology *versus* electrochemical potential gradient in the biochemical assays). To resolve this issue, further studies are required, e.g., studies using stopped-flow devices to measure transport activity, as was done for LacY [73]. Such studies can overcome the time resolution difference between biochemcal and electrophysiological experiments.

Cys scanning — i.e., systematic Cys replacement of each residue in a membrane protein and characterization of the mutations — is an effective experimental approach, used preferentially in Cys-less membrane proteins, to identify functionally important residues [74–76]. Importantly, Cys-less Ec-NhaA is as active as the wild-type antiporter [77], and the effects of the Cys replacements on activity and pH regulation have been analyzed as described above in this section and in Section 2.3.3. The Cys scanning data that are currently available for Ec-NhaA are comprehensive and include residues from TMs II [78], IV [79], VIII [80], IX [81], X [47], and XI [82]. For example, Cys scanning of TM IV highlighted the importance of this unwound TM in Ec-NhaA functionality [79]. Residues T132 and D133 change conformation with pH [79] and play a role in ligand binding [62].

Little information exists regarding the functional organization of Tt-NapA [35], ASBTs [36, 37], Pa-NhaP and Mj-NhaP1 [38, 43, 49]; the available data are mainly based on site-directed mutagenesis.

2.3.3 Where on the Crystal Structure Are the Dynamic Domains Located?

To understand the structure/function relationship in transport activity, it is critical to locate dynamic segments that change conformation during activity or regulation. For this purpose, it is essential to determine the atomic structure of all conformations, which is obviously very challenging. Accordingly, experimental and computational methods have been developed to localize dynamic segments in membrane proteins both *in vitro* and *in situ*.

Comparison of the crystal structures of different conformations enables mobile segments to be identified. The crystal structure of Ec-NhaA is only available at acidic pH; this structure corresponds to a down-regulated form of Ec-NhaA, in which the conformation of the transporter faces the cytoplasm (inward-open) [31]. This conformation does not bind Na^+/Li^+ because Ec-NhaA activates and binds its ligands only at alkaline pH [20, 62]. Nevertheless, this crystal structure provides a basis for computing other conformations. An outward-open conformation of EcNhaA has been modeled [46].

The crystal structure of Tt-NapA was determined in its outward-open, active, conformation at pH 7.8 [35]. On the basis of this conformation and the inward-open conformation of Ec-NhaA, a model of the antiporter activity has been suggested (see Section 2.5). The crystal structure of Nm-ASBT was captured in the inward-open conformation at pH 4.5 [36]. Modeling of the outward-facing conformation of Nm-ASBT was generated by using the symmetry of the inverted topology repeats. Yf-ASBT was crystallized in two conformations: the inward-open conformation of the wild-type antiporter at pH 5.5, and the outward-open conformation of a mutant (E254A) at pH 8.5 [37]. The two conformations enabled the structural basis of the transport mechanism to be modeled (see Section 2.5).

The crystal structure of Pa-NhaP was determined at two conformations, one at pH 4, and the other at pH 8. Both structures show inward-open conformations [38]. In contrast to the pH 8 structure, in the pH 4 structure the second narrow pathway is blocked by rearrangements of Ile151, Phe355, and Gly359. Small pH-induced changes were observed mainly at the dimer interface.

Remarkably, the outward-facing conformation of Mj-NhaP1 has been obtained by cryo-electron microscopy of 2D crystals grown at pH 4 without Na^+ [43]. The X-ray structure of Mj-NhaP1, previously determined at pH 8, was fitted manually to the EM map, and an atomic model was obtained. A calculated 3D difference map showed clear changes in the orientation of TMs; most of these changes were in the core domain: in TM6, and in the TM 5/TM 12 assembly. The core as a whole tilted by about 7 Å; on the cytoplasmic side the tilt was towards the dimer interface, and on the periplasmic side the tilt was away from the interface. This movement closes the cytoplasmic funnel and opens the periplasmic funnel. In the transition, the ion-binding site moves towards the extracellular side by about 5 Å.

The following biochemical approaches have been used to experimentally identify amino acid residues involved in conformational changes in Ec-NhaA and to characterize their structural localization: (i) measurement of the accessibility/reactivity of various residues in Ec-NhaA to trypsin [83] or mAb 1 F6 [84] as a function of pH; (ii) Cys-scanning, i.e., measurement of the accessibility of single-Cys replacements to various sulfhydryl (SH) reagents as a function of ligand concentration. This has become an almost routine procedure for locating ligand-induced conformational changes in many membrane proteins [74–76, 85]. This approach was used to identify pH-induced conformational changes in Ec-NhaA, preferentially using membrane-impermeant SH reagents similar in size and charge to Na^+

(2-sulfonatoethyl methanethiosulfonate (MTSES) or [2-(trimethyammonium) ethyl] methanethiosulfonate bromide (MTSET)) [79]. To assess whether a residue is exposed to the cation passage from the periplasm, intact cells, or right-side-out membrane vesicles, are used. Cytoplasmic accessibility is assessed using inverted membrane vesicles [79]. Comprehensive Cys scanning in Ec-NhaA has revealed many residues that change accessibility with pH (reviewed in [59]). These include: loop VIII-IX and K249 [83, 86], the N-terminus [84], TM IV [79] (an observation compatible with cryo-electron microscopy results [29]), E252C in TM IX [81, 85], and D65C on TM II [78]. An exciting recent experiment showed that the extended chain in TM IV changes conformation at alkaline pH [79].

Cys-scanning experiments done on Yf-ASBT used a membrane-impermeant probe (methoxypolyethelen glycol maleimide 5,000; mPEG-Mal-5 K, Sigma-Aldrich) to target Cys replacements in the highly-conserved crossover region [37]. The tests were conducted in intact cells (accessibility from the periplasm) and sonicated membrane fraction (accessibility from the cytoplasm and periplasm) [37]. Cys replacement of T106C showed that despite being inaccessible to the periplasm in the inward-facing conformation structure, the residue was accessible in the outward-facing conformation.

Changes in proximity between two sites within a protein can be assessed by specifically designing a pair of Cys replacements. Then, cross-linking is tested with bi-functional reagents of known distances *in situ*, as a function, e.g., of ligand concentration and/or pH. This approach has been used to identify pH-induced conformational change in TM X of Cys-less Ec-NhaA [47]. Similarly, ESR probes can be used.

It is important for all experiments based on Cys replacements and SH reagents to incorporate controls for the effects of the tested conditions on the reagents themselves. For example, the rate of the chemical modification (of SH groups) itself is pH-sensitive, and it may become less specific to SH groups above pH 8.5. Therefore, when the effect of a pH change is tested, Cys-less Ec-NhaA is used as a negative control for non-specific reactivity, and mutants that are accessible to the reagents in a pH-independent fashion are used as positive controls [80]. Other parameters such as stereochemistry can also affect the reactivity of the Cys replacements to any of the reagents. To exclude this possibility, the denatured protein is verified to react with the reagent [78]. Strict control reactions are also done at 4 °C to minimize thermal backbone motions.

2.3.4 Identification of Dynamic Areas with Fluorescent Reporters

Trp residues are fluorescent, and their fluorescence properties change as a function of the residues' environment in the protein. The use of Trp fluorescence to report conformational changes requires a functional Trp-less protein. To this end, a functional Trp-less Ec-NhaA was constructed, with phenylalanine residues replacing the eight native tryptophan residues [87]. The Trp-less Ec-NhaA was used to site-specifically insert a single tryptophan in order to monitor conformational changes of

TM IVp (W126) *versus* TM IVc (F136W) facing the pH sensor and TM XI (F399W), located in proximity to the active site. The F136W mutant revealed pH-induced conformational changes. The single-tryptophan mutant F399W revealed Li^+-induced conformational changes at the active site [87, 88]. This experimental system is currently being developed to investigate functional dynamics of Ec-NhaA by spectrofluorimetry combined with the stopped-flow technique.

2.3.5 Computational Approaches

Computational studies based on the crystal structure investigated the conformational alterations associated with pH changes and substrate binding [29, 65, 89, 90]. Arkin et. al. [65] conducted calculations in order to decipher the Na^+ and proton pathways. In another study, Schushan et al. [46] used the Ec-NhaA cytoplasm-facing structure to predict a periplasm-facing conformation. The prediction was based on two approaches: the first relied on pseudo-symmetric features of the crystal structure, and the second involved conducting Normal Mode Analysis using Elastic Network Models [46]. Importantly, the two approaches yielded similar results [46]. The predicted conformation was later shown to resemble the recently-determined outward-open conformation of the Tt-NapA transporter (RMSD of 3.4 Å over 288 C-alpha atoms).

2.4 Functional Dynamics; Determination of Kinetics of Conformational Changes in vitro and in situ in the Membrane and by in silico Predictions

In order to decipher the antiporter's turnover cycle and its regulation, the dynamics of the conformational changes need to be revealed. Crystal structures, despite offering invaluable atomic-level data, cannot provide information on the kinetics of conformational changes. Classic biochemical transport assays can only achieve a time resolution of seconds, and cannot be used to measure the fast turnover rate of the transporters, in particular that of Ec-NhaA, one of the fastest transporters (10^5 min^{-1}) [20]. The following complementary approaches fill this gap.

2.4.1 Site-Specific Tryptophan Fluorescence

As detailed in Section 2.3.4, different single Trp-Ec-NhaA were constructed to shed light on the kinetics of ligand/pH-induced conformational changes [87]. These experiments suggested that F136W (TM IVc) is located near a cluster of residues thought to be responsible for pH regulation (called the "pH sensor", see Section 2.6.1), and F339W (TM XI) is located near the active site. Stopped-flow spectrofluorimetry experiments are currently being carried out to determine the dynamics of these strategic sites.

An elegant approach has recently been developed to investigate the dynamics of conformational changes that are associated with substrate binding and transport of the bacterial leucine transporter, LeuT: Single-molecule fluorescence resonance energy transfer (smFRET) was used to quantify time-dependent changes in the LeuT structure, which might otherwise have been masked by ensemble-averaging in bulk measurements or suppressed through crystallographic conditions [91]. Computational methods were applied in parallel. This approach has recently been applied to the glutamate transporter [92–94]. As yet, it is not known whether this method can be applied to Ec-NhaA and similar transporters that are much faster than LeuT and Ph-Glt (Glt_{Ph}) [95].

2.4.2 Electrophysiology

Modern electrophysiology is a universal technique for the functional characterization of transport proteins [96]. However, most available transporter structures are of bacterial transporters, and, in general, cannot be investigated by conventional electrophysiology because of the cell wall and the small size of the bacterial cell. In these cases, SSM-based electrophysiology can be very useful [71, 72, 96]. This approach provides a means of measuring the transport activity of bacterial transporters under well-defined conditions at a time resolution of milliseconds, whereas the resolution of classic biochemical transport assays is around seconds. Using Na^+ or H^+ gradients as the driving force, translocation of charge by Ec-NhaA has been measured directly by monitoring transient currents generated in Ec-NhaA proteoliposomes/membrane vesicles absorbed on an SSM [71, 72]. These experiments yielded results that could not have been obtained otherwise. Specifically, as described in detail in Section 2.5.1, determine kinetic parameters of the antiport [72]. Because of the high time resolution of SSM-based electrophysiology, the method can be used to correlate conformational transitions and charge displacements and thus to gain insight into the transport process.

2.5 *The Na^+/H^+ Antiport Mechanism*

The function of the vast majority of secondary transporters is described using the alternate-access model [63] (see Section 2.3). Full exploration of the mechanism of transport requires determining all active conformations and also deciphering the dynamics and pathways between the conformations. Although we are far from having the entire picture, two models of antiport-mechanisms have been suggested.

2.5.1 Ec-NhaA

The Ec-NhaA crystal structure [31] revealed the delicate balance of electrostatic interactions in the region around the crossing of the extended chains of the TM IV/XI assembly. This delicate balance has been suggested to play an essential role in the highly rapid ion translocation of Ec-NhaA [31]. It is probable that sodium binds first to Asp164 (TM V), which causes a rearrangement of the TM IV/XI assembly, which is enough to translocate the small ions H^+/Na^+/Li^+. The conformational change of D133 in the middle of the assembly [79] supports this model.

To obtain information on the kinetics of EcNhaA transport activity, SSM-based electrophysiological experiments have been undertaken. This approach enables antiport activity to be measured at a resolution of milliseconds, under well-defined conditions on both sides of the membrane. Using Na^+ or H^+ gradients as the driving force translocation of charge by Ec-NhaA was measured directly by tracing transient currents generated in Ec-NhaA proteoliposomes, absorbed onto the SSM [72]. Forward (Na^+ excreted from the cytoplasm) and reverse (Na^+ taken up to the cytoplasm) transport directions were investigated using preparations of transporters with inside-out orientations (membrane vesicles) and with outside (right-side)-out orientations (proteoliposomes), respectively [72].

These experiments showed that Ec-NhaA is a symmetric transporter, and they contributed towards the development of a kinetic model of the Ec-NhaA translocation cycle [72]. This model supports the alternate accessibility mechanism of secondary transporters [63], wherein a single binding site alternates across the membrane. The translocation process includes an electrogenic rate-limiting step in which Na^+ is transferred across the membrane while bound to the aspartates of the binding site (Asp163, Asp164; TM V), and an electroneutral step transporting H^+ across the membrane while bound to these aspartates. H^+ and Na^+ compete for the single binding site, and the competition explains many phenomena of Ec-NhaA (see Section 2.3).

2.5.2 Tt-NapA, Nm- and Yf-ASBT: The Two-Domain Elevator Mechanism

A model of the operation mechanism of Tt-NapA was suggested on the basis of the antiporter's crystal structure in the outward-open conformation together with the Ec-NhaA crystal structure in the inward-open conformation [35]. A negatively-charged funnel is open to the periplasm in Tt-NapA. This funnel allows access to the strictly conserved Asp157 and Asp156 (equivalent to Asp163 and 164 in Ec-NhaA) believed to coordinate the ions. According to the crystal structures, the alternative access to this ion binding site, from the cytoplasm, requires large rotation of the core domain, about 20° against the dimerization interface (Figure 2). This large rotation of the core domain closes the periplasmic funnel seen in the Tt-NapA structure, and opens the cytoplasmic funnel, as observed in the Ec-NhaA structure. During this process, the two conserved, cation-binding, aspartate residues are shifted 10 Å toward the cytoplasm. Thus, a two-domain elevator mechanism model is suggested, which is similar to the model of the glutamate transporter (Glt_{Ph}) [95]. The model was supported by MD simulations and mutagenesis; in Tt-NapA, replacement of Asp156 and Asp157 with Ala [33] or Asn [35] resulted in complete loss of antiport activity. Of note, Ala replacement of Lys305 (homologue to Lys300 of Ec-NhaA) resulted in a compromised phenotype (an increase in apparent K_m for Na^+ (36 mM) and for Li^+ (17 mM)).

The two-domain elevator mechanism proposed for Tt-NapA was suggested to apply to both ASBTs. [36]. Using the internal symmetry of the topology-inverted motifs of Nm-ASBT according to [97], the outward-facing conformation was modeled [36]. In Nm-ASBT, sodium binding controls the conformation of the core domain, which in turn drives the movement of the panel domain (Figure 3b). This creates a large conformational change, which is required in order to alter the accessibility of the substrate binding site.

Two crystal structures of Yf-ASBT have been obtained in a lipid environment [37]: the wild-type was crystallized in an inward-facing conformation, and a mutant (E254A) was crystallized in an outward-facing conformation. Comparison between the two conformations revealed a large rigid-body rotation of the core domain that gives the crossover region accessibility to the other side of the membrane. This structural mechanism is very similar to that of Nm-ASBT and has been supported by mutagenesis and accessibility tests.

Remarkably, when transitioning between the cytoplasm-open and periplasm-open states, Tt-NapA, Nm-ASBT, and Yf-ASBT show similar rigid body movements of the core domain, despite substantial differences in the sizes of the substrates involved and the transport modes (antiport *versus* symport). The conformational change provides alternating access to the crossover region where Na^+ and protons are predicted to bind to a cluster of conserved acidic residues. However, the Na^+1 and Na^+2 sites are different. The deduced Na^+2 site in Tt-NapA (Arg331, Lys305, forming hydrogen bonds with the C-terminal ends of helices TM 4a and TM 11a) is replaced by two polar side chains in Yf-ASBT (Gln258, His71 forming hydrogen bonds with TM 4a and TM9a). In the structure of the carnitine transporter CaiT from *E. coli*, which is an Na^+-independent antiporter, an Arg mimics the bound Na^+ found in the Na^+-dependent symporters of the same fold [98]. It might be suggested that these residues in Tt-NapA have an analogous role to the bound Na^+ in Yf-ASBT and Nm-ASBT. This might provide insight into how the same fold and conformational changes act as scaffolding for highly distinct substrates and coupling mechanisms.

The question of whether the two-domain elevator mechanism is applicable to all Na^+/H^+ antiporters, including Ec-NhaA is still debatable [43, 49, 59], for the following reasons: (i) Ec-NhaA is one of the fastest transporters (turnover rate of 10^5 min^{-1} [20]). Notably, the rate of Tt-NapA is much slower (Table 1). (ii) Ec-NhaA shares this high turnover with the chloride/proton antiporter (ClC), which has a unique operation mechanism [99]. (iii) The elevator mechanism needs a rigid dimer interface for movement [35] (see Section 2.2.2). However, Ec-NhaA monomers have been shown to be functional [58]. (iv) The crystal structure of Mj-NhaP1 crystallized at pH 8 and the EM structure obtained at pH 4 resemble the structure of

Tt-NapA yet do not show the difference between the core and dimer domains suggested for Tt-NapA [43].

2.5.3 Pa-NhaP and Mj-NhaP1

In *Pyrococcus,* the Na^+ gradient required for ATP synthesis is maintained by Na^+/H^+ antiporters [100], most likely for pH homeostasis [101]. It has been suggested [38] that protons, probably in the form of hydronium ions, reach the Pa-NhaP binding site either through the cytoplasmic funnel or through the narrow pathway. Indeed, very slight rearrangement of the residues in either of these pathways enables H_3O^+ to pass. However, it is not yet clear which of the pathways operates to facilitate proton transfer, although it has been suggested that Na^+ is released from the cytoplasmic funnel and H^+ enters from the narrow pathway.

The transition of the Mj-NhaP1 structures from pH 4 to pH 8 was inferred to be induced by Na^+ [43]. The outward-open conformation resembles that of Tt-NapA. In both Mj-NhaP1 and ASBT, the core performs a rigid body rotation around an axis in the membrane (parallel to the dimer interface). In this transition the binding site moves upward and inward.

2.6 *Regulation of the Na^+/H^+ Antiporters*

2.6.1 Ec-NhaA

Similar to many prokaryotic and eukaryotic Na^+/H^+ antiporters [6, 21–23]), Ec-NhaA is drastically dependent on pH [20]; it is inactive below pH 6.5 and reaches maximal activity at pH 8.5. Comprehensive mutagenesis screening has been performed on this antiporter. When projected onto the structure, the mutagenesis map reveals a cluster of residues involved in transport activity, and another cluster that affects pH regulation (designated "pH sensor") (reviewed in [45, 59]). Two mechanisms of pH-regulation in Ec-NhaA have been proposed. The first is an allosteric mechanism [17], in which a "pH sensor", located far from the active site, accepts the pH signal and transduces it to activate conformational changes. This mechanism is supported by the following: (i) identification of clusters of amino acids distant from the active site that, when mutagenized, abrogate or change the pH profiles of Ec-NhaA as measured biochemically in inverted membrane vesicles and proteoliposomes [45]; (ii) identification of two conformational changes, one pH-induced and the other ligand-induced as determined by site-directed Trp fluorescence [87, 88]; (iii) prediction of a two-domain transport mechanism for the pH-induced conformational change of Ec-NhaA, on the basis of the two inverted repeats as well as data from elastic network modeling and biochemical cross-linking experiments ([46] and Section 2.3). Importantly, the allosteric mechanism model does not predict the movement of the ligand-binding domain but rather that of the pH sensor domain.

The second proposed mechanism is a direct competition of Na^+ and H^+ for the active site. Obviously, such a competition is expected to exist at the active site because the substrates of Ec-NhaA are H^+ and Na^+. However, the latter mechanism suggests that this competition also explains the pH regulation of Ec-NhaA, with no need for any allosteric effect of a "pH sensor." Results of electrophysiological experiments lend strong support to this proposition [72]. These experiments indicate that the decrease in transport activity at acidic pH values is due to an increase in K_m^{Na} and not due to a decrease in V_{max}. Furthermore, in the SSM system, even at pH 5, Ec-NhaA does not adopt an inactive conformation at either the cytoplasmic or periplasmic side.

In an attempt to reconcile the two proposed mechanisms, we have analyzed a number of different Ec-NhaA mutants, modulating the pH profile of the transporter using SSM-based electrophysiology [102]. The site-specific mutations affected Ec-NhaA activity in substantially different ways; residues as far apart as 15–20 Å from the binding site can have a significant impact on the dynamics of the conformational transitions or on the binding properties of Ec-NhaA [102]. As discussed below, the two mechanisms are not mutually exclusive.

2.6.2 Pa-NhaP and Mj-NhaP1

The pH-dependent transport activity of Pa-NhaP suggests a self-regulatory mechanism at pH 5, direct competition between Na^+ and H^+ for the binding site rather than regulation by a separate pH sensor as proposed for Ec-NhaA (Section 2.6.1). Nevertheless, at pH 6, the transport becomes cooperative, suggesting an allosteric mechanism related to the dimeric interface [38].

Mj-NhaP1 was shown to be active at pH 8, and although its crystal structure has not been determined, it has been suggested to be down-regulated at pH 4. Using SSM-based electrophysiology, a mechanism of direct competition of Na^+ and H^+ for the active site, similar to that proposed for Ec-NhaA, has been suggested for Mj-NhaP1 [49, 72, 103]. Accordingly, at low pH, a 10-fold higher K_m was observed. Nevertheless, Na^+ was shown to compete at pH 4 but not at pH 6 [103]. To explain the lack of competition at pH 6, the author concluded that there is another mechanism of pH regulation, e.g., side chain movements, and/or small helix movements that cannot be seen at low resolution.

In summary, there is currently no answer as to how the proteins of the NhaA fold are pH-regulated at the molecular level. The two proposed mechanisms of pH regulation described above obviously represent different mechanistic concepts, although they are not mutually exclusive, and the conformational transitions involved are not necessarily different. As stated above, given the ligand exchange mechanism, competition between the ligands, Na^+ and H^+, is expected to take place. The question remaining is whether in addition to the competition there is a pH-induced activation step.

It is important to note that the experiments supporting the allosteric mechanism were based on biochemical-biophysical approaches, whereas those supporting the direct competition mechanism were mainly based on SSM-based electrophysiology

[72, 104]. As discussed in Section 2.4.2, there are differences between the results obtained by electrophysiology and those obtained by classical transport assays. We believe that the differences in time resolution of the assays (electrophysiology: milliseconds; biochemistry: seconds) and driving forces (electrophysiology: chemical gradient; biochemistry: potential gradient) of the transport assays are the main factors driving the differences observed.

In conclusion, it is clear that capturing multiple distinct conformations of a given transporter is an essential step in elucidating the transport mechanism. Unfortunately, as yet, such a task has proven challenging to accomplish for many transporters. Instead, comparative analyses were performed in order to obtain mechanistic insight [105]. Additional structures and rapid assays of transport kinetics are needed in order to clarify by what means ion binding and release processes are coupled to structural changes and to driving forces, as well as to reveal how these systems are regulated.

3 Eukaryotic Na^+/H^+ Antiporter Proteins and Their Properties

Eukaryotic Na^+/H^+ antiporters have recently been the subject of several reviews, stressing their biological and clinical importance [3, 106–109].

3.1 The NHE Eukaryotic Na^+/H^+ Exchangers

The NHE Na^+/H^+ antiporters are ubiquitous ion transporters. They are present in all animal species, in the cytoplasmic membranes of cells and in the membranes of many organelles, and they have homologues in all kingdoms of life, from prokaryotes to eukaryotes. Currently, 13 evolutionarily conserved NHE isoforms are known in mammals [9, 21, 107, 110, 111]. Most species possess multiple Na^+/H^+ antiporters, which may serve redundant and/or specific functions.

The NHE Na^+/H^+ exchangers are encoded by the SLC9 gene family (Solute Carrier family 9) (HUGO nomenclature, http://www.genenames.org) and are a subgroup of the eukaryotic and prokaryotic monovalent CPA1 family (Transport Protein database http://www.tcdb.org) [9, 10, 21, 107]. The genes are dispersed throughout the genome, some of which are subject to alternative RNA splicing to yield multiple transcripts. The primary structures of the encoded proteins vary considerably in both sequence identity (13–68 % amino acid identity) and size (48–99 kDa).

On the basis of phylogeny, the SLC9 gene family is divided into three subgroups (http://www.bioparadigms.org). The SLC9A subgroup encompasses plasmalemmal isoforms NHE1-5 (SLC9A1-5) and the predominantly intracellular isoforms NHE6-9 (SLC9A6-9) [108]. However, NHE3 and NHE5 exist in a dynamic equilibrium with the recycling endosomal pool. The SLC9B subgroup consists of NHA1 (SLC9B1) and NHA2 (SLC9B2). The SLC9C subgroup consists of a sperm-specific plasmalemmal NHE (SLC9C1), as well as SLC9C2, a putative NHE for which functional data are currently lacking [107].

As yet, there are no available crystal structures for any of the NHEs, except for structures of short segments of NHE1 (see Section 3.1.1.2). Nevertheless, it is possible to conclude, on the basis of a large volume of biochemical data, that all NHEs are organized in a similar fashion: A very short N-terminus facing the cytoplasm followed by an N-terminal domain composed of 11–14 TMs (~450 residues) that carries out the Na^+/H^+ exchange, and an intracellular C-terminal domain (~125–440 residues, depending on the isoform) that is involved in regulation of the exchange activity. The C-terminal domain is suggested to associate with many regulatory factors and also with the intracellular aspects of the N-terminal domain as well as with the inner leaflet of the plasma membrane via interactions with negatively-charged phospholipids [112, 113]. All SLC9 members appear to exist as dimers, while transport function occurs in a monomer [114]. The dimerization is believed to provide stability.

Plasma membrane NHEs function by coupling to the Na^+ electrochemical gradient generated by the ubiquitous sodium pump (Na^+/K^+ ATPase), moving one Na^+ ion into the cytoplasm in exchange for removal of one H^+. The reported affinities of NHE1, –2, and –3 for extracellular Na^+ are in the range of 5–50 mM [115, 116]. The dependence of NHE1, –2, and –3 on [Na^+] shows simple Michaelis–Menten kinetics [106]. For NHE4, both Michaelis–Menten-type [Na^+] kinetics [117] and sigmoidal [Na^+] dependence have been reported [118]. In addition to Na^+ and H^+, Li^+, K^+, and NH_4^+ are also substrates of NHEs [106, 107, 119]; K^+ transport has been proposed to be a major function of the organellar NHE isoforms, NHE6, –7, and –9 [107].

Despite being bathed in neutral or slightly alkaline solutions, mammalian cells require active, continuous extrusion of protons to maintain their cytosolic pH within the physiological range. Excess acid equivalents are produced by a variety of metabolic pathways, and protons (equivalents) are driven into the cells through conductive pathways by the negative membrane potential. NHE1-5 participate in the regulation of cytosolic and organellar pH, cation composition, and cell volume, thereby contributing to the creation of environments suitable for cell function and survival. For example, in the intestine and kidney, NHEs, operating together with bicarbonate transporters, are critical for transepithelial movement of Na^+ and HCO_3^- and thus for whole body volume and acid–base homeostasis [106].

A bewildering collection of physiological agonists and antagonists regulates the NHEs, in most cases via the C-terminal domain; each NHE's regulation process will be described below (for review see [106, 120]). An interesting example is that many agents that stimulate NHE1 inhibit NHE3, and *vice versa*. NHE2 generally responds in a manner akin to NHE1 while the limited information available suggests that NHE5 shares regulatory properties with NHE3. The regulation can be targeted to changes in the proton set point, protein exchange activity, protein expression, redistribution in different membranes and protein synthesis and stability [121, 122]. For a summary of the various inhibitors of NHEs, turnover and trafficking, see a comprehensive recent review [3].

NHE6 and NHE9 are currently the only NHEs directly linked to human disease (Sects. 3.1.6 and 4.3); mutations in these genes cause neurological disease in humans. However, it is becoming increasingly apparent that members of SLC9 gene family contribute to the pathophysiology of multiple human diseases (see Section 4).

3.1.1 SLC9A1-NHE1

NHE1 is a plasma membrane protein that uses the chemical energy of the Na^+ gradient maintained by the Na^+/K^+ ATPase across the plasma membrane of eukaryotes for electro-neutral counter transport of H^+ [123–127]. It can operate reversibly [128, 129]. The Na^+/H^+ stoichiometry is 1/1 [130] or 2/2 [131]. Hence, NHE1 is electroneutral and does not disturb membrane potential during activity. This is particularly critical for excitable cells, in which the membrane potential must be stringently regulated. NHE1 can exchange Li^+ for Na^+, but other alkali cations are excluded.

NHE1 exists in the plasma membranes of almost all mammals [124], with some exceptions [132, 133]. It is the most abundant antiporter in the heart. It resides almost exclusively at the surface of cells, but preferentially accumulates in discrete microdomains within the plasma membrane, depending on the type and state of the cell [133–135]. In migrating fibroblasts, NHE1 concentrates at the leading edge of the cell along the border of lamellipodia [136]. NHE1 abundance at the plasma membrane is regulated by ubiquitination [122].

NHE1 is critical for regulation of intracellular pH, salt concentration and cell volume [3, 107–109, 137, 138]. As a result, NHE1 is often referred to as the "housekeeping" NHE isoform. By modifying the cytosolic pH, NHE1 also indirectly contributes to the regulation of the pH of the endoplasmic reticulum and of the nucleus, both of which seemingly lack independent pH homeostatic mechanisms [106], and whose membranes are freely permeable to H^+ equivalents. Notably, at the cytoplasmic membrane, NHE1 operates in conjunction with Na^+-coupled bicarbonate transporters [106].

NHE1 has numerous other indirect physiological roles that accompany or are a result of its pH regulatory function (see Section 3.1.1.3). NHE1 has long been a drug target because it has been suggested to play a role in heart pathology [109], hypertrophy [139], cardiac ischemia [140], and hypertension [141] (see Section 4.1).

3.1.1.1 Topology Model of NHE1

Hydropathy plots and extensive biochemical studies indicate that NHE1 has two functional modules [106, 142]: (i) the N-terminal ion translocation module, which is an integral membrane domain of 12–14 TMs (~500 amino acids), and (ii) the C-terminal regulatory module comprising ~300 cytoplasmic amino acids. The regulatory function is exerted by many agonists and antagonists through association with a number of signaling molecules (see Section 3.1.1.3). The NHE1 protein is both N- and O-glycosylated, although glycosylation is not required for transport function [143, 144].

A topology model of the translocation module of NHE1 has been suggested on the basis of hydropathy plots and Cys scanning accessibility tests [109, 145] (see Sects. 2.3.2 and 2.3.3). Intracellular- and extracellular-accessible Cys replacements were identified through the use of whole and permeabilized cells. The constructed model, called herein the Wakabayashi model, shows 12 TMs (designated here I-XII) with N^{in}-C^{in} topology and a glycosylation site at N75 between TMs I and II. TM X in the hydrophobicity plot has extracellularly-accessible residues on both ends, and therefore was proposed to be a re-entrant loop rather than a TM. Two intracellular loops, TMs IV-V and TMs VIII-IX, were also suggested to be re-reentrant loops. For further details see [126].

A three-dimensional homology model of the membrane part of NHE1 has been built [11] using the Ec-NhaA crystal structure [31, 51] as a template (Figure 6a), in spite of very low homology (sequence identity of ~10 %). According to this model, called herein the Landau model, the topology of NHE1 also contains 12 TMs (denoted here 1–12 to avoid complexity). The model has some similarities to and differences from the Wakabayashi model as follows: (a) the two first TMs (I and II) are absent in the Landau model because they are poorly conserved and are not important for functionality [146]. However, fully and partially glycosylated NHE1 (between Wakabayashi's TM 1 and 2) are often found in Western blots of NHE1, raising the possibility that there are 14 TMs in NHE1. (b) Excluding these first two TMs, the other first six helices (TMs 1–6 in the Landau model), and TMs III-VIII in the Wakabayashi model, have similar assignments in the two models. (c) TM IX of the Wakabayashi model is split and extended into two short helices in the Landau model (TMs 7–8). (d) The reentrant loop (between IX and X of the Wakabayashi model) is now reassigned as TM 9 in the Landau model. (e) The helix assignment of TM 10–12 is similar in both models. (f) Resembling the TM IV/XI assembly in Ec-NhaA (see Section 2.2.1), TM 4 and TM 11 are disrupted by extended chains in the Landau model (Figure 6).

The Landau model is supported by observations of conserved charged residues in the core of the protein, at the extended chain region [11]: D238 is thought to be equivalent to D133 in Ec-NhaA, compensating for the helix-end dipole charges. D163 and D164, which, in Ec-NhaA, are essential and bind ligands [62] (see Section 2.2.1), are replaced with N266 and D267 in NHE1 [11]. In addition, NHE1 contains conserved amino acid clusters facing extra-cellularly, which are important in pH regulation and inhibition of NHE.

The two models are reviewed in [126]. Another model has been suggested on the basis of the Ec-NhaA structure and a single-electron paramagnetic resonance measurement [147]. This model differs from the previous models [148]. Remarkably, strong support for the Landau model has recently been obtained: *in vitro* topology analysis has shown the presence of a signal sequence in NHE1, suggesting possible cleavage of the N-terminus [149]. Furthermore, recent amino acid sequencing analysis of NHE1 purified from fibroblasts or SF9 insect cells has revealed that cleavage indeed occurs between Gly36 and Leu37 just after TM I of the Wakabayashi model [109]. Therefore, the Wakabayashi topology model might have based on a minor fraction of NHE1 in which the N-tail was under forced expression.

According to the Landau model, TM 4 and TM 11 of NHE1 are in close proximity to each other and contain an unwound crossing region in the middle of the TMs, generating an extended antiparallel arrangement that is thought to form part of the ion permeation pathway [11]. This notion has been supported by mutational analyses of amino acids in TMs 4 and 11 [23, 150, 151] that altered affinities for Na^+ and H^+, respectively. These helices, along with TM 9 and TM 10 of the Landau model, also confer sensitivity

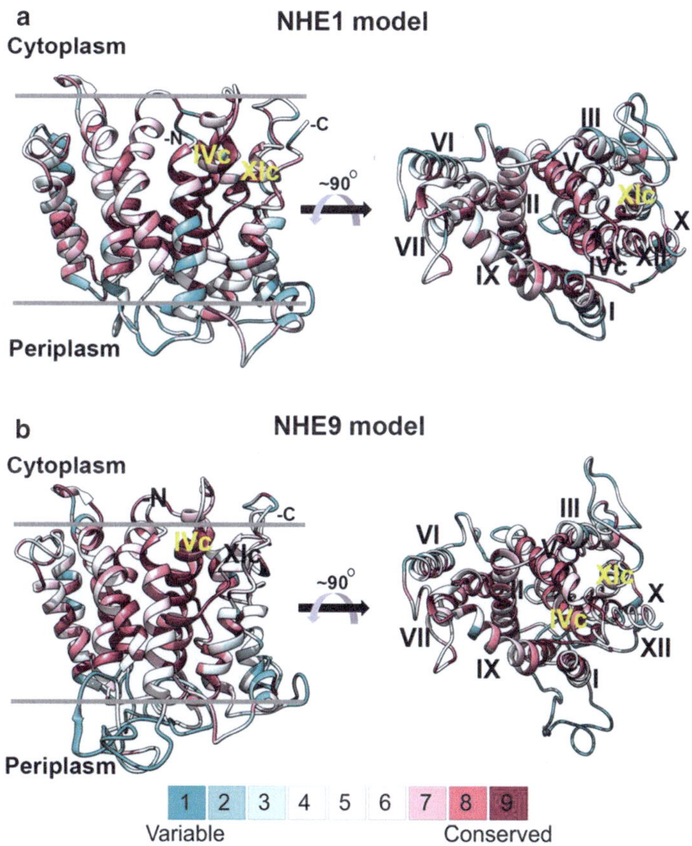

Figure 6 Evolutionary conservation scores projected on the modeled structures of the human antiporters NHE1 and NHE9. The modeled structures of human NHE1 [11] (**a**) and NHE9 [248], (**b**) based on the Ec-NhaA [31] crystal structure, are displayed in a ribbon representation. The ribbons are colored according to evolutionary conservation scores calculated with the ConSurf server with turquoise-through-maroon indicating variable-through-conserved [397–399]. The images on the *left side* are viewed in the membrane plane, with the intracellular side facing upward. Tentative boundaries of the membrane are shown. In the images on the *right side*, the molecule is rotated about 90°, and viewed from the cytoplasmic side into the membrane.

to competitive inhibitors of Na^+ transport (for a review see [106]), consistent with a prominent role for these regions in ion translocation. Hence, remarkably, the membrane domain structure of NHE1 shares the NhaA fold together with the prokaryotic Ec-NhaA [31], NhaP [38, 43, 57], ASBT [36, 37], and NapA [35].

NHE1 exists as a homodimer in the plasma membrane [106, 114, 152–154]. A dominant negative effect was observed [155] at neutral pH but not at acidic pH. This observation supports the view that dimerization of two active subunits is required for NHE1 to possess exchange activity in the neutral pH range, although each subunit is capable of catalyzing transport in the acidic pH range [131].

3.1.1.2 A Long Way Remains to an Atomic Structure of NHE1

Experimental structural information on NHE1 has been obtained only for segments of the protein. Fractions of its soluble domain were analyzed by nuclear magnetic resonance (NMR) [156] or by crystallography [157, 158]. The structures of single putative TMs of the membrane domain were modeled by NMR [127, 159]. While much of the required structural information is contained within the primary amino acid sequences of the TM segments, interactions between TM segments can obviously not be studied with single segments. Additionally, in some cases, the interactions between helices and the presence or absence of inter-helical loops can affect the conformation of TM segments [160]. Recently, the structure of a region spanning two transmembrane domains (TMs 4–5 in the Landau model) was determined using NMR [161]. Tryptophan scanning and site-directed mutagenesis of the first segment (TM 4 in the Landau model) supported the structure [161].

The C-terminus of NHE1 extends into the cytoplasm and is the least conserved region among the NHE family members (ranging in length from ~50–450 amino acids). This segment contains numerous putative as well as proven recognition sites for protein kinases and other ancillary factors that regulate the exchanger activity in response to diverse stimuli, though the molecular mechanisms by which these signals are transduced from the regulatory domain to the catalytic domain remain obscure. *In silico* analyses predict that intrinsic disorder exists in the distal part of the C-terminal tail [3]. Interestingly, in this part of the SLC9A tail, the intrinsic disorder is much more strongly conserved than the sequence is.

Three important crystal structures of parts of the C-terminus tail have paved the way to better understanding of NHE1 regulation [156, 158, 162] (see Section 3.1.1.3).

3.1.1.3 The Regulation of NHE1

At the N-terminal module, NHE1 is allosterically regulated by H^+ [115, 117, 119, 123, 128, 163–165]. As the cytosol becomes acidified, increased substrate $((H^+_{in}))$ availability tends to accelerate Na^+/H^+ exchange. However, this thermodynamic response is insufficient for accurate pH homeostasis. The concentration dependence of cytosolic H^+ on Na^+/H^+ exchange is much steeper, with a Hill coefficient of >2

[23, 116, 128]. Such pH_{in} dependence suggests the existence of two H^+ binding sites. One H^+ is bound at the active site as a substrate, while the other binding site uses cytosolic H^+ to activate the exchange activity (this site is called the H^+-modifier site or "H^+-sensor" or "set point"). The latter exerts an allosteric effect on the exchanger: Transport becomes markedly activated when the intracellular pH drops below a threshold level or "set point" [116, 165]. The existence of a modifier site has also been interpreted as a necessary feature to safeguard cells from excessive alkalinization, while effectively protecting them against acidosis. Not coincidentally, the set point of the housekeeping isoform NHE1 is similar to the resting physiological pH [166, 167]. It was also proposed that the stimulating effect of growth factors and other stimuli at the C-terminal module on NHE1 can be explained by these agents causing an increase in the apparent H^+ affinity of the allosteric H^+-binding site [109].

Notably, a recent study reports that H^+ activation of NHE1 is best explained by an allosteric model for a dimer transporter that does not include an additional H^+ sensor site [103, 168, 169]. Although a definitive conclusion for the mechanism underlying H^+ activation awaits detailed mapping of H^+-binding sites on the NHE1 molecule, several critical amino acid residues such as Glu131, Arg327, and Arg440 have been identified as important determinants of the pH_{in} dependence of NHE1 [151, 170].

NHE1 is highly sensitive to amiloride and is also sensitive to lipophilic amiloride derivatives including ethylisopropylamiloride (EIPA), benzoylguanidines HOE694 and capiroide [171]. Derivatives of the NHE1 inhibitor amiloride are known to interact with the N-terminal domain of NHE1 [172, 173]. Ec-NhaA is only sensitive to an amiloride derivative, 2-aminoperimidine (AP [174]), whose binding site has been identified on the Ec-NhaA crystal structure using the Landau NHE1 model and mutagenesis [11].

The regulation of NHE1 at the C-terminal module [175] involves phosphoinositides, postranslational modifications (phosphorylation, ubiquitylation), and binding proteins [3, 109]. The cytoplasmic region proximal to the membrane (~60 residues), termed the lipid-interacting domain (LID), is an important regulatory domain of NHE1. Modulation of NHE1 activity by various activators and inhibitors occurs through the direct binding of these molecules to the LID, which alters the association of the LID with the plasma membrane; ATP and the NHE1 activator phosphatidylinositol 4,5-bisphosphate bind competitively to the LID [109, 176].

The second messenger, Ca^{2+}, is involved in NHE1 regulation via four Ca^{2+} binding proteins, including calmoduline (CaM), which bind to two sites at the C-terminus of the regulatory module. A crystal structure (2.3 Å resolution) of the NHE1 binding region (amino acids 622–684) in complex with CaM (amino acids 5–48) and Ca^{2+} has been determined [158]. Calmodulin binds to a high-affinity site (amino acids 636–656) and to a low affinity site (amino acids 657–700). The structure sheds light on how Ca^{2+} might regulate NHE1 activity [158]. In the absence of calmodulin, the high-affinity binding site of NHE1 interacts with another region of NHE1, most likely the proton modifier site in the transmembrane domain, acting in an autoinhibitory manner. In the presence of increased intracellular Ca^{2+} and calmodulin, this interaction is released, and activation of NHE1 follows. In support of this model, deletion of the high-affinity calmodulin binding site on the C-terminus renders NHE1 constitutively active [158]. Association of NHE1 and CaM through endothelin-dependent signaling pathways has recently been shown *in vivo* [177]. For an extensive review see [109].

Calcineurin homologous proteins (CHPs) 1–3 constitute another well-studied group of NHE1-binding proteins (review in [109]). CHP1 is cytosolic and myristoylated, binds Ca^{2+} and interacts with a hydrophobic cluster of residues in NHE1 located between amino acids 510 and 530. CHP1 is an obligatory binding partner of NHE1 in the membrane (and also of NHE2 and NHE3), and its primary function in NHE1 appears to be stabilization of the NHE1 structure in the physiologically relevant conformation for optimal surface expression and exchange activity [178, 179]. Thus, NHE1 or CHP1 mutations that interfere with the interaction result in a dramatic loss of both NHE1 activity and NHE1 protein expression in the plasma membrane. Furthermore, CHP1 is co-purified with NHE1 from fibroblasts at nearly 1:1 stoichiometry [180]. CHP1 is phosphorylated in cells [181]; however, the role of this phosphorylation awaits further investigation. It is not clear whether or how CHP1 is involved in the external stimulus-induced activation of NHE1 [157, 178]. In cardiac muscles, both CHP1 and CHP3 are expressed; however, it is not yet known how cardiac NHE1 is differentially regulated via interaction with these distinct CHP isoforms.

CHP2 has 61 % amino acid identity to CHP1 and binds to the same juxtamembranous region on the NHE1 C-terminus. CHP2 is highly expressed only in the intestine and in tumor cells and protects cells from serum deprivation-induced cell death by activating NHE1 and increasing cell pH. The crystal structure of CHP2 in conjunction with the juxta-membranous portion of the NHE1 C-terminus has recently been solved [156, 162].

Evidence suggests that NHE1 binds ATP [182]. However, the need for ATP is largely the result of cellular depletion of PIP2 when ATP is depleted [183]. The inner leaflet of the plasma membrane is rich in PIP2, which at physiological pH bears between 3 and 4 negative charges on average. The high charge density probably attracts positively charged regions of the tail of NHE1, which have been shown to be essential for optimal NHE1 activity [106]. Such polycationic motifs are discernible in the cytosolic domains of NHE1 and NHE3 [184, 185]. NHE1 contains two C-terminal juxta-membrane stretches of positively charged amino acids that bind PIP2 *in vitro* [184]. Mutation of these two sites greatly reduces NHE1 activity and PIP2 binding. Accordingly, synthetic peptides mimicking the polycationic motifs of NHE3 detach from the membrane when PIP2 is depleted [186]. Because depletion of ATP is accompanied by net dephosphorylation of PIP2 into inositides of lower charge, detachment of the tail from the membrane could readily account for the observed inhibition of transport [184]. Indeed, NHE1 function in ATP-depleted cells can be restored by direct perfusion of PIP2 into the cell. Nevertheless, the underlying mechanisms by which PIP2 affects NHE1 activity remain unknown.

Sardet et al. first demonstrated that NHE1 is a phosphoprotein [187]. All relevant *in vivo* phosphorylation sites map to the NHE1 C-terminal 636–815 amino acids. However, deletion of these sites resulted in only a ~50 % loss of the growth factor stimulatory effect on NHE1 activity, suggesting that activation of NHE1 by growth factors occurs via additional mechanisms that do not require its direct phosphorylation. Several kinases have been shown to be involved in NHE1 phosphorylation [12, 22, 107, 188, 189]; other kinases and isoforms potentially involved in phosphorylation have been investigated poorly, if at all. Dephosphorylation of NHE1 is carried out by protein phosphatases PP1 and PP2a, and prevention of dephosphorylation by inhibition of phosphatases maintains NHE1 in an active state.

A recently identified NHE1 splice variant is an Na^+/Li^+ exchanger and not an Na^+/H^+ exchanger. The functional relevance of this protein has not been established. Notably, NHA2 also carries out Na^+/ Li^+ exchange (see Section 3.2).

3.1.2 SLC9A2-NHE2

The NHE2 protein [190, 191], an epithelial Na^+/H^+ exchanger, has 812 amino acids and is O-, but not N-linked glycosylated [192]. In contrast to the ubiquitous expression of NHE1, NHE2–5 show a restricted expression pattern: for example, NHE2–4 are predominantly expressed in the epithelial cells of the kidney, small intestine, and stomach, and NHE5 is mainly expressed in the brain [193–196]. When expressed in epithelial cells, NHE2 predominantly localizes to the apical membrane [193, 196]. For further details see [107, 190, 191, 193]. NHE2 has been suggested to be involved in intestinal and renal Na^+ absorption and possibly also in repair of epithelial damage [197]. Most recently, NHE2 expression was shown to be localized in the pituitary, with functional implications [198].

3.1.3 SLC9A3-NHE3

The NHE3 protein [190, 199] contains 834 amino acids in humans and is an example of the NHE isoforms that continually traffic between the recycling endosome system and the plasma membrane with their major function being in the plasma membrane [200, 201]. NHE3 is thus most related to NHE5 and is dissimilar to the intra-organellar NHEs, whose major function is in the organellar membranes [9] (Section 3.1.6). NHE3 is highly resistant to amiloride and its derivatives [202].

NHE3 is present in the Na^+ absorptive cells of the mammalian small intestine, colon, gall bladder, renal proximal tubule, and thick and thin limbs of the loop of Henle [193]. The apically-localized NHE3 is responsible for the majority of intestinal and renal Na^+ absorption [107, 203]. In these absorption processes, NHE3 is linked to members of the SLC26A family of Cl/HCO3 exchangers, particularly DRA (SLC26A3) [204]. The linkage is established indirectly via changes in pH_{in}, although there may also be a direct physical linkage between NHE3 and DRA.

In intestinal Na^+ absorptive cells, nutrient (D-glucose and D-galactose, L-amino acid) related Na^+ absorption and neutral NaCl absorption via NHE3 were once thought to be independent processes [203]. However, it is now known that D-glucose, acting via an SGLT1 transporter (SLC5A1), increases NHE3 trafficking to the brush border of intestinal Na^+ absorptive cells [205, 206]. Specifically, SGLT1 uptake of D-Glu initiates an apical signaling cascade to increase NHE3 exocytosis. This process, along with changes in tight junctional permeability, is likely to contribute to the stoichiometry of Na^+ to glucose absorption occasionally being higher than that of SGLT1 (2 Na^+:1 D-Glu). The involvement of NHE3 in respiration is not well defined [207].

Despite being classified as plasmalemmal, NHE3 and NHE5 are also present in internal membranes, specifically in recycling endosomes [208, 209]. Operation of the NHEs in their conventional (H^+ excretion) mode predicts that Na^+ trapped during endocytosis drives the translocation of cytosolic protons into the endosomal lumen, where these protons accumulate. Indeed, a modest yet significant acidification attributable to Na^+/H^+ exchange has been reported [208, 210]. Vesicular traffic along the endocytic pathway is known to be dependent on proper luminal acidification [211]. Accordingly, knocking out the NHE3 gene or inhibiting its activity have been associated with impairment of albumin delivery to sorting endosomes and lysosomes in renal cells, as well as with tubular proteinuria in whole animals [210, 212]. Despite these fascinating observations, it is not clear whether the plasmalemmal-type NHEs can contribute significantly to organellar pH homeostasis. Because they are highly selective for Na^+, these exchangers can only operate transiently, since their luminal substrate becomes rapidly depleted, unless replenished by active Na^+/K^+-ATPases. However, in the case of polarized epithelial cells, the apically derived vesicles bearing NHE3 are devoid of Na^+/K^+-ATPases.

It must be borne in mind that vacuolar-type H^+ pumps (V-ATPases) are generally acknowledged to be the primary determinants of luminal acidification in all the subcompartments of the endocytic pathway. Hence, it is unclear whether the limited contributions of NHE3 and similar exchangers have a measurable impact on the maintenance of the luminal pH of endosomes. It is also not clear how this function interacts with the organellar V-ATPase.

NHE3 is one of the most highly regulated transport proteins, with regulation occurring acutely, as well as on both an intermediate and a long-term basis, including circadian regulation [213, 214]. For detailed reviews of second-messenger and ligand regulation of NHE3, see [107, 200, 201, 203, 206, 213, 215–218].

The NHE3 C-terminus is necessary for NHE3 regulation. The NHE3 C-terminus acts as a scaffold binding multiple proteins and various lipids [188, 213, 219–224]. It has recently been shown that NHE3, when complexed with CHP1, undergoes a shift in pH_{in} sensitivity (0.4 units) toward the acidic side in comparison with NHE3 alone, as measured by oscillating pH electrodes combined with whole-cell patch clamping [225]. Thus, CHP1 interaction with NHE3 apparently establishes the exchanger set point for pH_{in}.

Structurally, the C-terminus of NHE3 is predicted to have multiple helical domains close to the N-terminal transport domain. One of these domains, NHE3 amino acids 473–497, is highly homologous to a domain in the NHE1 C-terminus (amino acids 516–540) for which the structure has been solved when bound to CHP2 [157].

3.1.4 SLC9A4-NHE4

Human NHE4 has 798 amino acids [107]. NHE4 expression is highest in the stomach and is also found, although at lower levels, in the kidney medulla, hippocampus, zymogen granules of the pancreas, and salivary gland.

NHE4 localizes to the basolateral membrane of epithelial cells. In the stomach, NHE4 is expressed in parietal cells. In the salivary gland, it is expressed in acinar and duct cells. However, NHE4 is not found in the small or large intestine. When expressed in fibroblasts, NHE4 is virtually inactive but can be activated by cell shrinkage or by an anion transporter inhibitor [195]. NHE4 is relatively resistant to amiloride and its analogs. NHE4-knockout mice have provided some insights into the protein's physiological and pathological functions [226, 227] (see Section 4.2).

3.1.5 SLC9A5-NHE5

Human NHE5 [228] has 896 amino acids. Its expression is highly restricted to the brain and sperm, although it is also observed in the spleen and skeletal muscle [228–230]. NHE5 found in multiple regions of the brain, suggesting that the protein might be neuron-specific.

NHE5 has high homology with NHE3 (50 % amino acid identity) and shares similar pharmacological, regulatory, and cell localization properties. Thus, NHE5 is located at both plasma membrane and recycling endosomes when expressed in mammalian cells, and it cycles between these two pools [209]. The internalization of NHE5 is clathrin-mediated, and the protein forms a complex with β-arrestins [231, 232]. Mutations within NHE5 amino acids 702–714 have been shown to abolish phosphorylation and binding of β-arrestin-2 *in vitro*. However, the mutant retained its ability to form a complex with β-arrestin-2 *in vivo*. Phosphorylation of NHE5 by the kinase CK2 is required for binding to β-arrestin-2 and for internalization of NHE5 to recycling endosomes [231, 233]. For further details see [107, 164, 233, 234].

NHE5 has recently been suggested to control dendritic spine growth via a pH-dependent negative-feedback mechanism [235]. A human brain cDNA library for NHE5-interacting proteins was screened, and a 2-catalytic subunit of AMP-activated protein kinase (AMPK) was identified as a putative partner [236]. AMPK is an evolutionarily conserved serine/threonine protein kinase that functions as a key sensor and master regulator of energy homeostasis at the cellular and organismal levels. Experimental results show that NHE5 forms a complex with both catalytic subunits of AMPK, and that activation of AMPK regulates hippocampal neuronal pH_{in} response to metabolic stress-induced acidosis by promoting cell surface accumulation of NHE5. This interaction represents a potentially novel mechanism for coupling energy metabolism to pH_{in} homeostasis in nervous tissue. It is puzzling that the allosteric effect of cytosolic H^+ was not detected in the case of NHE5 [237].

3.1.6 SLC9A6, SLC9A7, and SLC9A9-Intracellular NHEs: NHE6, NHE7, NHE9

NHE6, NHE7, and NHE9 [111, 238] are considered together as intracellular members of the NHE family derived evolutionarily from the *Saccharomyces cerevisiae* Nhx1, the first member of the NHE family to be molecularly identified [239]. Nhx1 can probably serve as a model for the function of the other intracellular NHEs, and is a K^+ and Na^+/H^+ antiporter. It is present in the yeast vacuolar membrane, where it serves to exchange intra-vacuolar H^+ for cytosolic K^+ and thus sets the vacuolar and cytosolic pH. It also appears to be involved in membrane trafficking from the late endosome to the vacuole and fusion in the vacuole [240]; these processes are dependent on intra-vacuolar pH. Abnormalities in these steps due to Nhx1 knockout can be reversed by correcting the vacuolar pH. Nhx1 has also been studied in a cell-free system [241].

In contrast to the transport of plasmalemmal NHEs, active transport by endosomal NHE is driven by the H^+ gradient generated by the V-type H^+-ATPase, resulting in cation (Na^+ or K^+) sequestration coupled to removal of protons from the compartmental lumen. Hence, the plasma membrane and endosomal subtypes have different ion selectivity, with the former being Na^+-selective, whereas most intracellular isoforms transport both K^+ and Na^+ ions [242]. The intra-organellar NHEs have similar V_{max} and K_m for Na^+ and K^+. The cation selectivity of endosomal NHEs has been related to the presence of multiple charged residues at the putative second extracellular loop.

In the endocytic pathway, organellar pH gradually decreases from early endosomes to lysosomes, with the latter exhibiting a pH of 4.5–5 [243]. Large pH gradients also exist within the recycling endosomal system of polarized cells [244]. Organellar pH is a critical regulator of enzyme activity, intracellular trafficking, membrane fusion, posttranslational modifications, dissociation reactions of receptor–ligand complexes, and uptake of neurotransmitters [242]. Despite the importance of tightly-controlled endomembrane pH, the molecular mechanisms maintaining pH in each organelle remain poorly understood. It is generally accepted that organellar pH is not set by regulation of V-ATPase activity per se but by proton-leak pathways (e.g., NHEs) and counter-ion conductance (e.g., Cl^- channels) [243, 245].

Compared with the extensive research done on plasma membrane isoforms, functional characterization of mammalian endosomal NHE has been conspicuously slow. This is in part due to methodological limitations. Standard assays of plasma membrane-NHE activity, which follow pH-dependent recovery, consistently fail to detect endosomal NHE function. Measurement of luminal pH uses compartment-specific pH-sensitive green fluorescent protein (GFP) constructs, which often show only modest changes, possibly, in part, because of the buffering role of GFP itself in small vesicles. In COS7 cells, an elevation of 0.4 pH units in recycling endosomes upon overexpression of NHE9 has been observed [246]. However, in the latter study, no effect of NHE6 overexpression was observed.

Nevertheless, in light of their distinct localization primarily along the endosomal and secretory pathways [242, 246], it is likely that intracellular NHEs play a critical part in the regulation of organellar pH. NHE6 is mostly considered a recycling endosomal protein [247], NHE7 is present in trans-Golgi [238], and NHE9 is present in late and recycling endosomes and late endosomes/lysosomes [246]. NHE8, which is discussed separately (see Section 3.1.7) is present in mid- to trans-Golgi, although it appears to be present in the brush border of some epithelial cells. Providing additional supportive evidence of the role of intracellular NHEs, studies have shown that overexpression or loss of these NHEs alters organellar pH [208, 246, 248, 249]. This notion is challenged, however, by studies in which knockdown or loss of individual organelle-specific NHEs had no effect on pH [250–252], although redundancy may theoretically explain the latter finding. Nevertheless, the spatial and temporal regulation of intracellular NHE expression in the brain may hold clues to the distinct and diverse phenotypes associated with these proteins, including autism and neurological disorders [111] (see Section 4).

In specialized tissues, NHE6, NHE8, and NHE9 appear, at least transiently, at the plasma membrane. NHE6 has been detected in the basolateral membrane of mineralizing osteoblasts, where it seemingly contributes to the maintenance of cytosolic pH [253]. NHE6 and NHE9 were also detected at the surface membrane of hair cell stereocilia of the inner ear [254]. In this system, the exchangers seem to exchange cytosolic H^+ for K^+, which is concentrated in the endolymph facing the extracellular surface of the hair cells.

Since NHEs exchange cations for protons, they affect not only intraluminal H^+ concentration but also the concentration of Na^+ and K^+, depending on ion selectivity. However, very little is known about the regulation of organellar Na^+ and K^+ by cation/proton antiporters. It is likely that organellar cations participate in the regulation of many processes, as does pH [246, 248, 253–255].

Regulation of the intracellular NHEs has not been comprehensively studied. NHE6 binds RACK1 (as do NHE7 and NHE9), using the NHE cytoplasmic C-terminus. Knocking down RACK1 leads to higher expression of intracellular NHE6 and less expression of plasma membrane NHE6 and is accompanied by higher pH in the recycling endosomes [242, 256]. This suggests a RACK1-dependent regulated trafficking of NHE6 between the plasma membrane and recycling endosomes. NHE9 also binds RACK1 via its cytoplasmic C-terminus, and this interaction seems to affect the steady-state distribution of NHE9 between endosomes and the plasma membrane. For additional reviews see [108, 111, 257].

NHE6 and NHE7 have multiple C-terminal binding partners (cytosolic proteins); however, the functions relating to regulation of these NHEs are poorly understood [258, 259]. NHE7 appears to be the NHE isoform present in the trans-Golgi, although it also traffics to the recycling system and to the plasma membrane [238]. Like other NHEs, it is partially localized to lipid rafts and associates with caveolin. The NHE7 C-terminus appears responsible for its trans-Golgi localization compared to NHE6, but NHE6 localization to the recycling system could not be attributed to the NHE6 C-terminus. An unusual feature of NHE7 is its insensitivity to amiloride, together with its inhibition by benzamil and quinine.

3.1.7 SLC9A8-NHE8

Human NHE8 [260] has 581 amino acids. On the basis of phylogenetic analysis, SLC9A8 has been predicted to encode an intracellular NHE and shares about 25 % homology with other members of the gene family [9]. Furthermore, the C-terminus of NHE8 is 50–100 amino acids shorter than that of other NHEs and shares no significant amino acid similarity, suggesting that the regulation of NHE8 is different from that of other family members. NHE8 is sensitive to amiloride [261].

NHE8 is ubiquitously expressed in human tissues, with the highest expression in the skeletal muscle and kidney. In the human kidney, NHE8 expression is restricted to the proximal tubule, whereas in the intestine, NHE8 is found in the stomach, duodenum, jejunum, ileum, and colon. NHE8 is also present ubiquitously in the trans-Golgi network and multivesicular bodies or late endosomes but is uniquely enriched in the apical membrane of some epithelial cells [261, 262]. It is unclear whether NHE8 is an intracellular or a plasma membrane NHE [107, 197, 246, 263–267].

NHE3 and NHE8 are developmentally regulated; NHE8 seems to be the major intestinal brush border NHE in neonates, and NHE3 is the predominant brush border NHE in adults [264, 268]. In renal proximal tubules, brush border NHE8 protein expression decreases with maturation, although the total amount of NHE8 in renal cortical membranes is higher in adults than in neonates. Furthermore, immunostaining of adult proximal tubules has revealed NHE8 in coated pit regions in addition to brush borders [251, 262].

Silencing NHE8 in HeLa-M cells results in perinuclear clustering of endosomes and lysosomes and perturbs endosomal protein trafficking, suggesting that NHE8 might be a negative regulator of inward vesiculation or that it might promote back fusion [269].

3.2 The NHA Eukaryotic Na^+/H^+ Exchangers

The SLC9B exchangers have recently been added to the SLC9 family [9, 240], and have also been classified as part of the CPA1 family. Compared with NHE isoforms 1–9, these newly identified antiporters have higher homology to prokaryotic Na^+/H^+ antiporters such as Ec-NhaA, and have thus been named NHA1 and NHA2, respectively. NHAs are predicted to contain 12 transmembrane domains, but unlike NHEs 1–9, seem to possess only short intracellular N- and C-termini.

NHA1 is 515 amino acids long [270] with ~50 kDa. According to RT-PCR experiments of human RNA tissue samples, it is expressed exclusively in the testis. The human biologic function of NHA1 remains unclear. The *Drosophila melanogaster* homologues of NHA1 and NHA2 (CG10806 and CG31052 or Dm-NHA1 and DM-NHA2) are localized in the apical plasma membrane of Malpighian (renal) tubules, together with the proton pumping V-ATPase [271, 272]. It has been proposed that they secrete cations into the tubular lumen, in conjunction with the V-ATPase. This close coupling of a primary V-ATPase with a secondarily active cation/H^+ exchange has been referred to as the Wieczorek model [271, 273]. Upon heterologous expression in NHE-deficient yeast (*Saccharomyces cerevisiae*), Dm-NHA1 localized to the plasma membrane and conferred protection against excess K^+. Heterologously overexpressed Dm-NHA2, on the other hand, was present on the vacuolar membrane, and transformed yeast displayed increased tolerance towards excess Na^+.

The subcellular localization of human NHA2 has been a matter of controversy [274–276]. Xiang and coworkers [276] showed that NHA2 is ubiquitously expressed in mouse tissues. According to immunofluorescence studies, pancreatic β-cells, distal tubular cells of the kidney (MDCK) and NHE-deficient yeast heterologously express human NHA2. NHA2 can also be found at the plasma membrane of native osteoclasts or in transfected mammalian cells. In addition, NHA2 has been found in endosomes at both functional and biochemical levels [250, 275].

When overexpressed in NHE-deficient yeast, NHA2 conferred tolerance to Li^+ and Na^+ but not K^+, in a pH-dependent manner [274, 276, 277]. NHA2 was inhibited by phloretin but not by the classical NHE-inhibitor amiloride, even at high concentrations. Mutation of two conserved aspartic acid residues (equivalent to Asp163 and Asp164 in Ec-NhaA) in the putative TM 5 of NHA2 (likely involved in cation binding and/or transport) led to loss of salt tolerance in transfected yeast [276].

Rao and colleagues [278] recently discovered an unconventional chemiosmotic coupling of NHA2 to a plasma membrane H^+ gradient, via V-ATPase in mammalian cells. They observed NHA2-mediated phloretin-sensitive Na^+/Li^+-counter transport in MDCK cells overexpressing NHA2, with important functional implications (see Section 4.4). While detailed kinetic studies are currently lacking, these studies clearly indicate that NHA2 is a *bona fide* NHE.

3.3 SLC9C, Sperm NHEs

A review of the physiology of Na^+/H^+ exchangers in the male reproductive tract has recently been published [279]. The Slc9c1 gene encoding sperm-specific NHE was originally identified in a mouse spermatid (haploid cell)-enriched cDNA library [280, 281]. Mammalian SLC9Cs lack distinct orthologues in non-mammalian genomes and are grouped separately from NHEs (SLC9A subfamily) and NHAs (SLC9B subfamily) in a distinct SLC9C subfamily.

SLC9Cs comprise an NHE-like N-terminal domain and a long non-conserved C-terminal part. with similarity to the Na^+-transporting carboxylic acid decarboxylase transporter family, a subfamily of the CPA superfamily [9, 240]. Northern blotting and dot blot analysis of a wide array of mouse tissues indicted that SLC9C1 mRNA expression is restricted to testicular tissue [280]. The putative open reading frame of mouse Slc9c1 encodes 1120 amino acids. Based on hydropathy analysis, the protein is proposed to contain 14 transmembrane domains, which include a unique voltage-sensing motif and a cyclic nucleotide-binding domain [280].

When transfected in NHE-null fibroblasts, full-length sperm NHE was poorly expressed. A chimeric construct in which the first transmembrane domain of sperm NHE was replaced with that of NHE1, however, exhibited improved expression at the plasma membrane and measurable NHE activity, suggesting that sperm NHE is indeed a functional Na^+/H^+ exchanger [278].

4 Na^+/H^+ Antiporters in Human Disease and as Drug Targets

Certain important functions in cell physiology are accomplished by cooperation of several NHEs [106]. Researchers should take this cooperation into account when seeking to identify specific NHEs as drug targets. Here we will relate separately to each antiporter. Mutations, knockout mice and use of NHE inhibitors are highly instructive in the study of the physio-pathological roles of a transporter. Furthermore, improvements in DNA sequencing technology have enabled large-scale sequencing of the human genome to be used to discover mutations associated with specific diseases or traits, including diseases or phenotypes not widely known or previously identified [282].

Nevertheless, although invaluable as a spotlight on the NHEs, genetic analyses and gene sequencing approaches alone do not provide definitive functional insight in the absence of structure-function information. Functional/structural evaluation of disease-associated NHE variants will be essential to predict clinical outcome, as a prerequisite to personalized therapy in patients with autism and other neurological diseases (see following subsections). Thus, a given alteration may be a harmless polymorphism or causal to the disease.

4.1 NHE1

Because NHE1 is critical for maintaining fundamental processes in cellular physiology, homeostasis of intracellular pH, salt concentration and cell volume [107, 109, 137, 138] (see Section 3.1.1), a change in its activity is detrimental to the cell. Furthermore, NHE1 has numerous indirect physiological roles that accompany or are a result of its pH regulatory function.

Two different Slc9a1-knockout mice have been reported. One is a spontaneous mutation that arose with a subsequent NHE-null phenotype [283]. The second knockout was engineered by homologous recombination resulting in deletion of NHE1 transmembrane domains 6 and 7 [284]. Compared with wild-type mice, NHE1 knockout mice showed lower steady-state pH_{in}, attenuated pH_{in} recovery from cell acidification (even in the presence of HCO_3^-) and increased expression and current density of voltage-gated Na^+ channels in hippocampal and cortical regions [285–287]. Loss of NHE1 was compatible with embryogenesis but was associated with a lower rate of postnatal growth and higher mortality [284]. In addition, knockout mice suffered from ataxia and epileptic seizures. This phenotype was associated with brain and other neural damage.

Ischemia-reperfusion induces cardiac injuries that are mediated by NHE1 [8, 288]. This conclusion is based on studies indicating that NHE1 inhibitors (review in [109]) afford significant protection against these injuries in many animal models and in patients undergoing coronary interventions [289, 290]. Evidence that NHE1-knockout mice are resistant to ischemia-reperfusion injury further supports the potential involvement of NHE1 in these injuries [291]. However, in a recent report, paradoxical resistance to myocardial ischemia-reperfusion injury was observed in transgenic mice overexpressing NHE1. This resistance may have been due to a stress response rather than NHE1 activation [292].

Overexpression of NHE1 also exaggerates the damage that occurs during myocardial ischemia and reperfusion. Enhanced NHE1 activity can cause substantial intracellular Na^+ accumulation that, through the activity of the plasmalemmal Na^+/Ca^{2+} exchanger, induces a deleterious increase of intracellular Ca^{2+} that ultimately leads to cell death [293]. Pharmacologic inhibition of NHE1 during episodes of ischemia–reperfusion has been shown to mitigate cardiac and neural injuries both *in vivo* and *in vitro* in rodents and pigs [294], with the mechanism presumably being lower accumulation of Na^+-driven intracellular Ca^{2+}.

NHE1 is also implicated in the cardiac hypertrophy and heart failure that often result from post-infarction remodeling. In addition to causing dysregulation of intracellular Ca^{2+} homeostasis, enhanced activities and/or expression levels of NHE1 are thought to be responsible for the cardiac remodeling observed in various animal models. In addition, treatment with NHE1 inhibitors reduced the pathological phenotypes of these animals [295–297] all of which exhibited pathological hypertrophy. The inhibition of Na^+/H^+ exchange prevented hypertrophy, fibrosis, and heart failure in β1-adrenergic receptor transgenic mice in which cardiac NHE1 was up-regulated [139]. A similar protective effect of cariporide was reported in other cases [298]. Furthermore, in both animal and human hearts, sarcolemmal NHE1 activity was demonstrated to be significantly elevated during chronic end-stage heart failure. For reviews of downstream signaling leading to cardiac hypertrophy and heart failure, see [109, 299–301].

Because many signal transduction pathways are activated during hypertrophy, it is not clear whether activation of NHE1 alone is sufficient to induce hypertrophy. This question was directly addressed by Wakabayashi and colleagues, who developed a transgenic mouse that selectively overexpressed a constitutively active NHE1 in the heart [179, 302]. This form of NHE1 lacks the CaM-binding autoinhibitory domain [163], which results in an alkaline shift of pH_{in}-dependence and hence mimics the pathological states in which hormonal stimulation occurs. Activation of NHE1 was found sufficient to initiate cardiac hypertrophy and heart failure via activation of Ca^{2+}-dependent prohypertrophic signaling pathways, and possibly via acceleration of Ca^{2+}-induced cell death.

Thus, NHE1 inhibitors have been promising in preclinical trials, protecting the myocardium from various diseases [303]. However, clinical trials in humans have not been promising, showing no overall benefit of NHE1 inhibition by cariporide in acute coronary syndromes and no benefits in patients with acute myocardial infarction. The failure to obtain practical benefit from NHE1 inhibitors in humans is largely due to a lack of inhibitor specificity [8, 109, 304]. Hence, increased knowledge of the atomic structure of NHE1 may facilitate the development of improved inhibitors for clinical use.

NHE1 also participates in cell migration [136, 305–307]. Inhibition or genetic ablation of NHE1 from fibroblasts significantly reduces migration speed and inhibits chemotaxis. Both ion translocation and anchoring of cytoskeletal proteins through the intracellular NHE1 C-terminus are required in order for migration to take place [136]. Involvement of NHE1 in cell migration is apparent *in vitro* in certain cell types; however, inhibition of NHE1 transport in granulocytes does not affect chemotaxis and chemokinesis [308].

NHE1 also plays a role in the pathogenesis of cancer (see [309] for a comprehensive review). In athymic nude mice, CCL39 hamster lung fibroblasts deficient in NHE1 caused tumors less frequently than did wild-type cells with functional NHE1 [310]. NHE1-dependent intracellular alkalinization seems to play an important role in the development of a transformed phenotype, which can be prevented by NHE1 inhibition [311]. In breast cancer and leukemic cells, inhibition of NHE1 exerts a protective effect against cancer, inducing apoptosis [312, 313]. As described above in this section, NHE1 has a fundamental role in cell migration. In breast cancer cells, serum deprivation activates NHE1 to induce cell motility and invasion [183]. The protons extruded by NHE1 at the cell front create an acidic environment optimal for the activity of proteinases involved in the degradation of the extracellular matrix. Low pH also enhances cell–matrix interactions and cell adhesion at the front [309, 314–317]. Thus, NHE1 seems to promote tumorigenesis at several levels, including cell proliferation, cell migration, invasion, metastasis, and suppression of apoptosis [318, 319]. However, there is no evidence that inhibition of NHE1 is useful in the treatment of cancer in humans.

Investigations of polymorphisms [282] in the SLC9A1 gene have shown that the Ser771Pro mutant is defective in expression and has unusual activation under sustained intracellular acidosis [161]. This is the first characterization of the effect of a human mutation on the SLC9A1 gene.

4.2 NHE2-NHE5 and NHE8

4.2.1 NHE2

Because the physiological role of NHE2 is not known (see Section 3.1.2), the pathological impact of this protein is also not clear [107]. NHE2-knockout mice display higher renal and plasma renin levels compared with wild-type mice [320]; furthermore, NHE2 shows a role in salivation responses [321]. Despite this, the knockout mice do not differ from their wild-type counterparts with respect to blood pressure, plasma aldosterone levels, renal sodium excretion or feedback responses, although significant NHE2 expression and activity has been observed in relation to these processes. It appears that the absence of NHE2 can be compensated for, most likely by NHE3 and NHE8 [322–328]. Most recently, NHE2 expression was localized to the pituitary, and its absence was shown to result in gross histological abnormalities, suggesting that the protein fulfills additional roles [198]. Ultimately, there is no known human disease ascribed to defects in this transporter [329].

4.2.2 NHE3

Linkage of SLC9A3 to human diseases is limited, although animal models suggest likely clinical implications: NHE3-knockout mice display metabolic acidosis and alkaline urine [330, 331] and absorptive defects of both sodium and water in both intestinal and renal tubular epithelia [325, 332, 333]. This results in volume depletion and hypotension despite increased renin expression [331]. Knocking out NHE3 in the intestine and conditionally knocking out renal NHE3 [330, 334] confirmed that both systems contribute to the significant decrease in circulating volume observed in NHE3-knockout mice. Given the large amounts of sodium and water that are not reabsorbed from the proximal tubule in the absence of NHE3, it is surprising that the NHE3-knockout mice survive at all [335, 336]. For further details see [107, 327, 331, 337].

NHE3 is also implicated in Ca^{2+} homeostasis. Specifically, it is likely that NHE3 provides the driving force for Ca^{2+} absorption from renal and intestinal epithelia [108, 338, 339]. NHE3 has also been found to play a role in intestinal inflammation [323, 340–343] and is a target for treatment of acute diarrhea [206].

4.2.3 NHE4

The physiological function of NHE4 is not clear, although NHE4-knockout mice have provided some insight into the protein's physiological and pathological functions [107, 226, 227].

4.2.4 NHE5

NHE5 is of particular interest in the nervous system, as its expression is largely confined to nervous tissue (see Section 3.1.5). Knockdown of SLC9A5, or expression of a dominant-negative mutant (E209 mutant), results in spontaneous spine outgrowth. It is therefore suggested that NHE5 controls dendritic spine growth via a pH-dependent negative-feedback mechanism [235]. Enhanced neuronal activity targets NHE5 from endosomes to plasma membranes of dendritic spines. The increased activity of NHE5 results in alkalinization of the dendritic spine and concomitant acidification of the synaptic cleft. This local acidification may serve as an autocrine feedback mechanism that regulates pH-sensitive proteins at the post-synapse membrane and possibly also at the pre-synapse membrane. Hence, NHE5 may be a very important pharmacological target.

4.2.5 NHE8

The pathophysiology of NHE8 is not clear. While brush-border NHE activity is diminished in patients with congenital Na^+ diarrhea (as compared with healthy individuals), there is no identified exonic mutation in SLC9A3 or SLC9A8 in these patients (the same applies to SLC9A1/2/5). Intestinal inflammations are associated with decreased expression of both NHE3 and NHE8, accounting for reduction of Na^+ absorption in inflammatory diarrhea. Somatostatin stimulates NHE8 but not NHE3 expression in the mouse intestine [333].

Loss of NHE8 has been shown to decrease mucus secretion and to increase susceptibility to gastric ulcers [344]. In addition, compared with wild-type mice, NHE8-knockout mice exhibit disorganized mucus layers and greater adhesion of bacteria to the distal colon, and are more susceptible to mucosal injury [345]. A combination of studies in mutant mice indicates that NHE8 plays an important role in proximal tubular Na^+ and HCO_3^- reclamation as well as in the protection of intestinal epithelia from bacterial infections. However, NHE8-knockout mice do not demonstrate metabolic acidosis and have unaltered blood pressure compared to wild-type mice. This is probably due to compensatory up-regulation of brush-border NHE3 in the kidney [346]. In support of this, NHE3/NHE8 double-knockout mice were shown to have lower blood pressure and lower proximal tubular NHE activity compared with NHE3-knockout mice.

4.3 NHE6, 7, 9

For a comprehensive review of NHE6, 7, and 9, see [111]. Notably, neuronal activity and synaptic transmission require precise control of ionic balance, as well as vesicular trafficking to ensure appropriate surface delivery and turnover of synaptic membrane receptors and transporters [347]. Therefore, the family of intracellular cation/proton exchangers (NHE6, 7, and 9) residing on endosomal and recycling compartments, regulating luminal pH [242, 246, 247, 252, 254, 348, 349] and controling vesicular trafficking, have been implicated in a range of neuropsychiatric disorders [9, 242, 251, 350–354].

Mutations in the NHE6 or NHE9 genes cause neurological disease in humans. These are currently the only NHEs directly linked to human disease, in keeping with the importance of maintenance of organellar pH [355]. NHE6 is encoded by the X-chromosome both in mice and in humans. Nonsense and missense mutations, as well as deletions in the SLC9A6 gene, have been observed [251, 308, 352]. These cause three phenotypes in humans: (a) X-linked Angelman syndrome, characterized by intellectual disability, microcephaly, epilepsy, ataxia, and behavioral abnormalities [251, 353]; (b) an Angelman-like syndrome known as Christianson syndrome [356]; and (c) a syndrome presenting with corticobasal degeneration with severe intellectual disability and autistic behavior [352]. The abnormal accumulation of glutamate in the brain of patients with NHE6 mutations is indicative of an underlying problem in glutamate clearance from the synapse [251]. Glutamate is excitotoxic at high concentrations, and aberrant glutamate levels are associated with neurological abnormalities of epilepsy and cerebellar degeneration, both characteristic symptoms of Angelman-like syndrome associated with NHE6 mutations [251]. The bulk of glutamate released into the neuronal synapse is cleared by rapid uptake into astrocytes, the single largest population of cells in the brain. Aside from the neurological phenotypes discussed above, little is known about extra-cranial consequences of SLC9A6 gene mutations [251, 356].

The neurological phenotype of NHE6-knockout mice has recently been described [357]. Mutant mice have no obvious phenotype, even at older ages. Extensive behavioral testing revealed that certain histological abnormalities result in mild motor hyperactivity and deficits in motor coordination in the mutant mice. Remarkably, studying the same knockout mouse model, Morrow and co-workers recently found that NHE6-deficient neurons exhibit over-acidified endosomes [358]. This over-acidification leads to disruption of endosomal signaling; this disruption leads to neuronal endolysosomal storage disease with cell death, as well as to more subtle alterations in endosomal signaling pathways that impair the wiring of neuronal circuits.

A possible function of NHE6 has also been demonstrated in hepatoma cells [257] expressing a splice variant of Slc9a6 (NHE6.1). Knockdown of NHE6.1 lowered recycling endosomal pH and surprisingly disrupted the apical canalicular plasma membrane, causing failure to traffic or maintain apical proteins; these effects were associated with reduced amounts of apical lipids and suggests an effect of NHE6 in regulating apical recycling that involves lipids as well as proteins [257]. However, in other cell models, knocking down NHE6 did not alter endosomal pH, while knocking down both NHE9 and NHE6 did lead to acidification of early endosomes, suggesting either protein is sufficient for pH regulation [252].

NHE7 has recently been identified as a novel target gene for Alzheimer's disease [359] and X-linked mental retardation contiguous gene syndromes [360]. It has been proposed that NHE7 works in a reverse mode compared with plasma membrane NHEs, by using cytosolic potassium in exchanging vesicular protons. A recent study of a fibroblast cell line that expresses wild-type NHE7 at the plasma membrane indeed shows that, contrary to the prevailing model, NHE7 functions as a proton-loading transporter in a manner additive to V-ATPases [361]. This function is compatible with the neuronal distribution of the NHE7 protein observed in the study.

Genetic approaches have identified NHE9 as a candidate gene in attention deficit hyperactivity disorder (ADHD), addiction, autism and mental retardation [283, 355, 358, 362–368]. Two other NHE9 mutations were identified in a rat model of ADHD, although the causal link to the disease was not established [368]. Therefore it could not be excluded that the variants associated with autism or ADHD represent benign polymorphisms [251, 283].

Remarkably, the functional consequences of human NHE9 missense mutations in astrocytes have recently been clarified [111, 248]. Overexpression of wild-type NHE9 but not of NHE9 mutants in astrocytes caused endosomal alkalinization and enhanced transferrin and glutamate uptake, indicating that the three missense mutations found in human NHE9 are loss-of-function mutations. As discussed above in this section, the abnormal accumulation of glutamate in the brain of patients with NHE6 mutations is indicative of characteristic symptoms of the Angelman-like syndrome [251]. Given the shared clinical symptoms in patients, it seems probable that, like NHE6, NHE9 could regulate glutamate clearance in the synapse. Altered gene dosage can significantly change vesicular pH and modulate cell surface expression and activity of glutamate transporters in murine cortical astrocytes, where these transporters have the potential to impact neurotransmission. Kondapalli and colleagues [111, 248] have developed a neurobiological model that establishes a role for NHE9.

4.4 NHA1 and NHA2

The physiological and pathological roles of NHA1 are not clear.

Regarding NHA2, Battaglino and coworkers first identified and cloned Slc9b2 from osteoclast precursor cells [369]. SiRNA-mediated knockdown significantly inhibited osteoclast differentiation and osteoclast function *in vitro*. Surprisingly, characterization of an NHA2-deficient mouse [275] showed normal development; in contrast to the published *in vitro* evidence, Slc9b2-deficient mice had normal bone density and their bones were characterized by normal structural parameters. These findings were recently confirmed independently [370]. Even when osteoclast differentiation and activation were stimulated by ovarectomy, no difference in bone loss could be observed between the two groups of mice up to 12 weeks after the intervention (Fuster, Hofstetter et al., unpublished observations). These findings suggest that NHA2 is dispensable for osteoclast differentiation and bone resorption in mice both *in vitro* and *in vivo*, at least under the conditions employed so far. The reason for the discrepancy between the *in vitro* and *in vivo* findings remains unclear.

On the basis of expression pattern, genomic localization and inhibitor characteristics (amiloride resistance and phloretin sensitivity), NHA2 has been proposed to be the Na^+/Li^+ counter transporter [276] originally described in the early 1980s at a functional level, linked to the development of arterial hypertension and diabetes [371–373]. In support of this hypothesis, Moe and coworkers found NHA2 to be expressed in the distal tubule of the rat kidney, a renal tubular segment that is paramount for Na^+ and blood pressure homeostasis in mammals [274].

Remarkably, Rao and colleagues [278] have recently established a mechanistically novel virtual H^+-coupled Na^+ (Li^+)-efflux pump in mammalian cells that mediates phloretin-sensitive Na^+-Li^+ counter-transport activity driven by V-ATPase [278, 374]. Although plasma membrane H^+ gradients have been observed in some specialized mammalian cells, the ubiquitous tissue distribution of NHA2 suggests that H^+-coupled transport is more widespread. The coexistence of Na^+ and H^+-driven chemiosmotic circuits has implications for salt and pH regulation in the kidney [278].

A recent study looked at the role of NHA2 in the endocrine pancreas in detail [108, 250, 375]. NHA2 was found to be expressed both in human as well as rodent β-cells and β-cell lines. Knockdown of NHA2 in the murine β-cell line Min6 reduced glucose- and sulfonylurea-induced insulin secretion. Simultaneous overexpression of wild-type but not functionally dead human NHA2 rescued the insulin secretion deficit induced by knockdown of endogenous NHA2. Cellular insulin or pro-insulin contents were unaltered in NHA2-deficient Min6 cells. Therefore, impaired insulin synthesis or maturation was not the cause of the insulin secretion deficit observed. Identical findings were observed *in vitro* when insulin secretion was studied with islets from two different NHA2-deficient mice or when islets were subjected to an NHA2 inhibitor. Interestingly, islets isolated from heterozygous mice were not normal but also exhibited a secretory deficit, suggesting haploinsufficiency. *In vivo*, both strains of NHA2-deficient mice displayed a pathological glucose tolerance with impaired insulin secretion but normal peripheral insulin sensitivity, compatible with the insulin secretion deficit observed *in vitro*. Loss of NHA2 may therefore affect insulin secretion indirectly, and much of the underlying mechanisms remain unknown [376–379].

In summary, strong evidence points to the involvement of NHA2 in essential hypertension and diabetes. These interesting results highlight NHA2 as an important drug target [375].

4.5 *Sperm NHEs*

Slc9c1-null male mice are completely infertile with severely diminished sperm motility but normal sperm numbers and sperm morphology [279, 280]. The corresponding protein was detected in testicular lysates of wild-type but not Slc9c1-knockout mice. Addition of ammonium chloride and cell-permeant cAMP analogues partially rescued the motility and fertility defects. Knockout of sperm-NHE1 results in a complete loss of full-length bicarbonate-sensitive soluble adenylate cyclase, and a substantial reduction of bicarbonate-stimulated soluble adenylyl cyclase activity [281].

SLC9C2 (also known as NHE11) is another member of the SLC9C family for which no functional data exist. EST databases indicate that the protein is expressed in the testis and is probably also present in the ovary, skin and connective tissue.

Despite being likely candidates for drug targets, sperm NHE and NHE11 have not been linked to male infertility in humans.

5 Bacterial Homologues of Human Na^+/H^+ Antiporters as Models for Drug Design

The crystal structure of a target protein that is involved in a disease is an important basis for drug design. However, although many transporters are now known to be involved in pathogenesis in humans, we are still far behind in obtaining crystal structures of mammalian transporters. GLUT1, the human glucose uniport, is the first human transporter to be determined [380]; its crystallization constitutes a great step towards understanding the mechanistic role of disease-causing mutations and their effects on the malfunction of the protein or its overexpression. The cause of the delay in structural determination of human transporters compared with bacterial ones stems from the great difficulty in overexpressing, purifying, and crystallizing eukaryotic membrane proteins.

Nevertheless, several of the human transporters that are involved in human diseases have bacterial homologues. Although bacterial integral membrane transporters are also difficult to crystallize, bacteria offer a great advantage in growth manipulation in addition to molecular tools for overexpression and purification of proteins. Thus, the crystal structures of several prokaryotic transporters that are homologues to clinically-important human transporters have been determined. Several examples include: LeuT from *Aquifex aeolicus,* a bacterial homologue for the neurotransmitter transporter that transports biogenic amines from the synapse to the cytoplasm of neurons and glia [61]; Glt_{ph} from *Pyrococcus horikoshii,* homologous to the eukaryotic glutamate transporter that is critical in glutaminergic synapse [381]; Ec-NhaA from *E. coli* [31], a homologue of the human NHA2 $Na^+/Li^+/H^+$ antiporter implicated in essential hypertension [278] and diabetes [250, 375] (see Section 4.4). Furthermore, archaeal and mammalian CPA1 antiporters are suggested to work essentially in the same way [274], given that they share several common features such as sequence homology, particularly in unwound stretches, as well as pH profile and transport kinetics. Thus, the novel Pa-NhaP and Mj-NhaP1 crystal structures can serve as models for the membrane part of NHE1 [38, 43]. Finally, Nm-ASBT from *Neisseria meningitidis* and Yf-ASBT from *Yersinia frederiksenii* are the bacterial homologues of the eukaryotic bile acid transporter (Section 2.1.3).

The Ec-NhaA structure has been used to model several structures of related human transporters of clinical importance. A model of NHE1 was used to predict the binding site of NHE1 inhibitors [11]; a model of NHA2 associated with essential human hypertension [276, 278] was used to guide mutagenic evaluation of transport function, ion selectivity, and pH dependence in yeast [13]. A model of NHE9 (Figure 6b), which is genetically linked to autism, was used to evaluate autism-associated mutations in the transmembrane domain [111, 248]. A model of a plant antiporter has guided mutagenesis [382].

Nm-ASBT and YF-ASBT, which share the NhaA fold, are homologues of the human bile acid transporter (see Section 2.2.3). Mutations in the human ASBT gene cause a condition of primary bile acid malabsorption [383]. ASBT is a pharmaceutical target for drugs aimed at lowering cholesterol [384] and type 2 diabetes [385]. Several ASBT inhibitors are effective in animal models [386, 387]. Because some drugs are poorly absorbed in the intestine or need to be targeted to the liver, ASBT and its close liver paralogue, NTCP, have also received attention as pro-drug carriers capable of transporting various compounds coupled to bile acid [388, 389]. The prokaryotic ASBT structure should aid the design of new inhibitors against ASBT with the goal of treating hypercholesterolaemia.

6 Na^+/H^+ Antiporters in Pathogenic Bacteria

There are several examples of the role of Na^+/H^+ antiporters in virulence and/or their importance in the adaptation of pathogenic bacteria to their ecological niches. These were revealed by deletions of the antiporter genes and testing the viability and virulence of the deleted mutants. Recently, the genome project has given rise to population genetics approaches for identifying virulent genes.

The *E. coli* species is divided into phylogenetic groups that differ in their virulence and commensal distribution. Strains belonging to the B2 group are involved in extra-intestinal pathologies but also appear to be more prevalent as commensals among human occidental populations. Recently, the genetic specificities of the B2 sub-group, using 128 sequenced genomes investigated [19]. The authors identified genes in the core genome that differ substantially from those in non-B2 genomes. Remarkably, the strongest divergence between B2 and non-B2 strains was observed in the antiporter gene *nha*A. Sequence analysis and experiments in a murine model of septicemia revealed that the *nha*A operon in the core genome has an important role in extra-intestinal virulence without strong effect on commensalism. Thus, *nha*A is a landmark of the B2 pathogenic strains. This important work shows that NhaA is a drug target for developing antibiotics against *E. coli* pathogens.

NhaA is widely spread in enterobacteria. The NhaA-type Na^+/H^+ antiporter of the Bubonic plague pathogen, *Yersinia pestis*, has recently been shown to be essential for its virulence [18]. The NhaA-type Na^+/H^+ antiporter of *Vibrio cholera* is important for the viability of the pathogen in its Na^+ rich biotope [390]. NhaP2, a $K^+/Na^+/H^+$ antiporter of *Vibrio cholera*, is important for its adaptation to acidic environment [391, 392].

There are several reasons why antiporters might serve as human drug targets and as targets for eradicating bacterial pathogens from an environment: (i) Although the major bacterial antiporters share a structural fold with the eukaryotic antiporters, their evolutionary conservation is low and different enough to design specific inhibitors, which may lead to new antibiotic drugs. (ii) Many pathogens live in environments characterized by high concentrations of salt and/or extreme pH levels, and therefore it is crucial to determine the role of their Na^+/H^+ antiporters in virulence or viability.

7 Perspective

A central aim in the structural research of prokaryotic Na^+/H^+ antiporters is to obtain the crystal structures of the active conformations of these proteins (e.g., Ec-NhaA). Nevertheless, crystal structures are snapshots, whereas antiporters such as Ec-NhaA are highly dynamic ‘nano-machines’. It is necessary to complement structural efforts with studies of structural and functional dynamics. The latter studies should use interdisciplinary approaches, including biochemistry, biophysics, and computation, in order to understand the mechanism of antiport activity and pH regulation.

The NhaA fold is shared by many Na^+/H^+ antiporters and secondary transporters, as revealed by the crystal structure alone without sequence homology. Comparative structural/functional biology can shed light on how a similar fold is manipulated to transport very different substrates and to couple differently to the proton motive force (stoichiometry of $1Na^+/1H^+$ or $1Na^+/2H^+$), and can provide insight into the regulation of these processes (activated at alkaline pH or at acidic pH).

Regarding eukaryotic Na^+/H^+ antiporters, a central aim is to obtain high-resolution structural information on the SLC9As and NHA2. All existing models are based on bacterial NhaAs, with the NhaA fold being a consensus. However, NhaAs and SLC9As are clearly also fundamentally different; they exhibit distinctly different stoichiometry and cooperativity, and Ec-NhaA is activated at alkaline pH, whereas NHE1 is activated at acidic pH.

Abbreviations and Definitions

ADHD	attention deficit hyperactivity disorder
Ala, A	alanine
AMP	adenosine 5′-monophosphate
AMPK	AMP-activated protein kinase
AP	2-aminopermidine
Arg, R	arginine
ASBT	apical sodium-dependent bile acid transporter
Asn, N	asparagine
Asp, D	aspartic acid
ATP	adenosine 5′-triphosphate
CaM	calmodulin
cAMP	cyclic adenosine monophosphate
CHPs	calcineurin homologous proteins
ClC	chloride/proton antiporter
CPA	cation proton antiporter superfamily
Cys, C	cysteine
D-Glu	D-glucose
DDM	n-dodecyl-β-D-maltoside
Ec-NhaA	Na^+/H^+ antiporter of *Escherichia coli*
EC_{50}	the concentration of an agent that gives half maximal response
EIPA	ethylisopropylamiloride
EM	electron microscopy
ESR	electron spin resonance
GFP	green fluorescent protein
Gln, Q	glutamine
GltPh	glutamate transporter
Glu, E	glutamic acid
GLUT1	human glucose uniporter

Gly, G	glycine
His, H	histidine
Ile, I	isoleucine
ITC	isothermal titration calorimetry
Leu, L	leucine
LeuT	leucine transporter
LID	lipid-interacting domain
Lys, K	lysine
MD	molecular dynamics
MDCK	MDCK cell line derived from a kidney
Met, M	methionine
MFS	major facilitator superfamily
Mj-NhaP1	homologue of human NHE1 from *Methanocaldococcus jannaschii*
MTSES	2-sulfonatoethyl methanethiosulfonate
MTSET	[2-(trimethylammonium)ethyl]methanethiosulfonate bromide
NaT-DC	Na^+-transporting carboxylic acid decarboxylase transporter family
NHA	eukaryotic Na^+/H^+ exchangers from the SLC9B family
NhaA	Na^+/H^+ antiporter
NHEs	the human Na^+/H^+ exchangers encoded by the SLC9 gene family
NHX	plant NHX cation/proton antiporters
Nm-ASBT	bacterial homologues of the human apical sodium-dependent bile acid transporter from *Neisseria meningitidis*
NMR	nuclear magnetic resonance
NTCP	a close liver paralogue of ASBT
Pa-NhaP	homologue of human NHE1from *Pyrococcus abyssi*
Phe, F	phenylalanine
PIP2	phosphatidylinositol 4,5-bisphosphate
PP	protein phosphatase
Pro, P	proline
RACK	a receptor for Activated Protein Kinase C
RMSD	root-mean-square deviation of atomic positionsreal-time PCR systems
RT-PCR	reverse transcription polymerase chain reaction
Ser, S	serine
SH	sulfhydryl
SLC10	sodium bile acid co-transporter family
SLC9 gene family	Solute Carrier family 9
smFRET	single-molecule fluorescence resonance energy transfer
SOS1	*Arabidopsis thaliana* Na^+/H^+ antiporter
SSM	solid support membrane
Thr, T	threonine

TM	transmembrane segment
Trp, W	tryptophan
Tt-NapA	Na^+/H^+ antiporter of *Thermus thermophilus*
Tyr, Y	tyrosine
V-ATPase	vacuolar ATPase
Val, V	valine
Yf-ASBT	bacterial homologues of the human apical sodium-dependent bile acid transporter from *Yersinia frederiksenii*
YvgP	a_monovalent_cation/H_antiporter from *B. subtilis* that exhibits homology to the cation:proton antiporter-1 (CPA-1) family.

Acknowledgment EP and ML thank the DIP (DFG) (German-Israeli Project Cooperation # LA3655/1-1 MI236/5-1). EP thanks the Israel Science Foundation (grant No. 284/12). ML thanks the Alon Fellowship from the Israeli Council for Higher Education, David and Inez Mayers Career Advancement Chair in Life Sciences, the I-CORE Program of the Planning and Budgeting Committee and The Israel Science Foundation, Center of Excellence in Integrated Structural Cell Biology; Grant No 1775/12, and the Support for training and career development of researchers (Marie Curie) CIG, Seventh framework program, Single Beneficiary.

References

1. E. Padan, M. Venturi, Y. Gerchman, N. Dover, *Biochim. Biophys. Acta* **2001**, *1505*, 144–157.
2. T. A. Krulwich, G. Sachs, E. Padan, *Nature Rev. Microbiol.* **2011**, *9*, 330–343.
3. R. Hendus-Altenburger, B. B. Kragelund, S. F. Pedersen, *Curr. Top. Membr.* **2014**, *73*, 69–148.
4. I. C. West, P. Mitchell, *Biochem. J.* **1974**, *144*, 87–90.
5. E. Padan, T. Tzubery, K. Herz, L. Kozachkov, A. Rimon, L. Galili, *Biochim. Biophys. Acta* **2004**, *1658*, 2–13.
6. J. Orlowski, S. Grinstein, *Curr. Opin. Cell Biol.* **2007**, *19*, 483–492.
7. L. Fliegel, *J. Mol. Cell Cardiol.* **2008**, *44*, 228–237.
8. M. Karmazyn, *J. Mol. Cell. Cardiol.* **2013**, *61*, 77–82.
9. C. L. Brett, M. Donowitz, R. Rao, *Am. J. Physiol. Cell. Physiol.* **2005**, *288*, C223–239.
10. M. H. Saier, Jr., V. S. Reddy, D. G. Tamang, A. Vastermark, *Nucleic Acids Res.* **2014**, *42*, D251–258.
11. M. Landau, K. Herz, E. Padan, N. Ben-Tal, *J. Biol. Chem.* **2007**, *282*, 37854–37863.
12. E. R. Slepkov, J. K. Rainey, B. D. Sykes, L. Fliegel, *Biochem. J.* **2007**, *401*, 623–633.
13. M. Schushan, M. Xiang, P. Bogomiakov, E. Padan, R. Rao, N. Ben-Tal, *J. Mol. Biol.* **2010**, *396*, 1181–1196.
14. J. S. Chen, V. Reddy, J. H. Chen, M. A. Shlykov, W. H. Zheng, J. Cho, M. R. Yen, M. H. Saier, Jr., *J. Mol. Microbiol. Biotechnol.* **2011**, *21*, 83–96.
15. D. Taglicht, E. Padan, S. Schuldiner, *J. Biol. Chem.* **1993**, *268*, 5382–5387.
16. G. Speelmans, B. Poolman, T. Abee, W. N. Konings, *Proc. Natl. Acad. Sci. USA* **1993**, *90*, 7975–7979.
17. E. Padan, E. Bibi, I. Masahiro, T. A. Krulwich, *Biochim. Biophys. Acta* **2005**, *1717*, 67–88.
18. Y. Minato, A. Ghosh, W. J. Faulkner, E. J. Lind, S. Schesser Bartra, G. V. Plano, C. O. Jarrett, B. J. Hinnebusch, J. Winogrodzki, P. Dibrov, C. C. Hase, *Infect. Immun.* **2013**, *81*, 3163–3172.

19. M. Lescat, F. Reibel, C. Pintard, S. Dion, J. Glodt, C. Gateau, A. Launay, A. Ledda, S. Cruvellier, J. Tourret, O. Tenallion, *PLOS ONE* **2014**, *9*, 1–11.
20. D. Taglicht, E. Padan, S. Schuldiner, *J. Biol. Chem.* **1991**, *266*, 11289–11294.
21. J. Orlowski, S. Grinstein, *Pflugers Arch.* **2004**, *447*, 549–565.
22. L. K. Putney, S. P. Denker, D. L. Barber, *Annu. Rev. Pharmacol. Toxicol.* **2002**, *42*, 527–552.
23. S. Wakabayashi, T. Hisamitsu, T. Pang, M. Shigekawa, *J. Biol. Chem.* **2003**, *278*, 43580–43585.
24. D. Alkoby, A. Rimon, M. Burdak, M. Patino-Ruiz, O. Călinescu, K. Fendler, E. Padan, *PLOS 1* **2014**, *9(4):e93200.*
25. Y. Gerchman, A. Rimon, M. Venturi, E. Padan, *Biochemistry* **2001**, *40*, 3403–3412.
26. D. Hilger, H. Jung, E. Padan, C. Wegener, K. P. Vogel, H. J. Steinhoff, G. Jeschke, *Biophys. J.* **2005**, *89*, 1328–1338.
27. D. Hilger, Y. Polyhach, E. Padan, H. Jung, G. Jeschke, *Biophys. J.* **2007**, *93*, 3675–3683.
28. K. A. Williams, U. Geldmacher-Kaufer, E. Padan, S. Schuldiner, W. Kuhlbrandt, *EMBO J.* **1999**, *18*, 3558–3563.
29. M. Appel, D. Hizlan, K. R. Vinothkumar, C. Ziegler, W. Kuhlbrandt, *J. Mol. Biol.* **2009**, *386*, 351–365.
30. E. Screpanti, E. Padan, A. Rimon, H. Michel, C. Hunte, *J. Mol. Biol.* **2006**, *362*, 192–202.
31. C. Hunte, E. Screpanti, M. Venturi, A. Rimon, E. Padan, H. Michel, *Nature* **2005**, *435*, 1197–1202.
32. C. Lee, S. Yashiro, D. L. Dotson, P. Uzdavinys, S. Iwata, M. S. Sansom, C. von Ballmoos, O. Beckstein, D. Drew, A. D. Cameron, *J. Gen. Physiol.* **2014**, *144*, 529–544.
33. E. M. Furrer, M. F. Ronchetti, F. Verrey, K. M. Pos, *FEBS Lett.* **2007**, *581*, 572–578.
34. E. Pinner, Y. Kotler, E. Padan, S. Schuldiner, *J. Biol. Chem.* **1993**, *268*, 1729–1734.
35. C. Lee, H.J. Kang, C. von Ballmoos, S. Newstead, P. Uzdavinys, D. L. Dotson, S. Iwata, O. Beckstein, A. D. Cameron, D. Drew, *Nature* **2013**, *501*, 573–577.
36. N. J. Hu, S. Iwata, A. D. Cameron, D. Drew, *Nature* **2011**, *478*, 408–411.
37. X. Zhou, E. J. Levin, Y. Pan, J. G. McCoy, R. Sharma, B. Kloss, R. Bruni, M. Quick, M. Zhou, *Nature* **2014**, *505*, 569–573.
38. D. Wöhlert, O. Yildiz, W. Kühlbrandt, *eLife* **2014**, *3*.
39. J. Hellmer, R. Patzold, C. Zeilinger, *FEBS Lett.* **2002**, *527*, 245–249.
40. J. Hellmer, A. Teubner, C. Zeilinger, *FEBS Lett.* **2003**, *542*, 32–36.
41. W. Jones, J. Leigh, F. Mayer, C. Woese, R. Wolfe, *Arch. Microbiol.* **1983**, *136*, 254–261
42. R. K. Thauer, A. K. Kaster, H. Seedorf, W. Buckel, R. Hedderich, *Nature Rev. Microbiol.* **2008**, *6*, 579–591.
43. C. Paulino, D. Wéhlert, E. Kapotova, O. Yildiz, W. Kühlbrandt, *eLife* **2014**, *3*.
44. E. Padan, D. Zilberstein, H. Rottenberg, *Eur. J. Biochem.* **1976**, *63*, 533–541.
45. E. Padan, *Trends Biochem. Sci.* **2008**, *33*, 435–443.
46. M. Schushan, A. Rimon, T. Haliloglu, L. R. Forrest, E. Padan, N. Ben-Tal, *J. Biol. Chem.* **2012**, *287*, 18249–18261.
47. L. Kozachkov, K. Herz, E. Padan, *Biochemistry* **2007**, *46*, 2419–3240.
48. K. A. Williams, *Nature* **2000**, *403*, 112–115.
49. C. Paulino, W. Kühlbrandt, *eLife* **2014**, *3:e01412*.
50. Y. Shi, *Annu. Rev. Biophys.* **2013**, *42*, 51–72.
51. E. Padan, L. Kozachkov, K. Herz, A. Rimon, *J. Exp. Biol.* **2009**, *212*, 1593–1603.
52. H. Krishnamurthy, C. L. Piscitelli, E. Gouaux, *Nature* **2009**, *459*, 347–355.
53. O. Boudker, G. Verdon, *Trends Pharmacol. Sci.* **2010**, *31*, 418–426.
54. L. R. Forrest, R. Kramer, C. Ziegler, *Biochim. Biophys. Acta* **2011**, *1807*, 167–188.
55. L. R. Forrest, *Science* **2013**, *339*, 399–401.
56. K. R. Vinothkumar, S. Raunser, H. Jung, W. Kuhlbrandt, *J. Biol. Chem.* **2006**, *281*, 4795–4801.
57. P. Goswami, C. Paulino, D. Hizlan, J. Vonck, O. Yildiz, W. Kuhlbrandt, *EMBO J.* **2011**, *30*, 439–449.

58. A. Rimon, T. Tzubery, E. Padan, *J. Biol. Chem.* **2007**, *282*, 26810–26821.
59. E. Padan, *Bioch. Biophys. Acta* **2014**, *1837*, 1047–1062.
60. E. Screpanti, C. Hunte, *J. Struct. Biol.* **2007**, *159*, 261–267.
61. A. Yamashita, S. K. Singh, T. Kawate, Y. Jin, E. Gouaux, *Nature* **2005**, *437*, 215–223.
62. M. Maes, A. Rimon, L. Kozachkov-Magrisso, A. Friedler, E. Padan, *J. Biol. Chem.* **2012**, *287*, 38150–38157.
63. O. Jardetzky, *Prog. Biophys. Mol. Biol.* **1996**, *65*, 171–219.
64. N. Kuwabara, H. Inoue, Y. Tsuboi, N. Nakamura, H. Kanazawa, *J. Biol. Chem.* **2004**, *279*, 40567–40575.
65. I. T. Arkin, H. Xu, M. O. Jensen, E. Arbely, E. R. Bennett, K. J. Bowers, E. Chow, R. O. Dror, M. P. Eastwood, R. Flitman-Tene, B. A. Gregersen, J. L. Klepeis, I. Kolossvary, Y. Shan, D. E. Shaw, *Science* **2007**, *317*, 799–803.
66. O. Boudker, R. M. Ryan, D. Yernool, K. Shimamoto, E. Gouaux, *Nature* **2007**, *445*, 387–393.
67. Z. Tao, A. Gameiro, C. Grewer, *Biochemistry* **2008**, *47*, 12923–12930.
68. V. Villeret, S. Huang, H. J. Fromm, W. N. Lipscomb, *Proc. Natl. Acad. Sci. USA* **1995**, *92*, 8916–8920.
69. B. Roux, S. Berneche, B. Egwolf, B. Lev, S. Y. Noskov, C. N. Rowley, H. Yu, *J. Gen. Physiol.* **2011**, *137*, 415–426.
70. E. B. Goldberg, T. Arbel, J. Chen, R. Karpel, G. A. Mackie, S. Schuldiner, E. Padan, *Proc. Natl. Acad. Sci. USA* **1987**, *84*, 2615–2619.
71. D. Zuber, R. Krause, M. Venturi, E. Padan, E. Bamberg, K. Fendler, *Biochim. Biophys. Acta* **2005**, *1709*, 240–250.
72. T. Mager, A. Rimon, E. Padan, K. Fendler, *J. Biol. Chem.* **2011**, *286*, 23570–23581.
73. I. Smirnova, V. Kasho, J. Sugihara, H. R. Kaback, *Proc. Natl. Acad. Sci. USA* **2013**, *110*, 8876–8881.
74. M. H. Akabas, C. Kaufmann, P. Archdeacon, A. Karlin, *Neuron* **1994**, *13*, 919–927.
75. M. H. Akabas, D. A. Stauffer, M. Xu, A. Karlin, *Science* **1992**, *258*, 307–310.
76. L. Guan, H. R. Kaback, *Nat. Protoc.* **2007**, *2*, 2012–2017.
77. Y. Olami, A. Rimon, Y. Gerchman, A. Rothman, E. Padan, *J. Biol. Chem.* **1997**, *272*, 1761–1768.
78. K. Herz, A. Rimon, E. Olkhova, L. Kozachkov, E. Padan, *J. Biol. Chem.*, *285*, 2211–2220.
79. A. Rimon, L. Kozachkov-Magrisso, E. Padan, *Biochemistry* **2012**, *51*, 9560–9569.
80. M. Diab, A. Rimon, T. Tzubery, E. Padan, *J. Mol. Biol.* **2011**, *413*, 604–614.
81. T. Tzubery, A. Rimon, E. Padan, *J. Biol. Chem.* **2008**, *283*, 15975–15987.
82. L. Galili, K. Herz, O. Dym, E. Padan, *J. Biol. Chem.* **2004**, *279*, 23104–23113.
83. Y. Gerchman, A. Rimon, E. Padan, *J. Biol. Chem.* **1999**, *274*, 24617–24624.
84. M. Venturi, A. Rimon, Y. Gerchman, C. Hunte, E. Padan, H. Michel, *J. Biol. Chem.* **2000**, *275*, 4734–4742.
85. T. Tzubery, A. Rimon, E. Padan, *J. Biol. Chem.* **2004**, *279*, 3265–3272.
86. A. Rothman, *Ph.D. Thesis, Hebrew University of Jerusalem* **1997**.
87. L. Kozachkov, E. Padan, *Proc. Natl. Acad. Sci. USA* **2011**, *108*, 15769–15774.
88. L. Kozachkov, E. Padan, *Mol. Mem. Biol.* **2012**, *30*, 90–100.
89. E. Olkhova, C. Hunte, E. Screpanti, E. Padan, H. Michel, *Proc. Natl. Acad. Sci. USA* **2006**, *103*, 2629–2634.
90. E. Olkhova, E. Padan, H. Michel, *Biophys. J.* **2007**, *92*, 3784–3791.
91. Y. Zhao, D. Terry, L. Shi, H. Weinstein, S. C. Blanchard, J. A. Javitch, *Nature* **2010**, *465*, 188–193.
92. N. Akyuz, R. B. Altman, S. C. Blanchard, O. Boudker, *Nature* **2013**, *502*, 114–118.
93. N. Reyes, S. Oh, O. Boudker, *Nat. Struct. Mol. Biol.* **2013**, *20*, 634–640.
94. I. Hanelt, D. Wunnicke, E. Bordignon, H. J. Steinhoff, D. J. Slotboom, *Nat. Struct. Mol. Biol.* **2013**, *20*, 210–214.
95. N. Reyes, C. Ginter, O. Boudker, *Nature* **2009**, *462*, 880–885.

96. P. Schulz, J. J. Garcia-Celma, K. Fendler, *Methods* **2008**, *46*, 97–103.
97. L. R. Forrest, Y. W. Zhang, M. T. Jacobs, J. Gesmonde, L. Xie, B. H. Honig, G. Rudnick, *Proc. Natl. Acad. Sci. USA* **2008**, *105*, 10338–10343.
98. S. Kalayil, S. Schulze, W. Kuhlbrandt, *Proc. Natl. Acad. Sci. USA* **2013**, *110*, 17296–172301.
99. A. Picollo, Y. Xu, N. Johner, S. Berneche, A. Accardi, *Nat. Struct. Mol. Biol.* **2012**, *19*, 525–531, S1.
100. P. M. McTernan, S. K. Chandrayan, C. H. Wu, B. J. Vaccaro, W. A. Lancaster, Q. Yang, D. Fu, G. L. Hura, J. A. Tainer, M. W. Adams, *J. Biol. Chem.* **2014**, *289*, 19364–19372.
101. G. N. Cohen, V. Barbe, D. Flament, M. Galperin, R. Heilig, O. Lecompte, O. Poch, D. Prieur, J. Querellou, R. Ripp, J. C. Thierry, J. Van der Oost, J. Weissenbach, Y. Zivanovic, P. Forterre, *Mol. Microbiol.* **2003**, *47*, 1495–1512.
102. T. Mager, M. Braner, B. Kubsch, L. Hatahet, D. Alkoby, A. Rimon, E. Padan, K. Fendler, *J. Biol. Chem.* **2013**, *288*, 24666–24675.
103. O. Calinescu, C. Paulino, W. Kuhlbrandt, K. Fendler, *J. Biol. Chem.* **2014**, *289*, 13168–13176.
104. O. Calinescu, C. Paulino, W. Kuhlbrandt, K. Fendler, *J. Biol. Chem.* **2014**, *289*, 13168–13176.
105. Y. Sonoda, S. Newstead, N. J. Hu, Y. Alguel, E. Nji, K. Beis, S. Yashiro, C. Lee, J. Leung, A. D. Cameron, B. Byrne, S. Iwata, D. Drew, *Structure* **2011**, *19*, 17–25.
106. J. Orlowski, S. Grinstein, *Compr. Physiol.* **2011**, *1*, 2083–2100.
107. M. Donowitz, C. Ming Tse, D. Fuster, *Mol. Aspects Med.* **2013**, *34*, 236–251.
108. D. G. Fuster, R. T. Alexander, *Pflugers Arch.* **2014**, *466*, 61–76.
109. S. Wakabayashi, T. Hisamitsu, T. Y. Nakamura, *J. Mol. Cell. Cardiol.* **2013**, *61*, 68–76.
110. S. H. Lee, T. Kim, E. S. Park, S. Yang, D. Jeong, Y. Choi, J. Rho, *Biochem. Biophys. Res. Commun.* **2008**, *369*, 320–326.
111. K. C. Kondapalli, H. Prasad, R. Rao, *Front. Cell. Neurosci.* **2014**, *8*, 172.
112. T. Ikeda, B. Schmitt, J. Pouyssegur, S. Wakabayashi, M. Shigekawa, *J. Biochem.* **1997**, *121*, 295–303.
113. R. T. Alexander, L. Fliegel, *Biochem. Cell Biol. (=Biochimie et Biologie Cellulaire)* **2011**, *89*, 85–86.
114. P. Fafournoux, J. Noel, J. Pouyssegur, *J. Biol. Chem.* **1994**, *269*, 2589–2596.
115. P. S. Aronson, *Annu. Rev. Physiol.* **1985**, *47*, 545–560.
116. S. A. Levine, M. H. Montrose, C. M. Tse, M. Donowitz, *J. Biol. Chem.* **1993**, *268*, 25527–25535.
117. R. Chambrey, P. L. St John, D. Eladari, F. Quentin, D. G. Warnock, D. R. Abrahamson, R. A. Podevin, M. Paillard, *Am. J. Physiol. Renal Physiol.* **2001**, *281*, F707–717.
118. C. Bookstein, M. W. Musch, A. DePaoli, Y. Xie, K. Rabenau, M. Villereal, M. C. Rao, E. B. Chang, *Am. J. Physiol.* **1996**, *271*, C1629–1638.
119. S. Wakabayashi, T. Ikeda, T. Iwamoto, J. Pouyssegur, M. Shigekawa, *Biochemistry* **1997**, *36*, 12854–12861.
120. I. A. Bobulescu, O. W. Moe, *Pflugers Arch.* **2009**, *458*, 5–21.
121. D. Rotin, O. Staub, *Pflugers Arch.* **2011**, *461*, 1–21.
122. A. Simonin, D. Fuster, *J. Biol. Chem.* **2010**, *285*, 38293–38303.
123. S. Wakabayashi, M. Shigekawa, J. Pouyssegur, *Physiol. Rev.* **1997**, *77*, 51–74.
124. C. Sardet, A. Franchi, J. Pouyssegur, *Cell* **1989**, *56*, 271–280.
125. L. Counillon, J. Pouyssegur, *J. Biol. Chem.* **2000**, *275*, 1–4.
126. G. Kemp, H. Young, L. Fliegel, *Channels* **2008**, *2*, 329–336.
127. B. L. Lee, B. D. Sykes, L. Fliegel, *J. Mol. Cell. Cardiol.* **2013**, *61*, 60–7.
128. P. S. Aronson, J. Nee, M. A. Suhm, *Nature* **1982**, *299*, 161–163.
129. S. F. Pedersen, M. E. O'Donnell, S. E. Anderson, P. M. Cala, *Am. J. Physiol. Regul. Integr. Comp. Physiol.* **2006**, *291*, R1–25.
130. S. Grinstein, J. D. Goetz, A. Rothstein, *J. Gen. Physiol.* **1984**, *84*, 565–584.
131. D. Fuster, O. W. Moe, D. W. Hilgemann, *J. Gen. Physiol.* **2008**, *132*, 465–480.
132. J. Peti-Peterdi, R. Chambrey, Z. Bebok, D. Biemesderfer, P. L. St John, D. R. Abrahamson, D. G. Warnock, P. D. Bell, *Am. J. Physiol. Renal Physiol.* **2000**, *278*, F452–463.

133. D. Biemesderfer, R. F. Reilly, M. Exner, P. Igarashi, P. S. Aronson, *Am. J. Physiol.* **1992**, *263*, F833–840.
134. S. Grinstein, M. Woodside, T. K. Waddell, G. P. Downey, J. Orlowski, J. Pouyssegur, D. C. Wong, J. K. Foskett, *EMBO J.* **1993**, *12*, 5209–5218.
135. K. Petrecca, R. Atanasiu, S. Grinstein, J. Orlowski, A. Shrier, *Am. J. Physiol.* **1999**, *276*, H709–717.
136. S. P. Denker, D. L. Barber, *J. Cell. Biol.* **2002**, *159*, 1087–1096.
137. J. Pouyssegur, A. Franchi, G. L'Allemain, S. Paris, *FEBS Lett.* **1985**, *190*, 115–119.
138. S. Grinstein, D. Rotin, M. J. Mason, *Biochim Biophys Acta* **1989**, *988*, 73–97.
139. S. Engelhardt, L. Hein, U. Keller, K. Klambt, M. J. Lohse, *Circulation Res.* **2002**, *90*, 814–819.
140. D. G. Allen, X. H. Xiao, *Cardiovasc. Res.* **2003**, *57*, 934–941.
141. V. N. Phan, M. Kusuhara, P. A. Lucchesi, B.C. Berk, *Hypertension* **1997**, *29*, 1265–1272.
142. S. Wakabayashi, P. Fafournoux, C. Sardet, J. Pouyssegur, *Proc. Natl. Acad. Sci. USA* **1992**, *89*, 2424–2428.
143. L. Counillon, J. Pouyssegur, R. A. Reithmeier, *Biochemistry* **1994**, *33*, 10463–10469.
144. R. S. Haworth, O. Frohlich, L. Fliegel, *Biochem. J.* **1993**, *289*, 637–640.
145. S. Wakabayashi, T. Pang, X. Su, M. Shigekawa, *J. Biol. Chem.* **2000**, *275*, 7942–7949.
146. L. D. Shrode, B. S. Gan, S. J. D'Souza, J. Orlowski, S. Grinstein, *Am. J. Physiol.* **1998**, *275*, C431–439.
147. E. B. Nygaard, J. O. Lagerstedt, G. Bjerre, B. Shi, M. Budamagunta, K. A. Poulsen, S. Meinild, R. R. Rigor, J. C. Voss, P. M. Cala, S. F. Pedersen, *J. Biol. Chem.* **2011**, *286*, 634–648.
148. M. Schushan, M. Landau, E. Padan, N. Ben-Tal, *J. Biol. Chem.* **2011**, *286*, le9; author reply Ie10.
149. E. Miyazaki, M. Sakaguchi, S. Wakabayashi, M. Shigekawa, K. Mihara, *J. Biol. Chem.* **2001**, *276*, 49221–49227.
150. N. Touret, P. Poujeol, L. Counillon, *Biochemistry* **2001**, *40*, 5095–5101.
151. S. Wakabayashi, T. Hisamitsu, T. Pang, M. Shigekawa, *J. Biol. Chem.* **2003**, *278*, 11828–11835.
152. L. Fliegel, R. S. Haworth, J. R. Dyck, *Biochem. J.* **1993**, *289*, 101–107.
153. T. Hisamitsu, T. Pang, M. Shigekawa, S. Wakabayashi, *Biochemistry* **2004**, *43*, 11135–11143.
154. K. Moncoq, G. Kemp, X. Li, L. Fliegel, H. S. Young, *J. Biol. Chem.* **2008**, *283*, 4145–4154.
155. T. Hisamitsu, Y. Ben Ammar, T. Y. Nakamura, S. Wakabayashi, *Biochemistry* **2006**, *45*, 13346–13355.
156. M. Mishima, S. Wakabayashi, C. Kojima, *J. Biol. Chem.* **2007**, *282*, 2741–2751.
157. Y.B. Ammar, S. Takeda, T. Hisamitsu, H. Mori, S. Wakabayashi, *EMBO J.* **2006**, *25*, 2315–2325.
158. S. Koster, T. Pavkov-Keller, W. Kuhlbrandt, O. Yildiz, *J. Biol. Chem.* **2011**, *286*, 40954–40961.
159. B. L. Lee, B. D. Sykes, L. Fliegel, *Biochem. Cell Biol. (= Biochimie et Biologie Cellulaire)* **2011**, *89*, 189–199.
160. J. F. Hunt, T. N. Earnest, O. Bousche, K. Kalghatgi, K. Reilly, C. Horvath, K. J. Rothschild, D. M. Engelman, *Biochemistry* **1997**, *36*, 15156–15176.
161. C. Alves, B. L. Lee, B. D. Sykes, L. Fliegel, *Biochemistry* **2014**, *53*, 3658–3670.
162. Y. Ben Ammar, S. Takeda, M. Sugawara, M. Miyano, H. Mori, S. Wakabayashi, *Acta Crystallogr. Sect. F Struct. Biol. Cryst. Commun.* **2005**, *61*, 956–958.
163. S. Wakabayashi, B. Bertrand, M. Shigekawa, P. Fafournoux, J. Pouyssegur, *J. Biol. Chem.* **1994**, *269*, 5583–5588.
164. S. Attaphitaya, K. Nehrke, J. E. Melvin, *Am. J. Physiol., Cell Physiol.* **2001**, *281*, C1146–1157.
165. A. Kapus, S. Grinstein, S. Wasan, R. Kandasamy, J. Orlowski, *J. Biol. Chem.* **1994**, *269*, 23544–23552.

166. J. R. Casey, S. Grinstein, J. Orlowski, *Nature Rev. Mol. Cell Biol.* **2010**, *11*, 50–61.
167. W. H. Moolenaar, L. G. Tertoolen, S. W. de Laat, *J. Biol. Chem.* **1984**, *259*, 7563–7569.
168. J. Lacroix, M. Poet, C. Maehrel, L. Counillon, *EMBO rep.* **2004**, *5*, 91–96.
169. J. Lacroix, M. Poet, L. Huc, V. Morello, N. Djerbi, M. Ragno, M. Rissel, X. Tekpli, P. Gounon, D. Lagadic-Gossmann, L. Counillon, *Biochemistry* **2008**, *47*, 13674–13685.
170. T. Hisamitsu, K. Yamada, T. Y. Nakamura, S. Wakabayashi, *FEBS J.* **2007**, *274*, 4326–4335.
171. B. Masereel, L. Pochet, D. Laeckmann, *Eur. J. Med. Chem.* **2003**, *38*, 547–554.
172. L. Counillon, A. Franchi, J. Pouyssegur, *Proc. Natl. Acad. Sci. USA* **1993**, *90*, 4508–4512.
173. J. Orlowski, R. A. Kandasamy, *J. Biol. Chem.* **1996**, *271*, 19922–19927.
174. P. Dibrov, A. Rimon, J. Dzioba, A. Winogrodzki, Y. Shalitin, E. Padan, *FEBS Lett.* **2005**, *579*, 373–378.
175. M. Baumgartner, H. Patel, D. L. Barber, *Am. J. Physiol., Cell Physiol.* **2004**, *287*, C844–850.
176. N. Shimada-Shimizu, T. Hisamitsu, T.Y. Nakamura, N. Hirayama, S. Wakabayashi, *Mol. Pharmacol.* **2014**, *85*, 18–28.
177. X. Li, D. Prins, M. Michalak, L. Fliegel, *Am. J. Physiol., Cell Physiol.* **2013**, *305*, C1161–1169.
178. T. Pang, T. Hisamitsu, H. Mori, M. Shigekawa, S. Wakabayashi, *Biochemistry* **2004**, *43*, 3628–3636.
179. T. Y. Nakamura, Y. Iwata, Y. Arai, K. Komamura, S. Wakabayashi, *Circulation Res.* **2008**, *103*, 891–899.
180. T. Hisamitsu, T. Y. Nakamura, S. Wakabayashi, *Mol. Cell. Biol.* **2012**, *32*, 3265–3280.
181. X. Lin, D. L. Barber, *Proc. Natl. Acad. Sci. USA* **1996**, *93*, 12631–12636.
182. N. Shimada-Shimizu, T. Hisamitsu, T. Y. Nakamura, S. Wakabayashi, *FEBS J.* **2013**, *280*, 1430–1442.
183. S. J. Reshkin, A. Bellizzi, V. Albarani, L. Guerra, M. Tommasino, A. Paradiso, V. Casavola, *J. Biol. Chem.* **2000**, *275*, 5361–5369.
184. O. Aharonovitz, H. C. Zaun, T. Balla, J.D. York, J. Orlowski, S. Grinstein, *J. Cell. Biol.* **2000**, *150*, 213–224.
185. J. Tzeng, B. L. Lee, B. D. Sykes, L. Fliegel, *J. Biol. Chem.* **2010**, *285*, 36656–65.
186. R. T. Alexander, V. Jaumouille, T. Yeung, W. Furuya, I. Peltekova, A. Boucher, M. Zasloff, J. Orlowski, S. Grinstein, *EMBO J.* **2011**, *30*, 679–691.
187. C. Sardet, L. Counillon, A. Franchi, J. Pouyssegur, *Science* **1990**, *247*, 723–726.
188. M. Donowitz, S. Mohan, C. X. Zhu, T. E. Chen, R. Lin, B. Cha, N. C. Zachos, R. Murtazina, R. Sarker, X. Li, *J. Exp. Biol.* **2009**, *212*, 1638–1646.
189. K. Kurashima, F. H. Yu, A. G. Cabado, E. Z. Szabo, S. Grinstein, J. Orlowski, *J. Biol. Chem.* **1997**, *272*, 28672–28679.
190. C. M. Tse, S. A. Levine, C. H. Yun, M. H. Montrose, P. J. Little, J. Pouyssegur, M. Donowitz, *J. Biol. Chem.* **1993**, *268*, 11917–11924.
191. Z. Wang, J. Orlowski, G. E. Shull, *J. Biol. Chem.* **1993**, *268*, 11925–11928.
192. C. M. Tse, S. A. Levine, C. H. Yun, S. Khurana, M. Donowitz, *Biochemistry* **1994**, *33*, 12954–12961.
193. W. A. Hoogerwerf, S. C. Tsao, O. Devuyst, S. A. Levine, C. H. Yun, J. W. Yip, M. E. Cohen, P. D. Wilson, A. J. Lazenby, C. M. Tse, M. Donowitz, *Am. J. Physiol.* **1996**, *270*, G29–41.
194. M. Repishti, D. L. Hogan, V. Pratha, L. Davydova, M. Donowitz, C. M. Tse, J. I. Isenberg, *Am. J. Physiol. Gastrointest. Liver Physiol.* **2001**, *281*, G159–163.
195. R. Chambrey, J. M. Achard, D. G. Warnock, *Am. J. Physiol.* **1997**, *272*, C90–98.
196. R. Chambrey, D. G. Warnock, R. A. Podevin, P. Bruneval, C. Mandet, M. F. Belair, J. Bariety, M. Paillard, *Am. J. Physiol.* **1998**, *275*, F379–386.
197. H. Xu, J. Li, R. Chen, B. Zhang, C. Wang, N. King, H. Chen, F. K. Ghishan, *Am. J. Physiol. Gastrointest. Liver Physiol.* **2011**, *300*, G647–653.
198. M. L. Miller, A. Andringa, P. J. Schultheis, G. E. Shull, *J. Biomed. Biotechnol.* **2011**, *2011*, 510827.
199. J. Orlowski, R. A. Kandasamy, G. E. Shull, *J. Biol. Chem.* **1992**, *267*, 9331–9339.

200. A. J. Janecki, M. H. Montrose, P. Zimniak, A. Zweibaum, C. M. Tse, S. Khurana, M. Donowitz, *J. Biol. Chem.* **1998**, *273*, 8790–8798.
201. C. W. Chow, S. Khurana, M. Woodside, S. Grinstein, J. Orlowski, *J. Biol. Chem.* **1999**, *274*, 37551–37558.
202. J. Orlowski, *J. Biol. Chem.* **1993**, *268*, 16369–16377.
203. N. C. Zachos, M. Tse, M. Donowitz, *Annu. Rev. Physiol.* **2005**, *67*, 411–443.
204. P. Jacob, H. Rossmann, G. Lamprecht, A. Kretz, C. Neff, E. Lin-Wu, M. Gregor, D. A. Groneberg, J. Kere, U. Seidler, *Gastroenterology* **2002**, *122*, 709–724.
205. J. R. Turner, E. D. Black, *Am. J. Physiol., Cell Physiol.* **2001**, *281*, C1533–1541.
206. R. Lin, R. Murtazina, B. Cha, M. Chakraborty, R. Sarker, T. E. Chen, Z. Lin, B. M. Hogema, H. R. de Jonge, U. Seidler, J. R. Turner, X. Li, O. Kovbasnjuk, M. Donowitz, *Gastroenterology* **2011**, *140*, 560–571.
207. M. Wiemann, J. R. Schwark, U. Bonnet, H. W. Jansen, S. Grinstein, R. E. Baker, H. J. Lang, K. Wirth, D. Bingmann, *Pflugers Arch.* **1999**, *438*, 255–262.
208. S. D'Souza, A. Garcia-Cabado, F. Yu, K. Teter, G. Lukacs, K. Skorecki, H. P. Moore, J. Orlowski, S. Grinstein, *J. Biol. Chem.* **1998**, *273*, 2035–2043.
209. K. Szaszi, A. Paulsen, E. Z. Szabo, M. Numata, S. Grinstein, J. Orlowski, *J. Biol. Chem.* **2002**, *277*, 42623–42632.
210. M. Gekle, K. Drumm, S. Mildenberger, R. Freudinger, B. Gassner, S. Silbernagl, *J. Physiol.* **1999**, *520 Pt 3*, 709–721.
211. O. A. Weisz, *Int. Rev. Cytol.* **2003**, *226*, 259–319.
212. M. Gekle, K. Volker, S. Mildenberger, R. Freudinger, G. E. Shull, M. Wiemann, *Am. J. Physiol. Renal Physiol.* **2004**, *287*, F469–473.
213. M. Donowitz, X. Li, *Physiol. Rev.* **2007**, *87*, 825–872.
214. M. Sotak, L. Polidarova, J. Musilkova, M. Hock, A. Sumova, J. Pacha, *Am. J. Physiol. Gastrointest. Liver Physiol.* **2011**, *301*, G1066–1074.
215. I. A. Bobulescu, O. W. Moe, *Seminars in Nephrology* **2006**, *26*, 334–344.
216. X. Li, T. Galli, S. Leu, J. B. Wade, E. J. Weinman, G. Leung, A. Cheong, D. Louvard, M. Donowitz, *J. Physiol.* **2001**, *537*, 537–552.
217. N. C. Zachos, O. Kovbasnjuk, M. Donowitz, *Ann. NY Acad. Sci.* **2009**, *1165*, 240–248.
218. X. Zhu, B. Cha, N. C. Zachos, R. Sarker, M. Chakraborty, T. E. Chen, O. Kovbasnjuk, M. Donowitz, *J. Biol. Chem.* **2011**, *286*, 34486–34496.
219. P. He, H. Zhang, C. C. Yun, *J. Biol. Chem.* **2008**, *283*, 33544–33553.
220. B. Cha, M. Tse, C. Yun, O. Kovbasnjuk, S. Mohan, A. Hubbard, M. Arpin, M. Donowitz, *Mol. Biol. Cell* **2006**, *17*, 2661–2673.
221. J. S. Lee, Y. M. Lee, J. Y. Kim, H. W. Park, S. Grinstein, J. Orlowski, E. Kim, K. H. Kim, M. G. Lee, *J. Biol. Chem.* **2010**, *285*, 8104–8113.
222. A. C. Girardi, B. C. Degray, T. Nagy, D. Biemesderfer, P. S. Aronson, *J. Biol. Chem.* **2001**, *276*, 46671–46677.
223. R. T. Alexander, A. Malevanets, A. M. Durkan, H. S. Kocinsky, P. S. Aronson, J. Orlowski, S. Grinstein, *J. Biol. Chem.* **2007**, *282*, 7376–7384.
224. H. Hayashi, K. Szaszi, N. Coady-Osberg, W. Furuya, A.P. Bretscher, J. Orlowski, S. Grinstein, *J. Gen. Physiol.* **2004**, *123*, 491–504.
225. V. Babich, K. Vadnagara, F. Di Sole, *FASEB J.* **2013**, *27*, 4646–4658.
226. S. Bourgeois, L. V. Meer, B. Wootla, M. Bloch-Faure, R. Chambrey, G. E. Shull, L. R. Gawenis, P. Houillier, *J. Clin. Invest.* **2010**, *120*, 1895–1904.
227. L. R. Gawenis, J. M. Greeb, V. Prasad, C. Grisham, L. P. Sanford, T. Doetschman, A. Andringa, M. L. Miller, G. E. Shull, *J. Biol. Chem.* **2005**, *280*, 12781–12789.
228. C. A. Klanke, Y. R. Su, D. F. Callen, Z. Wang, P. Meneton, N. Baird, R. A. Kandasamy, J. Orlowski, B. E. Otterud, M. Leppert, G. E. Shull, A. G. Menton, *Genomics* **1995**, *25*, 615–622.
229. S. Attaphitaya, K. Park, J. E. Melvin, *J. Biol. Chem.* **1999**, *274*, 4383–4388.
230. N. R. Baird, J. Orlowski, E. Z. Szabo, H. C. Zaun, P. J. Schultheis, A. G. Menon, G. E. Shull, *J. Biol. Chem.* **1999**, *274*, 4377–4382.

231. V. Lukashova, E. Z. Szabo, T. Jinadasa, A. Mokhov, D. W. Litchfield, J. Orlowski, *J. Biol. Chem.* **2011**, *286*, 11456–11468.
232. E. Z. Szabo, M. Numata, V. Lukashova, P. Iannuzzi, J. Orlowski, *Proc. Natl. Acad. Sci. USA* **2005**, *102*, 2790–2795.
233. G. H. Diering, J. Church, M. Numata, *J. Biol. Chem.* **2009**, *284*, 13892–13903.
234. I. Onishi, P. J. Lin, G. H. Diering, W. P. Williams, M. Numata, *Cell. Signal.* **2007**, *19*, 194–203.
235. G. H. Diering, F. Mills, S. X. Bamji, M. Numata, *Mol. Biol. Cell* **2011**, *22*, 2246–2257.
236. T. Jinadasa, E. Z. Szabo, M. Numata, J. Orlowski, *J. Biol. Chem.* **2014**, *289*, 20879–20897.
237. E. Z. Szabo, M. Numata, G. E. Shull, J. Orlowski, *J. Biol. Chem.* **2000**, *275*, 6302–6307.
238. M. Numata, J. Orlowski, *J. Biol. Chem.* **2001**, *276*, 17387–17394.
239. R. Nass, K. W. Cunningham, R. Rao, *J. Biol. Chem.* **1997**, *272*, 26145–26152.
240. C. L. Brett, D. N. Tukaye, S. Mukherjee, R. Rao, *Mol. Biol. Cell* **2005**, *16*, 1396–1405.
241. L. M. Kallay, C. L. Brett, D. N. Tukaye, M. A. Wemmer, A. Chyou, G. Odorizzi, R. Rao, *J. Biol. Chem.* **2011**, *286*, 44067–44077.
242. R. Ohgaki, S. C. van Ijzendoorn, M. Matsushita, D. Hoekstra, H. Kanazawa, *Biochemistry* **2011**, *50*, 443–450.
243. N. Demaurex, *News Physiol. Sci.* **2002**, *17*, 1–5.
244. E. Wang, P. S. Brown, B. Aroeti, S. J. Chapin, K. E. Mostov, K. W. Dunn, *Traffic* **2000**, *1*, 480–493.
245. B. E. Steinberg, K. K. Huynh, A. Brodovitch, S. Jabs, T. Stauber, T. J. Jentsch, S. Grinstein, *J. Cell. Biol.* **2010**, *189*, 1171–1186.
246. N. Nakamura, S. Tanaka, Y. Teko, K. Mitsui, H. Kanazawa, *J. Biol. Chem.* **2005**, *280*, 1561–1572.
247. C. L. Brett, Y. Wei, M. Donowitz, R. Rao, *Am. J. Physiol., Cell Physiol.* **2002**, *282*, C1031–1041.
248. K. C. Kondapalli, A. Hack, M. Schushan, M. Landau, N. Ben-Tal, R. Rao, *Nat. Commun.* **2013**, *4*, 2510.
249. Q. Ouyang, S. B. Lizarraga, M. Schmidt, U. Yang, J. Gong, D. Ellisor, J. A. Kauer, E. M. Morrow, *Neuron* **2013**, *80*, 97–112.
250. C. Deisl, A. Simonin, M. Anderegg, G. Albano, G. Kovacs, D. Ackermann, H. Moch, W. Dolci, B. Thorens, M. A. Hediger, D. G. Fuster, *Proc. Natl. Acad. Sci. USA* **2013**, *110*, 10004–10009.
251. G. D. Gilfillan, K. K. Selmer, I. Roxrud, R. Smith, M. Kyllerman, K. Eiklid, M. Kroken, M. Mattingsdal, T. Egeland, H. Stenmark, H. Sjoholm, A. Server, L. Samuelsson, A. Christianson, P. Tarpey, A. Whibley, M. R. Stratton, P. A. Futreal, J. Teague, S. Edkins, J. Gecz, G. Turner, F. L. Raymond, C. Schwartz, R. E. Stevenson, D. E. Undlien, P. Stromme, *Am. J. Hum. Genet.* **2008**, *82*, 1003–1010.
252. I. Roxrud, C. Raiborg, G. D. Gilfillan, P. Stromme, H. Stenmark, *Exp. Cell Res.* **2009**, *315*, 3014–3027.
253. L. Liu, P. H. Schlesinger, N. M. Slack, P. A. Friedman, H. C. Blair, *J. Cell. Physiol.* **2011**, *226*, 1702–1712.
254. J. K. Hill, C. L. Brett, A. Chyou, L. M. Kallay, M. Sakaguchi, R. Rao, P. G. Gillespie, *J. Neurosci.* **2006**, *26*, 9944–9955.
255. N. C. Rockwell, R. S. Fuller, *J. Biol. Chem.* **2002**, *277*, 17531–17537.
256. R. Ohgaki, N. Fukura, M. Matsushita, K. Mitsui, H. Kanazawa, *J. Biol. Chem.* **2008**, *283*, 4417–4429.
257. R. Ohgaki, M. Matsushita, H. Kanazawa, S. Ogihara, D. Hoekstra, S. C. van Ijzendoorn, *Mol. Biol. Cell* **2010**, *21*, 1293–1304.
258. P. J. Lin, W. P. Williams, Y. Luu, R. S. Molday, J. Orlowski, M. Numata, *J. Cell Sci.* **2005**, *118*, 1885–1897.
259. T. Kagami, S. Chen, P. Memar, M. Choi, L. J. Foster, M. Numata, *Mol. Membr. Biol.* **2008**, *25*, 436–447.
260. S. Goyal, G. Vanden Heuvel, P. S. Aronson, *Am. J. Physiol. Renal Physiol.* **2003**, *284*, F467–473.

261. J. Zhang, I. A. Bobulescu, S. Goyal, P. S. Aronson, M. G. Baum, O. W. Moe, *Am. J. Physiol. Renal Physiol.* **2007**, *293*, F761–766.
262. S. Goyal, S. Mentone, P. S. Aronson, *Am. J. Physiol. Renal Physiol.* **2005**, *288*, F530–538.
263. W. Kang'ethe, K. G. Aimanova, A. K. Pullikuth, S. S. Gill, *Am. J. Physiol. Renal Physiol.* **2007**, *292*, F1501–1512.
264. H. Xu, H. Chen, J. Dong, R. Lynch, F. K. Ghishan, *Cell. Physiol. Biochem.* **2008**, *21*, 109–116.
265. J. Y. Choi, M. Shah, M. G. Lee, P. J. Schultheis, G. E. Shull, S. Muallem, M. Baum, *The J. Clin. Invest.* **2000**, *105*, 1141–1146.
266. H. Xu, B. Zhang, J. Li, H. Chen, C. Wang, F. K. Ghishan, *Am. J. Physiol. Gastrointest. Liver Physiol.* **2010**, *299*, G921–927.
267. J. Gattineni, D. Sas, A. Dagan, V. Dwarakanath, M. Baum, *Am. J. Physiol. Renal Physiol.* **2008**, *294*, F198–204.
268. A. M. Becker, J. Zhang, S. Goyal, V. Dwarakanath, P. S. Aronson, O. W. Moe, M. Baum, *Am. J. Physiol. Renal Physiol.* **2007**, *293*, F255–2561.
269. S. P. Lawrence, N. A. Bright, J. P. Luzio, K. Bowers, *Mol. Biol. Cell* **2010**, *21*, 3540–3551.
270. G. Ye, C. Chen, D. Han, X. Xiong, Y. Kong, B. Wan, L. Yu, *Mol. Biol. Reports* **2006**, *33*, 175–180.
271. J. P. Day, S. Wan, A. K. Allan, L. Kean, S. A. Davies, J. V. Gray, J. A. Dow, *J. Cell Sci.* **2008**, *121*, 2612–2619.
272. V. R. Chintapalli, J. Wang, P. Herzyk, S. A. Davies, J. A. Dow, *BMC Genomics* **2013**, *14*, 518.
273. H. Wieczorek, D. Brown, S. Grinstein, J. Ehrenfeld, W. R. Harvey, *Bioessays* **1999**, *21*, 637–648.
274. D. G. Fuster, J. Zhang, M. Shi, I. A. Bobulescu, S. Andersson, O. W. Moe, *J. Am. Soc. Nephrol.* **2008**, *19*, 1547–1556.
275. W. Hofstetter, M. Siegrist, A. Simonin, O. Bonny, D. G. Fuster, *Bone* **2010**, *47*, 331–40.
276. M. Xiang, M. Feng, S. Muend, R. Rao, *Proc. Natl. Acad. Sci. USA* **2007**, *104*, 18677–18681.
277. X. Huang, L. R. Morse, Y. Xu, J. Zahradka, H. Sychrova, P. Stashenko, F. Fan, R. A. Battaglino, *Biochim. Biophys. Acta* **2010**, *1800*, 1241–1247.
278. K. C. Kondapalli, L. M. Kallay, M. Muszelik, R. Rao, *J. Biol. Chem.* **2012**, *287*, 36239–36250.
279. A. D. Martins, R. L. Bernardino, A. Neuhaus-Oliveira, M. Sousa, R. Sa, M. G. Alves, P. F. Oliveira, *Biol. Reprod.* **2014**, *91*, 11.
280. D. Wang, S. M. King, T. A. Quill, L. K. Doolittle, D. L. Garbers, *Nature Cell Biol.* **2003**, *5*, 1117–1122.
281. D. Wang, J. Hu, I. A. Bobulescu, T. A. Quill, P. McLeroy, O. W. Moe, D. L. Garbers, *Proc. Natl. Acad. Sci. USA* **2007**, *104*, 9325–9330.
282. G. R. Abecasis, D. Altshuler, A. Auton, L. D. Brooks, R. M. Durbin, R. A. Gibbs, M. E. Hurles, G. A. McVean, *Nature* **2010**, *467*, 1061–1073.
283. G. A. Cox, C. M. Lutz, C. L. Yang, D. Biemesderfer, R. T. Bronson, A. Fu, P. S. Aronson, J. L. Noebels, W. N. Frankel, *Cell* **1997**, *91*, 139–148.
284. S. M. Bell, C. M. Schreiner, P. J. Schultheis, M. L. Miller, R. L. Evans, C. V. Vorhees, G. E. Shull, W. J. Scott, *Am. J. Physiol.* **1999**, *276*, C788–795.
285. X. Q. Gu, H. Yao, G. G. Haddad, *Am. J. Physiol., Cell Physiol.* **2001**, *281*, C496–503.
286. Y. Xia, P. Zhao, J. Xue, X. Q. Gu, X. Sun, H. Yao, G. G. Haddad, *J. Neurophysiol.* **2003**, *89*, 229–236.
287. H. Yao, E. Ma, X. Q. Gu, G. G. Haddad, *J. Clin. Invest.* **1999**, *104*, 637–645.
288. M. Karmazyn, X. T. Gan, R. A. Humphreys, H. Yoshida, K. Kusumoto, *Circ. Res.* **1999**, *85*, 777–786.
289. M. Avkiran, G. Gross, M. Karmazyn, H. Klein, E. Murphy, K. Ytrehus, *Cardiovasc. Res.* **2001**, *50*, 162–166.
290. H. Clements-Jewery, F. J. Sutherland, M. C. Allen, W. R. Tracey, M. Avkiran, *Brit. J. Pharmacol.* **2004**, *142*, 57–66.

291. Y. Wang, J. W. Meyer, M. Ashraf, G. E. Shull, *Circ. Res.* **2003**, *93*, 776–782.
292. A. R. Cook, S. C. Bardswell, S. Pretheshan, K. Dighe, G. S. Kanaganayagam, R. I. Jabr, S. Merkle, M. S. Marber, S. Engelhardt, M. Avkıran, *J. Mol. Cell. Cardiol.* **2009**, *46*, 225–233.
293. M. Karmazyn, *Am. J. Physiol.* **1988**, *255*, H608–615.
294. M. Lazdunski, C. Frelin, P. Vigne, *J. Mol. Cell. Cardiol.* **1985**, *17*, 1029–1042.
295. N. G. Perez, B. V. Alvarez, M. C. Camilion de Hurtado, H. E. Cingolani, *Circ. Res.* **1995**, *77*, 1192–1200.
296. A. Darmellah, D. Baetz, F. Prunier, S. Tamareille, C. Rucker-Martin, D. Feuvray, *Diabetologia* **2007**, *50*, 1335–1344.
297. A. Kilic, A. Velic, L. J. De Windt, L. Fabritz, M. Voss, D. Mitko, M. Zwiener, H. A. Baba, M. van Eickels, E. Schlatter, M. Kuhn, *Circulation* **2005**, *112*, 2307–2317.
298. S. Aker, A. K. Snabaitis, I. Konietzka, A. Van De Sand, K. Bongler, M. Avkiran, G. Heusch, R. Schulz, *Cardiovasc. Res.* **2004**, *63*, 273–282.
299. R. L. White, J. E. Doeller, V. K. Verselis, B. A. Wittenberg, *J. Gen. Physiol.* **1990**, *95*, 1061–1075.
300. L. Xu, G. Mann, G. Meissner, *Circ. Res.* **1996**, *79*, 1100–1109.
301. A. B. Norholm, R. Hendus-Altenburger, G. Bjerre, M. Kjaergaard, S. F. Pedersen, B. B. Kragelund, *Biochemistry* **2011**, *50*, 3469–3480.
302. F. Mraiche, T. Oka, X. T. Gan, M. Karmazyn, L. Fliegel, *Basic Res. Cardiol.* **2011**, *106*, 603–616.
303. L. Fliegel, *J. Mol. Cell. Cardiol.* **2008**, *44*, 228–237.
304. E. Murphy, D. G. Allen, *J. Mol. Cell. Cardiol.* **2009**, *46*, 137–141.
305. M. Klein, P. Seeger, B. Schuricht, S. L. Alper, A. Schwab, *J. Gen. Physiol.* **2000**, *115*, 599–608.
306. A. Lagana, J. Vadnais, P. U. Le, T. N. Nguyen, R. Laprade, I. R. Nabi, J. Noel, *J. Cell Sci.* **2000**, *113*, 3649–3662.
307. L. Schneider, C. M. Stock, P. Dieterich, B. H. Jensen, L. B. Pedersen, P. Satir, A. Schwab, S. T. Christensen, S. F. Pedersen, *J. Cell. Biol.* **2009**, *185*, 163–176.
308. H. Hayashi, O. Aharonovitz, R. T. Alexander, N. Touret, W. Furuya, J. Orlowski, S. Grinstein, *Am. J. Physiol., Cell Physiol.* **2008**, *294*, C526–534.
309. S. R. Amith, L. Fliegel, *Cancer Res.* **2013**, *73*, 1259–1264.
310. A. E. Lagarde, A. J. Franchi, S. Paris, J. M. Pouyssegur, *J. Cell. Biochem.* **1988**, *36*, 249–260.
311. S. J. Reshkin, A. Bellizzi, S. Caldeira, V. Albarani, I. Malanchi, M. Poignee, M. Alunni-Fabbroni, V. Casavola, M. Tommasino, *FASEB J.* **2000**, *14*, 2185–2197.
312. S. J. Reshkin, A. Bellizzi, R. A. Cardone, M. Tommasino, V. Casavola, A. Paradiso, *Clin. Cancer Res.* **2003**, *9*, 2366–2373.
313. I. N. Rich, D. Worthington-White, O. A. Garden, P. Musk, *Blood* **2000**, *95*, 1427–1434.
314. G. Busco, R.A. Cardone, M. R. Greco, A. Bellizzi, M. Colella, E. Antelmi, M. T. Mancini, M. E. Dell'Aquila, V. Casavola, A. Paradiso, S. J. Reshkin, *FASEB J.* **2010**, *24*, 3903–3915.
315. C. Stock, R. A. Cardone, G. Busco, H. Krahling, A. Schwab, S. J. Reshkin, *Eur. J. Cell Biol.* **2008**, *87*, 591–599.
316. S. Harguindey, G. Orive, J. Luis Pedraz, A. Paradiso, S. J. Reshkin, *Biochim. Biophys. Acta* **2005**, *1756*, 1–24.
317. A. Paradiso, R. A. Cardone, A. Bellizzi, A. Bagorda, L. Guerra, M. Tommasino, V. Casavola, S. J. Reshkin, *Breast Cancer Res.* **2004**, *6*, R616–628.
318. S. J. Reshkin, M. R. Greco, R. A. Cardone, *Phil. Trans. R. Soc. London. Series B, Biol. Sci.* **2014**, *369*, 20130100.
319. E. Boedtkjer, L. Bunch, S.F. Pedersen, *Curr. Pharm. Design* **2012**, *18*, 1345–1371.
320. F. Hanner, R. Chambrey, S. Bourgeois, E. Meer, I. Mucsi, L. Rosivall, G. E. Shull, J. N. Lorenz, D. Eladari, J. Peti-Peterdi, *Am. J. Physiol. Renal Physiol.* **2008**, *294*, F937–944.
321. K. Park, R. L. Evans, G. E. Watson, K. Nehrke, L. Richardson, S. M. Bell, P. J. Schultheis, A. R. Hand, G. E. Shull, J. E. Melvin, *J. Biol. Chem.* **2001**, *276*, 27042–27050.

322. O. Bachmann, B. Riederer, H. Rossmann, S. Groos, P. J. Schultheis, G. E. Shull, M. Gregor, M. P. Manns, U. Seidler, *Am. J. Physiol. Gastrointest. Liver Physiol.* **2004**, *287*, G125–133.
323. L. R. Gawenis, X. Stien, G. E. Shull, P. J. Schultheis, A. L. Woo, N. M. Walker, L. L. Clarke, *Am. J. Physiol. Gastrointest. Liver Physiol.* **2002**, *282*, G776–784.
324. P. Hua, H. Xu, J. K. Uno, M. A. Lipko, J. Dong, P. R. Kiela, F. K. Ghishan, *Am. J. Physiol. Gastrointest. Liver Physiol.* **2007**, *293*, G146–153.
325. C. Ledoussal, J. N. Lorenz, M. L. Nieman, M. Soleimani, P. J. Schultheis, G. E. Shull, *Am. J. Physiol. Renal Physiol.* **2001**, *281*, F718–727.
326. J.N. Lorenz, I. Dostanic-Larson, G. E. Shull, J. B. Lingrel, *J. Am. Soc. Nephrol.* **2006**, *17*, 2457–2463.
327. P. J. Schultheis, L. L. Clarke, P. Meneton, M. Harline, G. P. Boivin, G. Stemmermann, J. J. Duffy, T. Doetschman, M. L. Miller, G. E. Shull, *J. Clin. Invest.* **1998**, *101*, 1243–1253.
328. M. Hatch, R. W. Freel, *Am. J. Physiol. Gastrointest. Liver Physiol.* **2008**, *295*, G348–356.
329. T. Muller, C. Wijmenga, A. D. Phillips, A. Janecke, R. H. Houwen, H. Fischer, H. Ellemunter, M. Fruhwirth, F. Offner, S. Hofer, W. Muller, I. W. Booth, P. Heinz-Erian, *Gastroenterology* **2000**, *119*, 1506–1513.
330. H. C. Li, Z. Du, S. Barone, I. Rubera, A. A. McDonough, M. Tauc, K. Zahedi, T. Wang, M. Soleimani, *J. Mol. Med.* **2013**, *91*, 951–963.
331. P. J. Schultheis, L. L. Clarke, P. Meneton, M. L. Miller, M. Soleimani, L. R. Gawenis, T. M. Riddle, J. J. Duffy, T. Doetschman, T. Wang, G. Giebisch, P. S. Aronson, J. N. Lorenz, G. E. Shull, *Nature Genetics* **1998**, *19*, 282–285.
332. C. Ledoussal, A. L. Woo, M. L. Miller, G. E. Shull, *Am. J. Physiol. Gastrointest. Liver Physiol.* **2001**, *281*, G1385–1396.
333. T. Wang, C. L. Yang, T. Abbiati, P. J. Schultheis, G. E. Shull, G. Giebisch, P. S. Aronson, *Am. J. Physiol.***1999**, *277*, F298–302.
334. W. T. Noonan, A. L. Woo, M. L. Nieman, V. Prasad, P. J. Schultheis, G. E. Shull, J. N. Lorenz, *Am. J. Physiol. Regul. Integr. Comp. Physiol.* **2005**, *288*, R685–691.
335. S. Nakamura, H. Amlal, P. J. Schultheis, J. H. Galla, G. E. Shull, M. Soleimani, *Am. J. Physiol.* **1999**, *276*, F914–921.
336. A. L. Woo, W. T. Noonan, P. J. Schultheis, J. C. Neumann, P. A. Manning, J. N. Lorenz, G. E. Shull, *Am. J. Physiol. Renal Physiol.* **2003**, *284*, F1190–1198.
337. J. L. Kinsella, P. S. Aronson, *Biochim. Biophys. Acta* **1982**, *689*, 161–164.
338. W. Pan, J. Borovac, Z. Spicer, J. G. Hoenderop, R. J. Bindels, G. E. Shull, M. R. Doschak, E. Cordat, R.T. Alexander, *Am. J. Physiol. Renal Physiol.* **2012**, *302*, F943–956.
339. J. Rievaj, W. Pan, E. Cordat, R. T. Alexander, *Am. J. Physiol. Gastrointest. Liver Physiol.* **2013**, *305*, G303–313.
340. M. R. Amin, J. Malakooti, R. Sandoval, P. K. Dudeja, K. Ramaswamy, *Am. J. Physiol., Cell Physiol.* **2006**, *291*, C887–896.
341. F. Rocha, M. W. Musch, L. Lishanskiy, C. Bookstein, K. Sugi, Y. Xie, E. B. Chang, *Am. J. Physiol., Cell Physiol.* **2001**, *280*, C1224–1232.
342. S. Sullivan, P. Alex, T. Dassopoulos, N. C. Zachos, C. Iacobuzio-Donahue, M. Donowitz, S. R. Brant, C. Cuffari, M. L. Harris, L. W. Datta, L. Conklin, Y. Chen, X. Li, *Inflam. Bowel Dis.* **2009**, *15*, 261–274.
343. S. Yeruva, K. Farkas, J. Hubricht, K. Rode, B. Riederer, O. Bachmann, A. Cinar, Z. Rakonczay, T. Molnar, F. Nagy, J. Wedemeyer, M. Manns, D. Raddatz, M. W. Musch, E. B. Chang, P. Hegyi, U. Seidler, *Inflam. Bowel Dis.* **2010**, *16*, 1149–1161.
344. I.A. Bobulescu, F. Di Sole, O. W. Moe, *Curr. Opin. Nephrol. Hypertens.* **2005**, *14*, 485–494.
345. C. Liu, H. Xu, B. Zhang, M. E. Johansson, J. Li, G. C. Hansson, F. K. Ghishan, *Am. J. Physiol., Cell Physiol.* **2013**, *305*, C121–128.
346. M. Baum, K. Twombley, J. Gattineni, C. Joseph, L. Wang, Q. Zhang, V. Dwarakanath, O. W. Moe, *Am. J. Physiol. Renal Physiol.* **2012**, *303*, F1495–502.
347. J. E. Melom, J. T. Littleton, *Curr. Opin. Genet. Dev.* **2011**, *21*, 256–261.
348. L. Xinhan, M. Matsushita, M. Numaza, A. Taguchi, K. Mitsui, H. Kanazawa, *Am. J. Physiol., Cell Physiol.* **2011**, *301*, C1431–1444.

349. E. C. Deane, A. E. Ilie, S. Sizdahkhani, M. Das Gupta, J. Orlowski, R. A. McKinney, *J. Neurosci.* **2013**, *33*, 595–610.
350. M. E. Malo, L. Fliegel, *Can. J. Physiol. Pharmacol.* **2006**, *84*, 1081–1095.
351. J. Luo, D. Sun, *Curr. Neurovasc. Res.* **2007**, *4*, 205–215.
352. J. Y. Garbern, M. Neumann, J. Q. Trojanowski, V. M. Lee, G. Feldman, J. W. Norris, M. J. Friez, C. E. Schwartz, R. Stevenson, A. A. Sima, *Brain J. Neurol.* **2010**, *133*, 1391–1402.
353. Y. Takahashi, K. Hosoki, M. Matsushita, M. Funatsuka, K. Saito, H. Kanazawa, Y. Goto, S. Saitoh, *Am. J. Med. Genetics. Part B, Neuropsychiatric Genetics* **2011**, *156B*, 799–807.
354. Y. Yasuda, R. Hashimoto, H. Yamamori, K. Ohi, M. Fukumoto, S. Umeda-Yano, I. Mohri, A. Ito, M. Taniike, M. Takeda, *Molecular Autism* **2011**, *2*, 9.
355. E. M. Morrow, S. Y. Yoo, S. W. Flavell, T. K. Kim, Y. Lin, R. S. Hill, N. M. Mukaddes, S. Balkhy, G. Gascon, A. Hashmi, S. Al-Saad, J. Ware, R. M. Joseph, R. Greenblatt, D. Gleason, J. A. Ertelt, K. A. Apse, A. Bodell, J. N. Partlow, B. Barry, H. Yao, K. Markianos, R. J. Ferland, M. E. Greenberg, C. A. Walsh, *Science* **2008**, *321*, 218–223.
356. A. L. Christianson, R. E. Stevenson, C. H. van der Meyden, J. Pelser, F. W. Theron, P. L. van Rensburg, M. Chandler, C. E. Schwartz, *J. Med. Genetics* **1999**, *36*, 759–766.
357. P. Stromme, K. Dobrenis, R. V. Sillitoe, M. Gulinello, N. F. Ali, C. Davidson, M. C. Micsenyi, G. Stephney, L. Ellevog, A. Klungland, S. U. Walkley, *Brain J. Neurol.* **2011**, *134*, 3369–3383.
358. M. Schwede, K. Garbett, K. Mirnics, D. H. Geschwind, E. M. Morrow, *Mol. Psychiatry* **2014**, *19*, 277–279.
359. S. A. Meda, B. Narayanan, J. Liu, N. I. Perrone-Bizzozero, M. C. Stevens, V. D. Calhoun, D. C. Glahn, L. Shen, S. L. Risacher, A. J. Saykin, G. D. Pearlson, *NeuroImage* **2012**, *60*, 1608–1621.
360. L. Zhang, T. Wang, A. F. Wright, M. Suri, C. E. Schwartz, R. E. Stevenson, D. Valle, *Am. J. Med. Genetics Part A* **2006**, *140*, 349–357.
361. N. Milosavljevic, M. Monet, I. Lena, F. Brau, S. Lacas-Gervais, S. Feliciangeli, L. Counillon, M. Poet, *Cell Rep.* **2014**, *7*, 689–696.
362. M. G. de Silva, K. Elliott, H. H. Dahl, E. Fitzpatrick, S. Wilcox, M. Delatycki, R. Williamson, D. Efron, M. Lynch, S. Forrest, *J. Med. Genetics* **2003**, *40*, 733–740.
363. J. Lasky-Su, R. J. Anney, B. M. Neale, B. Franke, K. Zhou, J. B. Maller, A. A. Vasquez, W. Chen, P. Asherson, J. Buitelaar, T. Banaschewski, R. Ebstein, M. Gill, A. Miranda, F. Mulas, R. D. Oades, H. Roeyers, A. Rothenberger, J. Sergeant, E. Sonuga-Barke, H. C. Steinhausen, E. Taylor, M. Daly, N. Laird, C. Lange, S. V. Faraone, *Am. J. Med. Genetics. Part B, Neuropsychiatric Genetics* **2008**, *147B*, 1355–1358.
364. J. M. Vink, A. B. Smit, E. J. de Geus, P. Sullivan, G. Willemsen, J. J. Hottenga, J. H. Smit, W. J. Hoogendijk, F. G. Zitman, L. Peltonen, J. Kaprio, N. L. Pedersen, P. K. Magnusson, T. D. Spector, K. O. Kyvik, K.I . Morley, A. C. Heath, N. G. Martin, R. G. Westendorp, P. E. Slagboom, H. Tiemeier, A. Hofman, A. G. Uitterlinden, Y. S. Aulchenko, N. Amin, C. van Duijn, B. W. Penninx, D. I. Boomsma, *Am. J. Hum. Genet.* **2009**, *84*, 367–379.
365. B. Franke, B. M. Neale, S. V. Faraone, *Hum. Genet.* **2009**, *126*, 13–50.
366. C. A. Markunas, K. S. Quinn, A. L. Collins, M. E. Garrett, A. M. Lachiewicz, J. L. Sommer, E. Morrissey-Kane, S. H. Kollins, A. D. Anastopoulos, A. E. Ashley-Koch, *Psych. Genetics* **2010**, *20*, 73–81.
367. E. Mick, A. Todorov, S. Smalley, X. Hu, S. Loo, R. D. Todd, J. Biederman, D. Byrne, B. Dechairo, A. Guiney, J. McCracken, J. McGough, S. F. Nelson, A. M. Reiersen, T.E. Wilens, J. Wozniak, B. M. Neale, S. V. Faraone, *J. Am. Acad. Child Adolesc. Psychiatry* **2010**, *49*, 898–905 e3.
368. Y. Zhang-James, T. DasBanerjee, T. Sagvolden, F. A. Middleton, S. V. Faraone, *Am. J. Med. Genetics. Part B, Neuropsychiatric Genetics* **2011**, *156B*, 835–843.
369. R. A. Battaglino, L. Pham, L. R. Morse, M. Vokes, A. Sharma, P. R. Odgren, M. Yang, H. Sasaki, P. Stashenko, *Bone* **2008**, *42*, 180–192.
370. J. F. Charles, F. Coury, R. Sulyanto, D. Sitara, J. Wu, N. Brady, K. Tsang, K. Sigrist, D. M. Tollefsen, L. He, D. Storm, A. O. Aliprantis, *Bone* **2012**, *51*, 902–912.

371. M. Canessa, N. Adragna, H. S. Solomon, T. M. Connolly, D. C. Tosteson, *New England J. Med.* **1980**, *302*, 772–776.
372. M. Canessa, G. Zerbini, L. M. Laffel, *J. Am. Soc. Nephrol.* **1992**, *3*, S41–49.
373. R. Mangili, J. J. Bending, G. Scott, L. K. Li, A. Gupta, G. Viberti, *New England J. Med.* **1988**, *318*, 146–150.
374. C. F. Verkoelen, B. G. van der Boom, D. J. Kok, A. B. Houtsmuller, P. Visser, F. H. Schroder, J. C. Romijn, *Kidney International* **1999**, *55*, 1426–1433.
375. C. Deisl, G. Albano, D. G. Fuster, *Curr. Opin. Nephrol. Hypertens.* **2014**, *23*, 406–10.
376. P. E. MacDonald, L. Eliasson, P. Rorsman, *J. Cell Sci.* **2005**, *118*, 5911–5920.
377. L. Orci, F. Malaisse-Lagae, M. Ravazzola, M. Amherdt, A. E. Renold, *Science* **1973**, *181*, 561–562.
378. T. Kimura, Y. Kaneko, S. Yamada, H. Ishihara, T. Senda, A. Iwamatsu, I. Niki, *J. Cell Sci.* **2008**, *121*, 3092–3098.
379. L. Min, Y. M. Leung, A. Tomas, R. T. Watson, H. Y. Gaisano, P. A. Halban, J. E. Pessin, J. C. Hou, *J. Biol. Chem.* **2007**, *282*, 33530–33536.
380. D. Deng, C. Xu, P. Sun, J. Wu, C. Yan, M. Hu, N. Yan, *Nature* **2014**,
381. D. Yernool, O. Boudker, Y. Jin, E. Gouaux, *Nature* **2004**, *431*, 811–818.
382. L. Wang, X. Feng, H. Zhao, L. An, Q. S. Qiu, *PloS one* **2014**, *9*, e104147.
383. P. Oelkers, L. C. Kirby, J. E. Heubi, P. A. Dawson, *J. Clin. Invest.* **1997**, *99*, 1880–1887.
384. W. Kramer, H. Glombik, *Curr. Med. Chem.* **2006**, *13*, 997–1016.
385. L. Chen, X. Yao, A. Young, J. McNulty, D. Anderson, Y. Liu, C. Nystrom, D. Croom, S. Ross, J. Collins, D. Rajpal, K. Hamlet, C. Smith, B. Gedulin, *Am. J. Physiol. Endocrin. Metabolism* **2012**, *302*, E68–76.
386. B. G. Bhat, S. R. Rapp, J. A. Beaudry, N. Napawan, D. N. Butteiger, K. A. Hall, C. L. Null, Y. Luo, B. T. Keller, *J. Lipid Res.* **2003**, *44*, 1614–1621.
387. M. C. Lewis, L. E. Brieaddy, C. Root, *J. Lipid Res.* **1995**, *36*, 1098–1105.
388. W. Kramer, G. Wess, *Eur. J. Clin. Invest.* **1996**, *26*, 715–732.
389. S. Tolle-Sander, K. A. Lentz, D. Y. Maeda, A. Coop, J. E. Polli, *Mol. Pharm.* **2004**, *1*, 40–48.
390. K. Herz, S. Vimont, E. Padan, P. Berche, *J. Bacteriol.* **2003**, *185*, 1236–1244.
391. C. T. Resch, J. L. Winogrodzki, C. T. Patterson, E. J. Lind, M. J. Quinn, P. Dibrov, C. C. Hase, *Biochemistry* **2010**, *49*, 2520–8.
392. E. J. Wiens, J. L. Winogrodzki, C. T. Resch, G. L. Orriss, J. Stetefeld, P. Dibrov, *Mol. Cell. Biochem.* **2014**, *389*, 51–58.
393. E. Padan, S. Schuldiner, *J. Bioenerg. Biomembr.* **1993**, *25*, 647–669.
394. L. Galili, A. Rothman, L. Kozachkov, A. Rimon, E. Padan, *Biochemistry* **2002**, *41*, 609–617.
395. K. R. Vinothkumar, S. H. Smits, W. Kuhlbrandt, *EMBO J.* **2005**, *24*, 2720–2729.
396. E. F. Pettersen, T. D. Goddard, C. C. Huang, G. S. Couch, D. M. Greenblatt, E. C. Meng, T. E. Ferrin, *J. Comput. Chem.* **2004**, *25*, 1605–12.
397. G. Celniker, G. Nimrod, H. Ashkenazy, F. Glaser, E. Marz, I. Mayrose, T. Pupko, N. Ben-Tal, *Isr. J. Chem.* **2013**, DOI: 10.1002/ijch.201200096.
398. M. Landau, I. Mayrose, Y. Rosenberg, F. Glaser, E. Martz, T. Pupko, N. Ben-Tal, *Nucleic Acids Res.* **2005**, *33*, W299–302.
399. H. Ashkenazy, E. Erez, E. Martz, T. Pupko, N. Ben-Tal, *Nucleic Acids Res.* **2010**, *38*, W529–533.

Chapter 13
Proton-Potassium (H^+/K^+) ATPases: Properties and Roles in Health and Diseases

Hideki Sakai, Takuto Fujii, and Noriaki Takeguchi

Contents

H. Sakai (✉) • T. Fujii
Department of Pharmaceutical Physiology, Graduate School of Medicine and Pharmaceutical Sciences, University of Toyama, 2630 Sugitani, Toyama City, Toyama 930-0194, Japan
e-mail: sakaih@pha.u-toyama.ac.jp

N. Takeguchi
Takeguchi Digestive Research, Toyama 930-0125, Japan
e-mail: qqze72p9@tiara.ocn.ne.jp

A. Sigel, H. Sigel, and R.K.O. Sigel (eds.), *The Alkali Metal Ions: Their Role for Life*, Metal Ions in Life Sciences 16, DOI 10.1007/978-3-319-21756-7_13

Abstract As a physiological phenomenon, acid secretion from the stomach was known already at least in the 17th century. But its mechanism was elucidated in more recent times only. At the end of the 20th century, gastric H^+/K^+-ATPase in the parietal cells was found to be responsible for a final step of H^+ secretion in these cells. In this century, several Cl^--transporting proteins for gastric acid (hydrochloric acid; HCl) secretion have been found. As inhibitors of gastric acid secretion, histamine H_2 receptor antagonists (H_2 blockers) were developed in the 1970's. This discovery brought a great benefit; that is, peptic ulcers became treatable by administration of a drug. In 1980's, proton pump inhibitors (PPIs) were developed. The target of PPIs is gastric H^+/K^+-ATPase and the PPIs exert generally more potent effects compared with H_2 blockers. Most recently, several K^+-competitive inhibitors of the ATPase are being developed. Here, we introduce gastric H^+/K^+-ATPase and its related proteins for gastric acid secretion, and several gastric diseases and their treatment by medicines.

Keywords Cl^- transporter • Gastric acid • K^+ channel • Parietal cell • PPI • Proton pump • Stomach

Please cite as: *Met. Ions Life Sci.* 16 (2016) 459–483

1 Introduction

In virtually all vertebrates, gastric acid secretion is a common physiological process, and it is essential for food digestion. The acid (H^+) secretion is mediated by gastric H^+/K^+-ATPase, which establishes a million-fold gradient of protons across the apical canalicular membrane of the gastric parietal cell. Interestingly, the earliest phylogenetic appearance of gastric acid secretion is in cartilaginous fishes and the expression of gastric H^+/K^+-ATPase was actually found in them [1].

Gastric H^+/K^+-ATPase is composed of α- and β-subunits. The α-subunit is a large catalytic subunit (around 1030 amino acids), whereas the β-subunit is a heavily glycosylated subunit (around 290 amino acids) and stabilizes expression of the α-subunit in the plasma membrane of the cells [2]. The α- and β-subunits have ten and one transmembrane domains, respectively. Primary sequence of the H^+/K^+-ATPase is highly conserved from fish to mammals: that is, the cDNA sequence of the α-subunit of Atlantic stingrays is over 80 % identical to that of mammals [3].

Gastric H^+/K^+-ATPase is the major therapeutic target in the treatment of acid-related diseases such as peptic ulcer and gastroesophageal reflux disease. So far, several proton pump inhibitors (PPIs) have been clinically used. The PPIs are the most effective among current acid-suppressive drugs. In addition, various K^+-competitive inhibitors of the ATPase are being developed, with the advantage of complete and rapid inhibition of acid secretion [4].

The main component of gastric acid is hydrochloric acid (HCl). H^+ is actively secreted by the H^+/K^+-ATPase. On the other hand, the mechanism of apical Cl^- transport for HCl secretion is not well established yet.

Apical K^+ recycling is essential for maintaining H^+/K^+-ATPase activity. So far, expression and function of a variety of K^+ channels and transporters have been reported. However, the K^+ recycling mechanism in the apical canalicular membrane may be more complex than we imagine.

In this review, we summarize molecular, biochemical, structural, and physiological aspects of gastric H^+/K^+-ATPase, the mechanism of gastric acid secretion, Cl^-- and K^+-transporting proteins associated with H^+/K^+-ATPase, gastric acid-related diseases, and the H^+/K^+-ATPase inhibitors.

2 Mechanism of Gastric Acid Secretion

2.1 Food Intake and Acid Secretion

The stomach secretes 1–2 L of gastric juice per day. Maximal speed of the secretion is 2.5 mL per minute at the stimulated phase. Cationic composition of the juice in this phase is different from that at the resting phase. At the stimulated phase, the juice contains more than 100 mM of H^+, 5–10 mM of K^+, and 10–20 mM of Na^+. In contrast, the juice at the resting phase contains 20–30 mM of H^+, 10–15 mM of K^+ and 60–80 mM of Na^+ [5]. The main anion involved in the gastric juice is Cl^-, and its concentration is 120–160 mM at the stimulated and resting phases [6].

The stomach has four anatomical regions: the cardia, fundus, corpus (body), and pylorus (Figure 1). The cardiac region is closest to the esophagus, and it contains cardiac glands which secrete mucus. The fundus and corpus regions are the largest parts of the stomach, and they contain gastric (fundic) glands which secrete hydrochloric acid (HCl), pepsinogen (a precursor of pepsin), gastric lipase, and mucus. The pyloric region is closest to the duodenum of the small intestine, and it contains pyloric glands, which secrete mucus and gastrin.

Regulation of gastric secretion associated with eating is divided into three stages called the cephalic, gastric, and intestinal phases. In the cephalic phase, the mere sight, smell, taste, or thought of food sends nervous impulses to the medulla oblongata. Vagal efferent activation promotes the secretion of entero-pancreatic hormones, including ghrelin, which increases food intakes and body weight [6]. Then, via the vagus nerve, parasympathetic neurons stimulate secretion of gastric juice containing HCl, pepsin which breaks down proteins, and lipase which breaks down fats. The parasympathetic stimulation also causes gastrin secretion from the pyloric region of the stomach. Gastrin travels through the bloodstream and further stimu-

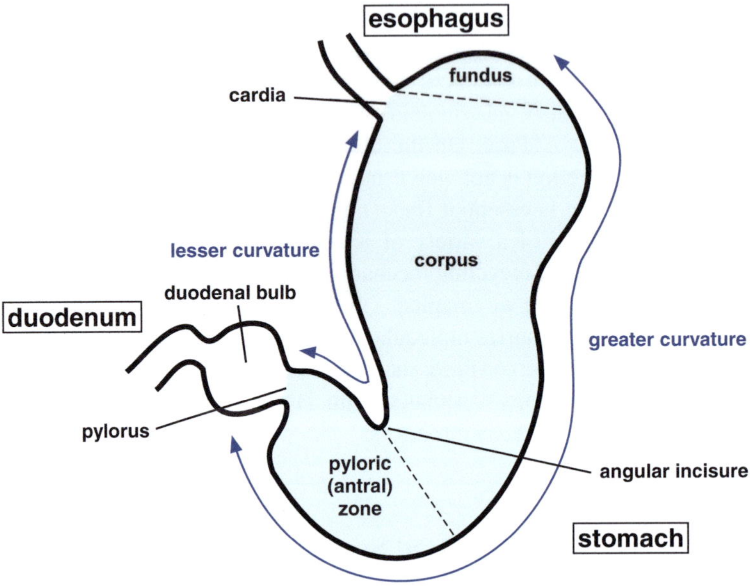

Figure 1 Outline of each region of the stomach.

lates the gastric secretion in the corpus venticuli of the stomach. In the gastric phase, food enters and distends the stomach. This distension activates a parasympathetic reflex via the medulla oblongata, and also has a direct stimulatory effect on the gastric glands. Then secretion of HCl and pepsinogen is activated and continues. In the intestinal phase, the stomach contents (chyme) enter the duodenum. The duodenum initially enhances gastric secretion, but soon inhibits it.

2.2 *Gastric Gland*

The gastric mucosa of the fundus and corpus regions forms invaginations called gastric pits. The lamina propria contains gastric (fundic) glands which open into the bases of the pits. The gastric glands contain several types of cells such as surface epithelial, mucous neck, parietal, chief, and enterochromaffin-like (ECL) cells (Figure 2).

Surface epithelial cells produce the mucus which protects the gastric mucosa from strong acid and enzymes in the lumen. The life of these cells is 4–6 days. Mucous neck cells are located in the isthmus and neck of gastric glands, and secrete mucus. Parietal cells are located in the isthmus, neck, and base of the glands. The cells have the largest diameter (20–30 μm) in the glands. They produce gastric acid (HCl) and intrinsic factor which is required for the absorption of vitamin B_{12} (cobalamin) in the small intestine. Chief cells are located in the base of gastric glands and secrete pepsinogen, a precursor of pepsin. ECL cells in the bases of the glands secrete histamine, an acid secretagogue.

Figure 2 An illustrated model for several types of cells in gastric glands.

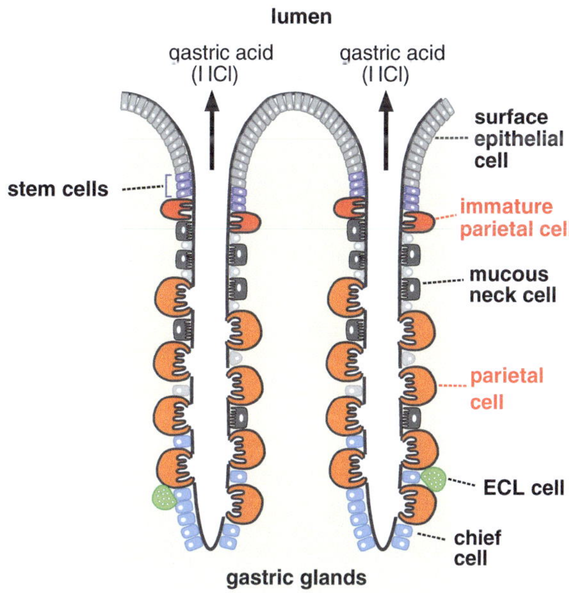

2.3 *Gastric Parietal Cell*

A higher concentration of HCl (160 mM at the stimulated phase) is secreted by gastric parietal cells in the stomach. The cells are rich in mitochondria which can produce a large amount of ATP for promoting the enzymatic activity of gastric H^+/K^+-ATPase. Gastric acid secretion is accompanied with dramatic morphological changes of the parietal cells (Figure 3). In resting parietal cells, tubulovesicles, which are rich in H^+/K^+-ATPase, are present in intracellular compartments underlying the apical membrane and forming a reticulated meshwork. Upon stimulation, tubulovesicles translocate and connect with the apical membrane, resulting in massive acid secretion [7, 8]. In this stimulated phase, the tubulovesicular and apical membranes do not mix but remain separate and distinct [9, 10] (Figure 3).

On the other hand, the parietal cell has the muscarinic M3 receptor, the cholecystokinin-2 (CCK-2) receptor, and the histamine H_2 receptor in its basolateral membrane. In this membrane, cytoprotective Cl^- channels are abundantly present (20,000–30,000 in a parietal cell) [11]. These Cl^- channels are activated by prostaglandin E_2 and cGMP [11, 12], and inhibited by interleukin-1β [13]. Opening of the channels can attenuate the ethanol-induced cell injury [14]. This cytoprotective function may explain why the parietal cells have a long lifespan (150–200 days) compared with that of surface epithelial cells (4–6 days).

The parietal cells migrate from the luminal to the basal region of gastric glands, and the (younger) luminal parietal cells much more actively secrete acid than do the (older) basal parietal cells [15–17].

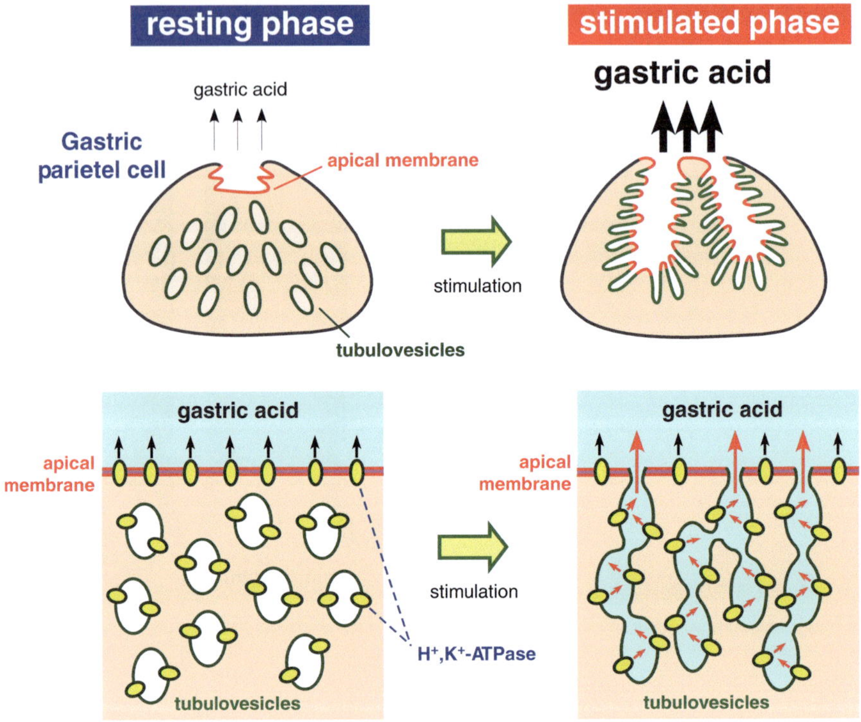

Figure 3 An illustrated model for the gastric parietal cell showing the basal acid secretion in the resting phase and the strong acid secretion in the stimulated phase.

2.4 Acid Secretagogues

Acetylcholine, gastrin, and histamine are the three major factors that stimulate gastric acid secretion in gastric parietal cells [18] (Figure 4). Acetylcholine is secreted by parasympathetic nerve fibers of both short and long reflex pathways, and it binds to the muscarinic M3 receptor in the parietal cell. Gastrin is a hormone produced by G cells in the pyloric glands, and it can directly bind to cholecystokinin-2 receptor in parietal cells. Gastrin also activates the parietal cell through an indirect pathway involving the release of histamine from ECL cells. The released histamine binds to histamine H_2 receptor of the parietal cell in a paracrine manner. In the gastric mucosa of the mammalian stomach, histamine is stored in ECL cells and in mucosal mast cells. At least 80 % of the gastric mucosal histamine resides in ECL cells [19].

In gastric parietal cells, the CCK-2 and muscarinic M3 receptors couple to the G-protein Gq, which activates phospholipase C to induce an increase in inositol trisphosphate (IP_3) and release of intracellular Ca^{2+} [20, 21]. The histamine H_2 receptor couples via the G-protein Gs to adenylate cyclase and stimulates cyclic AMP formation [22]. Intracellular Ca^{2+} and cAMP act as the second messenger for gastric acid secretion (Figure 4).

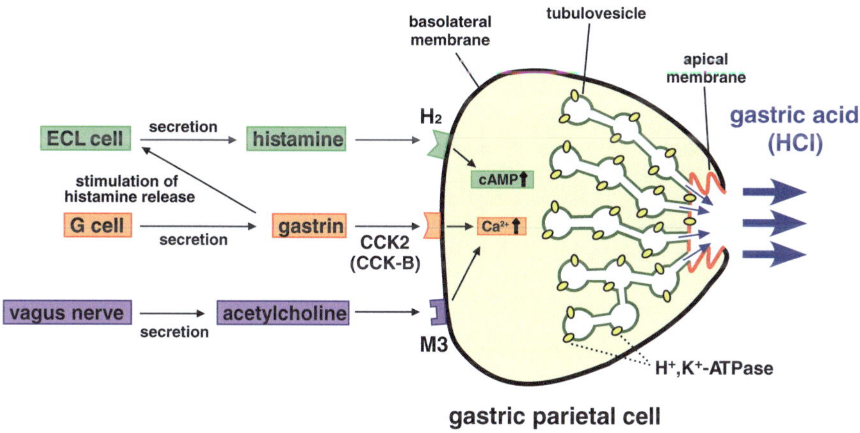

Figure 4 Secretagogues for gastric acid secretion and their receptors in the basolateral membrane of the parietal cell.

3 Properties of Gastric H^+/K^+-ATPase

3.1 Physiological and Biochemical Studies of Gastric H^+/K^+-ATPase

Physiological evidence that Na^+ is not essential but K^+ is the more important ion for acid secretion was obtained in 1963 from studies of frog gastric mucosa [23]. Later, the acid secretory rate from isolated frog gastric mucosa set in Ussing chambers was found to depend on the free K^+ concentration in the intracellular K^+ pool. The pool K^+ concentration was a function of the serosal medium K^+ concentration following a Michaelis-Menten equation and it was also linearly related with the mucosal medium K^+ concentration [24, 25].

The reason why K^+ is necessary for acid secretion was solved based on biochemical evidence. Ouabain-insensitive K^+-ATPase activities in isolated gastric microsomes were found in 1965 [26] and 1967 [27], which hinted at the presence of a Na^+/K^+-ATPase-like but ouabain-insensitive specific molecule for gastric secretion. In 1973, K^+-dependent ATPase activity in microsomes was found to be stimulated by the presence of valinomycin, a K^+ ionophore, in the medium [28]. But the acid secretion by isolated frog mucosa set in Ussing chambers was not stimulated by the addition of valinomycin in bathing solutions. This paradox was quickly solved in 1974 by the findings that gastric microsomes are inside-out closed vesicles, addition of ATP in the medium induces accumulation of H^+ in vesicles, valinomycin increases K^+ influx into vesicles, the amount of H^+ accumulation depends on the amount of K^+ in vesicles, and K^+ is actively transported from the inside to the outside of the vesicles [29]. The K^+-ATPase was found to be the H^+/K^+-ATPase. The H^+/K^+ exchange is non-electrogenic [30].

Figure 5 Reaction scheme of gastric H^+/K^+-ATPase. Cation binding sites of H^+/K^+-ATPase at the E_1 state face the cytoplasm and the ones at the E_2 state face the lumen. "In" and "out" indicate free cations in the cytoplasm and lumen, respectively.

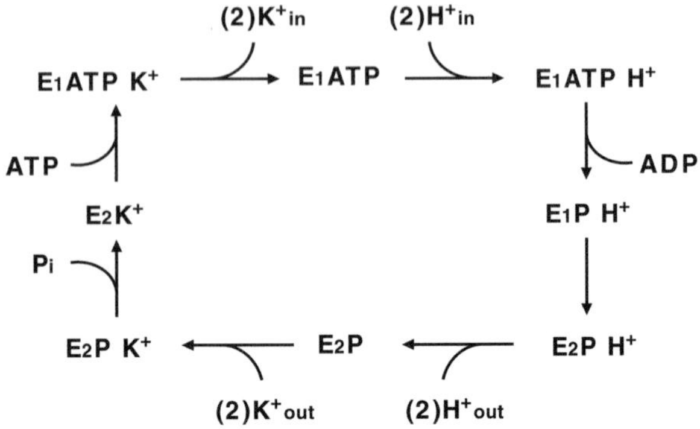

The biochemical reaction steps of H^+/K^+-ATPase consist of phosphorylation and dephosphorylation accompanying cyclic conformational changes as $E_1 \rightarrow E_2 \rightarrow E_1$, where cation binding sites of H^+/K^+-ATPase face the cytoplasm at the E_1 state or the extracellular surface at the E_2 state (Figure 5) [31–34]. Although these reaction steps resemble those of Na^+/K^+-ATPase, E_2K and E_1K states of H^+/K^+-ATPase are nearly equivalent energetically while E_2K is 3 orders of magnitude more stable than E_1K in Na^+/K^+-ATPase [35]. Na^+/K^+-ATPase can be phosphorylated from acetyl phosphate and dephosphorylated resulting in active ion transport, while H^+/K^+-ATPase can be also phosphorylated from acetyl phosphate but the conformational states of phosphorylated intermediates are not energy-rich resulting in non-active ion transport [36].

The sum of the H^+- and K^+-transport steps in H^+/K^+-ATPase is non-electrogenic because both H^+- and the K^+-transport steps are electrogenic, but oppositely directed [37, 38]. It is noted that Na^+/K^+-ATPase is an electrogenic pump with an electrogenic Na^+ step and an electroneutral K^+ step.

The complete inhibition of K^+-ATPase activity and proton uptake by a specific proton pump inhibitor rabeprazole (E3810) occurred when rabeprazole was bound to half of the H^+/K^+-ATPase α-subunit (Figure 6), suggesting that dimeric interaction between the α-subunits is involved in the enzyme activity and proton transport [39].

3.2 Molecular and Structural Studies of Gastric H^+/K^+-ATPase

Gastric H^+/K^+-ATPase consists of one catalytic α-subunit and one non-catalytic β-subunit. Molecular cloning of α- and β-subunits was first reported in 1986 and 1990, respectively [40, 41]. Figure 7 shows the schematic model for the N domain (nucleotide), the P domain (phosphorylation), and the A domain (actuator). Naming of N, P, and A was first proposed for sarcoplasmic reticulum Ca^{2+}-ATPase [42]. The α-subunit of H^+/K^+-ATPase in different species consists of 1031–1035 amino acids and has 10 transmembrane segments M1 – M10. The catalytic center resides in the large cytoplasmic domain between M4 and M5 segments. The β-subunit (290–299

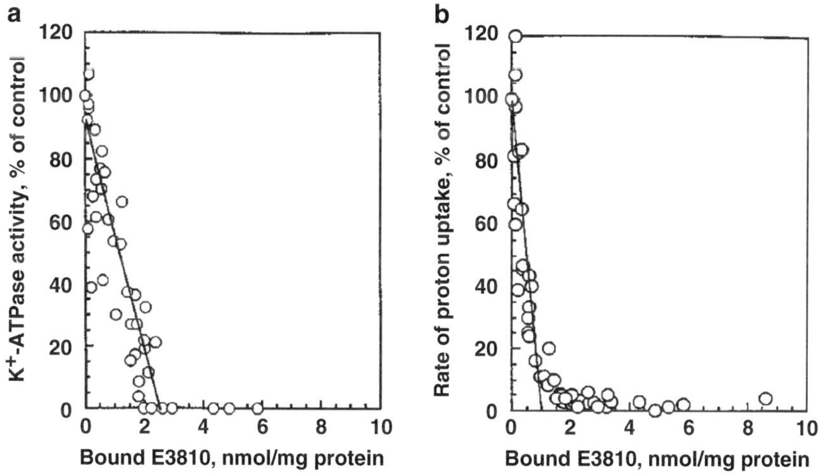

Figure 6 Relationship between the amount of bound rabeprazole (E3810) and the K^+-ATPase activity (**a**) and between the amount of bound rabeprazole and the rate of proton uptake (**b**) in gastric vesicles measured at 0.2 mM ATP. Taken with permission from Figures 6 and 7 in [39].

amino acids) consists of a short cytoplasmic domain, single transmembrane domain, and a large ectodomain which has 6–10 carbohydrates chains (Figure 7).

The transient functional expression of gastric H^+/K^+-ATPase α- and β-subunits in a HEK-293 cell line was first reported [43], which enabled to study site-directed mutagenesis of the putative cation binding site and catalytic center. Then, the stable expression of H^+/K^+-ATPase α- and β-subunits was reported [44].

Asp^{387} in the P domain is phosphorylated from ATP. Glu^{345} in the M4 segment, Glu^{797}, Lys^{793}, and Gln^{794} in the M5 segment [45, 46], and Lys^{819}, Glu^{822}, Asp^{826}, Ile^{827}, and Leu^{833} in the M6 segment [47, 48] are involved in cation binding (Figure 7). Fluorescein isothiocyanate (FITC) covalently binds with Lys^{519} in the N domain. K^+ binding in the cavity surrounded by transmembrane M4, M5, and M6 segments induces the gross conformational change, resulting in the local hydrophobicity change around Lys^{519} which is remote from the transmembrane domain (Figure 7).

Gastric H^+/K^+-ATPase transports 2 mol of H^+/mol of ATP hydrolysis in isolated hog gastric vesicles (Figure 5) [30]. There is an interesting question whether the H^+ transport mechanism is due to charge transfer and/or transfer of hydronium ion (H_3O^+). The measurement of transport of $[^{18}O]H_2O$ showed that 1.8 mol of water molecules/mol of ATP hydrolysis was actively transported [49]. The molecular dynamics simulation of the three-dimensional structure model of the H^+/K^+-ATPase α-subunit at E_1 conformation and site-directed mutagenesis studies predicted (i) the presence of a charge transfer pathway from the hydronium ion in cytosolic medium to Glu^{345} in the cation binding site 2 (H_3O^+-Lys^{164}-Gln^{161}-Glu^{345}) and (ii) the presence of a charge transfer from sites 2 to 1 via a water wire and a charge transfer pathway (H_3O^+-Asn^{794} -Glu^{797}). Figure 8 shows that protons are

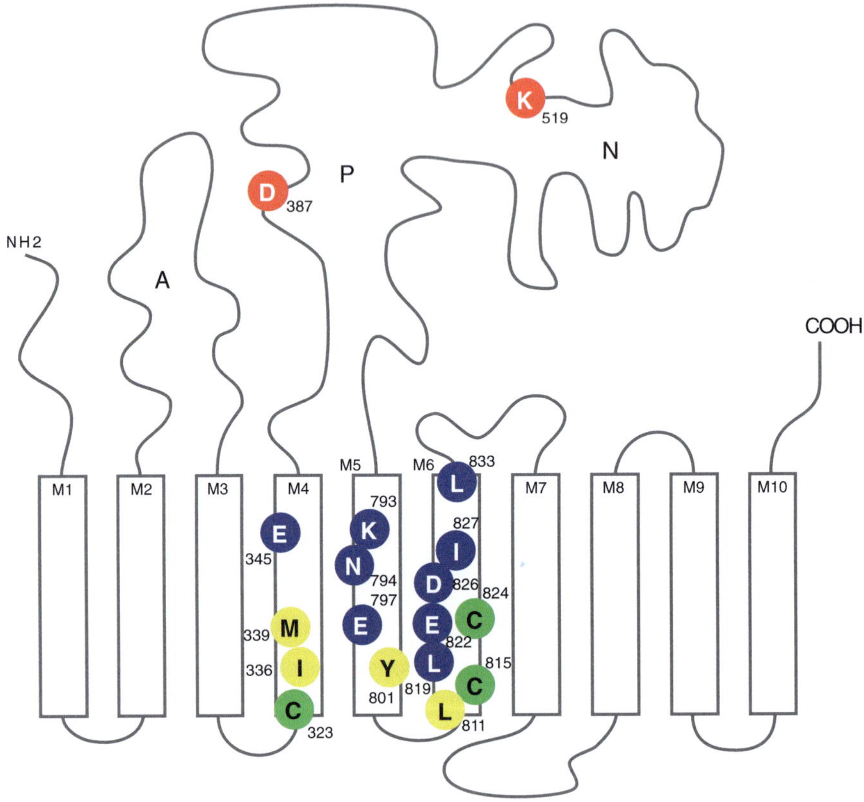

Figure 7 Schematic model of the gastric H^+/K^+-ATPase α-subunit. P = phosphorylation domain, N = nucleotide domain, and A = actuator domain. Blue circles show amino acid residues involved in cation binding, green circles show Cys residues involved in binding of proton pump inhibitors (PPIs), and yellow circles show amino acid residues involved in SCH28080 binding. K^{519} is the FITC-binding site and D^{387} is the phosphorylation site. Modified from Figure 3 in [34] with permission from the Pharmaceutical Society of Japan; © copyright 2004.

charge-transferred from the cytosolic side to H_2O in sites 2 and 1where the H_2O comes from the cytosolic medium, and H_3O^+ in the sites is transported into the lumen during the conformational transition from E_1P to E_2P [49].

The β-subunit is also involved in H^+/K^+-ATPase activity, formation of the αβ-subunit complex, and surface delivery of the complex. The β-subunit has six extracellular cysteine residues and form three disulfide bonds (Figure 9). Each of these S-S crosslinking is necessary for functions of the β-subunit [50]. The ectodomain of the rabbit β-subunit is modified with seven *N*-linked carbohydrate chains. Mutagenetic deletion of any one of seven carbohydrate chains did not affect the function of H^+/K^+-ATPase, but deletion of more than three carbohydrate chains

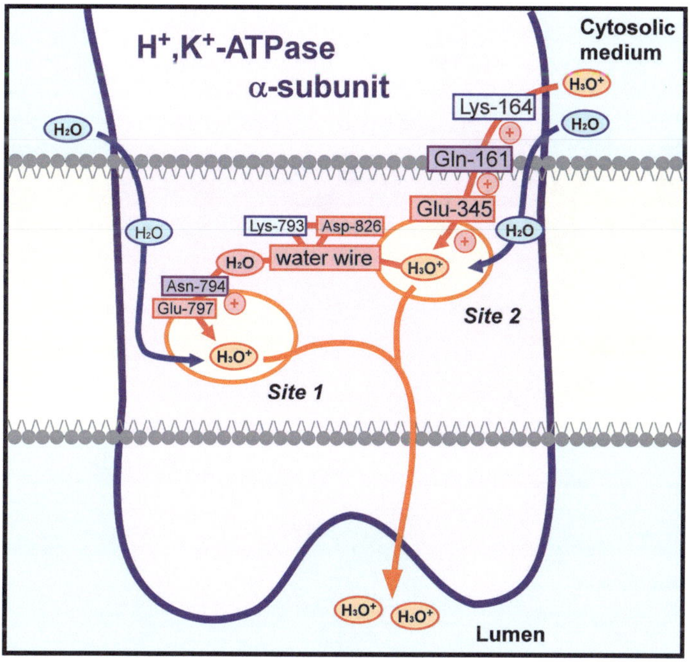

Figure 8 An illustrated model for proton transport in gastric H^+/K^+-ATPase. In the E_1 conformation, H_2O in the cytosolic medium is transported to cation binding sites 1 and 2. Charges of H_3O^+ in the cytosolic medium are transported to the H_2O molecule in site 2 *via* the charge transfer pathway H_3O^+-Lys^{164}-Gln^{161}-Glu^{345}-H_2O. Charges of H_3O^+ in site 2 are transported to the water molecule in site 1 *via* the water wire and the charge transfer pathway H_3O^+-Asn^{941}-Glu^{797}-H_2O. Finally, H_3O^+ in sites 2 and 1 is transported to the lumen during the conformational transition from E_1P to E_2P. Taken with permission from Figure 10 in [49].

accompanied decrease in β-subunit expression level, αβ complex formation, and H^+/K^+-ATPase activity [51].

3.3 *Gastric H^+/K^+-ATPase in Tissues Other than Stomach*

In mammals, gastric H^+/K^+-ATPase is predominantly expressed in parietal cells of the stomach, and is responsible for gastric acid secretion. Aside from the stomach, the same or similar ATPases have also been found to be functionally present in other tissues.

In the kidney, gastric H^+/K^+-ATPase mRNA [52] and protein [53, 54] are expressed. The ATPase is localized at the apical membrane of intercalated cells in the collecting duct. Acid secretion was impaired in the collecting duct of gastric H^+/K^+-ATPase-deficient mice, suggesting that the ATPase is responsible for the luminal acidification of the collecting duct [55]. On the other hand, the pump may

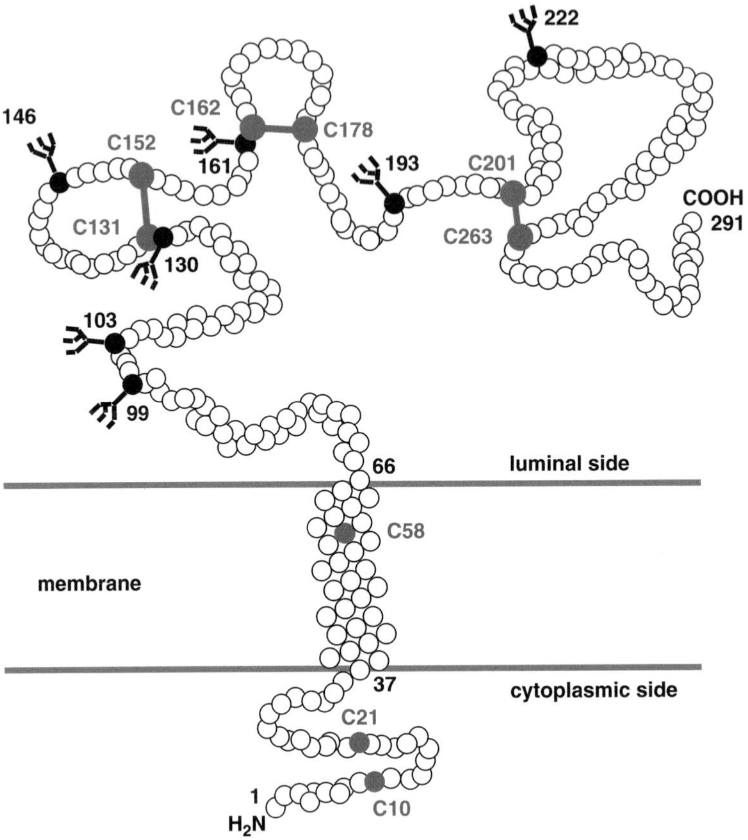

Figure 9 A schematic model of the β-subunit of gastric H^+/K^+-ATPase. Taken with permission from Figure 1 in [50].

not be involved in K^+ absorption of the collecting duct [56, 57]. Renal K^+ absorption is mediated by another type of H^+/K^+-ATPase (colonic H^+/K^+-ATPase or non-gastric H^+/K^+-ATPase) expressed in principal cells of the collecting duct [54, 57]. In the large intestine, colonic (non-gastric) H^+/K^+-ATPase is involved in K^+ absorption in the crypt cells of the distal colon and rectum [58, 59]. In the heart, mRNA and the protein of gastric ATPase are expressed, and the ATPase may be involved in the regulation of K^+ and H^+ homeostasis (pH regulation) in cardiac myocytes [60]. In the pancreas, mRNA and protein of gastric ATPase are expressed and may be involved in pancreatic secretion in ducts [61]. In the cochlear of the ear, mRNA and the protein of gastric ATPase are expressed, and it is involved in the formation of the endocochlear potential [62]. In the lung, an immunohistochemical study showed that gastric ATPase is expressed in mucous cells and ducts in the glands [63].

4 Ion Transporting Proteins Associated with H^+/K^+-ATPase

4.1 Cl^--Transporting Proteins

In gastric parietal cells, Cl^- efflux across the apical membrane is necessary for gastric acid secretion. It has been assumed that Cl^- moves passively down its electrochemical gradient through apical channels. Although the intracellular Cl^- concentration of the parietal cell has not been reported, it is speculated to be much lower than that of the luminally-secreted HCl. Thus, Cl^- transport through Cl^- channels would require a large electrical potential difference across the apical membrane (~60 mV). But the actual electrical potential difference across the apical membrane is estimated to be low, such as 20–25 mV [64], suggesting that the Cl^- secretory mechanism may be more complex than previously assumed.

The location and function of tubulovesicles, which are rich in H^+/K^+-ATPase, dramatically changes between resting and stimulated phases in the parietal cells (Figure 3). In the stimulated phase, the tubulovesicular and apical membranes do not mix but remain separate and distinct [9, 10]. So far, several Cl^- channels and transporters such as the K^+-Cl^- cotransporter-4 (KCC4) [9], the cystic fibrosis transmembrane conductance regulator (CFTR) Cl^- channel [65], SLC26A9 [66], CLIC-6 (parchorin) [67, 68], and ClC-5 [69] have been suggested as candidates that could be involved in the luminal Cl^- efflux for HCl secretion in tubulovesicles or the apical membrane.

KCC4 is predominantly expressed and associated with H^+/K^+-ATPase in the apical membrane of gastric parietal cells. In contrast, it is not significantly expressed in the tubulovesicles. KCC4 can increase the H^+/K^+-ATPase activity by effectively supplying K^+ to the luminal surface of the ATPase. Thus, KCC4 is an important molecule for maintaining H^+/K^+-ATPase activity in the apical membrane, and is suggested to contribute to basal HCl secretion in resting parietal cells [9].

The CFTR Cl^- channel is localized predominantly in the tubulovesicles [9]. Secretagogue-induced H^+/K^+-ATPase activity in isolated gastric glands from mice could be significantly reduced by an inhibitor of CFTR [65]. SLC26A9, which functions as a Cl^- channel and Cl^-/HCO_3^- exchanger, is also expressed in tubulovesicles and plays an essential role in HCl secretion by regulating Cl^- secretion and/or by affecting the viability of tubulovesicles/secretory canaliculi in parietal cells [66]. SLC26A9-knockout mice exhibited a significant decrease in gastric acid secretion during acid stimulation [66]. CLIC-6 is distributed throughout the cytosol [67]. ClC-5 is predominantly expressed in the tubulovesicles. ClC-5 and H^+/K^+-ATPase are functionally associated [69].

Figure 10 summarizes the putative mechanisms of HCl secretion in two different phases. In the resting phase, KCC4 together with H^+/K^+-ATPase that is present in the apical membrane is involved in the basal acid secretion. Upon stimulation, CFTR, SLC26A9, CLIC-6, ClC-5, and KCC4 are involved in massive gastric acid secretion.

4.2 K^+-Transporting Proteins

For maintaining H^+/K^+-ATPase activity in the luminal membrane of gastric parietal cells, K^+ recycling across the membrane is necessary. Several K^+ channels and transporters are expressed in the parietal cells and the properties of the apical K^+ recycling systems have been reported (Figure 10).

KCNQ1 and its subunit KCNE2 are highly expressed in parietal cells [70]. Knockout mice of heteromeric KCNQ1/KCNE2 exhibited a significant decrease in gastric acid secretion during acid stimulation [71, 72]. A highly selective inhibitor of the KCNQ1 channel did not inhibit the acid secretion in the resting phase, while it significantly inhibited the secretagogue-stimulated acid secretion [73]. The KCNQ1/KCNE2 channels act as the K^+ channels that provide K^+ to the external binding site of the H^+/K^+-ATPase in the apical canalicular membrane during acid secretion [71, 72].

A recent study using knockout mice of the Kir1.1 (ROMK) K^+ channel suggested that Kir1.1 plays a crucial role in gastric acid secretion and functions together with the KCNQ1/KCNE2 channel to recycle K^+ across the apical canalicular membrane of the parietal cells [74].

Kir4.1, an inwardly rectifying K^+ channel, is also expressed in parietal cells [73, 75]. An excellent study using knockout mice of Kir4.1 suggested that the physiological role of the Kir4.1 channel may be to provide a balance between rapid K^+ loss via the KCNQ1/KCNE2 channel and K^+ absorption via the slower action of the

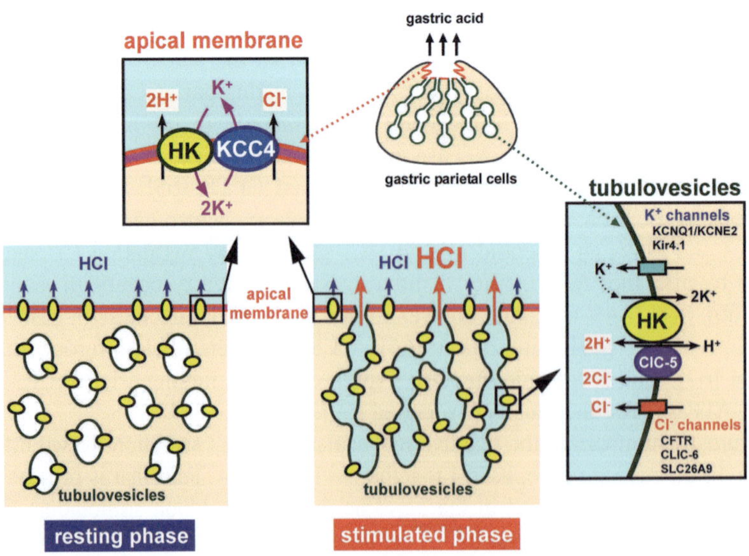

Figure 10 Molecular machineries involved in putative mechanisms of HCl secretion in the apical and tubulovesicular membranes of gastric parietal cells. HK, H^+/K^+-ATPase.

H^+/K^+-ATPase, and that Kir4.1 may influence the luminal K^+/H^+ concentration, membrane potential, and tubulovesicular membrane recycling [76].

Microarray analysis showed that the Kir4.2 channel is the most highly specific K^+ channel in the stomach, and a biochemical study suggested that the Kir4.2 channel is translocated from cytoplasm to the apical membrane upon stimulation and may be responsible for apical K^+ recycling [77].

5 Acid-Related Diseases

5.1 *Gastric and Duodenal Ulcers*

Normally, gastric cells are protected by the mucosal defense systems which are achieved by the mucus barrier against HCl and by secretion of bicarbonate (HCO_3^-) which neutralizes acid. Peptic ulcer is a defect in the mucosa of the stomach (gastric ulcer) and the duodenum (duodenal ulcer) that extends through the muscularis mucosae. *Helicobacter pylori* is one of the most common causes of peptic ulcer. This bacterium secretes urease which converts urea to ammonia. The production of ammonia (NH_3) neutralizes the acidity around *Helicobacter pylori*. Therefore, the bacterium can survive in the strong acidic environment of the stomach.

Taking aspirin and non-steroid anti-inflammatory drugs, which inhibit cyclooxygenase, is also a well-known risk factor for peptic ulcer. These drugs inhibit synthesis of prostaglandins which serve as an important function in maintaining normal gastric mucosal integrity [78]. Gastric hypersecretion and hyperacidity act as the attack factor against the mucus barrier.

Once mucosal defense systems are impaired due to imbalance between defensive and aggressive factors, the ulcers may be generated. As a therapy, taking proton pump inhibitors or histamine H_2 receptor blockers and eradication of *Helicobacter pylori* are effective.

5.2 *Gastroesophageal Reflux Disease*

A ring of muscle fibers in the lower esophagus prevents reflux of stomach contents into the esophagus. These muscle fibers are called the lower esophageal sphincter (LES). When the LES does not work well, the stomach contents reflux back up into the esophagus. Esophageal mucin secretion in patients with gastroesophageal reflux disease (GERD) was found to be significantly impaired [79]. Therefore, strong acids in the contents damage the lining of the esophagus and may cause inflammation and damage of the esophagus.

The risk factors for reflux include obesity, pregnancy, and smoking. Recently, an alternative pathogenic mechanism of GERD has been reported [80]. In this report, human esophageal squamous cells exposed to an acidic bile salt medium (similar in composition to gastric juice) secrete chemokines, including interleukin-8, which can induce the migration of inflammatory cells: that is, refluxed gastric juice does not directly damage the esophagus, but rather stimulates esophageal epithelial cells to secrete chemokines that mediate damage of esophageal tissue. PPIs can block chemokine (interleukin-8) production through mechanisms independent of their inhibitory effects on gastric acid secretion [81]. Some of the patients with GERD develop Barrett's esophagus, a risk factor for cancer of the esophagus.

5.3 Zollinger-Ellison Syndrome

The Zollinger-Ellison syndrome is characterized by gastrin-secreting tumors (gastrinomas) in the distal duodenum, proximal jejunum or pancreas. The gastrinomas secrete large amounts of gastrin which produces too much acid in the stomach. Then peptic ulcers are generated by excess acid [82]. Surgery is often required for treatment of this syndrome, although medical therapy with PPIs may be effective for reducing acid secretion.

6 H^+/K^+-ATPase as a Therapeutic Target

6.1 Proton Pump Inhibitors

6.1.1 Proton Pump Inhibitors and Their Different Binding Sites in the H^+/K^+-ATPase

Gastric H^+/K^+-ATPase that resides in the canaliculus membrane of parietal cells secretes acid. Commercially available proton pump inhibitors (PPIs) such as omeprazole, lansoprazole, rabeprazole (E3810), pantoprazole, esomeprazole have similar chemical structures (Figure 11).

PPIs are prodrugs. With a pK_a of 4.0 (omeprazole, lansoprazole, and pantoprazole) and 5.0 (rabeprazole), they accumulate in the acidic lumen of gastric parietal cells, are transformed into acid-activated active compounds, and bind with H^+/K^+-ATPase Cys residues that are accessible from the luminal side. The candidates are Cys^{323}, Cys^{815}, and Cys^{824} in the α-subunit of H^+/K^+-ATPase (Figure 7). Omeprazole binds Cys^{815} which is located near the K^+-binding site [83, 84].

Rabeprazole binds Cys^{323} at the luminal end of the α-subunit of H^+/K^+-ATPase [85]. Because rabeprazole has a long moiety of $OCH_2CH_2CH_2OCH_3$ at the R_2 position of the pyridine ring (Figure 11), it cannot penetrate into the K^+-binding site of H^+/K^+-ATPase. The conformation of H^+/K^+-ATPase changes as $E_1 \rightarrow E_2 \rightarrow E_1$ during the cyclic enzymatic reaction. The rabeprazole-bound H^+/K^+-ATPase is at the E_1

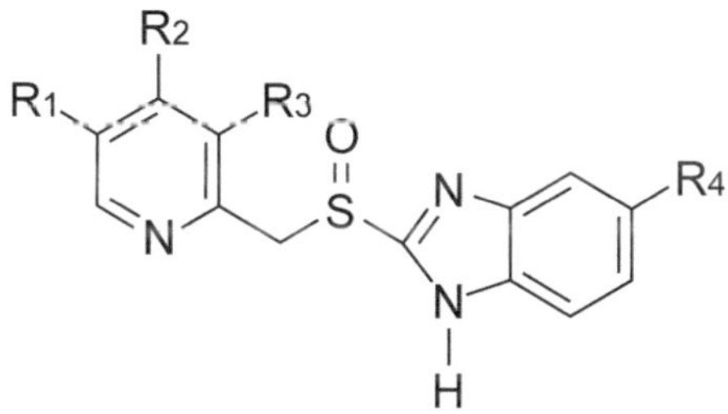

	R1	R2	R3	R4
Omeprazole	CH_3	OCH_3	CH_3	OCH_3
Lansoprazole	H	OCH_2CF_3	CH_3	H
Rabeprazole	H	$OCH_2CH_2CH_2OCH_3$	CH_3	H
Pantoprazole	H	OCH_3	OCH_3	$OCHF_3$
Esomeprazole	CH_3	OCH_3	CH_3	OCH_3

Figure 11 Chemical structures of proton pump inhibitors (PPIs).

state as determined by FITC fluorescence and the conformational state is not fixed: it can change to the E_2 state after the addition of K^+ in the medium, indicating that the binding of rabeprazole does not block K^+ penetration into the K^+-binding site (Figure 12) [86].

The omeprazole-bound H^+/K^+-ATPase is at the E_2 state, the addition of K^+ in the medium induces no further conformational change indicating that the binding of omeprazole blocks K^+ penetration into the K^+-binding site. The recovery from omeprazole-induced inhibition returns the conformational state to the E_1 state (Figure 12) [86]. The partial enzyme reactions that were most differently affected by these two inhibitors were the conformational change from E_2K^+ to E_1ATPK^+ for omeprazole and the luminal K^+-dependent dephosphorylation ($E_2PK^+ \rightarrow E_2\ K^+$) for rabeprazole [86].

When omeprazole first debuted in 1979, PPIs were generally considered to be irreversible inhibitors because of the chemical S-S crosslinking between H^+/K^+-ATPase and the PPIs. One important question was how the parietal cell reactivates acid secretion to normal level after the irreversible inhibition of H^+/K^+-ATPase. The normal half-life of H^+/K^+-ATPase is about 30–48 hrs so all of the H^+/K^+-ATPase is replaced every 72–96 hrs.

In 1991, rabeprazole was reported to restart acid secretion earlier than omeprazole in dogs [87]. This is due to the fact that the S-S crosslinking between rabeprazole and Cys^{323} of the H^+/K^+-ATPase α-subunit is more easily breakable by luminal endogenous glutathione compared with that between omeprazole and Cys^{815}. Soon, the reversal of the S-S crosslinking between omeprazole and H^+/K^+-ATPase by endogenous glutathione was also proposed [88]. Omeprazole induced a two-fold increase in the steady state mRNA level of the H^+/K^+-ATPase α-subunit in isolated

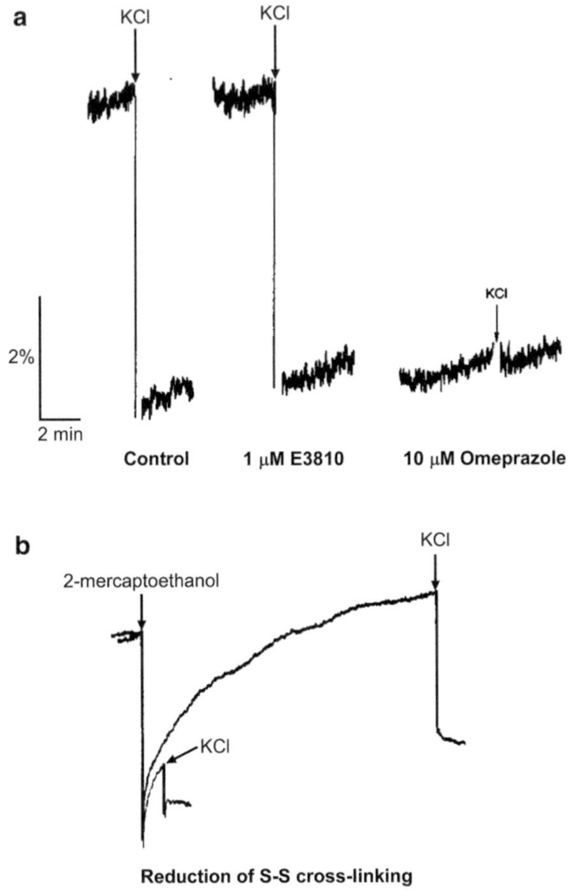

Figure 12 The conformational state of H^+/K^+-ATPase depends on the bound PPIs. (**a**) With the rabeprazole (E3810)-bound H^+/K^+-ATPase, the addition of K^+ in the medium induces the conformational change from the E_1 to the E_2 state as with uninhibited H^+/K^+-ATPase. (**b**) The conformational state of the omeprazole-bound H^+/K^+-ATPase is at the E_2 state and the addition of K^+ in the medium did not induce further conformational change. The recovery from omeprazole-induced inhibition returns the conformational state from E_2 to E_1. Taken with permission from Figures 3 and 5 in [86].

canine parietal cells [89]. Taken together, in long-term use, rabeprazole compared with omeprazole may have less stress on *de novo* synthesis of H^+/K^+-ATPase with less damage to the parietal cells which have a half-life of several months. Pantoprazole binds with Cys^{824} of the H^+/K^+-ATPase α-subunit and S-S crosslinking may not be easily breakable because endogenous glutathione cannot reach to the deep location of Cys^{824} (Figure 7) [90].

When rabeprazole or lansoprazole was singly administrated in gastric fistula rats, the complete arrest of acid secretion, which reflected the complete inhibition of H^+/K^+-ATPase in the secretory canaliculus membrane of parietal cells, occurred within 1–2 hrs after administration of both PPI while the complete inhibition of H^+/K^+-ATPase in intracellular tubulovesicles of gastric parietal cells occurred 4 hrs after administration of rabeprazole and more than 10 hrs after administration of lansoprazole [91]. These results suggest that inhibition of H^+/K^+-ATPase first occurs in the secretory canaliculus membrane, the inhibited H^+/K^+-ATPase in the secretory canaliculus is translocated into intracellular tubulovesicles, and the translocation speed of the inhibited H^+/K^+-ATPase by rabeprazole is faster than that by lansoprazole. Furthermore, 24 hrs after a single dose of rabeprazole in rats, H^+/K^+-ATPase activity in tubulovesicles recovered by about 50 % in food-stimulated rats but only slightly in fasting rats, indicating that food stimulates the translocation of H^+/K^+-ATPase between the apical canaliculus and intracellular tubulovesicles and the reactivation of H^+/K^+-ATPase from the inhibition occurs in the apical canaliculus [91].

At present, new PPIs that show a stronger acid suppression above the already existing PPIs are expected to result in an elevated clinical efficacy for acid-related diseases, which still stimulates development of new PPIs such as E3710 [92] and new potassium-competitive acid blockers described below.

6.1.2 Metabolism of Proton Pump Inhibitors by Liver Cytochrome Enzymes, the Polymorphism of the Enzymes, and Clinical Efficacy of Proton Pump Inhibitors

PPIs are metabolized by liver cytochrome enzymes such as CYP2C19 and CYP3A4. Polymorphism of CYP2C19 affects acid suppression because extensive metabolizers (EM) of CYP2C19 have lower serum concentrations of PPIs compared with poor metabolizers (PM). The prevalence of homoEM, heteroEM, and PM of CYP2C19 in Caucasians are 60–70 %, 27–37 %, and 3–6 %, which are in contrast with those in Asians of 28–42 %, 46–50 %, and 13–23 %, respectively. The standard doses of PPIs in Caucasians would reflect mainly homoEM and in part heteroEM. The standard doses in Asians would reflect mainly heteroEM and in part homoEM and PM, thus the standard doses of PPIs tend to be not enough for Asian homoEM patients.

For example, the recurrence rate during GERD maintenance therapy using 15 mg lansoprazole for 6 months in Japanese was reported to be 39 %, 22 %, and 0 % in CYP2C19 homoEM, heteroEM, and PM [93]. The metabolisms of omeprazole and lansoprazole mainly depend on CYP2C19 and CYP3A4, while the metabolism of rabeprazole by CYP2C19 and CYP3A4 contribute a small fraction to the overall metabolism. In fact, a recent study of effects of PPIs on CYP activity assessed by the [^{13}C]-aminopyrene CYP human breast test showed that omeprazole and lansoprazole inhibit CYP activity, while rabeprazole does not [94].

6.1.3 Interference Between Proton Pump Inhibitors and Other Drugs

PPIs are frequently, administrated concomitantly with other drugs, which requests special caution to select PPIs. Three examples are shown hereafter.

Dual antiplatelet therapy with aspirin and clopidogrel after percutaneous coronary intervention requests concomitant administration of PPIs to prevent aspirin-induced gastrointestinal bleeding. Clopidogrel is a prodrug and changes into the active metabolite by CYP2C19. A cross-over Japanese patient study showed that the antiplatelet effects of clopidogrel is weakened by omeprazole which is also metabolized by CYP2C19 but not by rabeprazole which is only slightly metabolized by CYP2C19 [95].

In CYP2C19 homoEM, omeprazole had no effect on clopidogrel-induced platelet inhibition, while in CYP2C19 heteroEM, omeprazole reduced the clopidogrel anti-platelet effect but rabeprazole did not [96].

Tacrolimus, an immunosuppressant, is metabolized by CYP3A4. The trough concentration of tacrolimus in CYP2C19 heteroEM patients, who received kidney transplantation, increased markedly after the introduction of lansoprazole which is metabolized by CYP2C19 and CYP3A4. The trough concentration returned to the therapeutic range after administration of lansoprazole was stopped [97].

6.2 Potassium-Competitive Acid Blockers

SCH28080 is a typical potassium-competitive acid blocker (PCAB) (Figure 13). H^+/K^+-ATPase in the E_2 state forms a SCH28080-binding cavity but not in the E_1 state. Amino acid residues involved in SCH28080 binding in M4, M5, and M6 segments of H^+/K^+-ATPase are shown in Figure 7. The proposed locations of the cavity for the two PCABs SCH28080 and SPI-447 are shown in Figure 14 [98].

SCH28080 inhibits H^+/K^+-ATPase but not Na^+/K^+-ATPase while ouabain inhibits Na^+/K^+-ATPase but not H^+/K^+-ATPase. Construction of the ouabain-binding cavity in gastric H^+/K^+-ATPase was possible by substitution of only seven amino acids of H^+/K^+-ATPase with corresponding amino acids of Na^+/K^+-ATPase [99]. The three dimensional structural study at 7 Å resolution showed that the shape and location of the SCH28080 cavity in H^+/K^+-ATPase roughly overlaps with those of the ouabain cavity in Na^+/K^+-ATPase [100].

AZD0865 (Figure 13) is a PCAB which underwent large-scale clinical trials. AZD0865 shares an imidazopyridine ring structure with SCH28080 and *in vitro* inhibition of H^+/K^+-ATPase by AZD9865 can be washed out. Although AZD0865 has superior pharmacodynamic properties of acid inhibition, the efficacy in treatment of GERD patients was similar to that of esomeprazole [101]. TAK-438 (vonoprazan) (Figure 13), another PCAB, has long-lasting anti-secretory effects because the binding of TAK-438 in its binding cavity is stabilized by additional hydrogen bonding [102]. Their acid-inhibitory effect is much more potent than that of lansoprazole. In a phase 2 clinical trial, treatment of GERD patients with TAK-438 was not inferior to lansoprazole, but TAK-438 accompanied much higher serum gastrin levels ± SD compared with lansoprazole [103].

7 Concluding Remarks and Future Directions

At present, the molecular mechanisms of Cl^- secretion and apical K^+ recycling for gastric acid secretion have been partly elucidated. Although several potent drugs for treating acid-related diseases have been developed, pathological conditions in some patients are still not improved by these medicines.

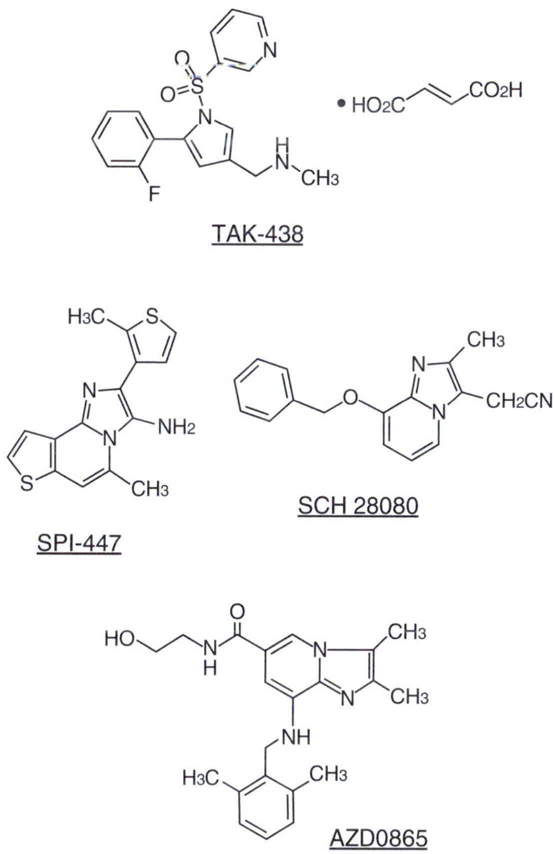

Figure 13 Chemical structures of potassium-competitive acid blockers (PCABs).

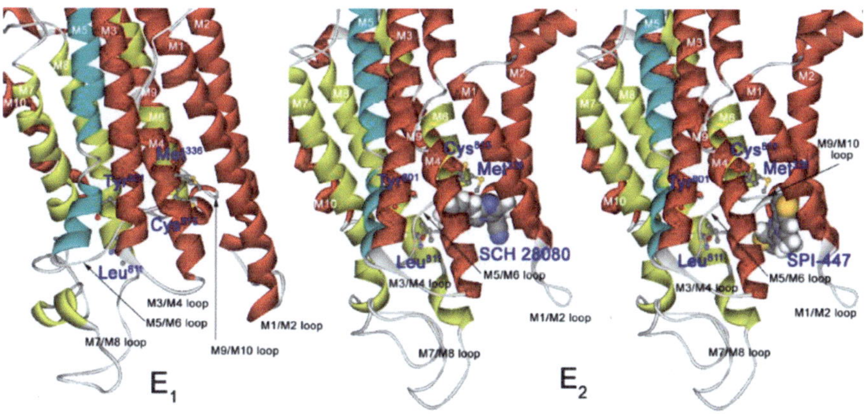

Figure 14 Binding sites of PCABs (SCH28080 and SPI-447) in the transmembrane domain of H^+/K^+-ATPase. Taken with permission from Figure 9 in [98].

For the future, one may expect that further mechanisms of Cl^- and K^+ transport will be elucidated and that more effective inhibitors of gastric H^+/K^+-ATPase will be introduced. The Cl^- and K^+ channels/transporters are also attractive targets for developing new drugs.

Abbreviations

ADP	adenosine 5′-diphosphate
Asn, N	asparagine
Asp, D	aspartic acid
ATP	adenosine 5′-triphosphate
cAMP	cyclic adenosine monophosphate
CCK-2	cholecystokinin-2
CFTR	cystic fibrosis transmembrane conductance regulator
cGMP	cyclic guanosine monophosphate
Cys, C	cysteine
ECL	enterochromaffin-like cells
EM	extensive metabolizer
FITC	fluorescein isothiocyanate
GERD	gastroesophageal reflux disease
Gln, Q	glutamine
Glu, E	glutamic acid
Ile, I	isoleucine
KCC4	K^+-Cl^- cotransporter-4
KCNE2	potassium channel, voltage gated subfamily E regulatory beta subunit 2
KCNQ1	potassium channel, voltage gated KQT-like subfamily Q, member 1
LES	lower esophageal sphincter
Leu, L	leucine
Lys, K	lysine
PCAB	potassium-competitive acid blocker
PM	poor metabolizer
PPI	proton pump inhibitor

Acknowledgment Our scientific studies described in this chapter were partly supported by Grants-in-aid for Scientific Research (KAKENHI) from the Japan Society for the Promotion of Science and Grants-in-aid the Ministry of Education, Culture, Sports, Science and Technology of Japan.

References

1. A. J. Smolka, E. R. Lacy, L. Luciano, E. Reale, *J. Histochem. Cytochem.* **1994,** *42*, 1323–1332.
2. G. Sachs, M. Besancon, J. M. Shin, F. Mercier, K. Munson, S. Hersey, *J. Bioenerg. Biomembr.* **1992,** *24*, 301–308.

3. K. P. Choe, J. W. Verlander, C. S. Wingo, D. H. Evans, *Am. J. Physiol. Regul. Integr. Comp. Physiol.* **2004,** *287*, R981–R991.
4. J. M. Shin, K. Munson, G. Sachs, *Compr. Physiol.* **2011**, *1*, 2141–2153.
5. D. G. McQuarrie, A. Eichenholz, A. S. Blumentals, J. A. Vennes, *Surgery* **1967,** *62*, 475–486.
6. P. Monteleone, C. Serritella, P. Scognamiglio, M. Maj, *Psychoneuroendocrinology* **2010,** *35*, 284–288.
7. C. T. Okamoto, J. G. Forte, *J. Physiol.* **2001,** *532*, 287–296.
8. X. Yao, J. G. Forte, *Annu. Rev. Physiol.* **2003,** *65*, 103–131.
9. T. Fujii, Y. Takahashi, A. Ikari, M. Morii, Y. Tabuchi, K. Tsukada, N. Takeguchi, H. Sakai, *J. Biol. Chem.* **2009,** *284*, 619–629.
10. M. Nishi, F. Aoyama, F, F. Kisa, H. Zhu, M. Sun, P. Lin, H. Ohta, B. Van, S. Yamamoto, S. Kakizawa, H. Sakai H, J. Ma, A. Sawaguchi, H. Takeshima, *J. Biol. Chem.* **2012,** *287*, 33523–33532.
11. H. Sakai, Y. Okada, M. Morii, N. Takeguchi, *J. Physiol.* **1992,** *448*, 293–306.
12. H. Sakai, E. Kumano, A. Ikari, N. Takeguchi, *J. Biol. Chem.* **1995,** *270*, 18781–18785.
13. H. Sakai, Y. Ohira, A. Tanaka, T. Suzuki, A. Ikari, M. Morii, N. Takeguchi, *J. Physiol.* **2003,** *551*, 207–217.
14. H. Sakai H, A. Ikari, T. Shimizu, T. Sato, N. Takeguchi, *Eur. J. Pharmacol.* **1998,** *361*, 109–117.
15. K. Bamberg, S. Nylander, K. G. Helander, L. G. Lundberg, G. Sachs, H. F. Helander, *Biochim. Biophys. Acta* **1994,** *1190*, 355–359.
16. S. M. Karam, X. Yao, J. G. Forte, *Am. J. Physiol.* **1997,** *272*, G161–G171.
17. G. Sachs, *Eur. J. Gastroenterol. Hepatol.* **2001,** *13 Suppl. 1*, S35–S41.
18. L. C. Samuelson, K. L. Hinkle, *Annu. Rev. Physiol.* **2003,** *65*, 383–400.
19. K. Andersson, D. Chen, H. Mattsson, F. Sundler, R. Håkanson, *Yale J. Biol. Med.* **1998,** *71*, 183–193.
20. J. L. Cabero, E. Grapengiesser, E. Gylfe, Z. Q. Li, S. Mårdh, *Biochem. Biophys. Res. Commun.* **1992,** *183*, 1097–1102.
21. L. Wang, I. Gantz, J. DelValle, *Am. J. Physiol.* **1996,** *271*, G613–G620.
22. S. J. Hill, *Agents Actions* **1991,** *Suppl. 33*, 145–159.
23. H. W. Davenport, *Am. J. Physiol.* **1963**, *204*, 213–216.
24. N. Takeguchi, I. Horikoshi, M. Hattori, *Am. J. Physiol.* **1977**, *232*, E294–E297.
25. N. Takeguchi, M. Hattori, A. Sano, I. Horikoshi, *Am. J. Physiol.* **1979**, *237*, E51–E55.
26. D. K. Kasbekar, R. P. Durbin, *Biochim. Biophys. Acta* **1965,** *105*, 472–482.
27. J. G. Forte, G. M. Forte, P. Saltman, *J. Cell. Physiol.* **1967**, *69*, 293–304.
28. A. L. Ganser, J. G. Forte, *Biochem. Biophys. Res. Commun.* **1973**, *54*, 690–696.
29. J. Lee, G. Simpson, P. Scholes, *Biochem. Biophys. Res. Commun.* **1974**, *60*, 825–832.
30. G. Sachs, H. H. Chang, E. Rabon, R. Schackman, M. Lewin, G. Saccomani, *J. Biol. Chem.* **1976**, *251*, 7690–7698.
31. B. Wallmark, H. B. Stewart, E. Rabon, G. Saccomani, G. Sachs, *J. Biol. Chem.* **1980**, *255*, 5313–5319.
32. J. J. Schrijen, W. A. van Groningen-Luyben, J. J. de Pont, S. L. Bonting, *Biochim. Biophys. Acta.* **1981**, *640*, 473–486.
33. M. Morii, N. Ishimura, N. Takeguchi, *Biochemistry* **1984**, *23*, 6816–6821.
34. S. Asano, M. Morii, N. Takeguchi, *Biol. Pharm. Bull.* **2004**, *27*, 1–12.
35. L. D. Faller, R. A. Diaz, G. Scheiner-Bobis, R. A. Farley, *Biochemistry* **1991**, *30*, 3503–3510.
36. S. Asano, S. Kamiya, N. Takeguchi, *J. Biol. Chem.* **1992**, *267*, 6590–6595.
37. P. Lorentzon, G. Sachs, B. Wallmark, *J. Biol. Chem.* **1988**, *263*, 10705–10710.
38. H. T. van der Hijden, E. Grell, J. J. de Pont, E. Bamberg, *J. Membr. Biol.* **1990,** *114*, 245–256.
39. M. Morii, Y. Hayata, K. Mizoguchi, N. Takeguchi, *J. Biol. Chem.* **1996**, *271*, 4068–4072.
40. G. E. Shull, J. B. Lingrel, *J. Biol. Chem.* **1986**, *261*, 16788–16791.
41. G. E. Shull, *J. Biol. Chem.* **1990**, *265*, 12123–12126.

42. C. Toyoshima, M. Nakasako, H. Nomura, H. Ogawa, *Nature* **2000,** *405*, 647–655.
43. S. Asano, Y. Tega, K. Konishi, M. Fujioka, N. Takeguchi, *J. Biol. Chem.* **1996**, *271*, 2740–2745.
44. N. Lambrecht, K. Munson, O. Vagin, G. Sachs, *J. Biol. Chem.* **2000,** *275*, 4041–4048.
45. H. P. Hermsen, J. B. Koenderink, H. G. Swarts, J. J. De Pont, *Biochemistry* **2000,** *39*, 1330–1337.
46. O. Vagin, K. Munson, N. Lambrecht, S. J. Karlish, G. Sachs, *Biochemistry* **2001**, *40*, 7480–7490.
47. H. G. Swarts, H. P. Hermsen, J. B. Koenderink, F. M. Schuurmans Stekhoven, J. J. De Pont, *EMBO J.* **1998**, *17*, 3029–3035.
48. S. Asano, T. Io, T. Kimura, S. Sakamoto, N. Takeguchi, *J. Biol. Chem.* **2001**, *276*, 31265–31273.
49. M. Morii, M. Yamauchi, T. Ichikawa, T. Fujii, Y. Takahashi, S. Asano, N. Takeguchi, H. Sakai, *J. Biol. Chem.* **2008**, *283,*16876–16884.
50. T. Kimura, Y. Tabuchi, N. Takeguchi, S. Asano, *J. Biol. Chem.* **2002**, *277*, 20671–20677.
51. S. Asano, K. Kawada, T. Kimura, A. V. Grishin, M. J. Caplan, N. Takeguchi, *J. Biol. Chem.* **2000**, *275*, 8324–8330.
52. S. Ono, J. Guntupalli, T. D. Dubose Jr., *Am. J. Physiol.* **1996,** *270*, F852–F861.
53. B. Bastani, *J. Am. Soc. Nephrol.* **1995,** *5*, 1476–1482.
54. J. A. Kraut, J. Hiura, M. Besancon, A Smolka, G. Sachs, D. Scott, *Am. J. Physiol.* **1997,** *272*, F744–F750.
55. I. J. Lynch, A. Rudin, S. L. Xia, L. R. Stow, G. E. Shull, I. D. Weiner, B. D. Cain, C. S. Wingo, *Am. J. Physiol. Renal Physiol.* **2008,** *294*, F621–F627.
56. O. Dherbecourt, L. Cheval, M. Bloch-Faure, P. Meneton, A. Doucet, *Pflügers Arch.* **2006,** *451*, 769–775.
57. W. H. Wang, G. Giebisch, *Pflügers Arch.* **2009,** *458*, 157–168.
58. M. Takeguchi, S. Asano, Y. Tabuchi, N. Takeguchi, *Gastroenterology* **1990,** *99*, 1339–1346.
59. Y. Tabuchi, M. Takeguchi, S. Asano, N. Takeguchi, *Jpn. J. Physiol.* **1992,** *42*, 577–589.
60. V. Beisvag, G. Falck, J. P. Loennechen, G. Qvigstad, P. Jynge, T. Skomedal, J. B. Osnes, A. K. Sandvik, Ø. Ellingsen, *Acta Physiol. Scand.* **2003,** *179*, 251–262.
61. I. Novak, J. Wang, K. L. Henriksen, K. A. Haanes, S. Krabbe, R. Nitschke, S. E. Hede, *J. Biol. Chem.* **2011,** *286*, 280–289.
62. T. Shibata, H. Hibino, K. Doi, T. Suzuki, Y. Hisa, Y. Kurachi, *Am. J. Physiol. Cell Physiol.* **2006,** *291*, C1038–C1048.
63. K. W. Altman, J. D. Waltonen, G. Tarjan, J. A. Radosevich, G. K. Haines III, *Ann. Otol. Rhinol. Laryngol.* **2007,** *116*, 229–234.
64. J. R. Demarest, D. D. F. Loo, *Annu. Rev. Physiol.* **1990,** *52*, 307–319.
65. S. M. Sidani, P. Kirchhoff, T. Socrates, L. Stelter, E. Ferreira, C. Caputo, K. E. Roberts, R. L. Bell, M. E. Egan, J. P. Geibel, *J. Biol. Chem.* **2007,** *282*, 6068–6074.
66. J. Xu, P. Song, M. L. Miller, F. Borgese, S. Barone, B. Riederer, Z. Wang, S. L. Alper, J. G. Forte, G. E. Shull, J. Ehrenfeld, U. Seidler, M. Soleimani, *Proc. Natl. Acad. Sci. USA* **2008,** *105*, 17955–17960.
67. T. Nishizawa, T. Nagao, T. Iwatsubo, J. G. Forte, T. Urushidani, *J. Biol. Chem.* **2000,** *275*, 11164–11173.
68. G. Sachs, J. M. Shin, O. Vagin, N. Lambrecht, I. Yakubov, K. Munson, *J. Clin. Gastroenterol.* **2007,** *41, Suppl. 2*, S226–S242.
69. Y. Takahashi, T. Fujii, K. Fujita, T. Shimizu, T. Higuchi, Y. Tabuchi, H. Sakamoto, I. Naito, K. Manabe, S. Uchida, S. Sasaki, A. Ikari, K. Tsukada, H. Sakai, *Biol. Open* **2014,** *3*, 12–21.
70. N. W. Lambrecht, I. Yakubov, D. Scott, G. Sachs, *Physiol. Genomics* **2005,** *21*, 81–91.
71. T. K. Roepke, A. Anantharam, P. Kirchhoff, S. M. Busque, J. B. Young, J. P. Geibel, D. J. Lerner, G. W. Abbott, *J. Biol. Chem.* **2006,** *281*, 23740–23747.
72. P. Song, S. Groos, B. Riederer, Z. Feng, A. Krabbenhöft, A. Smolka, U. Seidler, *J. Physiol.* **2009,** *587*, 3955–3965.
73. M.-A. Kaufhold, A. Krabbenhöft, P. Song, R. Engelhardt, B. Riederer, M. Fährmann, N. Klöcker, W. Beil, M. Manns, S. J. Hagen, U. Seidler, *Gastroenterology* **2008,** *134*, 1058–1069.

74. E. Vucic, T. Alfadda, G. G. MacGregor, K. Dong, T. Wang, J. P. Geibel, *Pflügers Arch.* **2014,** in press.
75. A. Fujita, Y. Horio, K. Higashi, T. Mouri, F. Hata, N. Takeguchi, Y. Kurachi, *J. Physiol.* **2002,** *540*, 85–92.
76. P. Song, S. Groos, B. Riederer, Z. Feng, A. Krabbenhöft, M. P. Manns, A. Smolka, S. J. Hagen, C. Neusch, U. Seidler, *J. Biol. Chem.* **2011,** *286*, 14120–14128.
77. W. He, W. Liu, C. S. Chew, S. S. Baker, R. D. Baker, J. G. Forte, L. Zhu, *Am. J. Physiol. Gastrointest. Liver Physiol.* **2011,** *301*, G591–G600.
78. T. A. Miller, *Am. J. Physiol.* **1983,** *245*, G601–G623.
79. I. Sarosiek, M. Olyaee, M. Majewski, E. Sidorenko, K. Roeser, S. Sostarich, G. Wallner, J. Sarosiek, *Dig. Dis. Sci.* **2009,** *54*, 2137–2142.
80. R. F. Souza, X. Huo, V. Mittal, C. M. Schuler, S. W. Carmack, H. Y. Zhang, X. Zhang, C. Yu, K. Hormi-Carver, R. M. Genta, S. J. Spechler, *Gastroenterology* **2009,** *137*, 1776–1784.
81. X. Huo, X. Zhang, C. Yu, Q. Zhang, E. Cheng, D. H. Wang, T. H. Pham, S. J. Spechler, R. F. Souza, *Gut* **2014,** *63*, 1042–1052.
82. I. Epelboym, H. Mazeh, *Oncologist* **2014,** *19*, 44–50.
83. G. Sachs, C. Prinz, D. Loo, K. Bamberg, M. Besancon, J. M. Shin, *Yale J. Biol. Med.* **1994,** *67*, 81–95.
84. J. M. Shin, G. Sachs, *Biochem. Pharmacol.* **2004**, *68*, 2117–2127.
85. M. Morii, K. Hamatani, and N. Takeguchi, *Biochem. Pharmacol.* **1995**, *49*, 1729–1734.
86. M. Morii, N. Takeguchi, *J. Biol. Chem.* **1993**, *268*, 21553–21559.
87. H. Fujisaki, H. Shibata, K. Oketani, M. Murakami, M. Fujimoto, T. Wakabayashi, I. Yamatsu, M. Yamaguchi, H. Sakai, N. Takeguchi, *Biochem. Pharmacol.* **1991**, *42*, 321–328.
88. K. Gedda, D. Scott, M. Besancon, P. Lorentzon, G. Sachs, *Gastroenterology* **1995**, *109*, 1134–1141.
89. V. W. Campbell, T. Yamada, *Am. J. Physiol.* **1991**, *260*, G434–G439.
90. J. M. Shin, G. Sachs, *Gastroenterology* **2003**, *123*, 1588–1597.
91. Y. Tomiyama Y, M. Morii M, N. Takeguchi, *Biochem. Pharmacol.* **1994**, *48,* 2049–2055.
92. K. Kodama, H. Fujisaki, H. Tonomura, M. Jindo, M. Watanabe, J. Nagakawa, N. Takeguchi, in *E3710, Long-Acting PPI as New Approach for the Treatment of Unmet Medical Needs for GERD, Gastroesophageal Reflux Disease*, Chapter 7, Ed. M. Bortolotti, InTech, Rijeka, Croatia, 2012, 129–150, available online, doi: 10.5772/29085.
93. M. Kawamura, S. Ohara, T. Koike, K. Iijima, H. Suzuki, S. Kayaba, K. Noguchi, S. Abe, M. Noguchi, T. Shimosegawa, *J. Gastroenterol. Hepatol.* **2007**, *22,* 222–226.
94. C. Kodaira, S. Uchida, M. Yamade, M. Nishino, M. Ikuma, N. Namiki, M. Sugimoto, H. Watanabe, A. Hishida, T. Furuta, *J. Clin. Pharmacol.* **2012**, *52*, 432–439.
95. K. Yamane, Y. Kato, J. Tazaki, T. Tada, T. Makiyama, M. Imai, T. Jinnai, T. Ikeda, R. Shirakawa, T. Kimura, H. Horiuchi, *J. Atheroscler. Thromb.* **2012**,*19*, 559–569.
96. S. Kenngott, R. Olze, M. Kollmer, H. Bottheim, A. Laner, E. Holinski-Feder, M. Gross, *J. Med. Res.* **2010**,*15,* 220–224.
97. K. Takahashi, H. Motohashi, A. Yonezawa, M. Okuda, N. Ito, S. Yamamoto, O. Ogawa, K. Inui, *Ann. Pharmacother.* **2004**, *38*, 791–794.
98. S. Asano, A. Yoshida, H. Yashiro, Y. Kobayashi, A. Morisato, H. Ogawa, N. Takeguchi, M. Morii, *J. Biol. Chem.* **2004**, *279*, 13968–13975.
99. L. Y. Qiu, E. Krieger, G. Schaftenaar, H. G. Swarts, P. H. Willems, J. J. De Pont, J. B. Koenderink, *J. Biol. Chem.* **2005**, *280*, 32349–32355..
100. K. Abe, K. Tani, Y. Fujiyoshi, *Nature Commun.* **2011**, *2*, 155.
101. P. J. Kahrilas, J. Dent, K. Lauritsen, P. Malfertheiner, H. Denison, S. Franzén, G. Hasselgren, *Clin. Gastroenterol. Hepatol.* **2007**, *5*, 1385–1591.
102. J. M. Shin, N. Inatomi, K. Munson, D. Strugatsky, E. Tokhtaeva, O. Vagin, G. Sachs, *J. Pharmacol. Exp. Ther.* **2011**, *339*, 412–420.
103. T. Chiba, Y. Sakurai, A. Nishimura, N. Hiramatsu, E. Umegaki, K. Iwakiri, K. Ashida, *Gastroenterology* **2013**, *144*, S564–S565.

Chapter 14
Bioinspired Artificial Sodium and Potassium Ion Channels

Nuria Rodríguez-Vázquez, Alberto Fuertes, Manuel Amorín, and Juan R. Granja

Contents

Abstract In Nature, all biological systems present a high level of compartmentalization in order to carry out a wide variety of functions in a very specific way. Hence, they need ways to be connected with the environment for communication, homeostasis equilibrium, nutrition, waste elimination, etc. The biological membranes carry out these functions; they consist of physical insulating barriers

N. Rodríguez-Vázquez • A. Fuertes • M. Amorín • J.R. Granja (✉)
Department of Organic Chemistry, Singular Research Centre in Chemical Biology and Molecular Materials (CIQUS), University of Santiago de Compostela (USC), E-15782 Santiago de Compostela, Spain
e-mail: juanr.granja@usc.es

A. Sigel, H. Sigel, and R.K.O. Sigel (eds.), *The Alkali Metal Ions: Their Role for Life*, Metal Ions in Life Sciences 16, DOI 10.1007/978-3-319-21756-7_14

constituted mainly by phospholipids. These amphipathic molecules spontaneously aggregate in water to form bilayers in which the polar groups are exposed to the aqueous media while the non-polar chains self-organize by aggregating to each other to stay away from the aqueous media. The insulating properties of membranes are due to the formation of a hydrophobic bilayer covered at both sides by the hydrophilic phosphate groups. Thus, lipophilic molecules can permeate the membrane freely, while the small charged or very hydrophilic molecules require the assistance of other membrane components in order to overcome the energetic cost implied in crossing the non-polar region of the bilayer. Most of the large polar species (such as oligosaccharides, polypeptides or nucleic acids) cross into and out of the cell via endocytosis and exocytosis, respectively. Nature has created a series of systems (carriers and pores) in order to control the balance of small hydrophilic molecules and ions. The most important structures to achieve these goals are the ionophoric proteins that include the channel proteins, such as the sodium and potassium channels, and ionic transporters, including the sodium/potassium pumps or calcium/sodium exchangers among others. Inspired by these, scientists have created non-natural synthetic transporting structures to mimic the natural systems. The progress in the last years has been remarkable regarding the efficient transport of Na^+ and K^+ ions, despite the fact that the selectivity and the ON/OFF state of the non-natural systems remain a present and future challenge.

Keywords Ion carriers • Ion channels • Ionophores • Ion selectivity • Potassium • Sodium • Transmembrane transport

Please cite as: *Met. Ions Life Sci.* 16 (2016) 485–556

1 Introduction

1.1 Natural Systems

Transmembrane ion transport and its equilibrium are vital in any living cell. In this regard, the two most studied and relevant ions are Na^+ and K^+, together with all the transporter systems like carriers, channels, pumps and receptors that permit their flow in the living organisms (Fig. 1) [1, 2]. These ions are fundamental for cell survival, thus membranes maintain the osmotic balance across the animal cell, keeping the intracellular concentration of sodium low [3, 4]. Therefore, other cations, mainly potassium, have to be pumped into the cell to balance the fixed pH and charge determined by the intracellular anion concentration. The cells create a membrane potential between the interior and the exterior of the cell that is critical for many vital physiological functions. For this purpose, membranes contain proteins embedded in the bilayer, called ionophoric proteins, to transport ions across the membrane.

These membrane proteins include the channel proteins, such as the sodium and potassium channels, and ionic transporters, including the sodium or potassium

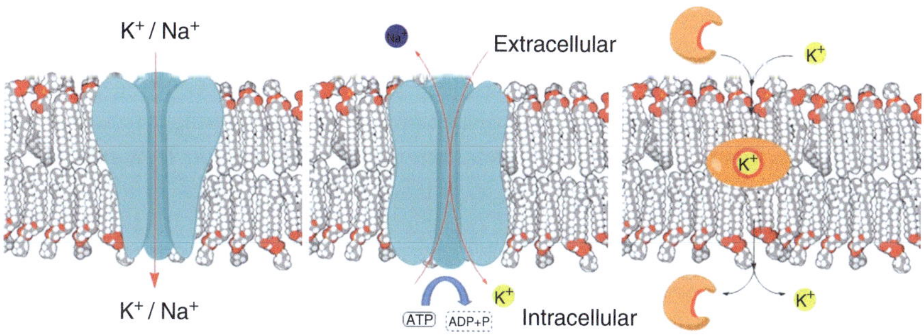

Figure 1 Models of mechanisms of sodium and/or potassium transport. **Left**: ion channel; **center**: ion pump; **right**: ion carrier.

pumps or calcium/sodium exchangers, among others. In addition, certain cells, called excitable cells, have the ability or the need to generate electrical signals. Several types of excitable cells are known, such as neurons, muscle cells and touch receptor cells, which convert chemical or mechanical messages into electrical signals through the use of ion channels. Biologically, the ion channels are responsible of maintaining the resting potential of the cells. For example, in excitable cells, such as neurons, the delayed counter flow of potassium ions shapes the action potential.

The sodium and potassium channels are ion transport membrane proteins whose main functional difference is its transport selective filter, which is caused by the residues located in the narrow part of the inner cavity of the protein. These kinds of channels are transmembrane proteins with an internal hydrophilic pore that extends from one side of the membrane to the other. Ion transport is considered to be passive as far as the ion flow is controlled solely by the concentration gradient, with transport rates close to diffusion in bulk water. In general, these proteins are in a closed state, in which no ions can flow. The channels are subsequently activated by an external stimulus, such as a neurotransmitter, a hormone or a voltage variation, which induces a protein conformational change that opens a pore allowing the ions to cross the membrane [2, 5, 6].

In general, almost any transmembrane forming nanopore species (leukocidins, gramicidins, hemolysins, etc.) can transport alkali metal ions, but the key feature of the sodium and potassium ion channels and pumps is their transport selectivity and the control over the open/closed states. Many nanopore proteins, like the hemolysins, do not require the control of the opening and closing state because their function demands being always in the open state. In general, these types of molecules are toxins whose function is to destabilize the cell ion balance by forming large holes in their membranes.

The function of potassium channels is to conduct this alkali metal ion directed by the electrochemical gradient. This transport is carried out both rapidly (close to its water diffusion rate) and selectively, excluding other ions, most notably, sodium. The discrimination between potassium and sodium ions, whose difference in radius is only in the sub-ångstrom range, is remarkable especially considering that this differentiation takes place at transport rates close to one hundred millions of ions per second. One of the biological functions of these channels is to set or reset the resting potential in the cells.

In excitable cells, such as cardiac muscle, potassium channels contribute to regulate the functional potential and are also involved in maintaining vascular tone. Thus, the malfunction of these channels may cause arrhythmias. In addition, the potassium channels are also present in irregular cellular processes such as the secretion of hormones, as for example the release of insulin from beta-cells in the pancreas, thus their malfunction can lead, among other diseases, to diabetes.

On the other hand, sodium channels are responsible in excitable cells (neurons, myocytes, or endocrine cells) for the rising phase of action potentials, the short-lasting event by means of which the electrical membrane potential of a cell rapidly rises and falls. Sodium channels also play a central role in rapidly transmitting the depolarizing impulses all through cells and cell networks, thereby enabling coordination of complex and multifunctional processes extending from locomotion to cognition.

Once the sodium or potassium channels are open and the corresponding ions move downhill across membranes, the living systems need to recover the original ion gradient. The ion pump proteins, such as the sodium or potassium pumps, are responsible of carrying out these functions. The process requires energy input to overcome the thermodynamic cost related with moving the ions against the concentration gradient.

The Na^+/K^+ pumps are membrane proteins formed by two subunits, alpha and beta, that self-assemble into tetrameric structures. They carry out active transport by consuming energy (ATP). These pumps can translocate three Na^+ outside the cell (extracellular medium) while two K^+ are transported into the cell (intracellular medium) [7]. They are responsible of maintaining the low Na^+ concentration (~5 mM) in the cell plasma compared to the extracellular medium (~145 mM; 10–30 times higher concentration). On the other hand, K^+ concentration inside the cell is very high (~140 mM) compared with the low concentration (~5 mM) in the external medium.

The extensive research carried out in the field of natural transport systems has also revealed that Na^+/K^+ pumps play a central role in an enormous number of signaling mechanisms in the cells, as a consequence, any minor malfunction of these channel proteins triggers the appearance of some illnesses known as channelopathies, such as fibromyalgia or cystic fibrosis [8, 9].

Nature also uses other transport systems to interfere in the ion balance of the cell. The ion carriers (Figure 1), which also transport ions passively, are perhaps the best-known systems. Despite the fact that just a few are voltage-dependent [10], all of them exhibit high ion selectivity. Generally, they are lipid-soluble molecules produced by a variety of microbes as a defense weapon against other (competing) microbes. One of the best known is valinomycin (Figure 2) [11], a cyclic depsipeptide with antibiotic properties that selectively recognizes and transports K^+ across the lipid membranes. Some of the carriers are small molecules that shuttle ions from one side of the membrane to the other, while in the case of the largest carriers, in

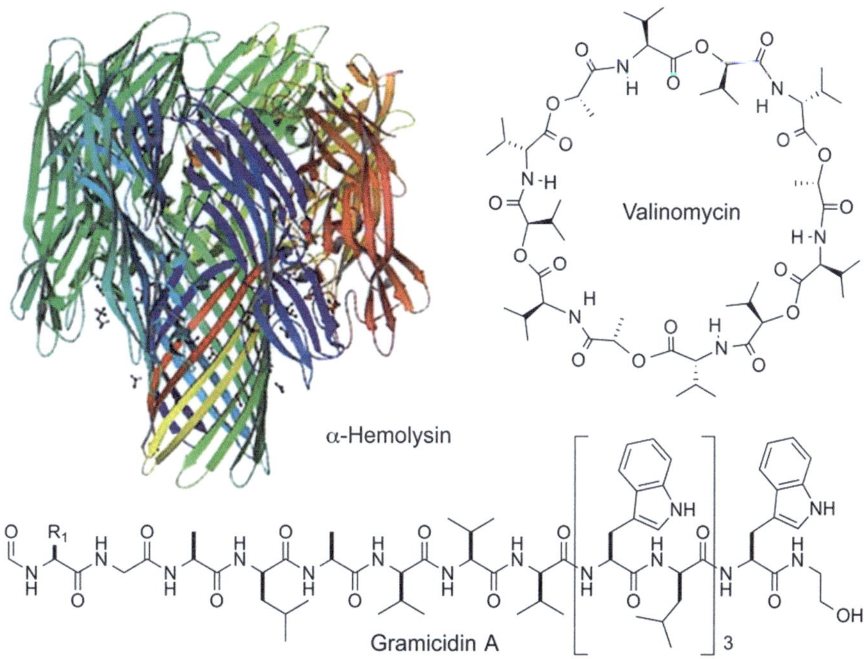

Figure 2 **Top left**: Structure of α-hemolysin (PDB code 3ANZ) formed by seven subunits (each color corresponds to a different monomer) that self-assemble through the formation of a β-barrel. In addition, the structures of the ion carrier valinomycin (**top right**) and the pentadecapeptide gramicidin A (**bottom**) are shown.

general, the transport system implies mechanisms that are not related with their diffusion, but the transporter undergoes conformational changes in the carrier structure.

The properties of all these natural transport systems (channels, pumps, carriers and ionophores) can be summed up in three characteristics: efficient ion mobility, high selectivity and externally (ligand or voltage) mediated transport.

In this review we will cover the most relevant synthetic transporting systems prepared to transport potassium and sodium ions. The chapter is divided into the previously mentioned systems, that are carriers and ion channels, but we have also included a third class, the membrane-spanning ionophores. This term includes a large family of molecular transporters that facilitate the transport of ions across the membranes despite the fact that they neither bind specifically the ions nor form structurally well defined and water-filled pores. These systems are generally amphiphilic molecules that can insert into the bilayer by increasing the permeability of the membranes, therefore granting the movement of ions across them in a passive and, sometimes, selective way, even though they do not possess a central channel *per se*.

1.2 Supramolecular Approaches

Natural systems have inspired scientists to develop synthetic systems which resemble the previously described carriers and channels, searching for the properties that Nature has mastered over the evolutionary process: transport efficiency, ion selectivity, and a tunable "ON/OFF" state control. Originally, these synthetic transporters were prepared seeking to mimic some of these properties but also looking for understanding them by means of simpler components. In the last years, authors were aiming to improve some of the natural properties or even to develop (create) functions not previously observed.

Due to the great complexity and size of most of the channels, carriers, and pumps, their synthesis by classical chemical methods based on covalent bond-forming processes are not feasible because they would require unapproachable long synthetic routes and a high need of manpower. Such large molecules would also suffer similar folding problems than natural proteins. In addition, the membrane transporters would require hydrophobic properties to facilitate their membrane insertion and hydrophilic characteristics to facilitate the ion transport, making their synthesis almost unaffordable.

Not surprisingly for a long time, one of the main drawbacks of membrane protein studies derived from the difficulties to get high-resolution crystal structures. Therefore, in the last decades chemists are searching for alternative strategies that rely on supramolecular chemistry. Supramolecular chemistry is based on appropriately designed basic components that interact with each other *via* weak forces (hydrogen bonding, π-stacking, dipole-dipole interaction, hydrophobic effect and so on) to eventually form super- or supra-molecules with the appropriate topology for carrying out a specific function. One of the main advantages of supramolecular chemistry is its efficiency and simplicity. Only relatively simple building blocks are required and synthesized and they have the ability to associate into more complex structures driven by the thermodynamics of the process. The final structure is achieved by different equilibria to form the thermodynamically most stable complex. Self-assembly and recognition processes are the two most important strategies for the supramolecular construction, the former one being the most relevant in the preparation of complex structures.

Inspired by the simple structure of the natural receptor valinomycin (Figure 2), the researchers have developed a variety of ion receptor carriers to mimic its natural function and selectivity. For the synthetic designs, the molecular recognition processes have played the most relevant role. For the synthesis of non-natural receptors, prior knowledge of the charge, the size, and the electronic properties of the ion allows the design of very selective receptors. The receptor/ion complex must form a lipophilic structure that facilitates the diffusion through lipid membranes.

On the other hand, in order to form transporters with an internal pore, natural systems like gramicidins, for their simplicity, or hemolysins, due to their robustness, have inspired the construction of non-natural channels (Figure 2) [12–14].

Gramicidins are linear pentadecapeptides made of alternating *L*- and *D*-amino acids that self-assemble on the membrane to form dimeric helical structures with an internal pore that is perpendicularly oriented to the membrane. The resulting channel is selective to alkali metal ions compared to divalent cations or anions but it can vaguely discriminate between alkali metal ions. Hemolysins are pore-forming toxins with a mushroom-shaped structure made of seven homo-oligomeric monomers that self-assemble into a β-barrel structure (Figure 2). This channel remains open indefinitely allowing the uncontrolled permeation of water, ions, and small molecules.

Based on these attractive structures a variety of synthetic helical or barrel-shaped constructions with the ability to insert in the lipid bilayer and form channels have been prepared. The creativity of the researchers has also derived in structures not previously observed in Nature. For example, the gramicidin A (gA) have inspired the design of peptide nanotubes made of cyclic molecules than can self-assemble on top of each other [15, 16]. A system in which small cyclic components adopt a flat conformation and stack on top of each other to form the nanotubes. In general, these synthetic approaches show a high transport rate, but other properties of natural ion systems, such as ion selectivity and fully controlled opening and closing, are not completely achieved. Finally, synthetic systems able to efficiently carry ion transport against the concentration gradient such as the Na^+/K^+ pumps remain elusive and will need more efforts.

Regarding the stoichiometry and transport direction, the ion translocation is classified as uniport, symport, and antiport (Figure 3). In the first type the conveyor translocates a single ion across a phospholipid membrane, while the last two refer to transporters that translocate simultaneously two different ions. The symport involves the movement of two or more ions, generally a cation and an anion, in the same direction while the antiporter (also called exchanger or counter-transporter) moves the ions across the membrane in opposite directions.

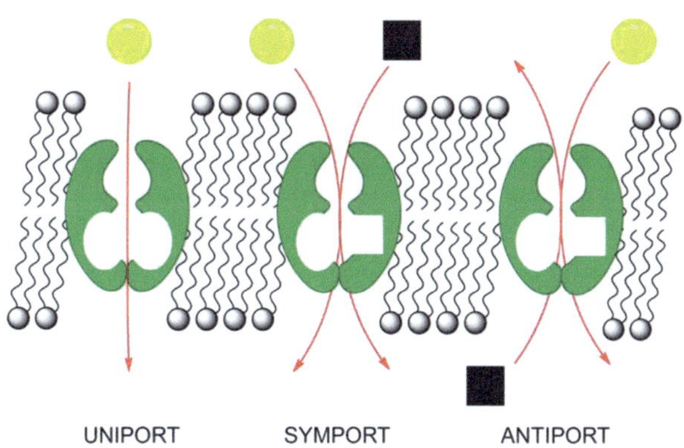

Figure 3 Models of transport mechanisms depending on stoichiometry and directionality of ions translocation: uniport, symport, and antiport.

2 Ion Carriers

Ion carriers are liposoluble compounds that facilitate ions to cross the membrane. In general, the carriers are ionophores that bind to a particular ion shielding it from the surrounding environment. Most of the carriers exhibit great selectivity and, generally, the transport is potential-dependent [17, 18]. Considering the transport mechanisms, they can be differentiated between the carriers that translocate from one side of the membrane to the other facilitating the diffusion of the ions and those transmembrane structures whose active transport mechanisms only require conformational changes to facilitate the ion movement. To differentiate both types of mechanisms we have divided the carriers in this chapter in two different groups: receptor-carriers and membrane-spanning ionophores.

2.1 Receptor-Carriers

Receptors-carriers bind ions with high affinity at the water-membrane interface. The resulting complex is lipophilic and can translocate between both sides of the lipid bilayer to release the ion in the most diluted aqueous solution.

Regarding their structure, in general these compounds have several polar groups (amide, ester or ether) oriented towards the central cavity (recognition center) in which the ion can be located. The resulting hydrophilic cages surrounded by a hydrophobic surface facilitate the immersion of the whole structure in the lipophilic media. Thus, this special arrangement produces lipophilic structures with highly hydrophilic cavities that can diffuse through the membrane bilayer shuttling the ions from one side to the opposite side. Generally, a conformational change at the membrane interface allows the release or capture of the ions.

The high selectivity achieved by some carriers is attributed to the size of the hydrophilic cage. Thus, only cations whose radius fits perfectly in the cavity are entrapped. Larger cations have to deform the cage while smaller ones do not find the appropriate coordination geometry. Regarding the structure there are two classes, the neutral and the ionic (carboxylate) carriers. The latter form neutral complexes with the cation because the carboxylic group is deprotonated and generally the transport properties are related to their ability to exchange protons and cations in an electroneutral process.

There are several natural compounds with receptor-carrier properties that interact with different ions to transport them across the biological membranes [17]. In general, the alkali metal ion transporters are oxygen-rich organic compounds as simple as oligoester chains with low molecular weight [19]. Many antibiotics, particularly the macrolide antibiotics, are carriers that exhibit high affinities for different ions such as valinomycin (Figure 2; binds selectively to potassium cations), calcimycin (divalent ion carrier), monensin (important role as Na^+/H^+ antiporter), enniatin (binds selectively to ammonium ions), beauvericin (alkaline earth metals), nigericin (acts as an H^+, K^+, Pb^{2+} transporter) among others (Figure 4) [20–22].

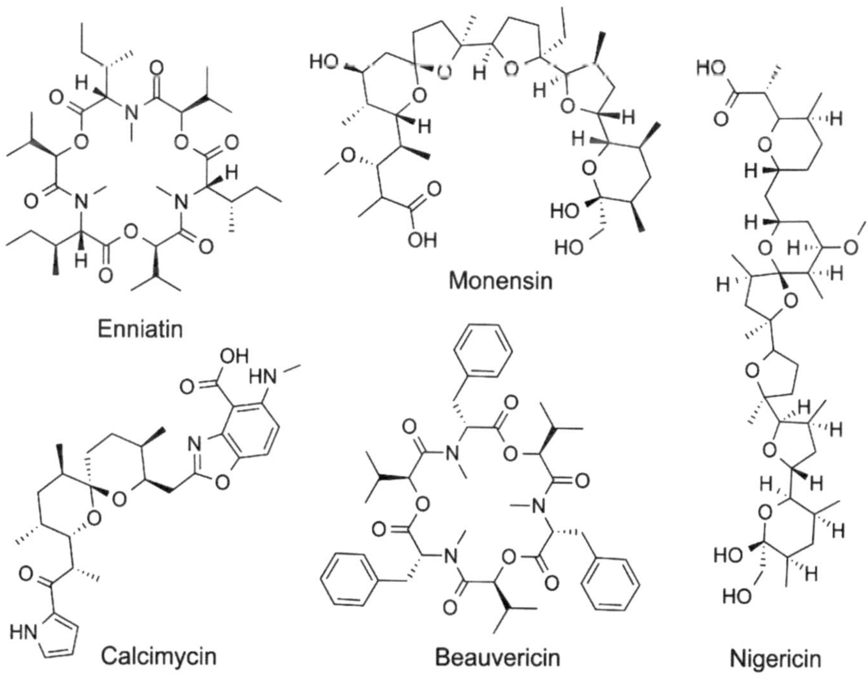

Figure 4 Structures of some natural receptor-carriers: enniatin, monensin, calcimycin, beauvericin, and nigericin.

Valinomycin is a well-known natural product that selectively binds a potassium ion, forming a complex that transports this cation across biological membranes [23]. It is a cyclic dodecadepsipeptide, *cyclo*-[(*D*-Val-*L*-Lac-*L*-Val-*D*-HyV)$_3$–] with alternating amide and ester bonds. This structure confers the conformational properties that result in its excellent cation-carrier activity [24, 25]. It binds numerous cations including those of the alkali, alkaline earth, and transition metals but presents a remarkably high selectivity to complex K^+ compared to Na^+.

The stability of the K^+ complex is between three and four orders of magnitude higher than that of the sodium complex. Apparently, its structure is so well designed that all stages of potassium transport have similar energetic barriers and no 'bottle necks' occur [26]. Therefore, in order to study the structure–function relationship, different analogues have been synthesized by changing their side chains, the configuration or the ring size [27]. Some of these analogues present a different ion selectivity, such as those in which the number of monomeric fragments, three instead of four, were modified such that they prefer to recognize bulky cations. Most of the analogues were less efficient in transporting ions than the natural compound.

Further studies were carried out with modified depsipeptides, such as those prepared by the group of Ranganathan [28]. They synthesized cyclic depsipeptide analogues in which a serine residue was used to link different adamantanebiscarboxylate groups to provide the desired membrane solubility and conformational constraints to facilitate the efficient ion transport in membranes (Figure 5). The resulting derivatives

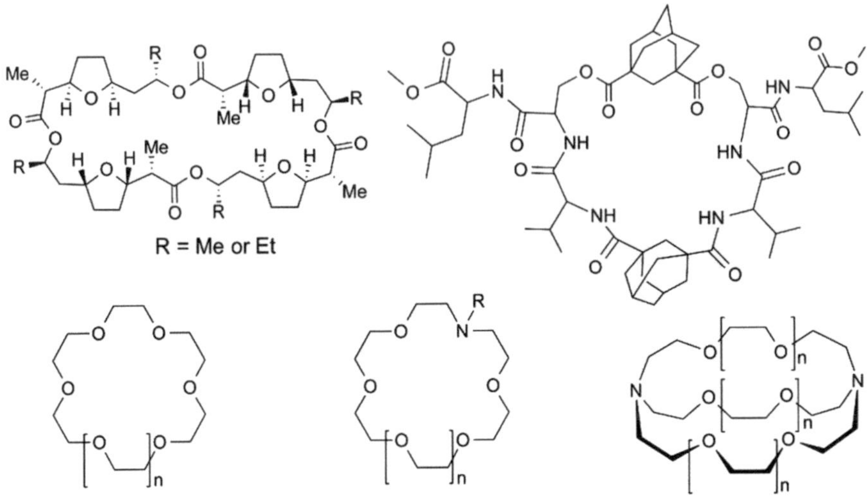

Figure 5 **Top left**: Structures of nonactin (R = Me), tetranactin (R = Et), and monactin, dinactin, trinactin if one, two or three methyl groups, respectively, are replaced by ethyl groups. **Top right**: Structure of a non-natural depsipeptide that mimics the ion carrier properties of valinomycin. **Bottom**: Examples of crown ether, lariat ether, and cryptand (2.2.2-cryptand, n = 1) structures.

were almost as efficient as valinomycin (with respect to the lipid:receptor ratio) in Na^+, Mg^{2+}, and Ca^{2+} cation transport, but unfortunately most of its selectivity was lost.

Enniatins (Figure 4) are also natural cyclic hexadepsipeptides whose ring is half the size of valinomycin [29–32]. In addition, the skeleton amide functionality is *N*-methylated, so no amide protons are involved in hydrogen bond stabilization of the ion/complex structure. The enniatins bind weaker to potassium, are less active, and the potassium/sodium selectivity is smaller than that of valinomycin. They can form different complexes with the ion [1:1, 2:1 (sandwich) or 3:2 (stacks)]. Biophysical studies showed that complex types change depending on the alkali metal ion, being 1:1 for sodium, 2:1 for potassium, and 3:1 for cesium.

Macrotetralide nactins (Figure 5) are a series of natural cyclic esters produced by various strains of *Actinomyces* [33–35]. They are called (nonactin, monactin, dinactin, trinactin, and tetranactin) by the number of ethyl radicals present in their structure instead of the methyl groups. The repeating units of nonactin (R = Me) and tetranactin (R = Et) are asymmetric, and both enantiomeric forms of the hydroxyacids can alternate in the macrotetralides. The ion/macrocycle complex was found to resemble the form of a molecular tennis ball. For the nactin series, antibiotic activity augmented with the increase of ethyl group substituents that is also related with the increase of affinity of complexation ratios between Na^+ and K^+.

Transport properties of these natural systems triggered the interest of chemists to develop *de novo* synthetic compounds that could facilitate the understanding of this process and also improve the observed transport properties. Original studies by Petersen, Lehn and Cram developed simple receptors like the crown ethers or cryptands that provided the first step towards the understanding of the structural characteristic of ion recognition (Figure 5) [36]. It was found that it was possible to differentiate the alkali metal ions with simple macrocycles by modifying the number of ethylene oxide units. Therefore, 12-crown-4 ethers recognized preferentially lithium ions while the 18-crown-6 ethers, like valinomycin, preferably bind potassium ions.

Another important concept in ion recognition introduced by Cram is the preorganization, in which the macrocycles based on six phenol rings bind Li^+ with a very large selectivity (Figure 6) as a consequence of the organization of the oxygen lone pairs towards a small area in which only a small ion like lithium can fit in [37, 38]. Further designs described the macrobicycle effects using cryptands, i.e., macrobicycle systems (Figure 5) made by three ethylene ether chains that showed even larger selectivities [39]. The substitution of one of the three ethyleneoxide tethers by an alkyl group, although it reduced the binding affinity, increased the selectivity between alkali ions and divalent cations like Ba^{2+} or Ca^{2+}.

These systems also allowed studying the differences between thermodynamic and kinetic effects, showing that an important characteristic of the cryptate formation process is the rate of cation exchange. Therefore, the most stable cryptate complexes release the ion very slowly [40, 41]. The dissociation rates are several orders of magnitude slower than those of macrocyclic or antibiotic complexes, and therefore an increase in ion release supposes a reduction of complex stability. Such effects were later used to explain the differences in the transport selectivity in favor of the less stable but more dynamic complexes. In general, the transport rates are often discussed in connection with models in which the rate-limiting step is the diffusion across the membrane of the metal complex while the complexation/decomplexation equilibria at the membrane interface is very fast.

The modified crown ethers, i.e., lariat ethers that are macrocyclic compounds with one or more pending side-arms (Figure 5) [42], have acceptable complexation/decomplexation dynamics coupled with the possibility of three-dimensional cation encapsulation by the crown ether unit and the side arm. In order to effectively transfer the metal ion in a separation process, the system must have a soluble counter anion in both the aqueous and the organic phases.

Surprisingly, some of these neutral molecules extract only the cation and do not require a counterion (uniport); its action is thus electrogenic, resulting in a change of potential across the membrane. However, the distribution coefficients of complexes formed with common anions, such as chlorides, sulfates, or nitrates, between the aqueous and the organic phases are too low to be useful. Attaching a proton-ionizable sidearm to the crown ether ring can eliminate the need to transfer anions from the aqueous into the organic phase. In addition, the presence of carboxylate groups increase the selectivity towards the cations of high charge density (Li^+ and Na^+) reducing the effects derived from the ring size.

Another advantage of proton-ionizable lariat ethers as ion carriers in liquid membrane transport processes is the coupling of metal ion transport from the aqueous source phase into the aqueous receiving phase with the back-transport of protons. Thus, a pH gradient provides the potential for metal ion transport [43]. In some cases, as for the macrocycle bearing the acetate group illustrated in Figure 6, very high sodium selectivity in polymer inclusion membranes is observed. The best

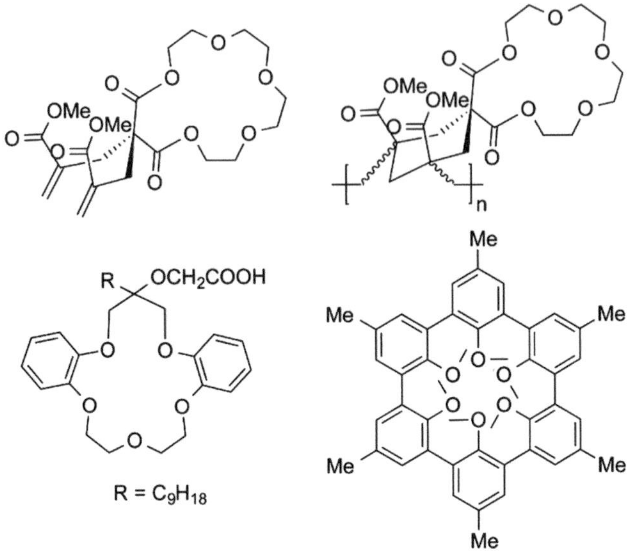

Figure 6 **Top**: Dialkylated malonate crown ether (left) used as monomeric model of corresponding polymer (right) [45]. **Bottom**: Lariat ether with a carboxylic acid for selective transport of sodium ions (left) and Cram spherand that selectively recognizes lithium ions (right).

transport was achieved with large alkyl groups ($R=C_9H_{18}$) with no lithium or potassium detected in the receiving phase.

Polymers containing crown ethers have also been prepared and their ion selectivity studied, they showed some cooperative effects between adjacent crown ethers [44]. Malonate-derived crown ethers were described, and the carbonyl group showed no particular sodium/potassium selectivity (Figure 6). Righetti et al. described their incorporation in polymers containing other carboxylic groups [45].

The transport behavior for the monomer and the polymer presented also some differences. For example, the 17-crown-5 ether has transport rates higher for the smaller cation (Na^+), probably because this ion fits better in its cavity. On the contrary, the selectivity is shifted toward the K^+ ion in the corresponding polymer. Additionally, the transport rate for the sodium ion was lower than in the corresponding monomer. This change in the sodium transport efficiency must be related to the polymer binding efficiency; it is possible that the conformation required for the selective binding of sodium in the monomer could not be reached in the polymer.

All these studies suggested that the ion transport based on this type of ion carriers could not exceed a maximal velocity. Thus, new alternatives to surpass this dilemma became very attractive in which the structure of the carriers could be reversibly interconverted between the strong-binding form (increase the interaction with the ion) and the less prone shape that would facilitate the ion release. With this idea light-driven transport systems were studied [46]. In this respect, azabis-crown ethers were prepared, in which the two macrocycles were attached by an azo group (Figure 7). This group isomerizes (*cis/trans*) upon selective light irradiation (ultraviolet or visible light). The

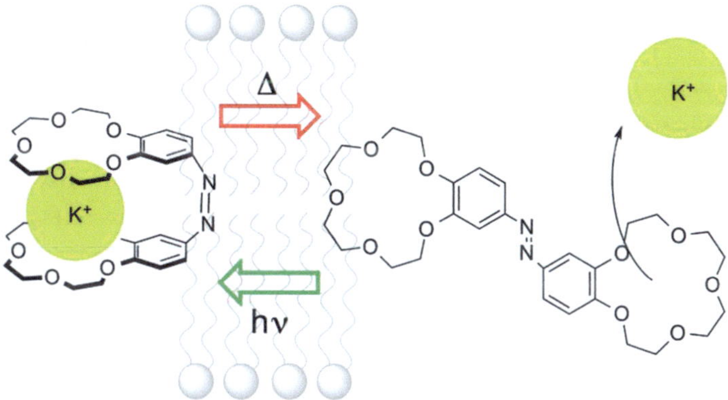

Figure 7 The *cis/trans* isomerization permits efficient K^+ complexation/liberation of the azabis-crown ether.

Z-isomers bring together both macrocycles and large cations can bind in a 1:1 ratio in a very tight way (sandwich-type complexes), while the *E*-form separates the rings facilitating the cation release.

Since these compounds can change their complexation properties on the basis of reversible isomerization, it is possible to control the uptake or the release velocity when light irradiation is applied to the transport system. An acceleration of the transport rate of alkali metals was observed. The results observed depend on the cation and the macrocycle size. For the less-stable complexes (Na^+) the transport rate was proportional to the concentration of the (*Z*)-formed carrier. For K^+ complexes it was found to depend on the counter-anion type and light used. Such behavior was explained on the basis of the rate-limiting step of the process, which was the cation extraction for hydrophilic anions, thus, the light increased the *cis*-form raising the cation affinity.

On the other hand, hydrophobic anions modified the rate-limiting step from the extraction step to the cation release, so the increase of the *cis*-form reduced the transport rate. These experiments carried out with hydrophobic anions showed that the transport was retarded upon UV irradiation, whereas with more hydrophilic counter-anions the transport was significantly increased. In this case, visible light irradiation, which efficiently mediates *cis/trans* isomerization, retarded the transport rate. As expected, alternate irradiation of the membrane by UV and visible light significantly accelerated the rate of K^+ translocation when hydrophobic counter-ions were used. In contrast, the same alternating irradiation protocol retarded the rate of Na^+ transport.

In general, salt transport studies through liquid membranes with these simple-molecular models could be related to the Gibbs' energies of the ion hydration and transfer in the membrane [47]. For example, the transport rates of potassium salts using 18-crown-6 ethers vary in eight orders of magnitude depending on counter-ion properties. The anions with smaller hydration free energies give faster cation transport.

The calixarene motif (Figure 8) is another chemical platform extensively studied as alkali metal ion carrier. The calix[n]arenes are cyclic oligomers derived from the condensation between a phenol and an aldehyde in which "n" denotes the number of repeating units in the macrocycle. These amphiphilic compounds have a hydrophilic ion-binding cavity with an external lipophilic surface that protects the polar ion from suffering strong repulsive interactions with the lipid membrane and hence allowing the ion flow. In fact, in 1996, Jin et al. synthesized the first selective artificial sodium transporter based on a calix[4]arene derivative [48].

The calixarenes are interesting compounds because of their pre-organized core of aromatic units, which provides the possibility of establishing interactions between the electron-rich aromatic ring and the cation (cation-π interaction). The combination of these mentioned characteristics is fundamental for its transport properties, demonstrating that the mentioned calix[4]arene ether (Figure 8) mediates selective Na^+ transport through phospholipid bilayer membranes. Despite the fact that the linear organization of two calix[4]arene could form a channel across the membrane, the studies showed that in this case the Na^+ transport was explained by the carrier mechanism, where the complex Na^+-calixarene diffuses through the membrane as a monomeric species. In particular, the calixcrowns in which an oligo(oxyethylene) chain links two oxygens of a non-adjacent phenolic ring of the macroarene framework, are important groups of these carriers due to their highly selective metal ion recognition.

The observed selectivity depends on the size of the oxyethylene chain, on the calixarene conformation and on the type and structure of substituents at the upper or lower rims (the wide and narrow part, respectively, of the characteristic three-dimensional basket structure of calixarenes) [49]. For example, it has been shown that the 1,3-alternating conformation of calix[4]arene-crown-6 derivatives bind Cs^+ ions with improved selectivity and strength than simple crown ethers [50–52]. Such selectivity was later implemented for Cs^+ transport in supported liquid membranes [53]. Calixarenes have also been used for studies of alkali metal cations through polymer inclusion membranes (PIMs) [54].

A dimeric tubular-shaped receptor has been designed by Beer's group that displays a high affinity for potassium ions compared to other alkali metals due to the presence of eight coordinating binding sites (Figure 8) [55]. These dimeric calixarenes could potenially work as membrane-spanning compounds but, based on the results reported in this article, because of their length and transport rates, they present properties that correspond to an ion carrier.

Larger calixarenes were also studied such as the pentamethyl ester of calix[5]arene derivatives bearing benzyl or *tert*-octyl moieties at the *para*-position. These studies by liquid-liquid extraction experiments showed a high selectivity within the alkali metal cation series [56]. This selectivity depends on different factors such as the macrocycle conformation, the substituents on the *para*-positions and the solvent. Another example of calixarene macrocycles is a dimeric channel in which diazobenzenes were placed at the membrane interface to achieve responsiveness to light by their incorporation in the larger rim of the macrocycle [57].

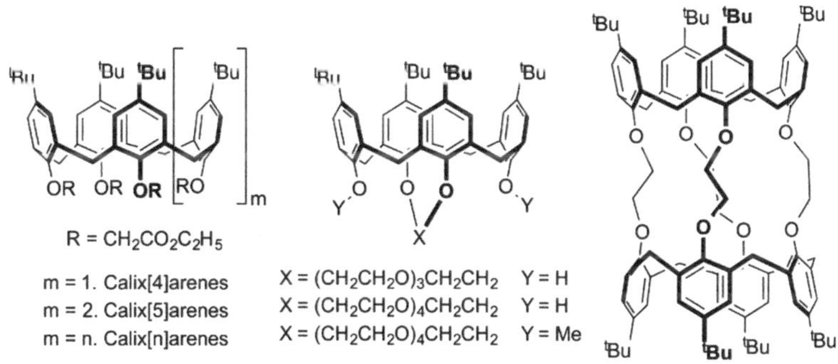

Figure 8 Structure of calix[n]arenes (**left**) and the calix[4]crown ether derivative (**center**) used in Na^+ transport through lipid membranes. **Right**: dimeric calix[4]arene that forms a tubular structure.

2.2 *Membrane-Spanning Ionophores*

Membrane-spanning ionophores are defined as molecular systems that insert in lipid membranes and modulate the ion permeability without a strong and selective binding to the cation, neither through the formation of a structurally well-defined nor a water-filled pore [19, 58, 59]. Because of their structural properties, they are not shuttling ions through the membrane by means of their membrane diffusion. These systems cannot be considered as molecular channels because they do not form structurally defined architectures but they can induce membrane permeability through a variety of mechanisms that include the formation of toroidal channels (Figure 9). The permeability of these compounds is usually voltage-dependent and some of the transport active structures give similar signals in planar lipid bilayer as the ion channels. These transporters can also give groove- or trench-like structures in which the molecules span the bilayer and the ions move through different binding sites of the transport system. It seems that there are, in general, various conducting structures.

Although some of them may be membrane-spanning, the most prevalent are those that involve the formation of transmembrane structures that provide the ion with a continuous pathway from one face to the other of the bilayer. To form the active structures, the transporter must start penetrating the bilayer. This process is occasionally described as the rate-limiting step and sometimes has the characteristics of the lipid flip-flop mechanism [60]. Once the monomeric transporter interacts and inserts on the membrane, it might adopt different conformations depending on the structure of the compound and the interactions with the membrane. The different parameters to consider are the flexibility of the ionophore, which might induce bending or folding, the segment length, the complementarity with the leaflet thickness, and so on. Some of these folded structures, depending on their depth, disrupt one of the layers generating a conducting pathway through some water penetration. These membrane perturbations can also give rise to aggregates and eventually to microphase separation. These initial loosely aggregates are also good candidates of some of the conducting structures.

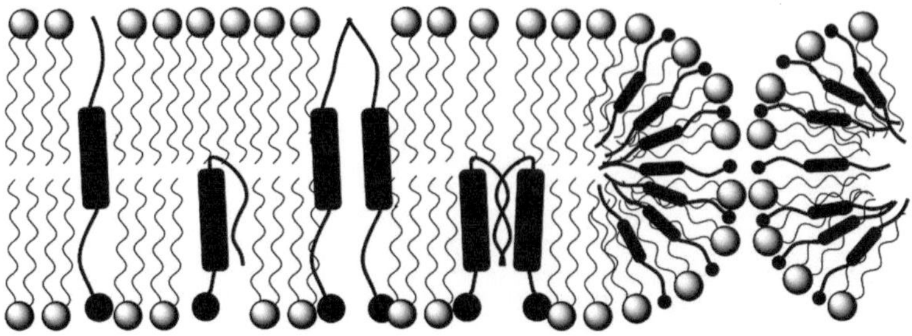

Figure 9 Different approaches for the ion transport mechanism of the amphiphilic molecules (see text).

In general, such structures would have relatively long-lived transporting events but with relatively low conductance. Such behavior is due to the stabilization by the phase behavior of the single monolayer [3]. Finally, the transporter could also uncoil to form a structure that spans the membrane. These extended forms could aggregate to form homo- or heteromeric species with transporting properties. These membrane structures are responsible of channel recording with large conductances but short-living and less regular conducting events. These transporting properties are a consequence of the stress derived from the large mechanical forces on the spanning structures that have independent motions compared to the bilayer. Finally, the transport properties of some of these ionophores correspond to the toroidal model in which the transporters are intercalated with the membrane phospholipids stabilizing the positive curvature perpendicular to the bilayer plane. These structures, in general, are responsible of long living and very high-conductance channels.

Despite the fact that for some of the transporters mentioned in this section the authors claimed the construction of membrane channels, the lack of rigidity and/or hydrophilicity to form water filled pores in the membrane and also due to the lack of experimental evidences of ion channel formation we decided to include them here.

The design of these synthetic ionophores requires the adequate organization of donor groups in the supramolecular architecture in order to ensure a highly efficient and selective ion transport. Here we have included also a variety of synthetic designs whose aim was to create components that spanned the lipid bilayer by itself or by association into dimeric components. Some of them include a macrocyclic component in which several pendant chains were attached. Depending on the position of the macrocycle on the membrane there are designs with the ring in the central part of the membrane whereas some others present it at the polar side of the membrane. Finally, simpler designs were also prepared; these were linear amphipathic molecules, which, depending on their chain length, could span the whole membrane bilayer or only one of the leaflets. Most of these compounds are derived from the idea of designing ion channels that are structurally similar to the amphotericin active form while using simple and synthetically accessible compounds.

Following the initial studies with crown ethers and lariat structures and with the aim of developing synthetic ion channels, Lehn and coworkers described one of the first structures incorporating a macrocycle in the central part of the membrane [61–63]. These compounds were called bouquet-like molecules or *chundle* approach (hybrid word of *chan*nel and bun*dle*). For that purpose they described crown ethers in which two of the ethylene groups were substituted by enantiomerically pure tartaric acid moieties (Figures 10 and 11). The carboxylate groups of the tartaric acid have a strong conformational preference for their anti-disposition, restricting the conformational freedom of the macrocycle and facilitating the functionalization of the ring at both sides. The incorporation of the PEG chains on the carboxylic groups (two per carboxylate) provided the active components [61–63]. Carboxylic groups are placed at the end of each PEG chains to ensure their positioning at the water interface.

Fyles and coworkers also exploited the same idea but using a core unit of polycarboxylate crown ethers made by a variable number of (*R*,*R*)-tartaric acid subunits [64–66]. Macrocyclic tetraesters of maleic acid were attached to the carboxylates to play the role of molecular walls on the channel. At the opposite side of the central ring the head groups were incorporated to ensure their location at the membrane interface. The macrocyclic tetraester and polycarboxylate crown ether hybrids were studied in bilayer vesicles by measuring proton and ion transport by NMR or fluorescence spectroscopy on liposome-entrapped dyes. In general, the active compounds were less active than gramicidin. Some of the derivatives, those with only two wall molecules, have carrier properties while the larger (tetra and hexa) derivatives were assigned channel-like properties. The results also suggest that these highly oxygenated synthetic transporters can migrate throughout the whole liposome population.

Cyclodextrins (CDs, Figure 10) are other macrocyclic structures commonly used in molecular recognition in water due to their hydrophobic cavity. Albeit the

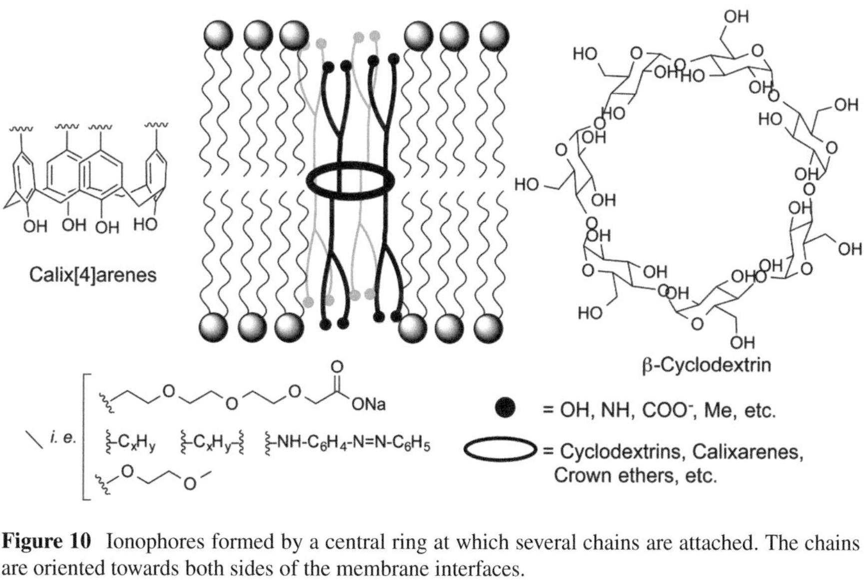

Figure 10 Ionophores formed by a central ring at which several chains are attached. The chains are oriented towards both sides of the membrane interfaces.

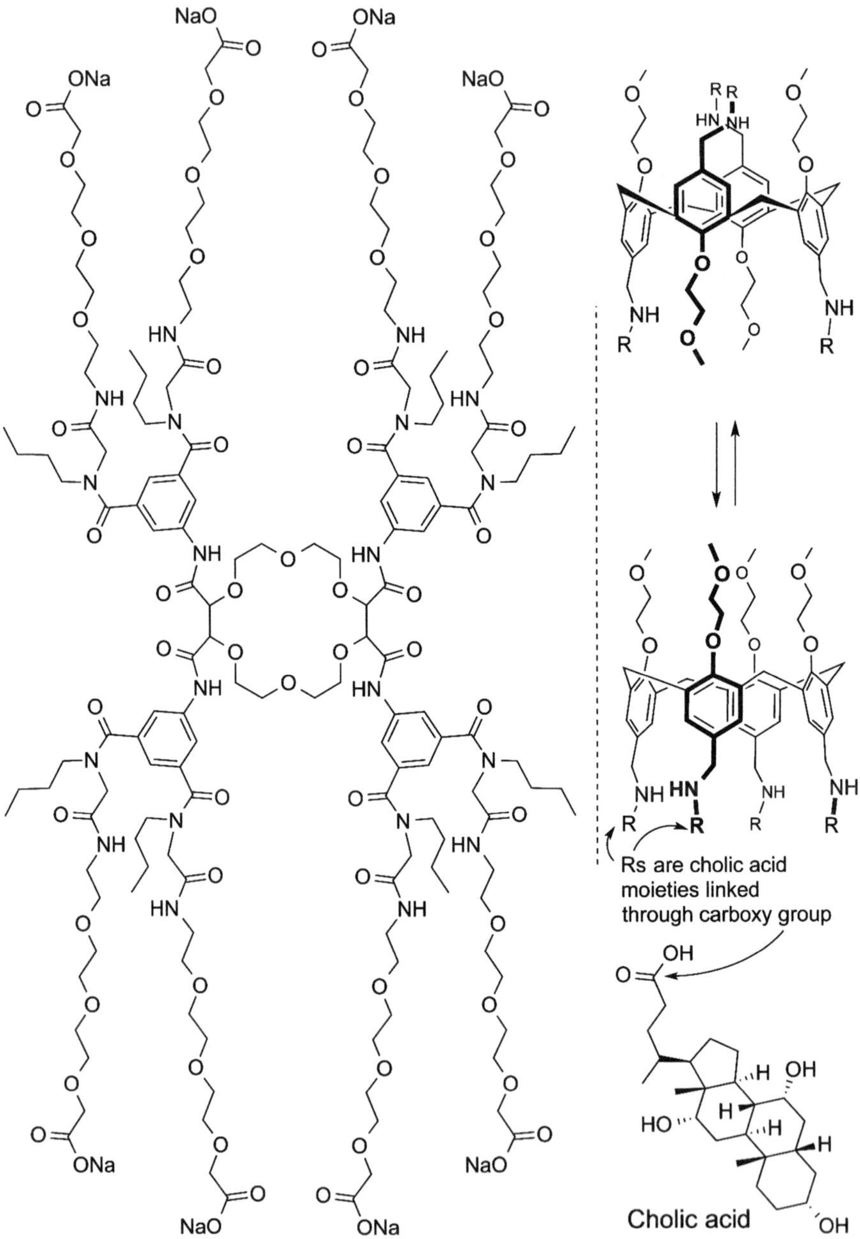

Figure 11 Example of ionophores containing a central macrocyclic molecule. **Left**: crown ether. **Right**: Calixarenes in which the two conformers, the extended (1,3 alternating) and the cone-shape, (**bottom**) are shown. The more active component is the extended one (**top**).

hydrophobic character, one of the first synthetic designs of ring-shaped transporters was based on CDs [67]. This initial work inspired other authors to use the CD motif in the design of other transport systems. For example, Lehn et al. designed another bouquet-like molecule that spans the lipid bilayer and mediates the alkali metal ion transport [68]. This structure has a central core made by the CD and again oligoethylene glycol chains were attached to both sides of the cyclic oligosaccharide to ensure the spanning of the membrane. These bouquet structures were found to cause a one-for-one antiport ion, sodium *versus* lithium, exchange. The transport studies were carried out by ^{23}Na NMR experiments and no channel mechanism was proven, although the observation that the sodium transport rates were similar both in fluid- and gel-state membranes suggests that ion migration does not take place through a carrier-type mechanism.

The calix[4]arene platform has also been used in combination with cholic acid in this type of designs (Figure 11) [69]. Cone-shaped and 1,3-alternate calixarene derivatives were prepared and they showed effective transport of sodium ions and protons across liposomal membranes. The extended, 1,3-alternated conformation, was more active than the cone-shaped. This, together with the observed unimolecular mechanism, suggests that the ion transporting properties depend on the overall length of the transporter. Active components are those whose length roughly matches the thickness of the hydrophobic core of the membrane.

The same macrocyclic moieties were used for the second group of membrane ionophores. In this case all the pending chains were attached at the same side of the ring forming a kind of molecular octopus-like transporters. Most of the transporters were prepared using chains long enough to span the phospholipid bilayer while the hydrophilic macrocycle would reside at the water/membrane interface. A second group of transporters contain chains inserted sufficiently deep in only one of the leaflets of the membrane. Such a structure would be active just by allowing the penetration of some water molecules that would open a conducting pathway. On the other hand, a flip-flop mechanism could facilitate the incorporation of the half-channel at both sides of the membrane, whose dimerization process would form the full transmembrane channel. Although some authors have attributed to this form the active species, mostly there is not enough driving force (molecular interactions) that can stabilize the dimers, especially considering the independent motion of the two layers of the membrane.

The first design of this type was the previously mentioned work of Tabushi et al. searching for novel synthetic ion channels [67]. This design consists of a β-cyclodextrin derivative, in which several alkyl chains were linked to the smaller base of the frustum, where the primary hydroxyl groups are projected. The alkyl chains had also amide groups at their ends to facilitate the interactions with the ions. The original idea behind this design was the formation of ion-channel mimics by associating two CDs into the channel (Figure 12). The cyclic oligosaccharide would reside at the membrane surface, with the polar hydroxyl groups in contact with water, while the hydrophobic chains dipped in the membrane. The transport rate for divalent ions was much faster than for the specific monovalent carriers like 18-azacrown-N6 and it was interpreted by the metal ion rapidly jumping from one binding site to another within the structure.

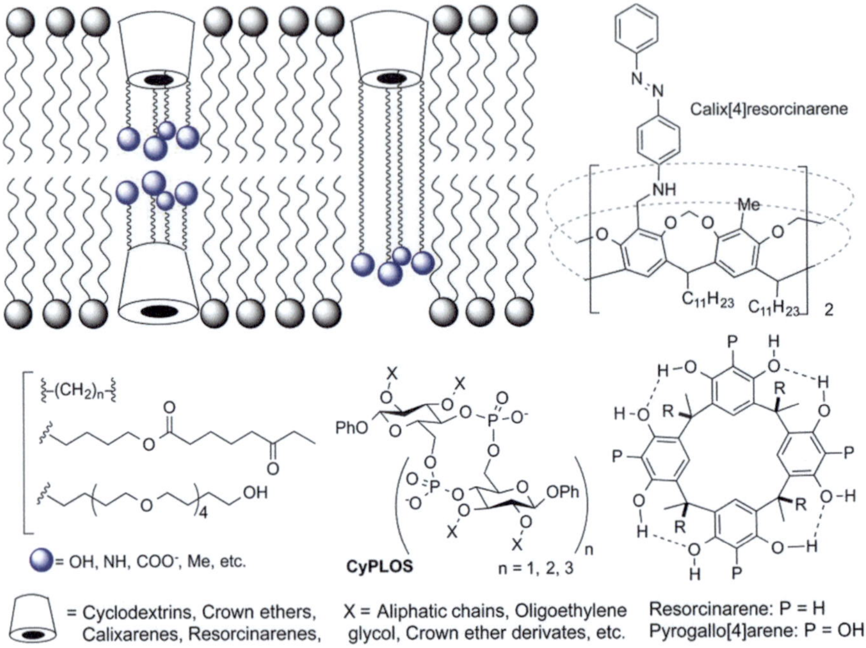

Figure 12 Structures of the membrane-spanning ionophores with a cyclic component (resorcinarenes, pyrogalloarenes, cavitands, and CyPLOS) that generate cavities at the lipid membrane surface.

Similarly, a CD with seven amino-5,10,15,20-tetraoxatetracosanol moieties instead of alkyl groups was also prepared [70, 71]. In principle, this ionophore would only require one molecule to span the bilayer, considering the length of the pentabutylenglycol pendants. The NMR studies suggested that the Na^+ transport rate in liposomes is about 36 % of that of gA activity. It should be mentioned that the liposome experiments only provide information on transporter formation and not about ion transport rate because the diffusion on the membrane and assembling or folding are the rate-limiting step in this process and not the ion diffusion. Probably this is one of the reasons why no difference in transport rates for other alkali metal ions was found. In any case, the transport rates depend on the type of counter-anion and, interestingly, present the selectivity ($I^- > Br^- > Cl^-$). Perhaps this is an after-effect of the basic amine groups at the chain-functionalized rim of the CD that can be protonated. Actually, the ion transport rates (both for cation and anion) increased upon augmentation of pH, probably as a consequence of the unique electrostatic properties of the multiple ammonium groups that line up the channel pore and the symporter properties of the channel.

Recently, Chui and Fyles carried out the most rigorous study of CD-based transporters [72]. They prepared a variety of synthetic semi-channels in which several different substituents were attached to the CD (α and β) at the small base of the frustum by means of the copper-catalyzed "click" cycloaddition reaction. In this way, several transporter precursors were prepared with different structural characteristics such as alkyl chains, perfluorated chains, cholic acid esters, and so on. All of them have in common the presence of chains with a polar group at their end. The activity of these CD derivatives was assessed using the bilayer clamp technique and analyzed using the activity grid analysis [73, 74].

The activity grid analysis is a method that provides quantitative information about conductance and opening duration in a five-level color code, representing qualitative information on the nature of the conductance time profiles. Analysis of the cumulative dataset suggests that the reported conductance data can reflect the structural features of the compounds prepared, but it also reflects the energetic landscape of the bilayer membrane in which synthetic ion channels function. Using this method they could detect the signal of the expected channel structure, together with the characteristic complexities of the small-molecule transporters. Thus, regarding the ion recording, the addition of this type of compounds to bilayers always gave a number of parallel pathways working simultaneously. In fact, defect-related channels were superior to the design expectations. Transport activity for the derivatives with shorter pendants was observed only after the addition of the transporter to both sides of the membrane, thus the inserted monomer must not be sufficient to facilitate ion migration (Figure 13). The CD derivatives with longer chains, such as steroidal moieties, can be active in the monomeric form, probably because they are able to interfere in the opposed bilayer leaflet.

Other proposed transporting species are the end-to-end dimers, in which the ions pass through the hydrophobic cavity of the CD, and large aggregates in which the ions travel through the contact regions of two or three CDs. The observed Eisenman I sequence indicates that ion dehydration must take place during the membrane crossing [75]. The activity was also studied by experiments of transient blockage of the channel through the addition of hydrophobic guests. Again these experiments were consistent with the formation of the dimeric channel but together with other competing and interconverting motifs.

A similar idea was developed and exploited by Kobuke et al. for the preparation of K^+-selective channels using resorcin[n]arenes instead of a CD [76, 77]. These macrocycles, like calixarenes, were obtained by condensation of resorcinol and aldehydes with long aliphatic chains whose principal characteristic is the presence of two hydroxylic groups per aromatic ring in the wide upper rim. The proposed structure is again a semi-channel that has alkyl chains with similar lengths to the lipids used in the study. The dihydroxylated aryl rings force the molecule to place the cup at the interface with the alkyl group penetrating the membrane. Therefore, the proposed active channel should be a dimeric structure. The transport studies were carried out in planar lipid bilayers, showing stable transitions between open and closed states with constant conductance at different voltages. The observed symmetry in the channel current suggests the existence of a symmetric channel structure. The ion permeability value between potassium and chloride ions was found to be 20, showing similar selectivity as other natural transporters like gA. Additionally, their analysis reported a remarkable K^+/Na^+ ratio of 3, representing one of the first examples of an artificial ion channel that displays selectivity for

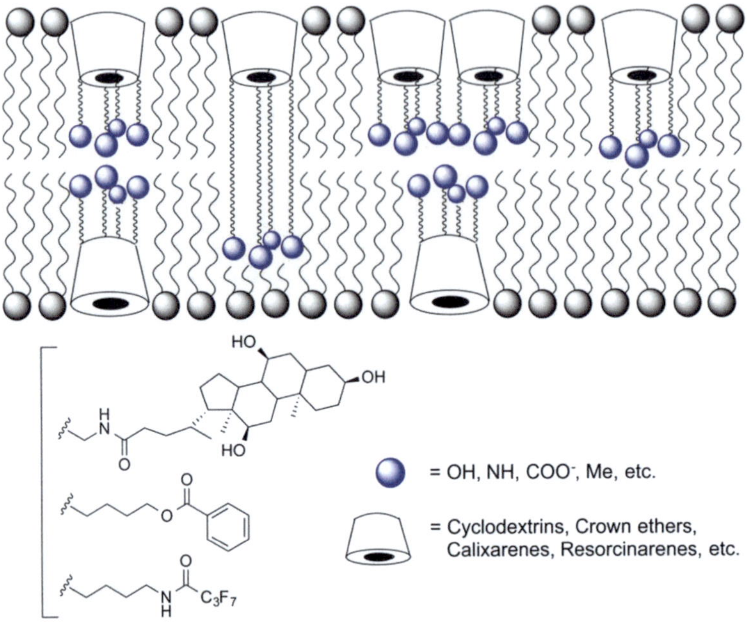

Figure 13 Schematic representation of several ion transport models systems based on cyclodextrins.

K^+ ions. Interestingly the incorporation of a nitrile group between both hydroxylic groups provided pH-depending conductivity changes. The reason of the observed variation of conductance values at pH 8.5 is not clear but it might be related to the change of acidity of the hydroxyl groups at the mouth of the channel that might interfere in the solvation and interaction with the transporting ion.

Another example is based on cavitands derived from resorcinarenes in which the phenolic groups of two adjacent resorcinols are covalently linked through a methylene bridge. In this study diazobenzenes were incorporated at the upper rim to place them at the membrane interface with the purpose of achieving a light-responsive transport process [78]. Successive short opening and closing states in transport experiments were obtained after incorporation of the thermodynamically favored EE isomer of 11,23-dimethyl-bis(5,17-*p*-phenylazophenyl-aminomethyl)-cavitand into the bilayer. The authors proposed a tail-to-tail dimer as the active component because channel activity was not observed when only resocinarene derivatives were added into one side of the lipid bilayer.

Pyrogallol[4]arenes are also calixarene-type structures derived from pyrogallol that were first exploited by Gokel's group (Figure 10). They prepared macrocycles formed from pyrogallol and dodecanal that could insert in phospholipid bilayers and form conducting pores that undergo reversible switching over a wide range of potentials [79]. The analogue with a short alkyl chain (propyl) was membrane-active but showed a very irregular behavior, suggesting some kind of membrane disruption. The derivative with long chains (dodecyl) showed current–time traces with fast open–close behavior. The estimated pore diameter, using the Hille equation [3], was around 11 Å, suggesting that the active form is derived from a bundle of transporter aggregates made of six molecules. Finally, derivatives containing branched chains formed supramolecular interlocked nanotubes composed by six monomers connected via an extensive hydrogen bond network [80]. These large nanotubes were also active in transport [81].

A very simple approach was the family of cyclic glycomimetics, called CyPLOS (Figure 12). They consist of a phosphodiester of β-phenyl-*D*-glucopyranoside with a variety of substituents linked to the hydroxyl groups at C2 and C3 [82–84]. The most active derivative contained a tetraethylenbenzyl ether chain (TEB). This ionophore appears to be active as a monomer. The proposed mechanism supposes that the polar anionic macrocyclic head remains on the surface of the membrane while TEB chains are inserted in the bilayer. The TEB chains are not able to span the membrane and must be located between the polar surface and the hydrocarbon core. Probably, this produces a destabilization of the leaflet of the bilayer, altering its permeability.

The idea of preparing ion-transport systems constructed on the basis of placing a macrocycle, azacrown ethers, at both bilayer interfaces was extended to other models that included also a ring in the central part of the membrane (Figure 14) [85–87]. The designed channels were called hydraphiles, because they are two-headed molecules that resemble the mythical hydra serpent. They are composed of three crown ethers attached initially through two alkyl chains facilitating the incorporation of two of the rings at both sides of the membrane. Other derivatives with different linkers were prepared later. The third ring would be in the center of the bilayer, chains that contain between 12 and 16 carbons being the optimal spacer length for the typical phospholipids. The transport studies were measured by ^{23}Na NMR in large unilamellar vesicles. To facilitate the incorporation of the transporter into the membrane, it was necessary to heat the mixture of vesicles and the trismacrocycle to 50 °C. These transport experiments showed that the synthetic channel exhibits first-order kinetics with a cation flux about 40-fold greater than that of the simple carrier but two orders of magnitude poorer than that of gramicidin. The selectivity for Na^+ over K^+ is 4:1. The authors proposed a transporting system with a groove shape similar to the one proposed in the Figure 14 whose transport mechanism is based on the ions relaying from ring to ring. The covalently linked macropentacycle is more active than the linear hydraphiles [88]. These derivatives might work more like an ion channel in which the ions are translocated in the hydrated state and therefore the transport enhancement observed for the cyclic tetra-crown ether can be due to the increase of donor atoms placed deep within the channel. Interestingly, these compounds have antimicrobial properties and their activity is correlated with their ion transport activity [89, 90].

From these initial studies different authors started to study less complex molecules finding also remarkable transport properties. Most of these designs were prepared with the idea of forming a barrel-stave structure in the membrane similar to the amphotericin (AmB) active form [91, 92]. In this way, some of the designs were constructed with a length similar to the natural product size, half-channel, although

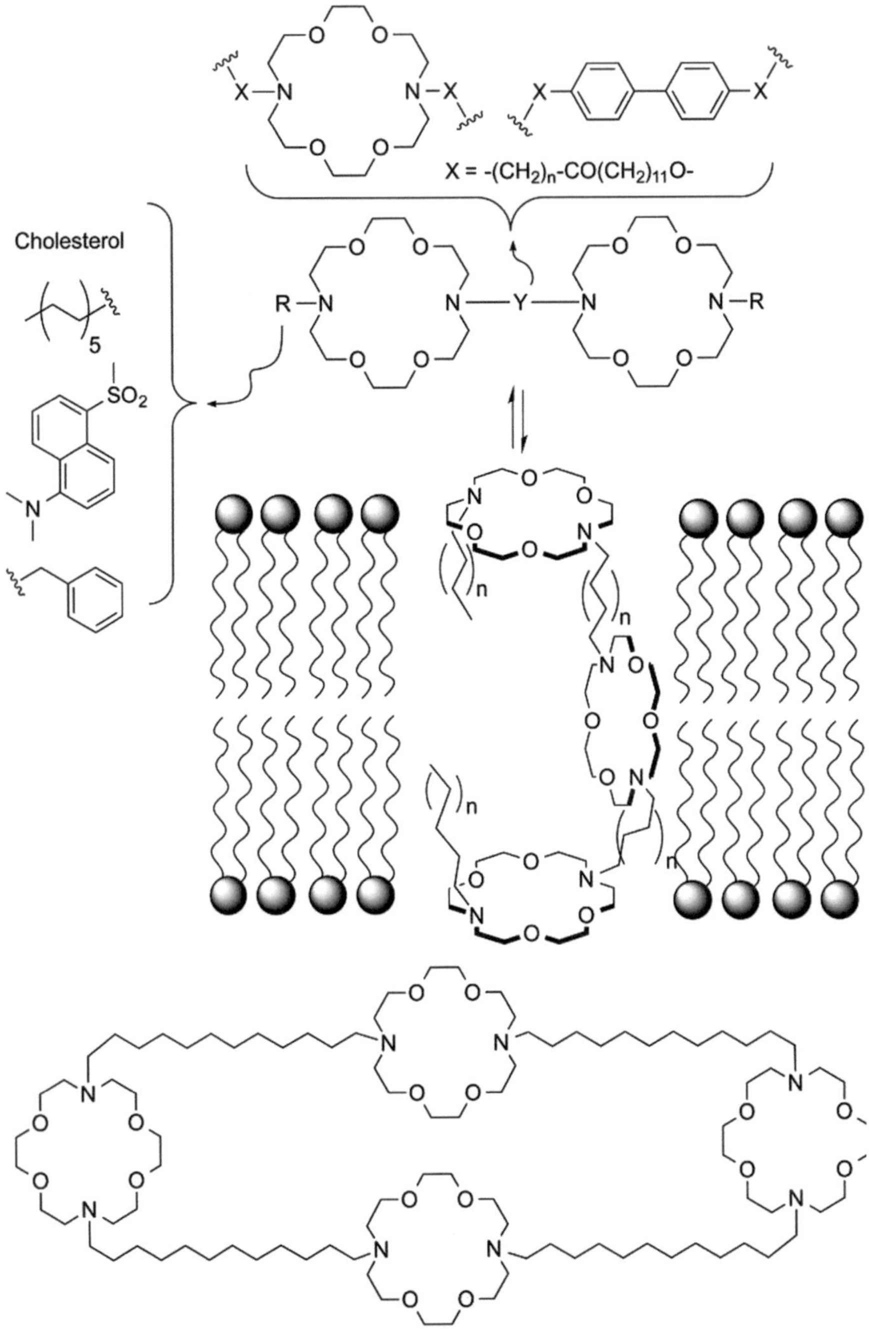

Figure 14 Hydraphile structures composed of two, three (**top**) or four (**bottom**) crown ethers and proposed model for the transport-active form in the lipid bilayers.

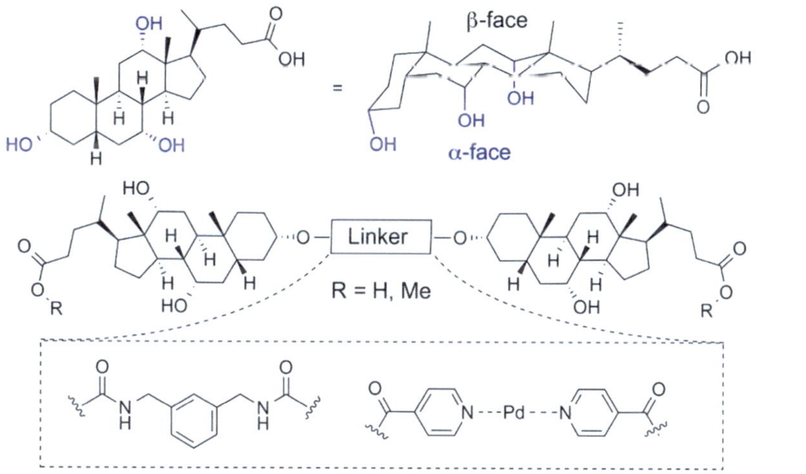

Figure 15 **Top**: Cholic acid showing the two amphipathic faces. **Bottom**: Cholic acid dimers connected by a covalent linker (biscarbamate) or a palladium complex through the coordination to a pyridine motif.

other derivatives were already prepared to span the full membrane thickness in order to reduce some of the entropic costs of forming large aggregates in the membrane. These types of molecules were called bolaamphiphiles, which are amphipathic molecules characterized by a long hydrophobic chain that contains hydrophilic groups at both ends [93]. In some cases a lipid-like structure was used as building block component in the preparation of these kinds of systems.

Cholic acid is a very well known steroid that presents two different faces on its topology. Because of its amphipathic behavior and rigidity, this compound was used very often in the preparation of ionophores. The hydrophilic groups are oriented towards its concave face, providing a conformation with a hydrophobic external surface and also forming a cavity in which small molecules can interact (Figure 15). The presence of the three hydroxyl groups in the hydrophilic face provides the appropriated environment for cation transport [94]. Other groups, such as ureas, were added to build cholic acid analogues suitable for anion transport [95]. Using this compound, Kobuke et al. synthesized transmembrane ionophores by attaching two molecules of the dimethyl ether of cholic acid through a biscarbamate link. The resulting dimeric compounds showed high cation selectivity even when either an anionic (carboxylate) or cationic (trimethylammonium) group was incorporated at the end of steroid side chains [96, 97].

The acid derivative showed permeability ratios for the potassium cation seventeen times faster than for the chloride anion. Interestingly, the incorporation of the trimethylglycine moiety provided also an ion transporter but with reduced potassium/chloride selectivity, reflecting the importance of the ionic character of the polar heads in transport selectivity. Both transporters showed a 3:1 selectivity for potassium over sodium. The authors have proposed a channel model in which three membrane-spanning derivatives self-assemble directed by the formation of hydrogen bonds between the hydroxyl or the methoxy groups of cholic acid, in this last case mediated by water molecules. Such aggregates form a hydrophilic cavity through which the cations can migrate. The larger conductance events observed for the tetramethoxybischolic acid derivative was explained by the larger cavity of the water-mediated trimer. The equilibrium between trimeric and tetrameric aggregates would explain the different channel events observed for the same transporter. The use of asymmetric anionic head groups on the bolaamphiphiles provided, upon bilayer incorporation, assemblies that exhibited rectification properties for the first time for non-peptidic transporters [98].

More recently, in 2008, Wilson and Webb reported the preparation of non-covalent analogues of Kobuke's compounds by replacing the carbamate linker with a metal complex [99]. With this aim, they incorporated a pyridine moiety at the C3 hydroxyl group. The coordination with Pd(II) forms a rigid framework that can also span the membrane and facilitate ion transport. The resulting complex showed an increased activity for transporting Na^+ and K^+ compared to Kobuke's carbamate-linked dimer. In contrast to this compound, the rate of K^+ transport was slower than that of Na^+. U-tube ion transport experiments allowed discarding the carrier mechanism. A remarkable aspect of this design is the inhibition of transport by a chemical signal. Thus, the addition of a palladium(II) chelating agent (hexathia-18-crown-6 ether) stops the ion flow probably by inducing the channel dissociation. Hence, it could be considered as a tunable transporter, as far as the flow can be controlled (switched on and off) by external stimulus such as the addition of a specific metal cation. In this sense, other authors used also the metal-mediated self-assembling approach to develop new synthetic ion transporters that will be discussed in the following section [100].

Another attractive approach by Kobuke used the idea of combining two components with different properties [77, 101]. This approach is based on the self-organization properties directed by salt bridge interactions between carboxylates and ammonium groups. Therefore, they used two components with different properties, one hydrophilic and the second one hydrophobic. The resulting couples have amphipathic properties that allow the interaction with the bilayer providing a hydrophilic environment that facilitates ion transport. Among others, an oligobutylene glycol bearing a carboxylic group and an ammonium derivative that contains two long alkyl chains were combined and showed to have transport activities similar to the ion channels. The planar lipid bilayer experiments showed the formation of three conductance levels that were frequently coexisting. The permeability ratio between potassium and chloride, calculated by the Goldman-Hodgkin-Katz equation, was approximately five [102–104]. These channels do not seem to discriminate among monovalent cations.

Very interesting results were obtained when an azobenzene moiety was incorporated in one of the components (the hydrophobic ammonium) (Figure 16). The resulting amphipathic couple gave photoswitched conductance responses [105]. The *trans-azo* isomer was active while the *cis* derivative did not transport. The UV light irradiation of the membrane causes isomerization of the azo group to the *cis* form that must

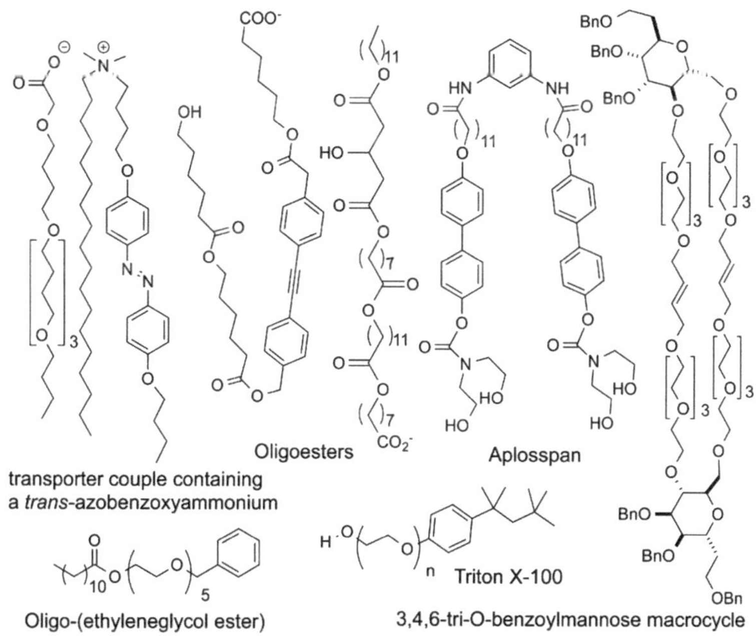

Figure 16 Examples of transporters based on linear amphiphilic molecules.

disrupt the structure responsible of the transport. Although the *trans*-azo compound showed a small conductivity, it provided a high K^+/Na^+ selectivity (six times more active), something quite uncommon for this type of supramolecular ion channels.

Simple bolaamphiphiles molecules were also studied and found to be membrane-active [93]. Very recently, Fyles has covered the ion transport properties of these simple materials [19]. The first compounds described were two macrocyclic tetraesters of maleic acid attached by two sulfides through an aromatic spacer [106, 107]. These synthetically simpler compounds preserved all the essential features of the activity of the more complex derivatives in which the same type of macrocycles were attached to crown ethers [64]. Clearly, the bis-macrocycles were capable of assembling to form transport-competent structures. Later, this group prepared even more simple structures such as the linear oligoester derivative illustrated in Figure 16.

These compounds provided also very active transport properties [74, 108, 109], with conductance traces that were similar to the more complex structures. The bilayer conductance analysis suggested, like previously mentioned for CDs, that typically several conducting structures are formed (Figure 9). Thus, the transport mechanism must include a number of different active species that work simultaneously. In some cases transport appears to be associated with first-formed and poorly structured species, rather than with well-defined and more complex structures. The formation of these aggregates must be induced by gathering and eventually microphase separation as driving force. In some cases, transport events can be derived from the formation of an open structure at one half of the bilayer that would perturb the opposing leaflet of the membrane thereby opening a conducting pathway. In addition, the flexible bolaamphiphiles can uncoil to form the spanning state that would aggregate and eventually give microphase separation. These aggregates are also responsible of some of the observed transporting properties.

Moreover, extremely simple amphiphilic structures, so-called *aplosspans* (Figure 16), can span the bilayer and transport ions in a very similar manner [110, 111]. Even simpler are the fatty acid esters of oligo-(ethylene glycol) that provided a surprising channel-like activity in which the cations have to strip off, at least partially, their hydration layer in order to pass through the membrane [112].

The family of polyether macrocycles that incorporate two residues of 3,4,6-tri-O-benzoyl mannose was also described as ion transporters (Figure 16) [113]. These systems were studied by ^{23}Na NMR techniques, showing differences in flux rates that correlate quite well with the length of the polyoxyethylene chains. The compounds with lengths larger than 26 Å were active in lipid bilayers, but analogues larger than 38 Å were less active. This suggests that the higher flexibility was in detriment of the sodium transport. Therefore the active transporter system must be formed once these molecules span the membrane and facilitate the ion transport.

Simple compounds such as pure membrane lipids have also been reported to form the so-called "lipid channels" [114, 115]. Thus, the pore mechanism has been used to describe a variety of transport processes as, for example, the proton permeation in membranes composed of pure lipids with short alkyl chains (less than 20 carbons) [116–118]. In the proposed mechanism the permeation across this type of phospholipid bilayer takes place through temporary pores generated by thermal fluctuations. This permeation avoids the thermodynamic cost associated to high-energy barriers required in partitioning and diffusion of the proton into the hydrophobic layer [119]. This latter mechanism, commonly described as the solubility-diffusion mechanism, is the one that apparently uses larger ions (e.g., potassium) or neutral molecules to cross the membranes in lipid vesicles [116, 120, 121]. It is worth mentioning that simple molecules like Triton X-100 can also transport ions by means of a pore mechanism that cannot be easily explained by a simple model based on some type of transmembrane structure (Figure 16) [122, 123]. It is clear that there is a fine line between ion channel formation and general membrane disruption. Triton X-100 is apparently capable of straddling the line under some circumstances.

3 Ion Channels

Ion channels are perhaps the most relevant and efficient group of proteins responsible for the selective transport of sodium and potassium ions. These transporters facilitate the passive diffusion of analytes through membranes by lowering the high activation energy required for the desolvation of the ions prior passing across the lipid bilayer (Figure 17).

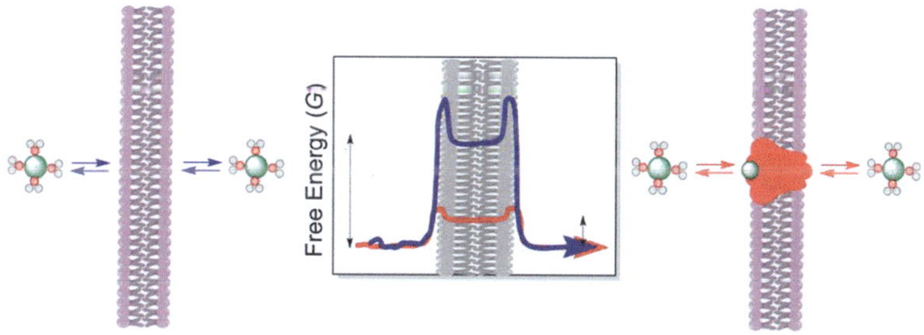

Figure 17 Migration of polar solutes (green sphere) across a lipid bilayer. In the central part the free energy comparison between the non-promoted translocation (blue, left) and channel assisted process (red, right) is depicted.

Their architecture is based on the presence of an inner cavity, normally filled with water molecules, that facilitates the flow of ions from one side of the membrane to the other, whereas it presents a hydrophobic outer surface that allows the insertion into the hydrocarbon layer of the membrane. Based on these simple design principles, researchers have developed a variety of strategies and structures to fabricate active synthetic membrane pores [124, 125].

3.1 Helical Structures

The pioneering studies by Montal in San Diego with different helical peptides that mimic the sequence of the putative transmembrane domain of natural ion channels established the molecular basis of transmembrane ion transport [126–130]. In his groundbreaking studies with short helical peptides that mimic the sequence of the transmembrane domains of the voltage-dependent sodium channel, he was able to identify the basic structural requirement to form ionic channels in lipid bilayers. The resulting structure was generated by the assembling of several helical peptides. Like the natural protein, it was cation-selective although it did not discriminate between Na^+ and K^+. In this respect, the *de novo* design of amphipathic-type peptides by DeGrado et al. allowed establishing the minimal system able to form transmembrane pores [131, 132].

The helical motif is present in many different natural channel-forming proteins with a barrel-stave type structure composed of a variable number of parallel-oriented helices. In addition, since the discovery of gramicidin D more than 70 years ago, designs based on a single helix have also attracted a great deal of attention [133, 134]. The gramicidins (Figure 18) are a group of short linear peptides with alternating *L*- and *D*-amino acids that adopt a helical structure, being the most abundant gramicidin A (gA).

It has been found that gramicidins mainly adopt two major folding motifs; the single-stranded helical dimer that has been proposed as the active channel form and the double-stranded helix [135, 136]. Interestingly, the preference of one form to the

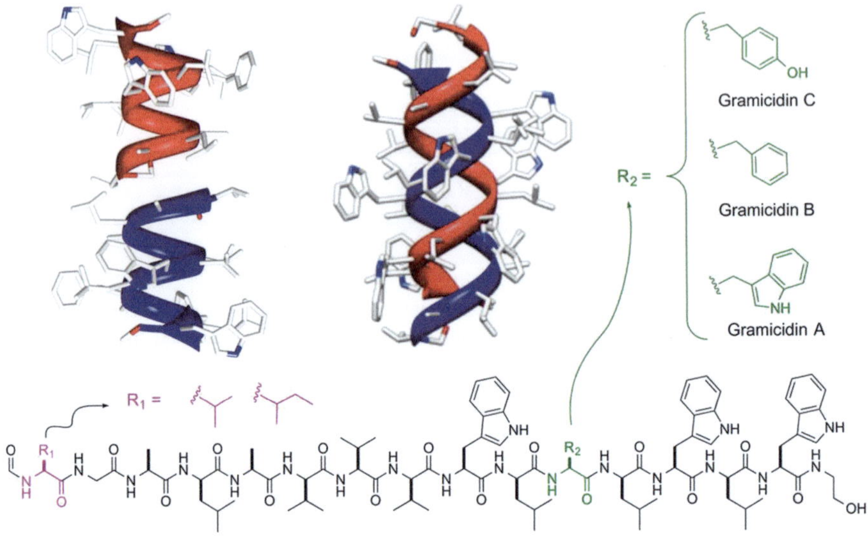

Figure 18 The natural mixture of gramicidins (gramicidin D) contains mostly gramicidin A (about 85 %) in which residue 11 is Trp (in green). In the other gramicidins (B and C) the Trp at position 11 is replaced by Phe or Tyr, respectively. Furthermore, Val at position 1 (in purple) can be replaced by Ile in about 5 to 20 % of molecules. There are two major folding motifs of gramicidins, i.e., a single-stranded helical dimer (PDB code 1GRM) (**top left**) and a double-stranded helix (**top right**) (PDB code 1BDW).

other depends on the media; the double-helical conformation is favored in organic solvents while the membrane-bound structure is the helical dimer [137]. Apparently the double-helical structure does not seem to be consistent with channel activity because its pore is too narrow for ions passing through. The active channel structure fits well with a head-to-head dimer of the initially denominated $\pi_{LD}{}^{6}$-helix [135], and nowadays called $\beta^{6.3}$-helix to better correlate the type of hydrogen-bonding pattern [138].

Due to its sequence in which *L*- and *D*-residues are alternated, all the amino acid side chains are projected outwards, in contact with neighboring lipid fatty chains. The interior of the channel, with a diameter size of ~4 Å, is decorated with the polar peptide backbone. The dimeric structure is held by fifteen intramolecular hydrogen-bonding interactions that stabilize the helical structure and six intermolecular hydrogen bonds that bring together the two-peptide subunits. As a result of the dynamics of the self-association process (monomer-dimer equilibrium), the channel presents the typical gated events, with a fast transition between the open and closed states. Hydrated ions have larger diameter than the gramicidin pore, so partial ion desolvation must take place prior to passing through the polypeptide orifice. The observed Eisenman I selectivity [75] for alkali metal ion is related with the increase of hydration energy from cesium to lithium. Presumably, the hydration shell is removed at the entrance of the channel mouth. Divalent cations are not transported as a consequence of their higher desolvation energies but, also, it has been suggested that these cations block the channel [139]. The ions, once in the pore, are coordinated to the amide bonds (π-orbital and oxygen lone pairs) and a couple of water molecules.

During the last decades several gramicidin modifications have been developed to better understand its transport and gating properties and to improve/modify the ion selectivity [140–145]. The strategies were mainly directed to introduce additional functional groups at the central part of the channel in order to increase the dimer stability by covalent strategies or assist the desolvation of the cations at the channel entrance. Although, the detailed analysis of all the wide variety of gA derivatives that have been prepared in the last few years is beyond the scope of this chapter, it is worth mentioning some clever synthetic designs. In this respect, the incorporation of an aza-[18]crown-6 ether linked to the gA through the side chains of residue twelve (Lys was used instead of Leu) and the C-terminus is a good example of a combination of the ion transport properties of gA with the ion selectivity properties of crown ethers [146]. The two-point attachment ensures the positioning of the crown ether precisely at the mouth of the channel. The resulting transporting channel showed a reversal in the potassium/cesium selectivity, transporting K^+ twice as fast as Cs^+. The partial desolvation of the potassium ion facilitated by the crown ether at the entrance of the channel resulted in an improved transport rate for this ion.

The incorporation of ionic groups at the channel mouth modifies the transport properties of gA [144, 147]. The anionic modified channel (phosphate or sulfonate groups attached at the C-terminus) presented increased single channel conductances as compared to the natural peptide, while a cationic group (trimethylamonium group) provided channels with reduced transporting properties. The preparation of asymmetric channels incorporating a cationic gA in one layer and the anionic derivative on the opposite layer provided heterodimeric assemblies with rectifying properties. Due to the difficulties of the ionic gA to cross the membrane the transport properties were stable for several hours.

The dimeric structure of the gA channel provides an additional place for channel modification. The central channel region allows the incorporation of new functional groups to increase the channel stabilization and so on. In this regard, the covalent link between both subunits through a cyclohexylaminoalcohol moiety modified the transport properties of the resulting transmembrane structure [148, 149]. The oxygen and nitrogen incorporated at the linker must be oriented towards the pore providing new properties. The covalently linked gA showed good conductivity properties for cesium or ammonium anions but only very poor activity for potassium and almost neglectable ones for sodium. This finding suggested that the incorporation of functional groups at the central part might create a filter area that improves the selectivity properties of the natural peptide. Some other pore modifications have also been reported, such as the use of δ-amino acids that contain a tetrahydrofuran moiety. Several units of the amino acid were linked in one of the gA subunits through a tartaric acid to provide a transmembrane channel with several coordination sites in the inner cavity. Unfortunately, the resulting channel did not provide any remarkable improved transport properties [150].

Inspired by the helical structure of the natural polymers, in the last few years synthetic oligomers called foldamers have also been designed [151, 152]. Some of these fold in a helical conformation generating an empty space, like the gramicidins, and they have been used in ion transport studies. The family of aromatic amide or hydrazide-made oligomers represent a remarkable example (Figure 19) [153]. These compounds adopt a variety of conformations and structures because of their rotation restriction around the C-N bond and the stabilization caused by aromatic ring substituents through the formation of hydrogen bonds with the NH groups. The conformational curved oligomers adopt helical structures whose internal diameter pore depends on the number of monomers per turn and also on the size and substituents of the aryl group.

Recently, a helical hydrazide oligomer with 1 nm internal diameter whose cavity can host alkyl saccharides has been described [154]. The incorporation of a phenylalanine tripeptide in the solvent-exposed surface of these foldamers facilitates insertion in the lipid bilayers [155]. These helical structures form channels that show NH_4^+/K^+ selectivity larger than the natural gA [156, 157]. This experimental observation was attributed to the differences in the ion desolvation step. In the proposed channel model the helical hydrazide foldamer was not able to span the lipid bilayer but the incorporation of a triphenylalanine peptide guides lateral pressure of the lipids to orient the helical foldamer parallel to its alkyl chains. Thus, peptide chains must be pushed by the lipids to contract the channel and increase its rigidity and hence prevent their collapse, facilitating the ion migration.

Cholic acid, as mentioned before, is an amphipathic steroid that has been used quite often as a building block in synthetic transport systems [158]. The first studies using this moiety comprised linear oligomers that formed nanopores with hydrophilic internal properties [159]. The transport experiments confirmed their capacity to translocate polar molecules across membranes although the channel mechanism

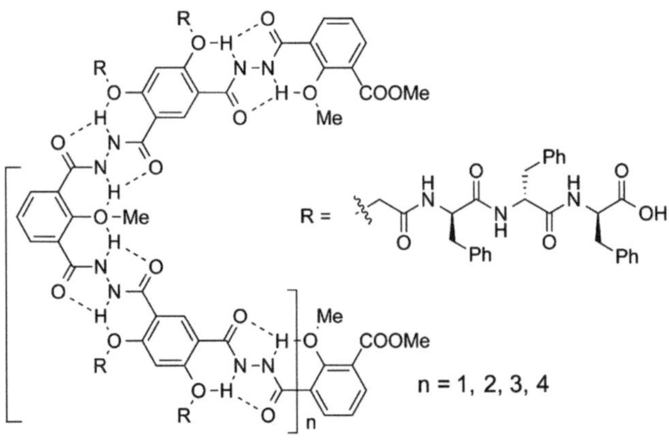

Figure 19 Helical aromatic hydrazide oligomers that form channels with better NH_4^+/K^+ selectivity than gramicidin A.

was not finally confirmed [160, 161]. Clearly, the foldamer studies suggest that it may be possible to design new helical systems in the future for transporting cations in a more specific way by incorporating appropriated functional groups into their cavity.

3.2 Barrel-Stave Methods

In addition to α-helical motifs described above, β-barrel is another protein architecture extensively found in natural transporters [162]. This structure is formed by the interaction of several peptide chains with a secondary β-sheet structure [163]. In this motif all hydrophilic side chains are pointing into the inner cavity of the barrel, while a layer of lipophilic amino acids forms the hydrophobic belt that favors the insertion into the lipid bilayer. β-Strands of these β-barrels are typically arranged in an antiparallel fashion. The barrels are classified depending on the number of strands that form the cylindrical structure. In general, they have wider pores than the α-helical proteins. For this reason, they are normally used to translocate relatively large molecules such as porins and other bacterial toxins [164]. While the design and synthesis of α-helical peptides and their aggregation properties have already been understood, the synthesis of peptide β-barrels have not been mastered yet. In this respect β-sheet aggregates have been suggested as being responsible of some intractable protein aggregation diseases, such as Alzheimer's, Huntington's, or Parkinson's disease as well as diabetes, among others [165–167]. A variety of different approaches has been developed to design soluble and structural stable β-sheet aggregates to understand better the β-sheet properties [168].

The existence of naturally occurring barrel staves has also been reported for small molecules, one of the best known examples is the transport-active structure of amphotericin B (AmB) [91]. The AmB is a polyenic macrolide antibiotic generally used in the treatment of systemic fungal infections (Figure 20). The selectivity against fungal cells is believed to derive from its association with biological membranes that contain ergosterol. Apparently, the macrocycle in the presence of the fungal sterol assembles into a transmembrane pore in the lipid bilayer. The barrel structure implies eight AmB molecules alternated with ergosterol to generate a half channel that dimerizes to form the active structure. The internal diameter of the channel lumen was estimated to be around 8 Å. Despite the fact that its biological action does not involve exclusively the transport of alkali metal ions, apparently it acts as an antimicrobial membrane-disrupting agent, as suggested by some recent studies in which the formation of the barrel-stave structure is not required for its antifungal activity [169, 170].

AmB channels transport monovalent ions, especially alkali metal ions, being Li^+ the most permeable one for AmB channels containing ergosterol [171]. The potassium, sodium, and chloride selectivity changes with the amphotericin concentration and type. The changes in selectivity observed with antibiotic concentration and time suggest the existence of different types of channels.

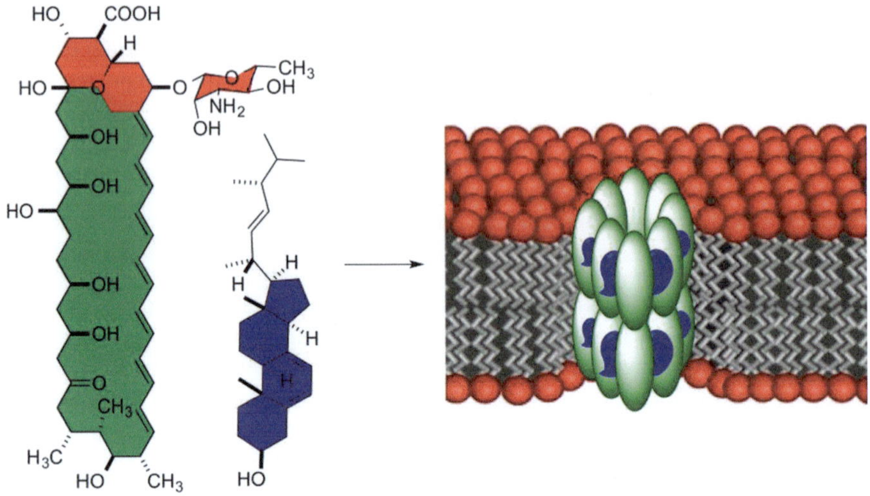

Figure 20 Structure of amphotericin B (in green) and ergosterol (in blue) associated for the formation of a barrel-stave structure (**right**) that disrupts the ionic balance of the cells.

The development of bio-inspired channels for ion transport based on this kind of barrel-stave approach has been carried out. One of the most remarkable examples are the transporter systems developed by Fyles' group, some of which were already mentioned in the previous section [19]. In his designs the entropic penalty to pay for the formation of the supramolecular channel was avoided by using rigid rod-shaped molecules that could span the membrane bilayer and self-assemble into the barrel structure. The previously mentioned use of polycarboxylate crown ethers made by a variable number of (*R*,*R*)-tartaric acid subunits represent one of the designs created to mimic an AmB channel model [64]. Macrocyclic tetraesters of maleic acid were attached to the carboxylates to play the role of molecular walls of the channel displayed by the polyene macrocycle of the natural antimicrobial.

Oligocholates of the type shown in Figure 21 with up to eleven moieties present membrane-active properties. The authors proposed a barrel-type channel model based on the formation of bundles in one leaflet that dimerize to form the transmembrane hydrophilic pore [172, 173]. The transport measured by ^{23}Na NMR experiments suggested different aggregation states, dimeric to monomeric, depending on the type of spacer used in the central part of the oligolysine core [174]. The change from unimolecular (rigid spacer) to supramolecular barrel-stave channels must be related to the reduction of the thermodynamic stability.

Another interesting example of this type of designs is the use of rigid *p-septiphenyl* and *optiphenyl* rods by Matile (Figure 21) [175]. In his initial studies, the oligophenyl rods were attached through the phenol group of each ring to a variety of diols that were able to transport protons by a two-step hop-and-turn mechanism [176–178]. The experiments carried out in vesicles suggested that the proton selectivity compared to that for potassium ion was more than 15 times larger. Other simple rigid-rod components were designed to study the effect of the

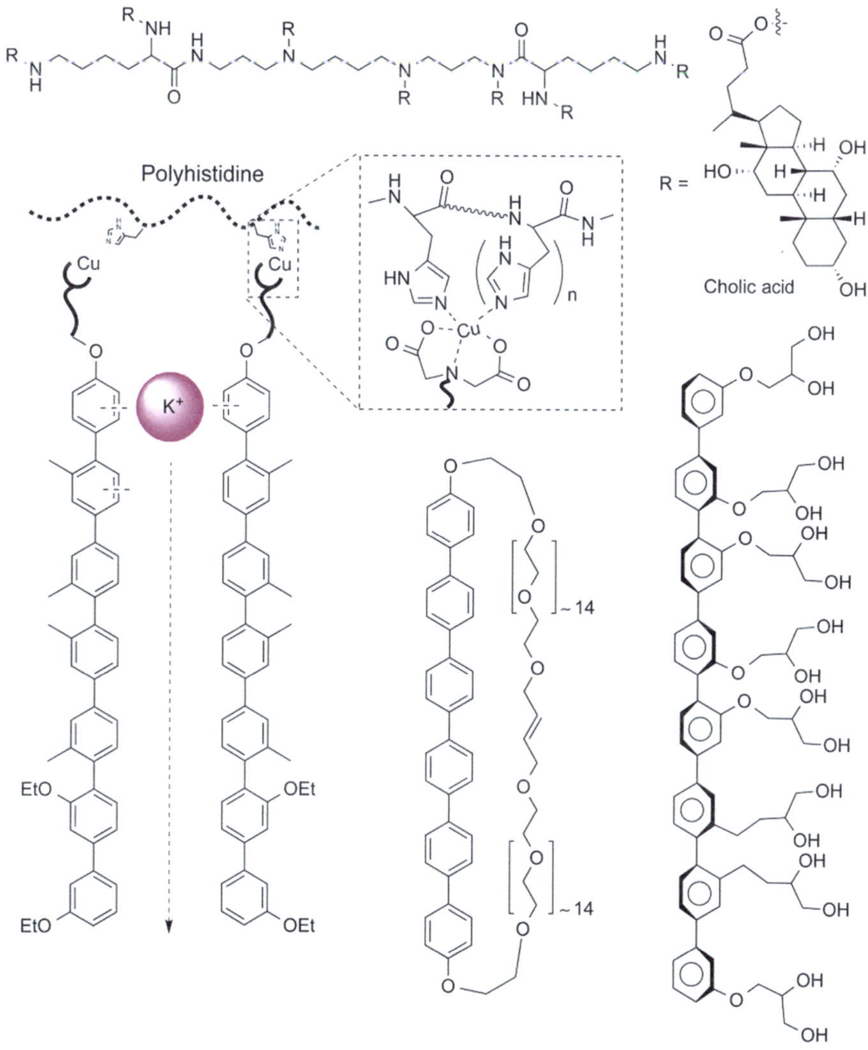

Figure 21 **Top**: Example of cholic acid oligomers used in ion channel formation whose proposed mechanism is the barrel-stave structure. **Bottom**: Examples of channel structures based on *p-septiphenyl* rods that promote the transport of K⁺ across the membranes. On the right, a polyhydroxylatedoctiphenyl rod precursor of ion channels is depicted.

cation-π interaction on the selectivity of ion transport [179]. In this case the oligophenyl rods were modified to increase its solubility in water with an iminodiacetate moiety that was exposed to the aqueous interface. Additionally, this group coordinates metals to facilitate the assembling process and the formation of the barrel structure. Thus, addition of a polyhistidine and Cu(II) induced the aggregation into flexible arene arrays of the rods in the lipid bilayer as confirmed by circular dichroism spectroscopy. The ion transport activity of the resulting ligand-receptor complex was carried out with fluoroscopic micelles and showed a value that was comparable to that of AmB.

These studies showed a biomimetic Eisenman IV topology ($K^+ > Rb^+ > Cs^+ > Na^+ > Li^+$) and biomimetic blockage with tetraethylammonium cations similar to the one found in natural ion channel proteins [75]. The nature of the substituents near to one end of the channel plays also an important role in the transport properties as a consequence of modifying the septiphenyl packing. The authors suggest a sort of ion hopping mechanism for the transmembrane ion translocation process. Although recent studies with natural proteins have ruled out the importance of cation-π interactions in the filter area of the channel in the ion selectivity, it provides a chemically attractive concept for the design of novel transport processes [179].

Based also on polybisphenols, but incorporating short peptides on each phenol hydroxyl group (Blue, Figure 22), Matile's group has prepared a more robust supramolecular channel for a wide range of applications [180]. The short peptides with an odd number of residues were attached in such a way that allows their interdigitation in an antiparallel fashion to form tetrameric aggregates. The pore diameter depends on the number of staves that form the channel but also on the number and type of amino acids of the peptides chains [181, 182]. Because of this design, one out of two amino acids, the ones at the even position, orientate their side chains towards the lumen of the channel allowing a simple tuning of their pore properties. The remaining residues have their side chains exposed towards the external surface allowing their incorporation or formation in different media.

This special design is also interesting for different applications, such as molecular transport systems [183, 184], biosensors [185], catalytic systems [174], etc. The β-barrels can recognize a variety of guests depending on their electronic and/or structural properties and the residues placed at even positions [181]. The barrel staves derived from the rods containing tripeptides (LHL) were more stable than those formed by pentapeptides (LHLHL) as can be inferred from the open-state lifetimes on planar lipid bilayers [186]. Interestingly, the first one presents an excellent ion channel activity while the last one acts as an esterase. The ion selectivity can be modulated depending on the peptide sequence and conditions. For example, the β-barrel that contains the pentapeptide LRLHL, in which side chains of Arg and His are projected into the channel cavity, reverts the potassium/chloride selectivity with the pH. While at acidic pH (<5.5) the channel is anion-selective, at higher pH it starts acting as potassium-selective channel [187].

On the other hand, the substitution of the Arg residue by Lys provided channels that give a high-conductance with a long life-time in planar bilayers at an applied voltage of 25 mV and ohmic behavior characteristic for symmetric pores [174]. The mean life-time of a single pore of this kind decreased upon increasing voltage. The life-time of high-conductance observed for this β-barrel at low voltage was beyond the transporting properties of gA. The observed conductance allowed estimating a pore diameter (d_{Hille}) of 12 Å using Hille's equation [2]. The versatility of this synthetic channel allowed the preparation of channels whose transport properties can be modulated (gated or blocked) by external signals [188, 189]. For that purpose the

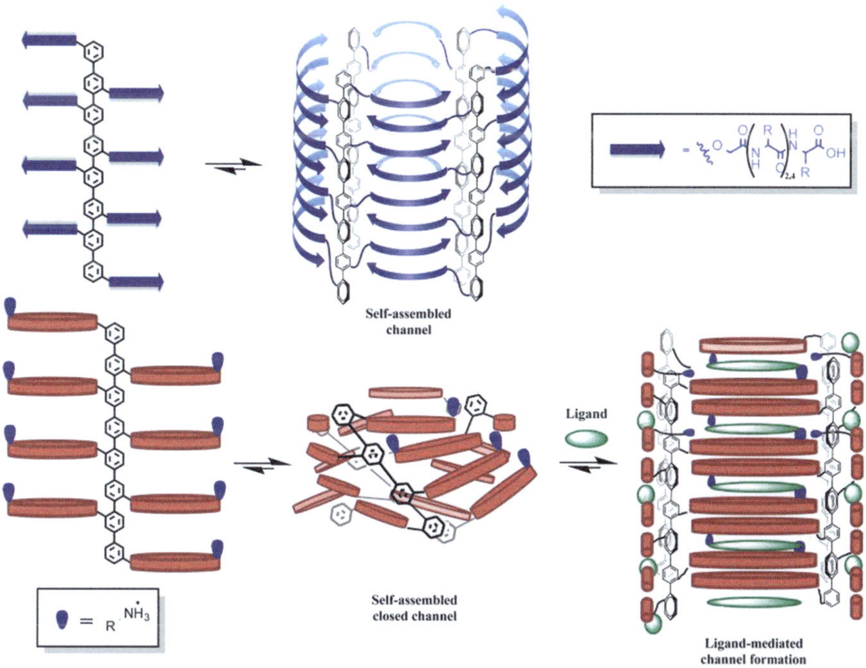

Figure 22 Barrel stave forming oligobisphenol models used in transmembrane transporting channels. **Top**: β-barrel formed by polybisphenols that incorporate short peptides (in blue). **Bottom**: model of ligand-gated barrel-stave ion channels prepared from the octaphenylene that contains NDI moieties (in red). This conjugate forms in the membrane a π-helix structure (closed state) that switches to the β-barrel form upon addition of an electron-rich aromatic compound (in green).

authors used pentapeptides (LHRHL) with one cationic residue exposed towards the external surface. Therefore, the resulting channel cannot form active pores in lipid membranes. Interaction with hydrophobic anionic ligands form lipophilic structures that dissolve in the membrane to generate transmembrane channels with efficient transport properties. If the ligand is added as an ester, then esterase enzymes are required to hydrolyze the ester and activate the channel. Finally the active channel was blocked by addition of polyglutamate that interacts strongly with the cationic cavity of the channel.

Push-pull systems were also designed using different substituents at the edge of the oligophenylen rods, allowing the preparation of asymmetric barrels [190, 191]. In this design, the authors incorporate a donor (methoxy) and an acceptor (sulfone) group in each side of the push-pull rod creating a macrodipole. The resulting push-pull β-barrel recognized polarized bilayer membranes at nanomolar concentration and amplified the selectivity and activity. Another clever modification was the incorporation of a short hydrophilic peptide chain at one of the ends of the rod [192]. The resulting nanopore can still operate at low concentration showing high selectivity because of the increased solubility of the barrel components. The tetralysine anchor was designed as an extension from the transmembrane β-barrel into the membrane water interface.

Moreover, the hydrophilic anchors are expected to enforce vectorial partitioning, and promote transmembrane orientation of the pore. This approach, in which the pendant is interacting with the membrane environment and interface, might open the opportunity to develop selective antimicrobial properties to these β-barrels. Substitution of the phenolic short peptides by naphthalenediimide units (NDIs) in the rigid rod template provided a nice example of ligand-gated barrel-stave channel (Figure 22) [184, 193].

The NDIs-octaphenylene conjugate allowed the construction of helical, π-acidic π-stack architectures through the combination of hydrogen bonds between the NDI chains and the aromatic (π-stacking) interactions between the NDIs themselves. This π-helix structure corresponds to the closed form of the channel that was designed to open, producing the formation of a β-barrel, by addition of electron-rich aromatic compounds in response to a charge-transfer complex formation. The transition from the helix to the barrel occurs because of the distance between the π-stacks, which was initially shorter than in the previous β-sheets. Thus, to satisfy NDI electronic properties, the structure collapses into the helical structure closing the aggregate pore. Although the barrel interior has cationic properties, this ligand-gated ion channel presents low anion selectivity, with a permeability ratio P_{Cl^-}/P_{K^+} of only 1.4 [102]. The transport studies showed a selectivity sequence of $Cl^- > K^+ > H^+$ as could be inferred from the experiments in which the channel was sensitive to the known proton carrier carbonylcyanide-*p*-trifluoromethoxyphenyl-hydrazone (FCCP) but not to valinomycin [194].

The construction of nanometer-size toroidal structures has also been reported using rigid/flexible amphiphilic small molecules that are composed of a rigid aromatic fragment (olygophenylens) and flexible coil segments (Figure 21) [195]. These supramolecular entities have also been used as ion channel models [196]. In these studies, a hexaphenylen rod was attached to a polyethyleneoxide chain at both ends of the oligoarene moiety. The resulting macrocycle assembled into barrel-like nanostructures directed by the parallel packing of the hexa-*p*-phenyl rods, which are surrounded both in their interior and exterior by the hydrophilic flexible segments. The incorporation of the resulting toroidal structure in fluorogenic vesicles produced a significant change in the fluorescence intensity, thereby indicating membrane activity. The channel structure might interact with hydrophilic head groups of the phospholipids and also with the hydrophilic flexible coil segment facilitating the penetration in the membrane and the formation of the membrane pore.

Lee's group also described an alternative strategy for the formation of transmembrane channels based on a different toroidal structure [197]. They used a more flexible curved component by designing a T-shaped hybrid building block (Figure 23) made of a β-sheet-forming peptide and an oligoethylenoxide dendron attached to the side chain of one of the central amino acids. The bulky dendron, because of the steric repulsion, induces curvature at the interface between building blocks. The resulting toroids formed by the antiparallel β-sheet organization of the undecapeptide have an interior decorated with the tryptophan side chain. Replacement of the oligoethyl-

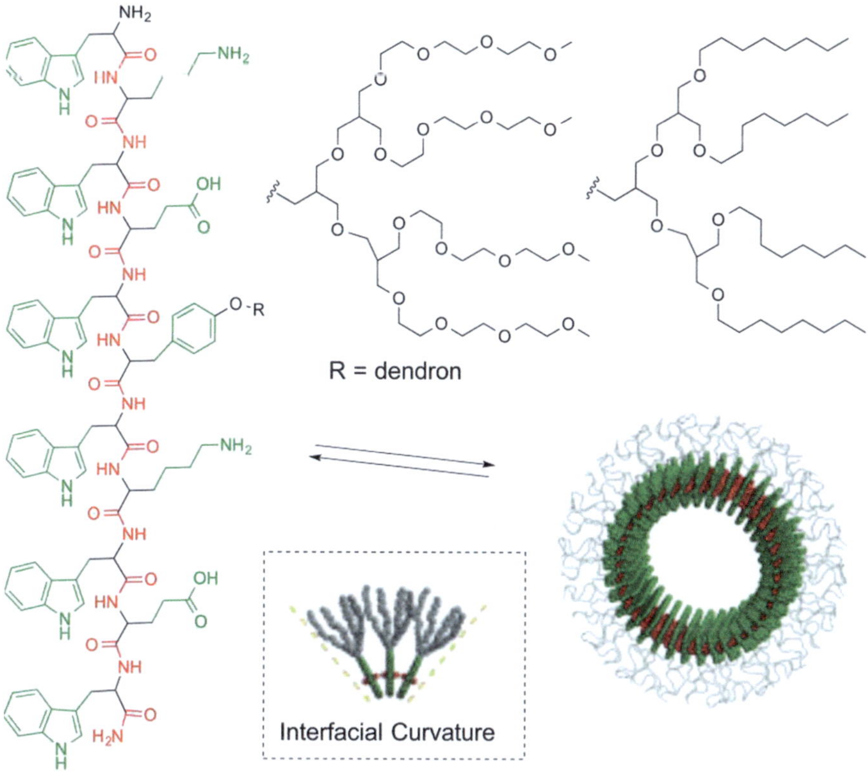

Figure 23 β-sheet forming T-shaped peptide that forms a toroidal structure thanks to the interfacial curvature generated by the β-sheet aggregate. The hydrophobic character of the dendron provided the lipophilic properties to solubilize it in the phospholipid membranes. The lower part is reproduced from [197] with permission from Wiley & Sons, © copyright 2011.

enoxide dendron by a hydrophobic one, provided a barrel structure that can be incorporated in the lipid membranes. Black lipid membrane experiments gave strong evidence of a single-ion channel despite their hydrophobic character. Hille's equation [2] gave an estimated internal diameter of 0.5 nm, smaller than the inner diameter (3 nm) of hydrophilic nanorings measured by TEM. This diameter contraction might derive from the further stabilization of the β-sheet structure induced by the membrane but also from the hydrophobic character of the nanoring pore.

A G-quadruplex is a supramolecular complex formed by stacks of four guanosine bases associated through Hoogsteen type hydrogen-bonding interactions (Figure 24). The four-guanines complex has a square planar structure called a G-quartet, and two or more guanine tetrads stack then on top of each other to form a G-quadruplex [198–200]. Generally, the quadruplexes are stabilized by alkali metal cations forming a sandwich type structure in which the cation is filling the central cavity between each pair of G-quartets. These structures are also formed in G-rich nucleotides (RNA and DNA) like telomeres, RNA packing sites or gene

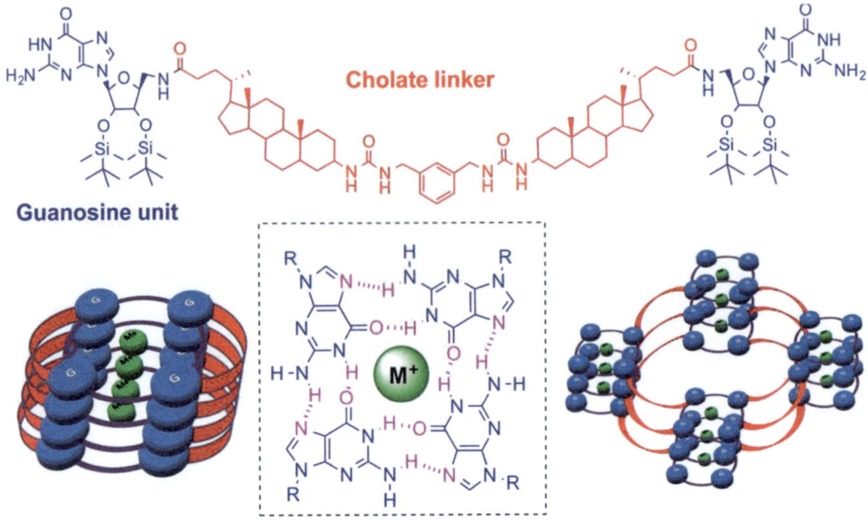

Figure 24 **Top**: Structure of a bis-guanosine derivative linked with a channel-forming cholate. **Bottom**: Proposed ion channel model based on stacking of G-quartets: at the left a single channel and on the right a cartoon model of tetrameric G-quartets stacks of a macropore is given. **Middle**: G-quartet structure generated by Hoogsteen-type hydrogen bonds (in purple).

promoter regions among others [201–203]. The role of the G-quadruplexes *in vivo* is still an area of intense debate; nevertheless it is quite obvious that these DNA and RNA structures are very relevant [204–207].

Nowadays intramolecular G-quadruplexes formed by human telomeres are attractive anticancer targets [208, 209]. The G-quadruplexes are implicated in cancer therapies as potential targets for new small molecules that would bind the quadruplexes interfering in telomerase activity and thereby reducing tumor growth. Other molecules similar to guanosine such as isoguanosine or folic acid also can form this kind of structures [210–212]. The incorporation of nonpolar protecting groups on the sugar hydroxyl groups of guanosine or analogues provided hydrophobic G-quartets that can dissolve in a membrane-like environment. The entrapped cation of G-quadruplexes formed by dimeric guanosine derivatives can stack on top of each other in the membrane lipids forming a transmembrane tubular structure that can transport alkali metal ions, as we will see in the next section [213]. Alkali metal ions (Na^+, K^+) can be translocated through films made of G-quadruplex polymers [214].

These supramolecular guanosine quartets represent another example of barrel-stave architecture used in ion transport [215]. In this strategy, and inspired by previous studies of Kobuke's group, guanosine dimers were linked through semi-rigid bis-cholic acid derivatives [98]. Using these ditopic guanosine-sterol derivatives they found two types of channels, in which the smaller conductance events may correspond to the ions moving through the central channel of the G-quadruplex. The estimated channel diameter, using Hille's equation [2], of around 2.6 Å fits quite well with the dimensions of central cavity of the quartets. The larger transport events correspond to larger channels for which the authors suggest the formation of very large metallopores in which the stacks of G-quadruplexes form columns that arrange into tetrameric or even larger aggregates through the bis-cholic linkers (Figure 24). Interestingly the channels are very stable, with seconds of "open" times, and cation-selective ($P_{K^+} > P_{Cl^-}$) [102]. Two different linkers were studied that differ in the connectors (carbamante or urea), providing the last linker channels of larger life times. These results support the proposed macrochannel structure that must be stabilized by interconnector hydrogen bonds [216].

In the last few years, DNA molecules have been also shown to be efficient platforms for building a variety of nanoparticles with different shapes [217–220]. In fact, DNA nanotubes have been prepared by different strategies, including barrel-stave or ring stacking [221–226]. In this respect, DNA-based barrel-stave synthetic channels have been described very recently (Figure 25) [227]. Despite the fact that

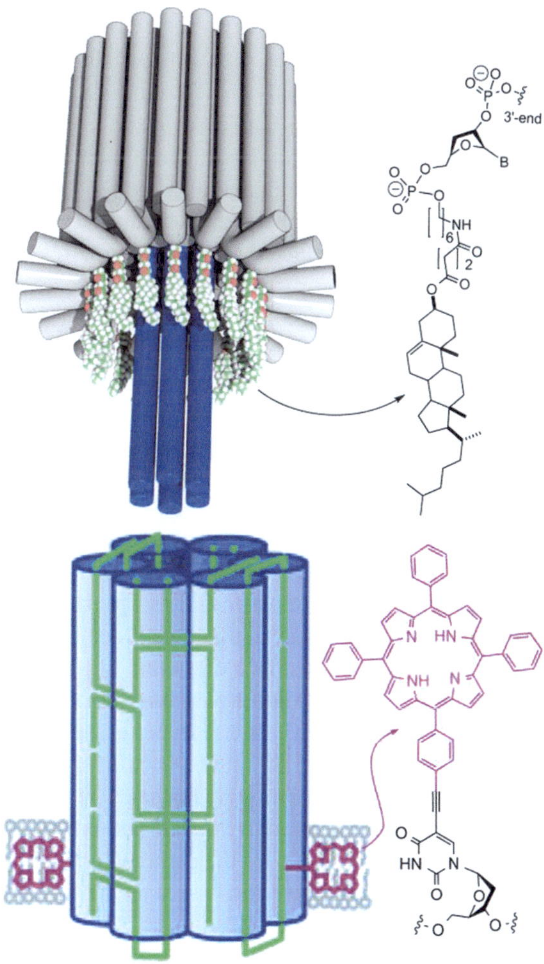

Figure 25 The first design of DNA-based nanotubes used a mushroom-shaped structure in which six double helices penetrate and span the lipid membrane acting as the stem (**top**). The second model used DNA modified with porphyrin moieties to facilitate the insertion in the lipid bilayer (**bottom**). The lower part is partially reproduced from [229] by permission of Wiley & Sons, © copyright 2013.

intuitively the polyanionic skeleton of this biomolecule does not appear to be suitable to insert in lipid bilayers, the incorporation of hydrophobic tails is enough to favor it. Inspired by the shape of naturally occurring α-hemolysin (Figure 2) (mushroom-shaped barrel-stave structure), Simmel's group designed a DNA channel formed by 54 double-helical DNA domains packed on a honeycomb structure [227]. The resulting transporting system has two different regions; the six double helices domain that penetrates and spans the lipid membrane acting as the stem and the upper cap that is responsible of its adhesion to the membrane.

Actually, the adhesion to the membrane is driven by the twenty-six cholesterol moieties that are conjugated to the barrel at the bottom part of the barrel-shaped cap, close to the membrane-spanning domain. The internal diameter of the resulting billycock-shaped structure is 2 nm and allows the transport of potassium ions up to values close to 1 nS (nanoSiemens). Remarkably, the pore size allowed also the translocation of short DNA hairpins, detected by the changes on potassium transport recordings. This property might have application in the development of DNA sequencing methods [228].

Almost in parallel, Howorka and collaborators designed a simpler transmembrane channel based on only six DNA double helices (Figure 25) [229]. To facilitate the incorporation of the nanotube in the lipid membrane initially they prepared DNA barrels using six ethyl-phosphorothioate-modified scaffold strands that form the hydrophobic bell around the assembly [224]. Later, in order to position the barrel in a directional manner, they also used oligonucleotides containing two tetraphenylporphyrins linked to deoxyuridine residues [230, 231]. The tetraphenylporphyrin was also useful in terms of fluorescence detection in the key inclusion step in the membrane. The DNA nanopore was prepared by typical heating and cooling procedures using an equimolar mixture of four regular DNA strands together with two strands containing the porphyrin. The incorporation of the assembly into the lipid membrane was visualized by fluorescence microscopy. The resulting DNA barrel was found to successfully span the membrane and it showed steady conductance events for K^+ with large transport events. The porphyrin-modified barrels were around 40 % less active than the original ethyl-phosphorothioate derivatives. The differences might derive from the way in which the nanotube is inserted into the membrane, such as the insertion at the initial part (porphyrin-conjuganted channel) or in the central part. This difference might modify the inner tube cross-section because of the differences in the lateral membrane pressure.

In concluding this section, we would like to mention the recent work published by Fyles et al. using a bolaamphiphile triazolic compound that assembles forming hollow columns that mimic the properties of gA (Figure 26) [232]. However, the proposed mechanism of action does not involve the same vertical disposition structure in the membrane than gA. Instead, the authors propose a horizontal pile of molecules, which forms an octagonal column in which eight different amphiphilic molecules expose to the pore the carbonyl and amino groups. This hydrophilic cavity has been demonstrated to possess the ability of forming water wires that easily facilitate protons diffusion. In addition, they have a cation/anion transport selectivity (K^+/Cl^-) larger than 2.

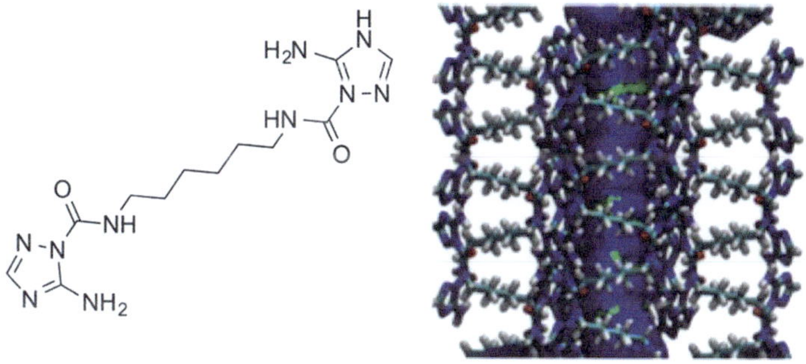

Figure 26 Triazol bolaamphiphile (**left**) and model of supramolecular channel filled with water molecules (**right**). The model at the right is reproduced from [232] by permission of Nature Publishing Group, © copyright 2014.

3.3 *Cyclic Structures*

One of the most promising strategies for the construction of synthetic ion channels is the stacking of cyclic molecules, despite the fact that this structural motif is not present in Nature *per se*. One of the advantages of the piling-up strategy is the precise control of the internal diameter of the nanotube. The diameter depends solely on the size of the stacking macrocycle used in the self-assembling process. Different types of cyclic compounds have been applied to the nanotube formation providing nanotubes with different properties and also extending and diversifying the strategies of nanotube preparation. Depending on the specific characteristics of the desired channel, the stacking between the macromolecules has been achieved by covalent bonding strategies that link the rings in an ordered and irreversible manner or by means of weak forces (supramolecular means). The later approach is more versatile and relevant nowadays because the process is self-regulated. The monomers contain all the information to recognize other molecules to assemble in a reversible manner, and, finally, the stacking can be controlled by media conditions or chemical signals [124].

3.3.1 Ring Stacking Induced by Covalent Bonds

Crown ethers have been shown to interact with different types of ions depending mainly on their ring size and number of donor atoms [36]. As mentioned before, they have been studied as ion transporting entities using several different designs and strategies. The high ion affinity and selectivity observed with crown ethers have led to large attempts of designing original strategies to align the macrocycles in the membrane environment. In this regard, the first approach to align crown ethers perpendicularly to the membrane was described by Nolte et al. using a polymeric template (Figure 27) [233, 234]. The polymer was an isocyanide structure that formed very regular and rigid helices with four repeating units per turn. The polymer contained benzo-18-crown-6 ether, as side chains, attached to the iminium group. As a consequence of the rigidity of the polymeric backbone, the macrorings are spatially disposed on top of each other at a distance of 4 Å. In this way, four channels are oriented parallel to the polymer axis. Transport experiments were carried out in bilayers of dihexadecyl phosphate vesicles, showing the transmembranous transport of cobalt ions. The determination of ion transport rates provided an Arrhenius activation energy parameter similar to gramicidin, which confirmed that

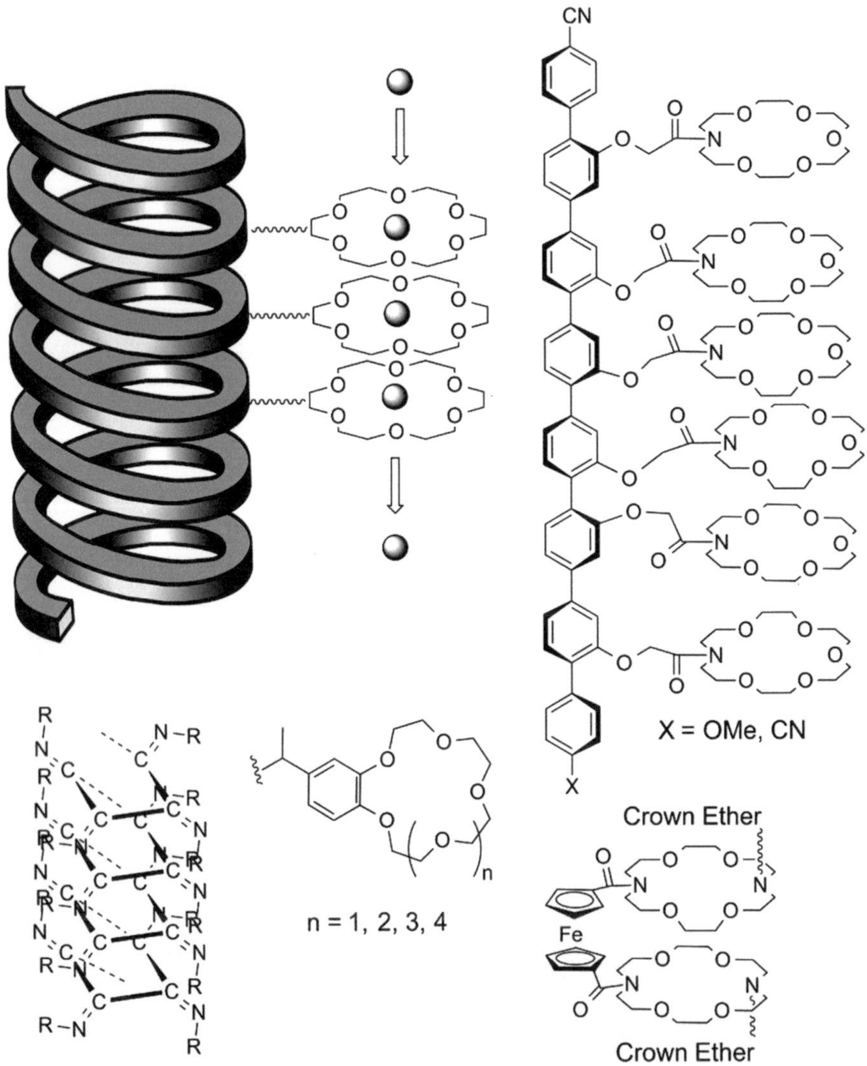

Figure 27 Schematic representation of covalently linked crown ethers to rigid templates, such as the α-helix (**top left**), rigid rods (**top right**) or isocyanate helical polymer (**bottom**), aligned on top of each other to form transmembrane channels.

the transport was mediated by a pore mechanism. The alkali metal transport properties were studied in liquid-liquid or solid-liquid extraction experiments, using polymers with benzo-crown ethers of different sizes [235]. These experiments confirmed their interaction with these ions and their ability to incorporate them to the lipophilic media.

A more structurally defined model was prepared by Voyer et al., who used a modified helical peptide structure that incorporated synthetic amino acids with different types of crown ethers (Figure 27) [236–239]. The crown ether residues were placed at strategic positions (i, i+4, i+7, i+11...) of the polypeptide chains in concordance with the helical structure to ensure their axial arrangement on top of each other. From these studies, they established the peptide length required to span the membrane bilayer, which was found to be 21 residues. The shorter ones (7 and 14 residues) were not able to actively transport ions. The crown ether size was also tested, and peptides bearing residues with 21-crown-7, 18-crown-6, 15-crown-5, and 13-crown-4 were prepared. As expected, the smallest crown ether was selective for Li^+ compared to other alkali metal ions, although it was not very active. Transport of the other alkali metal ions was almost negligible.

The 15-crown-5 presented some selectivity for Na^+ compared with K^+, although unfortunately no selectivity was observed between Li^+ and Na^+. Actually, all channels transported Li^+. On the other hand, Cs^+ was easily transported by 21-crown-7, but the transport decreased dramatically with the 18-crown-6 and was almost not observed for smaller crowns, in which a nonspecific hopping mechanism might be responsible for the observed transport. In comparison, K^+ is translocated through both 21- and 18-crown ethers very efficiently, being the last one selective for this ion. They also prepared peptide-helix hybrids with a different number of subunits and distances between crown ethers to establish the importance of the ring separation to act as ion relays [236–239]. These studies indicated that a distance of 17 Å is too long to achieve efficient transport. Interestingly, despite the fact that the optimal distance between two relays for sodium cation transport is 6 Å, a distance of 11 Å was tolerated. It should be remarked that these observations are in agreement with the results obtained with gA [240].

Similarly, Matile's group used the previously mentioned oligophenylen strategy to precisely attach several 18-azacrown-6 ethers and align them on top of each other (Figure 27) [241, 242]. In the initial studies, they used rigid push-pull rods, finding that the non-symmetric rod depolarized more efficiently the polarized bilayer membrane than the nearly identical symmetric derivatives. They further analyzed the transport properties through the incorporation of substituents with different electronic properties (cationic and neutral, with and without axial dipole). These studies were carried out in isoelectric, anionic, and polarized bilayer membranes under comparable conditions. The planar lipid bilayer studies reported that the higher activity observed corresponds to the asymmetric rod in polarized membranes as a consequence of their single channel rectification properties. The most relevant results were obtained from the cationic derivatives. Two model channels were prepared, one with an ammonium group near the negative end of the rod dipole, mimicking the bee toxin melittin, and the regioisomer in which the cationic group is near the positive end, simulating the structure of the natural antibiotic magainin. The first one showed higher activity in polarized membranes due to the constructive alignment of the molecular dipole with the potential. Such behavior might explain the low affinity of magainins for mammalian cells.

Simpler channels were also designed using the aza-crown ethers but linked linearly using the flexible alkyl chains mentioned previously [89, 243]. The original design contained three lariat aza-18-crown-6 ethers aiming to create a transmembrane channel by stacking co-facially all three aza-crown ethers (Figure 14). Later, studies with other derivatives suggest that the central ring must lay perpendicularly to the other two, being this central macrocycle essential for high activity. Addition of a larger number of crown ethers to form a macropentacycle provided channels three times more active than the original linear tricycle [88]. Interestingly, these hydraphiles are toxic to both Gram-positive and Gram-negative bacteria, and, apparently, this activity appears to be correlated with their ion channel activity [90]. The linear tricyclic compound was able to depolarize the membrane on cells as well.

The ferrocene moiety has also been applied to link two diaza-18-crown-6 ethers to develop a redox-active ion channel (Figure 27) [244, 245]. The ferrocene bridge provides rigidity within the channel structure and is necessarily placed at the central part of the bilayer. A second 18-crown-6 ether was attached to the other amino group through an alkyl chain, thus four-stacked rings form the transmembrane channel. Transport properties were confirmed by cation flux in unilamellar vesicles by ^{23}Na NMR spectroscopy and also with planar lipid bilayer measurements. Additionally, 'inside-out' clamp experiments with living cells tested K^+ transport across the membrane and showed that the ion transport was promoted by the artificial channel. Moreover, a reduction of the transmembrane current was observed as the interior of the cell changed from negative to positive potential. A channel derivative containing diaza-15-crown-5 ether instead of the 18-crown-6 at the central ring, in an attempt to create a selective filter, showed a 2–3-fold selective permeability for Na^+ over K^+, assuming that the channels were impermeable to chloride. Unfortunately, this selectivity was not consistent and some experiments reported on cation selectivity. Finally, the introduction of an oxidizing agent to both recording chambers cancelled the ion channel activity, so this ferrocene channels could be considered as an example of a switchable transporter, as far as the electronic nature of the media controls its ability to permeate ions on lipid bilayers.

3.3.2 Supramolecular Stacking

An alternative approach for tubular structure formation is piling macrocyclic compounds using non-covalent interactions to facilitate their appropriate alignment [125, 246]. In this strategy, simple molecules that adopt a flat conformation are held together through a variety of weak interactions such as hydrogen bonds, π-π stacking, and so on. Despite the fact that each individual contribution to the overall energy balance is very weak, the supramolecular approach combines, in general, a large number of these weak interactions to generate a total enthalpic contribution large enough to overcome the unfavorable entropic penalty derived from the ordered disposition. In addition, the dielectric properties of the lipid media reinforce the strength of some non-covalent interactions and contribute also to lipophilic interactions.

3.1.1.1 Nanotubes Based on Cyclic Peptides

One of the pioneer works in this field was the peptide nanotubes (Figure 28) [16]. These are biomaterial structures with tubular shape formed by β-sheet-type interactions between several cyclic peptides (CPs) that adopt a flat-shaped conformation. To ensure the required flat conformation, the original nanotubes were prepared using cyclic peptides with an even number of residues with alternating α-carbon chirality. These peptides arrange all the residue side chains projected in the ring/nanotube external surface, leaving an internal hollow cavity. In this pioneer work the incorporation of glutamic acid and glutamine residues allowed the use of a pH-triggered assembling process. Under appropriate conditions the peptides are soluble through the ionization (protonation/deprotonation) of their polar side chains, but the pH change reduced the peptide solubility inducing the assembling process and nanotube formation.

One of the advantages of this process is the precise control of the internal nanotube pore. In addition, it is possible to alter the nanotube diameter by modifying the ring size of the CP, simply through the increase or reduction of the number of amino acids. Therefore, changing the eight-residue CP, which formed nanotubes with an estimated internal diameter of 7.5 Å, with a dodecapeptide provided nanotubes with 13 Å diameter [247, 248].

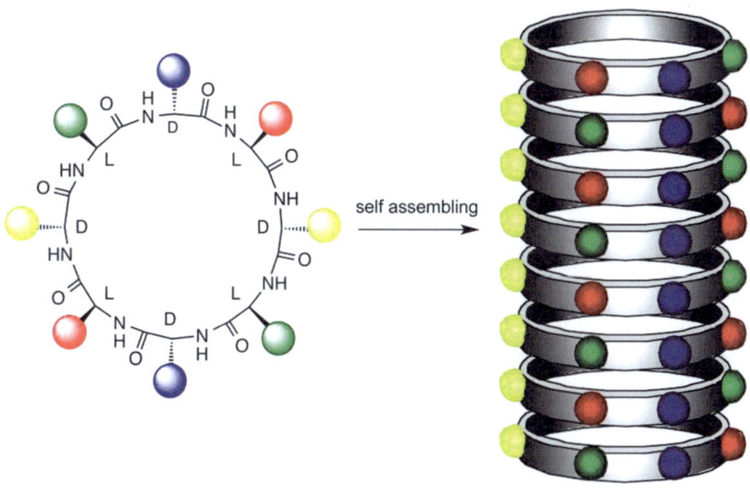

Figure 28 Scheme of a peptide nanotube formed by the self-assembly of cyclic peptides.

The radial orientation of all the side chains dresses the nanotubes, thus, depending on the selected sequence, it is possible to prepare ensembles with different external properties, such as the previously mentioned soluble in acidic or basic media nanotubes. The use of hydrophobic amino acids allowed the preparation of nanotubes soluble in lipid media (Figure 29) [249–251]. Thus, the cyclic octapeptide *c*-[*L*-Gln-(*D*-Leu-*L*-Trp)$_3$-*D*-Leu-] provided ion channels with a conductivity for alkali metal ions three times faster than the native gA. In general, most of these peptide channels do not transport divalent ions, such as Ca^{2+}, or anions like Cl^-. The rates of transport for the alkali metal ions follow their diffusion rates in water ($Cs^+ > K^+ > Na^+ > Li^+$, Eisenman I sequence) [75], in which the selectivity is related to the hydration energy. A characteristic feature of these channels are the different conductance levels observed in their current recordings. These levels were assigned to nanotubes with different numbers of stacked peptides. The length of the nanotube is restricted to multiples of five, considering an inter-subunit distance of 4.85 Å [252].

The Nernst–Planck equation [103, 104] allowed establishing the relationship between conductance and channel length, being composed the most common channels by six stacked subunits. This distance matches with the thickness of the hydrophobic part of the membrane. The nanotube formation was confirmed by infrared spectroscopy through the observation of the typical absorption bands of the antiparallel β-sheet interaction [249, 253–255]. This technique was also used to confirm the nanotube orientation on the membrane [256]. Thus, the parallel polarized grazing angle IR spectrum and analysis of the polarized ATR-IR indicated a nanotube tilt angle relative to the membrane plane of 7–15°, similar to the observed angle for the alkyl groups of the phospholipid bilayers. The relationship between ion transport properties and channel diameter of these systems was also studied with a series of CPs with different ring size (six, eight, ten and twelve residues). It was found that the channel dwell opening time (τ) was reduced upon increasing the nanotube diameter [250, 257, 258]. Interestingly, the channels made of cyclic decamers can transport hydrophilic molecules such as glucose or glutamic acid while the octamer was unable. This experiment confirmed that the transport was carried out by the supramolecular nanotube that is able to discriminate between different substrates according to their size.

The different microenvironmental properties of the CPs, which are at the edge of the nanotube, at the membrane interface where the phospholipid heads are, or the rest of the nanotubes, allowed the preparation of heteromeric nanotubes (Figure 29) [259]. Therefore, addition of two cyclic peptides with different properties, one hydrophilic and another hydrophobic, to the aqueous solutions that contain the lipid bilayers immediately formed the tubular structure. The transport properties of the resulting nanotube suggest the incorporation of both peptides to satisfy the optimal interactions with the membrane and the interphase. Some of these nanotubes presented non-ohmic current rectification derived from the electronic properties of the nanotube cap regulating the entry of ions into the pore. Thus, cyclic peptide caps bearing positive charged groups (ammonia) decreased the channel conductance values while the transport properties were enhanced with those that incorporated carboxylic groups.

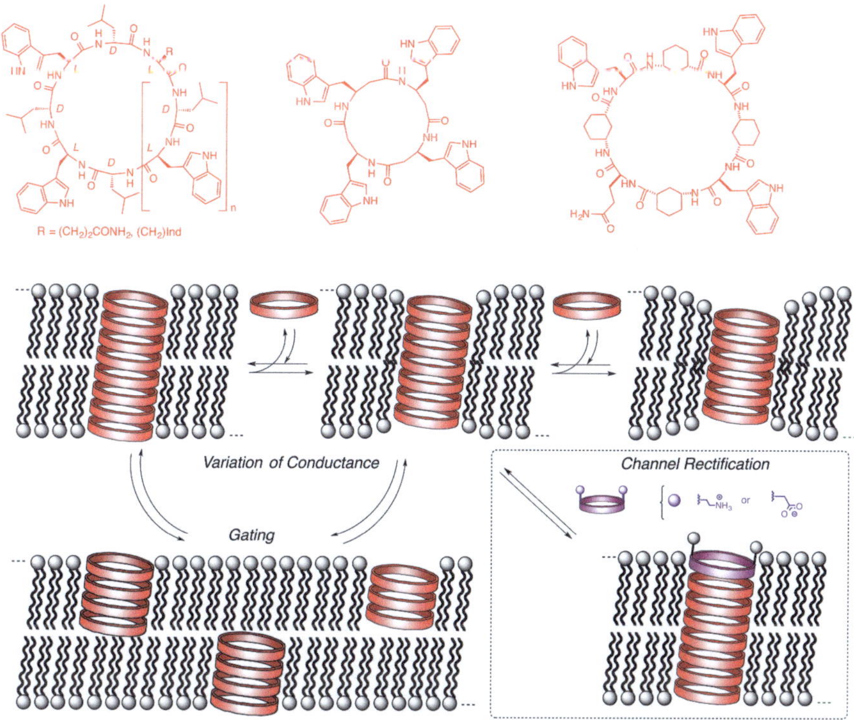

Figure 29 Peptide nanotubes inserted in a lipid membrane with the proposed mechanism for gating and current levels. **Top**: Different types of cyclic peptides studied in transmembrane ion transport. **Inset**: model of heterodimeric nanotubes used to explore the different membrane microenvironment.

The incorporation of polymeric chains linked to the peptides allowed the preparation of cyclic peptide nanotubes with control of their length and also properties (Figure 30) [260]. The incorporation of hydrophobic polymers (polyacrylates with different alkyl substituents) allowed the formation of hybrid derivatives (peptide nanotubes/polymers) that can be incorporated in membrane bilayers. The transport properties were studied using liposome assays containing pH-responsive fluorescent dyes, similar to those carried out with original nanotubes [249]. The proton transport activity confirmed the formation of transmembrane channels. The activity of these channels was related to the hydrophobic properties of the alkyl groups attached to the acrylates, confirming that the limiting step was the incorporation of the peptide nanotube to the membrane. No further experiments were carried out to confirm the ion transport activity but, considering the characteristic of the nanotube cavity, similar ion activity and selectivity should be expected. Perhaps longer open times could have been observed.

Very recently, Jolliffe et al. have reported the preparation of nanotubes with dual functionality in the form of either two faces or mixed polymeric corona through the incorporation of two different polymer types, two miscible for mixed polymers or two that give microphase separation for the non-mixed nanotubes [261]. The pore

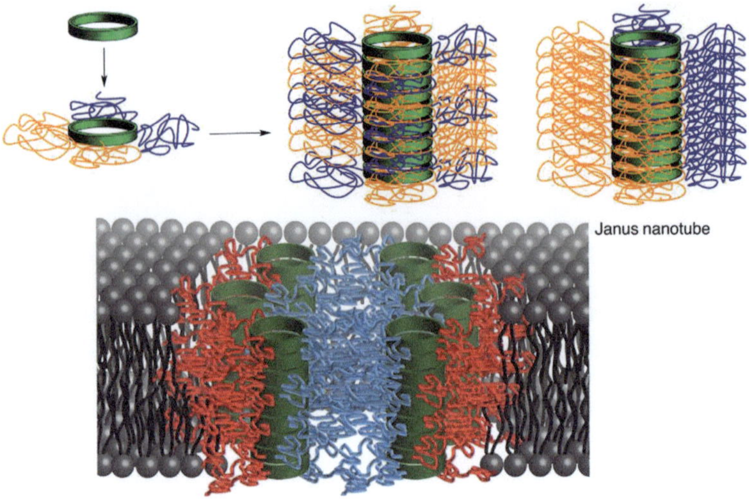

Figure 30 Cyclic peptide-polymer hybrids and model of the formation of bundles of Janus-type nanotubes. The lower part is reproduced from [263] by permission of Nature Publishing Group; © copyright 2013.

formation was proved by calcein release of the Janus-type nanotube (two faces) while no nanotube incorporation was observed in mixed nanotubes. The transport properties derived from the association of the nanotube hybrids to form a macropore. This nanotube bundle explained the release of the calcein dye, because this dye cannot pass through the nanotube pore as a consequence of its large size.

Other peptide platforms were prepared and shown to form nanotubes through similar mechanisms, e.g., using peptides made of β-, γ- and δ-amino acids that have different properties [262–266]. Additionally, peptide hybrids made of α- and β-Aas, α- and γ-Aas, or α- and ε-Aas have been reported to form different type of peptide nanotubes [267–271].

The β-peptide nanotubes were prepared using amino acids in which the side chains are in the β-carbon (HAa) such as *c*-[(β^3-HAla–)$_4$]. In the crystal structure, this peptide appears in a flat conformation forming four hydrogen bonds with each of the two neighboring subunits, leading to nanotubes whose internal diameter was 2.6–2.7 Å [262]. The greater flexibility of β-residues allows that both cyclic peptides of *R,S*-alternating amino acids or the homochiral ones adopt a flat conformation and therefore form nanotubes. A special feature of the resulting nanotubes, which are formed by CPs through a parallel-type interaction, is the resulting net macrodipole moment similar to the one observed in α-helices [272]. Almost simultaneously to this work, Ghadiri's group synthesized cyclic tetrapeptides composed of the hydrophobic β-amino acids homotryptophan (HTrp) and homoleucine (HLeu) and studied their ion transport properties (Figure 29) [273]. The cyclic peptide made of four HTrp residues has transport properties similar to the original nanotubes. The permanent macrodipole character of the channel, contrary to the initial expectations, did not influence their conductance through effects such as voltage gating or current rectification.

The use of α,γ-cyclic peptide hybrids brought new properties to peptide nanotubes (Figure 31) [274–277]. These peptides are made of an even number of residues with alternating α- and γ-amino acids. The γ-amino acids, *cis*-3-aminocycloalcanecarboxylic acids (γ-Acas), used in the nanotubes are cyclic derivatives possessing a *cis*-configuration [278]. The ring rigidity and the *cis*-configuration of the γ-Acas facilitate the peptides to adopt the required flat conformation and, consequently, their stacking to form nanotubes. In addition, in this conformation the β-methylene group of each γ-residue is projected towards the disk center. The resulting nanotube has an internal cavity with partial hydrophobic properties. In spite of these properties, computational studies showed that in water or membrane bilayer the nanotubes are immediately filled with partially ordered water molecules [279, 280]. Not surprisingly, the membrane nanotubes made of these cyclic peptides were able to efficiently facilitate the transmembrane transport of ions.

Transport studies in liposomes (H^+ migration) showed activity of both tryptophan-rich hexamers and octamers (Figure 29), while the planar lipid bilayer confirmed that only the last one has an orifice diameter large enough to facilitate the ion migration (Na^+, K^+, and Cs^+) [281]. The partial hydrophobic properties of the nanotube channels could provide microenvironment conditions appropriated for ion selectivity. In fact, this nanotube is selective for alkali metal ions, not observing transport of divalent cations or chloride. Among the alkali metal ions the transport rate follows the Eisenman I selectivity [75]. Transport rates are directly proportional to relative mobilities in the bulk solution, although sodium flux was 30–40 % greater than that of the heavier alkali metal ions compared to their relative bulk diffusion coefficients. Such differences might account for the influence of the ion dehydration on the transport rate.

The presence of the methylene group projected towards the nanotube lumen brings another key feature of these cyclic peptide hybrids: the modification of the

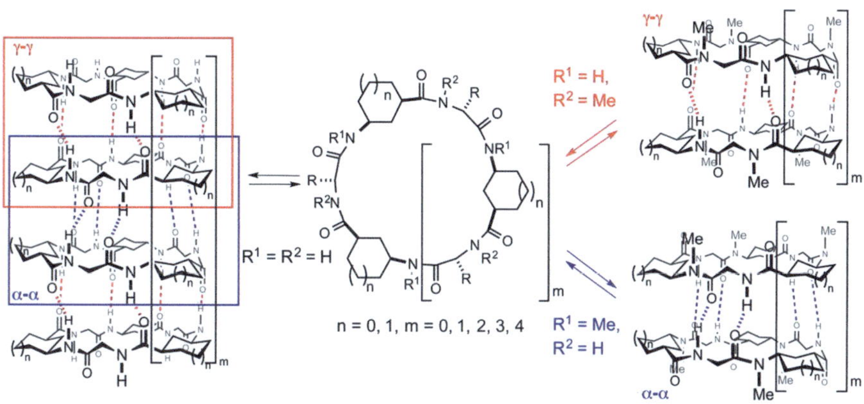

Figure 31 Representation of the self-assembling process of the *c*-[(*D*-α-Aa-γ-*L*-Aca-)$_n$] that leads to the nanotube formation (**left**) or dimers (**right**) depending on the presence and position of methyl groups (R^1 and R^2). The two nanotube interactions, the α,α-interaction (blue) and γ,γ-interaction (red) were studied in the dimers (**right**).

internal cavity properties through the functionalization of the β-carbon, something that is not easily affordable in other peptide nanotubes [282, 283]. The γ-amino acid that would play this role must contain the substituent on the β-carbon *trans*-oriented with respect to the carboxylic and amino groups. The first examples of the nanotube pore functionalization were carried out using the 4-amino-3-hydroxytetrahydrofuran-2-carboxylic acid (γ-Ahf), which was used in the CP dimeric model, *c*-[*D*-Leu-γ-Ahf-*D*-Tyr(Me)-^{Me}N-γ-Acp-], as a proof of concept (Figure 32).

The presence of the hydroxyl group in the cavity facilitates the assembling process towards the ensemble in which each cyclic peptide is hydrogen-bonded. Alternatively, the methylene group was substituted by oxygen using a glucuronic acid derivative with similar results (Figure 32) [264]. These amino acid modification strategies might allow the preparation of more selective ion channel models.

3.1.1.2 Other Nanotubes Based on Cyclic Molecules

Cyclic ureas and amide/urea hybrids can also form nanotubes by stacking of the cyclic components in a flat conformation through the formation of hydrogen-bonds between the carbonyl and NH groups that are lying perpendicularly to the plane of the ring [284–286]. Hydrophobic derivatives were also studied as ion transporters. In these studies the cycloureas were anion-selective due to the observed

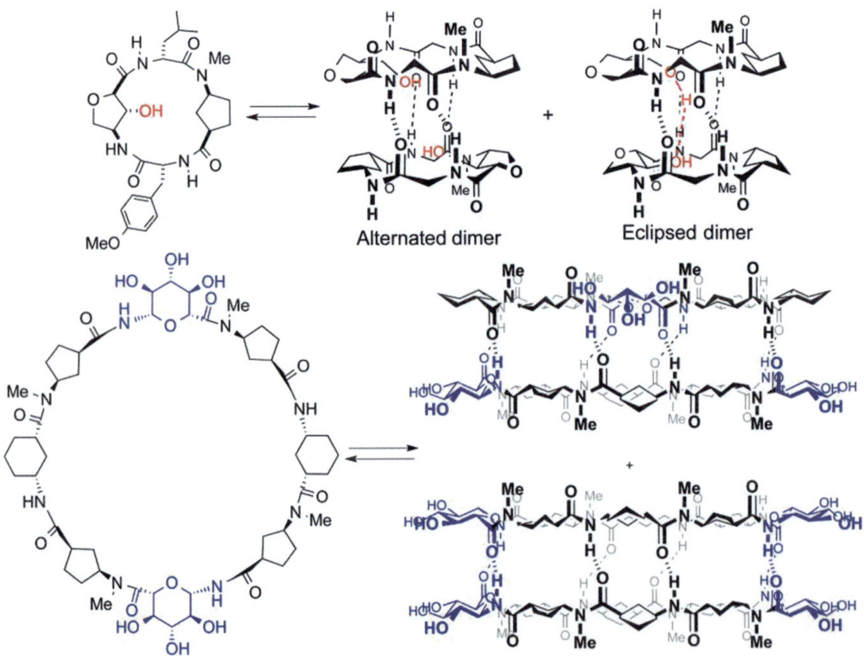

Figure 32 Designs of α,γ-cyclic peptides with functionalized inner cavity.

anion-macrodipole interactions but the mechanism is not related with the migration through the nanotube cavity [287].

Crown ethers were also studied in macroring self-assembling strategies (Figure 33). The strategy tried to substitute the previously mentioned synthetic rods (Figure 27), which were difficult to obtain, by rigid templates obtained by a self-forming process to direct the ring stacking. For that purpose 15-crown-5 or 18-crown-6 ether derivatives containing an aromatic ring functionalized with an urea group were prepared [288]. The crystal structure of one of the derivatives confirmed the formation of the urea ribbon in which the molecules are organized in an antiparallel dimeric fashion. The benzocrown ethers are interdigitated with the other substituent of the urea group (hexyl chain) in such a way that each alkyl group is asymmetrically sandwiched between two benzocrown ethers. Thus, two parallel crown ether channels are formed per urea ribbon. Most of these compounds showed membrane disruption properties in which the majority of opening events are spikes of very short duration and variable amplitude. In some cases at initial stage of time recording, some square-top openings of relatively long duration with spikes superimposed can be observed, being almost impossible to find openings without competing membrane disruption activity. Only the ureido-15-crown-5 ether with a hexyl chain presented typical channel openings at low concentrations in potassium transport experiments. Solution experiments had shown that they initially assemble to form dimers at low ureido-crown ether concentration. At higher concentration the dimers further assemble to form higher oligomers. For that reason, transport experiments were also carried out at higher ureido-crown ether concentrations. Surprisingly, the mentioned 15-crown did not show evidence of any regular behavior at higher concentration, meanwhile ureido-18-crown-6 ether displayed an extremely

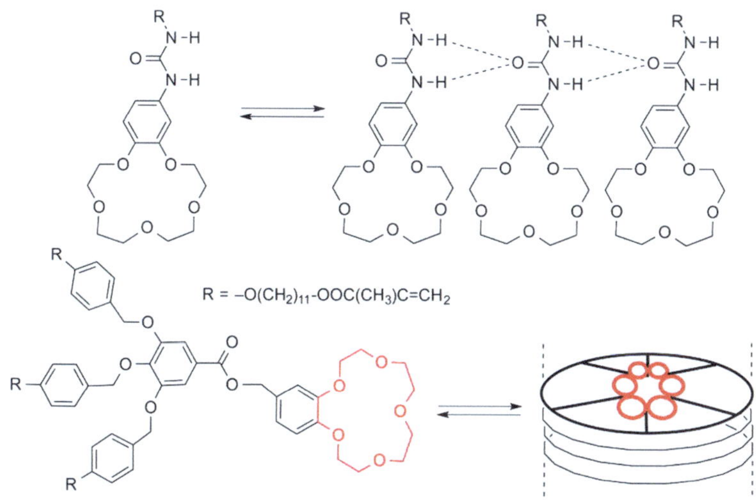

Figure 33 **Top**: Self-assembly of crown ethers bearing urea functional groups to form nanotubes. **Bottom**: The attaching of mesogenic dendrons to crown ethers induce them to stack parallel to the column axis.

wide range of conductance levels, both in terms of conductance values and lifetimes, from milliseconds to seconds.

Apparently, there are small current channels with a step-like behavior of relatively narrow levels. Additionally, there are also long-lived channels with very large currents. Each channel type was assigned to different supramolecular entities. While the shorter ones were assigned to an oligomeric stack of dimers similar to the one observed in the crystal structure, the larger channels were attributed to a possible supramolecular organization of oligomeric dimers forming a toroidal structure. The loss of one of the crown ethers from the stack reduced the channel length and therefore gave a higher transport level. On the other hand, the high crown ether concentration used in these experiments gave multiple copies of these double-barreled channels that produced different levels for each species. Each channel is gaining and losing the outermost crown ethers to account for the observed variety of transport events. Crown ether columnar channels orientated along a silica mesoporous pore were prepared that evolve over time to form the fittest ion channels improving the transport properties [289].

Additional supramolecular forces were used to align crown ethers, in this case the strategy was based on forming crown ether stacks within liquid crystalline phases. In this approach the benzo-15-crown-5 ether was attached to mesogenic dendrons that direct the formation of supramolecular columns (Figure 33) [290, 291]. These mesogenic forces drive the crown ether moieties to stack parallel to the column axis, forming ion channels. The material showed ionic conductivity in its mesophase with lithium ions passing through the membrane faster than sodium or potassium ones (Eisenman XI) [75].

Oligoamide macrocycles were also used in designing synthetic ion channels. For instance, aromatic cyclic oligoamides, similar to the previously shown helical channels, have demonstrated a good ion transport activity (Figure 34) [292, 293]. These macrocycles have pores ranging from 10 to 30 Å and a good K^+ translocation activity. Single channel recording carried out with similar cyclic hexameric oligoamides indicated that ion transport occurs through the cavities of these molecules. The observed conductances were similar to that of α-hemolysin. The use of Hille's equation [2], considering a 40 Å channel length, suggests a pore diameter of 8.5 Å.

The same authors reported the unusual mass-transport properties of self-assembling hexa(*m*-phenylene ethynylene) macrocycles (Figure 34) [294]. The incorporation of an alanine residue in each phenyl moiety allowed the alignment and stacking of the rings by reinforcing the π-π interaction of the aromatic rings with hydrogen bonds. The transport test established that the macrocycle ensemble mediates a highly selective transmembrane ion transport, one of the few examples of a synthetic nanopore, and also highly efficient transmembrane water permeability. Transport experiments with liposomes enclosing a pH-sensitive dye (HPTS), showed high proton selectivity (P_{H^+}/P_{Cl^-} ratio > 3,000) [102]. In addition, planar lipid bilayer studies carried out under high KCl concentration reported potassium permeability, although it was 2,000 times smaller than H^+ activity. No transport of Na^+ or Li^+ ions was observed, illustrating the striking selectivity of these channels.

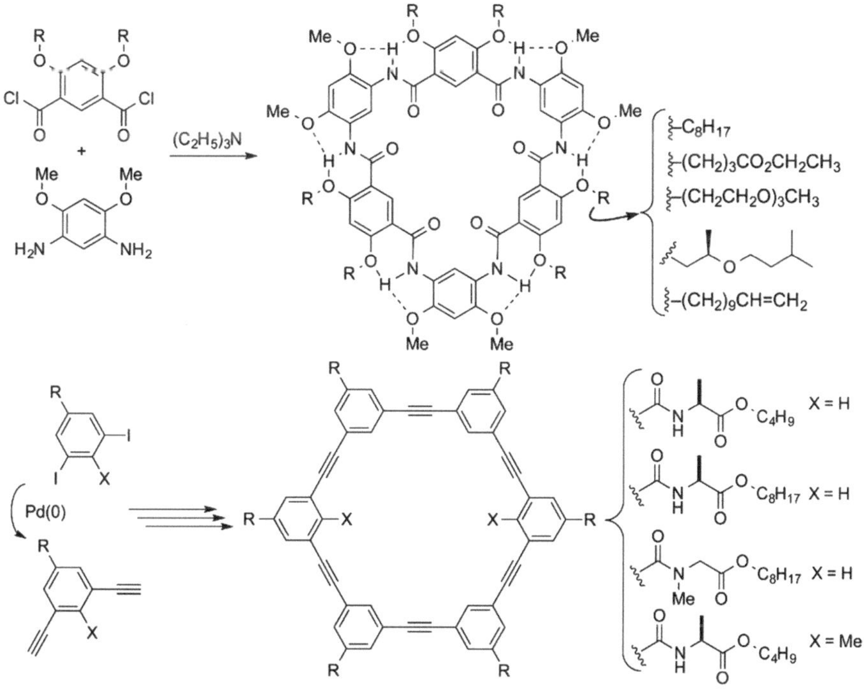

Figure 34 **Top**: Design of aromatic oligoamides that form ion channels. **Bottom**: Hexakis(*m*-pheneylene ethynylene) macrocycles that also form channels by ring stacking.

Analysis of the open/closed probability [131] gave a Hill's coefficient consistent with a functional transmembrane channel made of nine molecules.

Other examples of cyclic oligoamides for transporting ions are those that combine *m*-aminobenzoic acids with hydrophobic natural amino acids (Figure 35) [295, 296]. These type of peptides were described as receptors of a variety of ions and polar molecules [297–300]. The proposed structure resembles those of cyclodextrins or calixarenes, a toroidal basket with two rims of different sizes, creating a cavity decorated by the aromatic rings of the γ-amino acid. The incorporation of an additional substituent at the *meta* position of carboxylic and amino groups allowed increasing the membranephilicity and planarity of the ring. A variety of macrocycles made from six to ten residues were prepared and their transport properties studied. The planar lipid bilayer clearly showed for all of them well-defined single-step conductance changes with permeation ratios six times faster for potassium ion than chloride and almost twice if compared to sodium. The axial orientation of the alkyl groups and their projection towards the smaller opening of the toroidal structure prevents their stacking and nanotube formation. Instead, a channel model similar to those described in the previous section for CDs was proposed. Surprisingly, all the current recordings of hexamers, octamers, and decamers have almost the same conductance values, what might suggest that the ions do not flow through the interior of the CP. If so, larger voltage differences should be observed for the

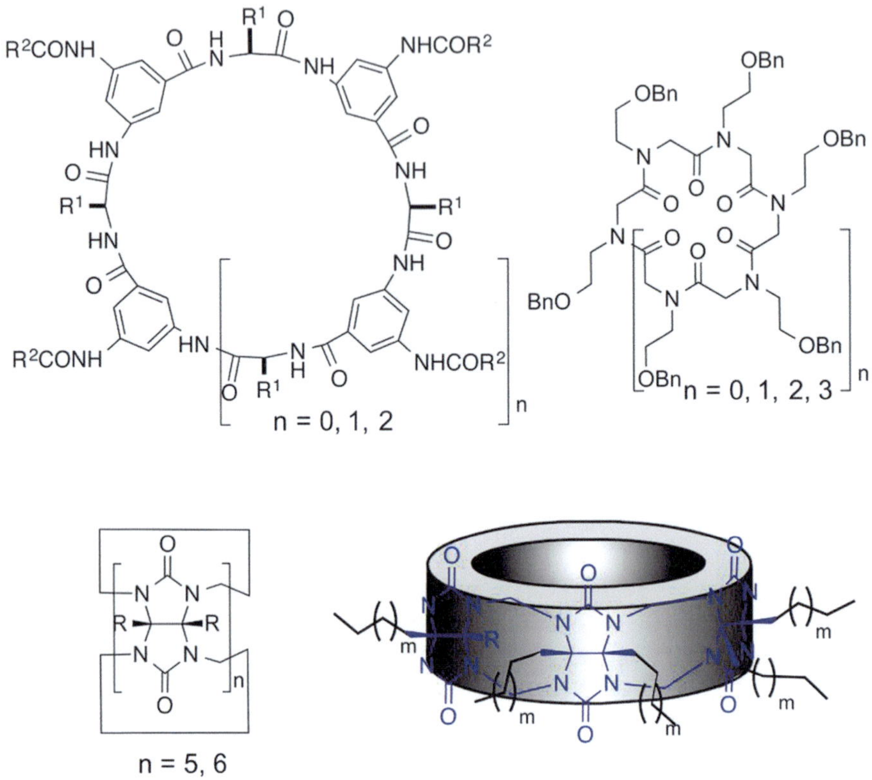

Figure 35 Model structure of 3-aminobenzoic acid CPs, cyclopeptoids (**top**) and cucurbiturils (**bottom**) used in alkali metal ion transport.

larger macrocycles. Another possible explanation is that the rate-limiting step is the movement through the alkyl chains attached to the aminobenzoic moiety.

Cyclopeptoids have also been used for cation transport properties [301]. The smaller macrocycles bind strongly to small ions (Li^+, Na^+, and K^+). Experiments carried out in vesicles showed that the hexapeptoid presented high sodium selectivity while the octapeptoid had a preference for Cs^+ transport. In this case the Hill plots suggested a carrier mechanism without formation of other supramolecular structures.

Cucurbit[6]urils (CB[6], Figure 35), a macrocycle made of six glycoluril monomers linked by methylene bridges, are cavitands whose entrance is flanked by six carbonyl groups and have a diameter of 3.9 Å. CB[6]s bearing hydrophobic chains, which are equatorially surrounding it, are membrane-soluble. It was found that this family of macromolecules could transport protons and alkali metal cations across phospholipid membranes in fluorometric vesicle assays involving pH-sensitive dyes [302]. The transport selectivity corresponds with a Eisenman I topology [75] with abnormal prevalence for lithium ion ($Li^+ > Cs^+ > Rb^+ > K^+ > Na^+$), which is opposite to the binding affinity of CB[6] toward alkali metal ions [303]. The contracted pentameric analog CB[5] exhibits also Li^+ over Na^+ selectivity but does not show transport of the larger alkali metal ions. Apparently, the macrocycle acts as a selectivity filter that retards the transport of the best-bound cations. Interestingly, acetylcholine, a well-known highly affinity guest, blocked the proton flux mediated by CB[6], supporting the hypothesis that ion transport takes place through its cavity. The planar lipid bilayer experiments exhibited short-lived, heterogeneous single-channel conductances what made their complete characterization unfeasable as previously reported for cyclodextrin analogues [72, 304]. The formation of unimolecular channels based on the formation of micellar domains in the bilayer around the macrocycle could be a possible transport model.

3.4 *Self-Assembling of Arc-Shaped Molecules*

In this section, we will consider the last approach used in the construction of synthetic channels, which consists in the design of arc-shaped molecules (sector molecules) that form cyclic structures *via* non-covalent or metal interactions. The resulting cyclic aggregates further assemble into nanotubes, following a pattern known as rosette model [305]. Alternatively, the arc-shaped molecules can form supramolecular helical structures with an internal pore.

3.4.1 Non-covalent Interactions to Form Cyclic Aggregates

As mentioned previously [210, 213], G-quartets are cyclic supramolecular aggregates generated by the formation of Hoogsteen-type hydrogen bond interactions between four guanine bases. In the presence of alkali metal ions they form sandwich-type structures. The incorporation of hydrophobic groups in the saccharide hydroxyl groups allowed their incorporation in lipidic media and their use as ion channel transporters. However, only the strategy based on the covalent oligomerization, through a ring-closing metathesis reaction (RCM), allowed the active transport of potassium ions (Figure 36) [213]. The metathesis reaction was carried out on the already preformed G-quadruplex to spawn an oligomeric complex whose molecular weight suggests that it is formed by sixteen 2,3′-isopropylidene guanosine units. This molecular weight correlates with four stacked G-quartets. These unimolecular G-quadruplexes oligomers can easily exchange sodium and potassium ions. ^{23}Na NMR experiments provided evidence of transmembrane transport and, importantly, no transport was observed when the monomeric precursor was used under identical conditions. The covalently linked G-quartets generate stable channels with a good K^+ selectivity in contrast to Cl^-.

Analogously to guanine, folate can also self-assemble to form similar quadruplexes (Figure 36) [211]. The incorporation of hydrophobic dendrons provides the self-assembling driving force for quadruplexes stacking without cation templation [306]. In addition, they can be incorporated in lipid bilayers forming hydrophilic

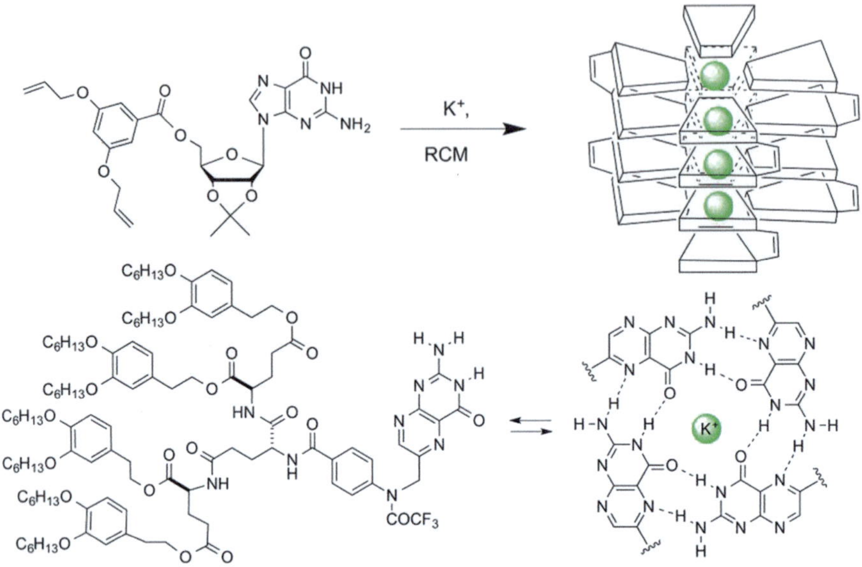

Figure 36 **Top**: Model of guanosine channel formation by initial assembling followed by RCM reaction to form covalently linked structures. **Bottom**: Folates can also form tetrameric structures, the incorporation of hydrophobic dendrons allows the formation of the membrane pore.

pores. These pores are based on planar cyclic tetramers that can coordinate cations with the carbonyl lone pairs pointing towards the center of the cavity. Sodium ion is the one that fits best [307]. In planar bilayer measurements, the dendritic folate rosettes gave remarkably homogeneous, long-lived single-channel currents. They also showed the expected preference for cations over anions (P_{K+}/P_{Cl^-} of 4.9). The observed Eisenman I sequence implies dominance of cation dehydration energies over binding selectivity [75]. The estimated Hille diameter (3.7 Å) was consistent with the size of the central cavity of the folate quartet [2].

The preparation of multiple channel structures with combined self-assembled helical rosette nanotubes and crown ethers (15-crown-5 or 18-crown-6) was described by Fenniri et al. [308]. The nanotube formation is based on a heteroaromatic bicycle system that possesses the Watson–Crick donor-donor-acceptor groups of guanine and the complementary acceptor-acceptor-donor of cytosine with a 120° arrangement. Under physiological conditions the basic heterocycle forms a six-membered aggregate held together by 18 H-bonds. The resulting supermacrocycle has an increased hydrophobic character that favors their stacking. The external properties of the nanotube can be modified by incorporating different groups at position 1 of the pyrido[4,3-*d*]pyrimidine-2,5(1*H*,6*H*)-dione system. In this case, benzo-crown ethers have been attached. Interestingly, dynamic light scattering and small-angle X-ray scattering studies confirmed that the presence of cations (Na^+/K^+) does not affect the 1-D organization, thereby reinforcing the potential exhibited by these ensembles in the preparation of selective ion channels.

Based on this design, several systems based on amphiphiles and peptides have been constructed, some of them have been included in Section 2.2 [195].

3.4.2 Metal-Assisted Channel Formation

This section is focused on the construction of synthetic channels by stacking of cyclic components through metal coordination to appropriated ligands [309]. In this respect, Fyles and Tong described a very simple design using square-shaped palladium complexes. The cyclic component was prepared by mixing the Pd(II) diamino complex that contains a long alkyl chain and a 4,4′-bipyridine ligand. Structural studies carried out on lipid membranes confirmed the ring formation and also the transport properties. In general, the current-time records showed both erratic, short-lived and stable, long-lived ion channel events. These results suggest that the transport mechanism is unlikely due to the "square" channel.

A similar approach was described using a porphyrin ring that contains two pyridine moieties [310]. The palladium complex forms, by itself, very large, high conductance and long-lived channels. The authors compared the channel activity of the palladium complex with that observed for ceramide [59]. They suggest the formation of a toroidal pore in the membrane by the stabilization of positive membrane curvature perpendicular to the plane of the lipid bilayer. A very similar approach but using a rhenium(I) complex to coordinate the pyridine moieties of the porphyrin was also described (Figure 37) [311]. The coordination properties of the Re(I) complex facilitate the formation of the square-shaped structure. The presence of carboxylic groups orthogonally oriented to the pyridines facilitates ring stacking through hydrogen bond formation between the four groups of each ring. The formation of nanopores, which is likely to be dimeric, was confirmed in liposomal membranes. The transport activity is intermediate between those of AmB and gA. The channel does not show selectivity toward first group cations or inorganic anions as could be expected for transport systems with a large pore diameter. The addition of G2 PAMAM dendrimer blocks the transport activity. The presence of amino groups that coordinates to the zinc of the porphyrins and the side of the dendrimer are fundamental to block the nanopore.

Kobuke's group created metallacycle units using three metal trisporphyrin components (Figure 38) [312]. The *meta* substitution of each of the phenyl moieties oriented the porphyrin rings with a 120° angle. The imidazole rings at both ends of the trisporphyrin ensure the oligomerization process. The central porphyrin contains two carboxylic groups perpendicular to the axis defined by the three porphyrins. The two *N*-methylimidazole rings are also aligned in this axis. The coordination between the zinc and the imidazole moiety induces the assembling process to form a trimeric cyclic component that was, finally, covalently trapped by RCM. The trimeric unit self-assembles, only two cyclic units are required, through the formation of hydrogen bonds between the carboxylic acids to form the transmembrane channel. The studies, as in the precedent example, do not show ion selectivity, observing ion current also for lithium and calcium ions, even though the large hydration energy of these ions

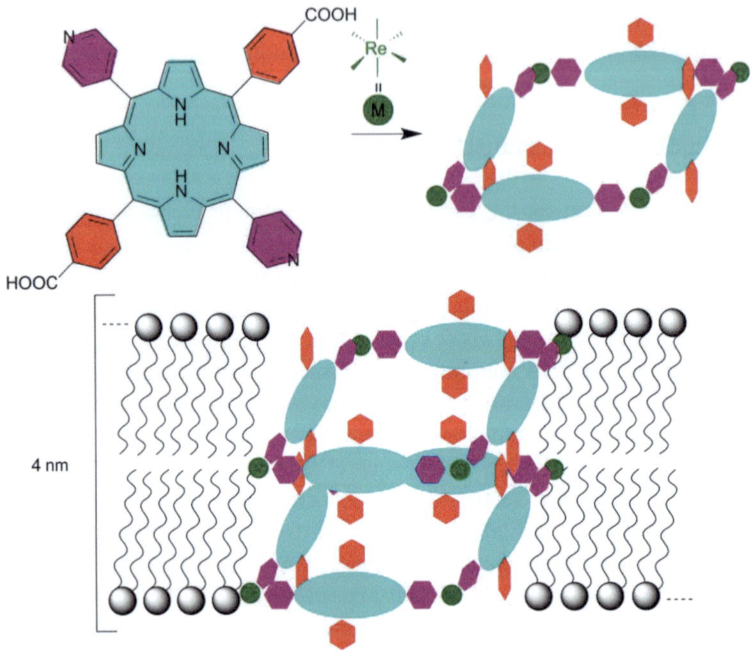

Figure 37 Nanotubes formed by cyclic molecules through metal bonds.

normally precludes them to pass through the channels. Fourth-generation PAMAM dendrimers were used to block the channel current. Reversing the applied potential and replacing the buffer solution allowed the recovering of the channel activity.

Three-dimensional (3D) metal-organic cages or polyhedra (MOP) are novel nanometric size materials consisting of metal ions or clusters coordinated to organic molecules that, generally, leave empty spaces in their interior [313–316]. These MOPs have been recently applied to the preparation of proton and alkali metal ion channels (Figure 38) [317]. For that purpose, the previously reported MOP-18 was used, which is a charge-neutral cuboctahedron core constructed by the self-assembling process of twelve dimeric copper complexes. These complexes form the so-called paddlewheel units, through the coordination with twenty-four molecules of 5-dodecoxybenzene-1,3-dicarboxylic acids. The dodecoxy groups provide a hydrophobic bell that surrounds the MOP structure providing the required lipophilic properties.

The complex has a hydrophilic cavity (13.8 Å diameter) that is accessible through triangular and square windows whose diameters are 3.8 Å and 6.6 Å, respectively [318, 319]. Fluorimetric vesicles and voltage-clamp methods confirmed the transporting properties. The studies showed the expected preference for cations over anions, with a K^+/Cl^- selectivity of 5.5, estimated by using the Goldman–Hodgkin–Katz equation [102]. Hille's equation provided a diameter of approximately 5.4 Å [2], therefore ion transport must take place through both windows of MOP complexes. Actually, a careful analysis of current measurements revealed the existence of two types of channels with different single-channel currents. The cation selectivity

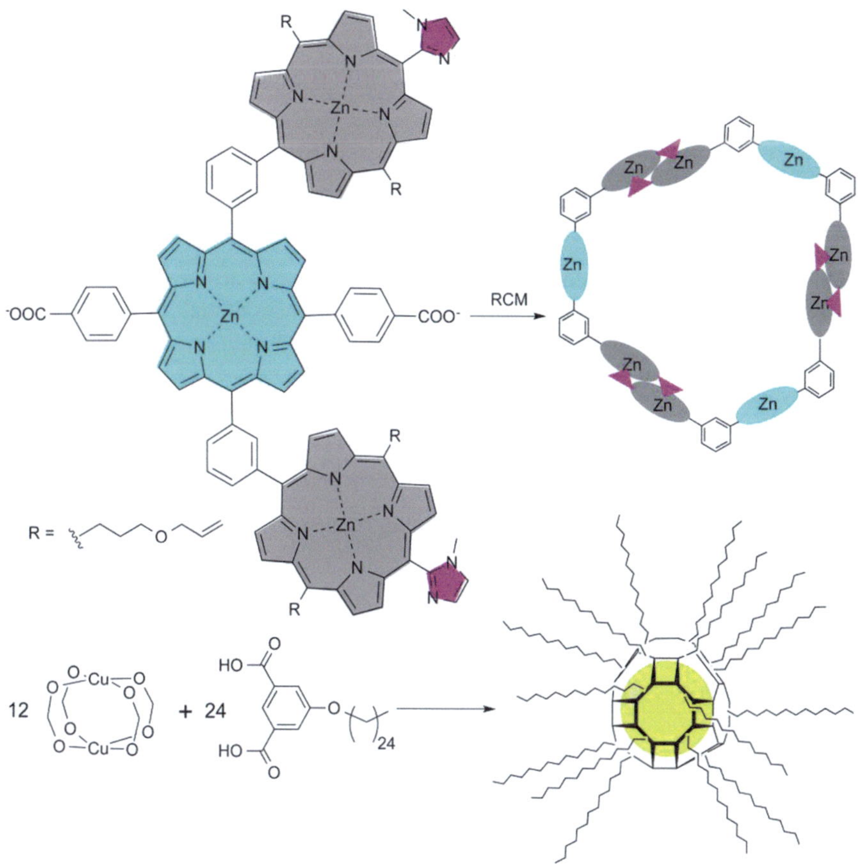

Figure 38 Trimeric structure derived from the zinc triporphyrin derivative through the zinc-imidazole coordination followed by RCM capture. Proposed MOP structure (MOP-18) of the cuboctahedron core constructed by assembling 5-dodecoxybenzene-1,3-dicarboxylic acids with a copper complex.

follows the order $Li^+ >> Na^+ > K^+ > Rb^+ > Cs^+$. This Eisenman XI sequence has a high preference for lithium ions [75]. This is quite unique, suggesting that the interactions between cations and the aromatic rings lining at the portals of the cavity must play a crucial role in the ion transport.

Similarly, C-alkylpyrogallol[4]arenes were also assembled in metallocapsules similar to MOP-18 by treatment with copper acetate [320, 321]. The incorporation of the long alkyl chains on the pyrogallarene facilitates its membrane insertion; the substitution of the undecenyl groups for ethyleneoxy units diminished the transport efficacy. One significant aspect is the greater stability of the open states upon voltage increase. The potassium *versus* chloride permeability ratio was smaller than the previously mentioned cuboctahedron (2.4:1). Cation selectivity follows the sequence $Na^+ > K^+ >> Cs^+$. A dimeric transmembrane model was proposed to explain the transport properties in which the movement of one molecule with respect to the other would explain the observed gating. Furthermore, at increasing voltages, the enhanced flow of ions through the two capsules forces the two components of the channel to stay aligned, explaining the observed stability of the open states.

4 Concluding Remarks and Future Directions

Nature uses the ionic gradient across cell membranes as a driving force for transmembrane interchange. To maintain the cell membrane potential, the intercellular media exhibits a low concentration of sodium ions and high levels of potassium. Although both ions are very similar, their radii differences are on the sub-ångstrom scale; cells have developed almost perfect membrane machines (ion channels, carriers or pumps) that transport selectively each ion in opposite directions. This selectivity is even more remarkable considering that channels can pass ions at up to diffusion-limited rates. Export of sodium ions from the cell provides the driving force for several secondary active transport proteins, which import glucose, amino acids, and other nutrients into the cell benefitting from the sodium gradient.

Not surprisingly, chemists have developed a variety of artificial transport systems that try to mimic the natural functions. The initial designs were based on ion carrier systems based on crown ether motifs. Some of them were very complex and able to switch from active to non-active systems. Later, synthetic ion channels have been created and most of them were programmed to self-assemble in the presence of biological membranes. In this direction, a variety of supramolecular designs were developed using different strategies, such as barrel, rosette or disk stacking. All this background allowed the preparation of systems that compete in efficiency and rates with the biological transporters. In addition, some of the synthetic systems are able to discriminate between different ions, and even some selectivity between sodium and potassium has been achieved. Finally, some of the synthetic models also present rectification and voltage or ligand gating in which the transport is controlled by external chemical stimuli.

Unfortunately, no synthetic system can compete with natural ion channels in rates and selectivity at the same time so far. Perhaps, the last challenge is the preparation of efficient ions pumps. Membrane transporter systems that work against the ionic gradient using an external stimulus are still a further challenge that will be achieved in the near future. Without any doubts, if chemists had access to these kinds of molecular machines, a new horizon would rise for novel technological applications.

Abbreviations

τ	dwell opening time
Aa	amino acid
Aca	3-aminocycloalcanecarboxylic acid

Acp	3-aminocyclopentanecarboxylic acid
Ahf	4-amino-3-hydroxytetrahydrofuran-2-carboxylic acid
AmB	amphotericin B
Arg	arginine
ATP	adenosine 5′-triphosphate
ATR-IR	attenuated total reflection infrared
CB[6]	cucurbit[6]urils
CD	cyclodextrin
CP	cyclic peptide
CyPLOS	cyclic phosphate-linked oligosaccharide
FCCP	carbonylcyanide-*p*-trifluoromethoxyphenyl-hydrazone
G2 PANAM	second generation poly(amidoamine) dendrimer
gA	gramicidin A
Gln	glutamine
HAa	homoamino acid
HAla	homoalanine
HLeu	homoleucine
HPTS	pyranine (8-hydroxypyrene-1,3,6-trisulfonic acid, trisodium salt)
HTrp	homotryptophan
HyV	2-hydroxy-3-methylbutanoic acid
Ile	isoleucine
IR	infrared
Lac	lactic acid
Leu	leucine
LHL	leucyl-histidyl-leucine (Leu-His-Leu)
LHLHL	leucyl-histidyl-leucyl-histidyl-leucine (Leu-His-Leu-His-Leu)
LHRHL	leucyl-histidyl-arginyl-histidyl-leucine (Leu-His-Arg-His-Leu)
LRLHL	leucyl-arginyl-leucyl-histidyl-leucine (Leu-Arg-Leu-His-Leu)
MOP	metal-organic cages or polyhedra
NDIs	naphthalenediimide units
NMR	nuclear magnetic resonance
nS	nanoSiemens
PEG	polyethylene glycol
Phe	phenylalanine
PIMs	polymer inclusion membranes
RCM	ring-closing metathesis
TEB	tetraethylenbenzyl ether chain
TEM	transmission electron microscopy
Trp	tryptophan
Tyr	tyrosine
Val	valine

Acknowledgments This work was supported by the Spanish Ministry of Economy and Competitivity (MEC) and the ERDF [CTQ2013-43264-R], by the Xunta de Galicia (GPC2013-039 and EM2014/011). N. R.-V. (FPU) and A. F. thanks the Spanish Ministry of Education and Competitivity (MEC) and Gil Dávila foundation for their fellowship grants, respectively.

References

1. W. D. Stein, *Channels, Carriers, and Pumps*, Academic Press, New York, 1990.
2. B. Hille, *Ion Channels of Excitable Membranes*, 3rd edn., Sinauer Associates, Inc., Sunderland, Mass., 2001.
3. R. L. Yeagle, *The Structure of Biological Membranes*, 2nd edn., CRC Press, Boca Raton, FL, USA, 2005.
4. M. Miaczynska, H. Stenmark, *J. Cell Biol.* **2008**, *180*, 7–11.
5. Y. Jiang, A. Lee, J. Chen, M. Cadene, B. T. Chait, R. MacKinnon, *Nature* **2002**, *417*, 515–522.
6. R. MacKinnon, *Nature* **2002**, *417*, 261–262.
7. J. P. Morth, B. P. Pedersen, M. S. Toustrup-Jensen, T. L.-M. Sørensen, J. Petersen, J. P. Andersen, B. Vilsen, P. Nissen, *Nature* **2007**, *450*, 1043–1049.
8. R. S. Kass, *J. Clin. Invest.* **2005**, *115*, 1986–1989.
9. S. Gilman, Ed., *Neurobiology of Disease,* Elsevier Academic Press, Maryland Heights, MO, USA, **2010**, pp. 319.
10. W. Chen, *Phys. Rev. E* **2006**, *73*, 021902.
11. K. Neupertlaves, M. Dobler, *Helv. Chim. Acta* **1975**, *58*, 432–442.
12. S. Weinstein, B. A. Wallace, E. R. Blout, J. S. Morrow, W. Veatch, *Proc. Natl. Acad. Sci. USA* **1979**, *76*, 4230–4234.
13. R. R. Ketchem, W. Hu, T. A. Cross, *Science* **1993**, *261*, 1457–1460.
14. L. Song, M. R. Hobaugh, C. Shustak, S. Cheley, H. Bayley, J. E. Gouaux, *Science* **1996**, *274*, 1859–1866.
15. P. DeSantis, S. Morosetti, R. Rizzo, *Macromolecules* **1974**, *7*, 52–58.
16. M. R. Ghadiri, J. R. Granja, R. A. Milligan, D. E. McRee, N. Khazanovich, *Nature* **1993**, *366*, 324–327.
17. A. Ovchinnikov, V. T. Ivanov, A. M. Skrob, *Membrane Active Complexones*, Elsevier, New York, 1974.
18. B. C. Pressman, *Annu. Rev. Biochem.* **1976**, *45*, 501–530.
19. T. M. Fyles, *Acc. Chem. Res.* **2013**, *46*, 2847–2855.
20. E. M. Scholar, W. B. Pratt, Eds., *The Antimicrobial Drugs*, 2nd edn., Oxford University Press, Oxford, 2000.
21. E. P. Bakker, *Ionophore Antibiotics* in *Mechanism of Action of Antibacterial Agents Antibiotics*, **1979**, *5*, 67–97.
22. A. Kart, A. Bilgili, *J. Anim. Vet. Adv.* **2008**, *7*, 748–751.
23. Y. A. Ovchinnikov, V. T. Ivanov, *Tetrahedron Report No.1*; Pergamon Press, New York, 1976.
24. R. M. Izatt, J. S. Bradshaw, S. A. Nielsen, J. D. Lamb, J. J. Christensen, *Chem. Rev.* **1985**, *85*, 271–339.
25. T. J. Marrone, K. M. Merz, Jr., *J. Am. Chem. Soc.* **1995**, *117*, 779–791.
26. Y. A. Ovchinnikov, *Eur. J. Biochem.* **1979**, *94*, 321–336.
27. D. G. Davis, B. F. Gisin, D. C. Tosteson, *Biochemistry* **1976**, *15*, 768–774.
28. D. Ranganathan, V. Haridas, K. P. Madhusudanan, R. Roy, R. Nagaraj, G. B. John, *J. Am. Chem. Soc.* **1997**, *119*, 11578–11584.
29. V. T. Ivanov, *Ann. N.Y. Acad. Sci.* **1975**, 264, 221–243.
30. M. M. Shemyakin, Y. A. Ovchinnikov, V. T. Ivanov, V. K. Antonov, A. M. Shkrob, I. I. Mikhaleva, A. V. Evstratov, G. G. Malenkov, *Biochem. Biophys. Res. Commun.* **1967**, 29, 834–841.
31. Y. A. Ovchinnikov, V. T. Ivanov, A. V. Evstratov, I. I. Mikhaleva, V. F. Bystrov, S. L. Portnova, T. A. Balashova, E. A. Meshcheryakova, V. M. Tulchinsky, *Int. J. Pept. Protein Res.* **1974**, 6, 465–498.
32. V. T. Ivanov, A. V. Evstratov, L. V. Sumskaya, E. I. Melnik, T. S. Chumburidze, S. L. Portnova, T. A. Balashova, Y. A. Ovchinnikov, *FEBS Lett.* **1973**, 36, 65–71.
33. J. Beck, H. Gerlach, V. Prelog, W. Voser, *Helv. Chim. Acta,* **1962**, 45, 620–630.

34. E. Meyers, F. E. Pansy, D. Perlman, D. A. Smith, F. L. Weisenborn, *J. Antibiot. Ser. A.* **1965**, *18*, 128–129.
35. S. N. Graven, H. A. Lardy, D. Johnson, A. Rutter, *Biochemistry* **1966**, *5*, 1729–1735.
36. B. A. Moyer, *Complexation and Transport*, in *Molecular Recognition: Receptors for Cationic Guests*, G. W. Gokel, Ed., in *Comprehensive Supramolecular Chemistry*, J. L. Atwood, J. E. D. Davies, D. D. MacNicol, F. Vögtle, J.-M. Lehn, Eds., Vol. 1, Pergamon, Elsevier, Oxford, 1996, pp. 377–416.
37. D. J. Cram, *Angew. Chem. Int. Ed.*, **1986**, *25*, 1039–1057.
38. D. J. Cram, H. P. deGrandpre, C. B. Knobler, K. N. Trueblood, *J. Am. Chem. Soc.*, **1984**, *106*, 3286–3292.
39. J.-M. Lehn, *Acc. Chem. Res.* **1978**, *11*, 49–57.
40. E. Graft, J. M. Lehn, *J. Am. Chem. Soc.*, **1975**, *97*, 5022–5024.
41. J. M. Lehn, *Pure Appl. Chem.* **1977**, *49*, 857–870.
42. G. W. Gokel, K. A. Arnold, M. Delgado, L. Echeverria, V. J. Gatto, D. A. Gustowski, J. Hernandez, A. Kaifer, S. R. Miller, L. Echegoyen, *Pure Appl. Chem.* **1988**, *60*, 461–465.
43. W. Walkowiak, C. A. Kozlowski, *Desalination* **2009**, *240*, 186–197.
44. J. Smid, *Pure Appl. Chem.* **1982**, *54*, 2129–2140.
45. E. Cagnoni, D. Pasini, A. Galbiati, M. Ricci, P. P. Righetti, *Macromol.* **2003**, *36*, 8894–8897.
46. M. Okahara, Y. Nakatsuji, *Topics in Current Chemistry: Biomimetic and Bioorganic Chemistry* **1985**, *128*, 37–59.
47. J. D. Lamb, J. J. Christensen, S. R. Izatt, K. Bedke, M. S. Astin, R. M. Izatt, *J. Am. Chem. Soc.* **1980**, *102*, 3399–3403.
48. T. Jin, M. Kinjo, T. Koyama, Y. Kobayashi, H. Harita, *Langmuir* **1996**, *12*, 2684–2689.
49. P. M. Marcos, S. Félix, J. R. Ascenso, M. A. Santos, M. A. P. Segurado, J. L. C. Pereira, *Tetrahedron* **2002**, *58*, 9223–9230.
50. A. Casnati, A. Pochini, R. Ungaro, F. Ugozzoli, F. Arnaud, S. Fanni, M. J. Schwing, R. J. M. Egberink, F. de Jong, D. N. Reinhoudt, *J. Am. Chem. Soc.* **1995**, *117*, 2767–2777.
51. F. Arnaud-Neu, Z. Asfari, B. Souley, J. Vicens, *New J. Chem.* **1996**, *20*, 453–463.
52. Z. Asfari, C. Naumann, J. Vicens, M. Nierlich, P. Thuéry, C. Bressot, V. Lamare and J.-F. Dozol, *New J. Chem.* **1996**, *20*, 1183–1194.
53. Z. Asfari, C. Bressot, J. Vicens, C. Hill, J.-F. Dozol, H. Rouquette, S. Eymard, V. Lamare, B. Tournois, *Anal. Chem.* **1995**, *67*, 3133–3139.
54. T. G. Levitskaia, D. M. Macdonald, J. D. Lamb, B. A. Moyer, *Phys. Chem. Chem. Phys.* **2000**, *2*, 1481–1491.
55. P. Schmitt, P. D. Beer, M. G. B. Drew, P. D. Sheen, *Angew. Chem., Int. Ed. Engl.* **1997**, *36*, 1840–1842.
56. F. Arnaud-Neu, Z. Asfari, B. Souley, J. Vicens, P. Thuéry, M. Nierlich, *J. Chem. Soc. Perkin Trans. 2*, **2000**, 495–499.
57. L. Husaru, M. Gruner, T. Wolff, W. D. Habicher, R. Salzer, *Tetrahedron Lett.* **2005**, *46*, 3377–3379.
58. T. M. Fyles, *Chem. Soc. Rev.* **2007**, *36*, 335–347.
59. L. Siskind, M. Colombini, *J. Biol. Chem.* **2000**, *275*, 38640–38644.
60. J. T. Davis, O. Okunola, R. Queseda, *Chem. Soc. Rev.* **2010**, *20*, 3843–3862.
61. J. P. Behr, J. M. Lehn, A. C. Dock, D. Moras, *Nature* **1982**, 295, 526–627.
62. J. P. Behr, J. M. Lehn, D. Moras, J. C. Thierry, *J. Am. Chem. Soc.* **1981**, *103*, 701–703.
63. L. Jullien, J.-M. Lehn, *Tetrahedron Lett.* **1988,** *29*, 3803–3806.
64. T. M. Fyles, T. D. James, K. C. Kaye, *J. Am. Chem. Soc.* **1993**, *115*, 12315–12321.
65. T. M. Fyles, T. D. James, A. Pryhitka, M. Zojaji, *J. Org. Chem.* **1993**, *58*, 7456–7468.
66. F. R. Fronczek, R. D. Gandour, T. M. Fyles, P. J. Hocking, S. J. McDermid, P. D. Wotton, *Can. J. Chem.* **1991**, *69*, 12–19.
67. I. Tabushi, Y. Kuroda, K. Yokota, *Tetrahedron Lett.* **1982**, *23*, 4601–4604.
68. M. J. Pregel, L. Jullien, J. Canceil, L. Lacombe, J.-M. Lehn, *J. Chem. Soc. Perkin Trans. 2*, **1995**, 417–426.

69. N. Maulucci, F. De Riccardis, C. B. Botta, A. Casapullo, E. Cressina, M. Fregonese, P. Tecilla, I. Izzo, *Chem. Commun.* **2005**, 1354–1356.
70. N. Madhavan, E. C. Robert, M. S. Gin, *Angew. Chem. Int. Ed.* **2005**, *44*, 7584–7587.
71. N. Madhavan, M. S. Gin, *ChemBioChem* **2007**, *8*, 1834–1840.
72. J. K. W. Chui, T. M. Fyles, *Org. Biomol. Chem.* **2014**, *12*, 3622–3634.
73. J. K. W. Chui, T. M. Fyles, *Chem. Soc. Rev.* **2011**, *41*, 148–175.
74. J. K. W. Chui, T. M. Fyles, H. Luong, *Beilstein J. Org. Chem.* **2011**, *7,* 1562–1569.
75. G. Eisenman, R. Horn, *J. Membr. Biol.* **1983**, *76*, 197–225.
76. Y. Tanaka, Y. Kobuke, M. Sokabe, *Angew. Chem. Int. Ed.* **1995**, *34*, 693–694.
77. W.-H. Chen, M. Nishikawa, S.-D. Tan, M. Yamamura, A. Satake, Y. Kobuke, *Chem. Commun.* **2004**, 872–873.
78. L. Husaru, R. Schulze, G. Steiner, T. Wolff, W. D. Habicher, R. Salzer, *Anal. Bioanal. Chem.*, **2005**, *382*, 1882–1888.
79. R. Li, O. V. Kulikov, G. W. Gokel, *Chem. Commun.* **2009**, 6092–6094.
80. O. V. Kulikov, M. M. Daschbach, C. R. Yamnitz, N. Rath, G. W. Gokel, *Chem. Commun.* **2009**, 7497–7499.
81. S. Negin, M. M. Daschbach, O. V. Kulikov, N. Rath, G. W. Gokel, *J. Am. Chem. Soc.* **2011**, *133*, 3234–3237.
82. D. Montesarchio, C. Coppola, M. Boccalon, P. Tecilla, *Carbohy. Res.*, **2012**, *356*, 62–74.
83. C. Coppola, A. Paciello, G. Mangiapia, S. Licen, M. Boccalon, L. De Napoli, L. Paduano, P. Tecilla, D. Montesarchio, *Chem. Eur. J.*, **2010**, *16*, 13757–13772.
84. G. Di Fabio, A. Randazzo, J. D'Onofrio, C. Ausìn, A. Grandas, E. Pedroso, L. De Napoli, D. Montesarchio, *J. Org. Chem.*, **2006**, *71*, 3395–3408.
85. A. Nakano, Q. Xie, J. V. Mallen, L. Echegoyen, G. W. Gokel, *J. Am. Chem. Soc.,* **1990**, *112*, 1287–1289.
86. G. W. Gokel, M. M. Daschbach, *Coor. Chem. Rev.*, **2008**, *252*, 886–902.
87. G. W. Gokel, *Acc. Chem. Res.*, **2013**, *46*, 2824–2833.
88. H. Shabany and G. W. Gokel, *Chem. Commun.*, **2000**, 2373–2374.
89. W. M. Leevy, J. E. Huettner, R. Pajewski, P. H. Schlesinger, G. W. Gokel, *J. Am. Chem. Soc.*, **2004**, *126*, 15747–15753.
90. W. M. Leevy, M. E. Weber, P. H. Schlesinger, G. W. Gokel, *Chem. Commun.*, **2005**, 89–91.
91. J. Bolard, *Biochim. Biophys. Acta* **1986**, *864*, 257–304.
92. D. M. Cereghetti, E. M. Carreira, *Synthesis* **2006**, 914–942.
93. N. Nuraje, H. Bai, K. Su, *Prog. Polym. Sci.* **2013**, *38*, 302–343.
94. F. De Riccardis, I. Izzo, D. Montesarchio, P. Tecilla, *Acc. Chem. Res.* **2013**, *46*, 2781–2790.
95. H. Valkenier, A. P. Davis, *Acc. Chem. Res.* **2013**, *46*, 2898–2909.
96. Y. Kobuke, T. Nagatani, *J. Org. Chem.* **2001**, *66*, 5094–5101.
97. M. Yoshii, M. Yamamura, A. Satake, Y. Kobuke, *Org. Biomol. Chem.* **2004**, *2*, 2619–2623.
98. C. Goto, M. Yamamura, A. Satake, Y. Kobuke, *J. Am. Chem. Soc.* **2001**, *123,* 12152–12159.
99. C. P. Wilson, S. J. Webb, *Chem. Commun.* **2008**, 4007–4009.
100. N. Sakai, S. Matile, *Angew. Chem. Int. Ed.* **2008**, *47*, 9603–9607.
101. Y. Kobuke, K. Ueda, M. Sokabe, *J. Am. Chem. Soc.* **1992**, *114*, 7618–7622.
102. D. E. Goldman, *J. Gen. Physiol.* **1943**, 27, 37–60.
103. A. L. Hodgkin, B. Katz, *J. Physiol.* **1949**, 108, 37–77.
104. R. B. Gunn, P. F. Curran, *Biophys. J.* **1971**, *11*, 559–571.
105. Y. Kobuke, A. Ohgoshi, *Colloids Surf. A* **2000**, *169*, 187–197.
106. T. M. Fyles, K. C. Kaye, T. D., James, D. W. M. Smiley, *Tetrahedron Lett.* **1990**, *31*, 1233–1236.
107. T. M. Fyles, D. Loock, W. F. Van Straaten-Nijenhuis, X. Zhou, *J. Org. Chem.* **1996**, *61*, 8866–8874.
108. T. M. Fyles, C. Hu, R. Knoy, *Org. Lett.* **2001**, *3*, 1335–1337.
109. H. Luong, T. M. Fyles, *Org. Biomol. Chem.* **2009**, *7*, 725–732.
110. W. Wang, R. Li, G. W. Gokel, *Chem. Commun.* **2009**, 911–913.
111. J. M. Moszynski, T. M. Fyles, *Org. Biomol. Chem.* **2010**, *8*, 5139–5149.

112. T. Renkes, H. J. Schafer, P. M. Siemens, E. Neumann, *Angew. Chem. Int. Ed.* **2000**, *39*, 2512–2516.
113. C. G. Espinola, R. Pérez, J. D. Martín, *Org. Lett.* **2000**, *2*, 3161–3164.
114. T. Heimberg, *Biophys. Chem.* **2010**, *150*, 2–22.
115. B. Wunderlich, C. Leirer, A. Idzko, U. F. Keyser, V. Myles, T. Heimburg, M. Schneider, *Biophys. J.* **2009**, *96*, 4592–4597.
116. M. Jansen, A. Blume, *Biophys. J.* ***1995***, *68*, 997–1008.
117. J. H. Ipsen, K. Jorgensen, O. G. Mouritsen, *Biophys. J.* **1990**, *58*, 1099–1107.
118. H. Hauser, D. Oldani, M. C. Phillips, *Biochemistry* **1973**, *12*, 4507–4517.
119. A. Parsegian, *Nature* **1969**, *221*, 844–846.
120. S. Paula, A. G. Volkov, A. N. Van Hoek, T. H. Haines, D. W. Deamer, *Biophys. J.* **1996**, *70*, 339–348.
121. X. D. Song, J. Perlstein, D. G. Whitten, *J. Am. Chem. Soc.* **1997**, *119*, 9144–9159.
122. T. K. Rostovtseva, C. L. Bashford, A. A. Lev, C. A. Pasternak, *J. Membr. Biol.* **1994**, *141*, 83–90.
123. P. Schlieper, E. De Robertis, *Arch. Biochem. Biophys.* **1977**, *184*, 204–208.
124. R. García-Fandiño, M. Amorín, J. R. Granja, *Synthesis of Supramolecular Nanotubes*, in *Supramolecular Chemistry: From Molecules to Nanomaterials*, P. A. Gale and J. W. Steed, Eds., Vol. 5, John Wiley & Sons Ltd, Chichester, UK, 2012, pp. 2149–2182.
125. D. T. Bong, T. D. Clark, J. R. Granja, M. R. Ghadiri, *Angew. Chem. Int. Ed.* **2001**, *40*, 988–1011.
126. M. Montal, *Curr. Opin. Struct. Biol.* **1996**, *6*, 499–510.
127. M. Montal, S. J. Opella, *Biochim. Biophys. Acta* **2002**, *1565*, 287–293.
128. S. Oiki, W. Danho, M. Montal *Proc. Natl. Acad. Sci. USA* **1988**, *85*, 2393–2397.
129. R. E. Greenblatt, Y. Blatt, M. Montal, *FEBS Lett.* **1985**, *193*, 125–134.
130. M. Montal, M. S. Montal, J. M. Tomich, *Proc. Natl. Acad. Sci. USA* **1990**, *87*, 6929–6933.
131. J. D. Lear, Z. R. Wasserman, W. F. DeGrado, *Science* **1988**, *240*, 1177–1181.
132. K. S. Akerfeldt, J. D. Lear, Z. R. Wasserman, L. A. Chung, W. F. DeGrado, *Acc. Chem. Res.* **1993**, *26*, 191–197.
133. R. J. Dubos, *J. Exp. Med.* **1939**, 70, 1–10.
134. D. A. Kelkar, A. Chattopadhyay, *Biochim. Biophys. Acta* **2007**, *1768*, 2011–2025.
135. D. W. Urry, *Proc. Natl. Acad. Sci. USA* **1971**, *68*, 672–676.
136. W. R. Veatch, E. T. Fossel, E. R. Blout, *Biochemistry* **1974**, *13*, 5249–5256.
137. B. A. Wallace, *Annu. Rev. Biophys. Chem.* **1990**, *19*, 127–157.
138. C. M. Venkatachalam, D. W. Urry, *J. Comp. Chem.* **1983**, *4*, 461–469.
139. V. B. Myers, D. A. Haydon, *Biochim. Biophys. Acta* **1972**, *274*, 313–322.
140. P. Läuger, *Angew. Chem. Int. Ed. Engl.* **1985**, *24*, 905–923.
141. C. J. Stankovich, S. H. Heinemann, J. M. Delfino, F. J. Sigworth, S. L. Schreiber, *Science* **1989**, *244*, 813–817.
142. R. E. Koeppe, II, O. S. Andersen, *Annu. Rev. Biophys. Biomol. Struct.* **1996**, *25*, 231–258.
143. L. Lien, D. C. J. Jaikaran, Z. Zhang, G. A. Wolley, *J. Am. Chem. Soc.* **1996**, *118*, 12222–12223.
144. M. X. Macrae, S. Blake, M. Mayer, J. Yang, *J. Am. Chem. Soc.* **2010**, *132*, 1766–1767.
145. P. Reiß, U. Koert, *Acc. Chem. Res.* **2013**, *46*, 2773–2780.
146. J. R. Pfeifer, P. Reiß, U. Koert, *Angew. Chem. Int. Ed.* **2006**, *45*, 501–504.
147. H. J. Apell, E. Bamberg, H. Alpes, P. Läuger, *J. Membr. Biol.* **1977**, *31*, 171–178.
148. H.-D. Arnot, A. Knoll, U. Koert, *Angew. Chem. Int. Ed.* **2001**, *40*, 2076–2078.
149. H.-D. Arndt, B. Ziemer, U. Koert, *Org. Lett.* **2004**, *6*, 3269–3272.
150. A. Schrey, A. Vescovi, A. Knoll, C. Rickert, U. Koert, *Angew. Chem. Int. Ed.* **2000**, *39*, 900–902.
151. G. Guichard, I. Huc, *Chem. Commun.* **2011**, *47*, 5933–5941.
152. T. A. Martinek, F. Fülöp, *Chem. Soc. Rev.* **2012**, *41*, 687–702.
153. D.-W. Zhang, Z. Zhao, J.-L. Hou, Z.-T. Li, *Chem. Rev.* **2012**, *112*, 5271–5316.

154. J.-L. Hou, X.-B. Shao, G.-J. Chen, Y.-X. Zhou, X.-K. Jiang, Z.-T. Li, *J. Am. Chem. Soc.* **2004**, *126*, 12386–12394.
155. L. Chen, W. Si, L. Zhang, G. Tang, Z.-T. Li, J.-L. Hou, *J. Am. Chem. Soc.* **2013**, *135*, 2152–2154.
156. P. Xin, P. Zhu, P. Su, J.-L. Hou, Z.-T. Li, *J. Am. Chem. Soc.* **2014**, *136*, 13078–13081.
157. O. S. Andersen, *Biophys. J.* **1983**, *41*, 119–133.
158. Y. Zhao, H. Cho, L. Widanapathirana, S. Zhang, *Acc. Chem. Res.* **2013**, *46*, 2763–2772.
159. Y. Zhao, Z. Zhong, *J. Am. Chem. Soc.* **2005**, *127*, 17894–17901.
160. S. Zhang, Y. Zhao, *Org. Biomol. Chem.* **2012**, *10*, 260–266.
161. S. Zhang, Y. Zhao, *Chem. Eur. J.* **2011**, *17*, 12444–12451.
162. M. Fioroni, T. Dworeck, F. Rodríguez-Ropero, *β-barrel Channel Proteins as Tools in Nanotechnology: Biology, Basic Science and Advanced Applications*, Springer Science & Business Media, Dordrecht, 2013.
163. W. C. Wimley, *Curr. Opin. Struct. Biol.* **2003**, *13*, 404–411.
164. M. Montoya, E. Gouaux, *Biochim. Biophys. Acta* **2003**, *1609*, 19–27.
165. F. Chiti, C. M. Dobson, *Annu. Rev. Biochem.* **2006**, 75, 333–366.
166. A. Aguzzi, T. O'Connor, *Nature Rev. Drug. Discov.* **2010**, 9, 237–248.
167. M. Bartolini, V. Andrisano, *ChemBioChem* **2010**, *11*, 1018–1035.
168. P.-N. Cheng, C. Liu, M. Zhao, D. Eisenberg, J. S. Nowick, *Nature Chem.* **2012**, *4*, 927–933.
169. D. S. Palacios, T. M. Anderson, M. D. Burke, *J. Am. Chem. Soc.* **2007**, *129*, 13804–13805.
170. M. Baginski, J. Czub, *Curr. Drug Metabol.* **2009**, *10*, 459–469.
171. S. C. Hartsel, S. K. Benz, W. Ayenew, J. Bolard, *Eur. Biophys. J.* **1994**, *23*, 125–132.
172. W.-H. Chen, S. L. Regen, *J. Am. Chem. Soc.* **2005**, *127*, 6538–6539.
173. W.-H. Chen, X.-B. Shao, S. L. Regen, *J. Am. Chem. Soc.* **2005**, *127*, 12727–12735.
174. S. Litvinchuk, G. Bollot, J. Mareda, A. Som, D. Ronan, M. R. Shah, P. Perrottet, N. Sakai, S. Matile, *J. Am. Chem. Soc.* **2004**, *126*, 10067–10075.
175. N. Sakai, J. Mareda, S. Matile, *Acc. Chem. Res.* **2005**, *38*, 79–87.
176. N. Sakai, K. C. Brennan, L. A. Weiss, S. Matile, *J. Am. Chem. Soc.* **1997**, *119*, 8726–8727.
177. L. A. Weiss, N. Sakai, B. Ghebremariam, C. Ni, S. Matile, *J. Am. Chem. Soc.* **1997**, *119*, 12142–12149.
178. C. Ni, S. Matile, *Chem. Commun.* **1998**, 755–756.
179. M. M. Tedesco, B. Ghebremariam, N. Sakai, S. Matile, *Angew. Chem. Int. Ed.* **1999**, *38*, 540–543.
180. N. Sakai, J. Mareda, S. Matile, *Acc. Chem. Res.* **2008**, *41*, 1354–1365.
181. B. Baumeister, S. Matile, *Chem. Eur. J.* **2000**, *6*, 1739–1749.
182. N. Sakai, B. Baumeister, S. Matile, *ChemBioChem* **2000**, *1*, 123–125.
183. N. Sakai, S. Matile, *Chem. Commun.* **2003**, *20*, 2514–2523.
184. P. Talukdar, G. Bollot, J. Mareda, N. Sakai, S. Matile, *J. Am. Chem. Soc.* **2005**, *127*, 6528–6529.
185. G. Das, P. Talukdar, S. Matile, *Science* **2002**, *298*, 1600–1602.
186. A. Som, N. Nakai, S. Matile, *Bioorg. Med. Chem.* **2003**, *11*, 1363–1369.
187. P. Talukdar, N. Sakai, N. Sordé, D. Gerard, V. M. F. Cardona, S. Matile, *Bioorg. Med. Chem.* **2004**, *12*, 1325–1336.
188. V. Gorteau, F. Perret, G. Bollot, J. Mareda, A. N. Lazar, A. W. Coleman, D.-H. Tran, N. Sakai, S. Matile, *J. Am. Chem. Soc.* **2004**, *126*, 13592–13593.
189. V. Gorteau, G. Bollot, J. Mareda, D. Pasini, D.-H. Tran, A. N. Lazar, A. W. Coleman, N. Nakai, S. Matile, *Bioorg. Med. Chem.* **2005**, *13*, 5171–5180.
190. N. Sakai, S. Matile, *J. Am. Chem. Soc.* **2002**, *124*, 1184–1185.
191. N. Sakai, D. Houdebert, S. Matile, *Chem. Eur. J.* **2003**, *9*, 223–232.
192. F. Mora, D.-H. Tran, N. Oudry, G. Hopfgartner, D. Jeannerat, N. Sakai, S. Matile, *Chem. Eur. J.* **2008,** *14*, 1947–1953.
193. P. Talukdar, G. Bollot, J. Mareda, N. Sakai, S. Matile, *Chem. Eur. J.* **2005**, *11*, 6525–6532.
194. N. Sakai, S. Matile, *J. Phys. Org. Chem.* **2006**, *19*, 452–460.
195. Y. Kim, W. Li, S. Shin, M. Lee, *Acc. Chem. Res.* **2013**, *46*, 2888–2897.

196. W. Y. Yang, J. H. Ahn, Y. S. Yoo, N. K. Oh, M. Lee, *Nat. Mater.* **2005**, *4*, 399–402.
197. I.-S. Park, Y.-R. Yoon, M. Jung, K. Kim, S. B. Park, S. Shin, Y.-B. Lim, M. Lee, *Chem. Asian J.* **2011**, *6*, 452–458.
198. J. T. Davis, *Angew. Chem. Int. Ed.* **2004**, *43*, 668–698.
199. G. N. Parkinson, M. P. H. Lee, S. Neidle, *Nature* **2002**, *417*, 876–880.
200. D. González-Rodríguez, J. L. J. Van Dongen, M. Lutz, A. L. Spek, A. P. H. J. Schenning, E. W. Meijer, *Nat. Chem.* **2009**, *1*, 154–155.
201. R. H. Shafer, I. Smirnov, *Biopolymers* **2001**, *56*, 209–227.
202. T. Simonsson, *Biol. Chem.* **2001**, *382*, 621–628.
203. H. Arthanari, P. H. Bolton, *Chem. Biol.* **2001**, *8*, 221–230.
204. N. Maizels, *Nat. Struct. Mol. Biol.* **2006**, *13*, 1055–1059.
205. L. A. Cahoon, H. S. Seifert, *Science* **2009**, *325*, 764–767.
206. E. Kruisselbrink, V. Guryev, K. Brouwer, D. B. Pontier, E. Cuppen, M. Tijsterman, *Curr. Biol.* **2008**, *18*, 900–905.
207. C. Ribeyre, J. Lopes, J.-B. Boulé, A. Piazza, A. Guédin, V. A. Zakian, J.-L. Mergny, A. Nicolas, *PLoS Genet.* **2009**, *5*, e1000475.
208. S. Balasubramanian, L. H. Hurley, S. Neidle, *Nature Rev. Drug Discovery* **2011**, *10*, 261–275.
209. T. A. Brooks, L. H. Hurley, *Genes & Cancer* **2010**, *1*, 641–649.
210. J. T. Davis, G. P. Spada, *Chem. Soc. Rev.* **2007**, *36*, 296–313.
211. F. Ciuchi, G. Di Nicola, H. Franz, G. Gottarelli, P. Mariani, M. G. P. Bossi, G. Spada, *J. Am. Chem. Soc.* **1994**, *116*, 7064–7071.
212. M. Cai, A. L. Marlow, J. C. Fettinger, D. Fabris, T. J. Haverlock, B. A. Moyer, J. T. Davis, *Angew. Chem. Int. Ed.* **2000**, *39*, 1283–1285.
213. M. S. Kaucher, W. A. Harrell, Jr., J. T. Davis, *J. Am. Chem. Soc.* **2006**, *128*, 38–39.
214. C. Arnel-Herault, A. Pasc, M. Michau, D. Cot, E. Petit, M. Barboiu, *Angew. Chem. Int. Ed.* **2007**, *46*, 8409–8413.
215. L. Ma, M. Melegari, M. Colombini, J. T. Davis, *J. Am. Chem. Soc.* **2008**, *130*, 2938–2939.
216. L. Ma, W. A. Harrell, J. T. Davis, *Org. Lett.* **2009**, *11*, 1599–1602.
217. K. Sanderson, *Nature* **2011**, *464*, 158–159.
218. P. W. K. Rothemund, *Nature* **2006**, *440*, 297–302.
219. N. C. Seeman, *Annu. Rev. Biochem.* **2010**, *79*, 65–87.
220. C. K. McLaughlin, G. D. Hamblin, H. F. Sleiman, *Chem. Soc. Rev.* **2011**, *40*, 5647–5656.
221. H. Lu, Y. Chen, Y. He, A. Ribbe, C. Mao, *Angew. Chem. Int. Ed.* **2006**, *45*, 1942–1945.
222. C. Lin, Y. Ke, Y. Liu, M. Mertig, J. Gu, H. Yan, *Angew. Chem. Int. Ed.* **2006**, *46*, 6089–6092.
223. T. L. Sobey, S. Renner, F. C. Simmel, *J. Phys., Condens. Matter* **2009**, *21*, 034112.
224. J. Burns, E. Stulz, S. Howorka, *Nano Lett.* **2013**, *13*, 2351–2356.
225. P. K. Lo, P. Karam, F. A. Aldaye, C. K. McLaughlin, G. D. Hamblin, G. Cosa, H. F. Sleiman, *Nat. Chem.* **2010**, *2*, 319–328.
226. P. K. Lo, F. Altvater, H. F. Sleiman, *J. Am. Chem. Soc.* **2010**, *132*, 10212–10214.
227. M. Langecker, V. Arnaut, T. G. Martin, J. List, S. Renner, M. Mayer, H. Dietz, F. C. Simmel, *Science* **2012**, *338*, 932–936.
228. J. E. Reiner, A. Balijepalli, J. W. F. Robertson, J. Campbell, J. Suehle, J. J. Kasianowicz, *Chem. Rev.* **2012**, *112*, 6431–6451.
229. J. R. Burns, K. Göpfrich, J. W. Wood, V. V. Thacker, E. Stulz, U. F. Keyser, S. Howorka, *Angew. Chem. Int. Ed.* **2013**, *52*, 12069–12072.
230. L.-A. Fendt, I. Bouamaied, S. Thöni, N. Amiot, E. Stulz, *J. Am. Chem. Soc.* **2007**, *129*, 15319–15329.
231. J. R. Burns, S. Preus, D. G. Singleton, E. Stulz, *Chem. Commun.* **2012**, *48*, 11088–11090.
232. M. Barboiu, Y. Le Duc, A. Gilles, P.-A. Cazade, M. Michau, Y. M. Legrand, A. Van der Lee, B. Coasne, P. Parvizi, J. Post, T. Fyles, *Nat. Commun.* **2014**, *5*, No. 4142.
233. A. J. M. Van Beijnen, R. J. M. Nolte, J. W. Zwikker, W. Drenth, *Recl. Trav. Chim. Pays-Bas* **1982**, *101*, 409–410.

234. J. G. Neevel, R. J. M. Nolte, *Tetrahedron Lett.* **1984**, *25*, 2263–2266.
235. M. F. M. Roks, R. J. M. Nolte, *Macromolecules* **1992**, *25*, 5398–5407.
236. F. Otis, M. Auger, N. Voyer, *Acc. Chem. Res.* **2013**, *46*, 2934–2943.
237. J. C. Meillon, N. Voyer, *Angew. Chem. Int. Ed. Engl.* **1997**, *36*, 967–969.
238. Y. R. Vandenburg, B. D. Smith, E. Biron, N. Voyer, *Chem. Commun.* **2002**, 1694–1695.
239. F. Otis, C. Racine-Berthiaume, N. Voyer, *J. Am. Chem. Soc.* **2011**, *133*, 6481–6483.
240. B. A. Wallace, *J. Struct. Biol.* **1998**, *121*, 123–141.
241. J.-Y. Winum, S. Matile, *J. Am. Chem. Soc.* **1999**, *121*, 7961–7962.
242. N. Sakai, D. Gerard, S. Matile, *J. Am. Chem. Soc.* **2001,** *123,* 2517–2524.
243. G. W. Gokel, R. Ferdani, J. Liu, R. Pajewski, H. Shabany, P. Uetrecht, *Chem. Eur. J.* **2001**, *7*, 33–39.
244. C. D. Hall, G. J. Kirkovits, A. C. Hall, *Chem. Commun.* **1999**, 1897–1898.
245. M. Tsikolia, A. C. Hall, C. Suarez, Z. O. Nylander, S. M. Wardlaw, M. E. Gibson, K. L. Valentine, L. N. Onyewadume, D. A. Ahove, M. Woodbury, M. M. Mongare, C. D. Hall, Z. Wang, B. Draghici, A. R. Katritzky, *Org. Biomol. Chem.* **2009**, *7*, 3862–3870.
246. R. J. Brea, C. Reiriz, J. R. Granja, *Chem. Soc. Rev.* **2010**, *39*, 1448–1456.
247. N. Khazanovich, J. R. Granja, D. E. McRee, R. A. Milligan, M. R. Ghadiri, *J. Am. Chem. Soc.* **1994**, *116*, 6011–6012.
248. J. D. Hartgerink, J. R. Granja, R. A. Milligan, M. R. Ghadiri, *J. Am. Chem. Soc.* **1996**, *118*, 43–50.
249. M. R. Ghadiri, J. R. Granja, L. K. Buehler, *Nature* **1994**, *369*, 301–304.
250. J. Montenegro, M. R. Ghadiri, J. R. Granja, *Acc. Chem. Res.* **2013**, *46*, 2955–2965.
251. N. Rodríguez-Vázquez, H. L. Ozores, A. Guerra, E. González-Freire, A. Fuertes, M. Panciera, J. M. Priegue, J. Outeiral, J. Montenegro, R. García-Fandiño, M. Amorín, J. R. Granja, *Curr. Top. Med. Chem.* **2014**, *14,* 2647–2661.
252. T. D. Clark, J. M. Buriak, K. Kobayashi, M. P. Isler, D. E. McRee, M. R. Ghadiri, *J. Am. Chem. Soc.* **1998**, *120*, 8949–8962.
253. P. I. Haris, D. Chapman, *Biopolymers* **1995**, *37*, 251–263.
254. S. Krimm, J. Bandekar, *Adv. Protein Chem.* **1986**, *38*, 181–364.
255. J. Bandekar, *Biochim. Biophys. Acta* **1992**, *1120*, 123–143.
256. H. S. Kim, J. D. Hartgerink, M. R. Ghadiri, *J. Am. Chem. Soc.* **1998**, *120*, 4417–4424.
257. J. R. Granja, M. R. Ghadiri, *J. Am. Chem. Soc.* **1994**, *116*, 10785–10786.
258. J. Sánchez-Quesada, H. S. Kim, M. R. Ghadiri, *Angew. Chem. Int. Ed.* **2001**, *40*, 2503–2506.
259. J. Sánchez-Quesada, M. P. Isler, M. R. Ghadiri, *J. Am. Chem. Soc.* **2002**, *124*, 10004–10005.
260. M. Danial, C. M.-N. Tran, K. A. Jolliffe, S. Perrier, *J. Am. Chem. Soc.* **2014**, *136*, 8018–8026.
261. M. Danial, C. M.-N. Tran, P. G. Young, S. Perrier, K. A. Jolliffe, *Nat. Commun.* **2013**, *4*, No. 2780.
262. D. Seebach, J. L. Matthews, A. Meden, T. Wessels, C. Baerlocher, L. B. McCusker, *Helv. Chim. Acta* **1997**, *80*, 173–182.
263. L. Li, H. Zhan, P. Duan, J. Liao, J. Quan, Y. Hu, Z. Chen, J. Zhu, M. Liu, Y.-D. Wu, J. Deng, *Adv. Funct. Mater.* **2012**, *22*, 3051–3056.
264. A. Guerra, R. J. Brea, M. Amorín, L. Castedo, J. R. Granja, *Org. Biomol. Chem.* **2012**, *10*, 8762–8766.
265. D. Gauthier, P. Baillargeon, M. Drouin, Y. L. Dory, *Angew. Chem. Int. Ed.* **2001**, *40*, 4635–4638.
266. S. Leclair, P. Baillargeon, R. Skouta, D. Gauthier, Y. Zhao, Y. L. Dory, *Angew. Chem. Int. Ed.* **2004**, *43*, 349–353.
267. I. L. Karle, B. K. Handa, C. H. Hassall, *Acta Cryst. Sect. B* **1975**, *31*, 555–560.
268. M. Amorín, L. Castedo, J. R. Granja, *J. Am. Chem. Soc.* **2003**, *125*, 2844–2845.

269. C. Reiriz, R. J. Brea, R. Arranz, J. L. Carrascosa, A. Garibotti, B. Manning, J. M. Valpuesta, R. Eritja, L. Castedo, J. R. Granja, *J. Am. Chem. Soc.* **2009**, *131*, 11335–11337.
270. J. Montenegro, C. Vázquez-Vázquez, A. Kalinin, K. E. Geckeler, J. R. Granja, *J. Am. Chem. Soc.* **2014**, *136*, 2484–2491.
271. W. S. Horne, C. D. Stout, M. R. Ghadiri, *J. Am. Chem. Soc.* **2003**, *125*, 9372–9376.
272. P. K. Kienker, W. F. Degrado, J. D. Lear, *Proc. Natl. Acad. Sci. USA* **1994**, *91*, 4859–4863.
273. T. D. Clark, L. K. Buehler, M. R. Ghadiri, *J. Am. Chem. Soc.* **1998**, *120*, 651–656.
274. R. J. Brea, M. Amorín, L. Castedo, J. R. Granja, *Angew. Chem. Int. Ed.* **2005**, *44*, 5710–5713.
275. M. Amorín, R. J. Brea, L. Castedo, J. R. Granja, *Org. Lett.* **2005**, *7*, 4681–4684.
276. M. Amorín, L. Castedo, J. R. Granja, *Chem. Eur. J.* **2005**, *11*, 6543–6551.
277. R. J. Brea, L. Castedo, J. R. Granja, *Chem. Commun.* **2007**, *31*, 3267–3269.
278. N. Rodríguez-Vázquez, S. Salzinger, L. F. Silva, M. Amorín, J. R. Granja, *Eur. J. Org. Chem.* **2013**, *17*, 3477–3493.
279. R. García-Fandiño, J. R. Granja, M. D'Abramo, M. Orozco, *J. Am. Chem. Soc.* **2009**, *131*, 15678–15686.
280. R. García-Fandiño, J. R. Granja, *J. Phys. Chem. C.* **2013**, *117*, 10143–10162.
281. R. García-Fandiño, M. Amorín, L. Castedo, J. R. Granja, *Chem. Sci.* **2012**, *3*, 3280–3285.
282. C. Reiriz, M. Amorín, R. García-Fandiño, L. Castedo, J. R. Granja, *Org. Biomol. Chem.* **2009**, *7*, 4358–4361.
283. For a clever approach for nanotube functionalization, see also: R. Hourani, C. Zhang, R. Van der Weegen, L. Ruiz, C. Li, S. Keten, B. A. Helms, T. Xu, *J. Am. Chem. Soc.* **2011**, *133*, 15296–15299.
284. V. Semetey, C. Didierjean, J.-P. Briand, A. Aubry, G. Guichard, *Angew. Chem. Int. Ed.* **2002**, *41*, 1895–1898.
285. L. Fischer, M. Decossas, J.-P. Briand, C. Didierjean, G. Guichard, *Angew. Chem. Int. Ed.* **2009**, *48*, 1625–1628.
286. J. L. López, E. M. Pérez, P. M. Viruela, R. Viruela, E. Ortí, N. Martín, *Org. Lett.* **2009**, *11*, 4524–4527.
287. A. Hennig, L. Fischer, G. Guichard, S. Matile, *J. Am. Chem. Soc.* **2009**, *131*, 16889–16895.
288. A. Cazacu, C. Tong, A. Van der Lee, T. M. Fyles, M. Barboiu, *J. Am. Chem. Soc.* **2006**, *128*, 9541–9548.
289. A. Cazacu, Y.-M. Legrand, A. Pasc, G. Nasr, A. Van der Lee, E. Mahon, M. Barboiu, *Proc. Natl. Acad. Sci. USA* **2009**, *106*, 8117–8122.
290. U. Beginn, G. Zipp, M. Möller, *Adv. Mater.* **2000**, *12*, 510–513.
291. V. Percec, G. Johansson, J. Heck, G. Ungarb, S. V. Battyb, *J. Chem. Soc. Perkin Trans. 1*, **1993**, 1411–1420.
292. B. Gong, Z. Shao, *Acc. Chem. Res.* **2013**, *46*, 2856–2866.
293. A. J. Helsel, A. L. Brown, K. Yamato, W. Feng, L. Yuan, A. J. Clements, S. V. Harding, Z. Shao, Z. F. Shao, B. Gong, *J. Am. Chem. Soc.* **2008**, *130*, 15784–15785.
294. X. Zhou, G. Liu, K. Yamato, Y. Shen, R. Cheng, X. Wei, W. Bai, Y. Gao, H. Li, F. Liu, F. T. Liu, D. M. Czajkowsky, J. Wang, M. J. Dabney, Z. Cai, J. Hu, F. V. Bright, L. He, X. C. Zeng, Z. Shao, B. Gong, *Nat. Commun.* **2012**, *3*, No. 949.
295. Z. Qi, M. Sokabe, K. Donowaki, H. Ishida, *Biophys. J.* **1999**, *76*, 631–641.
296. H. Ishida, Z. Qi, M. Sokabe, K. Donowaki, Y. Inoue, *J. Org. Chem.* **2001**, *66*, 2978–2989.
297. H. Ishida, M. Suga, K. Donowaki, K. Ohkubo, *J. Org. Chem.* **1995**, *60*, 5374–5375.
398. S. Kubic, *J. Am. Chem. Soc.* **1999**, *121*, 5846–5855.
399. S. Kubic, R. Goddard, *J. Org. Chem.* **1999**, *64*, 9475–9486.
300. J. Bitta, S. Kubic, *Org. Lett.* **2001**, *3*, 2637–2640.
301. C. De Cola, S. Liscen, D. Comegna, E. Cafaro, G. Bifulco, I. Izzo, P. Tecilla, F. De Riccardis, *Org. Biomol. Chem.*, **2009**, *7*, 2851–2854.
302. Y. J. Jeon, H. Kim, S. Jon, N. Selvapalam, D. Y. Oh, I. Seo, C.-S. Park, S. R. Jung, D.-S. Koh, K. Kim, *J. Am. Chem. Soc.* **2004**, *126*, 15944–15945.

303. X. X. Zhang, K. E. Krakowiak, G. Xue, J. S. Bradshaw, R. M. Izatt, *Ind. Eng. Chem. Res.* **2000**, *39*, 3516–3520.
304. J. K. W. Chui, T. M. Fyles, *Chem. Commun.* **2010**, *46*, 4169–4171.
305. H. Fenniri, P. Mathivanan, K. L. Vidale, D. M. Sherman, K. Hallenga, K. V. Wood, J. G. Stowell, *J. Am. Chem. Soc.* **2001**, *123*, 3854–3855.
306. T. Kato, T. Matsuoka, M. Nishii, Y. Kamikawa, K. Kanie, T. Nishimura, E. Yashima, S. Ujiie, *Angew. Chem. Int. Ed.* **2004**, *43*, 1969–1972.
307. N. Sakai, Y. Kamikawa, M. Nishii, T. Matsuoka, T. Kato, S. Matile, *J. Am. Chem. Soc.* **2006**, *128*, 2218–2219.
308. H. Fenniri, B.-L. Deng, A. E. Ribbe, K. Hallenga, J. Jacob, P. Thiyagarajan, *Proc. Natl. Acad. Sci. USA* **2002**, *99*, 6487–6492.
309. T. M. Fyles, C. C. Tong, *New J. Chem.* **2007**, *31*, 655–661.
310. U. Devi, J. R. D. Brown, A. Almond, S. J. Webb, *Langmuir* **2011**, *27*, 1448–1456.
311. M. Boccalon, E. Iengo, P. Tecilla, *J. Am. Chem. Soc.* **2012**, *134*, 20310–20313.
312. A. Satake, M. Yamamura, M. Oda, Y. Kobuke, *J. Am. Chem. Soc.* **2008**, *130*, 6314–6315.
313. M. Fujita, M. Tominaga, A. Hori, B. Therrien, *Acc. Chem. Res.* **2005**, *38*, 371–380.
314. S. R. Seidel, P. J. Stang, *Acc. Chem. Res.* **2002**, *35*, 972–983.
315. D. L. Caulder, K. N. Raymond, *Acc. Chem. Res.* **1999**, *32*, 975–982.
316. L. R. MacGillivray, J. L. Atwood, *Angew. Chem. Int. Ed.* **1999**, *38*, 1018–1033.
317. M. Jung, H. Kim, K. Baek, K. Kim, *Angew. Chem. Int. Ed.* **2008**, *47*, 5755–5757.
318. H. Furukawa, J. Kim, K. E. Plass, O. M. Yaghi, *J. Am. Chem. Soc.* **2006**, *128*, 8398–8399.
319. M. Eddaoudi, J. Kim, J. B. Wachter, H. K. Chae, M. O'Keeffe, O. M. Yaghi, *J. Am. Chem. Soc.* **2001**, *123*, 4368–4369.
320. O. V. Kulikov, R. Li, G. W. Gokel, *Angew. Chem. Int. Ed.* **2009**, *48*, 375–377.
321. S. J. Dalgarno, N. P. Power, J. E. Warren, J. L. Atwood, *Chem. Commun.* **2008**, 1539–1541

Chapter 15
Lithium in Medicine: Mechanisms of Action

Duarte Mota de Freitas, Brian D. Leverson, and Jesse L. Goossens

Contents

D. Mota de Freitas (✉) • B.D. Leverson • J.L. Goossens
Department of Chemistry and Biochemistry, Loyola University Chicago,
1068 West Sheridan Road, Chicago, IL 60660, USA
e-mail: dfreita@luc.edu; bleverson@luc.edu; jgoossens@luc.edu

A. Sigel, H. Sigel, and R.K.O. Sigel (eds.), *The Alkali Metal Ions: Their Role for Life*,
Metal Ions in Life Sciences 16, DOI 10.1007/978-3-319-21756-7_15

Abstract In this chapter, we review the mechanism of action of lithium salts from a chemical perspective. A description on how lithium salts are used to treat mental illnesses, in particular bipolar disorder, and other disease states is provided. Emphasis is not placed on the genetics and the psychopharmacology of the ailments for which lithium salts have proven to be beneficial. Rather we highlight the application of chemical methodologies for the characterization of the cellular targets of lithium salts and their distribution in tissues.

Keywords Bipolar disorder • Glycogen synthase kinase-3β • Lithium • Mechanisms of action

Please cite as: *Met. Ions Life Sci.* 16 (2016) 557–584

1 Introduction

Lithium (*lithos* for stone in Greek) was discovered in Sweden in the early part of the nineteenth century. It is an alkali metal and the third element on the Periodic Table. The electronic configuration of the lithium atom is $1s^2 2s^1$. The outer electron is easily removed thereby forming the lithium ion (Li^+), which has the stable electronic configuration, $1s^2$, of the noble gas helium. Its high reactivity explains why Li never exists by itself, thus accounting for its preferred occurrence as Li^+ in seawater and in minerals. In organolithium compounds, lithium forms a polar covalent bond with carbon by sharing its outer electron [1]. In non-aqueous media, organolithium compounds are a source of carbanions that are useful for the formation of carbon–carbon bonds [1]. Li and Li^+ have found important industrial applications, including the manufacture of lithium-ion batteries [2], ceramics and alloys [3], and as a source of tritium for nuclear usage [4]. The solvent in living systems is water. For lithium salts used in therapy, primarily lithium carbonate, the outer electron of Li is completely transferred to the anion to form the Li^+ cation. Henceforth, any referral to lithium signifies Li^+ and not Li, because it is Li^+ that is therapeutically relevant, and not the anion with which it forms a counterion.

Unlike other alkali and alkaline earth metal ions, such as Na^+, K^+, Mg^{2+}, and Ca^{2+}, Li^+ is only present in trace amounts in normal individuals. It therefore begs the question: How can Li^+ impact the levels of these four important ions in human physiology? General trends in ionic radii stem from their positions on the Periodic Table: (i) increases down a group due to population of additional electron shells; and (ii) decreases across a period because of a build-up in nuclear charge within an electron shell. It follows from these trends that ionic radii derived from elements that are positioned diagonally on the periodical table have similar ionic radii, the so-called *diagonal relationship*.

Table 1 Ionic radii of Li^+ and other biochemically-related radii [a,b]

Ion	r (CN=4)	r (CN=6)
Li^+	59	76
Mg^{2+}	57	72
Na^+	99	102
Ca^{2+}	not available	100

[a]r, ionic radius; CN, coordination number.
[b]Data collected from [5].

In Table 1, the listed coordination numbers and the r values (in pm), are from Shannon [5]. As it is apparent from Table 1, when six ligands are bound to the central metal ion, Li^+ and Mg^{2+} have similar ionic radii; the same relationship applies to coordination number four. Both Li^+ and Mg^{2+} are hard acids that prefer hard oxygen-containing side chains of amino acids, such as aspartate, glutamate, and asparagine. It is important to note that there is a two-fold difference in the charges of the two ions. In addition to stronger binding affinities for Mg^{2+}, the differences in charge density may account for distinct preferences in coordination number and geometries (4 and tetrahedral for Li^+ [6], and, for Mg^{2+}, 6 and octahedral [7]). More details on the inorganic chemistry of Li^+ can be found in our previously published review [8].

Frausto da Silva and Williams [9] first proposed that competition between Li^+ and Mg^{2+} bound to biological molecules may be the underlying theme in Li^+ interactions in cells. This hypothesis has since been demonstrated experimentally by us and others using a myriad of spectroscopic techniques [8].

Mg^{2+} is ubiquitous in the cell and its presence is essential for the function of many enzymes and other biological processes. If Li^+ were to dislodge Mg^{2+} from its native binding sites, the cell machinery would be disrupted. It is therefore feasible that, in an Mg^{2+}-containing protein that has both high-affinity and low-affinity sites, Li^+ would compete for the low-affinity Mg^{2+} binding site. Studies using density functional theory indicate that solvent accessibility, net charge, and ligand environment of the Mg^{2+} sites may all play a part in metal ion competition [10].

For adult patients, lithium carbonate doses of 900 to 1,200 mg/day result in steady-state lithium levels in the range of 0.5 mM to 1.2 mM, which are optimal for effective maintenance of bipolar patients [11]. For tissues in a 70 kg-adult, the total and free concentrations of Mg^{2+} are, respectively, in the range of 20 mM and 0.2 mM [12]. Depending on the geographical location, the highest reported Li^+ concentration in tap water is approximately 160 μg/L, which is three orders of magnitude less than that used in the treatment of bipolar disorder and depression. Although there are only trace amounts of Li^+ in drinking water, some studies have suggested that Li^+ may aid in suicide prevention both in patients with mood disorders and in normal individuals [13]. If indeed the putative correlation between suicide prevention and lithium levels in tap water is valid, a mechanism other than ion competition at low ratios of Li^+/bound Mg^{2+} must be present.

2 Uses of Lithium in Medicine

Originally used to treat bipolar disorder, the use of Li^+ has expanded to other illnesses such as Alzheimer's disease and thyroid cancer. Although lacking specificity, Li^+ used in tandem with other medications has proven to be effective.

2.1 *Bipolar Disorders*

Bipolar disorders (BD) are debilitating psychiatric mood disorders characterized by repeating cycles of depression and mania. Diagnosis of each type of BD is based on meeting a strict set of criteria describing: mania, hypomania, and major depression [14]. BD type I manifests itself as a manic state, which may be followed by hypomania or major depression. This differs from BD type II, which is characterized by both a hypomanic and major depressive state, while unipolar patients suffer solely from major depression. Approximately 2 % of the world population struggles with this disease while 10–12 % commit suicide to escape the crippling emotional roller coaster [15].

It has been well established that Li^+ is effective in treating manic states. Although there have been many drugs developed to treat BD, such as valproate and lurasidone, their effectiveness varies depending on several factors, including the type of BD and unique neurological profile of the patient. With regards to BD type I disorder, after an initial delay of 1–2 weeks, Li^+ has a response rate of 70–80 % and improvement continues until it reaches a plateau at 3–4 weeks [16]. These mood stabilizing effects can continue indefinitely without recurrences and is not immediately reversible after treatment is discontinued [17]. For treatment of BD type II disorder, Li^+ efficacy is seen at 6–8 weeks [18] and its effectiveness is the greatest soon after diagnosis [19]. Although anti-psychotic drugs, such as quetiapine and olanzapine, are very effective, Li^+ is still prescribed for treatment of BD type II disorder [20].

2.2 *Alzheimer's Disease*

Alzheimer's disease (AD) is a common chronic form of dementia characterized by loss of memory, change in behavior, and inability to concentrate. The exact cause of AD is still unknown but identifiable histopathological trademarks have been found, such as hyperphosphorylated neuronal microtubule-associated protein (tau), amyloid plaques, and a decrease in brain-delivered neurotropic factor (BDNF). Li^+ has the unique ability to repress each of these AD phenotypes. In a cross sectional study, older BD patients undergoing chronic Li^+ treatment had a significantly lower incidence of AD compared to those untreated or the general population (3 % *versus* 19 % respectively) [21]. These findings suggest that chronic Li^+ treatment of BD patients can protect against the development of dementia-type disorders, in particular AD [22].

The precise mechanism by which Li^+ delays the progression of AD has yet to be determined. An understanding of the involvement of tau in AD is emerging. Tau protein promotes and stabilizes microtubule formation, a process that is regulated by glycogen synthase kinase 3 (GSK-3). In AD patients, tau has been found to be hyperphosphorylated by GSK-3, causing the formation of paired helical fibers and the weakening of the cytoskeleton [23]. Li^+ reduces GSK-3-induced tau phosphorylation at multiple epitopes in both transfected cells and in neurons [24, 25]. The effects are both time- and concentration-dependent and effective at 1 mM and 3 mM. It was also demonstrated that Li^+-induced changes in tau phosphorylation are not the result of neurotoxicity, since even at 25 mM Li^+ hippocampal cell morphology was unaffected and the changes in tau phosphorylation were entirely reversible [24, 25].

Like tau proteins, amyloid β-peptides (Aβ) are thought to be one of the primary factors that contribute to AD. In recent years, the connection between Aβ plaques and AD has been called into question because the concentration of Aβ is not correlated to the degree of cognitive impairment observed in AD subjects. Preceding the onset of pathology, Li^+ decreased the rate of formation of toxic plaques, and prevented degeneration of memory in a PS1xAPP mouse model [26]. Another group found that inhibition of GSK-3 with Li^+ led to a decrease of the concentration of amyloid precursor protein by enhancing its degradation [27]. Although in both of these studies Li^+ appeared to act as an inhibitor, there were other investigations that contradicted these results [28, 29].

Due to the neuroprotective effects observed in BD patients, it has been proposed that Li^+ could be used to treat AD. Clinical trials have begun to investigate the effect Li^+ has on AD patients. Li^+ has been shown to increase BDNF, which has been linked to a delay and/or stagnation of the cognitive impairment symptoms in AD patients [30, 31].

2.3 *Thyroid Cancer Treatment*

Effects of Li^+ on patients with thyroid cancer are well documented and give insight on possible mechanistic pathways. Patients undergoing Li^+ treatment have a higher probability of developing hyperthyroidism. This side effect can be useful because Li^+ can inhibit thyroid hormones and increase retention of radioactive iodine. Therefore, Li^+ can be used as an adjunctive therapy in the treatment of hyperthyroidism and thyroid cancer [32, 33].

Li^+ is thought to inhibit coupling of iodotyrosine residues in the formation of thyroxine (T4) and triiodothyrosine (T3), which control the release of iodothyronines. Li^+ may also affect deiodinase activity, which originates from a group of enzymes that activates thyroid hormones by converting T4 to T3, or inactivate hormones by converting either T4 to reverse triiodothyronine or T3 into inactive diiodothyronine [32]. In addition, Li^+ can participate in a feedback loop mechanism by increasing intrathyroidal iodine, and inhibiting the release of T3 and T4.

The excretion of thyroid hormones leads to a decrease of T3 and T4 levels in serum during iodine therapy [33, 34]. Due to this feedback loop, iodine retention is increased without interfering with iodine uptake, increasing radioactive iodine concentration in normal and cancer thyroid cells [35].

3 Methods of Lithium Detection

3.1 Overview of Tools for Lithium Detection

Analytical methods available for the direct detection of Li^+ in blood and other tissues fall under two categories: (i) those that measure *total* Li^+ concentrations (atomic absorption or flame emission spectrometry, inductively-coupled plasma mass spectrometry, solution ^{7}Li NMR (nuclear magnetic resonance spectroscopy) and solution ^{7}Li MRSI (magnetic resonance with spectroscopic imaging)) and (ii) those that detect *free* Li^+ concentrations (optical or fluorescence indicators, ion-selective microelectrodes, solution ^{7}Li NMR and solid-state ^{7}Li NMR). If competition between Li^+ and bound Mg^{2+} were present, an increase in total Li^+ concentration, $[Li^+]_{total}$, would increase the concentration of free Mg^{2+}, $[Mg^{2+}]_{free}$, and decrease the concentration of bound Mg^{2+}, $[Mg^{2+}]_{bound}$, with the total Mg^{2+} concentration, $[Mg^{2+}]_{total}$, remaining intact.

The total concentrations of Li^+ in blood of bipolar patients on lithium therapy are well within the detection limits of atomic absorption or flame spectrometry, making them the methods of choice for assessing drug response and toxicity in the clinic, as well as patient compliance. The principles of atomic absorption and flame spectrometry are beyond the scope of this chapter, and for more details the reader is referred to an earlier review on these methodologies [36].

Li^+-selective chromophores, fluorophores, and microelectrodes have been designed based on the premise that, relative to other alkali and alkaline earth metal ions, some crown ethers and cryptands selectively bind Li^+ [37]. However, the large excess of Na^+ over Li^+ in biological fluids and the low Li^+/Na^+ selectivity ratio place limitations on the usefulness of these methods for measuring free Li^+ concentrations, $[Li^+]_{free}$.

The principles behind other more informative methods, such as ^{7}Li NMR spectroscopy and ^{7}Li MRSI, or indirect detection of Mg^{2+} with the aid of fluorophores, are explained below, and their applications are described thereafter in this chapter. Before continuing with the description of these spectroscopic methods, it is important to point out that no radioactive isotope of lithium exists. Also, Li^+ and Mg^{2+} have no unpaired d electrons, and therefore they are not amenable to study by UV-Vis spectroscopy and electron paramagnetic resonance. In addition, Li^+ only has two valence electrons and thus, Li^+ bound to proteins cannot be detected directly by X-ray crystallography. Indeed, no crystal structure with bound Li^+ has been deposited in the Protein Data Bank. In contrast, the higher electron density of Mg^{2+} enables the investigation by X-ray diffraction of Mg^{2+} binding sites in proteins, as shown in the crystallographic structures described in Section 5.

3.2 Lithium Nuclear Magnetic Resonance Spectroscopy

In a one-dimensional NMR experiment, a sample is placed in a magnetic field and transitions among nuclear states are induced by pulses of radiofrequency; between successive pulses, a delay is required to allow for decay of the net magnetization from excited states to the ground state, a process that is quantified by spin-lattice (T_1) and spin-spin (T_2) relaxation times [38]. NMR-active nuclei have a nuclear spin, I, greater than zero, but, when I is larger than ½, quadrupolar relaxation occurs, resulting in broad lines.

Two Li isotopes exist, ^{7}Li and ^{6}Li, with natural abundances of 92.6 % and 7.4 %; because their nuclear spin values are 3/2 and 1, respectively, either nucleus can be detected by NMR spectroscopy. The quadrupole moments of both nuclei are relatively small, resulting in relatively sharp peaks [38]. In a magnetic field of 11.7 Tesla (in which ^{1}H resonates at 500 MHz), the resonance frequency for ^{7}Li is 194.3 MHz and 73.6 MHz for ^{6}Li. These frequencies are easily attainable with multinuclear NMR probes. Receptivity, the product of the natural abundance and the inherent sensitivity of a nucleus, provides insight on the easiness for the detection of a specific nucleus; relative to ^{1}H, the receptivities of ^{7}Li and ^{6}Li are, respectively, 0.27 and 6.31×10^{-4}. The T_1 values for ^{7}Li are in the msec range whereas those for ^{6}Li can be as large as one sec.

Two useful parameters in NMR spectroscopy are the chemical shift and the spin-spin coupling constant. From the number and positions of the chemical shifts, which are reported relative to a reference compound and expressed in ppm, one can garner structural information. Multiplet patterns in one-dimensional NMR spectra or cross-peaks in two-dimensional NMR spectroscopy are indicative of interactions between spins of different nuclei.

For organolithium compounds, Li atoms in distinct chemical environments exhibit different chemical shift values [39]. Similarly, depending on the nature of the carbon substituents, Li-C covalent bonds exhibit multiplets or cross peaks that result from coupling between ^{6}Li (and less often with ^{7}Li) and ^{13}C. Because the T_1 values of ^{6}Li are longer than those of ^{7}Li, solution and solid-state ^{6}Li NMR is preferable to ^{7}Li NMR for the structural analysis of organolithium compounds [39].

As a result of the larger natural abundance of ^{7}Li relative to ^{6}Li, solution ^{7}Li NMR spectroscopy is the method of choice for probing Li^+ within the 0.5 mM to 1.2 mM range used in therapy. In the cytosol, Li exists as a tetra- or hexaaqualithium(I) ion. Once bound to biomacromolecules, such as proteins or cell membranes, it is primarily present as a partially hydrated Li^+ ion. Consequently, the Li^+ environment in biological aqueous media does not vary greatly, resulting in a narrow and, for the most part, uninformative scale of chemical shift values. For example, in a solution ^{7}Li NMR study of Li^+ binding to inositol monophosphatase (IMPase), a small downfield shift of approximately 0.1 ppm was observed at a [protein]/[Li^+] ratio of *ca.* 4 [40]. In contrast, a magic-angle spinning, solid-state ^{7}Li NMR study of Li^+ binding to lyophilized *E. coli* IMPase resolved $[Li^+]_{free}$, $[Li^+]_{bound}$, and non-specifically bound Li^+ [41]. Because of the lack of covalent bonding between Li^+ and water or other hard oxygen-containing ligands, spin-spin coupling between the quadrupolar ^{7}Li nucleus and ligand hydrogens is not observed in solution [42].

However, useful information can be obtained from chemical shifts. Figure 1A [43] shows that, in a red blood suspension, intracellular Li^+ cannot be distinguished from extracellular Li^+. Representative lanthanide shift reagents, such as dysprosium triphosphate, $[Dy(PPP)_2]^{7-}$, or thulium 1,4,7,10-tetraazacyclodecane-*N,′N″,N‴*-tetramethylenephosphonate, $[HTm(DOTP)]^{4-}$, are, as a virtue of their high negative charge, impermeable through the hydrophobic interior of a cell membrane. Therefore, these reagents only shift the extracellular pool of Li^+; see Figure 1B for an example of packed red blood cells (RBCs) suspended in an isotonic medium containing $[Dy(PPP)_2]^{7-}$. A control experiment using coaxial tubes confirm the assignments (Figure 1C).

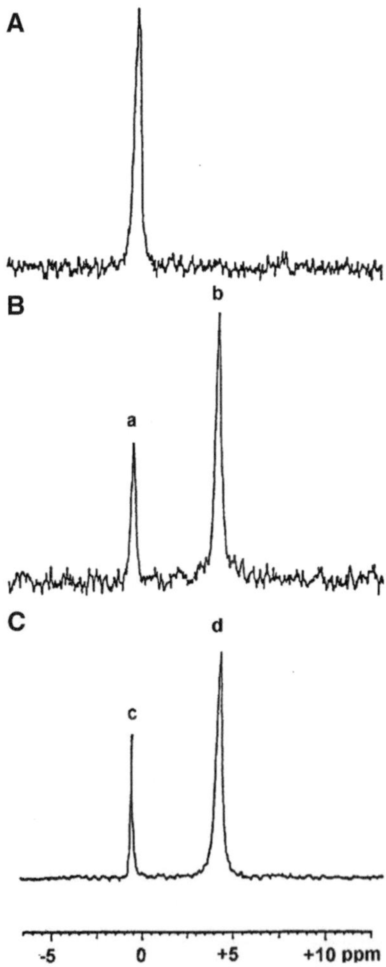

Figure 1 (**A**) ^{7}Li NMR spectrum (104.8 MHz, 37 °C) of gently packed RBCs (hematocrit was 13 %) in a medium containing 140 mM Na^+, 5 mM K^+, 3.5 mM Li^+, 10 mM glucose, and 50 mM HEPES at pH 7.5. Packed RBCs were incubated with 3.5 mM Li^+ at 37 °C for 12 h prior to NMR measurements. (**B**) ^{7}Li NMR spectrum of the same RBC suspension as in part A with the exception that 5 mM $Na_7Dy(PPP)_2$ was present in the medium instead of 50 mM NaC1. (**C**) ^{7}Li NMR spectrum of a concentric sample (5-mm inner tube inside a 10-mm outer tube) containing 20 mM Li^+ in the inner tube while the outer tube contained the same suspension medium as in part B. Labels a and c represent inner pools of Li^+ while b and d denote outer pools. Reproduced with permission from [43]; © copyright 1987, American Chemical Society.

The T_1/T_2 ratio is a measure of Li^+ immobilization. For $[Li^+]_{free}$, where the Li^+ rotational correlation time, τ_c (which is a scale of Li^+ motion) is short relative to the its NMR resonance frequency ω ($\omega\tau_c \ll 1$), $T_1 \approx T_2$. When Li^+ is immobilized in cells, τ_c is larger, and $\omega\tau_c \gg 1$, T_1 is longer than T_2 [44]. Determination of Li^+ binding constants, K_{Li}, to cellular fractions can be calculated as follows:

$$\frac{1}{\Delta R} = \frac{1}{(R_{obs} - R_{free})} = \frac{1}{K_{Li}[B](R_{bound} - R_{free})} + \frac{[Li^+]_{total}}{[B](R_{bound} - R_{free})} \quad (1)$$

where $[Li^+]_{total}$ and [B] are the concentrations of Li^+ and binding sites. R_{obs}, R_{free} and R_{bound} are, respectively, the values of the relaxation rates (the reciprocals of relaxation times), observed for the sample itself, with saturating $[Li^+]$, or with bound Li^+. From a James-Noggle plot [45], in which ΔR^{-1} is plotted against $[Li^+]_{total}$, K_{Li} can be calculated.

3.3 *Lithium Magnetic Resonance with Spectroscopy Imaging*

Unlike ^{7}Li NMR, ^{7}Li MR with spectroscopy imaging (MRSI) uses a fixed magnetic field and magnetic field gradients affording the opportunity to map distribution of Li^+ in organs, including human brains. T_1-weighted measurements enhance image resolution [46].

3.4 *Indirect Fluorescence Detection*

The levels of free Mg^{2+}, $[Mg^{2+}]_{free}$, can be measured by using Mg^{2+}-selective fluorescent dyes, such as furaptra [47]. As Figure 2 [48] indicates, indirect fluorescence detection of $[Mg^{2+}]_{free}$ can be used for probing competition between Li^+ and bound Mg^{2+} in biological samples, with $[Mg^{2+}]_{free}$ being calculated as per equation (2) [48]:

$$[Mg^{2+}]_{free} = \frac{K_d S_{min}(R - R_{min})}{S_{max}(R_{max} - R)} + [Li^+]_{free} \frac{K_d S'_{max}(R - R'_{max})}{K'_d S_{max}(R_{max} - R)} \quad (2)$$

where K_d and K'_d are, respectively, the dissociation constants of the furaptra complexes with Mg^{2+} and Li^+; S_{min}, S_{max} and S'_{max} are the fluorescence intensities at 374 nm in the absence and in the presence of saturating concentrations of Mg^{2+} and Li^+; and R, R_{min}, R_{max}, and R'_{max} are the fluorescence intensity ratios (342 nm/374 nm) observed for the sample, and in the absence and in the presence of saturating amounts of Mg^{2+} and Li^+. Competition between Li^+ and Mg^{2+} bound to a protein could be determined by adding a third term to equation (2) [49].

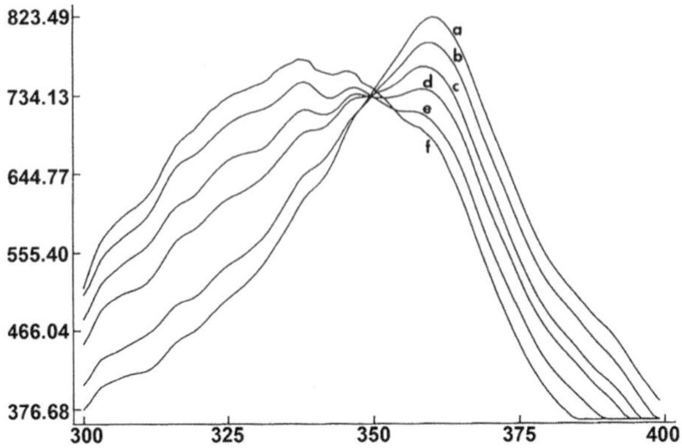

Figure 2 Fluorescence excitation spectra of 2 μM furaptra in a solution containing 5.0 mM ATP, 2.5 mM $MgCl_2$, 150 mM Tris-Cl (pH 7.4) and (a) 0 mM LiCl, (b) 5 mM LiCl, (c) 10 mM LiCl, (d) 20 mM LiCl, (e) 50 mM LiCl, and (f) 100 mM LiCl. Reproduced with permission from [48]; © copyright 1987, American Chemical Society.

4 Lithium in Living Systems

4.1 Lithium Transport Across Cell Membranes

Li^+ measures using atomic absorption and flame emission spectrophotometry may be error-prone because blood processing steps, such as centrifugation and cell lysis, may result in additional transmembrane Li^+ transport during sample preparation. ^{7}Li NMR studies are advantageous because cell suspensions can be investigated non-invasively. Figure 1 shows how RBC suspensions have been studied in this way. For nucleated cells (Figure 3 [50]), such as human neuroblastoma SHSY-5Y cells, continuous perfusion with aerated medium of gel-anchored cells is however required.

Li^+ uptake by RBCs takes place via a leak pathway, and through the transmembrane Cl^- exchanger or band 3 protein. The leak pathway contributes 70 % toward Li^+ influx and is thought to occur via water pores in the RBC membrane. In principle, a Li^+ cation could not use the Cl^- exchanger whose function is to transport Cl^- anions. However, CO_3^{2-} is believed to form an ion pair with Li^+, $[LiCO_3]^-$, resulting in net Li^+ influx in exchange with Cl^- efflux. Li^+ efflux from Li^+-loaded RBCs (for example, RBCs from blood of bipolar patients on Li^+ therapy), is mediated by the Na^+/Li^+ countertransporter. When the $[Na^+]$ gradient that it generated by active transport-mediated by Na^+/K^+-ATPase is partially dissipated, Li^+ efflux against a $[Li^+]$ gradient occurs, indicating that the Na^+/Li^+ countertransporter provides secondary active transport for Li^+. We later found out that Na^+/Li^+ countertransport is mediated by the Na^+/H^+ exchange protein located in the RBC membrane [51]. Mediation of Li^+ transport through other pathways, such as Na^+/K^+ cotransporter and Na^+/K^+-ATPase, has been reported; however, it does not occur at physiological concentrations of Na^+ and K^+.

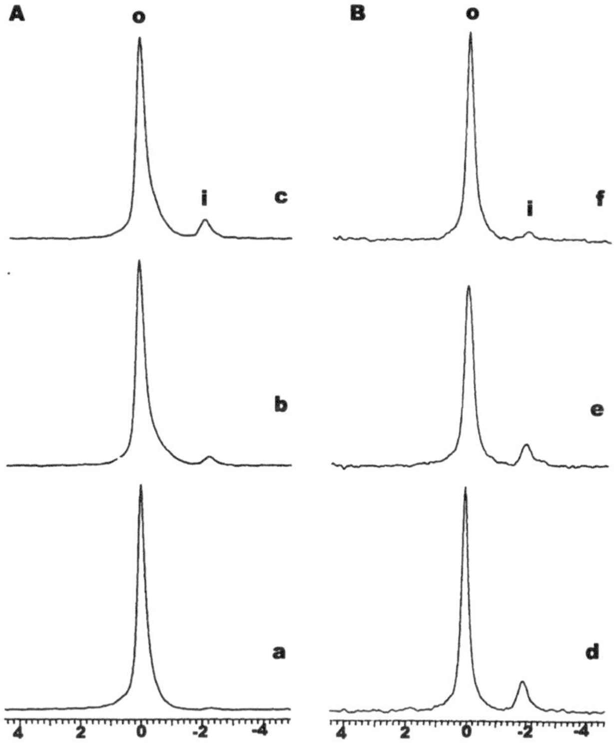

Figure 3 (**A**) ^{7}Li NMR spectra of Li^+ influx into human neuroblastoma SH-SY5Y cells embedded in agarose gel threads in a medium containing 3 mM $[HTmDOTP]^{4-}$ and 15 mM Li^+ at 11 min (spectrum a), 39 min (spectrum b) and 133 min (spectrum c) after starting the perfusion. (**B**) ^{7}Li NMR spectra of Li^+ efflux from Li^+-loaded cells at 75 min (d), 90 min (e), and 120 min (f). The symbols i and o denote the intracellular and extracellular ^{7}Li NMR signals. Reproduced with permission from [50]; © copyright 2002, John Wiley & Sons.

It was earlier speculated that the rates of Na^+/Li^+ countertransport in RBCs of bipolar patients undergoing Li^+ treatment were slower than for normal individuals, and that Li^+ ratios ($[Li^+]_{RBC}/[Li^+]_{plasma}$) were elevated for the patients. However, the rates of Na^+/Li^+ countertransport and of Li^+ ratios in RBCs of bipolar patients before and after undergoing chronic Li^+ therapy were not significantly different, indicating that the previously reported differences for RBCs of bipolar patients were not biological markers but a result of Li^+ treatment. Hypertensive patients have unequivocally shown to have significantly elevated rates of RBC Na^+/Li^+ countertransport relative to normotensive controls [52]. The cofounding effect of hypertension make earlier data from studies with RBCs of bipolar patients difficult to interpret. Details on RBC Li^+ transport pathways and their potential abnormalities have been previously described [36].

Instead of testing whether differences in rates of RBC Na^+/Li^+ countertransport and Li^+ ratios are biological markers, we studied RBCs from thirty Li^+-treated bipolar patients and found significant correlations between Li^+ affinity to the RBC membrane, as calculated from T_1 measures, and response to and toxicity of Li^+ therapy [53].

Figure 3 illustrates how Li^+ uptake and efflux can be monitored in eukaryotic cells, in this case human neuroblastoma SHSY-5Y cells [50]. Because oxygen is required for their viability, the cells, which were anchored on gel threads, were continuously perfused with an aerated medium. Addition to the medium of veratridine, a known inhibitor of voltage-sensitive Na^+ channels, reduced the rates of Li^+ uptake and efflux, suggesting that Li^+ transport may be mediated *via* Na^+ channels. A follow-up study indicated that 40 % of Li^+ uptake in the same cell line takes place *via* the Na^+/Ca^{2+} exchanger, with the leak pathway providing the remaining balance [54].

4.2 *Cellular Accumulation of Lithium*

We have used T_1/T_2 ratios to investigate Li^+ immobilization in intact human RBCs and in neuroblastoma SHSY-5Y cells, as well as in their cellular components [50, 55]. The cytosol and other subcellar fractions are viscous. Therefore, the viscosity of all fractions was adjusted to 5 cP (centipoise) by addition of glycerol. Under these conditions, a 3.0 mM viscosity-adjusted LiCl solution gave a T_1/T_2 ratio of approximately of 1. Table 2 indicates that the ratio is larger in intact SHSY-5Y cells than in RBCs, presumably because there are organelles present in the former that are absent in the latter. From the T_1/T_2 ratio values in subcellular fractions from SHSY-5Y cells, we concluded that the order of Li^+ immobilization is: plasma membrane ≫ microsomal membrane > nuclear membrane [50, 55].

We used the fluorescent dye furaptra to investigate Li^+/Mg^{2+} competition in human neuroblastoma SHSY-5Y cells. Chronic incubation of the cells for 24 to 72 hours with Li^+ resulted in a significant increase in $[Mg^{2+}]_{free}$, leading credence to the competition mechanism between Li^+ and bound Mg^{2+} in the therapeutic range (0.7 to 1.5 mM) [56].

Table 2 T_1/T_2 ratios for intact RBCs and SHSY-5Y cells as well as their subcellular fractions.[a]

Sample	$[Li^+]$ (mM)	T_1/T_2 ratio
RBCs	3.5	14
SHSY-5Y cells	2.9	100
RBC membrane	3.0	52
SHSY-5Y membrane	3.0	42
Cytosol-free homogenate	4.0	74
Cytosol-enriched fraction	4.0	6
Plasma-membrane fraction	4.0	106
Nuclear-membrane fraction	4.0	26
Mitochondrial-enriched fraction	4.0	14
Microsome-enriched fraction	4.0	52

[a]Data collected from [50] and [55].

4.3 Distribution of Lithium in Tissues

^{7}Li NMR has been used to study brains from patients with limited success; the images were however concentrated on wide brain slices or were unlocalized [57]. In contrast, with ^{7}Li MRSI, high-resolution images from brains of bipolar patients were obtained that allowed the determination of $[Li^+]_{total}$ and the observation of non-uniform Li^+ distribution [46]. Figure 4 [46] shows Li^+ brain maps from the same patient two weeks apart. In the first visit (Figure 4a), serum $[Li^+]$ was 0.3 mM lower than in the second visit for which the two sets of images were reproducible (Figures 4b and 4c). The detection of uneven Li^+ distribution in the brain of Li^+-treated bipolar patients is intriguing in that it may lead to the exploration of brain regions that are associated with emotion. This and similar studies show the promise on how Li^+ distribution can be investigated using Li^+ MRSI.

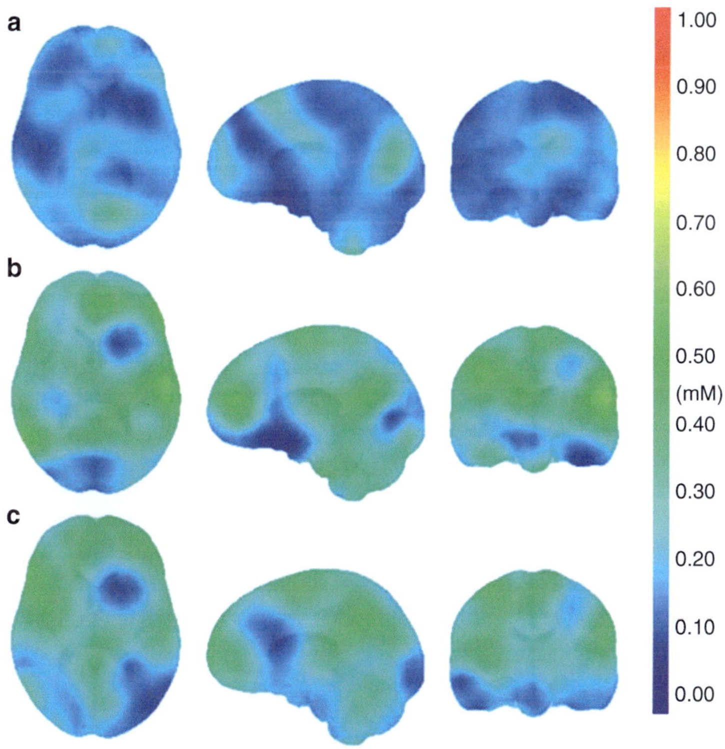

Figure 4 Comparison of Li^+ brain maps obtained from the same patient at visits two weeks apart. In the first visit (**a**), serum $[Li^+]$ was 0.5 mM, and, in the second visit, the two sets of images were at serum $[Li^+]$ of 0.8 mM (**b** and **c**). Reproduced with permission from [46]; © copyright 2012, John Wiley & Sons.

4.4 *Lithium Toxicity*

For most drugs, the concentration at which toxicity sets in is several orders of magnitude larger than the pharmacological level. In contrast, lithium treatment has to be monitored closely because at 2.0 mM significant side effects and toxicity occur, including, among others, weight gain, blurred vision, tremor [58], and, if left untreated, seizures and renal failure [59].

Weight gain impacts treatment compliance, and so do dermatologic side effects. Due to its high charge density, Li^+ has a very high hydration energy, which may account for the reported weight gain [60] and cutaneous adverse conditions [61].

5 Cellular Targets of Lithium

Considering the number of enzymes that utilize a magnesium cofactor, Li^+ acts on relatively few cellular targets. Curiously, Li^+ seems to mostly target enzymes associated with psychiatric disorders. This section provides an overview of the research into the mechanisms of action of Li^+ on some of the better studied Mg^{2+}-containing targets (Figure 5).

Li^+ inhibition is not limited to competition with the Mg^{2+} cofactor. Alternate mechanisms are thought to occur. This is the case for proteins such as the transcriptional factor cAMP response element-binding protein (CREB) [62, 63].

5.1 *Guanine Nucleotide Binding Proteins*

5.1.1 Mechanism and Li^+ Inhibition

Guanine nucleotide binding proteins (G-proteins) are regulatory membrane-bound proteins that play an indispensable role in transferring extracellular information across the cell membrane to affect intercellular functions. G-proteins are heterotrimeric in that they are composed of an α subunit (G_α), which regulates the activity of the effector protein, and a βγ subunit complex. G-proteins are attached to G-protein coupled receptors (GPCR), which are a type of hepta-helical serpentine transmembrane domain receptors. G-proteins regulate intracellular signaling cascades in response to specific GPCR activation by neurotransmitters or hormones. Once activated, GPCRs induce conformational changes in the G_α subunits, which contain the guanine nucleotide, resulting in exchange of GDP for GTP and dissociation of the βγ dimer. The dissociated α subunit then binds to its specific effector where it will stimulate or inhibit. This process is self-regulated with hydrolysis of bound GTP to GDP, effectively deactivating G_α and reforming the heterotrimeric GDP-bound G-protein conformation [64]. G_α subunits are separated into 4 primary classes based on their

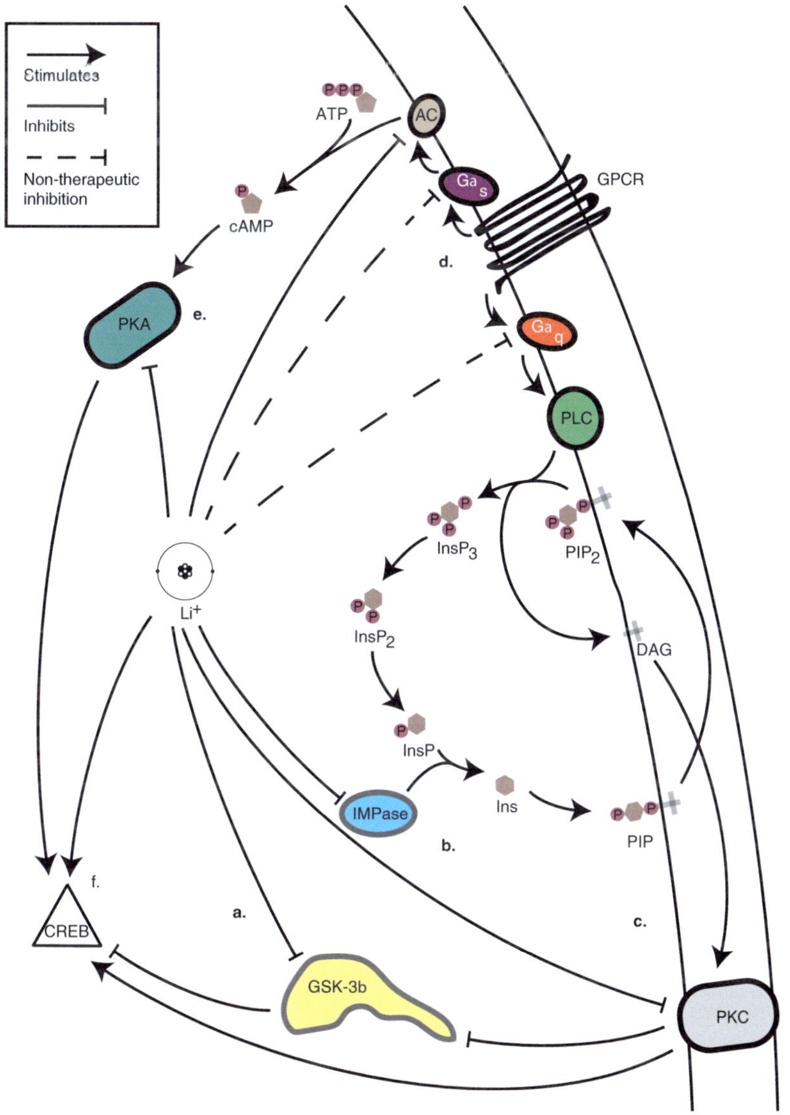

Figure 5 Li⁺ affects many cellular pathways. (**a**) Li⁺ is a noncompetitive inhibitor of GSK. (**b**) The PI signalling pathway is inhibited by IMPase, which prevents the cell from producing the secondary messengers DAG and PIP_2, further limiting the activity of downstream signaling proteins such as PKC. (**c**) The down regulation of MARCKS by Li^+ interrupts PKC signaling. (**d**) G-proteins $G_{\alpha s}$ and $G_{\alpha q}$ activate AC and PLC, respectively. (**e**) AC is inhibited both directly by Li^+ and indirectly by $G_{\alpha s}$. (**f**) The transcription factor CREB is activated by Li^+. AC, adenylyl cyclase, ATP, adenosine 5′-triphosphate, CREB, cAMP response element-binding protein, cAMP, cyclic AMP, DAG, 1,2-diacylglycerol, GPCR, G-protein coupled receptors, GSK-3, glycogen synthase kinase-3, IMPase, inositol monophosphatase, ($InsP_3$), 1,4,5-inositol triphosphate, ($InsP_2$), 4,5-inositol bisphosphate, (InsP), inositol phosphate, Ins, inositol, PI, phosphatidylinositol, (PIP_2), phosphatidylinositol 4,5-bisphosphate, PKA, protein kinase A, PKC, protein kinase C.

structure, G_i, G_s, G_q, and G_{11}. Each can be further separated into several different subgroups. The subunits G_s and G_i, which are involved in stimulation and inhibition of adenylyl cyclase, respectively, and G_q [64, 65], which stimulate IMPase.

In rat cortex membranes, adrenergic and cholinergic activation of G_s and $G_{i\alpha 1}$, respectively, was reported to be inhibited by Li^+ [65]. In bipolar patients, G_s was postulated to be associated with mania whereas $G_{i\alpha 1}$ with depression. Li^+ inhibition of G_s would result in a decrease in mania, while Li^+ inhibition of $G_{i\alpha 1}$ would alleviate depression. Because G_s and $G_{i\alpha 1}$ are Mg^{2+}-containing proteins, Li^+ competition for Mg^{2+} bound to these G-proteins might explain the effects of Li^+ on mania and depression.

Originally G-protein involvement in both manic and depressive cycles offered a simple and yet elegant explanation for Li^+ action in BD [66–68]. The hypothesis of Li^+ inhibition of G-proteins in BD was primarily based on cadaver and animal studies at therapeutic concentrations [65]. Studies have shown inhibition of $G_{i\alpha 1}$ to be negligible and not enough to produce the significant change observed in Li^+ treatment of BD [69]. However, these reports are in conflict with others [67, 70]. Several laboratories have examined deceased patients with mood disorders that were untreated or treated with Li^+. These studies focused on G-protein concentration and function. Conflicting data indicate increased levels of $G_{\alpha s}$, $G_{i\,\alpha 2}$, and $G_{\alpha 1/2}$ for patients with major depression [71] and increased $G_{\alpha s}$ and $G_{\alpha q}$ for BD [72], while many other studies showed no change in any G-protein concentration [68, 73]. Intrinsic variables, such as tissue storage, patient's age, and the inability to diagnose BD or major depression via tissue samples may account for these variances [65]. Furthermore, animal studies originally showed inhibition of G-proteins, but discrepancies and inconsistencies in the reported data have led to the conclusion that the results are not reproducible [74]. Lacking a universally accepted animal model for BD, makes drawing conclusions on therapeutic effects of Li^+ tenuous and the relevancy to BD debatable [74]. Although G-proteins are no longer the main focus of Li^+ inhibition in BD, they still remain a viable player in long-term treatment efficacy [75, 76].

G-proteins are Mg^{2+}-dependent and our recent fluorescence study on G_s and $G_{i\alpha 1}$ indicated that they contain two Mg^{2+} binding sites, one with high affinity and one with lower affinity. However, crystallographic reports of $G_{\alpha s}$ and $G_{i\alpha 1}$ only showed the presence of one Mg^{2+} binding site [77–82]. However, these studies were conducted at very high $[Li^+]$, which may have displaced Mg^{2+} from the low affinity Mg^{2+} site. Interestingly, for Ral, a monomeric G-protein, its crystal structure obtained in polyethylene glycol depicts two Mg^{2+} binding sites [83]. Regardless of their participation in Li^+ action in BP patients, G-proteins are therefore invaluable systems for studying Mg^{2+}/Li^+ competition due to being well characterized.

5.1.2 Li^+ Inhibition of Adenylyl Cyclase

Adenylyl cyclase (AC) is a membrane-bound enzyme responsible for converting adenosine triphosphate (ATP) to 3′,5′-cyclic adenosine monophosphate (cAMP), which serves as a secondary level of amplification of many signal transduction cycles. The increase of cAMP induces multiple effects, including opening of Ca^{2+}

channels and activation of protein kinase A (Figure 5). Li^+ is thought to inhibit AC in two ways, first by competing for its Mg^{2+} binding site or by inhibiting $G_{\alpha s}$, which is a known stimulator of cAMP formation [69]. The formation of forskolin-induced cAMP has been shown to be inhibited by Li^+ at therapeutic concentration in cerebral rat cortex [84]. In a study of euthymic BD patients undergoing Li^+ treatment, 0.5 mg of epinephrine, which is known to activate $G_{\alpha s}$ and stimulate the formation of cAMP, the plasma cAMP levels were lower when compared to untreated patients [85]. Similar studies in rodents showed a decrease in cAMP levels in cerebral cortex slices of rats treated with 2 mM of Li^+ [86] and at 1 mM Li^+ in guinea pig cortex [87, 88].

There are 10 isoforms of AC each having a different inhibition response to Li^+. Of these isoforms, AC-II, AC-V, and AC-VII are responsible for epinephrine-signal amplification and have been shown to be hypersensitive to Li^+ inhibition. Therefore, they have been implicated as being part of the array of the mechanisms of action for Li^+ treatment of BD [84, 89]. AC-V in particular is found in specific areas of the brain, including the nucleus accumbens and basal ganglia, which are associated with emotional responses. Further experiments using AC-V knock-out mice showed reduced mobility when performing the swim test, in which reduced mobility indicates BD-like behavior [84].

5.2 *Inositol Monophosphatase*

5.2.1 Phosphoinositide Turnover

Inositol phospholipids are the components of the phosphoinositol (PI) cycle. They represent an important class of secondary messengers, which are involved in triggering the formation and release of intracellular messengers that are vital for physiological responses within cells. Figure 5b outlines the PI cycle which is initiated by the activation of phospholipase C (PLC) by the α subunit of a specific G-protein, Gq, resulting in the conversion of phosphatidylinositol 4,5-bisphosphate (PIP_2) into 1,4,5-inositol triphosphate ($InsP_3$) and 1,2-diacylglycerol (DAG). DAG modulates the activity of protein kinase C (PKC) and $InsP_3$ induces the release of the secondary messenger Ca^{2+} [90]. The secondary messengers are then recycled and at the center is inositol monophosphatase (IMPase). IMPase catalyzes the hydrolysis of inositol derivatives and plays a crucial role in physiological response by maintaining the cellular levels of inositol (Ins), which is a precursor for many secondary messengers including DAG and PIP_2 [91].

5.2.2 Li^+ Inhibition of Inositol Monophosphatase

Mammalian IMPase is a highly conserved homodimer in which each 30 kDa subunit contains a key secondary structural feature of five alternating α-helices and β-sheets [92]. IMPase requires a trinuclear Mg^{2+} cofactor to function and is inhibited by Li^+ in an uncompetitive manner with an inhibitor constant (K_i) of 1.0 mM [40, 93]. Inhibition is thought to involve Li^+ binding to the enzyme-substrate complex and partial displacement of the Mg^{2+} cofactor [94]. As for G-proteins (see section 5.1), IMPase crystals collected from high-salt solutions showed only one Mg^{2+} site [95]. A subsequent study in which IMPase crystallization was conducted in polyethylene glycol demonstrated the presence of three Mg^{2+} binding sites [93], supporting the notion that Li^+ may compete in part for weakly bound Mg^{2+} in IMPase.

5.2.3 Inositol Depletion Hypothesis

Cellular concentrations of up to 10 mM Ins are found in the brain which reflect their importance in cell signaling [94]. One of the most appealing hypothesis for the mechanism of Li^+ action is the inositol depletion hypothesis. Ins represents the basis for two important signaling molecules, DAG and $InsP_3$. By inhibiting IMPase, Li^+ has been proposed to control the symptoms of bipolar disorder by depleting the cell of inositol, which in turn, would reduce the concentration of inositol-based secondary messengers such as DAG and PIP_2, preventing further signal transduction via PKC [96]. Several experiments with brain slices support this hypothesis [94].

The inositol depletion hypothesis does have some limitations. Klein and Melton demonstrated that Li^+ did not inhibit IMPase in *Xenopus* and proposed a new target for Li^+, glycogen synthase kinase-3(α) [97]. This has been further contradicted in genetic experiments which found that depletion of inositol had no effect on lithium-sensitive behavior [98].

5.3 Glycogen Synthase Kinase-3

GSK-3 is a constitutively active (no stimulation required) serine/threonine kinase. The name of the enzyme originates from a function that was early discovered: regulation of glycogen levels through inhibition of glycogen synthase. To regulate GSK-3 in this system, GSK is inhibited by insulin [99]. Extensive studies have provided evidence of GSK-3 functioning as an important regulatory kinase in a number of important roles, including inflammation, apoptosis, embryonic development, heart function, and synaptic transmission in neurons [100–105]. In fact, GSK-3 has now been shown to phosphorylate at least 50 different proteins, thus providing important regulation in an array of signaling pathways vital for homeostasis. Improper regulation on the part of GSK-3 can lead to a number of disease states. Indeed, hyperactive GSK-3 has been linked to several devastating illnesses including: Alzheimer's disease [101], diabetes [102], cancer [100, 106], Huntington's disease [107], and schizophrenia [108], to name a few.

GSK-3β can be found in the cytosol and in the nucleus as well as in mitochondria. It exists in two different splice variants GSK-3β1 and GSK-3β2. GSK-3β1 is a shorter variant and lacks exon 9 (a 13-residue exon), whereas GSK-3β2 is longer and does contain the 13-residue exon 9 in the catalytic domain [109]. While GSK-3β1 is expressed ubiquitously, GSK-3β2 is expressed almost exclusively in the central nervous system during development as well as into adulthood [109, 110]. The function of exon 9 in GSK-3β2 is still poorly understood. Unless otherwise stated, the rest of this section will be referring to GSK-3β1.

5.3.1 Mechanism of Glycogen Synthase Kinase-3

Serine/threonine kinases typically utilize a combination of alpha helices and beta sheets, which must be aligned in a specific orientation for enzymatic activity. GSK-3β does contain moieties with these unique secondary structural motifs. Most kinases use phosphorylated residues in an allosteric activation domain to achieve an active conformation. GSK-3 is more tightly regulated and phosphorylation may inhibit or activate it depending on the phosphorylation site. GSK-3 is inactivated when phosphorylated at the amine terminal serine (S) 9 in GSK-3β or at S21 in GSK-3α (Figure 6). When phosphorylated at these residues, a primed pseudo-substrate is formed in the active site, which acts as a competitive inhibitor [111]. In contrast, phosphorylation at tyrosine (Y) 216 in GSK-3β or Y279 in GSK-3α seems to promote activity, but because GSK-3 is a constitutively active enzyme it is unknown how important this phosphorylation actually is for activity [112].

Crystal structure analysis shows another unique trait of GSK-3β in that it prefers a phosphoserine on the substrate. More specifically, GSK-3 recognizes the sequence S/TXXXS, in which the P + 4 serine has been previously phosphorylated by another kinase. When the substrate has been "primed" by a prior phosphorylation event, GSK-3β is found to be exponentially more efficient [112, 113].

5.3.2 Li^+ Inhibition of Glycogen Synthase Kinase-3β

Like any important enzyme, GSK-3β activity must be closely regulated. Cells have four mechanisms for regulating GSK-3: phosphorylation of GSK-3β itself, phosphorylation of the substrate, subcellular localization and the formation of protein complexes [114]. For the native enzyme, the most effective mechanism of regulation is through phosphorylation of key residues in GSK-3. Inhibition of GSK-3β has therefore become a research topic of much significance. Li^+ has been shown to inhibit GSK-3β and is effective in the 1–2 mM range [97]; therefore, GSK-3β is significantly inhibited by Li^+ at therapeutically relevant concentrations.

There has been much debate as to the exact composition of the active site but what is known is that the active site of GSK-3β requires a Mg^{2+} cofactor to function. Li^+ is a noncompetitive inhibitor of GSK-3β with respect to substrate [97] but evidence has shown that Li^+ is a competitive inhibitor with respect to Mg^{2+} [115, 116]. This is a fairly unusual mechanism in that Li^+ binds the enzyme-substrate complex by partially displacing Mg^{2+} from the active site [97]. Because Li^+ and Mg^{2+} are

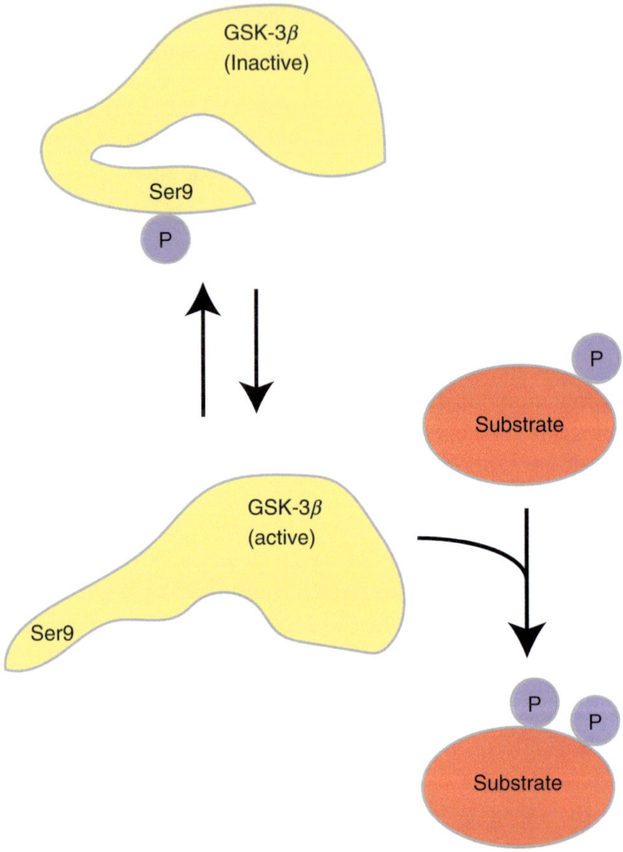

Figure 6 Roles of phosphorylation and dephosphorylation in the activity of GSK-3β. The N-terminus of GSK-3β acts as a pseudo-substrate and sterically hinders the active site when phosphorylated at Ser9. Upon dephosphorylation, the pseudo-substrate sways away from the active site making it accessible to the substrate. Active GSK-3β prefers to phosphorylate primed substrates.

similar in atomic radius, and both are positively charged, Li^+ is thought to be able to fit in the active site and displace, in part, the Mg^{2+} ion.

GSK-3 has been crystallized by several research groups. Figure 7 shows that, in the case of the ADP-bound structure the active site contains a mononuclear Mg^{2+} center in which a solvent-accessible cleft is lined with an aspartate and an asparagine residue [117]. In contrast, when an ATP analog is bound, an additional aspartate is present forming a binuclear Mg^{2+} center [118].

There has been much effort put into developing inhibitors of GSK-3 due to its involvement in a wide array of diseases. Design of such inhibitors has been met with challenges. Surrounding hydrophobic residues make accurately targeting the active site difficult [119].

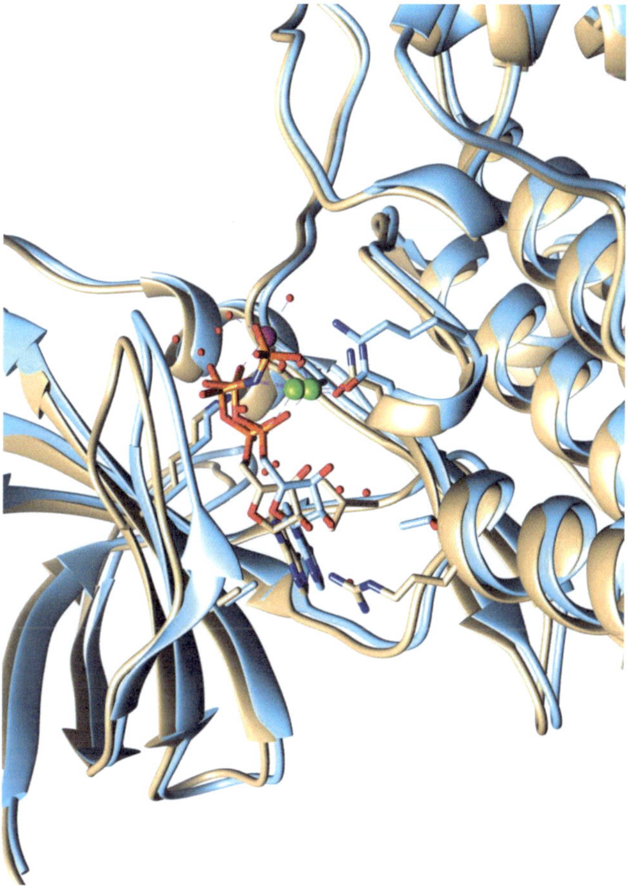

Figure 7 Crystallographic studies have shown GSK-3 to have mononuclear Mg^{2+} when ADP is bound in the active site (gold structure; Mg^{2+}, magenta). In contrast, when an ATP analog is bound, a dinuclear Mg^{2+} center (green) was observed (blue structure). Courtesy of the Protein Data Bank with the entries 1J1C (gold) [111], 1PYX (blue) [112]. Molecular graphics and analyses were performed with the UCSF Chimera package. Chimera is developed by the Resource for Biocomputing, Visualization, and Informatics at the University of California, San Francisco (supported by NIGMS P41-GM103311) [131].

5.4 Protein Kinase C and Myristoylated Alanine-Rich C-Kinase Substrate

The protein kinase C superfamily has been studied for many years and has greatly improved our knowledge of cell signaling and the use of secondary messengers within the cell. All isoforms of PKC contain a catalytic domain as well as a regulatory domain but their regulatory domains differ considerably in their primary structure. These differences dictate the activators that will bind to the mature enzyme.

PKC plays such a central role in cell signaling and is involved in many cellular functions. It is therefore not surprising that improper regulation of PKC has been observed in many diseases ranging from cancer [120] to BD [120, 121]. The first level of regulation is a maturation process that all PKC isoforms must undergo before they become active. Like GSK-3β, PKC also contains an autoinhibitory sequence (i.e., a pseudosubstrate) in the regulatory moiety that sterically hinders the active site, and must first be removed. This is accomplished by having a cofactor bind to the regulatory moiety [122]. At rest, PKC can be membrane-bound but is predominantly found in the cytosol [123]. PKC must also be phosphorylated at key residues and translocated to the membrane before it can be active [122, 124]. These post-translational modifications are necessary to correctly position key residues within the catalytic domain and to free the necessary space in the active site. PKC can be activated by certain upstream kinases including GPCRs, receptor tyrosine kinases, Ca^{2+}, and, depending on the isoform, DAG [120, 125]. Once activated, PKC can respond to secondary messengers and phosphorylate a number of specific downstream substrates, which, in turn, can regulate a variety of cell signaling pathways [122], one being the Wnt pathway of which is GSK-3β is a part [126, 127].

Myristoylated alanine-rich C-kinase substrate (MARCKS) is a highly conserved yet unstructured membrane-bound 32kD protein that is a common protein substrate for PKC and is myristoylated at the N-terminus. In addition to the N-terminal myristate, an acidic domain is crucial for MARCKS to bind negatively charged phospholipids in the membrane such as PIP_2. MARCKS and PKC colocalize on the membrane, allowing PKC to become active [128, 129]. Chronic Li^+ use has shown to inhibit PKC indirectly by altering the transcription levels of MARCKS, the key PKC substrate [130].

6 Concluding Remarks and Future Directions

With the exception of its general ability to compete with bound Mg^{2+} in Mg^{2+}-containing enzymes, Li^+ is a promiscuous inorganic ion that lacks binding specificity. However, recent theoretical studies have shed light on factors that may afford Li^+ affinity for cellular targets: (i) exposure of the Li^+ binding sites to solvent; (ii) the amino acids involved in metal ion coordination; and (iii) the net positive charge of a buried binding site [10]. Li^+ salts are, however, toxic drugs.

The understanding from Li^+ interactions with the cellular targets that have been identified may provide leads for the development of organic-based drugs with lower toxicity and higher specificity for GSK-3β and IMPase. Medication alone is not the only avenue that should be pursued when treating the chemical imbalance that is the hallmark of bipolar disorder. Patients with this illness must work closely with psychiatrists or psychologists in order to develop the tools for recognizing and seeking help for impending manic or depressive episodes [14].

Abbreviations

Aβ	amyloid β-peptides
AC	adenylyl cyclase
AD	Alzheimer's disease
ATP	adenosine 5′-triphosphate
[B]	concentration of binding sites
BD	bipolar disorder
BDNF	brain-delivered neurotropic factor
cAMP	3,5′-cyclic adenosine monophosphate
cP	centipoise, a unit of viscosity
CREB	cAMP response element-binding protein
DAG	1,2-diacylglycerol
$[Dy(PPP)_2]^{7-}$	dysprosium triphosphate
G_α	G-protein α subunit
$G_{\alpha s}$	G-protein αs subunit (AC stimulatory)
$G_{\alpha q}$	G-protein αq isoform (IMPase stimulatory)
$G_{i\alpha 1}$	G-protein iα1 isoform (AC inhibitor)
G-proteins	guanine nucleotide binding proteins
GDP	guanosine 5'-diphosphate
GPCR	G-protein coupled receptor
GSK-3	glycogen synthase kinase-3
GSK-3α	glycogen synthase kinase-3α
GSK-3β	glycogen synthase kinase-3β
GSK-3β1	glycogen synthase kinase-3β1
GSK-3β2	glycogen synthase kinase-3β2
GTP	guanosine 5' -triphosphate
HEPES	4-(2-hydroxyethyl)piperazine-1-ethanesulfonic acid
$HTm(DOTP)^{4-}$	thulium 1,4,7,10-tetraazacyclodecane-*N′*,*N″*,*N‴*-tetramethylene-phosphonate
IMPase	inositol monophosphatase
InsP	inositol phosphate
$InsP_2$	4,5-inositol bisphosphate
$InsP_3$	1,4,5-inositol triphosphate
K_{Li}	Li^+ binding constant
K_d	dissociation constant of the furaptra complex with Mg^{2+}
K_d'	dissociation constant of the furaptra complex with Li^+
K_i	inhibition constant
$[Li^+]_{bound}$	bound lithium concentration
$[Li^+]_{free}$	free lithium concentration
$[Li^+]_{total}$	total lithium concentration
MARCKS	myristoylated alanine-rich C-kinase
$[Mg^{2+}]_{bound}$	bound magnesium concentration
$[Mg^{2+}]_{free}$	free magnesium concentration

$[Mg^{2+}]_{total}$	total magnesium concentration
MHz	megahertz (10^9 s^{-1})
MRSI	magnetic resonance with spectroscopic imaging
Msec	millisecond (10^{-3} seconds)
NMR	nuclear magnetic resonance spectroscopy
PI	phosphatidylinositol
PIP_2	phosphatidylinositol 4,5-bisphosphate
PKA	protein kinase A
PKC	protein kinase C
PLC	phospholipase C
pm	picometer (10^{-12} meter)
r	ionic radius
R	fluorescence intensity ratios observed for the sample
RBC	red blood cell
R_{bound}	relaxation rate (reciprocal of T_1) for bound Li^+
R_{free}	relaxation rate (reciprocal of T_1) with saturating $[Li^+]_{total}$
R_{max}	fluorescence intensity ratio in the presence of saturating amounts of Mg^{2+}
R'_{max}	fluorescence intensity ratio in the presence of saturating concentrations of Li^+
R_{min}	fluorescence intensity ratios in the absence of Mg^{2+} and Li^+
R_{obs}	relaxation rate (reciprocal of T_1) observed with the NMR sample
Ser, S	serine
S_{min}	fluorescence intensity in the absence of Mg^{2+} and Li^+
S_{max}	fluorescence intensity in the presence of saturating concentrations of Mg^{2+}
S'_{max}	fluorescence intensity in the presence of saturating concentrations of Li^+
T_1	spin-lattice relaxation time
T_2	spin-spin relaxation time
T3	triiodothyrosine
T4	thyroxine
Tris	tris(hydroxymethyl)aminomethane
τ_c	correlation time
UV/Vis	ultra-violet/visible spectroscopy
ω	NMR resonance frequency
Tyr, Y	tyrosine

Acknowledgments Work in our laboratory is a culmination of a thirty-plus effort on the part of very capable graduate and undergraduate students, and postdoctoral fellows, and a long collaboration with Professor Margarida Castro and Professor Carlos Geraldes from the University of Coimbra, Portugal. Professor Mota de Freitas acknowledges that, without their talent and dedication, the projects described herein would not have come to fruition. We are grateful for funding from the Research Cooperation, the American Heart Association of Metropolitan Chicago, NATO (CRG921011), and the National Institute of Mental Health (MH45926).

References

1. Z. Rappoport, I. Marek, *The Chemistry of Organolithium Compounds,* John Wiley & Sons, 2004.
2. G. Eichinger, J. O. Besenhard, *J. Electroanal. Chem. Interfacial Electrochem.* **1976,** *72,* 1–31.
3. S. Huang, P. Cao, C. Wang, Z. Huang, W. Gao, *J. As. Cer. Soc.* **2013,** *1,* 46–52.
4. F. Barnaby, *How Nuclear Weapons Spread: Nuclear-Weapon Proliferation in the 1990s,* Routledge, London, 2012.
5. R. Shannon, *Acta Cryst.* **1976,** *A32,* 751–767.
6. J. Mahler, I. Persson, *Inorg. Chem.* **2012,** *51,* 425–438.
7. T. Dudev, J. A. Cowan, C. Lim, *J. Am. Chem. Soc.* **1999,** *121,* 7665–7673.
8. D. Mota de Freitas, M. M. Castro, C. F. Geraldes, *Acc. Chem. Res.* **2006,** *39,* 283–291.
9. J. J. Frausto da Silva, R. J. P. Williams, *Nature* **1976,** *263,* 237–239.
10. T. Dudev, C. Lim, *J. Am. Chem. Soc.* **2011,** *133,* 9506–9515.
11. G. S. Malhi, M. Tanious, D. Bargh, P. Das, M. Berk, *Aust Prescr.* **2013,** *36,* 18–21.
12. W. Jahnen-Dechent, M. Ketteler, *Clin. Kidney J.* **2012,** *5,* i3–i14.
13. H. Ohgami, T. Terao, I. Shiotsuki, N. Ishii, N. Iwata, *Br. J. Psychiatry* **2009,** *194,* 464–465; discussion 446.
14. American Psychiatric Association, *Diagnostic and Statistical Manual of Mental Disorders, (DSM-5),* American Psychiatric Pub., 2013.
15. E. Walker, R. E. McGee, B. G. Druss, *JAMA Psychiatry* **2015,** *72,* 334–341.
16. L. H. Price, G. R. Heninger, *N. Engl. J. Med.* **1994,** *331,* 591–598.
17. T. Suppes, R. J. Baldessarini, G. L. Faedda, M. Tohen, *Arch. Gen. Psychiatry* **1991,** *48,* 1082–1088.
18. K. N. Fountoulakis, E. Vieta, M. Siamouli, M. Valenti, S. Magiria, T. Oral, D. Fresno, P. Giannakopoulos, G. S. Kaprinis, *Ann. Gen. Psychiatry* **2007,** *6,* 27.
19. M. C. Wong, *Indian J. Psychol. Med.* **2011,** *33,* 18–28.
20. K. N. Fountoulakis, E. Vieta, *Imt. J. Neuropsychopharmacol.* **2008,** *11,* 999–1029.
21. L. V. Kessing, L. Sondergard, J. L. Forman, P. K. Andersen, *Arch. Gen. Psychiatry* **2008,** *65,* 1331–1335.
22. R. Nitrini, P. Caramelli, E. Herrera Jr, V. S. Bahia, L. F. Caixeta, M. Radanovic, R. Anghinah, H. Charchat-Fichman, C. S. Porto, M. T. Carthery, A. P. Hartmann, N. Huang, J. Smid, E. P. Lima, L. T. Takada, D. Y. Takahashi, *Alzheimer Dis. Assoc. Disord.* **2004,** *18,* 241–246.
23. M. Hong, D. C. Chen, P. S. Klein, V. M. Lee, *J. Biol. Chem.* **1997,** *272,* 25326–25332.
24. S. Lovestone, D. R. Davis, M. T. Webster, S. Kaech, J. P. Brion, A. Matus, B. H. Anderton, *Biol. Psychiatry* **1999,** *45,* 995–1003.
25. J. R. Munoz-Montano, F. J. Moreno, J. Avila, J. Diaz-Nido, *FEBS Lett.* **1997,** *411,* 183–188.
26. L. Trujillo-Estrada, S. Jimenez, V. De Castro, M. Torres, D. Baglietto-Vargas, I. Moreno-Gonzalez, V. Navarro, R. Sanchez-Varo, E. Sanchez-Mejias, J. C. Davila, M. Vizuete, A. Gutierrez, J. Vitorica, *Acta Neuropathol. Commun.* **2013,** *1,* 73.
27. C. Parr, R. Carzaniga, S. M. Gentleman, F. Van Leuven, J. Walter, M. Sastre, *Mol. Cell. Biol.* **2012,** *32,* 4410–4418.
28. C. Feyt, P. Kienlen-Campard, K. Leroy, F. N'Kuli, P. J. Courtoy, J. P. Brion, J. N. Octave, *J. Biol. Chem.* **2005,** *280,* 33220–33227.
29. A. Caccamo, S. Oddo, L. X. Tran, F. M. LaFerla, *Am. J. Pathol.* **2007,** *170,* 1669–1675.
30. M. A. Nunes, T. A. Viel, H. S. Buck, *Curr. Alzheimer Res.* **2013,** *10,* 104–107.
31. O. V. Forlenza, V. J. De-Paula, B. S. Diniz, *ACS Chem. Neurosci.* **2014,** *5,* 443–450.
32. J. Kohrle, *Mol. Cell. Endocrinol.* **1999,** *151,* 103–119.
33. S. C. Berens, R. S. Bernstein, J. Robbins, J. Wolff, *J. Clin. Invest.* **1970,** *49,* 1357–1367.
34. C. H. Emerson, A. J. Anderson, W. J. Howard, R. D. Utiger, *J. Clin. Endocrinol. Metab.* **1975,** *40,* 33–36.

35. D. Barbaro, G. Boni, G. Meucci, U. Simi, P. Lapi, P. Orsini, C. Pasquini, F. Piazza, M. Caciagli, G. Mariani, *J. Clin. Endocrinol. Metab.* **2003,** *88,* 4110–4115.
36. D. Mota de Freitas, M. Espanol, E. Dorus, *in Lithium Transport in Red Blood Cells of Bipolar Patients,* Vol. 4 of *Lithium and the Blood,* Eds F. Johnson, V. Gallicchio, Karger, Basel, 1991, pp. 96–120.
37. G. D. Christian, *J. Pharm. Biomed. Anal.* **1996,** *14,* 899–908.
38. C. J. Jameson, J. Mason, *in The Parameters of NMR Spectroscopy, Multinuclear NMR,* Ed J. Mason, Plenum Press, New York, 1987, pp. 3–50.
39. H. Günther, *in Advanced Applications of NMR to Organometallic Chemistry,* Eds B. Wrackmeyer, M. Gielen, R. Willem, Wiley, Chichester, UK, 1996, pp. 247–290.
40. V. Saudek, P. Vincendon, Q. T. Do, R. A. Atkinson, V. Sklenar, P. D. Pelton, F. Piriou, A. J. Ganzhorn, *Eur. J. Biochem.* **1996,** *240,* 288–291.
41. A. Haimovich, U. Eliav, A. Goldbourt, *J. Am. Chem. Soc.* **2012,** *134,* 5647–5651.
42. D. Mota de Freitas, *in Alkali Metal Nuclear Magnetic Resonance,* Vol. 227 of *Methods in Enzymology: Metallobiochemistry Part D: Physical and Spectroscopic Methods for Probing Metal Ion Environments in Metalloproteins,* Eds J. F. Riordan, B. L. Vallee, Academic Press, Waltham, USA, 1993, pp. 78–106.
43. M. C. Espanol, D. Mota de Freitas, *Inorg. Chem.* **1987,** *26,* 4356–4359.
44. Q. Rong, M. Espanol, D. Mota de Freitas, C. F. Geraldes, *Biochemistry* **1993,** *32,* 13490–13498.
45. T. L. James, J. H. Noggle, *Proc. Natl. Acad. Sci. USA* **1969,** *62,* 644–649.
46. J. Lee, C. Adler, M. Norris, W. Chu, E. M. Fugate, S. M. Strakowski, R. A. Komoroski, *Magn. Reson. Med.* **2012,** *68,* 363–368.
47. B. Raju, E. Murphy, L. A. Levy, R. D. Hall, R. E. London, *Am. J. Physiol.* **1989,** *256,* C540–C548.
48. D. Mota de Freitas, L. Amari, C. Srinivasan, Q. Rong, R. Ramasamy, A. Abraha, C. F. Geraldes, M. K. Boyd, *Biochemistry* **1994,** *33,* 4101–4110.
49. C. S. Malarkey, G. Wang, M. A. Ballicora, D. Mota de Freitas, *Biochem. Biophys. Res. Commun.* **2008,** *372,* 866–869.
50. J. Nikolakopoulos, C. Zachariah, D. Mota de Freitas, E. B. Stubbs Jr, R. Ramasamy, M. C. Castro, C. F. Geraldes, *J. Neurochem.* **1998,** *71,* 1676–1684.
51. Y. Chi, S. Mo, D. Mota de Freitas, *Biochemistry.* **1996,** *35,* 12433–12442.
52. M. Canessa, N. Adragna, H. S. Solomon, T. M. Connolly, D. C. Tosteson, *N. Engl. J. Med.* **1980,** *302,* 772–776.
53. B. T. Layden, N. Minadeo, J. Suhy, A. M. Abukhdeir, T. Metreger, K. Foley, G. Borge, J. W. Crayton, F. B. Bryant, D. M. d. Freitas, *Bipolar Disord.* **2004,** *6,* 53–61.
54. L. P. Montezinho, C. B Duarte, C. P. Fonseca, Y. Glinka, B. Layden, D. Mota de Freitas, C. F. Geraldes, M. M. Castro, *J. Neurochem.* **2004,** *90,* 920–930.
55. B. T. Layden, A. M. Abukhdeir, C. Malarkey, L. A. Oriti, W. Salah, C. Stigler, C. F. Geraldes, D. Mota de Freitas, *Biochim. Biophys. Acta* **2005,** *1741,* 339–349.
56. A. M. Abukhdeir, B. T. Layden, N. Minadeo, F. B. Bryant, E. B. Stubbs, D. Mota de Freitas, *Bipolar Disord.* **2003,** *5,* 6–13.
57. J. C. Soares, F. Boada, M. S. Keshavan, *Eur. Neuropsychopharmacol.* **2000,** *10,* 151–158.
58. J. Baek, G. Kinrys, A. Nierenberg, *Acta Psychiatr. Scand.* **2014,** *129,* 17–23.
59. L. Carter, M. Zolezzi, A. Lewczyk, *Can. J. Psychiatry* **2013,** *58,* 595–600.
60. E. D. Peselow, D. L. Dunner, R. R. Fieve, A. Lautin, *J. Affect. Disord.* **1980,** *2,* 303–310.
61. M. Jafferany, *Int. J. Dermatol.* **2008,** *47,* 1101–1111.
62. A. Nassar, A. N. Azab, *ACS Chem. Neurosci.* **2014**, *5*, 451–458.
63. P. R. Leeds, F. Yu, Z. Wang, C. T. Chiu, Y. Zhang, Y. Leng, G. R. Linares, D. M. Chuang, *ACS Chem. Neurosci.* **2014,** *5,* 422–433.
64. T. M. Cabrera-Vera, J. Vanhauwe, T. O. Thomas, M. Medkova, A. Preininger, M. R. Mazzoni, H. E. Hamm, *Endocr. Rev.* **2003,** *24,* 765–781.
65. J. Gonzalez-Maeso, J. J. Meana, *Curr. Neuropharmacol.* **2006,** *4,* 127–138.

66. S. Avissar, G. Schreiber, *Biol. Psychiatry* **1992,** *31,* 435–459.
67. S. Avissar, G. Schreiber, A. Danon, R. H. Belmaker, *Nature* **1988,** *331,* 440–442.
68. J. Gonzalez-Maeso, R. Rodriguez-Puertas, J. J. Meana, J. A. García-Sevilla, J. Guimon, *Mol. Psychiatry.* **2002,** *7,* 755–767.
69. J. Ellis, R. H. Lenox, *Lithium* **1991,** *2,* 141–147.
70. H. Y. Wang, E. Friedman, *Neuropharmacology* **1999,** *38,* 403–414.
71. M. A. Pacheco, C. Stockmeier, H. Y. Meltzer, J. C. Overholser, G. E. Dilley, R. S. Jope, *Brain Res.* **1996,** *723,* 37–45.
72. E. Friedman, H. Y. Wang, *J. Neurochem.* **1996,** *67,* 1145–1152.
73. D. Dowlatshahi, G. M. MacQueen, J. F. Wang, J. S. Reiach, L. T. Young, *J. Neurochem.* **1999,** *73,* 1121–1126.
74. M. L. Wong, J. Licinio, *Nat. Rev. Drug Discov.* **2004,** *3,* 136–151.
75. H. K. Manji, R. H. Lenox, *Biol. Psychiatry* **2000,** *48,* 518–530.
76. H. K. Manji, J. A. Quiroz, J. L. Payne, J. Singh, B. P. Lopes, J. S. Viegas, C. A. Zarate, *World Psychiatry* **2003,** *2,* 136–146.
77. R. K. Sunahara, J. J. Tesmer, A. G. Gilman, S. R. Sprang, *Science.* **1997,** *278,* 1943–1947.
78. J. J. Tesmer, R. K. Sunahara, A. G. Gilman, S. R. Sprang, *Science.* **1997,** *278,* 1907–1916.
79. D. E. Coleman, A. M. Berghuis, E. Lee, M. E. Linder, A. G. Gilman, S. R. Sprang, *Science* **1994,** *265,* 1405–1412.
80. D. E. Coleman, S. R. Sprang, *Biochemistry* **1998,** *37,* 14376–14385.
81. D. E. Coleman, S. R. Sprang, *Meth. Enzymol.* **1999,** *308,* 70–92.
82. D. E. Coleman, E. Lee, M. B. Mixon, M. E. Linder, A. M. Berghuis, A. G. Gilman, S. R. Sprang, *J. Mol. Biol.* **1994,** *238,* 630–634.
83. N. I. Nicely, J. Kosak, V. de Serrano, C. Mattos, *Structure* **2004,** *12,* 2025–2036.
84. L. Mann, E. Heldman, Y. Bersudsky, S. F. Vatner, Y. Ishikawa, O. Almog, R. H. Belmaker, G. Agam, *Bipolar Disord.* **2009,** *11,* 885–896.
85. R. H. Belmaker, M. Kon, R. P. Ebstein, H. Dasberg, *Biol. Psychiatry* **1980,** *15,* 3–8.
86. J. Forn, F. G. Valdecasas, *Biochem. Pharmacol.* **1971,** *20,* 2773–2779.
87. R. P. Ebstein, A. Reches, R. H. Belmaker, *J. Pharm. Pharmacol.* **1978,** *30,* 122–123.
88. A. Reches, R. P. Ebstein, R. H. Belmaker, *Psychopharmacology (Berl).* **1978,** *58,* 213–216.
89. L. Mann, E. Heldman, G. Shaltiel, R. H. Belmaker, G. Agam, *Int. J. Neuropsychopharmacol.* **2008,** *11,* 533–539.
90. R. Machado-Vieira, H. K. Manji, C. A. Zarate, *Bipolar Disord.* **2009,** *11, Suppl 2,* 92–109.
91. M. J. Berridge, C. P. Downes, M. R. Hanley, *Cell* **1989,** *59,* 411–419.
92. J. R. Atack, H. B. Broughton, S. J. Pollack, *FEBS Lett.* **1995,** *361,* 1–7.
93. R. Gill, F. Mohammed, R. Badyal, L. Coates, P. Erskine, D. Thompson, J. Cooper, M. Gore, S. Wood, *Acta Crystallogr. D Biol. Crystallogr.* **2005,** *61,* 545–555.
94. A. J. Harwood, *Mol. Psychiatry* **2005,** *10,* 117–126.
95. R. Bone, J. P. Springer, J. R. Atack, *Proc. Natl. Acad. Sci. USA* **1992,** *89,* 10031–10035.
96. C. J. Phiel, P. S. Klein, *Annu. Rev. Pharmacol. Toxicol.* **2001,** *41,* 789–813.
97. P. S. Klein, D. A. Melton, *Proc. Natl. Acad. Sci. USA* **1996,** *93,* 8455–8459.
98. A. Shaldubina, R. A. Johanson, W. T. O'Brien, R. Buccafusca, G. Agam, R. H. Belmaker, P. S. Klein, Y. Bersudsky, G. T. Berry, *Mol. Genet. Metab.* **2006,** *88,* 384–388.
99. N. Embi, D. B. Rylatt, P. Cohen, *Eur. J. Biochem.* **1980,** *107,* 519–527.
100. J. A. McCubrey, L. S. Steelman, F. E. Bertrand, N. M. Davis, M. Sokolosky, S. L. Abrams, G. Montalto, A. B. D'Assoro, M. Libra, F. Nicoletti, R. Maestro, J. Basecke, D. Rakus, A. Gizak, Z. N. Demidenko, L. Cocco, A. M. Martelli, M. Cervello, *Oncotarget* **2014,** *5,* 2881–2911.
101. L. Martin, X. Latypova, C. M. Wilson, A. Magnaudeix, M. L. Perrin, C. Yardin, F. Terro, *Ageing Res. Rev.* **2013,** *12,* 289–309.
102. S. Amar, R. H. Belmaker, G. Agam, *Curr. Pharm. Des.* **2011,** *17,* 2264–2277.
103. C. Gao, C. Holscher, Y. Liu, L. Li, *Rev. Neurosci.* **2011,** *23,* 1–11.
104. H. Lal, F. Ahmad, J. Woodgett, T. Force, *Circ. Res.* **2015,** *116,* 138–149.

105. M. Llorens-Martin, J. Jurado, F. Hernandez, J. Avila, *Front. Mol. Neurosci.* **2014,** *7,* 46.
106. U. Maurer, F. Preiss, P. Brauns-Schubert, L. Schlicher, C. Charvet, *J. Cell. Sci.* **2014,** *127,* 1369–1378.
107. D. A. Perez-Martinez, *Neurologia* **2009,** *24,* 143–146.
108. A. R. Cole, *FEBS J.* **2013,** *280,* 5213–5227.
109. A. Wood-Kaczmar, M. Kraus, K. Ishiguro, K. L. Philpott, P. R. Gordon-Weeks, *Mol. Cell. Neurosci.* **2009,** *42,* 184–194.
110. F. Mukai, K. Ishiguro, Y. Sano, S. C. Fujita, *J. Neurochem.* **2002,** *81,* 1073–1083.
111. P. Cohen, S. Frame, *Nat. Rev. Mol. Cell Biol.* **2001,** *2,* 769–776.
112. R. Dajani, E. Fraser, S. M. Roe, N. Young, V. Good, T. C. Dale, L. H. Pearl, *Cell* **2001,** *105,* 721–732.
113. E. ter Haar, J. T. Coll, D. A. Austen, H. M. Hsiao, L. Swenson, J. Jain, *Nat. Struct. Biol.* **2001,** *8,* 593–596.
114. R. S. Jope, C. J. Yuskaitis, E. Beurel, *Neurochem. Res.* **2007,** *32,* 577–595.
115. W. J. Ryves, A. J. Harwood, *Biochem. Biophys. Res. Commun.* **2001,** *280,* 720–725.
116. W. J. Ryves, R. Dajani, L. Pearl, A. J. Harwood, *Biochem. Biophys. Res. Commun.* **2002,** *290,* 967–972.
117. M. Aoki, T. Yokota, I. Sugiura, C. Sasaki, T. Hasegawa, C. Okumura, K. Ishiguro, T. Kohno, S. Sugio, T. Matsuzaki, *Acta Crystallogr. D Biol. Crystallogr.* **2004,** *60,* 439–446.
118. J. A. Bertrand, S. Thieffine, A. Vulpetti, C. Cristiani, B. Valsasina, S. Knapp, H. M. Kalisz, M. Flocco, *J. Mol. Biol.* **2003,** *333,* 393–407.
119. H. C. Zhang, L. V. Bonaga, H. Ye, C. K. Derian, B. P. Damiano, B. E. Maryanoff, *Bioorg. Med. Chem. Lett.* **2007,** *17,* 2863–2868.
120. C. A. Zarate, H. K. Manji, *CNS Drugs* **2009,** *23,* 569–582.
121. R. S. Jope, *Mol. Psychiatry* **1999,** *4,* 117–128.
122. A. C. Newton, *J. Biol. Chem.* **1995,** *270,* 28495–28498.
123. H. K. Manji, R. H. Lenox, *Synapse* **1994,** *16,* 11–28.
124. A. C. Newton, J. E. Johnson, *Biochim. Biophys. Acta.* **1998,** *1376,* 155–172.
125. M. Serova, A. Ghoul, K. A. Benhadji, E. Cvitkovic, S. Faivre, F. Calvo, F. Lokiec, E. Raymond, *Semin. Oncol.* **2006,** *33,* 466–478.
126. N. Goode, K. Hughes, J. R. Woodgett, P. J. Parker, *J. Biol. Chem.* **1992,** *267,* 16878–16882.
127. H. K. Manji, R. H. Lenox, *Biol. Psychiatry* **1999,** *46,* 1328–1351.
128. J. M. Graff, J. I. Gordon, P. J. Blackshear, *Science* **1989,** *246,* 503–506.
129. A. Rosen, K. F. Keenan, M. Thelen, A. C. Nairn, A. Aderem, *J. Exp. Med.* **1990,** *172,* 1211–1215.
130. L. Wang, X. Liu, R. H. Lenox, *J. Neurochem.* **2001,** *79,* 816–825.
131. E. F. Pettersen, T. D. Goddard, C. C. Huang, G. S. Couch, D. M. Greenblatt, E. C. Meng, T. E. Ferrin, *J. Comput. Chem.* **2004,** *25,* 1605–1612.

Chapter 16
Sodium and Potassium Relating to Parkinson's Disease and Traumatic Brain Injury

Yonghwang Ha, Jeong A Jeong, Youngsam Kim, and David G. Churchill

Contents

Abstract Alkali metals, especially sodium and potassium, are plentiful and vital in biological systems. They take on important roles in health and disease. Such roles include the regulation of homeostasis, osmosis, blood pressure, electrolytic equilibria, and electric current. However, there is a limit to our present understanding; the ions have a great ability and capacity for action in health and disease, much greater than our current understanding. For the regulation of physiological homeostasis, there is a crucial regulator (renin-angiotensin system, RAS), found at both peripheral and central levels. Misregulation of the Na^+-K^+ pump, and sodium channels in RAS are important

Y. Ha • J. A Jeong • Y. Kim • D.G. Churchill (✉)
Department of Chemistry, Korea Advanced Institute of Science and Technology (KAIST), 373-1 Guseong-dong, Yuseong-gu, Daejeon 305-701, Republic of Korea
e-mail: dchurchill@kaist.ac.kr

A. Sigel, H. Sigel, and R.K.O. Sigel (eds.), *The Alkali Metal Ions: Their Role for Life*, Metal Ions in Life Sciences 16, DOI 10.1007/978-3-319-21756-7_16

for the understanding of disease progression, hypertension, diabetes, and neurodegenerative diseases, etc. In particular, RAS displays direct or indirect interaction important to Parkinson's disease (PD). In this chapter, the relationship between the regulation of sodium/potassium concentration and PD was sought. In addition, some recent biochemical and clinical findings are also discussed that help describe sodium and potassium in the context of traumatic brain injury (TBI). TBI is caused from the heavy striking of the head; this strongly affects ion flux in the affected tissue (brain) and damages cellular regulation systems. Thus, inappropriate concentrations of ions (hyper- and hyponatremia, and hyper- and hypokalemia) will perturb homeostasis giving rise to important and far reaching effects. These changes also impact osmotic pressure and the concentration of other metal ions, such as the calcium(II) ion.

Keywords Ion flux • Na^+-K^+ pump • Renin-angiotensin system • Sodium channel • α-synuclein

Please cite as: *Met. Ions Life Sci.* 16 (2016) 585–601

1 Introduction

1.1 Scope and Interrelationships

The connection between metal ions and diseases stands as an important interrelationship that involves the concept of homeostasis, and has been elucidated at varying degrees in many studies. In living organisms, although the weight amount (or mole fraction) of metal ions is small, relative to, e.g., carbon-containing compounds (e.g., carbohydrates, proteins, lipids), they are very highly concentrated with respect to transition metals and are important in mediating various physiological processes. Alkali metals (especially sodium and potassium) are greatly important in the processes of osmotic control, electrolytic equilibria, and electric current. Free alkali metal ions (Na^+, K^+) are stable and nontoxic, relative to transition metal ions [1]. However, because they involve more electrolytic-type interactions, and rely less on covalent bonding (like *d*-block metals), the influence of their coordination chemistry is more difficult to pin down at the molecular level.

In this chapter, the effect of sodium and potassium on Parkinson's disease (PD) and traumatic brain injury (TBI) is discussed. Importantly, the effects of sodium and potassium on certain physiological regulations in human subjects are also presented. Additionally, sodium and potassium ions, related compounds, related proteins, their presence in vital organs, details of certain channels in biological membranes and neurons, and regarding relationships to neurodegenerative disease research (particularly PD) will be discussed.

There is a paucity of PD studies that relate to sodium and potassium ions directly. Rather, mechanisms of the homeostasis of sodium and potassium, in particular, e.g., the renin-angiotensin system (RAS), were explored. These well-elucidated, ion-dependent systems are a clear potential for therapeutic strategies for PD. Therefore, this chapter will focus more on homeostasis mechanisms of sodium and potassium than on discrete sodium or potassium coordination chemistry.

1.2 Parkinson's Disease: Alkali Metal Concentration and Interactions

1.2.1 Proteins Implicated in Parkinson's Disease

After Alzheimer's disease (AD), PD is the second-most common neurodegenerative disease characterized by three principal clinical symptoms, bradykinesia (slow movement), resting tremor, and muscle rigidity. Many medicinal compounds have been developed and investigated for potential therapeutic use with PD [2–5]. However, little remains certain about how PD progression manifests at the molecular level; long-lasting therapeutic avenues are not yet available [6–8]. Studies endeavor to start to link the clinical symptoms by endogenous or exogenous toxic materials or genetic (familial) links [9–12]. In particular, certain toxic materials, reactive oxygen species (ROS), and uncontrolled transition metal ion concentration can disturb the physiological mechanism and induce neurotoxic α-synuclein oligomers. These factors can be correlated well with PD progression [13, 14].

The concentration of transition metal ions is relatively small in the brain (μM ~ nM level) [15]. Because of lesser abundances of these species, cellular transport, and storage regulation is highly sensitive to change in the concentration of these ions in cells [16]. Alkali metal ions, however, are present in high amounts, internal or external to cells [17]. They take on important roles in the regulation of osmotic pressure, electric potential, and current [18].

For the regulation of sodium and potassium concentration, the RAS plays a central role using various cellular channels or pumps [19]. One investigative potential therapeutic strategy in PD is related to the regulation of channel functions [20, 21]. It has been known that there is a RAS version working independently in the brain. In particular, angiotensin II can interact with the T1 receptor found in the cell membrane, which is currently one main potential therapeutic target [22].

1.2.2 α-Synuclein Protein and Parkinson's Disease

α-Synuclein is the main hallmark protein in the progression of PD [7]. One main neuropathological symptom is the deposit of α-synuclein in the *substantia nigra* of the brain [13]. Studies focus on the nature of α-synuclein oligomers coincident during PD progression [23–25]; these are generated during PD progression and may induce neurotoxic effects thought to be critical in the manifestation of PD [26, 27]. Although there are many studies about α-synuclein oligomerization with transition metal ions [28, 29], there are still few reports about oligomerization due to *alkali* metal ions; there is no link to date between sodium and potassium ions (redox inactive metal ions) with α-synuclein oligomerization.

2 Parkinson's Disease and Alkali Metal Ions

2.1 Sodium

2.1.1 Endogenous Concentrations

There are various reports involving endogenous concentrations of alkali metal ions in Parkinson's disease subjects or model systems [30]. Frausto da Silva and Williams reported the concentration of alkali metals and biologically important anions of the cell and sea water; this data is presented in Figure 1 [1]. As shown in Figure 1, sodium concentration is higher than that of potassium in sea water, ($[Na^+]=460$ mmol L^{-1}; $[K^+]=10$ mmol L^{-1}); trends found for blood plasma are similar ($[Na^+]=160$ mmol L^{-1}; $[K^+]=10$ mmol L^{-1}). In red blood cells, however, the reverse is found ($[Na^+]=11$ mmol L^{-1}; $[K^+]=92$ mmol L^{-1}, Figure 1). This is due to the cellular gradient of metal ions using energy-consuming pumps or channels. Therefore, it is essential to consider channels and pumps of sodium and potassium that are further related to wider physiological mechanisms and in health and disease states [1].

Madelin et al. investigated the sodium concentration in the human brain. This *in vivo* study used sodium magnetic resonance imaging (MRI) techniques [31]. Na^+ levels of ca. 11 mM could be determined in both grey and white matter in the brain. A concentration gradient is seen to be maintained between the inside and outside of cells. The concentration of sodium within cells is ca. 10–14 mM; external to cells, the concentration is ca.140 mM [31].

2.1.2 Known Actions of Sodium

Sodium concentration is regulated by various factors in biology such as hormones, membrane channels, etc. One important regulation function of sodium concentration involves RAS, a hormone-related system (Scheme 1) [32, 33], which can be classified into two types: peripheral RAS and central RAS [34]. Peripheral RAS shows direct ionotropic effects on the heart and kidney. The RAS also affects concentrations of sodium by using various hormones related to the regulation of vascular resistance [22]. Llorens-Cortes and Mendelsohn reported that the central RAS is the brain-localized regulation system; this can be divided into two primary brain angiotensinogenic pathways: (i) the forebrain pathway and (ii) the connecting pathway between the hypothalamus and medulla oblongata [34]. Angiotensins (I–IV) in the brain are generated from the same protein origin, angiotensinogen. Angiotensinogen is cleaved by renin, an angiotensin-converting enzyme (ACE), aminopeptidase A, and aminopeptidase N, in this order [34].

In particular, angiotensin II can interact with the T1 receptor to help regulate the Na^+ channel [35, 36]. This channel maintains the sodium concentration within the human brain. Many studies support the interplay between angiotensin II and the T1 receptor. Factors that also exist are alcohol consumption [37], baroreceptor reflexes [38], cellular growth [39], depression [40], drinking water contamination [41, 42], gastric ulceration

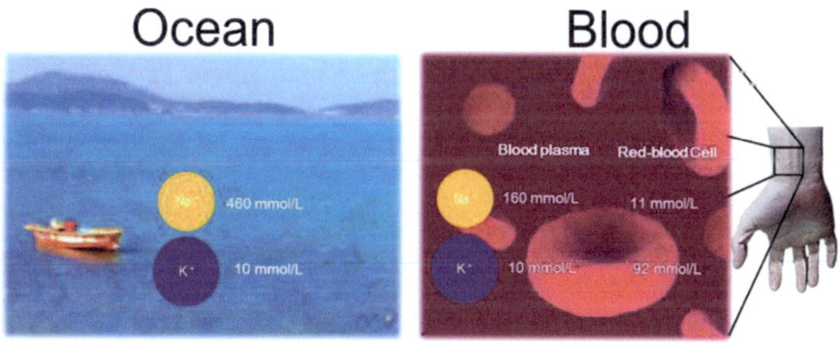

Figure 1 Concentrations of Na^+ and K^+ in living organisms and in the sea. Values taken from [1].

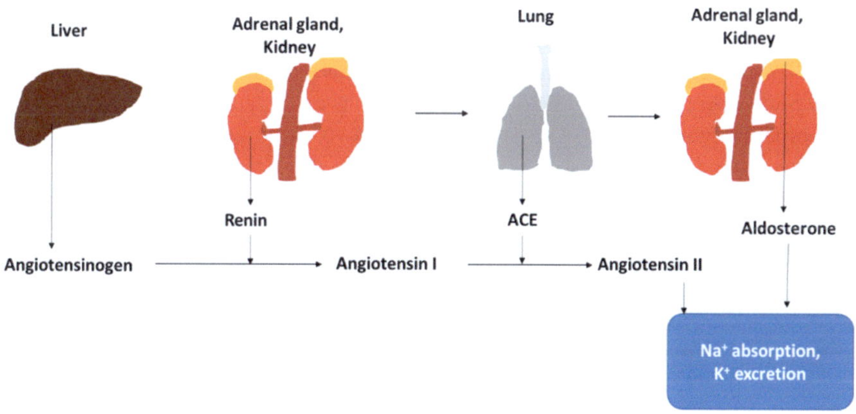

Scheme 1 Depiction of the regulation of Na^+ and K^+ concentration by the peripheral renin-angiotensin system. ACE, angiotensin-converting enzyme. Adapted from Ref. [22].

[43], the so-called "salt" appetite [44], seizure [45], sexual behavior [46], stress [47], and vasoconstriction [48]. Also, incidence of diabetes [49], PD, and AD exist [50].

The distribution and concentration of sodium ion is strongly correlated with physiological osmosis [51, 52] and the potential of a cellular membrane. This concentration gradient is sustained by the Na^+-K^+ pump, important for the transfer of electric potential. In particular, sodium takes on an important role for calcium regulation, cell volume control, glucose transport by sodium channels, the electric potential of membranes, pH regulation, and neurotransmission.

Kalupahan and Moustaid-Moussa reviewed studies about renin-angiotensin system-related obesity, inflammation, and insulin resistance [53]. The renin-angiotensin system is related to the ROS generation in neuronal cells. In particular, angiotensin II can activate NADPH oxidase which converts molecular oxygen to ROS (superoxide) [54]. Such ROS are sometimes important in normal function such as biosignalling pathways, but they mediate various cytotoxic pathways, and in excess, are deemed oxidative stress and are deleterious to any cell. These ROS are related to glucose metabolism, mitochondria function, and NF-κB release.

Although there was no direct mention of the relationship between the renin-angiotensin system and PD, reactive oxygen species are known as crucial factors for PD progression [53].

Mercer et al., reported Na_v 1.6 sodium channels in the globus pallidus which take on an important role for the regulation of speed and fast spiking in neuronal systems [55]. They mentioned that the Na_v 1.6 sodium channel in the globus pallidus is related to a prominent resurgent gating mode; the regulation of speed is not related with this mode, but fast spiking is affected by resurgence. The position of channels in the cell appeared as a primary factor. This result provides further support for macroscopic electrical stimulation as a potentially viable tool in Parkinson's disease therapy, not only understanding the cellular mechanisms underlying spiking in such neurons [55].

The Nishimura group reviewed the hypertension-related renin-angiotensin-aldosterone system (RAAS). High sodium uptake was found to increase sodium levels in cerebrospinal fluid (CSF). This stimulates epithelial sodium channels, RAAS, an endogenous digitalis-like factor, and the sympathetic nervous system, in this order. In particular, RAAS promotes oxidative stress in the brain and central nervous system [56].

A report by Zhou et al. involves glycosylation of the sodium channel β4 subunit. Relevant developments in mouse models were tracked; normally, glycosylation of β4 was found to be regulated. Alteration in glycosylation plays a role in morphological changes in the β4 subunit to increase the possibility for pathogenesis pathways suggested for PD [57].

Dopamine (DA) metabolism is vital for understanding PD etiology and PD progression. Characteristic of the disease is selective degeneration of DA neurons [9]. Some studies deal with the regulation of DA receptors D1, D2, and D5. [58, 59]. Gildea et al. [58] reported roles of DA receptors in the regulation of renal sodium transport. Using chemosensing methods including Förster resonance energy transfer studies, the stimulation of DA receptors was elucidated. Namely, D1R and D5R were determined to activate adenylyl cyclase and phospholipase C, which inhibited NHE3 and Na,K-ATPase activities that control sodium transport [58].

2.1.3 α-Synuclein and Sodium

To the best of our knowledge, there are no studies that correlate a sodium ion effect on α-synuclein structural changes. However, cellular concentration of sodium is related to inflammation and in generation of ROS in neuronal cells via RAS; this means, an irregular sodium concentration can indirectly induce toxic effects of e.g., α-synuclein (Scheme 2) [60]. In addition, their partial distribution on cellular membranes can generate electric potential related to the depolarization of the axon.

In vitro assays reported by the Standaert group discussed angiotensin II. This system was found to take a role of protection against α-synuclein toxicity and protein

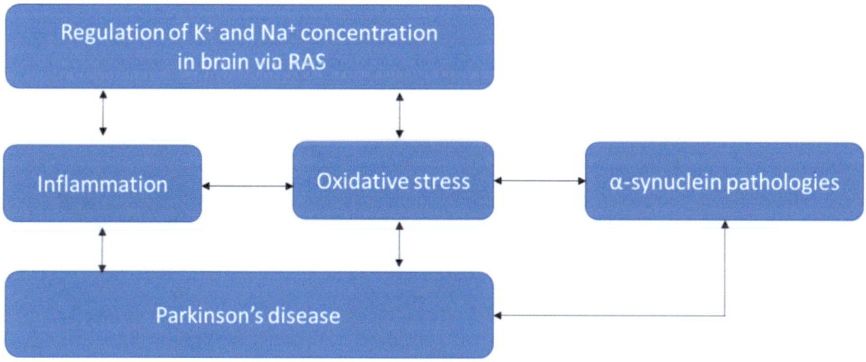

Scheme 2 Relationship between the concentrations of K^+, Na^+, and Parkinson's disease.

aggregation [61]. Angiotensin II was treated with Losartan (2-butyl-4-chloro-1-[[2′(H-tetrazol-5-yl)[1,1′-biphenyl]-4-yl]methyl-1H-imidazole-5-methanol), a AT1 antagonist. An 85 % reduction was found in α-synuclein-derived toxicity and a 19 % decrease in inclusion formation using α-synuclein overexpressed H4 cell line, a PD model. From these results, the regulation of RAS was found to be crucial for the possible prevention and treatment pathways for PD [61].

A report by Liss et al. discussed DA release in the midbrain [62]. In particular, *L*-type calcium channels and ATP-sensitive potassium channels in the midbrain display facile breakdown during neurodegeneration within PD. They underscored the need to better understand ATP-sensitive K^+ channel types. It was concluded that a complicated array of ion movement exists in both the substantia nigra [62] and ventral tegmental areas and in DA-producing neurons. In particular, they remarked that increasing Ca^{2+} levels in DA-releasing neurons relates to an interplay with sodium and potassium channels. Finally, strong links with high ROS levels in mitochondria exist, which may then be linked to PD (Figure 2).

A review by Edgerton and Jaeger from 2011 describes a model study to help explain cooperation between the dendrites and sodium channels using GENESIS 2.3 [63]. A simulation study of globus pallidus (model), which is from a combined electrophysiology and modeling database, was performed. It considered a mouse model and the amount of potassium that exists in four related systems [64]. The concentrations allowed for the analysis of normative values. Herein, relationships between the concentrations and incidences of myelinogenesis were formed. Freeze-dried samples from brain cells of rat models were analyzed through the use of proton-induced X-ray emission and Rutherford backscattering spectrometry. Four different genetic strains were sampled using these techniques (Figure 3).

In addition, a paper by Johnson et al. from 1992 described the role of *N*-methyl-*D*-aspartate in the activity of dopaminergic neurons [65]. DA neurons were discussed regarding the discharge of the neuron.

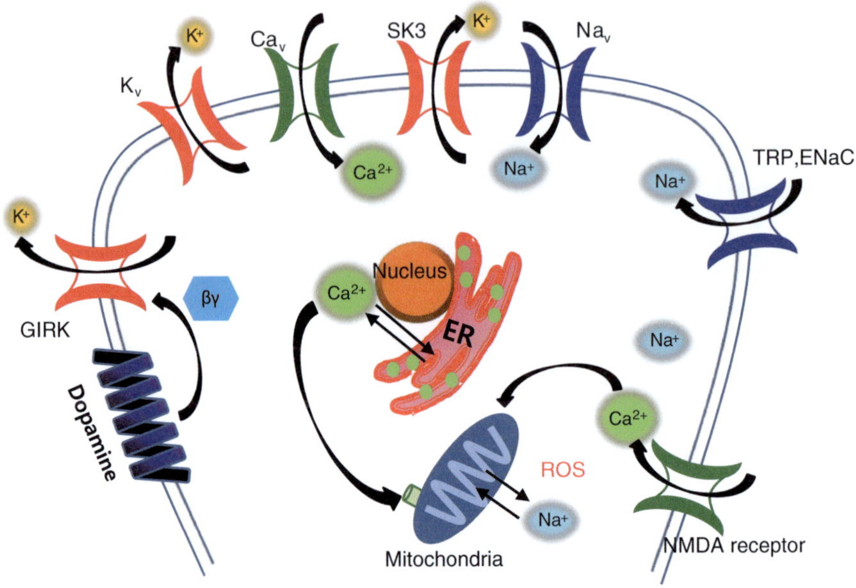

Figure 2 Schematic diagram of ion channels in the dopaminergic neurons of substantia nigra. Adapted from [62]. GIRK, G-protein-coupled inwardly-rectifying potassium channel; SK3, small conductance, calcium-sensitive potassium channel, NMDA: *N*-methyl-*D*-aspartate glutamate; TRP, transient receptor potential channel; ER, endoplasmic reticulum; ENaCs, epithelial sodium channels.

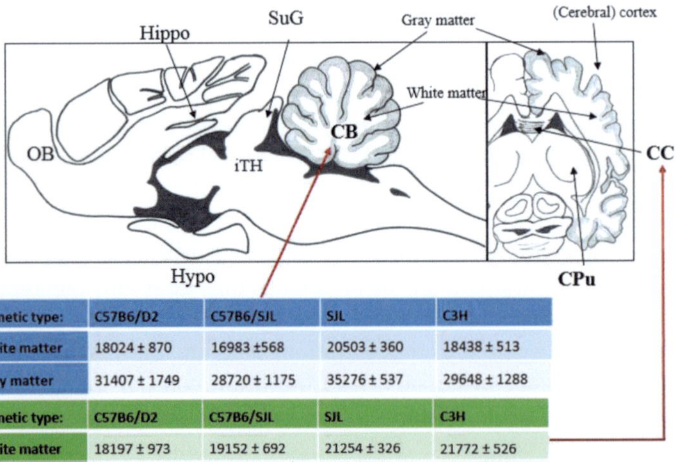

	Genetic type:	C57B6/D2	C57B6/SJL	SJL	C3H
CB	white matter	18024 ± 870	16983 ±568	20503 ± 360	18438 ± 513
	gray matter	31407 ± 1749	28720 ± 1175	35276 ± 537	29648 ± 1288
	Genetic type:	C57B6/D2	C57B6/SJL	SJL	C3H
CC	white matter	18197 ± 973	19152 ± 692	21254 ± 326	21772 ± 526
	Cortex	25704 ± 2075	15389 ± 1218	29746± 732	26107 ± 1031

Figure 3 Shown are the sagittal (**left**) and horizontal (**right**) views, and the amount of potassium concentration (μg/g in dry mass ± SEM) in different parts of mouse brain regarding different genetic types. SEM: standard error of the mean. Adapted from [64]. OB, olfactory bulb; Hypo, hypothalamus; Hippo, hippocampus; iTH, intralaminar thalamic nuclei; SuG, superficial grey layer of the superior colliculus; CB, cerebellum; CC, corpus callosum; CPu, caudate putamen.

2.2 *Parkinson's Disease and Potassium*

2.2.1 Endogenous Concentrations in Parkinson's Disease

The group of Santosh reported normative and disease-based concentrations of potassium [66]. Human serum was analyzed from 42 healthy controls and 45 diagnosed PD patients who were not treated by medicine (drug-naïve PD patients). The $[K^+]$ in these samples from PD patients was determined to be 4.17 ± 0.08 mM, relative to the normative value of 3.63 ± 0.2 mM [66]. The concentrations of intra-, and extracellular potassium were reported as 3,300 mM and $3.5 \sim 5$ mM, respectively. These values are affected by both dietary uptake and cellular pH [67].

2.2.2 Known Actions of Potassium

Potassium is critical in many different ways in biology. It is an important regulation element for osmotic pressure, electrolytic equilibria, and currents. Because of these roles, the action of potassium ion crossing cellular membranes is crucial. Potassium homeostasis in cells is regulated by cellular regulation proteins, potassium channels, and is related to both PD and RAS (Figure 3) [68–73]. Mertens et al. discussed if angiotensin II and pro-inflammatory compounds can affect the NADPH dependent oxidase of cellular membranes and can induce toxicity. It was found in these studies that ROS selectively degrade dopaminergic neuronal cells [73].

Potassium channels and their regulation and function exist as possible new therapeutic targets. The K^+-ATP channels consist of 2 subunits known as Kir6.2 and Sur1, a regulatory sulfonylurea receptor subunit [74]. Some studies have reported that selective degeneration of dopaminergic neurons in PD is strongly related with ATP-sensitive potassium channels [74, 75]. Liss et al. performed MPTP treatment of mouse models and found that substantia nigra dopaminergic neurons underwent degeneration; but in this study, the ventral tegmental area was not affected. This was achieved via hyperpolarization and electrical silencing [75]. Zheng et al., suggested that the regulation of potassium channels can be a promising therapeutic target for PD, in general [76].

Zheng et al. reported that Parkinsonian symptoms can be alleviated by the administration of Iptakalim (IPT, 2,3–dimethyl-N-(1-methylethyl)-2-butanamine•HCl) [76]. In this study, the symptoms, including catalepsy and hypolocomotion, were induced by haloperidol. IPT is a novel promotor of ATP-sensitive potassium channel (KATP) activity. Symptoms were monitored in a mouse model. From these results, they remarked that KATP correlated with incidence of Parkinson's disease. In this view, KATP may become a new therapeutic target [77]. Similarly, the group of Richter reported Retigabine, a Kv7 potassium channel activator, alleviated *L*-DOPA-induced dyskinesias in a lesioned rat model (6-OHDA, 6-hydroxydopamine). They found chronic treatment of Retigabine (also, Ezogabine; ethyl N-[2-amino-4-[(4-fluorophenyl)methylamino]phenyl]carbamate, GlaxoSmithKline, Valeant) can reduce the severity of abnormal involuntary movement. They mentioned Kv7 channels may be crucial for the pathophysiology of dyskinesias [78].

The Trudeau group tried to effect DA release through electrochemical methods (fast-scan cyclic voltammetry). These are related to potassium channel activation. Voltage-gating of the Kv1 family are important for regulating DA release. And, here, only Kv1.2, 1.3, and 1.6 were controlled by Kv subtype-selective blockers. Kv1 was found to be key in regulating DA release in the striatum [59]. The Chen group reviewed the role of potassium channels in the basal ganglia region [79]. Potassium channel functions are related importantly in cellular signaling, neurotransmitter release and neuronal excitability in basal ganglia. The authors suggested that the regulation of potassium channels in this brain region can provide insight for potential therapeutic targets in PD [79] .

2.2.3 α-Synuclein and Potassium

α-Synuclein is the hallmark protein for aggregation in PD. One characteristic is the deposition of such aggregated protein, not unlike how β-amyloid is deposited into senile plaques in AD. Aggregated α-synuclein has been correlated through *in vivo* studies with potassium concentration, or potassium channel activation [80–82].

Follett et al. reported that the depolarization of potassium can induce the increase of [Ca^{2+}]. This increase in turn induces α-synuclein aggregation [82]. When KCl was injected into neuroblastoma cell models (SH-SY5Y), they found an increase in intracellular Ca^{2+} concentration and an increase in the aggregation of α-synuclein relative to control samples. Thus, as described, suspected potassium effects on PD are related to the regulation of potassium channels [83].

3 Traumatic Brain Injury and Ischemia

TBI can lead to death or disability of the body, and can be classified into 3 types: penetrating TBI (pTBI), closed TBI (cTBI), and blast TBI (bTBI). Abnormal ion flux occurs after an accident or incident involving the striking of the head. Brain injuries can cause electrolyte disturbances [84]. Ischemia (*iskhaimous*: stopping of blood, Greek) is also related to conditions of stroke and brain damage. TBI is also accompanied by brain swelling which imparts higher levels of intracranial pressure (ICP).

3.1 Traumatic Brain Injury, Ischemia, and Sodium

Brain injuries are usually found coincident with disturbances in sodium ion concentration. The central nervous system (CNS) has the responsibility for the regulation of sodium/water homeostasis. Sodium, by concentration, is one of the major cations in biology (Figure 2), and it is closely related to osmosis. Ion concentration is regulated by the sodium-potassium ATPase (NKA) pump [85]. Since TBI relates to the ensuing osmotic pressure of tissue and fluids inside the skull, the concentrations of alkali metals are relevant to the discussion. Within this barometric perspective, the relationship between intracerebral Na^+ concentration and the content of water in the brain has been clearly demonstrated [86]. Also, these studies demonstrated that a decrease in extracellular Na^+ concentration induces the osmotic gradient so that the intracellular compartment is further filled with water. In a macroscopic perspective, this is characterized by levels of brain edema and ICP concentration [87]. *Hyponatremia* impairs incidence of edema in the injured brain [88].

There are two possible mechanisms that can be adapted with regard to hyponatremia in the brain [89, 90]; CSF flow hydrostatic pressure is increased by hyponatremia. Another mechanism is that the amount of intracellular osmolyte is decreased because of the lack of sodium ion. Species such as taurine, glutathione, and glutamate are found to be significantly important solutes for the survival of neurons. In particular, the role of glutamate is important for the adaptation of the cerebral environment, but the glutamate absorbed from extracellular space induces both the excitotoxicity of damaged neurons and an exacerbation of astrocyte damage in TBI with hyponatremia. Taurine (2-aminoethanesulfonic acid) acts as an antioxidant; it is redistributed by the neuron under hypo-osmotic stress in the brain. In this way, glutathione antioxidant is preserved. Neuron is more susceptible to TBI while the antioxidant is being reduced [90]. Therefore, severe hyponatremia induces specific processes in the brain and these undergo successive steps [89].

TBI is commonly accompanied by serious hyponatremia [91]. Recent experimental studies have involved rat models [92]. The volume of cortical contusion by TBI is different when comparing to cases in which hyponatremia precedes TBI. Brain acid buffering and the repairing system of intracellular acidosis are damaged by the occurrence of hyponatremia. Therefore, as a result of hyponatremia, the content of water is increased according to an osmotic gradient between the brain and blood so that neuronal swelling becomes significant, or severe. In brain stem tissue, in particular, the presence of both edema and additional increments of water content are observed. The brain stem is more sensitive to osmolarity changes [93]. Importantly, TBI makes neuronal elevation in intercellular sodium concentration cause cell swelling [94]. Intracellular sodium concentration is increased by the reuptake of glutamate; this causes swelling of the cell and intracellular acidosis. Extracellular fluid, in which $[Na^+]$ is decreased, indicates neuronal depolarization. Magnetic resonance imaging studies have revealed that a decline of intracellular pH in rat brain helps ameliorate TBI (fluid percussion) [95]. In astrocytic cells, Type 1 sodium-hydrogen antiporter (NHE-1), producing further Na^+ accumulation is activated by intracellular acidosis. Both (*i*) sodium-dependent glutamate transport and (*ii*) NHE-1 activation elevate Na^+ concentration. The elevation of $[Na^+]$ impacts other metal ion concentrations as well [96]. Increased ion levels could increase sodium-calcium exchanger (NCX) activation; this then initiates the import of toxic levels of calcium into the cell. These phenomena contribute to the abnormal function of astrocyte neurons [96], affect common astrocytic function, and lead to further neuronal excitotoxicity. However, (i) to cut off either the reverse mode of NCX, or (ii) to restrict glutamate transport may prohibit continuous damage [97].

The sodium gradient dominates both neuronal excitation and synaptic transmission. NKA is extremely reliant on Na^+ concentration to enable homeostasis in the cell. To allow for this activity, the brain expends half of its total daily amount of ATP. As considerable amounts of glutamate are released, extracellular glutamate is thus increased, and hyperactivity appears [98].

TBI patients residing in the intensive care unit have been found to present with hypernatremia upon treatment with hypertonic saline (HTS) solution. Hypernatremia, in extreme cases, leads to patient death; TBI patients generally have a greater possibility for morbidity due to hypernatremia [99]. About 29 % of TBI patients suffered from hypernatremia based on other experimental findings [100]. Due to HTS treatment, ICP lowering is observed; there are adverse effects as well, however, such as a reduction of the intravascular volume of the brain, and the potential failure of renal function [99].

3.2 *Traumatic Brain Injury, Ischemia, and Endogenous Potassium Concentrations*

TBI incidence also has correlations to potassium ion concentration. Potassium efflux may cause swelling in cytotoxic edema and neuronal astrocytes leading to uptake of an excess amount of K^+ [101]. Potassium channels have important roles in the regulation of the concentration of ion, and in the excitability of cells [102]. Potassium is forced to move oppositely against excitation through membrane action due to hyperpolarizability of the membrane. At nerve terminals, glutamate release will therefore be inhibited. Releasing glutamate may result in excitotoxicity to a cell [102]. In addition, when there is an incidence of contusion, the level of glutamate in patients is elevated [103]. Based on the characteristics of the ion, researchers have assumed that potassium channels and their action might be able to play a role in neuroprotection.

Namiranian et al. have revealed that the TREK-1 family of potassium channels, does not play an important role for protecting the nervous system after TBI, but it is neuroprotective with respect to ischemia [102]. TBI and ischemia have different pathological states that might be differently influenced by the presence or absence of TREK-1, even though the two disorders possess similar biochemical mechanisms [102].

Dyskalaemia is a common condition encountered by patients with TBI. Hypokalemia affects TBI patients more than other types of traumatic injury patients. Mild or moderate hypokalemia is not very life-threatening, but severe hypokalemia ($[K^+]<2.5$ mmol/L) may possibly lead to death in patients. When compared to mild and moderate hypokalemia, severe hypokalemia patients show related symtoms: high sodium ion and low phosphate ion levels. Also, higher mean ICP has been observed [84].

4 Conclusions

In this contribution, the direct or indirect relationships between alkali metal ion concentration and PD and TBI are presented. In almost all studies described here, a protein level, receptor or organelle, or organ-wide understanding exists. But, a detailed atomic level understanding is virtually absent. That is to say: What do characteristics of, e.g., hypernatremia and hyperkalemia look like at the molecular level?

Where possible, correlative and tabulated data is presented and explained herein, and trends in disease indeed relate back to the concentrations of the particular alkali metal ions, especially sodium and potassium. Clinical studies, as well as biochemical studies, were presented here and involve human subjects as well as animal models (e.g., rat). Attempts were made to connect a discussion of PD with that for TBI, and to show voids where efforts for future research may be placed.

Abbreviations

ACE	angiotensin-converting enzyme
AD	Alzheimer's disease
AT1	angiotensin receptor1
ATP	adenosine 5′-triphosphate
CNS	central nervous system
CSF	cerebrospinal fluid
DA	dopamine
GIRK	G-protein-coupled inwardly-rectifying potassium channel
HTS	hypertonic saline
IPT	iptakalim, 2,3-dimethyl-N-(1-methylethyl)-2-butanamine hydrochloride
ICP	intracranial pressure
KATP	ATP-sensitive potassium channel opener
Kir	inward-rectifier potassium ion channel
Kv	voltage-gated potassium channel
MPTP	1-methyl-4-phenyl-1,2,3,6-tetrahydropyridine
NADPH	reduced form of nicotinamide adenine dinucleotide phosphate
NCX	sodium-calcium exchanger
NHE-1	Type 1 sodium-hydrogen antiporter
NHE3	sodium–hydrogen exchanger 3
NKA	sodium-potassium ATPase
MPTP	1-methyl-4-phenyl-1,2,3,6-tetrahydropyridine
MRI	magnetic resonance imaging
NF-κB	nuclear factor kappa B
PD	Parkinson's disease
RAAS	renin-angiotensin-aldosterone system

RAS	renin angiotensin system
ROS	reactive oxygen species
SK3	small conductance, calcium-sensitive potassium channel
TBI	traumatic brain injury
TREK	potassium channel subfamily K member 2 (as known as KCNK2)
TRP	transient receptor potential channel

Acknowledgment Prof. D. G. Churchill (D.G.C.), Mr. Yonghwang Ha, and Ms. Jeong A Jeong acknowledge support from the NRF (National Research Foundation) of Korea (Grant 2011-0017280) for the operation of the Molecular Logic Gate Laboratory. D.G.C. acknowledges support from the Institute of Basic Science (IBS) of Korea.

References

1. J. J. R. Fraústo da Silva, R. J. P. Williams, *The Biological Chemistry of the Elements: The Inorganic Chemistry of Life*, Clarendon, Oxford, 1991.
2. M. Asanuma, I. Miyazaki, *Exp. Neurol.* **2007**, *206*, 172–178.
3. M. Asanuma, I. Miyazaki, *Curr. Pharm. Des.* **2008**, *14*, 1428–1434.
4. H. L. Chen, S. M. M. Zhang, M. A. Hernan, M. A. Schwarzschild, W. C. Willett, G. A. Colditz, F. E. Speizer, A. Ascherio, *Arch. Neurol.* **2003**, *60*, 1059–1064.
5. T. Mueller, *Expert Rev. Neurother.* **2013**, *13*, 969–977.
6. C. S. Chan, T. S. Gertler, D. J. Surmeier, *Mov. Disord.* **2010**, *25*, S63–S70.
7. K.-Y. Chau, H. L. Ching, A. H. V. Schapira, J. M. Cooper, *J. Neurochem.* **2009**, *110*, 1005–1013.
8. D. Sulzer, Y. Schmitz, *Neuron* **2007**, *55*, 8–10.
9. D. J. Moore, A. B. West, V. L. Dawson, T. M. Dawson, *Annu. Rev. Neurosci.* **2005**, *28*, 57–87.
10. H. Fujiwara, M. Hasegawa, N. Dohmae, A. Kawashima, E. Masliah, M. S. Goldberg, J. Shen, K. Takio, T. Iwatsubo, *Nat. Cell Biol.* **2002**, *4*, 160–164.
11. V. N. Uversky, *J. Neurochem.* **2007**, *103*, 17–37.
12. J. Lotharius, P. Brundin, *Nat. Rev. Neurosci.* **2002**, *3*, 932–942.
13. M. Goedert, *Nat. Rev. Neurosci.* **2001**, *2*, 492–501.
14. E. Gaggelli, H. Kozlowski, D. Valensin, G. Valensin, *Chem. Rev.* **2006**, *106*, 1995–2044.
15. J. D. Meeker, M. G. Rossano, B. Protas, M. P. Diamond, E. Puscheck, D. Daly, N. Paneth, J. J. Wirth, *Environ. Health Perspect.* **2008**, *116*, 1473–1479.
16. U. Krämer, I. N. Talke, M. Hanikenne, *FEBS Lett.* **2007**, *581*, 2263–2272.
17. H. H. Ussing, P. Kruhoffer, H. J. Thaysen, N. H. Thron, in *The Alkali Metal Ions in Biology*, Vol. 13, Springer, Berlin Heidelberg, 1959, pp. 1–195.
18. Y. Ha, O. G. Tsay, D. G. Churchill, *Monatsh. Chem.* **2011**, *142*, 385–398.
19. J. E. Sealey, *Hypertension*, **1990**, 1287–1317.
20. M. J. Hurley, B. Brandon, S. M. Gentleman, D. T. Dexter, *Brain* **2013**, *136*, 2077–2097.
21. M. J. Hurley, D. T. Dexter, *Pharmacol. Ther.* **2012**, *133*, 324–333.
22. J. W. Wright, J. W. Harding, *Pflugers Arch.* **2013**, *465*, 133–151.
23. Y. Ha, A. Yang, S. Lee, K. Kim, H. Liew, S. H. Lee, J. E. Lee, H. I. Lee, Y. H. Suh, H. S. Park, D. G. Churchill, *J. Neurosci. Res.* **2014**, *92*, 359–368.
24. Y. Ha, A. Yang, S. Lee, K. Kim, H. Liew, Y. H. Suh, H. S. Park, D. G. Churchill, *Biochem. Biophys. Res. Commun.* **2014**, *443*, 1085–1091.
25. Y. Ha, H. Liew, H. Y. Park, K. Kim, Y. H. Suh, D. G. Churchill, *J. Neurosci. Res.* **2011**, *71*, 168–177.
26. L. Chen, M. B. Feany, *Nat. Neurosci.* **2005**, *8*, 657–663.

27. J. Xu, S. Y. Kao, F. J. S. Lee, W. Song, L. W. Jin, B. A. Yankner, *Nat. Med.* **2002**, *8*, 600–606.
28. E. Carboni, P. Lingor, *Metallomics* **2015**, *7*, 395–404.
29. N. B. Cole, D. D. Murphy, J. Lebowitz, L. Di Noto, R. L. Levine, R. L. Nussbaum, *J. Biol. Chem.* **2005**, *280*, 9678–9690.
30. R. R. Crichton, D. T. Dexter, R. J. Ward, *Coord. Chem. Rev.* **2008**, *252*, 1189–1199.
31. G. Madelin, R. Kline, R. Walvick, R. R. Regatte, *Sci. Rep.* **2014**, *4*, 4763.
32. J. Rodriguez-Pallares, J. A. Parga, B. Joglar, M. J. Guerra, J. L. Labandeira-Garcia, *Age* **2012**, *34*, 863–880.
33. J. D. H. Slater, *Postgrad. Med. J.* **1964**, *40*, 479–496.
34. C. Llorens-Cortes, F. A. O. Mendelsohn, *J. Renin-Angiotensin-Aldosterone Syst.* **2002**, *3*, S39–S48.
35. M. Ohsawa, K. Tamura, H. Wakui, A. Maeda, T. Dejima, T. Kanaoka, K. Azushima, K. Uneda, Y. Tsurumi-Ikeya, R. Kobayashi, M. Matsuda, S. Uchida, Y. Toya, H. Kobori, A. Nishiyama, A. Yamashita, Y. Ishikawa, S. Umemura, *Kidney Int.* **2014**, *86*, 570–581.
36. M. Mamenko, O. Zaika, D. V. Ilatovskaya, A. Staruschenko, O. Pochynyuk, *J. Biol. Chem.* **2012**, *287*, 660–671.
37. B. Maul, W. Krause, K. Pankow, M. Becker, F. Gembardt, N. Alenina, T. Walther, M. Bader, W. E. Siems, *FASEB J.* **2005**, *19*, 1474–1481.
38. M. G. Garner, A. F. Phippard, P. J. Fletcher, J. M. Maclean, G. G. Duggin, J. S. Horvath, D. J. Tiller, *Hypertension* **1987**, *10*, 628–634.
39. C. M. Filipeanu, R. H. Henning, D. de Zeeuw, A. Nelemans, *Br. J. Pharmacol.* **2001**, *132*, 1590–1596.
40. J. M. Saavedra, E. Sánchez-Lemus, J. Benicky, *Psychoneuroendocrinology* **2011**, *36*, 1–18.
41. M. J. Wayner, A. D. Merkel, F. C. Barone, F. B. Jolicoeur, D. B. Rondeau, *Pharmacol. Biochem. Behav.* **1976**, *5*, 103–110.
42. C. M. Pawloski, G. D. Fink, *Am. J. Physiol.* **1990**, *259*, R531–538.
43. C. Bregonzio, I. Armando, H. Ando, M. Jezova, G. Baiardi, J. M. Saavedra, *Ann. N. Y. Acad. Sci.* **2004**, *1018*, 351–355.
44. J. T. Fitzsimons, *Physiol. Rev.* **1998**, *78*, 583–686.
45. J. Tchekalarova, V. Georgiev, *J. Physiol. Paris* **1999**, *93*, 191–197.
46. M. Nomura, H. Nishii, Y. Ozaki, N. Fujimoto, T. Matsumoto, *Physiol. Behav.* **2007**, *91*, 223–228.
47. J. M. Saavedra, J. Benicky, *Stress* **2007**, *10*, 185–193.
48. H. Kanaide, T. Ichiki, J. Nishimura, K. Hirano, *Circ. Res.* **2003**, *93*, 1015–1017.
49. D. J. Leehey, A. K. Singh, N. Alavi, R. Singh, *Kidney Int. Suppl.* **2000**, *77*, S93–98.
50. H. Brasch, L. Sieroslawski, P. Dominiak, *Hypertension* **1993**, *22*, 699–704.
51. D. Alvarez-Fischer, S. Guerreiro, S. Hunot, F. Saurini, M. Marien, P. Sokoloff, E. C. Hirsh, A. Hartmann, P. P. Michel, *J. Neurochem.* **2008**, *107*, 701–711.
52. J. Wardas, K. Kuter, W. Kolasiewicz, M. Zapala, K. Golembiowska, C. E. Muller, *Pharmacol. Rep.* **2009**, *61*, 1251–1252.
53. N. S. Kalupahana, N. Moustaid-Moussa, *Obes. Rev.* **2012**, *13*, 136–149.
54. K. K. Griendling, C. A. Minieri, J. D. Ollerenshaw, R. W. Alexander, *Circ. Res.* **1994**, *74*, 1141–1148.
55. J. N. Mercer, C. S. Chan, T. Tkatch, J. Held, D. J. Surmeier, *J. Neurosci.* **2007**, *27*, 13552–13566.
56. H. Takahashi, M. Yoshika, Y. Komiyama, M. Nishimura, *Hypertens. Res.* **2011**, *34*, 1147–1160.
57. T.-T. Zhou, Z.-W. Zhang, J. Liu, J.-P. Zhang, B.-H. Jiao, *Int. J. Biol. Sci.* **2012**, *8*, 630–639.
58. J. J. Gildea, I. T. Shah, R. E. Van Sciver, J. A. Israel, C. Enzensperger, H. E. McGrath, P. A. Jose, R. A. Felder, *Kidney Int.* **2014**, *86*, 118–126.
59. P. Martel, D. Leo, S. Fulton, M. Berard, L. E. Trudeau, *Plos One* **2011**, *6*.

60. S. Basak, G. V. R. K. Prasad, J. Varkey, K. Chattopadhyay, *ACS Chem. Neurosci.* **2015**, *6*, 239–246.
61. T. N. Grammatopoulos, T. F. Outeiro, B. T. Hyman, D. G. Standaert, *Biochem. Biophys. Res. Commun.* **2007**, *363*, 846–851.
62. E. Dragicevic, J. Schiemann, B. Liss, *Neuroscience* **2015**, *284*, 798–814.
63. J. R. Edgerton, D. Jaeger, *J. Neurosci.* **2011**, *31*, 10919–10936.
64. C. Sergeant, M. H. Vesvres, G. Devès, F. Guillou, *Nucl. Instrum. Methods Phys. Res. Sect. B*, **2005**, *231*, 234–238.
65. S. W. Johnson, V. Seutin, R. A. North, *Science* **1992**, *258*, 665–667.
66. S. Ahmed, W. Santosh, *Plos One* **2010**, *5*, e11252.
67. P. S. Aronson, G. Giebisch, *J. Am. Soc. Nephrol.* **2011**, *22*, 1981–1989.
68. D. Penton, S. Bandulik, F. Schweda, S. Haubs, P. Tauber, M. Reichold, L. D. Cong, A. El Wakil, T. Budde, F. Lesage, E. Lalli, M. C. Zennaro, R. Warth, J. Barhanin, *Endocrinology* **2012**, *153*, 4740–4748.
69. A. Kurtz, M. Hamann, K. Gotz, *Pflugers Arch.* **2000**, *440*, 889–895.
70. C. P. Ferrier, A. Kurtz, P. Lehner, S. G. Shaw, C. Pusterla, H. Saxenhofer, P. Weidmann, *Eur. J. Clin. Pharmacol.* **1989**, *36*, 443–447.
71. S. Bandulik, P. Tauber, D. Penton, F. Schweda, I. Tegtmeier, C. Sterner, E. Lalli, F. Lesage, M. Hartmann, J. Barhanin, R. Warth, *Endocrinology* **2013**, *154*, 2712–2722.
72. R. A. Abukishk, A. R. Noble, *J. Physiol. London* **1990**, *430*, P89–P89.
73. B. Mertens, P. Vanderheyden, Y. Michotte, S. Sarre, *J. Renin-Angiotensin-Aldosterone Syst.* **2010**, *11*, 49–56.
74. A. Y. Deutch, D. G. Winder, *Nat. Med.* **2006**, *12*, 17–18.
75. B. Liss, O. Haeckel, J. Wildmann, T. Miki, S. Seino, J. Roeper, *Nat. Neurosci.* **2005**, *8*, 1742–1751.
76. Y. Wang, P. L. Yang, J. F. Tang, J. F. Lin, X. H. Cai, X. T. Wang, G. Q. Zheng, *Med. Hypotheses* **2008**, *71*, 546–550.
77. S. Wang, L. F. Hu, Y. Yang, J. H. Ding, G. Hu, *Neuropharmacology* **2005**, *48*, 984–992.
78. S. E. Sander, C. Lemm, N. Lange, M. Hamann, A. Richter, *Neuropharmacology* **2012**, *62*, 1052–1061.
79. G. Wang, J. Zeng, R. J. Ren, S. D. Chen, *Front. Biosci.* **2008**, *13*, 3685–3698.
80. M. Subramaniam, D. Althof, S. Gispert, J. Schwenk, G. Auburger, A. Kulik, B. Fakler, J. Roeper, *J. Neurosci.* **2014**, *34*, 13586–13599.
81. M. Subramaniam, J. Roeper, *Acta Physiol.* **2014**, *210*, 108.
82. J. Follett, B. Darlow, M. B. Wong, J. Goodwin, D. L. Pountney, *Neurotoxic. Res.* **2013**, *23*, 378–392.
83. C. C. Shieh, M. Coghlan, J. P. Sullivan, M. Gopalakrishnan, *Pharmacol. Rev.* **2000**, *52*, 557–594.
84. X. Wu, X. Lu, X. Lu, J. Yu, Y. Sun, Z. Du, X. Wu, Y. Mao, L. Zhou, S. Wu, J. Hu, *Injury* **2015**, *46*, 35–41.
85. K. Bradshaw, M. Smith, *Contin. Educ. Anaesth. Crit. Care Pain* **2008**, *8*, 129–133.
86. B. R. Nathan, *Neurocrit. Care* **2007**, *6*, 72–78.
87. X. Di, B. G. Lyeth, R. J. Hamm, M. R. Bullock, *J. Neurotrauma* **1996**, *13*, 497–504.
88. J. C. Ayus, D. Armstrong, A. I. Arieff, *Kidney Int.* **2006**, *69*, 1319–1325.
89. C. Ke, W. S. Poon, H. K. Ng, F. M. M. Lai, N. L. S. Tang, J. C. S. Pang, *Exp. Neurol.* **2002**, *178*, 194–206.
90. S. W. Carlson, S. K. Madathil, D. M. Sama, X. Gao, J. Chen, K. E. Saatman, *J. Neuropathol. Exp. Neurol.* **2014**, *73*, 734–746.
91. N. Moro, Y. Katayama, T. Igarashi, T. Mori, T. Kawamata, J. Kojima, *Surg. Neurol.* **2007**, *68*, 387–393.
92. C. Ke, W. S. Poon, H. K. Ng, N. L. S. Tang, Y. Chan, J. Y. Wang, J. N. K. Hsiang, *Acta Neurochir.* **2000**, *76*, 405–408.
93. P. A. Walker, F. Jimenez, C. S. Cox, *Regener. Med.* **2010**, *5*, 65.

94. X. Zhao, F. A. Gorin, R. F. Berman, B. G. Lyeth, *J. Neurotrauma* **2008**, *25*, 1195–1205.
95. T. K. McIntosh, A. I. Faden, M. R. Bendall, R. Vink, *J. Neurochem.* **1987**, *49*, 1530–1540.
96. A. S. Filippidis, X. Liang, W. Wang, S. Parveen, C. M. Baumgarten, C. R. Marmarou, *J. Neurotrauma* **2014**, *31*, 1258–1267.
97. C. L. Floyd, F. A. Gorin, B. G. Lyeth, *Glia* **2005**, *51*, 35–46.
98. N. Khatri, H. Y. Man, *Front. Neurol.* **2013**, *4*, 1–11.
99. U. Maggiore, E, Picetti, E. Antonucci, E. Parenti, G. Regolisti, M. Mergoni, *Crit. Care* **2009**, *13*, R110.
100. M. Li, Y. H. Hu, G. Chen, *Injury* **2013**, *44*, 1213–1218.
101. H. S. Hatashita, J. T. Hoff, *Stroke* **1990**, *21*, 582–588.
102. K. Namiranian, C. D. Brink, J. C. Goodman, C. S. Robertson, Jr. R. M. Bryan, *J. Cereb. Blood Flow Metab.* **2011**, e1–e6.
103. M. Reinert, A. Khaldi, A. Zauner, E. Doppenberg, S. Choi, R. Bullock, *J. Neurosurg.* **2000**, *93*, 800–807.

Index

A. Sigel, H. Sigel, and R.K.O. Sigel (eds.), *The Alkali Metal Ions: Their Role for Life*, Metal Ions in Life Sciences 16, DOI 10.1007/978-3-319-21756-7

J

K

O

P

S

MIX
Papier aus verantwortungsvollen Quellen
Paper from responsible sources
FSC® C105338

FSC
www.fsc.org

If you have any concerns about our products,
you can contact us on
ProductSafety@springernature.com

In case Publisher is established outside the EU,
the EU authorized representative is:
Springer Nature Customer Service Center GmbH
Europaplatz 3, 69115 Heidelberg, Germany

Printed by Libri Plureos GmbH
in Hamburg, Germany